ADVANCED STRUCTURAL DYNAMICS

Advanced Structural Dynamics will appeal to a broad readership that includes both undergraduate and graduate engineering students, doctoral candidates, engineering scientists working in various technical disciplines, and practicing professionals in an engineering office. The book has broad applicability and draws examples from aeronautical, civil, earthquake, mechanical, and ocean engineering, and at times it even dabbles in issues of geophysics and seismology. The material presented is based on miscellaneous course and lecture notes offered by the author at the Massachusetts Institute of Technology for many years. The modular approach allows for a selective use of chapters, making it appropriate for use not only as an introductory textbook but later on functioning also as a treatise for an advanced course, covering materials not typically found in competing textbooks on the subject.

Professor Eduardo Kausel is a specialist in structural dynamics in the Department of Civil Engineering at the Massachusetts Institute of Technology. He is especially well known for two papers on the collapse of the Twin Towers on September 11, 2001. The first of this pair, published on the web at MIT only a few days after the terrorist act, attracted more readers around the world than all other works and publications on the subject combined. Professor Kausel is the author of the 2006 book *Fundamental Solutions in Elastodynamics* (Cambridge University Press).

Advanced Structural Dynamics

EDUARDO KAUSEL

Massachusetts Institute of Technology

CAMBRIDGE
UNIVERSITY PRESS

University Printing House, Cambridge CB2 8BS, United Kingdom

One Liberty Plaza, 20th Floor, New York, NY 10006, USA

477 Williamstown Road, Port Melbourne, VIC 3207, Australia

4843/24, 2nd Floor, Ansari Road, Daryaganj, Delhi – 110002, India

79 Anson Road, #06-04/06, Singapore 079906

Cambridge University Press is part of the University of Cambridge.

It furthers the University's mission by disseminating knowledge in the pursuit of
education, learning, and research at the highest international levels of excellence.

www.cambridge.org
Information on this title: www.cambridge.org/9781107171510
10.1017/9781316761403

First published 2017

A catalogue record for this publication is available from the British Library.

Library of Congress Cataloging-in-Publication Data
Names: Kausel, E.
Title: Advanced structural dynamics / by Eduardo Kausel, Massachusetts Institute of Technology.
Other titles: Structural dynamics
Description: Cambridge [England]: Cambridge University Press, 2017. |
Includes bibliographical references and index.
Identifiers: LCCN 2016028355 | ISBN 9781107171510 (hard back)
Subjects: LCSH: Structural dynamics – Textbooks. | Structural analysis (Engineering) – Textbooks.
Classification: LCC TA654.K276 2016 | DDC 624.1/71–dc23
LC record available at https://lccn.loc.gov/2016028355

ISBN 978-1-107-17151-0 Hardback

To my former graduate student and dear Guardian Angel Hyangly Lee,
in everlasting gratitude for her continued support of my work at MIT.

Contents

Contents

Preface

The material in this book slowly accumulated, accreted, and grew out of the many lectures on structural dynamics, soil dynamics, earthquake engineering, and structural mechanics that I gave at MIT in the course of several decades of teaching. At first these constituted mere handouts to the students, meant to clarify further the material covered in the lectures, but soon the notes transcended the class environment and began steadily growing in size and content as well as complication. Eventually, the size was such that I decided that it might be worthwhile for these voluminous class notes to see the light as a regular textbook, but the sheer effort required to clean out and polish the text so as to bring it up to publication standards demanded too much of my time and entailed sacrifices elsewhere in my busy schedule that I simply couldn't afford. Or expressing it in MIT-speak, I applied the *Principle of Selective Neglect*. But after years (and even decades) of procrastination, eventually I finally managed to break the vicious cycle of writer's block and brought this necessary task to completion.

Make no mistake: the material covered in this book far exceeds what can be taught in any one-semester graduate course in structural dynamics or mechanical vibration, and indeed, even in a sequence of two such courses. Still, it exhaustively covers the fundamentals in vibration theory, and then goes on well beyond the standard fare in – and conventional treatment of – a graduate course in structural dynamics, as a result of which most can (and should) be excluded from an introductory course outline, even if it can still be used for that purpose. Given the sheer volume of material, the text is admittedly terse and at times rather sparse in explanations, but that is deliberate, for otherwise the book would have been unduly long, not to mention tedious to read and follow. Thus, the reader is expected to have some background in the mechanical sciences such that he or she need not be taken by the hand. Still, when used in the classroom for a first graduate course, it would suffice to jump over advanced sections, and do so without sacrifices in the clarity and self-sufficiency of the retained material.

In a typical semester, I would start by reviewing the basic principles of dynamics, namely Newton's laws, impulse and conservation of linear and angular momenta, D'Alembert's principle, the concept of point masses obtained by means of mass lumping and tributary areas, and most importantly, explicating the difference between static and dynamic degrees of freedom (or master–slave DOF), all while assuming small displacements and skipping initially over the section that deals with Lagrange's equations. From there on I would move on to cover the theory of single-DOF systems and devote just about half of the semester to that topic, inasmuch as multi-DOF systems and continuous systems can largely be regarded as generalizations of those more simple systems. In the lectures, I often interspersed demonstration experiments to illustrate basic concepts and made use of brief Matlab® models to demonstrate the application of the concepts being learned. I also devoted a good number of lectures to explain harmonic analysis and the use of complex Fourier series, which in my view is one of the most important yet difficult concepts for students to comprehend and assimilate properly. For that purpose, I usually started by explaining the concepts of amplitude and phase by considering a simple complex number of the form $z = x + \mathrm{i}\,y$, and then moving on to see what those quantities would be for products and ratios of complex numbers of the form $z = z_1 z_2$, $z = z_1/z_2 = |z_1|/|z_2| e^{\mathrm{i}(\phi_1 - \phi_2)}$, and in particular $z = 1/z_2 = e^{-\mathrm{i}\phi_2}/|z_2|$. I completely omitted the use of sine and cosine Fourier series, and considered solely the complex exponential form of Fourier series and the Fourier transform, which I used in the context of periodic loads, and then in the limit of an infinite period, namely a transient load. From there the relationship between impulse response function and transfer functions arose naturally. In the context of harmonic analysis, I would also demonstrate the great effectiveness of the (virtually unknown) *Exponential Window Method* (in essence, a numerical implementation of the Laplace Transform) for the solution of lightly damped system via complex frequencies, which simultaneously disposes of the problems of added trailing zeroes and undesired periodicity of the response function, and thus ultimately of the "wraparound" problem, that is, causality.

Discrete systems would then take me some two thirds of the second half of the semester, focusing on classical modal analysis and harmonic analysis, and concluding with some lectures on the vibration absorber. This left me just about one third of the half semester (i.e., some two to three weeks) for the treatment of continuous systems, at which time I would introduce the use of Lagrange's equations as a tool to solve continuous media by discretizing those systems via the *Assumed Modes Method*.

In the early version of the class lecture notes I included support motions and ground response spectra as part of the single-DOF lectures. However, as the material dealing with earthquake engineering grew in size and extent, in due time I moved that material out to a separate section, even if I continued to make seamless use of parts of those in my classes.

Beyond lecture materials for the classroom, this book contains extensive materials not included in competing books on structural dynamics, of which there already exist a plethora of excellent choices, and this was the main reason why I decided it was worthwhile to publish it. For this reason, I also expect this book to serve as a valuable reference for practicing engineers, and perhaps just as importantly, to aspiring young PhD graduates with academic aspirations in the fields of structural dynamics, soil dynamics, earthquake engineering, or mechanical vibration.

Last but not least, I wish to acknowledge my significant indebtedness and gratitude to Prof. José Manuel Roësset, now retired from the Texas A&M University, for his most invaluable advice and wisdom over all of the years that have spanned my academic career at MIT. It was while I was a student and José a tenured professor here that I learned with him mechanics and dynamics beyond my wildest expectations and dreams, and it could well be said that everything I know and acquired expertise in is ultimately due to him, and that in a very real sense he has been the ghost writer and coauthor of this book.

> In problems relating to vibrations, nature has provided us with a range of mysteries which for their elucidation require the exercise of a certain amount of mathematical dexterity. In many directions of engineering practice, that vague commodity known as common sense will carry one a long way, but no ordinary mortal is endowed with an inborn instinct for vibrations; mechanical vibrations in general are too rapid for the utilization of our sense of sight, and common sense applied to these phenomena is too common to be other than a source of danger.
>
> C. E. Inglis, FRS, James Forrest Lecture, 1944

Notation and Symbols

Although we may from time to time change the meaning of certain symbols and deviate temporarily from the definitions given in this list, by and large we shall adopt in this book the notation given herein, and we shall do so always in the context of an upright, right-handed coordinate system.

Vectors and matrices: we use **boldface** symbols, while non-boldface symbols (in italics) are scalars. Capital letters denote matrices, and lowercase letters are vectors. (Equivalence with blackboard symbols: $\underset{\sim}{q}$ is the same as $\mathbf{q}$, and $\underline{M}$ is the same as $\mathbf{M}$).

Special Constants (non-italic)

e   Natural base of logarithms = 2.71828182845905...
i   Imaginary unit = $\sqrt{-1}$
π   3.14159265358979...

Roman Symbols

a	Acceleration
$\mathbf{a}$	Acceleration vector
A	Amplitude of a transfer function or a wave; also area or cross section
A_s	Shear area
b	Body load, $b = b(\mathbf{x},t)$
$\mathbf{b}$	Vector of body loads, $\mathbf{b} = \mathbf{b}(\mathbf{x},t)$
c	Viscous damping (dashpot) constant
C_1, C_2	Constants of integration
C_S	Shear wave velocity $\left(\sqrt{G/\rho}\right)$
C_r	Rod wave velocity $\left(\sqrt{E/\rho}\right)$
C_f	Flexural wave velocity $\left(\sqrt{C_r R \omega}\right)$
$\mathbf{C}$	Viscous damping matrix
$\mathbb{C}$	Modally transformed, diagonal damping matrix $(\mathbf{\Phi}^T \mathbf{C} \mathbf{\Phi})$
D	Diameter
f	Frequency in Hz; it may also denote a flexibility

f_d	Damped natural frequency, in Hz
f_n	Natural frequency, in Hz
$\hat{\mathbf{e}}$	Cartesian, unit base vector $\left(\hat{\mathbf{e}}_1, \hat{\mathbf{e}}_2, \hat{\mathbf{e}}_3 \equiv \hat{\mathbf{i}}, \hat{\mathbf{j}}, \hat{\mathbf{k}}\right)$
E	Young's modulus, $E = 2G(1+v)$
E_d	Energy dissipated
E_s	Elastic energy stored
$\hat{\mathbf{g}}$	Curvilinear base vector $\left(\hat{\mathbf{g}}_1, \hat{\mathbf{g}}_2, \hat{\mathbf{g}}_3\right)$
g	Acceleration of gravity
$g(t)$	Unit step-load response function
G	Shear modulus
h	Depth or thickness of beam, element, or plate
$h(t)$	Impulse response function
H	Height
$H(\omega)$	Transfer function (frequency response function for a unit input)
I	Area moment of inertia
j	Most often an index for a generic mode
J	Mass moment of inertia
k	Usually stiffness, but sometimes a wavenumber
k_c	Complex stiffness or impedance
K	Kinetic energy
$\mathbf{K}$	Stiffness matrix
L	Length of string, rod, beam, member, or element
m	Mass
$\mathbf{M}$	Mass matrix
n	Abbreviation for *natural*; also, generic degree of freedom
N	Total number of degrees of freedom
$p(t)$	Applied external force
$\tilde{p}(\omega)$	Fourier transform of $p(t)$, i.e., load in the frequency domain
p_0	Force magnitude
$\mathbf{p}$	External force vector, $\mathbf{p} = \mathbf{p}(t)$
$q(t)$	Generalized coordinate, or modal coordinate
$\mathbf{q}(t)$	Vector of generalized coordinates
r	Tuning ratio $r = \omega / \omega_n$; radial coordinate
$\mathbf{r}$	Radial position vector
R	Radius of gyration or geometric radius
S_a	Ground response spectrum for absolute acceleration (pseudo-acceleration)
S_d	Ground response spectrum for relative displacements
S_v	Ground response spectrum for relative pseudo- velocity
t	Time
t_d	Time duration of load
t_p	Period of repetition of load
T	Period $(= 1/f)$, or duration
T_d	Damped natural period

T_n	Natural period
$u(t)$	Absolute displacement. In general, $u = u(\mathbf{x},t) = u(x,y,z,t)$
$\tilde{u}(\omega)$	Fourier transform of $u(t)$; frequency response function
u_0	Initial displacement, or maximum displacement
$\dot{u}_0$	Initial velocity
u_g	Ground displacement
u_h	Homogeneous solution (free vibration)
u_p	Particular solution
u_{p0}	Initial displacement *value* (not condition!) of particular solution
$\dot{u}_{p0}$	Initial velocity *value* (not condition!) of particular solution
$\mathbf{u}$	Absolute displacement vector
$\dot{\mathbf{u}}$	Absolute velocity vector
$\ddot{\mathbf{u}}$	Absolute acceleration vector
v	Relative displacement (scalar)
$\mathbf{v}$	Relative displacement vector
V	Potential energy; also, magnitude of velocity
V_{ph}	Phase velocity
x,y,z	Cartesian spatial coordinates
$\mathbf{x}$	Position vector
Z	Dynamic stiffness or impedance (ratio of complex force to complex displacement)
$\mathbf{Z}$	Impedance matrix

Greek Symbols

α	Angular acceleration
$\boldsymbol{\alpha}$	Angular acceleration vector
γ	Specific weight; direction cosines; participation factors
$\delta(t)$	Dirac-delta function (singularity function)
Δ	Determinant, or when used as a prefix, finite increment such as Δt
ε	Accidental eccentricity
λ	Lamé constant $\lambda = 2Gv/(1-2v)$; also wavelength $\lambda = V_{ph}/f$
ϕ_{ij}	ith component of jth mode of vibration
ϕ_j	Generic, jth mode of vibration, with components $\phi_j = \{\phi_{ij}\}$
φ	Rotational displacement or degree of freedom
Φ	Modal matrix, $\Phi = \{\phi_j\} = \{\phi_{ij}\}$
θ	Azimuth; rotational displacement, or rotation angle
ρ	Mass density
ρ_w	Mass density of water
ξ	Fraction of critical damping; occasionally dimensionless coordinate
μ	Mass ratio
τ	Time, usually as dummy variable of integration
v	Poisson's ratio

ω Driving (operational) frequency, in radians/second

ω_d Damped natural frequency

ω_n Natural frequency, in rad/s

ω_j Generic jth modal frequency, in rad/s, or generic Fourier frequency

$\boldsymbol{\omega}$ Rotational velocity vector

$\boldsymbol{\Omega}$ Spectral matrix (i.e., matrix of natural frequencies), $\boldsymbol{\Omega} = \left\{ \omega_j \right\}$

Derivatives, Integrals, Operators, and Functions

Temporal derivatives $\dfrac{\partial u}{\partial t} = \dot{u} , \qquad \dfrac{\partial^2 u}{\partial t^2} = \ddot{u}$

Spatial derivatives $\dfrac{\partial u}{\partial x} = u', \qquad \dfrac{\partial^2 u}{\partial x^2} = u''$

Convolution $f * g \equiv f(t) * g(t) = \int_0^T f(\tau) g(t - \tau) d\tau = \int_0^T f(t - \tau) g(\tau) d\tau$

Real and imaginary parts: If $z = x + i\,y$ then $x = \mathrm{Re}(z), \quad y = \mathrm{Im}(z)$.

(Observe that the imaginary part does *not* include the imaginary unit!).

Signum function $\mathrm{sgn}(x - a) = \begin{cases} 1 & x > a \\ 0 & x = a \\ -1 & x < a \end{cases}$

Step load function $\mathcal{H}(t - t_0) = \begin{cases} 1 & t > t_0 \\ \frac{1}{2} & t = t_0 \\ 0 & t < t_0 \end{cases}$

Dirac-delta function $\delta(t - t_0) = \begin{cases} 0 & t > t_0 \\ \infty & t = t_0, \\ 0 & t < t_0 \end{cases} \qquad \int_{t_0 - \varepsilon}^{t_0 + \varepsilon} \delta(t - t_0)\, dt = 1, \qquad \varepsilon > 0$

Kronecker delta $\delta_{ij} = \begin{cases} 1 & i = j \\ 0 & i \neq j \end{cases}$

Split summation $\displaystyle\sum_{j=m}^{n} a_j = \tfrac{1}{2} a_m + a_{m+1} + \cdots + a_{n-1} + \tfrac{1}{2} a_n$ (first and last element halved)

Unit Conversions

Fundamental Units

	Metric				English	
Length	Mass	Time		Length	Force	Time
(m)	(kg)	(s)		(ft)	(lb)	(s)

Length

Distance

1 m = 100 cm = 1000 mm

1 dm = 10 cm = 0.1 m

1 ft	= 12 in.
1 yd	= 3' = 0.9144 m
1 mile	= 5280 ft = 1609.344 m

1 in.	=	2.54 cm
1 ft	=	30.48 cm

Volume

$1 \ dm^3 = 1 \ [l]$

Until 1964, the liter (or *litre*) was defined as the volume occupied by 1 kg of water at 4°C = 1.000028 dm^3. Currently, it is defined as being *exactly* 1 dm^3.

1 gallon	=	231 in^3 (exact!)
1 pint	=	1/8 gallon
	=	½ quart
1 cu-ft	=	1728 cu-in.
	=	7.48052 gallon
1 quart	=	2 pints = 0.03342 ft^3
	=	0.946353 dm^3 (liters)

1 gallon	=	3.785412 dm^3
1 cu-ft	=	28.31685 dm^3
1 pint	=	0.473176 dm^3

Mass

1 (kg)	=	1000 g		1 slug	=	32.174 lb-mass
1 (t)	=	1000 kg (metric ton)			=	14.594 kg
	=	1 Mg				

1 lb-mass	=	0.45359237 kg (exact!)
	=	453.59237 g
1 kg	=	2.2046226 lb-mass

Time

Second (s), also (sec)

Derived Units

Acceleration of Gravity

G	=	9.80665 m/s^2	(exact normal value!)	G	=	32.174 ft/s^2
	=	980.665 cm/s^2	(gals)		=	386.09 in./s^2

Useful approximation: g (in m/s^2) $\approx \pi^2 = 9.8696 \approx 10$

Density and Specific Weight

1 kg/dm^3	=	1000 kg/m^3	=	62.428 lb/ft^3
			=	8.345 lb/gal
			=	1.043 lb/pint

1 ounce/ft^3 = 1.0012 kg/m^3 (an interesting near coincidence!)

Some specific weights and densities (approximate values):

		Spec. weight		Density
Steel	=	490 lb/ft^3	=	7850 kg/m^3
Concrete	=	150 lb/ft^3	=	2400 kg/m^3
Water	=	62.4 lb/ft^3	=	1000 kg/m^3
Air	=	0.0765 lb/ft^3	=	1.226 kg/m^3

Force

1 N [Newton] = force required to accelerate 1 kg by 1 m/s^2

9.81 N	=	1 kg-force	= 1 kp ("kilopond"; widely used in Europe in the past, it is a metric, non-SSI unit!)
1 lb	=	4.44822 N	(related unit: 1 Mp =1 "megapond" = 1 metric ton)
		0.45359 kg-force	

Pressure

1 Pa	=	1 N/m^2	Normal atmospheric pressure	= 1.01325 bar (15°C, sea level)
1 kPa	=	10^3 Pa		= 101.325 kPa (exact!)
1 bar	=	10^5 Pa		= 14.696 lb/in^2 = 0.014696 ksi
1 ksi	=	6.89476 MPa		= 2116.22 lb/ft^2

Power

1 kW = 1000 W = 1 kN-m/s
= 0.948 BTU/s

1 HP = 550 lb-ft/s = 0.707 BTU/s = 0.7457 kW
1 BTU/s = 778.3 lb-ft/s = 1.055 KW
1 CV = 75 kp × m/s = 0.7355 kW "Cheval Vapeur"

Temperature

$T(°F) = 32 + 9/5\ T(°C)$ (some exact values: –40°F = –40°C, 32°F = 0°C and 50°F = 10°C)

1 Fundamental Principles

1.1 Classification of Problems in Structural Dynamics

As indicated in the list that follows, the study of structural dynamics – and books about the subject – can be organized and classified according to various criteria. This book follows largely the first of these classifications, and with the exceptions of nonlinear systems, addresses all of these topics.

(a) By the number of degrees of freedom:

$$
\begin{cases}
\textbf{Single DOF} \\
\textbf{Multiple DOFs} \begin{cases} \text{lumped mass (discrete) system (finite DOF)} \\ \text{continuous systems (infinitely many DOF)} \end{cases}
\end{cases}
$$

Discrete systems are characterized by systems of ordinary differential equations, while continuous systems are described by systems of partial differential equations.

(b) By the linearity of the governing equations:

$$
\begin{cases}
\textbf{Linear systems}\,(\text{linear elasticity, small motions assumption}) \\
\textbf{Nonlinear systems} \begin{cases} \text{conservative (elastic) systems} \\ \text{nonconservative (inelastic) systems} \end{cases}
\end{cases}
$$

(c) By the type of excitation:

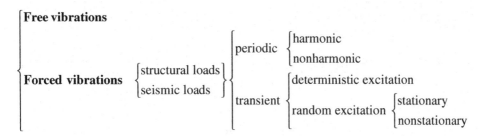

(d) By the type of mathematical problem:

$\begin{bmatrix}\textbf{Static} \rightarrow \text{boundary value problems} \\ \textbf{Dynamic} \begin{cases}\text{eigenvalue problems (free vibrations)} \\ \text{initial value problem, propagation problem (waves)}\end{cases}\end{bmatrix}$

(e) By the presence of energy dissipating mechanisms:

$\begin{bmatrix}\textbf{Undamped vibrations} \\ \textbf{Damped vibrations} \begin{cases}\text{viscous damping} \\ \text{hysteretic damping} \\ \text{Coulomb damping} \\ \text{etc.}\end{cases}\end{bmatrix}$

1.2 Stress–Strain Relationships

1.2.1 Three-Dimensional State of Stress–Strain

$$\begin{Bmatrix}\varepsilon_x \\ \varepsilon_y \\ \varepsilon_z\end{Bmatrix} = \frac{1}{E}\begin{bmatrix}1 & -v & -v \\ -v & 1 & -v \\ -v & -v & 1\end{bmatrix}\begin{Bmatrix}\sigma_x \\ \sigma_y \\ \sigma_z\end{Bmatrix}, \qquad \begin{aligned}E &= 2G(1+v) = \text{ Young's modulus} \\ v &= \text{Poisson's ratio}\end{aligned} \tag{1.1}$$

$$\begin{Bmatrix}\sigma_x \\ \sigma_y \\ \sigma_z\end{Bmatrix} = \frac{2G}{1-2v}\begin{bmatrix}1-v & v & v \\ v & 1-v & v \\ v & v & 1-v\end{bmatrix}\begin{Bmatrix}\varepsilon_x \\ \varepsilon_y \\ \varepsilon_z\end{Bmatrix}, \qquad \lambda = \frac{2Gv}{1-2v} = \text{ Lamé constant} \tag{1.2}$$

$$\tau_{xy} = G\gamma_{xy}, \qquad \tau_{xz} = G\gamma_{xz}, \qquad \tau_{yz} = G\gamma_{yz} \tag{1.3}$$

1.2.2 Plane Strain

$$\begin{Bmatrix}\varepsilon_x \\ \varepsilon_y\end{Bmatrix} = \frac{1}{2G}\begin{bmatrix}1-v & -v \\ -v & 1-v\end{bmatrix}\begin{Bmatrix}\sigma_x \\ \sigma_y\end{Bmatrix} \tag{1.4}$$

$$\begin{Bmatrix}\sigma_x \\ \sigma_y\end{Bmatrix} = \frac{2G}{1-2v}\begin{bmatrix}1-v & v \\ v & 1-v\end{bmatrix}\begin{Bmatrix}\varepsilon_x \\ \varepsilon_y\end{Bmatrix} \tag{1.5}$$

$$\varepsilon_z = \gamma_{xz} = \gamma_{yz} = 0 \qquad \sigma_z = v(\sigma_x + \sigma_y) = \lambda(\varepsilon_x + \varepsilon_y), \qquad \tau_{xy} = G\gamma_{xy} \tag{1.6}$$

1.2.3 Plane Stress

$$\begin{Bmatrix}\varepsilon_x \\ \varepsilon_y\end{Bmatrix} = \frac{1}{E}\begin{bmatrix}1 & -v \\ -v & 1\end{bmatrix}\begin{Bmatrix}\sigma_x \\ \sigma_y\end{Bmatrix} = \frac{1}{2G(1+v)}\begin{bmatrix}1 & -v \\ -v & 1\end{bmatrix}\begin{Bmatrix}\sigma_x \\ \sigma_y\end{Bmatrix} \tag{1.7}$$

$$\begin{Bmatrix} \sigma_x \\ \sigma_y \end{Bmatrix} = \frac{2G}{1-v} \begin{Bmatrix} 1 & v \\ v & 1 \end{Bmatrix} \begin{Bmatrix} \varepsilon_x \\ \varepsilon_y \end{Bmatrix} \tag{1.8}$$

$$\sigma_z = \tau_{xz} = \tau_{yz} = 0, \qquad \varepsilon_z = -\frac{v}{1-v}\left(\varepsilon_x + \varepsilon_y\right) = -\frac{v}{E}\left(\sigma_x + \sigma_y\right), \qquad \tau_{xy} = G\gamma_{xy} \tag{1.9}$$

1.2.4 Plane Stress versus Plane Strain: Equivalent Poisson's Ratio

We explore here the possibility of defining a *plane strain* system with Poisson's ratio $\tilde{v}$ such that it is equivalent to a *plane stress* system with Poisson's ratio v while having the same shear modulus G. Comparing the stress–strain equations for plane stress and plane strain, we see that this would require the simultaneous satisfaction of the two equations

$$\frac{1-\tilde{v}}{1-2\tilde{v}} = \frac{1}{1-v} \quad \text{and} \quad \frac{\tilde{v}}{1-2\tilde{v}} = \frac{v}{1-v} \tag{1.10}$$

which are indeed satisfied if

$$\boxed{\tilde{v} = \frac{v}{1+v}} \quad \text{plane strain ratio } \tilde{v} \text{ that is equivalent to the plane-stress ratio } v \tag{1.11}$$

Hence, it is always possible to map a plane-stress problem into a plane-strain one.

1.3 Stiffnesses of Some Typical Linear Systems

Notation

E = Young's modulus
G = shear modulus
v = Poisson's ratio
A = cross section
A_s = shear area
I = area moment of inertia about bending axis
L = length

Linear Spring

Longitudinal spring $\quad k_x$

Helical (torsional) spring $\quad k = \dfrac{G\,d^4}{8\,n\,D^3}$

where

$\quad\quad d$ = wire diameter

$\quad\quad D$ = mean coil diameter

$\quad\quad n$ = number of turns

Figure 1.1

Member stiffness matrix = $\mathbf{K} = \begin{Bmatrix} k & -k \\ -k & k \end{Bmatrix}$

Rotational Spring

Rotational stiffness $\qquad k_\theta$

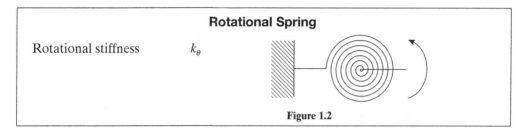

Figure 1.2

Floating Body

Buoyancy stiffness $\quad k_z = \rho_w\, g\, A_w \qquad$ Heaving (up and down)

$\qquad\qquad\qquad\quad k_{\theta x} = \rho_w\, g\, I_w \qquad$ Rolling about x-axis (small rotations)

$\qquad\qquad\qquad\quad k_{x,y} = 0 \qquad\qquad$ Lateral motion

in which ρ_w = mass density of water; g = acceleration of gravity; A_w = horizontal cross section of the floating body at the level of the water line; and I_w= area moment of inertia of A_w with respect to the rolling axis (x here). If the floating body's lateral walls in contact with the water are not vertical, then the heaving stiffness is valid only for small vertical displacements (i.e., displacements that cause only small changes in A_w). In addition, the rolling stiffness is just an approximation for small rotations, even with vertical walls. Thus, the buoyancy stiffnesses given earlier should be interpreted as tangent stiffnesses or incremental stiffnesses.

Observe that the weight of a floating body equals that of the water displaced.

Cantilever Shear Beam

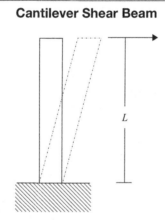

Figure 1.3

A shear beam is infinitely stiff in rotation, which means that no rotational deformation exists. However, a free (i.e., unrestrained) shear beam may rotate as a rigid body. After deformation, sections remain parallel. If the cantilever beam is subjected to a moment at its free end, the beam will remain undeformed. The moment is resisted by an equal and opposite moment at the base. If a force P acts at an elevation $a \leq L$ above the base, the lateral displacement increases linearly from zero to $u = Pa / GA_s$ and remains constant above that elevation.

Typical shear areas : Rectangular section $A_s = \frac{5}{6} A$

Ring $A_s = \frac{1}{2} A$

Stiffness as perceived at the top:

Axial stiffness $k_z = \dfrac{EA}{L}$

Transverse stiffness $k_x = \dfrac{GA_s}{L}$

Rotational stiffness $k_\varphi = \infty$

Cantilever Bending Beam

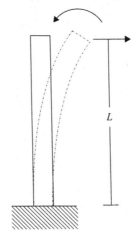

L

Figure 1.4

Stiffnesses as perceived at the top:

Axial stiffness $k_z = \dfrac{EA}{L}$

Transverse stiffness (rotation positive counterclockwise):

(a) Free end $k_x = \dfrac{3EI}{L^3}$ (rotation unrestrained)

$k_\phi = \dfrac{EI}{L}$ (translation unrestrained)

(b) Constrained end $k_{xx} = \dfrac{12EI}{L^3}$ $k_{x\varphi} = \dfrac{6EI}{L^2}$

$$k_{\varphi x} = \frac{6EI}{L^2} \qquad k_{\varphi\varphi} = \frac{4EI}{L}$$

$$\mathbf{K}_{BB} = \frac{EI}{L^3} \begin{Bmatrix} 12 & 6L \\ 6L & 4L^2 \end{Bmatrix}$$

Notice that carrying out the static condensations $k_x = k_{xx} - k_{x\varphi}^2 / k_{\varphi\varphi}$ and $k_\varphi = k_{\varphi\varphi} - k_{x\varphi}^2 / k_{xx}$ we recover the stiffness k_x, k_φ for the two cases when the loaded end is free to rotate or translate.

Transverse Flexibility of Cantilever Bending Beam

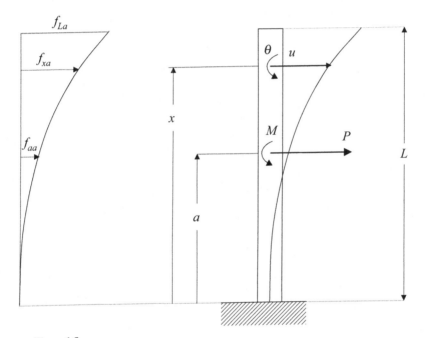

Figure 1.5

A lateral load P and a *counterclockwise* moment M are applied simultaneously to a cantilever beam at some arbitrary distance a from the support (Figure 1.5). These cause in turn a transverse displacement u and a counterclockwise rotation θ at some other distance x from the support. These are

$$u(x) = \frac{P}{6EI} x^2 (3a - x) - \frac{M}{2EI} x^2 \qquad x \le a$$

$$\theta(x) = -\frac{P}{2EI} x(2a - x) + \frac{M}{EI} x \qquad x \le a$$

and

$$u(x) = \frac{P}{6EI}a^2(3x-a) - \frac{M}{2EI}a(2x-a) \qquad x \geq a$$

$$\theta(\xi) = -\frac{P}{2EI}a^2 + \frac{M}{EI}a \qquad\qquad x \geq a$$

Transverse Flexibility of Simply Supported Bending Beam

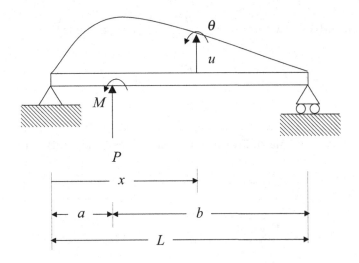

Figure 1.6

A transverse load P and a counterclockwise moment M are applied to a simply supported beam at some arbitrary distance a from the support (Figure 1.6). These cause in turn a transverse displacement u and rotation θ (positive up and counterclockwise, respectively) at some other distance x from the support. Defining the dimensionless coordinates $\alpha = a/L$, $\beta = 1-\alpha$ and $\xi = x/L$, the observed displacement and rotation are:

$$u(x) = \frac{PL^3}{6EI}\beta\xi(1-\beta^2-\xi^2) + \frac{ML^2}{6EI}\xi(\xi^2+3\beta^2-1), \qquad\qquad \xi \leq \alpha$$

$$\theta(x) = \frac{PL^2}{6EI}\beta(1-\beta^2-3\xi^2) + \frac{ML}{6EI}(3\xi^2+3\beta^2-1), \qquad\qquad \xi \leq \alpha$$

and

$$u(x) = \frac{PL^3}{6EI}\alpha(1-\xi)\left[1-\alpha^2-(1-\xi)^2\right] + \frac{ML^2}{6EI}(1-\xi)\left[1-3\alpha^2-(1-\xi)^2\right] \qquad \xi \geq \alpha$$

$$\theta(x) = \frac{PL^2}{6EI}\alpha\left[3\left(1-\xi\right)^2 + \alpha^2 - 1\right] + \frac{ML}{6EI}\left[3\left(1-\xi\right)^2 + 3\alpha^2 - 1\right] \qquad \xi \ge \alpha$$

In particular, at $\xi = \alpha$

$$u(\alpha) = \frac{PL^3}{3EI}\alpha^2\beta^2 + \frac{ML^2}{3EI}\alpha\beta(\beta - \alpha)$$

$$\theta(\alpha) = \frac{PL^2}{3EI}\alpha\beta(\beta - \alpha) + \frac{ML}{3EI}(1 - 3\alpha\beta)$$

Transverse stiffness at $x = a$ $\qquad k = \dfrac{3EIL}{a^2 b^2}$ $\qquad$ (rotation permitted)

Special case : $a = b = L/2$ $\qquad k = \dfrac{48EI}{L^3}$ $\qquad$ (rotation permitted)

Stiffness and Inertia of Free Beam with Shear Deformation Included

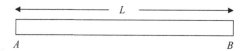

Figure 1.7

Consider a free, homogeneous bending beam AB of length L that lies horizontally in the vertical plane $x - y$ and deforms in that plane, as shown in Figure 1.7. It has mass density ρ, Poison's ratio v, shear modulus G, Young's modulus $E = 2G(1+v)$, area-moment of inertia I_z about the horizontal bending axis z (i.e., bending in the plane), cross section A_x, and shear area A_{sy} for shearing in the transverse y direction. In the absence of further information about shear deformation, one can choose either $A_{sy} = A_x$ or $A_{sy} = \infty$. Define

$$m = \rho A_x L, \quad j_z = \rho I_z / L = m\left(\frac{R_z}{L}\right)^2, \quad R_z = \sqrt{\frac{I_z}{A_x}}, \quad \phi_z = \frac{12EI_z}{GA_{sy}L^2} = 24(1+v)\frac{I_z}{A_{sy}L^2}$$

then from the theory of finite elements we obtain the *bending* stiffness matrix $\mathbf{K}_B$ for a bending beam with shear deformation included, together with the consistent bending mass matrix $\mathbf{M}_B$, which accounts for both translational as well as rotational inertia (rotations positive counterclockwise):

$$\mathbf{K}_B = \frac{EI_z}{(1+\phi_z)L^3}\begin{Bmatrix} 12 & 6L & -12 & 6L \\ 6L & (4+\phi_z)L^2 & -6L & (2-\phi_z)L^2 \\ -12 & -6L & 12 & -6L \\ 6L & (2-\phi_z)L^2 & -6L & (4+\phi_z)L^2 \end{Bmatrix}$$

$$\mathbf{M}_B = \frac{m}{420} \begin{bmatrix} 156 & 22L & 54 & -13L \\ 22L & 4L^2 & 13L & -3L^2 \\ 54 & 13L & 156 & -22L \\ -13L & -3L^2 & -22L & 4L^2 \end{bmatrix} + \frac{j_z}{30} \begin{bmatrix} 36 & 3L & -36 & 3L \\ 3L & 4L^2 & -3L & -L^2 \\ -36 & -3L & 36 & -3L \\ 3L & -L^2 & -3L & 4L^2 \end{bmatrix}$$

On the other hand, the axial degrees of freedom (when the member acts as a column) have *axial* stiffness and mass matrices

$$\mathbf{K}_A = \frac{EA_x}{L} \begin{Bmatrix} 1 & -1 \\ -1 & 1 \end{Bmatrix}, \qquad \mathbf{M}_A = \frac{\rho ab}{6} \begin{Bmatrix} 2 & 1 \\ 1 & 2 \end{Bmatrix}$$

The *local* stiffness and mass matrices $\mathbf{K}_L, \mathbf{M}_L$ of a beam column are constructed by appropriate combinations of the bending and axial stiffness and mass matrices. These must be rotated appropriately when the members have an arbitrary orientation, after which we obtain the *global* stiffness and mass matrices.

Inhomogeneous, Cantilever Bending Beam

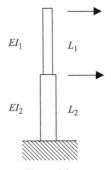

Figure 1.8

Active DOF are the two transverse displacements at the top ($j = 1$) and the junction ($j = 2$). Rotations are passive (slave) DOF.

Define the member stiffnesses $\quad S_j = \dfrac{3EI_j}{L_j^3}, \quad j = 1,2$

The elements $k_{11}, k_{12,} k_{21}, k_{22}$ of the lateral stiffness matrix are then

$$k_{11} = \frac{S_1}{1 + \dfrac{3S_1}{4S_2}\left(\dfrac{L_1}{L_2}\right)^2} \qquad k_{12} = -k_{11}\left(1 + \frac{3}{2}\frac{L_1}{L_2}\right)$$

$$k_{21} = -k_{11}\left(1 + \frac{3}{2}\frac{L_1}{L_2}\right) \qquad k_{22} = k_{11}\left[1 + \frac{S_2}{S_1} + 3\frac{L_1}{L_2} + 3\left(\frac{L_1}{L_2}\right)^2\right]$$

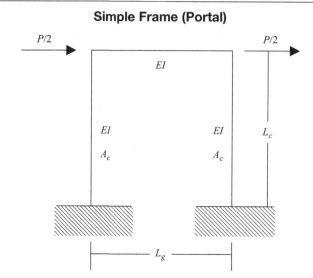

Simple Frame (Portal)

Figure 1.9

Active DOF is lateral displacement of girder.
The lateral stiffness of the frame is

$$k = \frac{24\,EI_c}{L_c^3}\left\{\frac{1+\frac{1}{6}\frac{I_c}{I_g}\frac{L_g}{L_c}+4\frac{I_c}{A_cL_g^2}}{1+\frac{2}{3}\frac{I_c}{I_g}\frac{L_g}{L_c}+16\frac{I_c}{A_cL_g^2}}\right\}$$

Includes axial deformation of columns. The girder is axially rigid.

Rigid Circular Plate on Elastic Foundation

To a first approximation and for sufficiently high excitation frequencies, a rigid circular plate on an elastic foundation behaves like a set of springs and dashpots. Here, $C_s = \sqrt{G/\rho}$ = shear wave velocity; R = radius of foundation; G = shear modulus of soil; v = Poisson's ratio of soil; and Rocking = rotation about a horizontal axis.

Table 1.1   Stiffness and damping for circular foundation

Direction	Stiffness	Dashpot
Horizontal	$k_x = \frac{8GR}{2-v}$	$c_x = 0.6\frac{k_xR}{C_s}$
Vertical	$k_z = \frac{4GR}{1-v}$	$c_z = 0.8\frac{k_zR}{C_s}$
Rocking	$k_r = \frac{8GR^3}{3(1-v)}$	$c_r = 0.3\frac{k_rR}{C_s}$
Torsion	$k_t = \frac{16GR^3}{3}$	$c_t = 0.3\frac{k_tR}{C_s}$

1.4 Rigid Body Condition of Stiffness Matrix

We show herein how the stiffness matrix of a one-dimensional member AB, such as a beam, can be inferred from the stiffness submatrix of the ending node B by means of the rigid body condition. Let $\mathbf{K}$ be the stiffness matrix of the complete member whose displacements (or degrees of freedom) $\mathbf{u}_A, \mathbf{u}_B$ are defined at the two ends A, B with coordinates $\mathbf{x}_A, \mathbf{x}_B$. The stiffness matrix of this element will be of the form

$$\mathbf{K} = \begin{Bmatrix} \mathbf{K}_{AA} & \mathbf{K}_{AB} \\ \mathbf{K}_{BA} & \mathbf{K}_{BB} \end{Bmatrix} \tag{1.12}$$

Now, define a set of six rigid body displacements relative to any one end, say B, which consists of three rotations and three translations. The displacements at the other end are then

$$\mathbf{u}_A = \mathbf{R}\mathbf{u}_B \tag{1.13}$$

where

$$\mathbf{R} = \begin{Bmatrix} 1 & 0 & 0 & 0 & +\Delta z & -\Delta y \\ 0 & 1 & 0 & -\Delta z & 0 & +\Delta x \\ 0 & 0 & 1 & +\Delta y & -\Delta x & 0 \\ 0 & 0 & 0 & 1 & 0 & 0 \\ 0 & 0 & 0 & 0 & 1 & 0 \\ 0 & 0 & 0 & 0 & 0 & 1 \end{Bmatrix}, \qquad \mathbf{R}^{-1} = \begin{Bmatrix} 1 & 0 & 0 & 0 & -\Delta z & +\Delta y \\ 0 & 1 & 0 & +\Delta z & 0 & -\Delta x \\ 0 & 0 & 1 & -\Delta y & +\Delta x & 0 \\ 0 & 0 & 0 & 1 & 0 & 0 \\ 0 & 0 & 0 & 0 & 1 & 0 \\ 0 & 0 & 0 & 0 & 0 & 1 \end{Bmatrix}, \tag{1.14}$$

and

$$\Delta x = x_A - x_B, \qquad \Delta y = y_A - y_B, \qquad \Delta z = z_A - z_B \tag{1.15}$$

On the other hand, a rigid body motion of a free body requires no external forces, so

$$\begin{Bmatrix} \mathbf{K}_{AA} & \mathbf{K}_{AB} \\ \mathbf{K}_{BA} & \mathbf{K}_{BB} \end{Bmatrix} \begin{Bmatrix} \mathbf{R} \\ \mathbf{I} \end{Bmatrix} = \begin{Bmatrix} \mathbf{O} \\ \mathbf{O} \end{Bmatrix} \tag{1.16}$$

Hence

$$\boxed{\mathbf{K}_{AB} = -\mathbf{K}_{AA}\mathbf{R}} \tag{1.17}$$

Also,

$$\mathbf{K}_{BA} = \mathbf{K}_{AB}^T = -\mathbf{R}^T\mathbf{K}_{AA}^T = -\mathbf{R}^T\mathbf{K}_{AA}, \qquad \mathbf{K}_{BB} = -\mathbf{K}_{BA}\mathbf{R} \tag{1.18}$$

so

$$\boxed{\mathbf{K}_{BB} = \mathbf{R}^T\mathbf{K}_{AA}\mathbf{R}} \tag{1.19}$$

For an upright member with A straight above B, then $\Delta x = \Delta y = 0$, $\Delta z = h =$ height of floor. Thus, the submatrices $\mathbf{K}_{AB}, \mathbf{K}_{BA}, \mathbf{K}_{BB}$ depend explicitly on the elements of $\mathbf{K}_{AA}$.

1.5 **Mass Properties of Rigid, Homogeneous Bodies**

Mass density is denoted by ρ, mass by m, mass moments of inertia (rotational inertia) by J, and subscripts identify axes. Unless otherwise stated, *all mass moments of inertia are relative to principal, centroidal axes!* (i.e., products of inertia are assumed to be zero).

Arbitrary Body

$$m = \int_{\text{Vol}} \rho \, dV$$

$$J_{xx} = \int_{\text{Vol}} \rho\left(y^2 + z^2\right) dV \qquad J_{yy} = \int_{\text{Vol}} \rho\left(x^2 + z^2\right) dV \qquad J_{zz} = \int_{\text{Vol}} \rho\left(x^2 + y^2\right) dV$$

Parallel Axis Theorem (Steiner's Theorem)

$$J_{\text{parallel}} = J_{\text{centroid}} + md^2$$

in which d = distance from centroidal axis to parallel axis. It follows that the mass moments of inertia about the centroid are the smallest possible.

Cylinder

$$m = \rho \pi R^2 H$$

$$J_{xx} = J_{yy} = m\left(\frac{R^2}{4} + \frac{H^2}{12}\right)$$

$$J_{zz} = m\frac{R^2}{2}$$

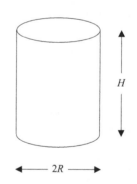

Figure 1.10

Rectangular Prism

$$m = \rho \, abc$$

$$J_{xx} = \tfrac{1}{12}m\left(b^2 + c^2\right)$$

$$J_{yy} = \tfrac{1}{12}m\left(a^2 + c^2\right)$$

$$J_{zz} = \tfrac{1}{12}m\left(a^2 + b^2\right)$$

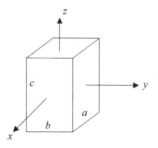

Figure 1.11

Thin Disk

$$J_{xx} = J_{yy} = m\frac{R^2}{4}$$

$$J_{zz} = m\frac{R^2}{2}$$

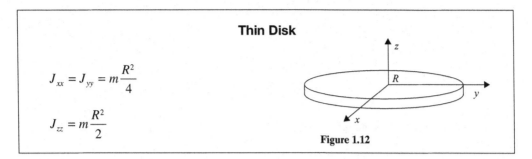

Figure 1.12

Thin Bar

$$J_{xx} \approx 0$$

$$J_{yy} = J_{zz} = \tfrac{1}{12}mL^2$$

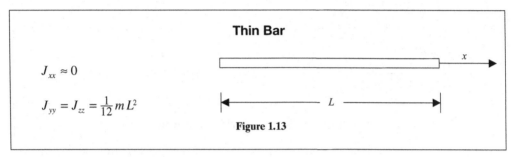

Figure 1.13

Sphere

$$V = \tfrac{4}{3}\pi R^3$$

$$A_{surf} = 4\pi R^2$$

$$J_{xx} = J_{yy} = J_{zz} = \tfrac{2}{5}mR^2$$

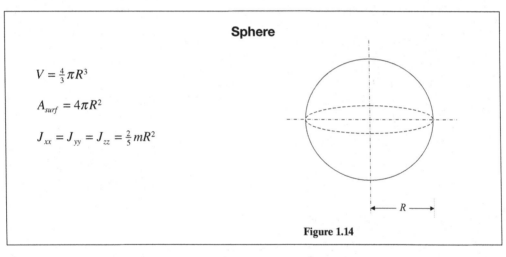

Figure 1.14

Ellipsoid

$$V = \tfrac{4}{3}\pi abc$$

$$J_{xx} = \tfrac{1}{5}m\left(b^2 + c^2\right)$$

$$J_{yy} = \tfrac{1}{5}m\left(a^2 + c^2\right)$$

$$J_{zz} = \tfrac{1}{5}m\left(a^2 + b^2\right)$$

$$J_{xy} = J_{yz} = J_{zx} = 0$$

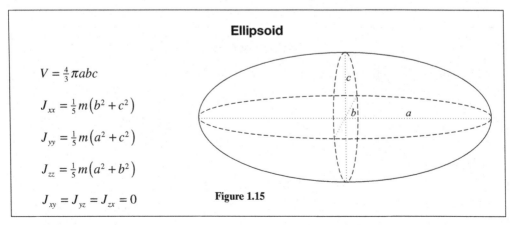

Figure 1.15

Cone

$$V = \frac{1}{3}\pi R^2 H$$

$e = \frac{1}{4}H$ = height of center of gravity

$$J_x = J_y = \frac{3}{80}m\left(4R^2 + H^2\right)$$

$$J_z = \frac{3}{10}mR^2$$

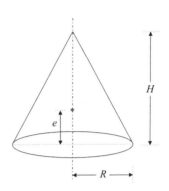

Figure 1.16

Prismatic Solids of Various Shapes

Let A, I_{xx}, I_{yy}, be the area and moments of inertia of the cross section, and H the height of the prism. The mass and mass-moments of inertia are then

$$m = \rho A H$$

$$J_{xx} = m\left(\frac{I_{xx}}{A} + \frac{H^2}{12}\right)$$

$$J_{yy} = m\left(\frac{I_{yy}}{A} + \frac{H^2}{12}\right)$$

$$J_{zz} = m\left(\frac{I_{xx} + I_{yy}}{A}\right)$$

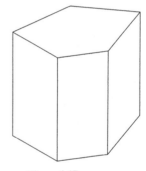

Figure 1.17

The values of I_{xx}, I_{yy}, I_{zz} for various shapes are as follows:

Triangle (Isosceles)

Height of center of gravity above base of triangle = $h/3$

$$A = \tfrac{1}{2}a^2 h$$

$$I_{xx} = \tfrac{1}{18}Ah^2$$

$$I_{yy} = \tfrac{1}{8}Aa^2$$

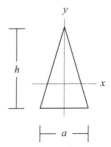

Figure 1.18

Circular Segment

$A = \frac{1}{2} R^2 (\alpha - \sin \alpha)$

$e = \dfrac{c^3}{12A}$

$c = 2R \sin \frac{1}{2}\alpha$

$I_{xx} = \frac{1}{16} R^4 (2\alpha - \sin 2\alpha) - Ae^2$

$I_{yy} = \frac{1}{48} R^4 (6\alpha - 8 \sin \alpha + \sin 2\alpha)$

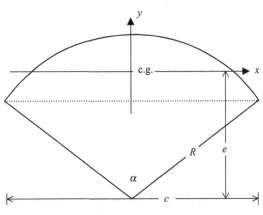

Figure 1.19

Circular Arc

t = thickness of arc

$A = \alpha t R$

$e = R \dfrac{\sin \frac{1}{2}\alpha}{\frac{1}{2}\alpha}$

$I_{xx} = \frac{1}{2} A R^2 \left(1 + \dfrac{\sin \alpha}{\alpha}\right) - Ae^2$

$I_{yy} = \frac{1}{2} A R^2 \left(1 - \dfrac{\sin \alpha}{\alpha}\right)$

For ring, set $\alpha = 2\pi$, $\sin \alpha = 0$.

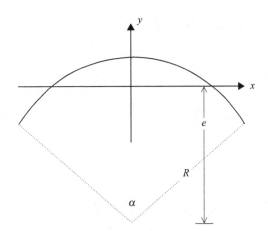

Figure 1.20

Arbitrary, Closed Polygon

Properties given are for arbitrary axes. Use Steiner's theorem to transfer to center of gravity. The polygon has n vertices, numbered in *counterclockwise* direction.
   The vertices have coordinates (x_1, y_1) through (x_n, y_n). By definition,

$x_{n+1} \equiv x_1 \qquad y_{n+1} \equiv y_1$

$x_0 \equiv x_n \qquad y_0 \equiv y_n$

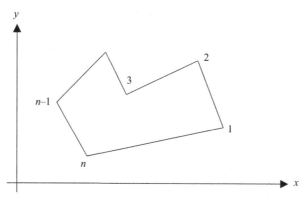

Figure 1.21

The polygon need not be convex. To construct a polygon with one or more interior holes, connect the exterior boundary with the interior, and number the interior nodes in *clockwise* direction.

$$A = \tfrac{1}{2} \sum_{i=1}^{n} x_i \left(y_{i+1} - y_{i-1} \right) = \tfrac{1}{2} \sum_{i=1}^{n} y_i \left(x_{i-1} - x_{i+1} \right)$$

$$x_{CG} = \frac{1}{6A} \sum_{i=1}^{n} y_i \left(x_{i-1} - x_{i+1} \right) \left(x_{i-1} + x_i + x_{i+1} \right)$$

$$y_{CG} = \frac{1}{6A} \sum_{i=1}^{n} x_i \left(y_{i-1} - y_{i+1} \right) \left(y_{i-1} + y_i + y_{i+1} \right)$$

$$I_{xx} = \frac{1}{24} \sum_{i=1}^{n} x_i \left(y_{i+1} - y_{i-1} \right) \left[\left(y_{i-1} + y_i + y_{i+1} \right)^2 + y_{i-1}^2 + y_i^2 + y_{i+1}^2 \right]$$

$$I_{yy} = \frac{1}{24} \sum_{i=1}^{n} y_i \left(x_{i-1} - x_{i+1} \right) \left[\left(x_{i-1} + x_i + x_{i+1} \right)^2 + x_{i-1}^2 + x_i^2 + x_{i+1}^2 \right]$$

$$I_{xy} = \frac{1}{24} \sum_{i=1}^{n} \left(x_i y_{i+1} - x_{i+1} y_i \right) \left[\left(2x_i + x_{i+1} \right) y_i + \left(x_i + 2x_{i+1} \right) y_{i+1} \right]$$

Solids of Revolution Having a Polygonal Cross Section

Consider a homogeneous solid that is obtained by rotating a polygonal area about the x-axis. Using the same notation as in the previous section, and defining ρ as the mass density of the solid, the volume, the x-coordinate of the centroid, and the mass moments inertia of the solid are as indicated below. Again, the polygon need not be convex, and holes in the body can be accomplished with interior nodes numbered in the clockwise direction.

$$V = \tfrac{\pi}{3} \sum_{i=1}^{n} x_i \left(y_{i-1} - y_{i+1} \right) \left(y_{i-1} + y_i + y_{i+1} \right)$$

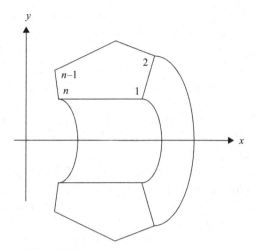

Figure 1.22

$$x_{CG} = \frac{\pi}{12V} \sum_{i=1}^{n} (x_i y_{i+1} - x_{i+1} y_i)\left[(2x_i + x_{i+1})y_i + (x_i + 2x_{i+1})y_{i+1}\right]$$

$$J_{xx} = \rho \frac{\pi}{10} \sum_{i=1}^{n} (x_i y_{i+1} - x_{i+1} y_i)(y_i + y_{i+1})(y_i^2 + y_{i+1}^2)$$

$$J_{yy} = \frac{1}{2}J_{xx} + \rho\frac{\pi}{30} \sum_{i=1}^{n} (x_i y_{i+1} - x_{i+1} y_i)\left[(x_i + x_{i+1})^2(y_i + y_{i+1}) + 2(x_i^2 y_i + x_{i+1}^2 y_{i+1})\right]$$

1.6 Estimation of Miscellaneous Masses

1.6.1 Estimating the Weight (or Mass) of a Building

Although the actual weight of a building depends on a number of factors, such as the geometry, the materials used and the physical dimensions, there are still simple rules that can be used to estimate the weight of a building, or at least arrive at an order of magnitude for what that weight should be, and this without using any objective information such as blueprints.

First of all, consider the fact that about 90% of a building's volume is just air, or else it would not be very useful. Thus, the solid part (floors, walls, etc.) occupies only some 10% by volume. Also, the building material is a mix of concrete floors, steel columns, glass panes, light architectural elements, furniture, people, and so on that together and on average weigh some 2.5 ton/m³ (150 pcf). Accounting for the fact that the solid part is only 10%, that gives an average density for the building of $\rho_{av} = 0.25$ ton/m³ ($\gamma_{av} = 15$ pcf), which is the density we need to use in our estimation.

Second, the typical high-rise building has an inter-story height of about $h = 4$ m (~13 ft), so if N is the total number of stories, then the building has a height of $H = Nh$. For example, the Prudential Building (PB) in Boston has $N = 52$ stories, so its height is $H = 52 \times 4 = 208$ meters (alternatively, $H = 60 \times 13 = 780$ ft; its actual height is given below).

Third, although you know only the number of stories, but not necessarily the base dimensions, you could still arrive at an excellent guess by observing the building of interest from the distance. Indeed, just by looking at it you can estimate by eye its *aspect ratio*, that is, how much taller than wider it is. Again, for the Prudential (or the John Hancock) Building in Boston, you would see that it is some five times taller than it is wide. Thus, the base dimension is $A = H / (\text{aspect})$, which in our example of the PB would be $A = 208 / 5 = 41.6$ m (~136 ft). Proceeding similarly for the other base dimension B, you will know (or have estimated) the length and width A, B as well as the height H. The volume is then simply $V = ABH$.

The fourth and last step is to estimate the mass or weight: You multiply the volume times the average density, and presto, there you have it. Mass: $M = \rho_{av}V = \rho_{av}ABH$. Continuing with our example of the PB, and realizing that the PB is nearly square in plane view, then $A = B$, in which case we estimate its weight as $W = 0.25 \times 41.6 \times 41.6 \times 208 = 89,989 \sim 100,000$ ton. Yes, the estimated weight may well be off from the true value by some 20% or even 30%, but the order of magnitude is correct, that is, the weight of the PB is indeed on the order of 100,000 ton.

For the sake of completeness: the actual height of the PB is 228 m (749 ft), which is close to our estimate. However, the aspect ratio, as crudely inferred from photos, is about 4.5, and not quite 5, as we had surmised by simple visual inspection from the distance.

And being on the subject of tall buildings, we mention in passing that a widely used rule of thumb (among other more complex ones) to predict the fundamental period of vibration of steel-frame buildings in the United States is $T = 0.1N$, that is, 10% of the number of stories. Hence, the John Hancock building in Boston, which has $N = 60$ stories, would have an estimated period of vibration of about 6 s (its true period is somewhat longer). Other types of buildings, especially masonry and concrete shear-wall buildings, are stiffer and tend to have shorter periods. All of these formulas are very rough and exhibit considerable scatter when compared to experimentally measured periods, say ambient vibrations caused by wind, so they should not be taken too seriously. Nonetheless, in the absence of other information they are very useful indeed.

1.6.2 Added Mass of Fluid for Fully Submerged Tubular Sections

Table 1.2 gives the added (or participating) mass of water per unit length and in lateral (here horizontal) motion for various fully submerged tubular sections. The formulas given herein constitute only a low-frequency, first-order approximation of the mass of water that moves in tandem with the tubular sections. In essence, it is for a cylinder of water with diameter equal to the widest dimension of the tubular section transverse to the direction of motion.

Table 1.2. Added mass of fluid for fully submerged tubular sections. The mass density of the fluid is ρ_w

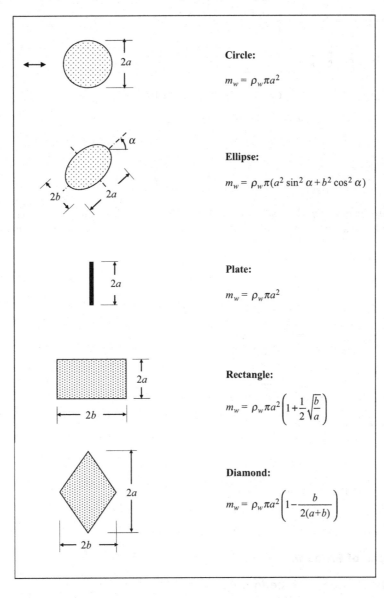

Circle:

$$m_w = \rho_w \pi a^2$$

Ellipse:

$$m_w = \rho_w \pi (a^2 \sin^2 \alpha + b^2 \cos^2 \alpha)$$

Plate:

$$m_w = \rho_w \pi a^2$$

Rectangle:

$$m_w = \rho_w \pi a^2 \left(1 + \frac{1}{2}\sqrt{\frac{b}{a}}\right)$$

Diamond:

$$m_w = \rho_w \pi a^2 \left(1 - \frac{b}{2(a+b)}\right)$$

1.6.3 Added Fluid Mass and Damping for Bodies Floating in Deep Water

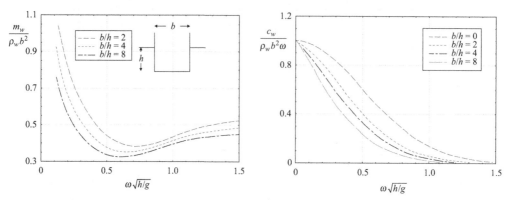

Figure 1.23. Added mass (*left*) and radiation damping (*right*) per unit length for a ship of rectangular cross section (2-D) heaving in deep water (i.e., vertical oscillations), as function of oscillation frequency ω (in rad/s).

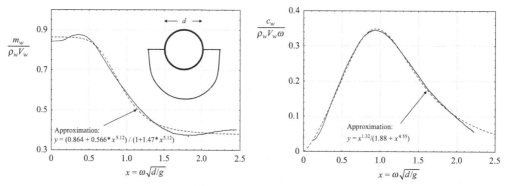

Figure 1.24. Added mass (*left*) and radiation damping (*right*) for semi-submerged spherical buoy heaving in deep water. $V_w = \pi d^2/12$ is the volume of water displaced.

1.7 Degrees of Freedom

1.7.1 Static Degrees of Freedom

Consider an ideally massless structural system subjected to time-varying loads. The number of *static* degrees of freedom in this structure is the number of parameters required to define its deformation state completely, which in turn depends on the number of independent *deformation modes* that it possesses. For continuous systems, the number of possible deformation modes is generally infinite. However, if we restrict application of the loads to some discrete locations only, or if we prescribe the spatial variation (i.e., shape) that the loads can take, then the number of possible deformation modes will be finite. In general,

Static DOF ≤ No. of independent load parameters ≤ No. of deformation modes

For example, consider a simply supported beam. In principle, this beam could be deformed in an infinite number of ways by appropriate application of loads. However, if we insist that the load acting on the beam can be only uniformly distributed, then only one independent parameter exists, namely the intensity of the load. If so, this system would have only one static degree of freedom, and all physical response quantities would depend uniquely on this parameter.

Alternatively, if we say that the beam is acted upon only by two concentrated loads p_1, p_2 at fixed locations x_1, x_2, then this structure will have only two static degrees of freedom. In this case, the entire deformational state of the system can be written as

$$u(x,t) = p_1(t) f(x,x_1) + p_2(t) f(x,x_2) \tag{1.20}$$

in which the $f(x,x_j)$ are the flexibility functions (= modes of deformation). Notice that all physical response parameters, such as reactions, bending moments, shears, displacements, or rotations, will be solely a function of the two independent load parameters. In particular, the two displacements at the location of the loads are

$$\begin{Bmatrix} u_1(t) \\ u_2(t) \end{Bmatrix} = \begin{bmatrix} f(x_1,x_1) & f(x_1,x_2) \\ f(x_2,x_1) & f(x_2,x_2) \end{bmatrix} \begin{Bmatrix} p_1(t) \\ p_2(t) \end{Bmatrix} \tag{1.21}$$

Furthermore, if we define u_1, u_2 as the *master* degrees of freedom, then all other displacements and rotations anywhere else in the structure, that is, the *slave* degrees of freedom, will be completely defined by u_1, u_2.

1.7.2 Dynamic Degrees of Freedom

When a structure with discrete (i.e., lumped) masses is set in motion or vibration, then by D'Alembert's Principle, every free (i.e., unrestrained) point at which a mass is attached experiences inertial loads opposing the motion. Every location, and each direction, for which such dynamic loads can take place will give rise to an independent *dynamic degree of freedom* in the structure. In general, each discrete mass point can have up to six dynamic degrees of freedom, namely three translations (associated each with the same inertial mass m) and three rotations (associated with three usually unequal rotational mass moments of inertia J_{xx}, J_{yy}, J_{zz}). Loosely speaking, the total number of dynamic degrees of freedom is then equal to the number of points and directions that can move and have an inertia property associated with them. Clearly, a continuous system with distributed mass has infinitely many such points, so continuous systems have infinitely many dynamic degrees of freedom.

Consider the following examples involving a massless, inextensible, simply supported beam with lumped masses as shown (motion in plane of drawing only). Notice that because the beam is assumed to be axially rigid, the masses cannot move horizontally:

1 DOF = vertical displacement of mass:

Figure 1.25

2 DOF = vertical displacement + rotation of mass

Figure 1.26

2 DOF = rotation of the masses. Even though two translational masses exist, there is no degree of freedom associated with them, because the supports prevent the motion.

Figure 1.27

By way of contrast, consider next an *inextensible* arch with a single lumped mass, as shown. This system has now 3 DOF because, unlike the beam in the second example above, the bending deformation of the arch will elicit not only vertical and rotational motion but also horizontal displacements.

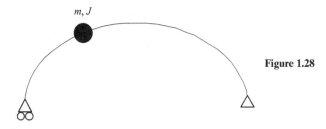

Figure 1.28

1.8 Modeling Structural Systems

1.8.1 Levels of Abstraction

The art of modeling a structural system for dynamic loads relates to the process by which we abstract the essential properties of an actual physical facility into an idealized mathematical model that is amenable to analysis. To understand this process, it is instructive for us to consider the development of the model in terms of steps or *levels of abstraction*.

To illustrate these concepts, consider a two-story frame subjected to lateral wind loads, as shown in Figures 1.29 to 1.32. As we shall see, the models for this frame in the first through fourth levels of abstraction will be progressively reduced from infinitely many dynamic degrees of freedom, to 12, 8, and 2 DOF, respectively.

In the first level of abstraction, the principal elements of the actual system are identified, quantified, and idealized, namely its distribution of mass, stiffness, and damping; the support conditions; and the spatial-temporal characteristics of the loads. This preliminary model is a continuous system with distributed mass and stiffness, it spans

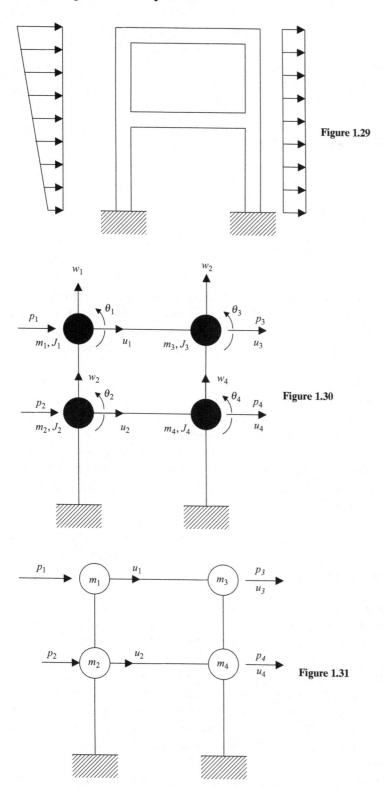

Figure 1.29

Figure 1.30

Figure 1.31

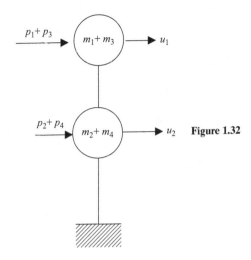

Figure 1.32

the full three-dimensional space, and it has infinitely many degrees of freedom. Of course, in virtually all cases, such a model cannot be analyzed without further simplifications. However, it is a useful conceptual starting point for the development of systems that can be analyzed, using some of the techniques that will be described in the next section.

In the second level of abstraction, we transform the continuous system into a discrete system by lumping masses at the nodes, and/or representing the structural elements with ideal massless elements, such as beams, plates, finite elements, and so forth. We also condense distributed dynamic loads, such as wind pressures, into concentrated, equivalent forces that act on the nodes alone. We initially allow the structural elements to deform in a most general fashion; for example, we assume that beams may deform axially, in bending and in shear. Also, the nodal masses have both translational as well as rotational inertia, so that nodes could typically have up to six degrees of freedom each. In general, this system will have a finite, but large number of degrees of freedom. Inasmuch as nodal coupling occurs only when two or more nodes are connected by a structural element, the stiffness and damping matrices will be either narrowly banded or sparse. Such a system is said to be *closely coupled*. In the example shown, the system has 12 DOF: 8 translational and 4 rotational DOF.

In a third level of abstraction, we may neglect vertical inertia forces as well as rotational inertias. Note carefully that this does not imply that the vertical motions or rotations vanish. Instead, these become static degrees of freedom, and thus depend linearly on the lateral translations, that is, they become slave DOF to the lateral translations, which are the master DOF. While the number of dynamic DOF of this model is now less than in the original one, we pay a price: the stiffness and damping matrices will now be fully populated, and the system becomes *far coupled*. In our example, the structure has now four dynamic degrees of freedom. The process of reducing the number of DOF as a result of neglecting rotational and translational inertias can formally be achieved by matrix manipulations referred to as *static condensation*.

In a fourth and last level of abstraction, we introduce further simplifications by assuming that certain structural elements are ideally rigid, and cannot deform. For example, we can usually neglect the axial deformations of beams (i.e., floors), an assumption that establishes a kinematic constraint between the axial components of motion at the two ends of the beam. In our example, this means that the horizontal motions at each elevation are uniform, that is, $u_1 = u_3$, $u_2 = u_4$. The system has now only 2 DOF. The formal process by which this is accomplished through matrix manipulations is referred to as *kinematic condensation*.

Once the actual motions u_1, u_2 in this last model have been determined, it is possible to undo both the static and dynamic condensations, and determine any arbitrary component of motion or force, say the rotations of the masses, the axial forces in the columns, or the support reactions.

1.8.2 Transforming Continuous Systems into Discrete Ones

Most continuous systems cannot be analyzed as such, but must first be cast in the form of discrete systems with a finite number of DOF. We can use two basic approaches to transform a continuous system into a discrete one. These are

- Physical approximations (*Heuristic* approach): Use common sense to lump masses, then basic methods to obtain the required stiffnesses.
- Mathematical methods: *Weighted residuals* family of methods, the *Rayleigh–Ritz* method, and the energy method based on the use of *Lagrange's equations*. We cite also the method of *finite differences*, which consists in transforming the differential equations into difference equations, but we shall not consider it in this work.

We succinctly describe the heuristic method in the following, but postpone the powerful mathematical methods to Chapter 6, Sections 6.2–6.4 since these methods are quite abstract and require some familiarity with the methods of structural dynamics.

Heuristic Method

Heuristics is the art of inventing or discovering. As the definition implies, in the heuristic method we use informal, common sense strategies for solving problems.

To develop a discrete model for a problem in structural dynamics, we use the following simple steps:

- Idealize the structure as an assembly of structural elements (beams, plates, etc.), abstracting the essential qualities of the physical system. This includes making decisions as to which elements can be considered infinitely rigid (e.g., inextensional beams or columns, rigid floor diaphragms, etc.), and also what the boundary conditions should be.
- Idealize the loading, in both space and time.
- On the basis of both the structural geometry and the spatial distribution of the loads, decide on the number and location of the discrete mass points or *nodes*. These will define the active degrees of freedom.
- Using common sense, *lump* (or concentrate) the translational and rotational masses at these nodes. For this purpose, consider all the mass distributed in the vicinity of the active nodes, utilizing the concept of *tributary areas* (or volume). For example, in

the simplest possible model for a beam, we would divide the beam into four equal segments, and lump the mass and rotational inertia of the two mid-segments at the center (i.e., half of the beam), and that of the lateral segments at each support (one quarter each). In the case of a frame, we would probably lump the masses at the intersection of the beams and columns.

- Lump the distributed loads at the nodes, using again the concept of tributary area.
- Obtain the global stiffness matrix for this model as in a Tinkertoy, that is, constructing the structure by assembly and overlap of the individual member stiffness matrices, or by either the *direct stiffness* approach or the *flexibility* approach. The latter is usually more convenient in the case of statically determinate systems, but this is not always the case.
- Solve the discrete equations of motion.

1.8.3 Direct Superposition Method

As the name implies, in this method the global stiffness matrix is obtained by assembling the complete system by superimposing the stiffness matrices for each of the structural elements. This entails overlapping the stiffness matrices for the components at locations in the global matrix that depend on how the members are connected together. Inasmuch as the orientation of the members does not generally coincide with the global directions, it is necessary to first rotate the element matrices prior to overlapping, so as to map the local displacements into the global coordinate system. You are forewarned that this method should never be used for hand computations, since even for the simplest of structures this formal method will require an inordinate effort. However, it is the standard method in computer applications, especially in the *finite element method*.

1.8.4 Direct Stiffness Approach

To obtain the stiffness matrix for the structural system using the direct stiffness approach, we use the following steps:

- Identify the active DOF.
- Constrain all active DOFs (and *only* the active DOF!). Thus, in a sense, they become "supports."
- Impose, one at a time, unit displacements (or rotations) at each and every one of the constrained DOF. The force (or moment) required to impose each of these displacements together with the "reaction" forces (or moments) at the remaining constrained nodes, are numerically equal to the terms of the respective column of the desired stiffness matrix.

This method usually requires considerable effort, unless the structure is initially so highly constrained that few DOF remain. An example is the case of a one-story frame with many columns (perhaps each with different ending conditions) that are tied together by an infinitely rigid girder. Despite the large number of members, such a system has only one lateral DOF, in which case the direct stiffness approach is the method of choice.

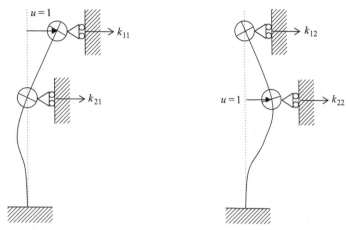

Figure 1.33. Direct stiffness method.

Let us illustrate this method by means of an example. Consider a structure composed of a bending beam with two lumped masses, as shown in Figure 1.33. While the masses can and do rotate, there are no rotational inertias associated with them, so they are simply static degrees of freedom. Hence, this system has a total of only two dynamic DOF, namely the lateral translation of the two masses. To determine the elements of the 2×2 stiffness matrix, we consider the two test problems shown below, and determine by some standard method, such as the *slope-deflection method*, the reactions at the fictitious supports. Notice that the masses are allowed to rotate freely, since they are not active, dynamic DOF.

$$\begin{Bmatrix} k_{11} & k_{12} \\ k_{21} & k_{22} \end{Bmatrix} \begin{Bmatrix} 1 \\ 0 \end{Bmatrix} = \begin{Bmatrix} k_{11} \\ k_{21} \end{Bmatrix} \qquad \begin{Bmatrix} k_{11} & k_{12} \\ k_{21} & k_{22} \end{Bmatrix} \begin{Bmatrix} 0 \\ 1 \end{Bmatrix} = \begin{Bmatrix} k_{12} \\ k_{22} \end{Bmatrix} \qquad (1.22)$$

1.8.5 Flexibility Approach

The flexibility approach is basically the converse of the direct stiffness approach, and leads to the inverse of the stiffness matrix, which is the flexibility matrix. It is particularly well suited for statically determinate structures, but may require considerable effort if the structure is indeterminate.

The discrete flexibility matrix is obtained as follows:

- Identify the active DOF.
- Apply, one at a time, a unit force (or moment) at each and every active DOF. The observed displacements (or rotations) at all active DOF are then numerically equal to the columns of the flexibility matrix. The stiffness matrix, if desired, must be obtained by inversion of the flexibility matrix.

To illustrate this method, consider once more the example of the previous section. Without constraining the structure, we now solve for the displacements produced by the

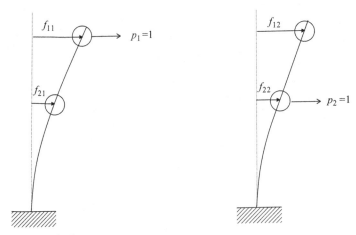

Figure 1.34. Flexibility method.

two test loadings shown in Figure 1.34. These displacements are numerically equal to the elements of the flexibility matrix. Since the cantilever column is statically determinate, this problem is particularly simple to solve, either from tabulated formulae (e.g., see the section "Stiffnesses of Some Typical Linear Systems"), or by standard methods such as the *conjugate beam method*.

$$\begin{bmatrix} f_{11} & f_{12} \\ f_{21} & f_{22} \end{bmatrix} \begin{Bmatrix} 1 \\ 0 \end{Bmatrix} = \begin{Bmatrix} f_{11} \\ f_{21} \end{Bmatrix} \qquad \begin{bmatrix} f_{11} & f_{12} \\ f_{21} & f_{22} \end{bmatrix} \begin{Bmatrix} 0 \\ 1 \end{Bmatrix} = \begin{Bmatrix} f_{12} \\ f_{22} \end{Bmatrix} \qquad (1.23)$$

Example 1

Consider a simply supported, homogeneous bending beam subjected to a dynamic load at the center, as shown in Figure 1.35. Although this structure can be analyzed rigorously as a continuous system, we shall pretend ignorance of this fact, and attempt instead to represent it as a discrete system. Clearly, the absolute simplest representation is a massless beam with a concentrated mass lumped at the location of the load equal to mass of the shaded (tributary) area, that is, $m = \frac{1}{2}\rho AL$. The remaining mass may be lumped at the supports, but because they do not move, they have no effect on the discrete model. This assumes that the temporal variation of the load changes no faster than the beam's fundamental period.

From the flexibility approach, a unit static load at the center causes a displacement $f = \frac{1}{48}L^3/EI$, so the stiffness of this 1-DOF system is $k = 1/f = 48EI/L^3$.

Example 2

By way of contrast, consider next the same problem subjected to two dynamic loads, as shown in Figure 1.36. A satisfactory dynamic model would involve now two or more masses, particularly if the loads are negatively correlated, that is, if $p_2(t) \sim -p_1(t)$.

Each of the three lumped masses now equals one-fourth of the total beam mass, while the elements of the 3×3 flexibility matrix can readily be obtained from the formula for deflections listed in the section on typical linear systems.

Figure 1.35. Mass lumping.

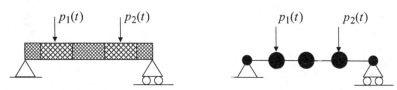

Figure 1.36

Example 3

As a last example, consider a horizontally layered soil subjected to vertically propagating shear waves (i.e., SH-waves). Inasmuch as motions in horizontal planes are uniform, it suffices to consider a column of soil of unit width and subjected to pure shear. We begin by discretizing the soil column into elements $j = 1, 2, 3\ldots$ of unit cross section ($A_j = 1$) whose thickness h_j is small in comparison to the typical wavelengths in the seismic motion.

Using the heuristic approach, we obtain lumped masses and discrete springs whose values are

$$m_j = \tfrac{1}{2}(\rho_{j-1} h_{j-1} + \rho_j h_j) \tag{1.24}$$

$$k_j = \frac{G_j}{h_j} = \frac{\rho_j C_{sj}^2}{h_j} \tag{1.25}$$

The mass and stiffness matrices are then

$$\mathbf{M} = \mathrm{diag}\{m_1 \quad m_2 \quad \cdots \quad m_n\}, \qquad \mathbf{K} = \begin{Bmatrix} k_1 & -k_1 & & \\ -k_1 & k_1 + k_2 & -k_2 & \\ & -k_2 & \ddots & -k_{n-1} \\ & & -k_{n-1} & k_{n-1} + k_n \end{Bmatrix}$$

1.8.6 Viscous Damping Matrix

Damping forces in a mechanical system result from various phenomena that lead to internal energy dissipation in the form of heat. These arise as a result of nonrecoverable processes associated with material hysteresis, drag, or friction. A particularly simple dissipation mechanism is in the form of linear, viscous damping, in which damping forces are proportional to the rate of material deformation, that is, to the velocity of deformation. Although most structural systems exhibit damping forces that are *not* of a viscous nature, it is nevertheless often convenient to approximate these as being of the viscous type.

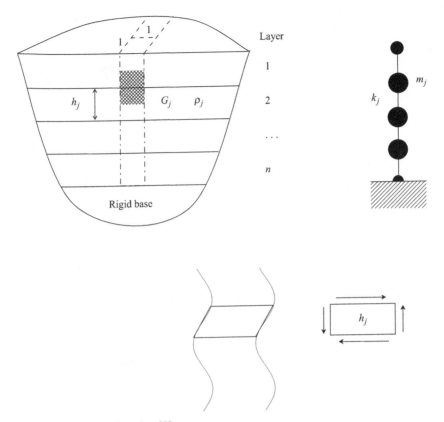

Figure 1.37. Column of soil subjected to SH waves.

The motivation for this lies in the great mathematical simplification that linear viscous damping confers to the idealized structural system, and the good approximations that can still be achieved whatever the actual damping type. Thus, in a way the main reason for using viscous damping is that "we can get away" with it, and much less so because it is convenient. Here we show how to construct damping matrices for an assembly of viscous dashpots, and shall defer the treatment of nonviscous damping and the formulation of appropriate nonviscous damping models to later sections.

A viscous dashpot (or shock absorber) is a mechanical device in which the force necessary to produce a deformation is proportional to the velocity of deformation. Damping forces that are proportional to the velocity of deformation are said to be viscous. In general, the assembly of damping elements follows the same formation rules as those for stiffness elements. Hence, viscous damping matrices have a structure that is similar to that of stiffness matrices. Consider, for example, a system consisting of two spring–dashpot systems in series subjected to forces p_1, p_2 that produce time-varying displacements u_1, u_2. The stiffness and damping matrices are

$$\mathbf{K} = \begin{Bmatrix} k_1 & -k_1 \\ -k_1 & k_1 + k_2 \end{Bmatrix} \tag{1.26}$$

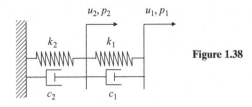

Figure 1.38

$$C = \begin{Bmatrix} c_1 & -c_1 \\ -c_1 & c_1 + c_2 \end{Bmatrix} \tag{1.27}$$

and the force-deformation equations is

$$p(t) = C\dot{u} + Ku \qquad u = \begin{Bmatrix} u_1 \\ u_2 \end{Bmatrix} \qquad p = \begin{Bmatrix} p_1 \\ p_2 \end{Bmatrix} \tag{1.28}$$

Then again, most structures do *not* have dashpots, in which case C cannot be assembled a priori.

1.9 Fundamental Dynamic Principles for a Rigid Body

1.9.1 Inertial Reference Frames

Consider a free, rigid body whose motion is being observed in a global or Newtonian *inertial reference frame* with unit base vectors $\hat{e}_1, \hat{e}_2, \hat{e}_3$. Fixed to this body at its *center of mass* is a set of *local* coordinate axes with unit base vectors $\hat{g}_1, \hat{g}_2, \hat{g}_3$ at which the motions (translations and rotations) elicited by an external force f and torque t are measured. The body has total mass m and a symmetric rotational inertia tensor $J = J_{ij}\,\hat{g}_i\,\hat{g}_j$ (summation over repeated indices implied), whose components are

$$J_{ij} = \delta_{ij} \int_{Vol} r^2 \rho\, dV - \int_{Vol} x_i x_j \rho\, dV \tag{1.29}$$

in which δ_{ij} is the Kronecker delta, $r^2 = x_1^2 + x_2^2 + x_3^2$, and the coordinates x_i are relative to the center of mass. The diagonal terms are the mass moments of inertia, while the off-diagonal terms are the products of inertia; notice the negative sign in front of the second integral. If the axes are principal, then the products of inertia vanish, and only the diagonal terms remain.

1.9.2 Kinematics of Motion

Consider a *rigid* body that moves and rotates in some arbitrary fashion, and within this body, choose as reference point the center of mass, which moves with velocity v and about which the body rotates with instantaneous angular velocity ω. We also choose another arbitrary observation point p on the body whose position relative to the reference point is $r = r(t)$, and whose velocity is v_p. From vector mechanics, the following holds:

$$\tfrac{d}{dt} r = \omega \times r \tag{1.30}$$

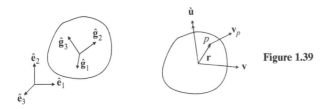

Figure 1.39

$$\mathbf{v}_p = \mathbf{v} + \boldsymbol{\omega} \times \mathbf{r} \qquad (1.31)$$

$$\mathbf{a}_p = \mathbf{a} + \boldsymbol{\alpha} \times \mathbf{r} + \boldsymbol{\omega} \times (\boldsymbol{\omega} \times \mathbf{r}) \qquad (1.32)$$

in which $\boldsymbol{\alpha} = \dot{\boldsymbol{\omega}}$ is the angular acceleration, $\boldsymbol{\omega} \times (\boldsymbol{\omega} \times \mathbf{r}) = \boldsymbol{\omega}(\boldsymbol{\omega} \bullet \mathbf{r}) - \omega^2 \mathbf{r}$ and $\omega^2 = \boldsymbol{\omega} \bullet \boldsymbol{\omega}$. Now, the position vector $\mathbf{r}$ can be expressed either in local (rotating) or global (nonrotating) coordinates with unit orthogonal basis vectors $\hat{\mathbf{g}}_i$ and $\hat{\mathbf{e}}_i$, respectively, that is, $\mathbf{r} = x_i' \hat{\mathbf{g}}_i = x_i \hat{\mathbf{e}}_i$. In particular, if we choose one of the rotating basis vectors $\mathbf{r} = \hat{\mathbf{g}}_i = \gamma_{ij} \hat{\mathbf{e}}_j$, in which the γ_{ij} are the direction cosines of the rotated basis with respect to the global basis, then

$$\tfrac{d}{dt} \hat{\mathbf{g}}_i = \boldsymbol{\omega} \times \hat{\mathbf{g}}_i = \tfrac{d}{dt} \gamma_{ij} \hat{\mathbf{e}}_j = \dot{\gamma}_{ij} \gamma_{kj} \hat{\mathbf{g}}_k \qquad (1.33)$$

or in matrix format

$$\tfrac{d}{dt} \begin{Bmatrix} \hat{\mathbf{g}}_1 \\ \hat{\mathbf{g}}_2 \\ \hat{\mathbf{g}}_3 \end{Bmatrix} = \begin{bmatrix} 0 & \omega_3' & -\omega_2' \\ -\omega_3' & 0 & \omega_1' \\ \omega_2' & -\omega_1' & 0 \end{bmatrix} \begin{Bmatrix} \hat{\mathbf{g}}_1 \\ \hat{\mathbf{g}}_2 \\ \hat{\mathbf{g}}_3 \end{Bmatrix} = \begin{bmatrix} \dot{\gamma}_{11} & \dot{\gamma}_{12} & \dot{\gamma}_{13} \\ \dot{\gamma}_{21} & \dot{\gamma}_{22} & \dot{\gamma}_{23} \\ \dot{\gamma}_{31} & \dot{\gamma}_{32} & \dot{\gamma}_{33} \end{bmatrix} \begin{bmatrix} \gamma_{11} & \gamma_{21} & \gamma_{31} \\ \gamma_{12} & \gamma_{22} & \gamma_{32} \\ \gamma_{13} & \gamma_{23} & \gamma_{33} \end{bmatrix} \begin{Bmatrix} \hat{\mathbf{g}}_1 \\ \hat{\mathbf{g}}_2 \\ \hat{\mathbf{g}}_3 \end{Bmatrix} = \dot{\mathbf{R}} \mathbf{R}^T \mathbf{G} \qquad (1.34)$$

The local components of the angular velocity vector are then

$$\omega_1' = \dot{\gamma}_{21} \gamma_{31} + \dot{\gamma}_{22} \gamma_{32} + \dot{\gamma}_{23} \gamma_{33} \qquad (1.35)$$

$$\omega_2' = \dot{\gamma}_{31} \gamma_{11} + \dot{\gamma}_{32} \gamma_{12} + \dot{\gamma}_{33} \gamma_{13} \qquad (1.36)$$

$$\omega_3' = \dot{\gamma}_{11} \gamma_{21} + \dot{\gamma}_{12} \gamma_{22} + \dot{\gamma}_{13} \gamma_{23} \qquad (1.37)$$

Cardanian Rotation

The rotation of a rigid body rotation in 3-D space can be described in terms of three independent, right-handed rotations about the local coordinate axes or *Cardanian angles* $\theta_1, \theta_2, \theta_3$ whose sequence must remain inviolate. Of various possible choices, we choose the following rule:

- Rotate about the global x-axis by an angle θ_1, which moves the axes to, say, position $\bar{x}, \bar{y}, \bar{z}$.
- Rotate about the local $\bar{y}$ by an angle θ_2, which moves the axes to position $\tilde{x}, \tilde{y}, \tilde{z}$.
- Rotate about the local $\tilde{z}$ axis by an angle θ_3 which moves the axes to their final position x', y', z'.

The three rotation matrices needed to accomplish these three sequential elementary rotations are

$$\mathbf{R}_1 = \begin{Bmatrix} 1 & 0 & 0 \\ 0 & \cos\theta_1 & \sin\theta_1 \\ 0 & -\sin\theta_1 & \cos\theta_1 \end{Bmatrix}, \quad \mathbf{R}_2 = \begin{Bmatrix} \cos\theta_2 & 0 & -\sin\theta_2 \\ 0 & 1 & 0 \\ \sin\theta_2 & 0 & \cos\theta_2 \end{Bmatrix}, \quad \mathbf{R}_3 = \begin{Bmatrix} \cos\theta_3 & \sin\theta_3 & 0 \\ -\sin\theta_3 & \cos\theta_3 & 0 \\ 0 & 0 & 1 \end{Bmatrix}$$
(1.38)

Observe the different signs of the off-diagonal terms for $\mathbf{R}_2$. With the shorthand $c_j = \cos\theta_j$, $s_j = \sin\theta_j$ we obtain the complete rotation matrix as

$$\mathbf{R} = \mathbf{R}_3\mathbf{R}_2\mathbf{R}_1 = \begin{Bmatrix} c_2c_3 & c_1s_3 + s_1s_2c_3 & s_1s_3 - c_1s_2c_3 \\ -c_2s_3 & c_1c_3 - s_1s_2s_3 & s_1c_3 + c_1s_2s_3 \\ s_2 & -s_1c_2 & c_1c_2 \end{Bmatrix} = \{\gamma_{ij}\}, \quad \mathbf{x}' = \mathbf{R}\mathbf{x}, \quad \mathbf{x} = \mathbf{R}^T\mathbf{x}'$$ (1.39)

This matrix contains the direction cosines γ_{ij} of the local axes with respect to the global axes. Using the preceding expressions together with MATLAB®'s symbolic manipulation capability, it can be shown that the components of the angular velocity vector are given by

$$\omega_1' = \dot\theta_1 \cos\theta_2 \cos\theta_3 + \dot\theta_2 \sin\theta_3$$ (1.40)

$$\omega_2' = \dot\theta_2 \cos\theta_3 - \dot\theta_1 \cos\theta_2 \sin\theta_3$$ (1.41)

$$\omega_3' = \dot\theta_3 + \dot\theta_1 \sin\theta_2$$ (1.42)

Eulerian Rotation

The most common alternative approach to describe rotations is by means of the so-called Eulerian angles, which differ from the Cardanian rotations in that one local axis is used twice. These angles can be visualized as follows. Assume that initially, the local axes with unit base vectors $\hat{\mathbf{g}}_1, \hat{\mathbf{g}}_2, \hat{\mathbf{g}}_3$ and the global axes with unit base vectors $\hat{\mathbf{e}}_1, \hat{\mathbf{e}}_2, \hat{\mathbf{e}}_3$ coincide in space. Thereafter, carry out the following three rotations (right-hand rule applies, and the sequence must remain inviolate):

- Rotate the local frame about the vertical axis $\hat{\mathbf{g}}_3$ by a horizontal angle (azimuth, or longitude) $0 \le \phi \le 2\pi$ until the plane x,z (i.e., $\hat{\mathbf{g}}_1, \hat{\mathbf{g}}_3$) contains the line from the center to the final north pole. Of the two possible rotations, which differ by 180 degrees, choose the one for which the angular distance (about $\hat{\mathbf{g}}_2$) from the initial to the final north poles satisfies $0 \le \theta \le \pi$.
- Rotate next the local frame about $\hat{\mathbf{g}}_2$ through a right-handed angle $0 \le \theta \le \pi$. This brings $\hat{\mathbf{g}}_3$ to its final position, so $\cos\theta = \hat{\mathbf{g}}_3 \cdot \hat{\mathbf{e}}_3$.
- Finally, rotate again the local frame about $\hat{\mathbf{g}}_3$ by an angle $0 \le \psi \le 2\pi$. This brings the local system to its ultimate configuration.

If $\mathbf{G} = \{\gamma_{ij}\} = \{\hat{\mathbf{g}}_i \cdot \hat{\mathbf{e}}_j\}$ is the matrix of direction cosines for the three local axes, then its elements γ_{ij} in terms of the Eulerian angles are as shown in Table 1.3.[1]

In *local* body coordinates, the components of the rotational velocity vector $\boldsymbol{\omega}$ are then

$$\omega_1 = \dot\theta \sin\psi - \dot\phi \sin\theta \cos\psi$$ (1.43)

---

[1] J. L. Synge and B. A. Griffith, *Principles of Mechanics* (New York: McGraw-Hill, 1959), 261.

Table 1.3. Direction cosines of local axes in terms of Eulerian angles

$\cos\theta\cos\phi\cos\psi - \sin\theta\sin\psi$	$\cos\phi\sin\psi + \cos\theta\sin\phi\cos\psi$	$-\sin\theta\cos\psi$
$-\cos\theta\cos\phi\sin\psi - \sin\phi\cos\psi$	$\cos\phi\cos\psi - \cos\theta\sin\phi\sin\psi$	$\sin\theta\sin\psi$
$\sin\theta\cos\phi$	$\sin\theta\sin\phi$	$\cos\theta$

$$\omega_2 = \dot\theta\cos\psi + \dot\phi\sin\theta\sin\psi \tag{1.44}$$

$$\omega_3 = \dot\phi\cos\theta + \dot\psi \tag{1.45}$$

1.9.3 Rotational Inertia Forces

Let's now examine the rotational inertia forces in a body. For this purpose, let $\mathbf{J}$ denote the rotational inertia tensor, which can be written in local or global coordinates (**Note**: Pairs of juxtaposed basis vectors constitutes a dyad, and repeated indices imply summation.)

$$\mathbf{J} = J'_{ij}\,\hat{\mathbf{g}}_i\,\hat{\mathbf{g}}_j = J_{ij}\,\hat{\mathbf{e}}_i\,\hat{\mathbf{e}}_j \tag{1.46}$$

In local coordinates, the components J'_{ij} of $\mathbf{J}$ are time invariant, so

$$\begin{aligned}
\tfrac{d}{dt}\mathbf{J} &= J'_{ij}\left(\tfrac{d}{dt}\hat{\mathbf{g}}_i\,\hat{\mathbf{g}}_j + \hat{\mathbf{g}}_i\,\tfrac{d}{dt}\hat{\mathbf{g}}_j\right) = J'_{ij}\left(\boldsymbol{\omega}\times\hat{\mathbf{g}}_i\,\hat{\mathbf{g}}_j + \hat{\mathbf{g}}_i\,\boldsymbol{\omega}\times\hat{\mathbf{g}}_j\right)\\
&= \boldsymbol{\omega}\times J'_{ij}\,\hat{\mathbf{g}}_i\,\hat{\mathbf{g}}_j - J'_{ij}\,\hat{\mathbf{g}}_i\,\hat{\mathbf{g}}_j\times\boldsymbol{\omega}\\
&= \boldsymbol{\omega}\times\mathbf{J} - \mathbf{J}\times\boldsymbol{\omega}
\end{aligned} \tag{1.47}$$

Now, the principle of conservation of angular momentum states that the external moment (or torque) $\mathbf{t}$ is given by the rate of change of the angular momentum $\mathbf{h} = \mathbf{J}\bullet\mathbf{E}$, that is,

$$\begin{aligned}
\mathbf{t} &= \tfrac{d}{dt}\mathbf{h} = \tfrac{d}{dt}(\mathbf{J}\bullet\boldsymbol{\omega}) = \tfrac{d}{dt}(\mathbf{J})\bullet\boldsymbol{\omega} + \mathbf{J}\bullet\tfrac{d}{dt}\boldsymbol{\omega} = (\boldsymbol{\omega}\times\mathbf{J} - \mathbf{J}\times\boldsymbol{\omega})\bullet\boldsymbol{\omega} + \mathbf{J}\bullet\boldsymbol{\alpha}\\
&= \mathbf{J}\bullet\boldsymbol{\alpha} + \boldsymbol{\omega}\times\mathbf{J}\bullet\boldsymbol{\omega}
\end{aligned} \tag{1.48}$$

This is because $\mathbf{J}\times\boldsymbol{\omega}\bullet\boldsymbol{\omega} = \mathbf{0}$. Notice also that $\mathbf{h} = \mathbf{J}\bullet\boldsymbol{\omega} = \boldsymbol{\omega}\bullet\mathbf{J}$, because $\mathbf{J}$ is symmetric. Expressing next the various terms in the expression above by means of the local (moving) unit vectors, we obtain

$$\mathbf{J}\bullet\boldsymbol{\omega} = J'_{ij}\,\hat{\mathbf{g}}_i\,\hat{\mathbf{g}}_j\bullet\omega'_k\,\hat{\mathbf{g}}_k = J'_{ij}\omega'_j\,\hat{\mathbf{g}}_i, \qquad \mathbf{J}\bullet\boldsymbol{\alpha} = J'_{ij}\,\alpha'_j\,\hat{\mathbf{g}}_i = J'_{ij}\,\dot\omega'_j\,\hat{\mathbf{g}}_i \tag{1.49}$$

Furthermore, if the moving axes attached to the rigid body are principal axes, then only the diagonal terms of the inertia tensor exist, so $\mathbf{J}\bullet\boldsymbol{\omega} = J'_{11}\omega'_1\,\hat{\mathbf{g}}_1 + J'_{22}\omega'_2\,\hat{\mathbf{g}}_1 + J'_{33}\omega'_3\,\hat{\mathbf{g}}_3$. Hence,

$$\begin{aligned}
\boldsymbol{\omega}\times(\mathbf{J}\bullet\boldsymbol{\omega}) &= \begin{vmatrix} \hat{\mathbf{g}}_1 & \hat{\mathbf{g}}_2 & \hat{\mathbf{g}}_3 \\ \omega'_1 & \omega'_2 & \omega'_3 \\ J'_{11}\omega'_1 & J'_{22}\omega'_2 & J'_{33}\omega'_3 \end{vmatrix}\\
&= (J'_{33} - J'_{22})\,\omega'_2\omega'_3\,\hat{\mathbf{g}}_1 + (J'_{11} - J'_{33})\,\omega'_3\omega'_1\,\hat{\mathbf{g}}_2 + (J'_{22} - J'_{11})\,\omega'_1\omega'_2\,\hat{\mathbf{g}}_3
\end{aligned} \tag{1.50}$$

$$\mathbf{t} = \left[J'_{11}\alpha'_1 + (J'_{33} - J'_{22})\,\omega'_2\omega'_3\right]\hat{\mathbf{g}}_1 + \left[J'_{22}\alpha'_2 + (J'_{11} - J'_{33})\,\omega'_3\omega'_1\right]\hat{\mathbf{g}}_2 + \left[J'_{33}\alpha'_3 + (J'_{22} - J'_{11})\,\omega'_1\omega'_2\right]\hat{\mathbf{g}}_3 \tag{1.51}$$

In matrix form and in local (rotating) coordinates, Eq. 1.51 can be written as

$$\begin{Bmatrix} T_1' \\ T_2' \\ T_3' \end{Bmatrix} = \begin{bmatrix} J_{11}' & 0 & 0 \\ 0 & J_{22}' & 0 \\ 0 & 0 & J_{33}' \end{bmatrix} \begin{Bmatrix} \dot{\omega}_1' \\ \dot{\omega}_2' \\ \dot{\omega}_3' \end{Bmatrix} + \begin{bmatrix} J_{33}' - J_{22}' & 0 & 0 \\ 0 & J_{11}' - J_{33}' & 0 \\ 0 & 0 & J_{22}' - J_{11}' \end{bmatrix} \begin{Bmatrix} \omega_2' \omega_3' \\ \omega_3' \omega_1' \\ \omega_1' \omega_2' \end{Bmatrix} \qquad (1.52)$$

This expression is valid for *arbitrarily large rotations*.

Finally, if we assume small rotations, we can neglect all double products of angular velocities, in which case there is no difference between the local and global reference frames. If so, the angular velocities can be defined in terms of the rotations as $\omega_j = \frac{d}{dt}\theta_j = \dot{\theta}_j$, and thus we recover the classical equation of linear structural dynamics relating applied torques with angular accelerations

$$\begin{Bmatrix} T_1 \\ T_2 \\ T_3 \end{Bmatrix} = \begin{bmatrix} J_{11} & 0 & 0 \\ 0 & J_{22} & 0 \\ 0 & 0 & J_{33} \end{bmatrix} \begin{Bmatrix} \dot{\omega}_1 \\ \dot{\omega}_2 \\ \dot{\omega}_3 \end{Bmatrix} = \begin{bmatrix} J_{11} & 0 & 0 \\ 0 & J_{22} & 0 \\ 0 & 0 & J_{33} \end{bmatrix} \begin{Bmatrix} \ddot{\theta}_1 \\ \ddot{\theta}_2 \\ \ddot{\theta}_3 \end{Bmatrix} = \mathbf{J} \cdot \boldsymbol{\alpha} \qquad (1.53)$$

1.9.4 Newton's Laws

Newton's famous three laws are

- A rigid body at rest or in uniform motion will continue in that state until an external force is applied.
- The force needed to impart some acceleration to a body is proportional to the mass of the body and coincides in direction with the acceleration.
- To each action (i.e., force) there is an equal and opposite reaction.

These laws are valid only in an inertial reference frame. The first law is a verbal description of the principle of conservation of linear momentum, the second law is a statement about dynamic equilibrium and defines the concept of mass, while the third is a statement about the forces between bodies that interact with one another. For example, when we push a car with some force, the car pushes back *on us* with that same force, and we in turn transfer that force to the ground through our feet. Also, the gravitational force exerted on some body by the earth is equal and opposite to the force with which that body attracts the earth.

The formulation of Newton's laws for rigid bodies undergoing rectilinear motion is straightforward, but the rotational case is complicated by the fact that the local axes change orientation in space as the body rotates, as we have already seen.

(a) Rectilinear Motion

The external force $\mathbf{f}$ required to impart onto the body an acceleration $\mathbf{a} = \dot{\mathbf{v}}$ is

$$\boxed{\mathbf{f} = m\mathbf{a}} \qquad (1.54)$$

This is Newton's second law. If $\mathbf{f} = \mathbf{0}$, the body moves with constant velocity along a rectilinear path or remains at rest, and this constitutes the first law.

(b) Rotational Motion

When an external torque **t** is applied to the body, it imparts on the body an angular acceleration $\alpha = \frac{d}{dt}\boldsymbol{\omega}$. In the *local* frame moving and rotating with the body, the elements of **J** are constant, but not so the base vectors, because as the body's axes rotate in space, it changes its global orientation. Thus, the time derivative of **J** does not vanish. In this moving system, the angular acceleration relates to the torque and the instantaneous angular velocity $\boldsymbol{\omega}$ through the expression

$$\mathbf{t} = \frac{d\mathbf{h}}{dt} = \frac{d(\mathbf{J}\cdot\boldsymbol{\omega})}{dt} = \mathbf{J}\cdot\boldsymbol{\alpha} + \boldsymbol{\omega}\times\mathbf{J}\cdot\boldsymbol{\omega} \tag{1.55}$$

where $\mathbf{h} = \mathbf{J}\cdot\boldsymbol{\omega}$ is the angular momentum. More generally, if one of the principal axes of the body is an axis of symmetry, then this body has at least two identical moments of inertia. If so, the *elements* of **J** are also constant with respect to a moving reference frame $\hat{\mathbf{g}}_i$ that does *not* rotate together with the body about that local axis, say $\hat{\mathbf{g}}_1$. In such reference frame, the base rotates with angular velocity $\boldsymbol{\Omega} = \Omega_1\,\hat{\mathbf{g}}_2 + \omega_2\,\hat{\mathbf{g}}_2 + \omega_3\,\hat{\mathbf{g}}_3$ that differs from that of the body, namely $\boldsymbol{\omega} = \omega_1\,\hat{\mathbf{g}}_1 + \omega_2\,\hat{\mathbf{g}}_2 + \omega_3\,\hat{\mathbf{g}}_3$, with $\Omega_1 \neq \omega_1$. The component Ω_1 can be defined to be zero, or deduced from constraints imposed on the other base vectors (e.g., that **h** be contained in the plane of $\hat{\mathbf{g}}_2, \hat{\mathbf{g}}_3$, etc.). Either way, it follows that $d\hat{\mathbf{g}}_i / dt = \boldsymbol{\Omega}\times\hat{\mathbf{g}}_i$ and

$$\mathbf{t} = \frac{d(\mathbf{J}\cdot\boldsymbol{\omega})}{dt} = \mathbf{J}\cdot\boldsymbol{\alpha} + \boldsymbol{\Omega}\times\mathbf{J}\cdot\boldsymbol{\omega} \tag{1.56}$$

This expression is particularly useful for vibration problems involving rotating machinery.

If the rotations remain *small* at all times, then we can indeed use the rotation angles about the global axes to measure the orientation, that is, $\boldsymbol{\omega} \approx \dot{\boldsymbol{\theta}}$. In that case, we can also neglect the quadratic term in Eq. 1.56, which linearizes the relationship between external torque and angular acceleration, an assumption we shall use throughout the remainder of this book. In that case,

$$\boxed{\mathbf{t} \approx \mathbf{J}\cdot\boldsymbol{\alpha} = \mathbf{J}\cdot\ddot{\boldsymbol{\theta}}} \tag{1.57}$$

1.9.5 Kinetic Energy

The total kinetic energy of a body is the sum of its translational and rotational kinetic energies (summation over repeated indices implied):

$$\begin{aligned} K &= \tfrac{1}{2}m\,\mathbf{v}\cdot\mathbf{v} + \tfrac{1}{2}\boldsymbol{\omega}\cdot\mathbf{J}\cdot\boldsymbol{\omega} \\ &= \tfrac{1}{2}m\,v_i v_i + \tfrac{1}{2}J_{ij}\omega_i\omega_j \end{aligned} \tag{1.58}$$

1.9.6 Conservation of Linear and Angular Momentum

If **v** and $\boldsymbol{\omega}$ are the instantaneous linear and angular velocity vectors, then the products $m\mathbf{v}$ and $\mathbf{J}\cdot\boldsymbol{\omega}$ are, respectively, referred to as the *linear momentum* and *angular momentum* of the rigid body. On the other hand, the *linear impulse* and the *angular impulse* are defined as the integrals in time of the external force and torque vectors, respectively. The following principles then apply.

(a) Rectilinear Motion

The change in linear momentum $\mathbf{p} = m\mathbf{v}$ between two arbitrary times t_1, t_2 equals the total external linear impulse applied, that is,

$$\Delta\mathbf{p} = m\mathbf{v}_2 - m\mathbf{v}_1 = \int_{t_1}^{t_2} \mathbf{f}\, dt \qquad (1.59)$$

If the external force vector is zero, then the linear momentum $\mathbf{p} = m\mathbf{v}$ remains constant, in which case the body's center of mass continues to move with unchanging speed and direction, or remains altogether at rest. This is the *principle of conservation of linear momentum.*

(b) Rotational Motion

The change in angular momentum $\mathbf{h} = \mathbf{J} \cdot \boldsymbol{\omega}$ between two instants t_1, t_2 equals the total external angular impulse applied, that is,

$$\Delta\mathbf{h} = \mathbf{J}_2\boldsymbol{\omega}_2 - \mathbf{J}_1\boldsymbol{\omega}_1 = \int_{t_1}^{t_2} \mathbf{t}\, dt \qquad (1.60)$$

Notice that if the inertia tensor $\mathbf{J}$ is defined with respect to the global coordinate system, then its components must necessarily be a function of time, since the body continually changes its orientation. On the other hand, in a local system moving and rotating with the body, the elements J_{ij} of $\mathbf{J}$ (but not the base vectors $\hat{\mathbf{g}}_i$) remain constant, but then $\mathbf{t}$ and $\boldsymbol{\omega}$ must be measured with respect to those moving axes.

   If the external torque vector is zero, then the angular momentum $\mathbf{h} = \mathbf{J} \cdot \boldsymbol{\omega}$ must remain constant in both magnitude and direction. This is the *principle of conservation of angular momentum.* However, the physical interpretation of the principle is obscured by the fact that $\mathbf{J}$ changes with the orientation of the local axes, so $\boldsymbol{\omega}$ is *not* constant. Hence, a freely rotating body may appear to wobble in space. For instance, try imagining the motion of a thin rod after you have imparted onto it a fast spin about its longitudinal axis, and slow rotation about a perpendicular axis.

   You might also wish to ponder the problem of how a space-walking astronaut with no initial angular momentum relative to the spaceship may rotate his body without the help of any external support points (i.e., without external torques). A related problem would be for you to elucidate how a free-falling cat in belly-up position manages to turn around in midair and land on its feet.

1.9.7 D'Alembert's Principle

This principle is a very useful alternative interpretation of Newton's laws, which is arrived at by rewriting the force–acceleration equations as

$$\mathbf{f} - m\mathbf{a} = \mathbf{0} \qquad (1.61)$$

and

$$\mathbf{t} - \mathbf{J} \cdot \boldsymbol{\alpha} = 0 \qquad \text{(assuming again small rotations).} \qquad (1.62)$$

In this new form, these two equations can be interpreted in the same way as the force and moment equilibrium equations in statics. It suffices to interpret $-m\mathbf{a}$ and $-\mathbf{J} \cdot \boldsymbol{\alpha}$ as fictitious external forces acting in direction opposite to the positive directions of acceleration, and from this point on treat these as static forces. Hence, the above equations can be referred to as the *dynamic equilibrium equations*.

1.9.8 Extension of Principles to System of Particles and Deformable Bodies

The dynamic principles previously stated for a rigid body also apply to a system of particles or to an elastic, deformable body. It suffices to carry out a sum over all particles, or integrate over the volume of the elastic body. For example, the principles of linear and angular momentum for a system of N particles (ideal point masses) are now

$$\sum_{i=1}^{N} m_i \mathbf{v}_i = \int_{t_1}^{t_2} \mathbf{f}\, dt \quad \text{and} \quad \sum_{i=1}^{N} \mathbf{r}_i \times (m_i\, \mathbf{v}_i) = \int_{t_1}^{t_2} \mathbf{t}\, dt \qquad (1.63)$$

in which $\mathbf{f}, \mathbf{t}$ are the net external force and torque, and $\mathbf{r}_i$ is the instantaneous position vector for the ith particle. Internal forces between the particles, if any, or elastic forces within the body, do not contribute to these expressions, because they will always appear in equal and opposite pairs (i.e., an action on one particle will be neutralized by an equal and opposite reaction on another particle). If there is no net external force and torque acting on the system of particles, then the total linear and angular momenta are conserved.

1.9.9 Conservation of Momentum versus Conservation of Energy

One should be aware of a very important point: the conservation of linear and/or angular momentum in a system of particles or in a deformable body has nothing to do with conservation of energy in that system. In fact, mechanical energy, such as kinetic or elastic energy, need not be conserved, even when the momentum is conserved. To illustrate this concept, consider the impact of a bullet onto a rigid body at rest. The bullet after impact remains embedded in, and moves with the body. If m_b and v_b are the mass and travel velocity of the bullet, and m, v are the mass of the body and its velocity after the impact, then from the principle of conservation of linear momentum, we have

$$m_b v_b = (m_b + m)\, v \qquad \text{so that} \qquad v = \frac{m_b v_b}{m_b + m} \qquad (1.64)$$

The ratio of kinetic energy *after* the impact to that *before* the impact is then

$$\varepsilon = \frac{\frac{1}{2}(m_b + m)v^2}{\frac{1}{2} m_b v_b^2} = \frac{(m_b + m)}{m_b v_b^2}\left[\frac{m_b v_b}{m_b + m}\right]^2 = \frac{m_b}{m_b + m} < 1 \qquad (1.65)$$

As can be seen, the kinetic energy is not conserved, but is partially transformed into heat as a result of the plastic deformation and internal rupture caused by the bullet.

If we knew instead that the kinetic energy was conserved during the collision of two bodies, then we could use the kinetic energy equation to determine the travel velocities of these bodies after the collision, provided these are rigid. By contrast, if the two colliding bodies are elastic, then some of the mechanical energy will be converted into vibrational energy (i.e., waves) within these bodies, in which case the total kinetic energy equation could not be used without much ado to decide on the final velocities of the two bodies after impact.

1.9.10 Instability of Rigid Body Spinning Freely in Space

A rigid body spinning freely in space conserves both its kinetic energy as well as its angular momentum. Expressing these quantities in terms of local axes $\hat{\mathbf{g}}_i$ that spin together with the body and assuming these to be principal axes of inertia, then

$$K = \tfrac{1}{2}\boldsymbol{\omega}\cdot\mathbf{J}\cdot\boldsymbol{\omega} = \tfrac{1}{2}J_{ij}\omega_i\omega_j = \tfrac{1}{2}\left(J_{11}\omega_1^2 + J_{22}\omega_2^2 + J_{33}\omega_3^2\right) \tag{1.66}$$

and

$$\mathbf{h} = \mathbf{J}\cdot\boldsymbol{\omega} = J_{ij}\omega_j\hat{\mathbf{g}}_i = J_{11}\omega_1\hat{\mathbf{g}}_1 + J_{22}\omega_2\hat{\mathbf{g}}_2 + J_{33}\omega_3\hat{\mathbf{g}}_3 \tag{1.67}$$

Although $\mathbf{h}$ has both constant magnitude and constant direction with respect to an inertial reference frame, it changes its relative orientation with respect to the body as the latter spins in space. The constant magnitude is

$$H = |\mathbf{h}| = \sqrt{\mathbf{h}\bullet\mathbf{h}} = \sqrt{\left(J_{11}\omega_1\right)^2 + \left(J_{22}\omega_2\right)^2 + \left(J_{33}\omega_3\right)^2} \tag{1.68}$$

These two physical quantities define the ellipsoids

$$\frac{\omega_1^2}{\left(\sqrt{\frac{2K}{J_{11}}}\right)^2} + \frac{\omega_2^2}{\left(\sqrt{\frac{2K}{J_{22}}}\right)^2} + \frac{\omega_3^2}{\left(\sqrt{\frac{2K}{J_{33}}}\right)^2} = 1 \quad \text{and} \quad \frac{\omega_1^2}{\left(\frac{H}{J_{11}}\right)^2} + \frac{\omega_2^2}{\left(\frac{H}{J_{22}}\right)^2} + \frac{\omega_3^2}{\left(\frac{H}{J_{33}}\right)^2} = 1 \tag{1.69}$$

For given initial conditions the constants K, H define two ellipsoids with minor, intermediate, and major axes given by $\sqrt{2K/J_{jj}}$ and H/J_{jj}. The intersections of the ω_j axes with the ellipsoids define three polar regions A, B, C while the ellipsoids themselves intersect along a curve that defines the allowable states of spin. As shown by Hugh Hunt of Cambridge University,[2] the intersection near the intermediate pole constitutes a saddle point, so a spin around that axis is unstable, whereas spins about the other two axes are stable. However, he goes on to argue that the spin about the small axis is not stable either: As the body spins about that minor axis in a state of high energy, it loses some of that energy by internal friction even if the angular momentum is conserved, so as the K-ellipsoid changes in time the allowable trajectory spirals around the surface until it settles around the lowest energy state, which is a spin about the major axis.

1.10 Elements of Analytical Mechanics

At the heart of the formal methods in analytical mechanics and dynamics lie the equations of Lagrange, which constitute a powerful means for deriving the equations of

---

[2] www.eng.cam.ac.uk/~hemh

motion of complicated mechanical and structural systems. These equations are based on abstract mathematical manipulations of various forms of mechanical energy in a system, such as kinetic and elastic energies, and allow these to be expressed in terms of generalized displacement coordinates. Because kinetic and elastic energies are scalar quantities, the methods based on Lagrange's equations have the great advantage of accomplishing their goal without recourse to either vector mechanics or considerations of free body equilibrium in the deformed and displaced configuration. Thus, it is not necessary to decompose forces or displacements along coordinate directions to find the equations of motion. The approach is particularly powerful when displacements and rotations are large.

1.10.1 Generalized Coordinates and Its Derivatives

Assume that the displacement vector $\mathbf{u}$ at any point $\mathbf{x}$ of a dynamic system can be completely defined by n independent *generalized coordinates* (or parameters), each of which depends on time only:

$$\mathbf{u}(\mathbf{x},t) = \mathbf{u}(\mathbf{x},q_1,q_2,\cdots q_n) = \mathbf{u}(\mathbf{x},\mathbf{q}), \qquad \mathbf{x} = \mathbf{x}(\mathbf{q}), \qquad \mathbf{q} = \mathbf{q}(t) \tag{1.70}$$

$$q_i = q_i(t) \qquad i = 1,2\cdots n \tag{1.71}$$

Clearly, such a system has only n degrees of freedom. Notice also that in the expression of the displacement vector in terms of the generalized coordinates, we do *not* include t as an explicit parameter, since the dependence of $\mathbf{u}$ on t is completely captured by the coordinates themselves. From this definition, it follows immediately that

$$\dot{q}_i = \frac{dq_i}{dt} = \frac{\partial q_i}{\partial t} \tag{1.73}$$

and

$$\dot{\mathbf{u}} = \frac{d\mathbf{u}}{dt} = \sum_{i=1}^{n} \frac{\partial \mathbf{u}}{\partial q_i}\frac{\partial q_i}{\partial t} + \frac{\partial \mathbf{u}}{\partial t} = \sum_{i=1}^{n} \frac{\partial \mathbf{u}}{\partial q_i}\dot{q}_i \tag{1.74}$$

That is,

$$\boxed{\dot{\mathbf{u}} = \sum_{i=1}^{n} \frac{\partial \mathbf{u}}{\partial q_i}\dot{q}_i} \tag{1.75}$$

In the last expression, the partial derivative with respect to time was dropped because $\mathbf{u}$ does not depend explicitly on t. Also,

$$\frac{\partial \dot{\mathbf{u}}}{\partial \dot{q}_i} = \frac{\partial}{\partial \dot{q}_i}\left[\frac{\partial \mathbf{u}}{\partial q_1}\dot{q}_1 + \frac{\partial \mathbf{u}}{\partial q_2}\dot{q}_2 + \cdots \frac{\partial \mathbf{u}}{\partial q_n}\dot{q}_n\right] = \frac{\partial \mathbf{u}}{\partial q_i} \tag{1.76}$$

which again results from the fact that the partial derivatives of $\mathbf{u}$ with respect to q_i do not depend on $\dot{q}_i$. Hence, we have obtained a most important rule, which is known as the *cancellation of the dots*:

$$\boxed{\frac{\partial \dot{\mathbf{u}}}{\partial \dot{q}_i} = \frac{\partial \mathbf{u}}{\partial q_i}} \tag{1.77}$$

Finally,

$$\frac{d}{dt}\frac{\partial \mathbf{u}}{\partial q_i} = \sum_{j=1}^{n}\frac{\partial}{\partial q_j}\left(\frac{\partial \mathbf{u}}{\partial q_i}\right)\dot{q}_j = \frac{\partial}{\partial q_i}\sum_{j=1}^{n}\frac{\partial \mathbf{u}}{\partial q_j}\dot{q}_j = \frac{\partial \dot{\mathbf{u}}}{\partial q_i} \tag{1.78}$$

That is,

$$\boxed{\frac{d}{dt}\frac{\partial \mathbf{u}}{\partial q_i} = \frac{\partial \dot{\mathbf{u}}}{\partial q_i}} \tag{1.79}$$

If we define the matrix of partial derivatives and the generalized coordinates vector as

$$\frac{\partial \mathbf{u}}{\partial \mathbf{q}} \equiv \begin{Bmatrix} \dfrac{\partial \mathbf{u}}{\partial q_1} \\[2mm] \dfrac{\partial \mathbf{u}}{\partial q_2} \\[2mm] \vdots \\[2mm] \dfrac{\partial \mathbf{u}}{\partial q_n} \end{Bmatrix} \quad \text{and} \quad \mathbf{q} = \begin{Bmatrix} q_1 \\ q_2 \\ \vdots \\ q_n \end{Bmatrix} \tag{1.80}$$

then we can write the previous boxed expressions in compact form as

$$\boxed{\dot{\mathbf{u}} = \dot{\mathbf{q}}^T \frac{\partial \mathbf{u}}{\partial \mathbf{q}}, \qquad \frac{\partial \dot{\mathbf{u}}}{\partial \dot{\mathbf{q}}} = \frac{\partial \mathbf{u}}{\partial \mathbf{q}}, \qquad \frac{d}{dt}\frac{\partial \mathbf{u}}{\partial \mathbf{q}} = \frac{\partial \dot{\mathbf{u}}}{\partial \mathbf{q}}} \tag{1.81}$$

In terms of these definitions, a perturbation (or complete differential) $\delta \mathbf{u}$ can be written symbolically as

$$\delta \mathbf{u} = \delta \mathbf{q}^T \frac{\partial \mathbf{u}}{\partial \mathbf{q}} \quad \text{and / or} \quad \delta \mathbf{u}^T = \left(\frac{\partial \mathbf{u}}{\partial \mathbf{q}}\right)^T \delta \mathbf{q} \equiv \delta \mathbf{q}^T \frac{\partial \mathbf{u}^T}{\partial \mathbf{q}} \tag{1.82}$$

This shorthand is also useful in the evaluation of quadratic forms. For example, consider a symmetric matrix $\mathbf{A}$ forming the scalar function f

$$f = \tfrac{1}{2}\mathbf{q}^T \mathbf{A}\mathbf{q} \tag{1.83}$$

The partial derivative of f with respect to q_j (the result of which is a scalar!) is

$$\begin{aligned}
\frac{\partial f}{\partial q_j} &= \frac{\partial}{\partial q_j}\left(\tfrac{1}{2}\mathbf{q}^T \mathbf{A}\mathbf{q}\right) = \tfrac{1}{2}\frac{\partial \mathbf{q}^T}{\partial q_j}\mathbf{A}\mathbf{q} + \tfrac{1}{2}\mathbf{q}^T \mathbf{A}\frac{\partial \mathbf{q}}{\partial q_j} \\[2mm]
&= \tfrac{1}{2}\frac{\partial \mathbf{q}^T}{\partial q_j}\left(\mathbf{A} + \mathbf{A}^T\right)\mathbf{q} \\[2mm]
&= \frac{\partial \mathbf{q}^T}{\partial q_j}\mathbf{A}\mathbf{q} = \mathbf{e}_j^T \mathbf{A}\mathbf{q}
\end{aligned} \tag{1.84}$$

in which $\mathbf{e}_j$ is a vector whose jth element is 1, and all others are zero. The second line above follows because the second term on the first line is a scalar, and the transpose of a scalar equals the scalar itself. In addition, $\mathbf{A} + \mathbf{A}^T = 2\mathbf{A}$ because the matrix is symmetric. Writing all of the partial derivatives in matrix form, we obtain

$$\frac{\partial f}{\partial \mathbf{q}} = \frac{1}{2} \frac{\partial \mathbf{q}^T \mathbf{A} \mathbf{q}}{\partial \mathbf{q}} = \left\{ \frac{\partial f}{\partial q_j} \right\} = \left\{ \mathbf{e}_j^T \right\} \mathbf{A} \mathbf{q} = \mathbf{I} \mathbf{A} \mathbf{q} \equiv \mathbf{A} \mathbf{q} \tag{1.85}$$

1.10.2 Lagrange's Equations

We proceed next to derive Lagrange's equations. For this purpose, we start from the principle of virtual work, which states: A system is in dynamic equilibrium if the virtual work done by the external forces, including the work done by the inertia or D'Alembert forces (i.e., the change in kinetic energy), equals the increase in strain (or deformation) energy plus the energy dissipated by damping. Clearly, this is simply a statement that no energy gets lost, only transformed (i.e., conservation of energy).

$$\delta W_{\text{elastic}} + \delta W_{\text{damping}} = \delta W_{\text{loads}} + \delta W_{\text{inertia}} \tag{1.86}$$

in which the δ symbol denotes a *virtual* quantity (i.e., an imagined perturbation or *variation*).

(a) Elastic Forces

We begin on the left-hand side with the virtual work done by elastic forces, and to focus ideas, we assume at first and without loss of generality that the elastic forces originate from a number springs of stiffnesses k_j that undergo instantaneous elongations $e_j = e_j(q_1 \cdots q_n)$ along directions that generally change with time. Thus, the elastic forces along the instantaneous local directions of these members are

$$\mathbf{p}_{\text{elastic}} = \sum_{\text{springs}} k_j e_j = \mathbf{K} \mathbf{e} \tag{1.87}$$

in which $\mathbf{K} = \text{diag}(k_j)$ is the constant (time invariant) matrix of spring stiffnesses. Applying virtual elongations $\delta \mathbf{e}$ along the direction of the forces, we conclude that they perform an amount of work given by

$$\delta W_{\text{elastic}} = \delta \mathbf{e}^T \mathbf{p}_{\text{elastic}} = \delta \mathbf{e}^T \mathbf{K} \mathbf{e} \tag{1.88}$$

But

$$\delta \mathbf{e}^T = \sum_{i=1}^{N} \delta q_i \frac{\partial \mathbf{e}^T}{\partial q_i} \equiv \delta \mathbf{q}^T \frac{\partial \mathbf{e}^T}{\partial \mathbf{q}} \tag{1.89}$$

so

$$\delta W_{\text{elastic}} = \delta \mathbf{q}^T \frac{\partial \mathbf{e}^T}{\partial \mathbf{q}} \mathbf{p}_{\text{elastic}} = \delta \mathbf{q}^T \mathbf{f}_{\text{elastic}} = \delta \mathbf{q}^T \frac{\partial V}{\partial \mathbf{q}} = \delta \left(\tfrac{1}{2} \mathbf{e}^T \mathbf{K} \mathbf{e} \right) = \delta V \tag{1.90}$$

in which[3] $V = \frac{1}{2}\mathbf{e}^T\mathbf{K}\mathbf{e}$ is the *elastic potential*, or also the potential energy function of the elastic system. More generally, in the case of continuous solid media, the strain (i.e., potential) energy function is of the form

$$V = \frac{1}{2}\iiint_{Vol} \boldsymbol{\varepsilon}^T\mathbf{E}\,\boldsymbol{\varepsilon}\,dVol \tag{1.91}$$

with $\boldsymbol{\varepsilon}$ and $\mathbf{E}$ being, respectively, the strain vector at a point and the constitutive matrix (i.e., the matrix of elastic constants), and *Vol* is the volume of the body in question. Similar expressions can be written for other systems, such as beams, plates, or shells. Thus, we conclude that the generalized elastic forces are

$$\mathbf{f}_{\text{elastic}} = \frac{\partial\mathbf{e}^T}{\partial\mathbf{q}}\mathbf{K}\mathbf{e} = \frac{\partial V}{\partial\mathbf{q}} \tag{1.92}$$

(b) Damping Forces

Next, we consider the virtual work dissipated by the generalized damping forces. For *viscous damping*, these forces are solely a function of the instantaneous rate of deformation of the members, and not of their current position or past deformation history. Hence, their virtual work can be obtained from a *damping potential D* in ways that perfectly parallel those of the elastic forces *V*. Indeed, consider a set of viscous dashpots c_i oriented along some generally time-varying directions that are subjected to instantaneous rates of deformation $\dot{e}_i$ along those same directions:

$$\mathbf{p}_{\text{damping}} = \sum_{\text{dampers}} c_j\,\dot{e}_j = \mathbf{C}\dot{\mathbf{e}}, \qquad \mathbf{C} = \text{diag}(c_1, c_2, \ldots), \qquad \dot{\mathbf{e}} = \begin{bmatrix} \dot{e}_1 & \dot{e}_2 & \cdots \end{bmatrix}^T = \dot{\mathbf{e}}(\mathbf{q}, \dot{\mathbf{q}}) \tag{1.93}$$

The virtual work done by these forces is then

$$\delta W_{\text{damping}} = \delta\mathbf{e}^T\mathbf{p}_{\text{damping}} = \delta\mathbf{e}^T\mathbf{C}\dot{\mathbf{e}} \tag{1.94}$$

But

$$\delta\mathbf{e}^T = \delta\mathbf{q}^T\frac{\partial\mathbf{e}^T}{\partial\mathbf{q}} \equiv \delta\mathbf{q}^T\frac{\partial\dot{\mathbf{e}}^T}{\partial\dot{\mathbf{q}}} \tag{1.95}$$

The last identity is due to the cancellation of the dots rule. Hence

$$\delta W_{\text{damping}} = \delta\mathbf{q}^T\mathbf{f}_{\text{damping}} = \delta\mathbf{q}^T\frac{\partial\dot{\mathbf{e}}}{\partial\dot{\mathbf{q}}}\mathbf{p}_{\text{damping}} = \delta\mathbf{q}^T\frac{\partial\dot{\mathbf{e}}^T}{\partial\dot{\mathbf{q}}}\mathbf{C}\dot{\mathbf{e}} \tag{1.96}$$

Hence, the generalized viscous damping forces are

$$\mathbf{f}_{\text{damping}} = \frac{\partial\dot{\mathbf{e}}^T}{\partial\dot{\mathbf{q}}}\mathbf{C}\dot{\mathbf{e}} = \frac{1}{2}\left(\frac{\partial\dot{\mathbf{e}}^T}{\partial\dot{\mathbf{q}}}\mathbf{C}\dot{\mathbf{e}} + \dot{\mathbf{e}}^T\mathbf{C}\frac{\partial\dot{\mathbf{e}}}{\partial\dot{\mathbf{q}}}\right) = \frac{\partial}{\partial\dot{\mathbf{q}}}\left(\frac{1}{2}\dot{\mathbf{e}}^T\mathbf{C}\dot{\mathbf{e}}\right) = \frac{\partial D}{\partial\dot{\mathbf{q}}} \tag{1.97}$$

---

[3] $\delta V = \delta\left(\frac{1}{2}\mathbf{e}^T\mathbf{K}\mathbf{e}\right) = \frac{1}{2}\left(\delta\mathbf{e}^T\mathbf{K}\mathbf{e} + \mathbf{e}^T\mathbf{K}\,\delta\mathbf{e}\right) = \frac{1}{2}\left(\delta\mathbf{e}^T\mathbf{K}\mathbf{e} + \delta\mathbf{e}^T\mathbf{K}^T\mathbf{e}\right) = \delta\mathbf{e}^T\mathbf{K}\mathbf{e}$

in which $D = \frac{1}{2} \sum\limits_{\text{dashpots}} c_j \dot{e}_j^2 = \frac{1}{2} \dot{\mathbf{e}}^T \mathbf{C} \dot{\mathbf{e}}$ is the "viscous damping potential," which has dimen-

sions of instantaneous power dissipation, and not of energy. In the case of continuous media and small strain, the damping potential is of the form

$$D = \frac{1}{2} \iiint\limits_{Vol} \dot{\boldsymbol{\varepsilon}}^T \mathbf{D} \, \dot{\boldsymbol{\varepsilon}} \, dVol \tag{1.98}$$

with $\dot{\boldsymbol{\varepsilon}}$ and $\mathbf{D}$ being, respectively, the vector of instantaneous rate of strain at a point and the viscosity matrix (i.e., the matrix of viscous constants).

(c) External Loads

We consider next the work done by the external loads. If $\mathbf{p}_j = \mathbf{p}_j(\mathbf{x}, t)$ are the external body forces acting on the system at discrete points $\mathbf{x}_j$, then

$$\delta W_{\text{loads}} = \sum\limits_{\text{loads}} \delta \mathbf{u}_j^T \mathbf{p}_j = \delta \mathbf{q}^T \sum\limits_{\text{loads}} \frac{\partial \mathbf{u}_j^T}{\partial \mathbf{q}} \mathbf{p}_j = \delta \mathbf{q}^T \mathbf{f}_e \tag{1.99}$$

in which $\mathbf{f}_e$ is the *generalized load vector*

$$\mathbf{f}_e = \sum\limits_{\text{loads}} \frac{\partial \mathbf{u}_j^T}{\partial \mathbf{q}} \mathbf{p}_j \tag{1.100}$$

In particular, if the loads do not depend on the displacements (i.e., on the generalized coordinates), then

$$\mathbf{f}_e = \frac{\partial}{\partial \mathbf{q}} \sum\limits_{\text{loads}} (\mathbf{u}_j^T \mathbf{p}_j) = \frac{\partial W_e}{\partial \mathbf{q}} \tag{1.101}$$

in which W_e is the external work function. Alternatively, if the loads can be derived from a potential V_e, then $W_e = -V_e$. In the case of continuous solid media subjected to body loads $\mathbf{b}(\mathbf{x}, t)$, the corresponding formulae are

$$\delta W_{\text{loads}} = \iiint\limits_{Vol} \delta \mathbf{u}^T \mathbf{b} \, dVol = \iiint\limits_{Vol} \left(\delta \mathbf{q}^T \frac{\partial \mathbf{u}^T}{\partial \mathbf{q}} \right) \mathbf{b} \, dVol = \delta \mathbf{q}^T \iiint\limits_{Vol} \frac{\partial \mathbf{u}^T}{\partial \mathbf{q}} \mathbf{b} \, dVol = \delta \mathbf{q}^T \mathbf{f}_e \tag{1.102}$$

with

$$\mathbf{f}_e = \iiint\limits_{Vol} \frac{\partial \mathbf{u}^T}{\partial \mathbf{q}} \mathbf{b} \, dVol \tag{1.103}$$

Again, if the actual body loads are not a function of the generalized coordinates, or if they can be derived from a load potential V_e, then

$$\mathbf{f}_e = \frac{\partial}{\partial \mathbf{q}} \iiint\limits_{Vol} \mathbf{u}^T \mathbf{b} \, dVol = \frac{\partial W_e}{\partial \mathbf{q}} = -\frac{\partial V_e}{\partial \mathbf{q}} \tag{1.104}$$

(d) Inertia Forces

Finally, we consider the inertia or D'Alembert forces. Their virtual work is

$$\delta W_{\text{inertia}} = \iiint_{Vol} \delta \mathbf{u}^T \left(-\rho \ddot{\mathbf{u}} \right) dVol = -\delta \mathbf{q}^T \iiint_V \frac{\partial \mathbf{u}^T}{\partial \mathbf{q}} \ddot{\mathbf{u}} \, dm \tag{1.105}$$

in which $dm = \rho \, dVol$ is the elementary mass. But

$$\begin{aligned}
\frac{d}{dt} \left(\frac{\partial \mathbf{u}^T}{\partial \mathbf{q}} \dot{\mathbf{u}} \right) &= \frac{d}{dt} \left(\frac{\partial \mathbf{u}^T}{\partial \mathbf{q}} \right) \dot{\mathbf{u}} + \frac{\partial \mathbf{u}^T}{\partial \mathbf{q}} \ddot{\mathbf{u}} \\
&= \frac{\partial \dot{\mathbf{u}}^T}{\partial \mathbf{q}} \dot{\mathbf{u}} + \frac{\partial \mathbf{u}^T}{\partial \mathbf{q}} \ddot{\mathbf{u}}
\end{aligned} \tag{1.106}$$

Hence

$$\frac{\partial \mathbf{u}^T}{\partial \mathbf{q}} \ddot{\mathbf{u}} = \frac{d}{dt} \left(\frac{\partial \mathbf{u}^T}{\partial \mathbf{q}} \dot{\mathbf{u}} \right) - \frac{\partial \dot{\mathbf{u}}^T}{\partial \mathbf{q}} \dot{\mathbf{u}} \tag{1.107}$$

and

$$\delta W_{\text{inertia}} = -\delta \mathbf{q}^T \iiint_V \left[\frac{d}{dt} \left(\frac{\partial \mathbf{u}^T}{\partial \mathbf{q}} \dot{\mathbf{u}} \right) - \frac{\partial \dot{\mathbf{u}}^T}{\partial \mathbf{q}} \dot{\mathbf{u}} \right] dm \tag{1.108}$$

We introduce at this point the *kinetic energy* function

$$K = \tfrac{1}{2} \iiint_V \dot{\mathbf{u}}^T \dot{\mathbf{u}} \, dm \tag{1.109}$$

which ostensibly satisfies

$$\frac{\partial K}{\partial q_i} = \tfrac{1}{2} \iiint_V \left[\frac{\partial \dot{\mathbf{u}}^T}{\partial q_i} \dot{\mathbf{u}} + \dot{\mathbf{u}}^T \frac{\partial \dot{\mathbf{u}}}{\partial q_i} \right] dm = \iiint_V \frac{\partial \dot{\mathbf{u}}^T}{\partial q_i} \dot{\mathbf{u}} \, dm \tag{1.110}$$

and

$$\frac{\partial K}{\partial \dot{q}_i} = \tfrac{1}{2} \iiint_V \left[\frac{\partial \dot{\mathbf{u}}^T}{\partial \dot{q}_i} \dot{\mathbf{u}} + \dot{\mathbf{u}}^T \frac{\partial \dot{\mathbf{u}}}{\partial \dot{q}_i} \right] dm = \iiint_V \frac{\partial \dot{\mathbf{u}}^T}{\partial \dot{q}_i} \dot{\mathbf{u}} \, dm \equiv \iiint_V \frac{\partial \mathbf{u}^T}{\partial q_i} \dot{\mathbf{u}} \, dm \tag{1.111}$$

(The last identity is due to the cancellation of dots rule.) Hence

$$\delta W_{\text{inertia}} = -\delta \mathbf{q}^T \left[\frac{d}{dt} \frac{\partial K}{\partial \dot{\mathbf{q}}} - \frac{\partial K}{\partial \mathbf{q}} \right] \tag{1.112}$$

(e) Combined Virtual Work

Combining all four terms and factoring out the virtual variation of generalized coordinates, we obtain

$$\delta \mathbf{q}^T \left[\frac{d}{dt} \frac{\partial K}{\partial \dot{\mathbf{q}}} - \frac{\partial K}{\partial \mathbf{q}} + \frac{\partial V}{\partial \mathbf{q}} + \frac{\partial D}{\partial \dot{\mathbf{q}}} - \mathbf{f}_e \right] = 0 \tag{1.113}$$

and since this must be true for arbitrary variations of the generalized coordinates $\delta\mathbf{q}$, then

$$\frac{d}{dt}\frac{\partial K}{\partial \dot{\mathbf{q}}} - \frac{\partial K}{\partial \mathbf{q}} + \frac{\partial V}{\partial \mathbf{q}} + \frac{\partial D}{\partial \dot{\mathbf{q}}} = \mathbf{f}_e \qquad \mathbf{f}_e = \sum_{\text{loads}} \frac{\partial \mathbf{u}_j^T}{\partial \mathbf{q}} \mathbf{p}_j \tag{1.114}$$

or in terms of the individual components

$$\boxed{\frac{d}{dt}\frac{\partial K}{\partial \dot{q}_i} - \frac{\partial K}{\partial q_i} + \frac{\partial V}{\partial q_i} + \frac{\partial D}{\partial \dot{q}_i} = f_{ei}} \qquad \boxed{f_{ei} = \sum_{\text{loads}} \frac{\partial \mathbf{u}_j^T}{\partial q_i} \mathbf{p}_j} \tag{1.115}$$

In most cases, the strain energy function V does not depend on the $\dot{q}_i$, so $\dfrac{\partial V}{\partial \dot{q}} = 0$, and the loads can be derived from a potential. If so, we can write the previous expression as

$$\boxed{\frac{d}{dt}\frac{\partial L}{\partial \dot{q}_i} - \frac{\partial L}{\partial q_i} + \frac{\partial D}{\partial \dot{q}_i} = \frac{\partial W_e}{\partial q_i} = -\frac{\partial V_e}{\partial q_i}} \qquad \boxed{W_e = \sum_{\text{loads}} (\mathbf{u}_j^T \mathbf{p}_j)} \tag{1.116}$$

in which

$$L = K - V \tag{1.117}$$

is the *Lagrangian*. We demonstrate the use of these equations with some examples.

Example 1: Cylinder Rolling on Surface

Consider a cylinder of radius R, mass m and mass moment of inertia $J = \frac{1}{2}mR^2$ that rolls without friction or slipping on a flat surface, as shown below. A harness of mass m_1 attached to the axis of the cylinder is connected to a spring k_1 in line with the axis. A second spring k_2 is connected to a plate of mass m_2 that rests on the cylinder and is dragged along laterally without slipping Determine the equation of motion.

Let u_1, u_2 be the lateral displacements of the plates at the center and top of the cylinder, respectively, and let θ be the rotation of the cylinder. Since the system rotates instantaneously about the contact point without sliding, this point has zero instantaneous velocity. Hence,

$$u_1 = \theta R \qquad u_2 = 2\theta R \tag{1.118}$$

Thus, this system has only one generalized coordinate (or degree of freedom), which is the rotation θ. Assuming the springs to be unstressed in the equilibrium position, then the elastic and kinetic energies are

$$V = \tfrac{1}{2}k_1 u_1^2 + \tfrac{1}{2}k_2 u_2^2 = \tfrac{1}{2}(k_1 + 4k_2)R^2\theta^2 \tag{1.119}$$

$$K = \tfrac{1}{2}(m + m_1)\dot{u}_1^2 + \tfrac{1}{2}m_2\dot{u}_2^2 + \tfrac{1}{2}J\dot{\theta}_2^2 = \tfrac{1}{2}\left(\tfrac{3}{2}m + m_1 + 4m_2\right)R^2\dot{\theta}^2 \tag{1.120}$$

Hence

$$\frac{\partial V}{\partial \theta} = (k_1 + 4k_2)R^2\theta \tag{1.121}$$

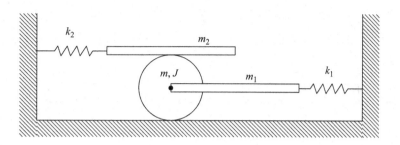

Figure 1.40

$$\frac{\partial K}{\partial \dot{\theta}} = \left(\tfrac{3}{2}m + m_1 + 4m_2\right)R^2\dot{\theta} \tag{1.122}$$

$$\frac{d}{dt}\frac{\partial K}{\partial \dot{\theta}} = \left(\tfrac{3}{2}m + m_1 + 4m_2\right)R^2\ddot{\theta} \tag{1.123}$$

and

$$\left(\tfrac{3}{2}m + m_1 + 4m_2\right)R^2\ddot{\theta} + \left(k_1 + 4k_2\right)R^2\theta = 0 \tag{1.124}$$

That is,

$$\left(\tfrac{3}{2}m + m_1 + 4m_2\right)\ddot{\theta} + \left(k_1 + 4k_2\right)\theta = 0 \tag{1.125}$$

Observe that a derivation using equilibrium concepts and free body diagrams would have required considerably more work.

Example 2: Rigid Bar on Springs and Dashpots

Consider a perfectly rigid bar of length $4L$, mounted on springs and dashpots, which is subjected to external time-varying forces, as in Figure 1.41. The bar has total mass m and mass moment of inertia J about the center of mass, which is known to be at the geometric center. The system is constrained to only translate vertically and/or rotate about some point in the plane. Derive the equations of motion for small rotations.

Clearly, this is a 2-DOF system. We choose the translation and rotation of the center of mass as the generalized coordinates, that is,

$$q_1 \equiv u \qquad \text{and} \qquad q_2 = \theta \tag{1.126}$$

In terms of these generalized coordinates, the displacements at the points of attachment of the springs, force, and dashpot are, respectively

$$u_\alpha = u - 2L\theta \qquad u_\beta = u + L\theta \qquad u_\gamma = u - L\theta \qquad u_\delta = u + 2L\theta \tag{1.127}$$

which are positive in the upward direction. In term of these parameters, we have

Kinetic energy :
$$K = \tfrac{1}{2}(m\dot{u}^2 + J\dot{\theta}^2) \tag{1.128}$$

Damping potential :
$$D = \tfrac{1}{2}c\dot{u}_\delta^2 = \tfrac{1}{2}c(\dot{u} + 2L\dot{\theta})^2 \tag{1.129}$$

Elastic potential :
$$V = \tfrac{1}{2}\left(k_\alpha u_\alpha^2 + k_\beta u_\beta^2\right) = \tfrac{1}{2}\left[k_\alpha(u - 2L\theta)^2 + k_\beta(u + L\theta)^2\right] \tag{1.130}$$

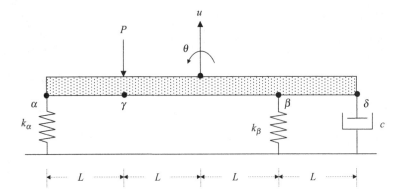

Figure 1.41

External work function : $W_e = (-p)u_\gamma = -p(u - L\theta)$

(negative, because force acts downward)
$$(1.131)$$

The required partial derivatives are

$$\frac{\partial K}{\partial \dot{u}} = m\dot{u} \quad \rightarrow \quad \frac{d}{dt}\frac{\partial K}{\partial \dot{u}} = m\ddot{u} \tag{1.132}$$

$$\frac{\partial K}{\partial \dot{\theta}} = J\dot{\theta} \quad \rightarrow \quad \frac{d}{dt}\frac{\partial K}{\partial \dot{\theta}} = J\ddot{\theta} \tag{1.133}$$

$$\frac{\partial D}{\partial \dot{u}} = c\left(\dot{u} + 2L\dot{\theta}\right) \qquad \frac{\partial D}{\partial \dot{\theta}} = c\left(\dot{u} + 2L\dot{\theta}\right)(2L) \tag{1.134}$$

$$\frac{\partial V}{\partial u} = k_\alpha\left(u - 2L\theta\right)(1) + k_\beta\left(u + L\theta\right)(1) = \left(k_\alpha + k_\beta\right)u + \left(k_\beta - 2k_\alpha\right)L\theta \tag{1.135}$$

$$\frac{\partial V}{\partial \theta} = k_\alpha\left(u - 2L\theta\right)(-2L) + k_\beta\left(u + L\theta\right)(L) = \left(k_\beta - 2k_\alpha\right)Lu + \left(4k_\alpha + k_\beta\right)L^2\theta \tag{1.136}$$

$$f_1 = \frac{\partial W_e}{\partial u} = -p \qquad f_2 = \frac{\partial W_e}{\partial \theta} = pL \tag{1.137}$$

Hence, the two Lagrange equations are

$$m\ddot{u} + c\left(\dot{u} + 2L\dot{\theta}\right) + \left(k_\alpha + k_\beta\right)u + \left(k_\beta - 2k_\alpha\right)L\theta = -p \tag{1.138}$$

and

$$J\ddot{\theta} + 2Lc\dot{u} + 4cL^2\dot{\theta} + \left(k_2 - 2k_1\right)Lu + \left(4k_1 + k_2\right)L^2\theta = pL \tag{1.139}$$

or in matrix form

$$\begin{bmatrix} m & 0 \\ 0 & J \end{bmatrix}\begin{Bmatrix} \ddot{u} \\ \ddot{\theta} \end{Bmatrix} + \begin{bmatrix} c & 2Lc \\ 2Lc & 4cL^2 \end{bmatrix}\begin{Bmatrix} \dot{u} \\ \dot{\theta} \end{Bmatrix} + \begin{bmatrix} k_\alpha + k_\beta & \left(k_\beta - 2k_\alpha\right)L \\ \left(k_\beta - 2k_\alpha\right)L & \left(4k_\alpha + k_\beta\right)L^2 \end{bmatrix}\begin{Bmatrix} u \\ \theta \end{Bmatrix} = \begin{Bmatrix} -p \\ pL \end{Bmatrix} \tag{1.140}$$

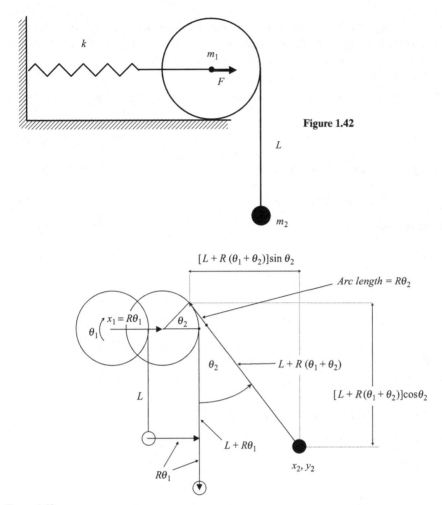

Figure 1.42

Figure 1.43

Example 3: Hanging Mass on Wheel that Rolls without Slippage

Consider a wheel of radius R, total mass m_1 and mass moment of inertia $J = \frac{1}{2}m_1 R^2$ that rolls without slipping on a flat horizontal surface. The axis of the wheel is attached via a harness to a horizontal spring of stiffness k, as shown. In addition, a second mass hangs from a string that is wound on the wheel. The length of this string is L when the wheel is in the starting position and the second mass hangs vertically. The axis of the wheel is subjected to a horizontal force $F(t)$. Considering the effect of gravity and neglecting any frictional losses, formulate the equations of motions using Lagrange's equations.

To formulate this problem, we begin by setting the origin of coordinates $x_1 = 0$, $y_1 = 0$ at the center of the wheel when the spring is *not* deformed, as shown in Figure 1.42, and denote the horizontal position of the wheel at later times by the coordinate x_1 of this center. Also, the second mass has an instantaneous position x_2, y_2 and oscillates as a pendulum, as shown in Figure 1.43, which also illustrates the various kinematic (i.e., motion) conditions. As the mass moves back and forth, the point at which it is tangent to the wheel changes with time.

We denote the angle by which the wheel rotates as θ_1 (positive clockwise), and the angle that the string forms with the vertical as θ_2 (positive *counter*clockwise!). Since the string remains taut as it oscillates (an assumption!), this is also the angle formed with the horizontal by the radius from the center of the wheel to the point at which the string is tangent. From Figure 1.43, the instantaneous positions of the two masses are as follows:

$$\text{Mass}\,1 : x_1 = R\theta_1, \qquad y_1 = 0 \tag{1.141}$$

$$\text{Mass}\,2 : x_2 = R\theta_1 + R\cos\theta_2 + \left[L + R\left(\theta_1 + \theta_2\right)\right]\sin\theta_2 \tag{1.142}$$

$$y_2 = R\sin\theta_2 - \left[L + R\left(\theta_1 + \theta_2\right)\right]\cos\theta_2 \tag{1.143}$$

Also, the chain rule gives

$$\frac{\partial \sin\theta_2}{\partial t} = \frac{\partial \theta_2}{\partial t}\frac{\partial \sin\theta_2}{\partial \theta_2} = \dot{\theta}_2\cos\theta_2, \qquad \frac{\partial \cos\theta_2}{\partial t} = \frac{\partial \theta_2}{\partial t}\frac{\partial \cos\theta_2}{\partial \theta_2} = -\dot{\theta}_2\sin\theta_2 \tag{1.144}$$

Hence,

$$\dot{x}_1 = R\dot{\theta}_1, \quad \dot{y}_1 = 0 \tag{1.145}$$

$$
\begin{aligned}
\dot{x}_2 &= R\dot{\theta}_1 - R\dot{\theta}_2\sin\theta_2 + R\left(\dot{\theta}_1 + \dot{\theta}_2\right)\sin\theta_2 + \left[L + R\left(\theta_1 + \theta_2\right)\right]\dot{\theta}_2\cos\theta_2 \\
&= R\dot{\theta}_1\left(1 + \sin\theta_2\right) + \left[L + R\left(\theta_1 + \theta_2\right)\right]\dot{\theta}_2\cos\theta_2
\end{aligned}
\tag{1.146}
$$

$$
\begin{aligned}
\dot{y}_2 &= R\dot{\theta}_2\cos\theta_2 - R\left(\dot{\theta}_1 + \dot{\theta}_2\right)\cos\theta_2 + \left[L + R\left(\theta_1 + \theta_2\right)\right]\dot{\theta}_2\sin\theta_2 \\
&= -R\dot{\theta}_1\cos\theta_2 + \left[L + R\left(\theta_1 + \theta_2\right)\right]\dot{\theta}_2\sin\theta_2
\end{aligned}
\tag{1.147}
$$

On the other hand, the kinetic energy is

$$
\begin{aligned}
K &= \tfrac{1}{2}\left[m_1\left(\dot{x}_1^2 + \dot{y}_1^2\right) + m_2\left(\dot{x}_2^2 + \dot{y}_2^2\right) + J\dot{\theta}_1^2\right] \\
&= \tfrac{1}{2}\left[\tfrac{3}{2}m_1 R^2\dot{\theta}_1^2 + m_2\left(\dot{x}_2^2 + \dot{y}_2^2\right)\right]
\end{aligned}
\tag{1.148}
$$

with the second term expressed in terms of Eqs. 1.146 and 1.147. Also, the potential energy is the sum of the energy in the spring and the gravitational energy:

$$
\begin{aligned}
V &= \tfrac{1}{2}kx_1^2 + \left(m_1 y_1 + m_2 y_2\right)g \\
&= \tfrac{1}{2}kR^2\theta_1^2 + m_2 y_2 g
\end{aligned}
\tag{1.149}
$$

Choosing generalized coordinates $q_1 \equiv \theta_1$ and $q_2 \equiv \theta_2$, then the various partial derivatives needed are as follows:

$$\frac{\partial \dot{x}_1}{\partial \theta_1} = 0, \qquad \frac{\partial \dot{x}_1}{\partial \theta_2} = 0 \qquad \left(\text{from Eq. 1.145}\right)$$

$$\frac{\partial \dot{x}_1}{\partial \dot{\theta}_1} = R, \qquad \frac{\partial \dot{x}_1}{\partial \dot{\theta}_2} = 0 \qquad \left(\text{from Eq. 1.145}\right)$$

$$\frac{\partial \dot{x}_2}{\partial \theta_1} = R\dot{\theta}_2\cos\theta_2 \qquad \left(\text{from Eq. 1.146}\right)$$

$$\frac{\partial \dot{x}_2}{\partial \theta_2} = R\left(\dot{\theta}_1 + \dot{\theta}_2\right)\cos\theta_2 - \left[L + R\left(\theta_1 + \theta_2\right)\right]\dot{\theta}_2\sin\theta_2 \qquad \text{(from Eq. 1.146)}$$

$$\frac{\partial \dot{x}_2}{\partial \dot{\theta}_1} = R\left(1 + \sin\theta_2\right), \qquad \frac{\partial \dot{x}_2}{\partial \dot{\theta}_2} = \left[L + R\left(\theta_1 + \theta_2\right)\right]\cos\theta_2 \qquad \text{(from Eq. 1.146)}$$

$$\frac{\partial \dot{y}_2}{\partial \theta_1} = R\cos\theta_2, \qquad \frac{\partial \dot{y}_2}{\partial \theta_2} = \left[L + R\left(\theta_1 + \theta_2\right)\right]\sin\theta_2 \qquad \text{(from Eq. 1.142)}$$

$$\frac{\partial \dot{y}_2}{\partial \theta_1} = R\dot{\theta}_2\sin\theta_2$$

$$\frac{\partial \dot{y}_2}{\partial \theta_2} = R\left(\dot{\theta}_1 + \dot{\theta}_2\right)\sin\theta_2 + \left[L + R\left(\theta_1 + \theta_2\right)\right]\dot{\theta}_2\cos\theta_2 \qquad \text{(from Eq. 1.147)}$$

$$\frac{\partial \dot{y}_2}{\partial \dot{\theta}_1} = -R\cos\theta_2, \qquad \frac{\partial \dot{y}_2}{\partial \dot{\theta}_2} = \left[L + R\left(\theta_1 + \theta_2\right)\right]\sin\theta_2 \qquad \text{(from Eq. 1.147)}$$

The *total* time derivatives that will be needed are then

$$\frac{d}{dt}\left(\frac{\partial \dot{x}_1}{\partial \dot{\theta}_1}\right) = 0, \qquad \frac{d}{dt}\left(\frac{\partial \dot{x}_1}{\partial \dot{\theta}_2}\right) = 0 \qquad (1.150)$$

$$\frac{d}{dt}\left(\frac{\partial \dot{x}_2}{\partial \dot{\theta}_1}\right) = R\dot{\theta}_2\cos\theta_2, \qquad \frac{d}{dt}\left(\frac{\partial \dot{x}_2}{\partial \dot{\theta}_2}\right) = R\left(\dot{\theta}_1 + \dot{\theta}_2\right)\cos\theta_2 - \left[L + R\left(\theta_1 + \theta_2\right)\right]\dot{\theta}_2\sin\theta_2 \qquad (1.151)$$

$$\frac{d}{dt}\left(\frac{\partial \dot{y}_2}{\partial \dot{\theta}_1}\right) = R\dot{\theta}_2\sin\theta_2, \qquad \frac{d}{dt}\left(\frac{\partial \dot{y}_2}{\partial \dot{\theta}_2}\right) = R\left(\dot{\theta}_1 + \dot{\theta}_2\right)\sin\theta_2 + \left[L + R\left(\theta_1 + \theta_2\right)\right]\dot{\theta}_2\cos\theta_2 \qquad (1.152)$$

Also,

$$\frac{\partial K}{\partial \theta_1} = m_2\left(\dot{x}_2\frac{\partial \dot{x}_2}{\partial \theta_1} + \dot{y}_2\frac{\partial \dot{y}_2}{\partial \theta_1}\right), \qquad \frac{\partial K}{\partial \dot{\theta}_1} = \tfrac{3}{2}m_1R^2\dot{\theta}_1 + m_2\left(\dot{x}_2\frac{\partial \dot{x}_2}{\partial \dot{\theta}_1} + \dot{y}_2\frac{\partial \dot{y}_2}{\partial \dot{\theta}_1}\right) \qquad (1.153)$$

$$\frac{\partial K}{\partial \theta_2} = m_2\left(\dot{x}_2\frac{\partial \dot{x}_2}{\partial \theta_2} + \dot{y}_2\frac{\partial \dot{y}_2}{\partial \theta_2}\right), \qquad \frac{\partial K}{\partial \dot{\theta}_2} = m_2\left(\dot{x}_2\frac{\partial \dot{x}_2}{\partial \dot{\theta}_2} + \dot{y}_2\frac{\partial \dot{y}_2}{\partial \dot{\theta}_2}\right) \qquad (1.154)$$

and

$$\frac{d}{dt}\left(\frac{\partial K}{\partial \dot{\theta}_1}\right) = \tfrac{3}{2}m_1R^2\ddot{\theta}_1 + m_2\left[\ddot{x}_2\frac{\partial \dot{x}_2}{\partial \dot{\theta}_1} + \ddot{y}_2\frac{\partial \dot{y}_2}{\partial \dot{\theta}_1} + \dot{x}_2\frac{d}{dt}\left(\frac{\partial \dot{x}_2}{\partial \dot{\theta}_1}\right) + \dot{y}_2\frac{d}{dt}\left(\frac{\partial \dot{y}_2}{\partial \dot{\theta}_1}\right)\right] \qquad (1.155)$$

$$\frac{d}{dt}\left(\frac{\partial K}{\partial \dot{\theta}_2}\right) = m_2\left[\ddot{x}_2\frac{\partial \dot{x}_2}{\partial \dot{\theta}_2} + \ddot{y}_2\frac{\partial \dot{y}_2}{\partial \dot{\theta}_2} + \dot{x}_2\frac{d}{dt}\left(\frac{\partial \dot{x}_2}{\partial \dot{\theta}_2}\right) + \dot{y}_2\frac{d}{dt}\left(\frac{\partial \dot{y}_2}{\partial \dot{\theta}_2}\right)\right] \qquad (1.156)$$

for which we have already expressions for all of the derivatives. On the other hand, the partial derivatives of the potential energy term is

$$\frac{\partial V}{\partial \theta_1} = kR^2\theta_1 + m_2g\frac{\partial y_2}{\partial \theta_1}, \qquad \frac{\partial V}{\partial \theta_2} = m_2g\frac{\partial y_2}{\partial \theta_2} \qquad (1.157)$$

Also,

$$W_e = Fx_1 = FR\theta_1, \qquad \frac{\partial W_e}{\partial \theta_1} = FR, \qquad \frac{\partial W_e}{\partial \theta_2} = 0 \tag{1.158}$$

Finally,

$$\frac{d}{dt}\left(\frac{\partial K}{\partial \dot{\theta}_j}\right) - \frac{\partial K}{\partial \theta_j} + \frac{\partial V}{\partial \theta_j} = \frac{\partial W_e}{\partial \theta_j}, \qquad j = 1,2 \tag{1.159}$$

which is formed with the expressions already found.

After solving the two non-linear equations of motion (which can be done only numerically), one must verify that the tension in the string is always nonnegative (i.e., zero or positive). This will depend in turn on the initial conditions and the actual variation of the force with time, and of course, on the material properties and physical dimensions. A linear solution for small oscillations can be obtained by setting $\sin\theta_2 \approx \theta_2, \cos\theta_2 \approx 1$ and neglecting powers and products of motions, for example, $\theta_1\theta_2 \approx 0$, and so on.

Example 4: Pendulum with Multiple Masses

A pendulum consists of a chain of n masses m_j $(j = 1, 2, ...n)$ that are connected by rigid, massless links of length L_j that are pivoted to one another, as shown below. Determine the equations of free vibration for arbitrarily large oscillations.

If θ_j are the rotations of each link with respect to the vertical, then the horizontal (i.e., left to right) and vertical displacements (i.e., upward from the equilibrium position) are

$$x_j = \sum_{i=1}^{j} L_i \sin\theta_i \tag{1.160}$$

$$y_j = \sum_{i=1}^{j} L_i \left(1 - \cos\theta_i\right) \tag{1.161}$$

Hence, the velocities are

$$\dot{x}_j = \sum_{i=1}^{j} L_i \dot{\theta}_i \cos\theta_i \tag{1.162}$$

$$\dot{y}_j = \sum_{i=1}^{j} L_i \dot{\theta}_i \sin\theta_i \tag{1.163}$$

The increase in gravitational energy from the rest position and the kinetic energy are then

$$V = \sum_{j=1}^{n} m_j\, g\, y_j = g\sum_{j=1}^{n} m_j \sum_{i=1}^{j} L_i \left(1 - \cos\theta_i\right) \tag{1.164}$$

$$K = \tfrac{1}{2}\sum_{j=1}^{n} m_j \left\{\left[\sum_{i=1}^{j} L_i \dot{\theta}_i \cos\theta_i\right]^2 + \left[\sum_{i=1}^{j} L_i \dot{\theta}_i \sin\theta_i\right]^2\right\} \tag{1.165}$$

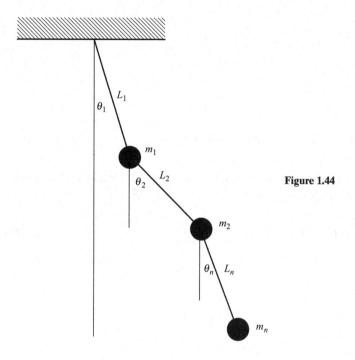

Figure 1.44

Hence

$$\frac{\partial V}{\partial \theta_k} = g \sum_{j=k}^{n} m_j L_k \sin \theta_k = g\, L_k \sin \theta_k \sum_{j=k}^{n} m_j \tag{1.166}$$

$$\frac{\partial K}{\partial \dot{\theta}_k} = L_k \sum_{j=k}^{n} m_j \sum_{i=1}^{j} L_i\, \dot{\theta}_i \left(\cos \theta_i \; \cos \theta_k + \sin \theta_i \; \sin \theta_k \right) \tag{1.167}$$

(Observe that $\dfrac{\partial K}{\partial \dot{\theta}_k} = 0$ for $k > j$). Hence,

$$\frac{\partial K}{\partial \dot{\theta}_k} = L_k \sum_{j=k}^{n} m_j \sum_{i=1}^{j} L_i\, \dot{\theta}_i \cos \left(\theta_k - \theta_i \right) \tag{1.168}$$

and

$$\frac{d}{dt} \frac{\partial K}{\partial \dot{\theta}_k} = L_k \sum_{j=k}^{n} m_j \sum_{i=1}^{j} L_i \left\{ \ddot{\theta}_i \cos \left(\theta_k - \theta_i \right) - \dot{\theta}_i \left(\dot{\theta}_k - \dot{\theta}_i \right) \sin \left(\theta_k - \theta_i \right) \right\} \tag{1.169}$$

Also

$$\begin{aligned}
\frac{\partial K}{\partial \theta_k} &= L_k \sum_{j=k}^{n} m_j \sum_{i=1}^{j} L_i\, \dot{\theta}_i\, \dot{\theta}_k \left\{ -\cos \theta_i \sin \theta_k + \sin \theta_i \cos \theta_k \right) \right\} \\
&= -L_k \sum_{j=k}^{n} m_j \sum_{i=1}^{j} L_i\, \dot{\theta}_i\, \dot{\theta}_k \sin \left(\theta_k - \theta_i \right)
\end{aligned} \tag{1.170}$$

Finally,

$$\frac{d}{dt}\frac{\partial K}{\partial \dot{\theta}_k} - \frac{\partial K}{\partial \theta_k} + \frac{\partial V}{\partial \theta_k} = 0 \tag{1.171}$$

That is,

$$\boxed{\sum_{j=k}^{n} m_j \sum_{i=1}^{j} L_i \left\{ \ddot{\theta}_i \cos(\theta_k - \theta_i) + \dot{\theta}_i^2 \sin(\theta_k - \theta_i) \right\} + g \sin\theta_k \sum_{j=k}^{n} m_j = 0} \tag{1.172}$$

If rotations remain small, then these equations can be linearized by neglecting quadratic terms and approximating the trigonometric functions, which results in

$$\sum_{j=k}^{n} m_j \sum_{i=1}^{j} L_i \ddot{\theta}_i + g \theta_k \sum_{j=k}^{n} m_j = 0 \qquad k = 1,\ 2, \ldots n \tag{1.173}$$

2 Single Degree of Freedom Systems

2.1 The Damped SDOF Oscillator

A single degree of freedom (SDOF) system is one for which the behavior of interest can be fully described at any time by the value of a single state variable such as the displacement. No structure is ever an SDOF system, but this simple model is often sufficient in many practical cases to provide a reasonable estimate of global response parameters, say the maximum deformation exhibited by the more complex physical system at hand. In addition, as we shall see later, multiple degree of freedom systems or continuous systems can be analyzed as the combination of SDOF systems (i.e., modal analysis). Thus, understanding the dynamic characteristics of SDOF systems is essential to estimate the potential importance of dynamic effects in a physical structure and assess the need for more elaborate analyses and the proper interpretation of the results provided by the added sophistication.

Let m, c, and k be the mass, damping, and stiffness of the SDOF system, $p(t)$ the time-varying external force applied onto the mass, and $u(t)$ the resulting displacement, as shown in Figure 2.1. Considering the mass as a free body in space, it must satisfy the dynamic equilibrium condition

$$m\ddot{u} + c\dot{u} + ku = p(t) \tag{2.1}$$

which is a *linear equation*, that is, the response to a linear combination of two different external loads is the linear combination of the responses induced by each of the loads acting alone. As is well known, the solution to this *second-order differential equation* is of the general form

$$u(t) = u_h(t) + u_p(t) \tag{2.2}$$

in which the first term is referred to as the *homogeneous solution*, while the second term is the *particular solution*. The homogeneous part is the solution to the dynamic equilibrium equation with zero right-hand side, so it must represent a *free vibration* (i.e., without external excitation). The particular solution or *forced vibration* part, on the other hand, is *any* solution satisfying the nonhomogeneous differential equation. We shall consider both solutions in the ensuing.

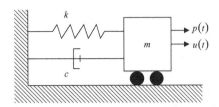

Figure 2.1. SDOF system.

2.1.1 Free Vibration: Homogeneous Solution

Consider the homogeneous equation

$$m\ddot{u}_h + c\dot{u}_h + ku_h = 0 \tag{2.3}$$

Dividing by m (which we assume to be nonzero),

$$\ddot{u}_h + \left(\frac{c}{m}\right)\dot{u}_h + \left(\frac{k}{m}\right)u_h = 0 \tag{2.4}$$

For the sake of greater generality, we shall temporarily allow m, c, k to be arbitrary functions of time, although they may *not* depend on the response. Next, we define the *integrating factor*

$$I(t) = \frac{1}{2}\int \frac{c(t)}{m(t)}dt \tag{2.5}$$

which can be used to express the homogeneous solution as

$$u_h = e^{-I(t)}y \tag{2.6}$$

involving the as yet unknown function $y = y(t)$. The time derivatives of this expression are

$$\dot{u}_h = e^{-I(t)}\left\{\dot{y} - \frac{c}{2m}y\right\} \tag{2.7}$$

$$\ddot{u}_h = e^{-I(t)}\left\{\ddot{y} - \frac{c}{m}\dot{y} + \left[\left(\frac{c}{2m}\right)^2 - \frac{d}{dt}\left(\frac{c}{2m}\right)\right]y\right\} \tag{2.8}$$

Substituting these three terms into the differential equation, we obtain

$$e^{-I(t)}\left\{\ddot{y} + \left[\frac{k}{m} - \left(\frac{c}{2m}\right)^2 - \frac{d}{dt}\left(\frac{c}{2m}\right)\right]y\right\} = 0 \tag{2.9}$$

which requires the term in braces to be zero:

$$\ddot{y} + \left[\frac{k}{m} - \left(\frac{c}{2m}\right)^2 - \frac{d}{dt}\left(\frac{c}{2m}\right)\right]y = 0 \tag{2.10}$$

This transformed homogeneous differential equation is simpler than the original equation, because it lacks the first time derivative of y.

We consider now the special case of **constant coefficients** m, c, and k. We define

$$\boxed{\omega_n = \sqrt{\frac{k}{m}}} \qquad \text{Angular frequency} \qquad (2.11)$$

$$\boxed{\xi = \frac{c}{2m\omega_n} = \frac{c}{2\sqrt{km}}} \qquad \text{Fraction of critical damping} \qquad (2.12)$$

Assuming temporarily that $\xi < 1$ (the *underdamped* case), we have then

$$\boxed{\omega_d = \omega_n \sqrt{1 - \xi^2}} \qquad \text{Damped angular frequency} \qquad (2.13)$$

In terms of these parameters, the integrating factor and the transformed differential equation are

$$I(t) = \xi\omega_n t \qquad (2.14)$$

$$\ddot{y} + \omega_d^2 y = 0 \qquad (2.15)$$

whose solution is

$$y = A\cos\omega_d t + B\sin\omega_d t \qquad (2.16)$$

where A and B are constants of integration. The free vibration solution (i.e., the homogeneous solution) is then

$$u_h = e^{-\xi\omega_n t}\left[A\cos\omega_d t + B\sin\omega_d t\right] \qquad (2.17)$$

The integration constants can be found by imposing initial conditions on displacement and velocity. The results are as follows.

Underdamped Case ($\xi < 1$)

$$\boxed{u_h = e^{-\xi\omega_n t}\left\{u_0\cos\omega_d t + \frac{\dot{u}_0 + \xi\omega_n u_0}{\omega_d}\sin\omega_d t\right\}} \qquad (2.18)$$

This is an oscillatory solution whose amplitude decays with time, as depicted in Figure 2.2.
Alternatively, the free vibration equation can be written as

$$u_h = Ce^{-\xi\omega_n t}\cos(\omega_d t - \varphi) \qquad (2.19)$$

in which

$$C = \sqrt{u_o^2 + \left(\frac{\dot{u}_o + \xi\omega_n u_o}{\omega_d}\right)^2} \qquad (2.20)$$

$$\varphi = \arctan\frac{\dot{u}_o + \xi\omega_n u_o}{\omega_d u_o} \qquad (2.21)$$

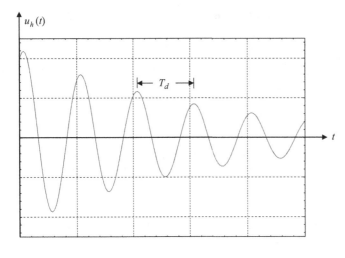

Figure 2.2. Free vibration, subcritical damping.

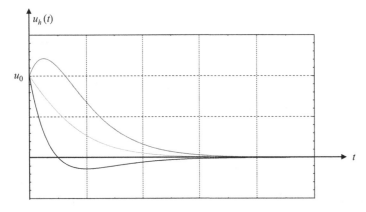

Figure 2.3. Free vibration, critical damping.

where C is the amplitude and φ is the phase angle. Observe that the latter can also be defined in terms of the time t_0 to the maximum response, that is, $\omega_d t_0 - \varphi = 0$, that is, $t_0 = \varphi / \omega_d$.

Critically Damped Case ($\xi = 1$)

The solution for this situation can be obtained from the previous one by considering the limit in which ξ approaches 1, which causes the damped angular frequency to approach zero:

$$\lim_{\omega_d \to 0} \frac{\sin \omega_d t}{\omega_d} = t \qquad \lim_{\omega_d \to 0} \cos \omega_d t = 1 \tag{2.22}$$

Hence

$$\boxed{u_h = e^{-\omega_n t} \left\{ u_0 + \left(\dot{u}_0 + \omega_n u_0 \right) t \right\}} \tag{2.23}$$

This solution may change sign at most once (which occurs when $u_0 > 0$ and $\dot{u}_0 + \omega_n u_0 < 0$, or vice versa), and thereafter it decays exponentially to zero. These various cases are illustrated in Figure 2.3.

Overdamped Case (ξ > 1)

We first need to introduce a new definition:

$$\boxed{\omega_o = \omega_n\sqrt{\xi^2-1}}\qquad\text{Overdamped angular frequency}\tag{2.24}$$

in terms of which the damped frequency can be written as

$$\omega_d = i\,\omega_o\tag{2.25}$$

in which i = √−1. Hence

$$\cos\omega_d t = \cos i\,\omega_o t = \cosh\omega_o t\tag{2.26}$$

$$\frac{\sin\omega_d t}{\omega_d} = \frac{\sin i\,\omega_o t}{i\,\omega_o} = \frac{i\sinh\omega_o t}{i\,\omega_o} = \frac{\sinh\omega_o t}{\omega_o}\tag{2.27}$$

It follows that

$$u_h = e^{-\xi\omega_n t}\left\{u_0\cosh\omega_o t + \frac{\dot{u}_0+\xi\omega_n u_0}{\omega_o}\sinh\omega_o t\right\}\tag{2.28}$$

This form is, however, computationally unstable. To change it into a stable form, we use the definitions of the hyperbolic functions, which leads us to

$$\boxed{u_h = \tfrac{1}{2}\left\{\left(u_0+\frac{\dot{u}_0+\xi\omega_n u_0}{\omega_o}\right)e^{-(\xi\omega_n-\omega_o)t} + \left(u_0-\frac{\dot{u}_0+\xi\omega_n u_0}{\omega_o}\right)e^{-(\xi\omega_n+\omega_o)t}\right\}}\tag{2.29}$$

When expressed in this form, the solution involves only decaying exponentials. Again, depending on the values of the initial conditions, this solution may reverse sign at most once. The vibration pattern is similar to that for critical damping.

2.1.2 Response Parameters

As seen in the preceding sections, the two main parameters affecting the free vibration of an SDOF system is the natural frequency ω_n (or natural period T_n) and the fraction of critical damping ξ. As we shall see later on, these are also the two parameters controlling the response to forced excitations. The natural frequency represents the way that the structure would like to vibrate on its own, and the fraction of damping defines how fast those vibrations decay with time. Most structural systems exhibit light damping, typically less than 10% (i.e., ξ < 0.1), and very frequently even less than 1% (ξ < 0.01), in which case there is negligible difference between the undamped natural frequency ω_n and the damped natural frequency ω_d. Indeed, even for 10% damping we would have $\omega_d = \omega_n\sqrt{1-0.1^2} = 0.995\,\omega_n$, that is, the damped frequency would still be merely 0.5% smaller than the undamped frequency. In those cases we can then assume that $\omega_d \approx \omega_n$.

In correspondence to the damped frequency ω_d, we also define the following two parameters:

$$f_d = f_n\sqrt{1-\xi^2}\qquad\text{Damped natural frequency in cycles per second}\tag{2.30}$$

Table 2.1. Relationship between angular frequency, natural frequency, and period

	Angular frequency (rad/s)	Natural frequency (Hz)	Natural period (s)
Angular frequency ω_n	ω_n	$2\pi f_n$	$\dfrac{2\pi}{T_n}$
Natural frequency f_n	$\dfrac{\omega_n}{2\pi}$	f_n	$\dfrac{1}{T_n}$
Natural period T_n	$\dfrac{2\pi}{\omega_n}$	$\dfrac{1}{f_n}$	T_n

$$T_d = \frac{1}{f_d} \qquad \text{Damped natural period} \tag{2.31}$$

The angular frequency is measured in radians per second, the natural frequency in cycles per second or Hertz (Hz), while the natural period is measured in seconds. The relationship between these various quantities is given in Table 2.1.

2.1.3 Homogeneous Solution via Complex Frequencies: System Poles

Consider once more the differential equation for the free vibration of an SDOF system

$$m\ddot{u} + c\dot{u} + ku = 0 \tag{2.32}$$

As we already have seen, after division by the mass m this equation can be expressed as

$$\ddot{u} + 2\xi\omega_n\,\dot{u} + \omega_n^2\,u = 0 \tag{2.33}$$

An alternative method for obtaining the solution to this equation consists in seeking a trial solution of the form

$$u = C\,e^{i\omega_c t} \tag{2.34}$$

with a generally complex frequency ω_c and a nontrivial integration constant C. Its derivatives are simply

$$\dot{u} = i\omega_c\,C\,e^{i\omega_c t}, \qquad \ddot{u} = -\omega_c^2\,C\,e^{i\omega_c t} \tag{2.35}$$

Substituting these expressions into the differential equation, factoring out common terms, and canceling the nonzero exponential term as well as the constant, we obtain

$$-\omega_c^2 + 2\xi\omega_n\omega_c + \omega_n^2 = 0 \tag{2.36}$$

This is a quadratic equation in ω_c. Its roots are

$$\omega_c = \pm\omega_d + i\xi\omega_n = \omega_n\left(\pm\sqrt{1-\xi^2} + i\xi\right) \tag{2.37}$$

These two complex roots are referred to as the *poles* of the SDOF system at hand, as depicted in Figure 2.4. Notice that the absolute value of each of the roots equals the undamped frequency, and that both have a nonnegative imaginary part. Thus, the poles are said to lie in the upper complex half-plane.

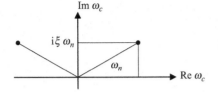

Figure 2.4. Complex poles.

The general solution is then a linear combination of the two roots, that is

$$
\begin{aligned}
u &= C_1 e^{i\omega_{c1} t} + C_2 e^{i\omega_{c2} t} \\
&= e^{-\xi\omega_n t} \left[C_1 e^{i\omega_d t} + C_2 e^{-i\omega_d t} \right] \\
&= e^{-\xi\omega_n t} \left[A \cos \omega_d t + B \sin \omega_d t \right]
\end{aligned}
\tag{2.38}
$$

which is identical to the free vibration solution obtained before. For this solution to represent actual vibrations, the integration constants A and B must be real. This implies that C_1 and C_2 must be complex conjugates of each other. If so, then $A = 2 \operatorname{Re} C_1$ and $B = -2 \operatorname{Im} C_1$.

2.1.4 Free Vibration of an SDOF System with Time-Varying Mass

To illustrate the behavior of a linear system with time-varying physical parameters, we present here an example related to the free vibration of an SDOF system whose mass decreases continuously with time.

Consider an undamped, SDOF system consisting of a tank filled with water, as shown in Figure 2.5. The tank rests on wheels, and is restrained laterally by an elastic spring. We assume also that the tank has a device (say a membrane) that prevents any sloshing of the water surface. The tank has an opening at the bottom through which the water flows out perpendicular to the direction of oscillations at a rate that depends on the water head. If we also neglect the pressure gradients that arise in the water from the lateral oscillations of the tank, then the water flow is given by

$$
Q(t) = A_0 \sqrt{2gh}
\tag{2.39}
$$

in which h is the height of the water in the tank, g is the acceleration of gravity, and A_0 is the cross section of the opening. If A denotes the cross section of the tank, then the rate of change of h is $\dot{h} = -Q(t)/A$, which leads to the differential equation for the water height,

$$
dt = -\frac{A}{A_0\sqrt{2g}} \frac{dh}{\sqrt{h}}
\tag{2.40}
$$

Integrating this equation, we obtain

$$
t - t_0 = -\frac{A}{A_0} \sqrt{\frac{2h_0}{g}} \left(1 - \sqrt{\frac{h}{h_0}} \right)
\tag{2.41}
$$

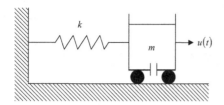

Figure 2.5. System with time-varying mass.

Beginning the observation at $t_0 = 0$, when the height of the water is h_0, and defining the parameter

$$a = \frac{A_0}{A}\sqrt{\frac{g}{2h_0}} \tag{2.42}$$

then

$$\sqrt{\frac{h}{h_0}} = 1 - at \tag{2.43}$$

Now, the mass of the water is proportional to the height of the water in the tank, so if m_T is the mass of the empty tank and m_W is the initial mass of the water, then the instantaneous mass of the system is

$$m = m_T + m_W (1 - at)^2 \qquad \text{Subjected to } at \leq 1 \tag{2.44}$$

To obtain the dynamic equation for the spring mass system, we apply next the principle of conservation of momentum to the system as a free body:

$$\frac{d(m\dot{u})}{dt} = -ku \tag{2.45}$$

That is, the rate of change in momentum equals the external force acting on the mass. It follows that

$$\frac{d}{dt}\left\{\left[m_T + m_W (1 - at)^2\right]\dot{u}\right\} + ku = 0 \tag{2.46}$$

To simplify matters further, we shall neglect the mass of the tank in comparison to that of the water. This means that our solution will be meaningful only for early times, while enough water remains in the tank, that is, while $at \ll 1$. The dynamic equilibrium equation then reduces to

$$\frac{d}{dt}\left[m_W (1 - at)^2 \dot{u}\right] + ku = 0 \tag{2.47}$$

This equation admits a solution of the form

$$u = (1 - at)^\lambda, \tag{2.48}$$

whose time derivatives are

$$\dot{u} = -a\lambda(1 - at)^{\lambda - 1}, \qquad \frac{d}{dt}\left[(1 - at)^2 \dot{u}\right] = a^2\lambda(\lambda + 1)(1 - at)^\lambda \tag{2.49}$$

Hence,

$$m_W a^2 \lambda (\lambda + 1)(1 - at)^\lambda + k(1 - at)^\lambda = 0 \tag{2.50}$$

If we define the initial natural frequency as $\omega_0 = \sqrt{k/m_W}$, and consider the fact that $1 - at \neq 0$ for $t > 0$, we obtain

$$a^2 \lambda(\lambda + 1) + \omega_0^2 = 0 \tag{2.51}$$

which yields

$$\lambda = \frac{-1 \pm \sqrt{1 - (2\omega_0/a)^2}}{2} \quad \text{if } \frac{\omega_0}{a} \leq \frac{1}{2} \qquad \text{Fast rate of mass loss} \tag{2.52}$$

$$\lambda = \frac{-1 \pm i\sqrt{(2\omega_0/a)^2 - 1}}{2} \quad \text{if } \frac{\omega_0}{a} > \frac{1}{2} \qquad \text{Slow rate of mass loss} \tag{2.53}$$

with D standing for either $D_{fast} = \sqrt{1 - (2\omega_0/a)^2}$ or $D_{slow} = \sqrt{(2\omega_0/a)^2 - 1}$, then

$$u = (1 - at)^\lambda = (1 - at)^{-\frac{1}{2}(1 \mp D)} = \frac{(1 - at)^{\pm\frac{1}{2}D}}{\sqrt{1 - at}} \tag{2.54}$$

Hence, the solutions for these two cases are then (with C_1, C_2 = constants of integration)

$$u = \frac{C_1 e^{\theta(t)} + C_2 e^{-\theta(t)}}{\sqrt{1 - at}}, \qquad \theta(t) = -\tfrac{1}{2}\ln(1 - at)D_{fast} \qquad \text{Fast rate} \tag{2.55}$$

$$u = \frac{C_1 \cos\theta(t) + C_2 \sin\theta(t)}{\sqrt{1 - at}}, \qquad \theta(t) = -\tfrac{1}{2}\ln(1 - at)D_{slow} \qquad \text{Slow rate} \tag{2.56}$$

2.1.5 Free Vibration of SDOF System with Frictional Damping

We consider herein the free vibration of a 1-DOF system that is affected by frictional resistance to motion. The resisting force is typically proportional to some weight mg times a coefficient of friction μ, that is, $F = \mu mg$, and the process will dissipate energy in the form of heat. This means that mechanical energy will be lost to the system through friction, a phenomenon that is referred to as *Coulomb damping*.

Consider an otherwise undamped SDOF system resting on a rough surface that can slide laterally and let F be the magnitude of the frictional force opposing the motion (Figure 2.6). The equation of motion and the initial conditions are then

$$m\ddot{u} + ku = -F\,\text{sgn}(\dot{u}), \qquad u\big|_{t=0} = u_0, \qquad \dot{u}\big|_{t=0} = \dot{u}_0 \tag{2.57}$$

where the sign function preceded by the negative sign indicates that the direction of the force is opposite to the instantaneous velocity. For convenience, we denote also

$$u_s = \frac{F}{k} \tag{2.58}$$

as the yield or *sliding* limit, which has units of displacement. Clearly, for the system to begin moving at all it behooves that

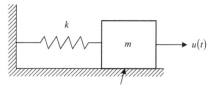

Figure 2.6. SDOF system with frictional damping.

Frictional surface with sliding resistance F

$$|ku_0| > F, \quad \text{i.e.,} \quad |u_0| > u_s \tag{2.59}$$

If this condition were not to be met, then the system would remain at rest in the initially displaced position.

We shall first consider the special case of positive initial displacement with no initial velocity, and thereafter we shall generalize for any arbitrary combination of these two.

(a) System Subjected to Initial Displacement

If we give an initial positive displacement $u_0 > u_s \equiv F / k, \dot{u}_0 = 0$, then the system will begin a recovery toward the rest position with $\dot{u} < 0$, in which case the initial equation of motion is

$$m\ddot{u} + ku = F = ku_s \tag{2.60}$$

the solution of which is

$$u = (u_0 - u_s)\cos\omega_n t + u_s \tag{2.61}$$

The system will continue this recovery until its velocity vanishes at $t_1 = \frac{1}{2}T_n, \omega_n t_1 = \pi$ while the displacement attains the maximum negative value $u_1 = u|_{t=t_1}$, that is,

$$u_1 = (u_0 - u_s)(-1) + u_s = 2u_s - u_0 = -(u_0 - 2u_s) \tag{2.62}$$

From this point on, if $|u_1| = u_0 - 2u_s > u_s$, that is, $u_0 > 3u_s$ the system will begin a recovery with positive velocity, during which the differential equation changes into

$$m\ddot{u} + ku = -F = -ku_s \tag{2.63}$$

for which the solution in delayed time is

$$\begin{aligned} u &= (u_1 + u_s)\cos\omega_n(t - t_1) - u_s \\ &= (3u_s - u_0)\cos\omega_n(t - t_1) - u_s \end{aligned} \tag{2.64}$$

This motion will attain a positive maximum at $t_2 = T, \omega_n t_2 = 2\pi, \cos\omega_n(t_2 - t_1) = \cos(\pi) = -1$ which is

$$u_2 = (3u_s - u_0)(-1) - u_s = u_0 - 3u_s \tag{2.65}$$

It now becomes clear that the system will experience successive minima and maxima at times $t_j = \frac{1}{2}jT_n, j = 1, 2, \cdots$ whose magnitudes are

$$|u_j| = u_0 - (2j - 1)u_s \tag{2.66}$$

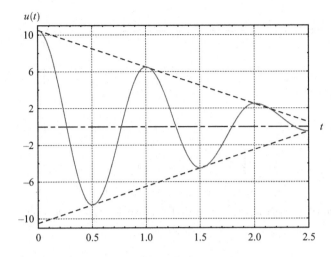

Figure 2.7. Coulomb damping response with $u_0 = 10.5$, $v_0 = \dot{u}_0 = 0$, $T_n = 1$, $u_s = 1$.

and this process will continue as long as $k\left|u_j\right| > F$, that is, $\left|u_j\right| = u_0 - (2j-1)u_s > u_s$, which implies

$$j < \tfrac{1}{2} u_0 / u_s = \tfrac{1}{2} k u_0 / F \tag{2.67}$$

Once this limit is either reached or first exceeded, the system comes to a complete stop and moves no further, that is, the motion ceases at

$$t_{\max} = \tfrac{1}{2} T_n \text{ round}\left[\tfrac{1}{2}\frac{u_o}{u_s}\right] \tag{2.68}$$

where "round" is the integer rounding function such that if $n = N + \varepsilon$, where N is an integer and $0 < \varepsilon < 1$, then

$$\text{round}\,(n) = \text{round}\,(N + \varepsilon) = \begin{cases} N & \varepsilon \le 0.5 \\ N+1 & \varepsilon > 0.5 \end{cases} \tag{2.69}$$

For example, if $\tfrac{1}{2}u_0/u_s = 5.875$ then $t_{\max} = \tfrac{1}{2}T_n \text{ round}[5.875] = \tfrac{1}{2}T_n \times 6 = 3T_n$, while $\tfrac{1}{2}u_0/u_s = 5.5$ yields $t_{\max} = \tfrac{1}{2}T_n \text{ ceil}[5.5] = \tfrac{1}{2}T_n \times 5 = 2.5T_n$. This is shown in Figures 2.7 and 2.8.
We observe that in all cases, the maxima decrease linearly and lie on the straight lines

$$y = \pm\left(u_0 - 4u_s \frac{t}{T_n}\right) \tag{2.70}$$

The system comes to a stop either at the intersection of these two enveloping lines, or shortly thereafter, namely at the next intersection of one of the two straight envelopes with the response curve. In the latter case, when the system stops it exhibits a permanent, residual dislocation, as it does in Figure 2.7.

(b) Arbitrary Initial Conditions

We now consider arbitrary initial conditions, but inasmuch as the response functions for negative initial conditions are simply the reversed (mirror image) response functions for

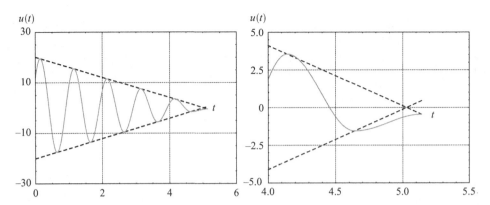

Figure 2.8. Coulomb damping response with $u_0 = 12, v_0 = \dot{u}_0 = 100, T_n = 1. u_s = 1$. The figure on the right shows the detail when motion ceases.

positive conditions, then it suffices to consider simply the case $u_0 \geq 0$ with either positive or negative initial velocity.

If $u_0 \geq 0$ and $\dot{u}_0 > 0$, then these initial conditions are associated with a resisting force $-F$ that will elicit the response functions

$$u(t) = (u_0 + u_s)\cos\omega_n t + \frac{\dot{u}_0}{\omega_n}\sin\omega_n t - u_s \tag{2.71}$$

$$\dot{u}(t) = \omega_n\left[-(u_0 + u_s)\sin\omega_n t + \frac{\dot{u}_0}{\omega_n}\cos\omega_n t\right] \tag{2.72}$$

The elongation will be maximum when $\dot{u}(t_1) = \dot{u}_1 = 0$, that is,

$$u_1 = \sqrt{\frac{\dot{u}_0^2}{\omega_n^2} + (u_0 + u_s)^2} - u_s, \qquad \omega_n t_1 = \arctan\frac{\dot{u}_0}{\omega_n(u_0 + u_s)} \tag{2.73}$$

From this point on, the solution is identical to that already found, with the trivial modifications $u_0 \rightarrow u_1, t \rightarrow t - t_1$, see Figure 2.8.

If $u_0 \geq 0$ and $\dot{u}_0 < 0$, then the response functions are

$$u(t) = (u_0 - u_s)\cos\omega_n t + \frac{\dot{u}_0}{\omega_n}\sin\omega_n t + u_s \tag{2.74}$$

$$\dot{u}(t) = \omega_n\left[-(u_0 - u_s)\sin\omega_n t + \frac{\dot{u}_0}{\omega_n}\cos\omega_n t\right] \tag{2.75}$$

The elongation will be minimum when $\dot{u}(t_1) = \dot{u}_1 = 0$, that is

$$u_1 = u_s - \sqrt{\frac{\dot{u}_0^2}{\omega_n^2} + (u_s - u_0)^2}, \qquad 0 < \omega_n t_1 = \arctan\frac{-\dot{u}_0}{\omega_n(u_s - u_0)} \leq \pi \tag{2.76}$$

From this point on, the solution is again like the solution for nonzero initial displacement, with the replacement $u_0 \rightarrow u_1$ and $t \rightarrow t - t_1$. If $u_1 > 0$, or if $u_1 < 0$ but $|u_1| \leq u_s$, then the system comes to a rest at that point. In other words, the negative initial velocity must be large

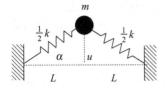

Figure 2.9. Nonlinear oscillator.

enough to cause the system to extend beyond $u = -u_s$, otherwise it comes to a complete stop. If $|u_1| > u_s$, the system behaves from that time on exactly the same as for a pure initial *negative* displacement. In most cases, the system will exhibit a final residual dislocation $0 \le |u_{\text{final}}| \le u_s$.

2.2 Phase Portrait: Another Way to View Systems

The phase portrait is an alternative method of representing the response of a vibrating system. In essence, the phase portrait is a plot of the instantaneous velocity versus position of the vibrating mass. The curves or trajectories in this plot are referred to as the *phase lines*. The phase portrait is particularly useful for nonlinear systems, especially when the solution to the differential equation of motion is unknown, or cannot be found.

2.2.1 Preliminaries

Consider a spring–mass system with two springs each of stiffness $k/2$ and length L to which a mass m is attached. Each spring is subjected to an initial tension T_0. The mass is given an initial lateral disturbance and is constrained to move in a direction perpendicular to the initial direction of the springs, as shown in Figure 2.9. If $u(t)$ is the instantaneous lateral (i.e., vertical) motion of the mass, then this produces an elongation in each spring

$$\Delta L = L\left[\frac{1}{\cos\alpha} - 1\right] \tag{2.77}$$

Disregarding gravity effects, the total force in each spring is then

$$F = F_0 + \tfrac{1}{2}k\,\Delta L \tag{2.78}$$

Taken together, both springs produce a net vertical force

$$F_v = 2F\sin\alpha = kL(\tan\alpha - \sin\alpha) + 2F_0\sin\alpha \tag{2.79}$$

But $\tan\alpha = u/L$. Hence

$$F_v = ku\left\{1 - \frac{1}{\sqrt{1+\left(\frac{u}{L}\right)^2}}\right\} + \frac{2F_0\,u}{L\sqrt{1+\left(\frac{u}{L}\right)^2}} = f(u) \tag{2.80}$$

By D'Alembert's principle, the dynamic equilibrium equation for free vibration is then

$$m\ddot{u} + f(u) = 0 \tag{2.81}$$

which is a homogeneous, nonlinear equation, that is, there is no external excitation, and the restoring force is nonlinear. For *small* amplitudes of vibration, the terms in the square roots can be expanded in Taylor series, which leads to

$$f(u) = 2F_0\left(\frac{u}{L}\right) + \left(\tfrac{1}{2}kL - F_0\right)\left(\frac{u}{L}\right)^3 + \cdots \tag{2.82}$$

The slope of the restoring force with respect to displacement is the tangent stiffness. For the nonlinear restoring force considered here, the tangent stiffness at the rest position is

$$k_t = \left.\frac{df(u)}{du}\right|_{u=0} = \frac{2F_0}{L} \tag{2.83}$$

As can be seen, if the initial tension in the springs were zero, the restoring force would be cubic in the lateral displacements, even when the displacements remain small. Thus, such a system would be inherently nonlinear, and it would have no initial tangent stiffness. By contrast, when the initial tension is high and we consider only small vibrations, the linear term dominates. We have then once more a linear system characterized by an equivalent spring k_t. Notice the difference with the case in which the mass is constrained to vibrate in a horizontal direction (i.e., in the direction of the springs). In such a case, the initial tension of the springs would play no role whatsoever, and the system would remain always linear.

To obtain the phase portrait, we begin by multiplying the equation of motion by the velocity

$$m\ddot{u}\dot{u} + f(u)\dot{u} = 0 \tag{2.84}$$

But $\ddot{u}\dot{u} = \tfrac{1}{2}\frac{d}{dt}\dot{u}^2$, so when we integrate with respect to time, we obtain

$$\tfrac{1}{2}m\dot{u}^2 + \int f(u)\,du = V_1 \tag{2.85}$$

We recognize the two terms on the left-hand side as the kinetic and strain energies K and V in the system, respectively, while the integration constant on the right-hand side is the initial energy in the system. Thus, this equation is nothing but a statement of the principle of conservation of energy. Hence

$$\dot{u} = \pm\sqrt{\frac{2}{m}\left(V_1 - V\right)} = \dot{u}(u) \tag{2.86}$$

which provides an expression with which the phase portrait can be evaluated. This equation implies that for this problem the phase portrait is symmetric with respect to the horizontal axis (u). From the expression for the restoring force, we obtain

$$V = \int\left[ku\left\{1 - \frac{1}{\sqrt{1+\left(\frac{u}{L}\right)^2}}\right\} + \frac{2F_0 u}{L\sqrt{1+\left(\frac{u}{L}\right)^2}}\right]du = \tfrac{1}{2}ku^2 + (2F_0 - kL)L\left[\sqrt{1+\left(\frac{u}{L}\right)^2} - 1\right] \tag{2.87}$$

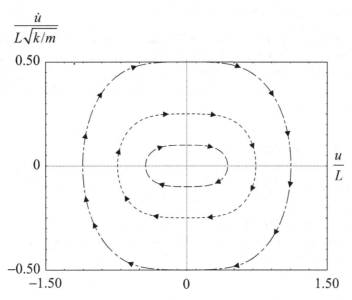

Figure 2.10. Phase lines for laterally oscillating mass without pre-tension of spring.

If we measure the initial energy of the system in the rest position of the springs (i.e., when $u = 0$), then only the mass has kinetic energy $E_0 = \frac{1}{2} m \dot{u}_0^2$. In that case,

$$\dot{u} = \sqrt{\dot{u}_0^2 - \frac{2}{m} \left[\tfrac{1}{2} k u^2 + (2F_0 - kL)L \left(\sqrt{1 + \left(\tfrac{u}{L}\right)^2} - 1 \right) \right]} \tag{2.88}$$

Clearly, the maximum elongation u_m occurs when the velocity is zero, that is, when

$$\tfrac{1}{2} k u_m^2 + (2F_0 - kL)L \left(\sqrt{1 + \left(\tfrac{u_m}{L}\right)^2} - 1 \right) = \tfrac{1}{2} m \dot{u}_0^2 \tag{2.89}$$

This expression can be reduced to a biquadratic equation with four roots. Of these, only two are proper roots, while the other two are spurious roots that arise from the rationalization, and do not satisfy Eq. 2.89. The proper roots have the same numerical value but opposite sign, so the phase lines for this system happen to be symmetric with respect to the vertical axis.

A plot of the phase lines for the case of no axial force ($F_0 = 0$) and for three values of the initial velocity is shown in Figure 2.10. As can be seen, the phase lines for different initial conditions are ellipsoidal. In fact, they are true ellipses when the system is linear. The arrows indicate the trajectories, that is, the direction in which time increases. The motions are periodic, and one period is the time it takes to travel once around.

2.2.2 Fundamental Properties of Phase Lines

Trajectory Arrows

The trajectory arrows *must* always point from left to right in the region above the horizontal axis, because if the velocity is positive, the displacements can only increase. For a similar reason, the arrows must point from right to left in the region below the axis.

Intersection of Phase Lines with Horizontal Axis

The tangents to the phase lines at points where they intersect the horizontal axis are always *perpendicular* to that axis, except perhaps at *singular points* as considered below. To see that this is so, we write the equation of motion as

$$m\frac{d\dot{u}}{dt} + f(u) = m\frac{d\dot{u}}{du}\frac{du}{dt} + f(u) = 0 \tag{2.90}$$

from which it follows immediately that

$$\frac{d\dot{u}}{du} = -\frac{1}{m\dot{u}}f(u) \tag{2.91}$$

Clearly, if $\dot{u} = 0$ and $f(u) \equiv f(u_m) \neq 0$ then we have a *regular point*, the slope is infinitely large, and the tangent is indeed vertical. The zero-crossings of the phase lines represent then extreme values for the displacements.

Asymptotic Behavior at Singular Points and Separatrix

Points at which both the velocity and the restoring force vanish are referred to as *singular points*. These points represent a (perhaps unstable) equilibrium position at which the system is at rest, so once there, the system does not move any further. However, a singular point may be approached only asymptotically, that is, it takes an infinite time to reach such a point. To prove this, assume that in the local neighborhood of a singular point, the equation for the phase line can be approximated by the tangent at that point, and that this tangent (inclined as shown in Figure 2.11) has an assumed nonvertical, finite slope c. Hence,

$$\dot{u} = \frac{du}{dt} = c(u_s - u) \tag{2.92}$$

in which u_s is the position of the singular point. From here

$$\int dt = \int \frac{du}{c(u_s - u)} \tag{2.93}$$

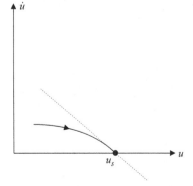

Figure 2.11. Phase line approaching a singular point.

that is

$$t - t' = -\frac{1}{c} \ln(u - u_s)\big|_{u'}^{u} = -\frac{1}{c} \ln \frac{u - u_s}{u' - u_s} \tag{2.94}$$

with primed quantities denoting an arbitrary starting position and time. Solving for the displacements, we finally find

$$u_s - u = (u_s - u')e^{-ct} \tag{2.95}$$

Since the exponential term becomes zero only when t approaches infinity, u can approach u_s only in an asymptotic fashion, as stated earlier. The tangent to the phase lines at a singular point is called the *separatrix*, which the phase lines themselves cannot cross.

Period of Oscillation

In the case of closed trajectories (closed phase lines), the period of oscillation can be obtained from the definition

$$\dot{u} = \frac{du}{dt} \rightarrow \quad dt = \frac{du}{\dot{u}(u)} \tag{2.96}$$

The period is then twice the time it takes to travel between extreme positions, that is,

$$T = 2 \int_{u_{\min}}^{u_{\max}} \frac{du}{\sqrt{\frac{2}{m}(V_0 - V)}} \tag{2.97}$$

with $V_0 = V(u_{\min}), V = V(u)$ being the initial and instantaneous potential energies.

2.2.3 Examples of Application

Phase Lines of a Linear SDOF System

The phase lines for a linear undamped system are ellipses, while those of damped systems are open lines that converge toward the origin. The origin itself is a singular point that can be reached only asymptotically. In the case of lightly damped systems ($\xi < 1$), the phase portrait is a spiral that has infinitely many loops, while for heavy damping ($\xi > 1$), the phase portrait consists of vortex-like curves, as shown in Figure 2.12.

Ball Rolling on a Smooth Slope

An interesting example is that of a small ball sliding downhill and without friction on a smooth slope. We assume that it is permissible to neglect the rotational inertia of the ball, and that at no time does the ball lose contact with the surface. Let the topography of the hill be described by the single-valued function $y = f(x)$, and let $\dot{u}$ be the instantaneous velocity of the ball along the tangent to a point. If the ball starts rolling from a height h above an arbitrary level (the origin of y) and without any initial velocity, then by

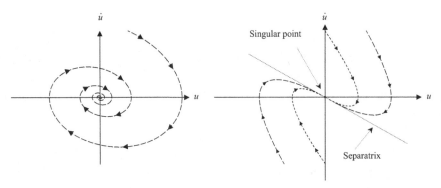

Figure 2.12. Phase lines for lightly damped system (left) versus heavily damped system (right).

conservation of energy, the net gain in kinetic energy on descent must equal the loss of potential energy. Hence

$$\dot{u} = \sqrt{2g(h-y)} = \sqrt{2g[h-f(x)]} \tag{2.98}$$

This very simple equation describes the phase lines in terms of the horizontal position x of the ball, as illustrated in Figure 2.13, which shows a drawing of a hill with an equation $y = 2x^3 - 3x^2$, evaluated in the range $-1.6 < x < 0.7$. The hill is shown in the upper part of the drawing, while the corresponding phase lines are in the lower part. To gain an understanding of these lines, consider, for example, the upper left side of the topmost curve. Assume that the ball had been given an appropriately large initial velocity in the *uphill* direction starting from a position on the far left. The ball would have rolled up the hill, past the crest, then down into the valley, and finally, uphill again until it came to a stop at the position at which the ball is drawn. At this point, the ball would have reversed direction, and repeated the motion in the opposite direction. It would finally have rolled past the starting position and disappeared into the abyss.

If the initial uphill velocity imparted on the ball had been much smaller, the ball would not have been able to reach the crest. Its motion would then have been described by a phase line similar to the leftmost branch. Suppose, however, that the ball had been given just the right uphill velocity so as to match the potential energy at rest at the hilltop (i.e., consistent with the second phase line). The ball would have then crept toward the hilltop, never quite making it, yet never starting to roll back. In fact, the ball would have reached the top of the hill only after an infinitely long time. Thus, the hilltop is a *singular point* in this case. If the ball had been left motionless at that point, it would have remained there for an indefinite period (even if unstably so).

Finally, the closed loops on the right correspond to stable oscillation within the valley portion of the slope.

The phase lines can also be expressed in terms of the horizontal component of the ball's velocity versus its horizontal position:

$$\dot{x} = \frac{\dot{u}}{\sqrt{1 + \left[\frac{d}{dx} f(x)\right]^2}} \tag{2.99}$$

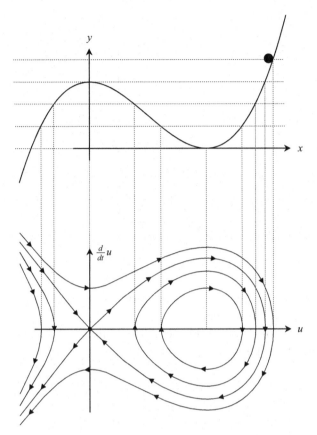

Figure 2.13. Phase lines for a ball rolling down on a smooth slope.

From here, we could, for example, obtain the travel time between two positions. Since $\dot{x} = \dfrac{dx}{dt}$, then

$$t = t_0 + \int_{x_1}^{x_2} \sqrt{\frac{1 + \left[\frac{d}{dx} f(x)\right]^2}{2g \left[h - f(x)\right]}}\, dx, \qquad x_2 > x_1 \tag{2.100}$$

2.3 Measures of Damping

From Section 2.1.1 we already know that when a viscously damped SDOF system is allowed to vibrate freely after some initial disturbance, that vibration dies out exponentially as $e^{-\xi \omega_n t}$, that is, it decreases at an exponential rate that depends on the natural frequency and the fraction of critical damping. In this section, we make use of these characteristics and assess damping by measuring how fast the vibration decays in time, and more specifically, by measuring the number of cycles until the vibration amplitude has dropped by 50%. We examine first the linearly viscous case and then make some comments on other forms of damping.

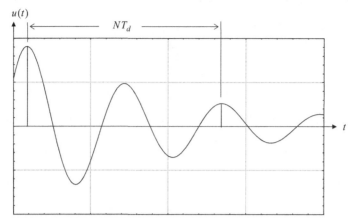

Figure 2.14. Logarithmic decrement: Measuring damping through amplitude decay.

2.3.1 Logarithmic Decrement

The *logarithmic decrement* is a quantity that can be used to determine the amount of damping in an oscillator by comparing the maximum free vibration responses at two instants in time. Consider again the free vibration of a subcritically damped oscillator shown in Figure 2.14. The response at an instant t is

$$u(t) = e^{-\xi\omega_n t}\left[A\cos\omega_d t + B\sin\omega_d t\right] \tag{2.101}$$

The motion at a later time $t + N T_d$ separated from t by an integer number N of damped periods T_d, is then

$$\begin{aligned}
u(t + NT_d) &= e^{-\xi\omega_n(t+NT_d)}\left[A\cos\omega_d(t+NT_d) + B\sin\omega_d(t+NT_d)\right]\\
&= e^{-\xi\omega_n(t+NT_d)}\left[A\cos\omega_d t + B\sin\omega_d t\right]
\end{aligned} \tag{2.102}$$

Taking the ratio of the displacements at the two instants, the terms in brackets cancel out, so that

$$\frac{u(t)}{u(t+NT_d)} = \frac{e^{-\xi\omega_n t}}{e^{-\xi\omega_n(t+NT_d)}} = e^{\xi\omega_n NT_d} \tag{2.103}$$

Taking the natural logarithm on both sides, we obtain

$$\Delta_N = \ln\frac{u(t)}{u(t+NT_d)} = \xi\omega_n NT_d \tag{2.104}$$

But

$$T_d = \frac{2\pi}{\omega_d} = \frac{2\pi}{\omega_n\sqrt{1-\xi^2}} \tag{2.105}$$

so that

$$\boxed{\Delta_N = \ln\frac{u(t)}{u(t+NT_d)}} = 2\pi N\frac{\xi}{\sqrt{1-\xi^2}} \qquad \text{Logarithmic decrement} \tag{2.106}$$

Solving for the fraction of critical damping,

$$\xi = \frac{\frac{1}{2\pi N}\Delta_N}{\sqrt{1+\left(\frac{1}{2\pi N}\Delta_N\right)^2}} \tag{2.107}$$

which for lightly damped systems can be approximated as

$$\boxed{\xi = \frac{\Delta_N}{2\pi N}} \tag{2.108}$$

This equation can be used to assess the fraction of critical damping in an SDOF system.

2.3.2 Number of Cycles to 50% Amplitude

We show next that by determining experimentally the number of cycles $N_{50\%}$ required for the amplitudes to wane to half of their initial values, we can easily estimate the damping without having to determine the logarithmic decrement. By definition of 50% reduction, we obviously have

$$\Delta_N = \ln\frac{u(t)}{u(t+N_{50\%}T_d)} = \ln 2 \tag{2.109}$$

which means that the damping must be

$$\xi = \frac{\ln 2}{2\pi N_{50\%}} = \frac{0.11}{N_{50\%}} \tag{2.110}$$

which can be approximated as

$$\boxed{\xi = \frac{1}{10 N_{50\%}}} \tag{2.111}$$

This is a very simple formula to remember. An interesting corollary of this result is that in a system with 10% damping, it takes only a single cycle for the amplitude to drop by 50%.

More generally, after $m = nN_{50\%}$ cycles, the amplitude of vibration will have decayed by a factor 2^n. This is so because if $A_0, A_1, A_2, \ldots$ are successive amplitudes of damped *free* vibration, the ratio of the amplitudes of oscillations at two arbitrary instants separated by $m = nN_{50\%}$ cycles is

$$\frac{A_0}{A_m} = \frac{A_0}{A_{N_{50\%}}}\frac{A_{N_{50\%}}}{A_{2N_{50\%}}}\cdots\frac{A_{(n-1)N_{50\%}}}{A_{nN_{50\%}}} = \left(e^{\xi\omega_n N_{50\%}T_d}\right)^n = 2^n \tag{2.112}$$

In particular, we choose $m = nN_{50\%}$ to be the number of cycles required for the oscillations to decay to a negligible fraction, that is, $A = A_0\,2^{-n}$. For instance, let $N_{50\%} = 3$ cycles (which corresponds to a fraction of damping $\xi = 0.037$). The number of cycles required for the vibration to decrease to a fraction $2^{-7} = 1/128 = 0.0078$ of the initial amplitude is then $n = 7$ and that decay occurs in just $7 \times 3 = 21$ cycles.

2.3.3 Other Forms of Damping

Linear viscous damping with a resisting force directly proportional to the velocity of deformation is commonly used in structural dynamics because of its mathematical simplicity, which leads to linear, ordinary differential equations with constant coefficients. However, most structures dissipate energy through mechanisms other than viscous damping. Then again a common characteristic of viscous damping with all of the other forms of damping is that the resisting force is always against the motion, that is, that damping forces always act in a direction opposite to the velocity. Also, for most structural systems without physical dashpots the rate of decay of vibrations is either moderate or small, in which case the error incurred by the adoption of a viscous model may not be very significant, that is, the number or cycles to 50% amplitude may still be a physically meaningful quantity.

As we shall see later on in Section 2.9, when an SDOF system is forced to vibrate harmonically at its natural frequency and at some constant maximum amplitude Δ, the ratio of energy E_d dissipated per cycle of motion to the total strain and kinetic energy E_s stored in the system is a measure of the effective damping, that is,

$$\xi_{\text{eff}} = \frac{1}{4\pi} \frac{E_d}{E_s} \qquad (2.113)$$

In the case of viscous damping ξ_v, it is easy to demonstrate that $\xi_{\text{eff}} = \xi_v$, which is independent of the maximum amplitude of motion. Thus, the fraction energy lost per cycle is constant from cycle to cycle, and this fact is consistent with the exponential rate decay of the vibration, that is, the ratio of two consecutive peaks $(e^{-\xi\omega_n T_d})$ is constant. But this is not so when damping is other than viscous. In fact, even when viscous losses exist, the damping need not be linear, as is the case of drag forces or of hydrodynamic damping, for which the resisting forces are proportional to the square of the relative velocity of the mass point with respect to the surrounding fluid. For an SDOF system such forces would be of the form $f_{\text{damp}} \sim \dot{u}^2$. In that case, one finds that $\xi_{\text{eff}} \sim \Delta$, where Δ is the amplitude of motion. Hence, a system subjected to drag forces at first decays rather fast, but as the velocity drops, so does the rate of decay associated with drag forces, which slows to less than exponential. Thus, the 50% amplitude rule would need to be adjusted so as to test for amplitudes comparable to the expected levels of motion.

At the other extreme is energy dissipation through friction (Section 2.1.5), which does not depend at all on the velocity, but only on the magnitude of frictional slipping Δ. In that case, the effective damping $\xi_{\text{eff}} \sim \Delta^{-1}$ increases as the vibration attenuates. This causes the system to come to a stop in a finite number of cycles, as we have already seen.

2.4 Forced Vibrations

2.4.1 Forced Vibrations: Particular Solution

In the introduction to Section 2.1, we gave the somewhat tautological definition of particular solution as *any solution satisfying the nonhomogeneous differential equation* (i.e., with nonvanishing right-hand side). However, we gave no hint as to how one may actually find such a solution. While in Section 2.4.2 we will give a general formula for obtaining particular solutions for any arbitrary loading $p(t)$ via *convolution* integrals, we begin with a simple *heuristic* method, which we follow with the more formal (and far more general) *variations of parameters* method.

(a) Heuristic Method

A simple heuristic rule, which is very useful for hand calculations, is

> *"Try a particular solution that looks like the right-hand side."*

If the right-hand side is

- a polynomial of order n, try a *complete* polynomial of that same order.
- an exponential times a polynomial, try the same exponential times a complete polynomial.
- an exponential times a sine or a cosine, try a combination of sine and cosine times the exponential.

We illustrate this strategy with an example. Consider the equation

$$m\ddot{u} + c\dot{u} + ku = at^2 \tag{2.114}$$

We then try a particular solution in the form of a complete second-order polynomial:

$$u_p = A + Bt + Ct^2 \tag{2.115}$$

whose derivatives are

$$\dot{u}_p = B + 2Ct \tag{2.116}$$

$$\ddot{u}_p = 2C \tag{2.117}$$

Substituting these expressions into the differential equation, we obtain

$$m(2C) + c(B + 2Ct) + k(A + Bt + Ct^2) = at^2 \tag{2.118}$$

Collecting terms in powers of t,

$$(2mC + cB + kA) + (2cC + kB)t + kCt^2 = at^2 \tag{2.119}$$

Since this expression must be valid at all times t, comparison with the right-hand side indicates that

$$2mC + cB + kA = 0 \tag{2.120}$$

$$2cC + kB = 0 \tag{2.121}$$

$$kC = a \tag{2.122}$$

This is a system of three equations in the three unknown integration constants A, B, and C, which can now easily be solved. We leave this straightforward task to the readers.

(b) Variation of Parameters Method

We next present a general, formal method to find particular solutions to any linear, second-order differential equation with either time-dependent or constant coefficients. Readers are forewarned, however, that this method generally requires considerable effort, even for very simple forcing functions acting on systems with constant coefficients.

Consider once more the differential equation for an SDOF system:

$$m(t)\ddot{u} + c(t)\dot{u} + k(t)u = p(t) \tag{2.123}$$

in which the mass $m(t) \neq 0$. Dividing by this term, we obtain an equation of the form

$$\ddot{u} + f(t)\dot{u} + g(t)u = r(t) \tag{2.124}$$

in which we assume $f(t), g(t)$ to be analytic functions of t in at least some temporal neighborhood. As shown in treatises on differential equations,[1] a particular solution can be obtained that has the form

$$u_p(t) = w_1 u_{h1} + w_2 u_{h2} \tag{2.125}$$

in which $u_{h1} = u_{h1}(t)$ and $u_{h2} = u_{h2}(t)$ are the two *known* solutions to the *homogeneous* equation, and $w_1 = w_1(t)$ and $w_2 = w_2(t)$ are as yet unknown functions of time. Requiring these to satisfy the arbitrary subsidiary condition

$$\dot{w}_1 u_{h1} + \dot{w}_2 u_{h2} = 0 \tag{2.126}$$

it then follows from the differential equation that (see Kreyszig, op. cit., p. 49).

$$\dot{w}_1 \dot{u}_{h1} + \dot{w}_2 \dot{u}_{h2} = r \tag{2.127}$$

Combining these two equations, we obtain

$$w_1(t) = -\int \frac{r\,u_{h2}}{W}\,dt \quad \text{and} \quad w_2(t) = \int \frac{r\,u_{h1}}{W}\,dt \tag{2.128}$$

in which

$$W = u_{h1}\dot{u}_{h2} - \dot{u}_{h1}u_{h2} \tag{2.129}$$

is the *Wronskian*.

As an illustration, consider once more the example in the previous section, which for positive times has a forcing function $p(t) = at^2$, so that $r = at^2 / m$. The two homogeneous solutions are in this case

$$u_{h1} = A\,e^{-\xi\omega_n t}\cos\omega_d t \qquad \dot{u}_{h1} = -A\,e^{-\xi\omega_n t}\left(\omega_d \sin\omega_d t + \xi\omega_n \cos\omega_d t\right) \tag{2.130}$$

$$u_{h2} = B\,e^{-\xi\omega_n t}\sin\omega_d t \qquad \dot{u}_{h2} = B\,e^{-\xi\omega_n t}\left(\omega_d \cos\omega_d t - \xi\omega_n \sin\omega_d t\right) \tag{2.131}$$

---

[1] See, for example, Erwin Kreyszig, *Advanced Engineering Mathematics,* 9th ed. (Hoboken, NJ: John Wiley & Sons, 2005).

After some algebra, the Wronskian evaluates to

$$W = \omega_d AB e^{-2\xi\omega_n t} \tag{2.132}$$

so that

$$u_p = \frac{a}{\omega_d} e^{-\xi\omega_n t} \left[-\cos\omega_d t \int_0^t \tau^2 e^{\xi\omega_n \tau} \sin\omega_d \tau \, d\tau + \sin\omega_d t \int_0^t \tau^2 e^{\xi\omega_n \tau} \cos\omega_d \tau \, d\tau \right] \tag{2.133}$$

This expression can be evaluated via integration by parts. The required integrals are found to be

$$\int t^2 e^{\alpha t} \sin\beta t \, dt = F(t)\sin\beta t - G(t)\cos\beta t \tag{2.134}$$

$$\int t^2 e^{\alpha t} \cos\beta t \, dt = G(t)\sin\beta t + F(t)\cos\beta t \tag{2.135}$$

with

$$F(t) = \frac{e^{\alpha t}}{\alpha^2 + \beta^2} \left[\alpha t^2 - 2\frac{\alpha^2 - \beta^2}{\alpha^2 + \beta^2} t + 2\alpha \frac{\alpha^2 - 3\beta^2}{(\alpha^2 + \beta^2)^2} \right] \tag{2.136}$$

$$G(t) = \frac{e^{\alpha t}}{\alpha^2 + \beta^2} \left[\beta t^2 - 4\frac{\alpha\beta}{\alpha^2 + \beta^2} t + 2\beta \frac{3\alpha^2 - \beta^2}{(\alpha^2 + \beta^2)^2} \right] \tag{2.137}$$

Making use of these integrals, the particular solution reduces, after some algebra, to

$$u_p(t) = \frac{a}{m}\left[(\omega t)^2 - 4\xi\omega t - 2(1 - 4\xi^2) \right] \tag{2.138}$$

This expression coincides with that obtained by the heuristic method, but requires an inordinate amount of effort. However, it has the advantage of being a rigorous method.

2.4.2 Forced Vibrations: General Solution

The complete solution to the differential equation is then the summation of the free vibration and the particular solution:

$$\boxed{u(t) = e^{-\xi\omega_n t} \left\{ (u_0 - u_{p0})\cos\omega_d t + \frac{(\dot{u}_0 - \dot{u}_{p0}) + \xi\omega_n(u_0 - u_{p0})}{\omega_d}\sin\omega_d t \right\} + u_p(t)} \tag{2.139}$$

with $u_{p0}, \dot{u}_{p0}$ being the initial values of the particular solution. It should be noted that these are *not* initial conditions being imposed on the particular solution, but simply the values that result from substituting $t = 0$ into that solution. Notice also that although these initial values are subtracted from the true initial conditions in the expressions within the braces, they are automatically added back by the last term. Hence, the complete expression has the system's correct initial conditions, namely $u_0, \dot{u}_0$. Later on, we shall also need the first derivative of this expression, which is

$$\boxed{\dot{u}(t) = e^{-\xi\omega_n t}\left\{ (\dot{u}_0 - \dot{u}_{p0})\cos\omega_d t - \frac{\omega_n}{\omega_d}\left[\xi(\dot{u}_0 - \dot{u}_{p0}) + \omega_n(u_0 - u_{p0}) \right]\sin\omega_d t \right\} + \dot{u}_p(t)} \tag{2.140}$$

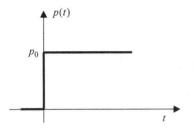

Figure 2.15. Step load.

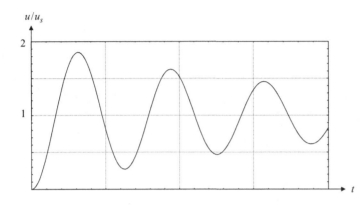

Figure 2.16. Step-load response function.

We apply next this general solution to the case of a suddenly applied load of constant amplitude.

2.4.3 Step Load of Infinite Duration

At $t = 0$, we apply a load of amplitude p_0, which we then maintain unchanged. Hence, the load is (Figure 2.15).

$$p(t) = p_0 \, \mathcal{H}(t) = \begin{cases} p_0 & \text{if} \quad t > 0 \\ 0 & \text{if} \quad t < 0 \end{cases} \tag{2.141}$$

Since the load is constant (i.e., a polynomial of zero degree), we try a particular solution that is also constant

$$u_p = \frac{p_0}{k} \equiv u_s \qquad \text{Static deflection} \tag{2.142}$$

Clearly, since this expression is constant, its initial values must be $u_{p0} = u_s$ and $\dot{u}_{p0} = 0$. Hence, for a system starting at rest, the general solution is (Figure 2.16)

$$u = u_s \left\{ 1 - e^{-\xi \omega_n t} \left(\cos \omega_d t + \frac{\xi}{\sqrt{1 - \xi^2}} \sin \omega_d t \right) \right\} \qquad t > 0 \tag{2.143}$$

The ratio between the dynamic solution and the static deflection (i.e., the term in braces) is often referred to as the *dynamic load factor*. In the particular case being considered here, this load factor is at most 2, but for other load types, it can be substantially greater.

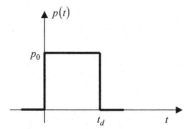

Figure 2.17. Box function.

We can express the response to a suddenly applied step load more compactly as

$$u(t) = p_0\, g(t) \tag{2.144}$$

in which

$$g(t) = \begin{cases} \dfrac{1}{k}\left\{1 - e^{-\xi\omega_n t}\left(\cos\omega_d t + \dfrac{\xi}{\sqrt{1-\xi^2}}\sin\omega_d t\right)\right\} & \text{if} \quad t > 0 \\[2ex] 0 & \text{if} \quad t < 0 \end{cases} \tag{2.145}$$

will be referred to as the *unit step-load response function*.

2.4.4 Step Load of Finite Duration (Rectangular Load, or Box Load)

From the foregoing, it is clear that if we delay the application of the step load by some time t_d, the response must also be delayed by that same amount. On the other hand, a step load of finite duration t_d is equivalent to the summation of two step loads of infinite duration, one at $t = 0$ and an equal and opposite one at $t = t_d$, which can be written as $p(t) = p_0\left[\mathcal{H}(t) - \mathcal{H}(t - t_d)\right]$ (Figure 2.17). Since the system is linear, superposition applies, so the response must be of the form

$$u = p_0\left[g(t) - g(t - t_d)\right] \tag{2.146}$$

where the functions must be understood as being zero if their arguments are negative.

2.4.5 Impulse Response Function

The total impulse of a rectangular load of amplitude p_0 and duration t_d is simply the product $p_0\, t_d$. Thus, the load amplitude required to produce a *unit* impulse must be

$$p_0 = \frac{1}{t_d} \tag{2.147}$$

which in turn produces a response

$$u = \frac{g(t) - g(t - t_d)}{t_d} \tag{2.148}$$

In the limit when the load duration shrinks to zero and the load amplitude grows to infinity, the load transforms into a singularity referred to as the Dirac delta function $\delta(t)$

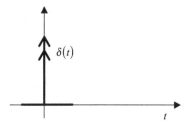

Figure 2.18. Unit impulse (Dirac delta).

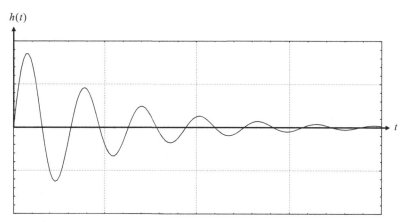

Figure 2.19. Impulse response function.

or unit impulse (Figure 2.18). The response produced by such a unit impulse follows from the limit

$$u = \lim_{t_d} \frac{g(t) - g(t - t_d)}{t_d} = \dot{g}(t) = \frac{1}{m\omega_d} e^{-\xi\omega_n t} \sin \omega_d t \qquad (2.149)$$

which is known as the impulse response function. This function, which is shown in Figure 2.19, is so important in structural dynamics that we reserve a special symbol for it:

$$\boxed{h(t) = \frac{1}{m\omega_d} e^{-\xi\omega_n t} \sin \omega_d t} \qquad \text{Impulse response function} \qquad (2.150)$$

Again, this function must be understood as being zero when the argument is negative.

A more direct and simpler derivation of this function could have been obtained by means of the *principle of conservation of momentum*, which states that the impulse imparted on a mass m abruptly changes its velocity from zero to $V = 1/m$. Since after the unit impulse the system is free from external loads, it is clear that the response must be the same as that of a free vibration with initial condition $u_0 = 0, \dot{u}_0 = 1/m$, which leads immediately to the same expression for $h(t)$ given previously.

The first two derivatives of the impulse response function are

$$\dot{h}(t) = \frac{1}{m} e^{-\xi\omega_n t} \left[\cos \omega_d t - \frac{\xi}{\sqrt{1-\xi^2}} \sin \omega_d t \right] \qquad (2.151)$$

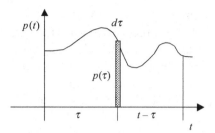

Figure 2.20. Duhamel's integral.

$$\ddot{h}(t) = \frac{1}{m}\left\{\delta(t) - \omega_n e^{-\xi\omega_n t}\left[2\xi\cos\omega_d t + \frac{1-2\xi^2}{\sqrt{1-\xi^2}}\sin\omega_d t\right]\right\} \qquad (2.152)$$

Observe that at $t = 0$, the second derivative exhibits a Dirac delta singularity.

2.4.6 Arbitrary Forcing Function: Convolution

Convolution Integral

The heuristic method for obtaining particular solutions presented previously is adequate for relatively simple excitations, but it is not appropriate for implementation in a computer, and it fails in the general case of arbitrary excitations. To obviate this difficulty, we will present a more general method, which is based on the use of the impulse response function.

Let $p(t)$ be a forcing function with some arbitrary variation in time. However, we shall restrict this excitation to begin no sooner than $t = 0$, that is, to be zero for negative times. A function satisfying this property is said to be *causal*.

Assume that we want to compute the response at some arbitrary time $t > 0$, as shown in Figure 2.20. Consider also a prior instant $\tau < t$, at which time we cut out a slice of width $d\tau$ in the forcing function. The elementary impulse associated with that slice is simply $p(\tau)\,d\tau$, that is, it is the area of the slice. Clearly, the elementary response du induced by *that* impulse at a time $t - \tau$ later must be the product of the impulse and the response function for a unit impulse, that is,

$$du(t) = h(t - \tau)p(\tau)\,d\tau$$

Since the system is linear, superposition applies, which means that the total response at time t must be the sum of all elementary responses:

$$u(t) = \int_0^t h(t-\tau)p(\tau)\,d\tau \qquad (2.153)$$

a result that is known as Duhamel's integral or the *convolution* integral. Notice that the upper limit of this integral is a variable parameter, namely the time at which the response is being computed. Making a simple change of variables $t' = t - \tau$, it is easy to show that this integral can also be expressed as

$$u(t) = \int_0^t h(\tau)p(t-\tau)\,d\tau \qquad (2.154)$$

That is, the arguments of the two functions can be interchanged. [It should be noted, however, that the meaning of τ in these two expressions is not the same; in the second, it is really t' in disguise, which on account of being a *dummy* variable of integration, we relabeled it into τ.] Because of the symmetry with respect of the argument of integration, one should choose whichever form is more convenient, usually the second. For this reason, a new operation symbol is reserved for the convolution, namely the star (*):

$$u = p * h = h * p \tag{2.155}$$

That is, the convolution is commutative.

Time Derivatives of the Convolution Integral

In applications, it is often desirable to obtain also the response velocity and/or acceleration. These can be derived from the convolution integral, but the task is complicated by the fact that the upper limit of the integral is the very variable with respect to which we take derivatives. As shown in books on calculus, the following formula applies for derivatives with respect to a variable that appears as an argument both in an integral and in the limits:

$$\frac{\partial}{\partial t} \int_{a(t)}^{b(t)} f(t,\tau) g(\tau) \, d\tau = \int_{a(t)}^{b(t)} \frac{\partial f(t,\tau)}{\partial t} g(\tau) \, d\tau + f(t,b) g(b) \frac{db}{dt} - f(t,a) g(a) \frac{da}{dt} \tag{2.156}$$

If we apply this formula to the two forms of the convolution integral (with $a = 0$, $b(t) = t$ and $\dot{h}(0) = 1/m$), we obtain the following results:

$$u = p * h \qquad \text{or} \qquad u = h * p \tag{2.157}$$

$$\dot{u} = p * \dot{h} \qquad \text{or} \qquad u = \dot{p} * h + p(0) h(t)$$

$$\ddot{u} = p * \ddot{h} + p(t)/m \quad \text{or} \quad \ddot{u} = \ddot{p} * h + \dot{p}(0) h(t) + p(0) \dot{h}(t) \tag{2.158}$$

Convolution as a Particular Solution

Clearly, since the convolution integral is a solution to the inhomogeneous differential equation, it must also be a valid particular solution, even if a very special one. Considering the fact that the convolutions for displacement and velocity at time $t = 0$ are empty integrals (i.e., limits 0 to 0), it follows from the preceding that *the convolution is a particular solution with zero initial conditions*. Hence, the general solution is

$$u(t) = e^{-\xi \omega_n t} \left\{ u_0 \cos \omega_d t + \frac{\dot{u}_0 + \xi \omega_n u_0}{\omega_d} \sin \omega_d t \right\} + p * h \tag{2.159}$$

$$\dot{u}(t) = e^{-\xi \omega_n t} \left\{ \dot{u}_0 \cos \omega_d t - \frac{\omega_n}{\omega_d} (\xi \dot{u}_0 + \omega_n u_0) \sin \omega_d t \right\} + p * \dot{h} \tag{2.160}$$

in which the frequency ratio could be written as $\frac{\omega_d}{\omega_n} = \sqrt{1-\xi^2}$. This equation should be contrasted with the one given earlier, in which the initial values of the particular solution were subtracted from the actual initial conditions to compensate for the contribution to those conditions provided by $u_p(t)$. For convenience, we list below some formulas that are useful in the evaluation of convolution integrals:

$$\int_0^t e^{-\xi\omega_n\tau} \sin \omega_d\tau \, d\tau = \frac{1}{\omega_n}\left\{\frac{\omega_d}{\omega_n} - e^{-\xi\omega_n t}\left[\frac{\omega_d}{\omega_n}\cos\omega_d t + \xi\sin\omega_d t\right]\right\} \tag{2.161}$$

$$\int_0^t e^{-\xi\omega_n\tau} \cos \omega_d\tau \, d\tau = \frac{1}{\omega_n}\left\{\xi - e^{-\xi\omega_n t}\left[\xi\cos\omega_d t - \frac{\omega_d}{\omega_n}\sin\omega_d t\right]\right\} \tag{2.162}$$

$$\int_0^t \tau e^{-\xi\omega_n\tau} \sin \omega_d\tau \, d\tau = \frac{1}{\omega_n^2}\left\{2\xi\frac{\omega_d}{\omega_n} - e^{-\xi\omega_n t}\left[(2\xi\frac{\omega_d}{\omega_n} + \omega_d t)\cos\omega_d t - (1 - 2\xi^2 - \xi\omega_n t)\sin\omega_d t\right]\right\}$$
$$\tag{2.163}$$

$$\int_0^t \tau e^{-\xi\omega_n\tau} \cos \omega_d\tau \, d\tau = \frac{-1}{\omega_n^2}\left\{1 - 2\xi^2 - e^{-\xi\omega_n t}\left[(1 - 2\xi^2 - \xi\omega_n t)\cos\omega_d t + (2\xi\frac{\omega_d}{\omega_n} + \omega_d t)\sin\omega_d t\right]\right\}$$
$$\tag{2.164}$$

2.5 Support Motion in SDOF Systems

2.5.1 General Considerations

Evaluation of the dynamic effects elicited by motions of the support point underneath a structure are of interest not only in connection with earthquakes, but also for other vibrating systems, such as a car traveling at some speed on a bumpy road, or a ship acted upon by rough seas.

Consider an SDOF system subjected to a support motion, as shown in Figure 2.21. We define

$u_g(t)$ = Ground displacement (assumed to be known)
$u(t)$ = *Absolute* displacement of the mass
$v(t)$ = *Relative* displacement of mass

From these definitions, these three displacements are related through

$$v = u - u_g \qquad \text{or} \qquad u = v + u_g \tag{2.165}$$

The inertia forces are proportional to the absolute acceleration, while the forces in the spring and dashpot are proportional to the deformation of these elements, that is, to the relative displacement and relative velocity. Hence, the dynamic equilibrium equation is

$$m\ddot{u} + c\dot{v} + kv = 0 \tag{2.166}$$

Substituting the relationship between relative and absolute motions, we obtain the two alternative equations

$$\boxed{m\ddot{v} + c\dot{v} + kv = -m\ddot{u}_g} \tag{2.167}$$

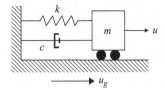

Figure 2.21. SDOF system subjected to support motion.

$$m\ddot{u} + c\dot{u} + ku = c\dot{u}_g + ku_g$$ (2.168)

Both equations are of the same form as that of an SDOF system subjected to a dynamic force. The first equation is the preferred choice in traditional earthquake engineering, because most seismic records are obtained in terms of ground accelerations. By contrast, the determination of both the ground velocity and the displacement time histories is an error-prone task, which requires numerical integration of the acceleration records and is sensitive to the assumed initial conditions. To minimize the accumulation of errors, the records are subjected prior to integration to modifications that are usually referred to as *instrument correction* and *baseline correction*. On the other hand, the second equation is to be preferred whenever the support motion is known in terms of velocity or displacements, or when the solution is obtained in the frequency domain. As written, however, this equation has the inconvenience of involving two excitation terms, but later on we shall circumvent this shortcoming.

The formal solution to the previous two equations is

$$v = e^{-\xi\omega_n t}\left[v_0\cos\omega_d t + \frac{\dot{v}_0 + \xi\omega_n v_0}{\omega_d}\sin\omega_d t\right] - m\ddot{u}_g * h$$ (2.169)

$$u = e^{-\xi\omega_n t}\left[u_0\cos\omega_d t + \frac{\dot{u}_0 + \xi\omega_n u_0}{\omega_d}\sin\omega_d t\right] + (c\dot{u}_g + ku_g) * h$$ (2.170)

As can be seen, the solutions are similar to those for forced excitations. Notice, however, that these two forms involve *different* initial conditions, and that these conditions are not independent. In fact, they are related through the initial *values* of the ground velocity and displacement. While the initial conditions for earthquakes are arguably zero, it often happens in practice that the initial portion of the record is imperfectly known, for example, when the instrument triggered only after the ground acceleration exceeded some threshold value, or when there is ambient noise present in the signal. A fact often ignored is that those nonzero initial values may have arisen naturally as a result of the instrument and baseline correction referred to previously, both of which tend to filter out very low and very high frequency components. Hence, assuming zero values for the ground motion's starting values may lead to invalid response computations, particularly for very soft or very stiff systems. The former are best handled by assuming zero absolute initial motions, while for the latter the preferred choice may be zero relative initial motions.

Returning briefly to the second solution, which expressed the response in terms of absolute motions, we shall now consolidate the two excitation terms. Using integration by parts, it can be shown that $\dot{u}_g * h = u_g * \dot{h}$. Hence, the convolution part of the absolute response can be written as

$$c\ddot{u}_g * h + k u_g * h = (c\dot{h} + kh) * u_g = \frac{k}{\omega_n}(2\xi\dot{h} + \omega_n h) * u_g \tag{2.171}$$

If we define

$$h_{u|u_g} = \frac{k}{\omega_n}(2\xi\dot{h} + \omega_n h) \tag{2.172}$$

as the *seismic impulse response function for absolute displacements due to absolute ground displacement*, and introduce into this definition the expressions for $h(t)$ previously presented, we obtain

$$h_{u|u_g}(t) = \omega_n e^{-\xi\omega_n t}\left\{2\xi\cos\omega_d t + \frac{\omega_n}{\omega_d}(1 - 2\xi^2)\sin\omega_d t\right\} \tag{2.173}$$

The absolute seismic response can then be written as

$$u = e^{-\xi\omega_n t}\left[u_0\cos\omega_d t + \frac{\dot{u}_0 + \xi\omega_n u_0}{\omega_d}\sin\omega_d t\right] + u_g * h_{u|u_g} \tag{2.174}$$

In particular, if the absolute initial conditions are zero (i.e., system starting at rest), then

$$u = u_g * h_{u|u_g} \tag{2.175}$$

An important advantage of the impulse response function just described is that it is invariant under a change of input–output type, that is, $h_{u|u_g} = h_{\dot{u}|\dot{u}_g} = h_{\ddot{u}|\ddot{u}_g}$. Hence

$$\dot{u} = \dot{u}_g * h_{u|u_g}, \qquad \ddot{u} = \ddot{u}_g * h_{u|u_g} \tag{2.176}$$

In most cases, it is the absolute acceleration response that is of interest in engineering applications, so the last equation can be used to evaluate this quantity in a direct fashion.

As we shall see later on in connection with harmonic response analyses, and more generally with *frequency-domain* analyses, the *Fourier transform* of the impulse response function – the *transfer function* – plays a fundamental role in structural dynamics. In the case of the seismic impulse response function $h_{u|u_g}$, this transform can be evaluated without much difficulty. Using integration by parts, the transfer function can be shown to be given by

$$H_{u|u_g}(\omega) = \int_0^\infty h_{u|u_g}e^{-i\omega t}dt = \frac{1 + 2i\xi\frac{\omega}{\omega_n}}{1 - (\frac{\omega}{\omega_n})^2 + 2i\xi\frac{\omega}{\omega_n}} \tag{2.177}$$

This expression agrees with the well-known *transfer function for absolute displacement due to ground displacement* (or velocity to velocity, or acceleration to acceleration). Later on we shall encounter this transfer function again when considering the topic of harmonic seismic response.

2.5.2 Response Spectrum

The seismic response spectrum is a plot depicting the maximum response of an SDOF system for a given fraction of critical damping to some earthquake excitation. It is plotted either as a function of the natural frequency or of the natural period of that system. The principal application of response spectra lies in engineering design, because when a structure is modeled as an SDOF system, all internal forces are synchronous with, and proportional to, the response. Hence, the maximum values of those internal forces, which are needed to design the structure, are attained when the response is maximum. A full description of this topic can be found in Chapter 7, Section 7.3. A brief summary follows.

The ground response spectrum is defined in terms of the following three quantities:

$$S_d \equiv S_d(\omega_n, \xi) = |v_{max}| \quad \text{Spectral displacement (max. relative displacement)} \tag{2.178}$$

$$S_v = \omega_n S_d \quad \text{Pseudo-velocity} \tag{2.179}$$

$$S_a = \omega_n^2 S_d = \omega_n S_v \quad \text{Pseudo-acceleration (spectral acceleration)} \tag{2.180}$$

The spectral displacement is a plot of the actual maximum *relative displacement* for the SDOF system at hand. At very low frequencies the spectral displacement approaches asymptotically the maximum ground displacement. This is so because a very flexible structure has no time to respond to the ground motion and basically remains stationary, with the ground moving back and forth underneath. On the other hand, the pseudo-acceleration is a close approximation to the peak absolute acceleration. At very high frequencies, the spectral accelerations approach asymptotically the maximum ground accelerations. This is because a very stiff structure will follow the ground motion at all times.

The pseudo-velocity is only weakly related to the maximum relative velocity. Indeed, at low frequencies the true peak relative velocity tends to the ground velocity and not to zero as S_v does, while at high frequencies the true peak relative velocity drops to zero with the square of the frequency as $\ddot{u}_{max}/\omega_n^2$; this is much faster than the pseudo-velocity, which decays only linearly with frequency, that is, $S_v \to \ddot{u}_{max}/\omega_n$. Thus, when drawn as distinct lines on a tripartite spectrum (see the next section), these two parameters must intersect at intermediate frequencies at which both roughly agree.

Tripartite Spectrum

The response spectrum can be displayed in many different formats, some which we review in more detail in Chapter 7. One of these is the so-called *tripartite spectrum* displayed in terms of either the natural oscillator frequency or the natural period, as shown in Figure 2.22. This particular trilogarithmic format of the response spectrum displays the frequency (or period) on the horizontal axis, the so-called pseudo-velocity on the vertical axis, then a very close approximation to the peak absolute acceleration referred to as the pseudo-acceleration (for practical purposes, this is the peak acceleration), and the peak relative displacement. Both of the latter are read along axes that are inclined at 45 degrees to the horizontal. The spectrum shown is for some specific earthquake with a peak acceleration of about 0.5g and a peak ground displacement of about 16 cm (0.16 m), namely

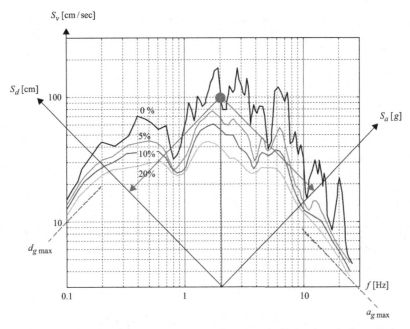

Figure 2.22. Pseudo-velocity tripartite response spectrum for some specific earthquake, here displayed in terms of frequency, and not period. Fractions of damping are 0.00, 0.05, 0.10, and 0.20.

the asymptotic value to which the accelerations tend to at very high frequencies (or short periods), and to which the relative displacements tend to at low frequencies (or long periods). If the spectrum shown had been plotted versus period, then the positions of the acceleration and relative displacement axes would have been flipped with respect to a vertical axis, and so would also have been the asymptotes. The spectrum usually plots various response spectra curves for different values of critical damping. The more highly damped, the lower the spectrum. In the present case, the spectrum includes four curves for different values of damping, namely 0% (top curve), 5%, 10%, and 20% (bottom curve). **Note:** For clarity of the figure, the rotated grid has been omitted.

As a specific example, consider an oscillator (structure) with a natural period of 0.5 second (i.e., natural frequency of 2 Hz) and damping of 0% (first curve from the top), the relative displacement for the earthquake represented by the spectrum shown is approximately 7.9 cm (0.079 m) (the value pointed at by the arrow to the left). Also, the maximum absolute acceleration is approximately 1.3g (the value pointed to by the arrow to the right). This agrees with the computation $a_{max} = (2\pi \times 2)^2 \times 0.079 / 9.81 = 1.272$ g.

2.5.3 Ship on Rough Seas, or Car on Bumpy Road

Consider a vehicle that travels with speed V on an uneven surface, such as a car on a bumpy road or a ship navigating in rough seas, as shown in Figure 2.23. In the first case, the road roughness and undulations y do not change with time, but depend solely on the position of the car, while in the second, the surface of the water changes in both space and time. We consider the latter, more general case first, which we then specialize to the case of a car.

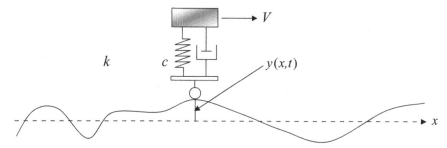

Figure 2.23. Car on bumpy road and a ship on rough seas.

Let $y(x,t)$ be a mathematical description of the vertical position of the point on the surface in the vertical travel plane at which the vehicle makes contact with the surface. At a small time Δt later, the vehicle moves to position $x + \Delta x$, so that the apparent vertical velocity of the support point, as perceived by an observer moving horizontally in tandem with the vehicle, is

$$\frac{dy}{dt} = \lim_{\Delta t \to 0} \frac{y(x + \Delta x, t + \Delta t) - y(x,t)}{\Delta t} = \lim_{\Delta t \to 0} \frac{1}{\Delta t}\left[\Delta x \frac{\partial y}{\partial x} + \Delta t \frac{\partial y}{\partial t}\right] = V\frac{\partial y}{\partial x} + \frac{\partial y}{\partial t}$$

$$= V(t)\,y'(x,t) + \dot{y}(x,t) \tag{2.181}$$

To a first coarse approximation, a ship heaving in water (or a car oscillating vertically on its shock absorbers) can be idealized as an SDOF system subjected to a support motion y. Hence

$$\boxed{m\ddot{u} + c\dot{u} + ku = c\frac{dy}{dt} + ky = c\left[\dot{y}(x,t) + V(t)\,y'(x,t)\right] + ky(x,t)} \tag{2.182}$$

in which $u(t)$ is the absolute *vertical* position of the vehicle with respect to an inertial reference frame, say the average ocean or road surface. In the case of a *constant* horizontal vehicle speed V, we can assume without loss of generality that $x = Vt$, in which case

$$m\ddot{u} + c\dot{u} + ku = c\left[\dot{y}(Vt,t) + V\,y'(Vt,t)\right] + k\,y(Vt,t) \qquad \text{Ship} \tag{2.183}$$

and by specialization with $\dot{y} = 0$,

$$m\ddot{u} + c\dot{u} + ku = cV\,y'(Vt) + k\,y(Vt) \qquad\qquad \text{Car} \tag{2.184}$$

We'll provide an example of a car and defer a moving ship for the section on harmonic analysis.

Example
A car that weighs 2000 kg travels on a bumpy road with constant velocity of 72 km/h. To a first approximation, it can be modeled as an SDOF system. The car is known to have an undamped natural frequency of 1 Hz (i.e., undamped period $T = 1$ s), and a fraction of critical damping of 50%. The road is flat, except for a ramp of length $L = 10$ m and a height $h = 0.10$ m. Evaluate the response.

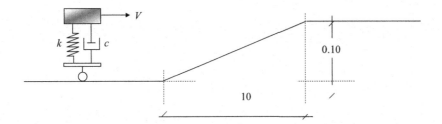

Expressed in metric units, the speed of the car is $V = 20$ m/s. When the car travels up the ramp, it covers an effective distance $L = \sqrt{10^2 + 0.1^2} \approx 10$ m, that is, the effect of the inclination on the distance traveled and on the horizontal velocity is negligible. The total time t_d needed to traverse the ramp is then

$$t_d = \frac{L}{V} = \frac{10 \,[\text{m}]}{20 \,[\text{m/s}]} = 0.5 \,[\text{s}] \tag{2.185}$$

that is, it traverses the slope in about half a period of vertical oscillation. Choosing as vertical reference level $y = 0$ to be the lower flat region and placing the origin $x = 0$ at the beginning of the ramp, and assuming without loss of generality that the car reaches the origin at time $t = 0$, then the equations of the road $y(x) = y(Vt)$ is given by

$$y = \begin{cases} 0 & t < 0 \\ \dfrac{V h}{L} t & 0 \le t \le t_d \\ h & t > t_d \end{cases} \tag{2.186}$$

which implies the apparent ground velocity

$$\dot{y} = \begin{cases} 0 & t < 0 \\ V \dfrac{h}{L} & 0 \le t \le t_d \equiv V \tfrac{h}{L} \big[\mathcal{H}(t) - \mathcal{H}(t - t_d) \big], \\ 0 & t > t_d \end{cases} \tag{2.187}$$

where $\mathcal{H}(t) \equiv \mathcal{H}_0(t)$ is the Heaviside (or unit step) function (see Chapter 9, Section 9.1). Hence, the apparent vertical ground acceleration is

$$\ddot{y} = V \tfrac{h}{L} \big[\delta(t) - \delta(t - t_d) \big] \tag{2.188}$$

because the derivative of the Heaviside function is the Dirac delta function. Expressing the solution in terms of relative displacements, and with the analogy $y \leftrightarrow u_g$, the equation of motion is then

$$m \ddot{v} + c \dot{v} + k v = -m \ddot{y} = -m V \tfrac{h}{L} \big[\delta(t) - \delta(t - t_d) \big] \tag{2.189}$$

in which case the solution is trivially obtained by simple inspection, namely

$$v = -m V \tfrac{h}{L} \big[h(t) - h(t - t_d) \big] \tag{2.190}$$

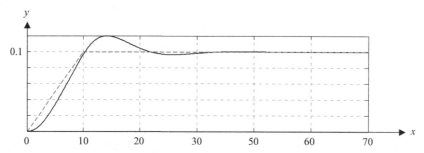

Figure 2.24. Car traveling at constant speed and climbing a ramp of finite length.

with $h(t)$ being the impulse response function. The *absolute* displacement is then

$$u = y + v = V\frac{h}{L}\left\{\left[\mathcal{R}(t) - \mathcal{R}(t - t_d)\right] - m\left[h(t) - h(t - t_d)\right]\right\} \qquad (2.191)$$

where

$$\mathcal{R}(t) \equiv \mathcal{H}_1(t) = \begin{cases} 0 & t \le 0 \\ t & t > 0 \end{cases} \qquad (2.192)$$

is the unit ramp function. This response function is easily computed and evaluated using MATLAB®. At first the car presses against the shock absorbers (i.e., the road) and after passing the transition to the flat part at $t_d = \frac{1}{2}T$, it reaches its maximum elongation at about $t = \frac{3}{4}T$, after which it bounces back and oscillates for a brief time before resuming its quiescent travel, as shown in Figure 2.24. (**Note**: vertical scale is highly magnified.)

2.6 Harmonic Excitation: Steady-State Response

The important case of harmonic (sinusoidal) excitation is of particular interest in structural dynamics, not only for its immediate practical applications, but also because of its intimate relationship through Fourier synthesis of arbitrary loadings. Moreover, mastery of the so-called frequency-domain analyses allows access to the powerful methods in signal processing. Indeed, it can safely be affirmed that the concepts of harmonic response play a central role in mechanical vibrations, and constitute a cornerstone in vibration theory.

2.6.1 Transfer Function Due to Harmonic Force

Consider an SDOF system subjected to a load that varies in time like a cosine or a sine function, with an arbitrary frequency ω (which should not be confused with the natural frequency ω_n)

$$m\ddot{u}_1 + c\dot{u}_1 + ku_1 = p_0 \cos \omega t \qquad (2.193)$$

$$m\ddot{u}_2 + c\dot{u}_2 + ku_2 = p_0 \sin \omega t \qquad (2.194)$$

Clearly, we could find a *particular solution* to either equation by considering a combination of a cosine and a sine function of this same frequency together with integration constants to be determined, but we refrain from doing so here for reasons that will become apparent. Instead, we present next an alternative strategy that allows us to solve for both excitations at once, and that has the added benefit of providing certain expressions of great importance in structural dynamics.

To solve these two equations, we employ an elegant trick: we multiply the second equation by $i = \sqrt{-1}$ and add the two equations together. After doing so, we obtain

$$m(\ddot{u}_1 + i\ddot{u}_2) + c(\dot{u}_1 + i\dot{u}_2) + k(u_1 + iu_2) = p_0(\cos\omega t + i\sin\omega t) = p_0 e^{i\omega t} \tag{2.195}$$

It should be noted that since both u_1 and u_2 are real quantities, the imaginary unit keeps them separate, and no mix-up between the two can occur on summation. That issue settled, we proceed to define the complex response function

$$u = u_1 + iu_2 \tag{2.196}$$

from which we can extract u_1 and u_2 by taking the real and imaginary parts

$$u_1 = \text{Re}(u) \tag{2.197}$$

$$u_2 = \text{Im}(u) \tag{2.198}$$

Also, we recognize Euler's formula $\cos\omega t + i\sin\omega t = e^{i\omega t}$. With these preliminaries, we can write

$$m\ddot{u} + c\dot{u} + ku = p_0 e^{i\omega t} \tag{2.199}$$

This equation allows a very simple *particular* solution of the form

$$u(t) = \tilde{u}\, e^{i\omega t} \tag{2.200}$$

with an as yet unknown *frequency response function* $\tilde{u} = \tilde{u}(\omega)$. This particular solution is periodic, and is referred to as the *steady-state* response (i.e., the response that will remain after any free vibration has died out). The time derivatives of this solution are

$$\dot{u}(t) = i\omega\tilde{u}\, e^{i\omega t} \tag{2.201}$$

$$\ddot{u}(t) = (i\omega)^2 \tilde{u}\, e^{i\omega t} = -\omega^2 \tilde{u}\, e^{i\omega t} \tag{2.202}$$

Substituting these expressions into the differential equation, factoring out the common term $\tilde{u}$, and canceling the common (nonzero) exponential term on either side, we obtain

$$(-\omega^2 m + i\omega c + k)\tilde{u} = p_0 \tag{2.203}$$

Finally, solving for $\tilde{u}$,

$$\tilde{u}(\omega) = p_0 \left[\frac{1}{k - \omega^2 m + i\omega c} \right] \equiv p_0\, H(\omega) \tag{2.204}$$

In the preceding expression, $k - \omega^2 m + i\omega c$ is the dynamic stiffness of the SDOF system, and

$$\boxed{H(\omega) = \frac{1}{k - \omega^2 m + i\omega c} \equiv \frac{1/k}{1 - \left(\frac{\omega}{\omega_n}\right)^2 + 2i\xi\frac{\omega}{\omega_n}}} \tag{2.205}$$

is the response function for unit harmonic load ($p_0 = 1$), which is known as the *transfer function*. In general, transfer functions in linear systems always express a complex-amplitude relationship between some input and some output, and this fact is often stated explicitly by writing the transfer function in a form such as

$$H(\omega) = H_{u|p}(\omega) \qquad (2.206)$$

However, if the input–output relationship is obvious from the context (and especially, if there is only one transfer function H), it may be preferable to exclude the sub-indices, because by doing so the algebra with transfer functions remains more transparent and readable. We adopt this strategy here, and will resort to sub-indices only when needed to avoid confusion.

Before proceeding further, consider first the polar representation of a ratio of two complex numbers:

$$\frac{z_1}{z_2} = \frac{x_1 + i\,y_1}{x_2 + i\,y_2} = \frac{r_1\,e^{i\varphi_1}}{r_2\,e^{i\varphi_2}} = \frac{\sqrt{x_1^2 + y_1^2}}{\sqrt{x_2^2 + y_2^2}}\,e^{i(\varphi_1 - \varphi_2)} \qquad (2.207)$$

in which r_1, r_2 are the *absolute values*, and $\varphi_i = \arctan(y_i / x_i)$ ($i = 1,2$) are the *phase angles*. We apply this representation to the transfer function, yielding:

$$H(\omega) = \frac{1/k}{\sqrt{\left[1 - \left(\frac{\omega}{\omega_n}\right)^2\right]^2 + 4\xi^2\left(\frac{\omega}{\omega_n}\right)^2}}\,e^{-i\varphi} \qquad (2.208)$$

in which the phase angle is that of the denominator alone (i.e., $\varphi \equiv \varphi_2$, since the numerator is a real, positive number, with $\varphi_1 = 0$):

$$\varphi = \arctan \frac{2\xi \frac{\omega}{\omega_n}}{1 - \left(\frac{\omega}{\omega_n}\right)^2} \qquad (2.209)$$

Hence, the response function $\tilde{u}$ is of the form

$$\tilde{u}(\omega) = \frac{p_0}{k}\frac{1}{\sqrt{\left[1 - \left(\frac{\omega}{\omega_n}\right)^2\right]^2 + 4\xi^2\left(\frac{\omega}{\omega_n}\right)^2}}\,e^{-i\varphi} = u_s A(\omega)e^{-i\varphi} \qquad (2.210)$$

with

$$u_s = \frac{p_0}{k} \qquad \text{Static deflection} \qquad (2.211)$$

and

$$A(\omega) = \frac{1}{\sqrt{\left[1 - \left(\frac{\omega}{\omega_n}\right)^2\right]^2 + 4\xi^2\left(\frac{\omega}{\omega_n}\right)^2}} \qquad \text{Dynamic amplification function} \qquad (2.212)$$

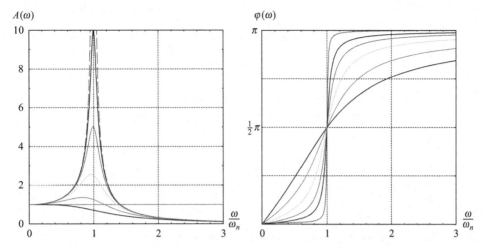

Figure 2.25. Dynamic amplification function (left) and phase angle (right) for an SDOF system subjected to harmonic load. Damping fractions used are $\xi = 0.01, 0.05, 0.10, 0.20, 0.40,$ and $0.707.$

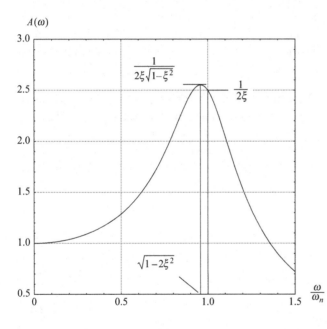

Figure 2.26. Peak response versus resonant response: The difference is usually negligible.

The complex response is then

$$u(t) = \tilde{u}e^{i\omega t} = u_s A e^{i(\omega t - \varphi)} \tag{2.213}$$

Taking the real and imaginary parts from this expression, it follows immediately that

$$\boxed{u_1 = u_s A \cos(\omega t - \varphi)} \qquad \boxed{u_2 = u_s A \sin(\omega t - \varphi)} \tag{2.214}$$

which are the actual (real) steady-state response functions to cosine loads and sine loads, respectively. Clearly, the dynamic amplification factor relates the amplitudes of dynamic and static responses, while the phase angle relates to the delay in time between the response and the forcing function.

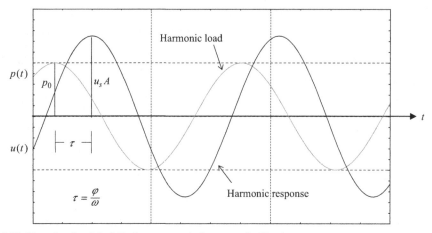

Figure 2.27. Phase lag (or delay) τ of response relative to applied load.

The peak amplification occurs at a tuning ratio $r = \sqrt{1 - 2\xi^2}$ and has a value $\left(2\xi\sqrt{1 - \xi^2}\right)^{-1}$. At $\omega = \omega_n$ ($r = 1$), the amplification is $1/2\xi$, which is an easy value to remember. Notice that when $\xi \geq \sqrt{0.5} = 0.71$, the peak occurs at the origin (zero frequency). The damped frequency lies nearly halfway between the peak and the undamped frequency, and has an amplitude $\left[2\xi\sqrt{1 - \frac{3}{4}\xi^2}\right]^{-1}$.

It is interesting also to consider the transfer function for velocity response, which is obtained from the displacement response function by simple multiplication by $i\omega$. The reason we mention it here is that when the absolute value of this function is plotted versus the logarithm of the tuning ratio $r = \omega/\omega_n$, the plot is symmetric with respect to the resonant frequency. In terms of the tuning ratio, the transfer function for velocity due to an applied force is

$$H_{\dot{u}|p} = \frac{1}{\sqrt{km}}\left[\frac{i\,r}{1 - r^2 + 2\,i\,\xi r}\right] \tag{2.215}$$

whose absolute value is

$$\left|H_{\dot{u}|p}\right| = \frac{1}{\sqrt{km}}\left[\frac{r}{\sqrt{\left(1 - r^2\right)^2 + 4\xi^2 r^2}}\right] \tag{2.216}$$

Notice that the value of this function remains the same when r is replaced by $1/r$. Hence, the function is symmetric when plotted versus $\log r$.

2.6.2 Transfer Function Due to Harmonic Support Motion

The equation for the *absolute* response of an SDOF system subjected to support motion is

$$m\ddot{u} + c\dot{u} + ku = c\dot{u}_g + ku_g \tag{2.217}$$

In the case of harmonic ground *displacement* of amplitude u_{g0}

$$u_g = u_{g0}\, e^{i\omega t} \tag{2.218}$$

the harmonic absolute response is of the form

$$u = u_{g0}\, H_{u|u_g}(\omega)\, e^{i\omega t} \tag{2.219}$$

in which $H_{u|u_g}(\omega)$ is the *transfer function from ground displacement to absolute displacement*. The velocities and accelerations are then

$$\dot{u} = i\,\omega u \qquad \ddot{u} = (i\,\omega)^2\, u = -\omega^2\, u \tag{2.220}$$

$$\dot{u}_g = i\,\omega u_g \qquad \ddot{u}_g = (i\,\omega)^2\, u_g = -\omega^2\, u_g \tag{2.221}$$

Substituting, as before, these expressions into the differential equation and solving for the transfer function, we obtain

$$H_{u|u_g}(\omega) = \frac{k + i\,\omega c}{k - \omega^2 m + i\,\omega c} \tag{2.222}$$

or dividing by k

$$\boxed{H_{u|u_g}(\omega) = \frac{1 + 2 i \xi \frac{\omega}{\omega_n}}{1 - \left(\frac{\omega}{\omega_n}\right)^2 + 2 i \xi \frac{\omega}{\omega_n}}} \tag{2.223}$$

If we repeat this analysis assuming that the support motion is specified in terms of either ground velocity or ground acceleration, and compute correspondingly the response in terms of absolute velocity and absolute acceleration, we will find that the transfer functions for these two cases are identical to the transfer function just found. In other words,

$$H(\omega) = H_{u|u_g}(\omega) \equiv H_{\dot{u}|\dot{u}_g}(\omega) \equiv H_{\ddot{u}|\ddot{u}_g}(\omega) \tag{2.224}$$

That is, the transfer functions for absolute motion are all equal, provided that the input–output relationship refers to motions of the same type (ground displacement to absolute displacement, ground velocity to absolute velocity, or ground acceleration to absolute acceleration). A similar identity would hold if we were to change the type output from absolute to relative displacements. In that case, the transfer functions would change into $H_{v|u_g} = H_{u|u_g} - 1$, such that

$$H_{v|u_g}(\omega) \equiv H_{\dot{v}|\dot{u}_g}(\omega) \equiv H_{\ddot{v}|\ddot{u}_g}(\omega) \tag{2.225}$$

In terms of absolute value and phase angle, the transfer function can be written as

$$H(\omega) = |H(\omega)|\, e^{i(\varphi_1 - \varphi_2)} = |H(\omega)|\, e^{-i\varphi} \tag{2.226}$$

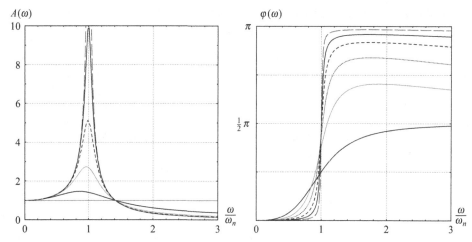

Figure 2.28. Dynamic amplification function (left) and phase angle (right) for an SDOF system subjected to unit support motion. Damping fractions used are $0.01, 0.05, 0.10, 0.20$, and 0.50.

in which the absolute value is the ratio of the absolute values of numerator and denominator, and the phase angle $\varphi = \varphi_2 - \varphi_1$ is *defined* as the difference between the phase angle of the denominator and that of the numerator (so as to have a positive phase angle). The results are

$$|H(\omega)| = \sqrt{\frac{1+4\xi^2(\frac{\omega}{\omega_n})^2}{\left[1-\left(\frac{\omega}{\omega_n}\right)^2\right]^2 + 4\xi^2(\frac{\omega}{\omega_n})^2}} \tag{2.227}$$

$$\tan \varphi_1 = 2\xi\frac{\omega}{\omega_n} \qquad \tan\varphi_2 = \frac{2\xi\frac{\omega}{\omega_n}}{1-(\frac{\omega}{\omega_n})^2} \qquad \tan(\varphi_2 - \varphi_1) = \frac{\tan \varphi_2 - \tan\varphi_1}{1+\tan\varphi_1\tan\varphi_2} \tag{2.228}$$

from which we can solve for the phase angle

$$\varphi = \arctan\frac{2\xi(\frac{\omega}{\omega_n})^3}{1-(1-4\xi^2)(\frac{\omega}{\omega_n})^2} \tag{2.229}$$

Notice how all amplification curves have unit value at $\sqrt{2}$ times the natural frequency. Beyond this point, an increase in damping *increases*, not decreases the response. The peak amplification occurs at a tuning ratio $\frac{\omega}{\omega_n} = \frac{1}{2\xi}\sqrt{\sqrt{1+8\xi^2}-1} \approx \frac{1}{2\xi}\sqrt{4\xi^2} \approx 1$; the shift away from resonance is less than in the nonseismic case. Observe also that the phase angle is no longer exactly $90°$ at resonance.

Example: Ship Heaving in Heavy Seas
Consider ideally harmonic ocean waves of amplitude A, period T and wavelength λ in an infinitely deep body of water. The dominant period of the waves relates to the sea conditions, and to a first very rough approximation can be estimated as indicated in Table 2.2.

Table 2.2. Dominant period of waves as a function of sea conditions

	T (sec)
Light seas	5
Moderate seas	10
Heavy seas	15

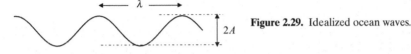

Figure 2.29. Idealized ocean waves.

These waves are known to satisfy

$$\lambda = \frac{g}{2\pi}T^2, \qquad k_w = \frac{2\pi}{\lambda} = \frac{1}{g}\left(\frac{2\pi}{T}\right)^2 = \frac{\omega^2}{g}, \qquad V_w = \frac{\omega}{k_w} = \frac{g}{\omega} = \frac{gT}{2\pi} = \text{velocity of waves} \quad (2.230)$$

in which g is the acceleration of gravity, and $\omega = 2\pi/T$. We mention in passing that the vertical amplitude of these ocean waves with respect to the mid-surface usually does not exceed the limit

$$\frac{2A}{\lambda} \cong \frac{1}{7} \tag{2.231}$$

because if $2A > \frac{1}{7}\lambda$, the waves would break. Also, from the preceding, it follows that,

$$y = A\sin\left(\omega t \mp k_w x\right) = A\sin\left[\omega\left(t \mp \frac{Vt}{V_w}\right)\right] = A\sin\left[\omega t\left(1 \mp \frac{V}{V_w}\right)\right] \tag{2.232}$$

We must use the upper (negative) sign when the waves move in the positive x direction (i.e., in the same direction as the ship), and the lower (positive) sign when they travel in the negative x direction. Next, we define

$$\omega_w = \omega\left(1 \mp \frac{V}{V_w}\right) = \text{effective frequency of water waves, as perceived by the ship} \tag{2.233}$$

Hence

$$y = A\sin\omega_w t \qquad \frac{dy}{dt} = A\omega_w \cos\omega_w t \tag{2.234}$$

and

$$m\ddot{u} + c\dot{u} + ku = A\left[c\omega_w \cos\omega_w t + k\sin\omega_w t\right] \tag{2.235}$$

from which the *steady-state* response can readily be computed:

$$u(t) = A\sqrt{\frac{1+\left(2\xi\frac{\omega_w}{\omega_n}\right)^2}{\left[1-\left(\frac{\omega_w}{\omega_n}\right)^2\right]^2 + \left(2\xi\frac{\omega_w}{\omega_n}\right)^2}}\sin\left(\omega_w t - \phi\right) \tag{2.236}$$

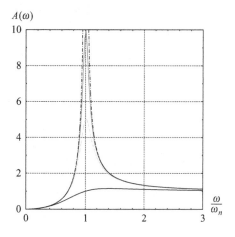

Figure 2.30. Amplification function for eccentric mass vibrator.

$$\phi = \arctan \frac{2\xi \left(\dfrac{\omega_w}{\omega_n}\right)^3}{1 - (1 - 4\xi^2)\left(\dfrac{\omega_w}{\omega_n}\right)^2} = \text{phase delay} \tag{2.237}$$

in which ξ is the fraction of critical damping. Observe that when the ship travels in the same direction and at the same speed as the water waves, then $\omega_w = 0$, that is, there is no dynamic effect.

2.6.3 Eccentric Mass Vibrator

An eccentric mass vibrator is a mechanical device used to induce harmonic forces in a dynamic system. In some cases, imperfectly balanced machines, such as fans and engines, may also act as eccentric mass vibrators and be the cause of accidental vibrations in the systems that support them. It should be noted that in those cases, the machine has a fixed operating frequency (typically 1800 rpm, or 30 Hz), in which case the driving frequency $\omega = \omega_o$ is fixed.

A mass vibrator may consist of two identical masses mounted eccentrically on parallel wheels that rotate in opposite directions. The reason for using two eccentric masses instead of just one is to create a harmonic force along a single coordinate direction. The effect of the two masses is designed to be additive along one axis of the device, and to cancel out in the direction perpendicular to it.

Consider a mass vibrator consisting of two small masses whose value is $0.5m_v$ mounted each with identical eccentricity ε and rotating with frequency ω around an axis in a rigid frame firmly attached to an SDOF system of mass m_s. The motion of each of the eccentric masses can be described as the sum of the translation u of the axis (i.e. the oscillator) in the direction of vibration, say the horizontal, and the component contributed by the rotation along that direction, that is

$$u_v = u + \varepsilon \cos \omega t \tag{2.238}$$

This motion induces an inertia force in the vibrator $m_v(\ddot{u} - \omega^2 \varepsilon \cos \omega t)$, which is transferred to the oscillator. The dynamic equilibrium equation of the oscillator is then

$$m_s \ddot{u} + m_v(\ddot{u} - \omega^2 \varepsilon \cos \omega t) + c\dot{u} + ku = 0 \tag{2.239}$$

that is

$$(m_s + m_v)\ddot{u} + c\dot{u} + ku = \omega^2 m_v \varepsilon \cos \omega t \tag{2.240}$$

Defining $m = m_s + m_v$, the equation of motion can be written as

$$m\ddot{u} + c\dot{u} + ku = \omega^2 m_v \varepsilon \cos \omega t \tag{2.241}$$

which is of the same form as that of a forced harmonic excitation encountered earlier, except that now the force amplitude $p_0 = \omega^2 m_v \varepsilon$ depends also on the excitation frequency. The response can then be written in the following two alternative ways:

$$u = \left(\frac{m_v \varepsilon \omega_n^2}{k} \right) \frac{\left(\frac{\omega}{\omega_n} \right)^2}{\sqrt{\left[1 - \left(\frac{\omega}{\omega_n} \right)^2 \right]^2 + 4\xi^2 \left(\frac{\omega}{\omega_n} \right)^2}} \cos(\omega t - \varphi) \tag{2.242}$$

or

$$u = \left(\frac{m_v \varepsilon \omega^2}{k} \right) \frac{\left(\frac{\omega_n}{\omega} \right)^2}{\sqrt{\left[1 - \left(\frac{\omega_n}{\omega} \right)^2 \right]^2 + 4\xi^2 \left(\frac{\omega_n}{\omega} \right)^2}} \cos(\omega t - \varphi) \tag{2.243}$$

where in both cases

$$\varphi = \arctan \frac{2\xi \frac{\omega}{\omega_n}}{1 - (\frac{\omega}{\omega_n})^2} \tag{2.244}$$

Equations 2.242 and 2.243 are fully equivalent, but are meant for different purposes. In principle, they differ solely in a convenient simultaneous multiplication and division by ω_n^2. In Eq. 2.242, it is assumed that the natural frequency will be constant and what changes is the operational frequency ω of the eccentric mass. Thus, the second fraction in Eq. 2.242 is the amplification function $A(\omega)$ for that case, and this is illustrated in Figure 2.30. Equation 2.243, on the other hand, is meant for use when the operational frequency ω is a constant, say the speed of a fan or motor (typically 25 Hz, 30 Hz, 50 Hz or 60 Hz), and because that equipment is unbalanced, it elicits vibrations in the supporting structure or floor. In that case, a simple remedy might be to attach some mass to the motor or fan, and thus lower the resonant frequency ω_n of the motor on the supporting system. Observe that in that case, the amplification function is defined in terms of the reciprocal of the tuning ratio ω_n / ω because it is ω_n which is now variable. In addition, the fraction of critical damping $\xi = \frac{1}{2} \frac{c}{\sqrt{km}}$ is variable too because m changes, and that must be taken into account as well. See Example 1 in Section 3.10.3 for further details.

Experimental Observation

When the preceding eccentric mass vibrator is demonstrated in class by means of a simple *crazy ballpoint pen* (a children's toy constructed with a small motor with an eccentric mass at its upper end, which makes writing very difficult or even impossible), it will generally be found that the oscillations are not harmonic, and this is despite the fact that the preceding equations involved no approximations whatsoever (i.e. the oscillations need not be small!). Why is this? Because in the derivation of the equations we assumed that the motor was powerful enough to deliver a torque able to overcome any dynamic feedback from the oscillations, that is, that it can sustain a constant rate of rotation. But the small motor in the crazy pen is weak, and is indeed affected by dynamic feedback, so its rate of rotation is not constant, and therefore, the driving frequency is not constant, which means that the response of the oscillator cannot be harmonic either.

2.6.4 Response to Suddenly Applied Sinusoidal Load

Consider a system subjected to a sinusoidal force $p = p_0 \sin \omega t$ that begins at $t = 0$. The response is then given by the superposition of the particular solution (= the steady-state response) and the homogeneous solution with initial conditions $u_0 - u_{p0}, \dot{u}_0 - \dot{u}_{p0}$, that is,

$$u = u_p + e^{-\xi \omega_n t} \left[\left(u_0 - u_{p0} \right) \cos \omega_d t + \frac{\left(\dot{u}_0 - \dot{u}_{p0} \right) + \xi \omega_n \left(u_0 - u_{p0} \right)}{\omega_d} \sin \omega_d t \right] \qquad (2.245)$$

with particular solution

$$u_p = u_s A \sin \left(\omega t - \phi \right), \qquad \dot{u}_p = \omega u_s A \cos \left(\omega t - \phi \right) \qquad (2.246)$$

where $u_s = p_0 / k$ is the static deflection and the amplification function A and phase angle ϕ are as before. Hence, at $t = 0$

$$u_{p0} = -u_s A \sin \phi, \qquad \dot{u}_{p0} = \omega u_s A \cos \phi \qquad (2.247)$$

so for a system that starts at rest with $u_0 = 0$ and $\dot{u}_0 = 0$

$$u = u_s A \left\{ \sin \left(\omega t - \phi \right) - e^{-\xi \omega_n t} \left[-\sin \phi \cos \omega_d t + \frac{\omega \cos \phi - \xi \omega_n \sin \phi}{\omega_d} \sin \omega_d t \right] \right\} \qquad (2.248)$$

As can be seen, the transient part decays exponentially, and after a short time, only the steady state remains. This is illustrated in Figure 2.31 at the top, specialized for a resonant condition $\omega = \omega_n$.

A special situation occurs when damping is zero, in which case $\xi = 0$, $A = \left(1 - r^2 \right)^{-1}$, $r = \omega / \omega_n$, and with the definition $\tau = \omega_n t$, we obtain

$$u = u_s \frac{\sin r\tau - r \sin \tau}{1 - r^2} \qquad (2.249)$$

which for $r \neq 1$ (nonresonant condition) oscillates and attains some finite, maximum value (middle drawing in Figure 2.31). For the resonant condition, $r \to 1$, we obtain an indeterminate form that can be evaluated with L'Hôspital's rule:

$$\boxed{u = \tfrac{1}{2} u_s \left(\sin \tau - \tau \cos \tau \right)}, \qquad \xi = 0, \qquad \omega = \omega_n \qquad (2.250)$$

The amplitude of this response function grows linearly with time, as shown at the bottom of Figure 2.31.

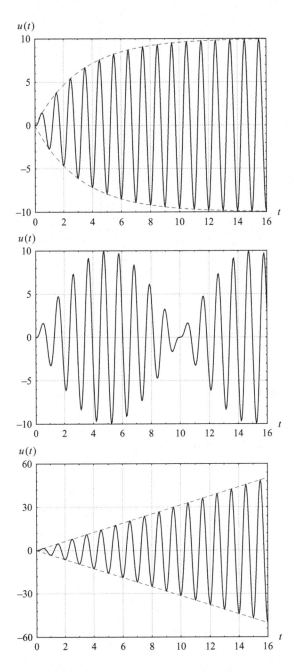

Figure 2.31. Response to transient sine load near or at resonance.

Top : $\xi = 0.05, \omega = \omega_n$

Middle : $\xi = 0, \ \omega = 0.9\,\omega_n$

Bottom : $\xi = 0, \ \omega = \omega_n$

2.6.5 Half-Power Bandwidth Method

The half-power bandwidth method is a procedure for determining experimentally the damping ratio in lightly damped structures. It consists in exciting the dynamic system near resonance and carefully monitoring the amplitude of harmonic response. In particular, we record the peak A_{max} and then measure the frequencies ω_1, ω_2 (or f_1, f_2, in Hz) at which the response is $A_1 = A_2 = A_{max}/\sqrt{2}$ (i.e., 70.7% of the peak) These frequencies are referred to as the half-power points. The fraction of damping is then given by

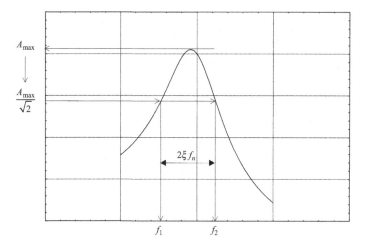

Figure 2.32. Half-power bandwidth method.

$$\xi = \frac{\omega_2 - \omega_1}{\omega_2 + \omega_1} \quad \text{or} \quad \boxed{\xi = \frac{f_2 - f_1}{f_2 + f_1}} \quad \boxed{f_n = \frac{f_1 + f_2}{2}} \tag{2.251}$$

The procedure is as follows:

- Determine experimentally the amplification curve near resonance.
- Draw a horizontal line tangent to the peak.
- Draw a parallel line at a level √2 lower than the peak.
- From the intersection of the previous line with the amplification curve, drop verticals to the frequency axis, and determine the half-power frequencies.
- Apply the damping formula in Eq. 2.251.

Notice that this procedure does not require estimating the resonant frequency itself, which can be known only imprecisely.

The proof is as follows. From the previous section, the peak response is

$$A_{\max} = \frac{1}{2\xi\sqrt{1-\xi^2}} \tag{2.252}$$

so that

$$A_i = \frac{1}{2\xi\sqrt{2}\sqrt{1-\xi^2}} = \frac{1}{\sqrt{\left[1-\left(\frac{\omega_i}{\omega_n}\right)^2\right]^2 + 4\xi^2\left(\frac{\omega_i}{\omega_n}\right)^2}} \qquad i = 1,2 \tag{2.253}$$

It follows that

$$\left[1-\left(\frac{\omega_i}{\omega_n}\right)^2\right]^2 + 4\xi^2\left(\frac{\omega_i}{\omega_n}\right)^2 = 8\xi^2(1-\xi^2) \tag{2.254}$$

which is a biquadratic equation in ω_i. Its solution is

$$\frac{\omega_i}{\omega_n} = \sqrt{1 - 2\xi^2 \mp 2\xi\sqrt{1-\xi^2}} \tag{2.255}$$

Expanding the square root according to the binomial theorem,

$$\frac{\omega_i}{\omega_n} = 1 - \xi\left(\xi \mp \sqrt{1-\xi^2}\right) + \frac{(2\xi)^2}{8}\left(\xi \mp \sqrt{1-\xi^2}\right)^2 + \cdots \tag{2.256}$$

which in turn can be expanded into

$$\frac{\omega_i}{\omega_n} = 1 \pm \xi - \tfrac{1}{2}\xi^2 + \cdots O(\xi^3) \tag{2.257}$$

Hence

$$\omega_2 - \omega_1 = \omega_n(1 + \xi - \tfrac{1}{2}\xi^2) - \omega_n(1 - \xi - \tfrac{1}{2}\xi^2) + \cdots O(\xi^3) \approx 2\xi\omega_n \tag{2.258}$$

Also

$$\omega_2 + \omega_1 = \omega_n(1 + \xi - \tfrac{1}{2}\xi^2) + \omega_n(1 - \xi - \tfrac{1}{2}\xi^2) + \cdots \approx 2\omega_n \tag{2.259}$$

Taking the ratio of these two expressions, we obtain finally

$$\boxed{\xi = \frac{\omega_2 - \omega_1}{\omega_2 + \omega_1} = \frac{f_2 - f_1}{f_2 + f_1}} \tag{2.260}$$

Although in engineering practice it is often assumed that this formula is exact, as we have seen it is only an approximation, even if a good one. Still, it deteriorates as damping becomes heavier. For example, specifying a known, true value of damping and using the formula to obtain an estimated damping, we obtain:

$$\text{For} \quad \xi_{\text{true}} = 0.10, \quad \xi_{\text{est}} = 0.102, \text{which is close, but} \tag{2.261}$$

$$\text{for} \quad \xi_{\text{true}} = 0.30, \quad \xi_{\text{est}} = 0.405, \text{which is not close.} \tag{2.262}$$

Application of Half-Power Bandwidth Method

Let $A_0, A_1, A_2, \ldots$ be the successive amplitudes of damped *free* vibration for *any* damping ξ. Then, from the above we have that

$$\frac{A_0}{A_N} = \frac{A_N}{A_{2N}} = \cdots \frac{A_{mN}}{A_{(m+1)N}} = e^{\xi\omega_n N T_d} \qquad (m, N = \text{integers}) \tag{2.263}$$

that is, the ratios of any two amplitudes of oscillations separated by *any* fixed integer number of cycles N is *constant*, because all factors in the exponent of the term on the right are constant. In particular, it is also constant for $N = N_{50\%}$, so

$$\frac{A_0}{A_{N_{50\%}}} = \frac{A_{N_{50\%}}}{A_{2N_{50\%}}} = \cdots \frac{A_{mN50\%}}{A_{(m+1)N_{50\%}}} = e^{\xi\omega_n N_{50\%} T_d} = 2 \qquad (\text{the 2 is by definition of } N_{50\%})$$

Raising each term to some arbitrary nth power, we obtain

$$\left(\frac{A_0}{A_{N_{50\%}}}\right)^n = \left(\frac{A_{N_{50\%}}}{A_{2N_{50\%}}}\right)^n = \cdots = \left(e^{\xi\omega_n N_{50\%} T_d}\right)^n \equiv e^{\xi\omega_n (nN_{50\%}) T_d} = 2^n \tag{2.264}$$

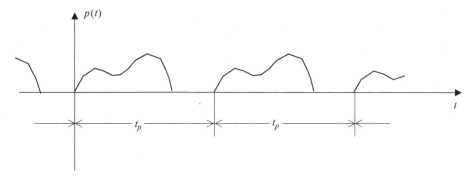

Figure 2.33. Periodic loading via Fourier series.

that is, $n\,N_{50\%}$ is the number of cycles required for the oscillations to decay to a (presumably negligible) fraction $A = A_0\,2^{-n}$. For example, if the number of cycles to 50% amplitude is $N_{50\%} = 3$ (this corresponds to $\xi = 0.037$) and we ask how many cycles are required for the vibration to decrease to a fraction $2^{-7} = 1/128 = 0.0078$ of the initial amplitude, this number is $7 \times 3 = 21$.

2.7 Response to Periodic Loading

2.7.1 Periodic Load Cast in Terms of Fourier Series

We have just learned how to obtain the steady-state response to a harmonic load with amplitude $\tilde{p}$ and frequency ω:

$$m\ddot{u} + c\dot{u} + ku = \tilde{p}\,e^{i\omega t} \tag{2.265}$$

$$u(t) = \tilde{p}\,H(\omega)\,e^{i\omega t} \tag{2.266}$$

with H being the complex-valued transfer function.

Consider now a periodic loading with arbitrary variation $p(t)$ and temporal period t_p.

From calculus, we know that any periodic function, and $p(t)$ in particular, can be expressed in terms of a (complex) Fourier series of the form (for further details, see Chapter 6, Section 6.6):

$$\boxed{p(t) = \frac{\Delta\omega}{2\pi} \sum_{j=-\infty}^{\infty} \tilde{p}_j\,e^{i\omega_j t}} \qquad \text{The Fourier series} \tag{2.267}$$

with

$$\boxed{\tilde{p}_j = \tilde{p}(\omega_j) = \int_0^{t_p} p(t)e^{-i\omega_j t}\,dt} \qquad \text{The complex Fourier load coefficients} \tag{2.268}$$

$$\omega_j = j\frac{2\pi}{t_p} = j\,\Delta\omega \qquad\qquad \text{The discrete frequencies (where } j \text{ is an integer)} \tag{2.269}$$

$$\Delta\omega = \frac{2\pi}{t_p} \qquad\qquad\qquad \text{The frequency step} \tag{2.270}$$

Hence, the differential equation for the periodic loading $p(t)$ can be expressed as

$$m\ddot{u} + c\dot{u} + ku = \frac{\Delta\omega}{2\pi} \sum_{j=-\infty}^{\infty} \tilde{p}_j \, e^{i\omega_j t} \tag{2.271}$$

Clearly, since each component of the load in the Fourier series is harmonic, the response to each of those components is also harmonic, and proportional to both the load component and the transfer function with that frequency:

$$H_j \equiv H(\omega_j) = \frac{1/k}{1 - \left(\dfrac{\omega_j}{\omega_n}\right)^2 + 2\,i\,\xi\dfrac{\omega_j}{\omega_n}} \tag{2.272}$$

On the other hand, since the system is linear, superposition applies, so the complete *steady-state* response must be the summation of all contributions (i.e., the Fourier series)

$$\boxed{u(t) = \frac{\Delta\omega}{2\pi} \sum_{j=-\infty}^{\infty} H_j \, \tilde{p}_j \, e^{i\omega_j t}} \tag{2.273}$$

It should be noted that this summation extends over both positive and negative values of the integer index j, that is, over both positive and "negative" frequencies ω_j. Observe also that all three factors in the summation exhibit complex-conjugate symmetry with respect to the frequency, that is,

$$H_{-j} = \text{conj}\left(H_j\right) = H_j^* \qquad \tilde{p}_{-j} = \text{conj}\left(\tilde{p}_j\right) = \tilde{p}_j^* \qquad e^{-i\omega_j t} = \text{conj}\,(e^{i\omega_j t}) \tag{2.274}$$

Also, the product of three conjugates is the conjugate of the product. Hence, when the terms for a given negative frequency and its matching positive frequency are added, their imaginary parts cancel out in pairs, and the result is real. Since this is true for every component, we arrive at the important result that the Fourier series representation of the steady-state response is *real*. Also, since all Fourier series are periodic, we conclude that *the steady-state response is periodic.*

2.7.2 Nonperiodic Load as Limit of Load with Infinite Period

Consider next the special case of a load that is periodic, but within each period has finite duration $t_d < t_p$. If the period is doubled without changing the duration of the load, we observe from the integral defining the Fourier series coefficients $\tilde{p}_j$ that these are not affected by the change in period. The only difference is that now the frequency step is half as large as it was before, so that there are more of these coefficients – indeed, twice as many. Thus, the *Fourier spectrum*, that is, the collection of all $\tilde{p}_j$ values expressed in terms of the frequencies, is denser, but its shape does not change. Imagine now that this process of doubling the period is repeated over and over again, until the load finally ceases to be periodic, or more precisely, until its period is infinitely large. If so, we can imagine a load occurring at minus infinity, another one at $0 < t < t_d$, and a third one at plus infinity. Since an infinite time elapses between the load at minus infinity and the load starting at zero, any vibration that was induced by the former will already have died out by the time the latter begins. Hence, the initial conditions at $t = 0$ must be zero displacement and velocity.

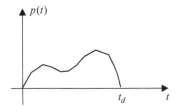

Figure 2.34. Nonperiodic loading via Fourier series.

Also, we can see that as the period grows larger, the frequency spacing becomes infinitesimally small, the discrete frequencies become continuous frequencies, and the summation converges to an integral, that is

$$t_p \to \infty, \qquad \Delta\omega \to d\omega, \qquad \omega_j \to \omega, \qquad \sum_{j=-\infty}^{\infty} f(\omega_j)\Delta\omega \to \int_{-\infty}^{\infty} f(\omega)\, d\omega \qquad (2.275)$$

Hence, we conclude that in the limit of a nonperiodic load, the response is nonperiodic and converges to

$$u(t) = \frac{1}{2\pi} \int_{-\infty}^{\infty} H(\omega)\, \tilde{p}(\omega)\, \mathrm{e}^{\mathrm{i}\omega t}\, d\omega \qquad (2.276)$$

with

$$\tilde{p}(\omega) = \int_{0}^{t_d} p(t)\, \mathrm{e}^{-\mathrm{i}\omega t}\, dt \qquad (2.277)$$

or since the duration is arbitrary, we can replace this equation with the more general formula

$$\tilde{p}(\omega) = \int_{-\infty}^{\infty} p(t)\, e^{-\mathrm{i}\omega t}\, dt \qquad \text{Fourier transform of load} \qquad (2.278)$$

which has the inverse

$$p(t) = \frac{1}{2\pi} \int_{-\infty}^{\infty} \tilde{p}(\omega)\, \mathrm{e}^{\mathrm{i}\omega t}\, d\omega \qquad \text{Inverse Fourier transform} \qquad (2.279)$$

$p(t)$ and $\tilde{p}(\omega)$ are said to be *Fourier transform pairs*.

Clearly, the response predicted by Eqs. 2.278 and 2.279 must be an alternative representation of the convolution integral, since it provides the nonperiodic response to a nonperiodic load, and it has zero initial conditions. Thus, we may say that the Fourier integral represents the solution in the *frequency-domain*, whereas the convolution provides the solution in the *time domain*. Hence,

$$p*h = \int_{-\infty}^{\infty} p(\tau)h(t-\tau)\, d\tau = \frac{1}{2\pi}\int_{-\infty}^{\infty} \tilde{p}(\omega)\, H(\omega)\, \mathrm{e}^{\mathrm{i}\omega t}\, d\omega \qquad (2.280)$$

Example: Sequence of Rectangular Pulses

Consider a periodic load with period t_p consisting of rectangular pulses of amplitude p_0 and duration t_d.

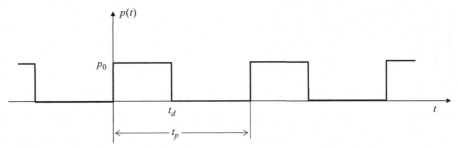

Figure 2.35. Rectangular wave.

The Fourier transform of *one* cycle of the load is then

$$\tilde{p}(\omega) = p_0 \int_0^{t_d} e^{-i\omega t} dt = p_0 \frac{1 - e^{-i\omega t_d}}{i\omega} = p_0 \frac{e^{i\omega t_d/2} - e^{-i\omega t_d/2}}{i\omega} e^{-i\omega t_d/2} \qquad (2.281)$$

With the definition,

$$\theta = \frac{\omega t_d}{2} \qquad (2.282)$$

the previous expression can be written as

$$\tilde{p}(\omega) = p_0 t_d \frac{\sin\theta}{\theta} e^{-i\theta} \qquad (2.283)$$

which has absolute value

$$|\tilde{p}(\omega)| = p_0 t_d \left| \frac{\sin\theta}{\theta} \right| \qquad (2.284)$$

and phase angle (with k being any integer)

$$\varphi(\omega) = \begin{cases} 2k\pi - \theta & \text{if } \sin\theta > 0 \\ (2k+1)\pi - \theta & \text{if } \sin\theta < 0 \end{cases} \qquad (2.285)$$

The bars shown in the drawing correspond to the values of $\tilde{p}_j = \tilde{p}(\omega_j)$ when $t_p = 10 t_d$, that is, when

$$\Delta\theta = \Delta\omega t_d / 2 = (2\pi / t_p) t_d / 2 = \pi / 5 \qquad (2.286)$$

Doubling this period would double the number of bars shown.

2.7.3 System Subjected to Periodic Loading: Solution in the Time Domain

While Fourier series are normally used to compute the response of general dynamic systems to periodic loads, in the case of SDOF systems it is also possible to obtain such solutions directly in the time domain. This goal is achieved by taking into account the periodicity of the response.

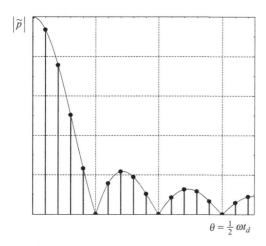

$$\theta = \tfrac{1}{2}\omega t_d$$

Figure 2.36. Fourier spectrum for rectangular wave.

Consider an SDOF system subjected to a periodic load of period t_p, and focus attention to the response during an individual time window, say $0 \leq t \leq t_p$. Within that window, the response can be written as the sum of a *free vibration* term with as yet unknown initial conditions u_0, $\dot{u}_0$, and a *convolution integral* with the current cycle of the load, which has zero initial conditions. In other words,

$$u(t) = \mathrm{e}^{-\xi\omega_n t}\left\{ u_0 \cos\omega_d t + \frac{\dot{u}_0 + \xi\omega_n u_0}{\omega_d}\sin\omega_d t \right\} + p * h \tag{2.287}$$

$$\dot{u}(t) = \mathrm{e}^{-\xi\omega_n t}\left\{ \dot{u}_0 \cos\omega_d t - \frac{\omega_n}{\omega_d}(\xi\dot{u}_0 + \omega_n u_0)\sin\omega_d t \right\} + p * \dot{h} \tag{2.288}$$

Since the response to the periodic loading must be periodic, this means that the final conditions at the end of the current time window must be the same as the initial conditions, that is,

$$u(t_p) = u(0) \equiv u_0 \tag{2.289}$$

$$\dot{u}(t_p) = \dot{u}(0) \equiv \dot{u}_0 \tag{2.290}$$

Hence

$$u_0 = \mathrm{e}^{-\xi\omega_n t_p}\left\{ u_0 \cos\omega_d t_p + \frac{\dot{u}_0 + \xi\omega_n u_0}{\omega_d}\sin\omega_d t_p \right\} + p * h\big|_{t_p} \tag{2.291}$$

$$\dot{u}_0 = \mathrm{e}^{-\xi\omega_n t_p}\left\{ \dot{u}_0 \cos\omega_d t_p - \frac{\omega_n}{\omega_d}(\xi\dot{u}_0 + \omega_n u_0)\sin\omega_d t_p \right\} + p * \dot{h}\big|_{t_p} \tag{2.292}$$

in which the last terms are the convolutions evaluated at the end of the period. With the abbreviations

$$C = \mathrm{e}^{-\xi\omega_n t_p}\cos\omega_d t_p \tag{2.293}$$

$$S = \frac{\omega_n}{\omega_d}\mathrm{e}^{-\xi\omega_n t_p}\sin\omega_d t_p \tag{2.294}$$

$$a = p * h\big|_{t_p} = \int_0^{t_p} p(\tau) h(t-\tau) d\tau \qquad (2.295)$$

$$b = p * \dot{h}\big|_{t_p} = \int_0^{t_p} p(\tau) \dot{h}(t-\tau) d\tau \qquad (2.296)$$

the two equations for u_0, $\dot{u}_0$ can be expressed as

$$(C + \xi S) u_0 + \frac{1}{\omega_n} S \dot{u}_0 + a = u_0 \qquad (2.297)$$

$$-\omega_n S u_0 + (C - \xi S) \dot{u}_0 + b = \dot{u}_0 \qquad (2.298)$$

This is a system of two equations in two unknowns. Its solution is

$$\begin{Bmatrix} u_0 \\ \dot{u}_0 \end{Bmatrix} = \frac{1}{(1-C)^2 + S^2(1-\xi^2)} \begin{Bmatrix} 1 - C + \xi S & \frac{1}{\omega_n} S \\ -\omega_n S & 1 - C - \xi S \end{Bmatrix} \begin{Bmatrix} a \\ b \end{Bmatrix} \qquad (2.299)$$

which can readily be evaluated. Once we have the requisite initial conditions, we can proceed to evaluate the sum of the free vibration and convolution at any arbitrary instant during the first time window.

2.7.4 Transfer Function versus Impulse Response Function

Consider next the special case of a unit, impulsive load, that is, $p(t) = \delta(t)$ (the Dirac-delta function). As we know, the response elicited by this load is the impulse response function $h(t)$. On the other hand, from the previous section, the Fourier transform of this load is

$$\tilde{p}(\omega) = \int_{-\infty}^{\infty} \delta(t) e^{-i\omega t} dt = e^0 = 1 \qquad (2.300)$$

which states that the unit impulse (or *shot noise*) has a constant Fourier spectrum, that is, it has energy at all frequencies. Hence, from the frequency-domain representation of the convolution, we arrive at the important conclusion that

$$\boxed{h(t) = \frac{1}{2\pi} \int_{-\infty}^{\infty} H(\omega) e^{i\omega t} d\omega} \qquad (2.301)$$

$$\boxed{H(\omega) = \int_{-\infty}^{\infty} h(t) e^{-i\omega t} dt} \qquad (2.302)$$

that is, *the transfer function and the impulse response functions are Fourier transform pairs*.

2.7.5 Fourier Inversion of Transfer Function by Contour Integration

Having demonstrated that the impulse response function and the transfer function are Fourier transform pairs, it remains for us to show that a Fourier inversion of the transfer

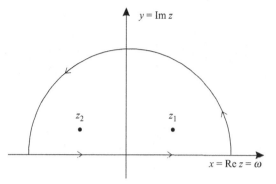

Figure 2.37. Contour integration for an SDOF system.

function does indeed recover the impulse response function. For this purpose, consider once more an SDOF system subjected to a unit impulse, that is,

$$m\ddot{h} + c\dot{h} + kh = \delta(t) \tag{2.303}$$

in which $h(t)$ is the impulse response function. A Fourier transform into the frequency domain yields

$$(-\omega^2 m + \mathrm{i}\,\omega c + k)\,H(\omega) = 1 \tag{2.304}$$

from which we can solve for the transfer function as

$$H(\omega) = \frac{1}{-\omega^2 m + \mathrm{i}\,\omega c + k} = \frac{-1}{m(\omega - z_1)(\omega - z_2)} \tag{2.305}$$

in which z_1, z_2 are the *poles* of the system, that is, the zeros of the transfer function's denominator:

$$-mz^2 + \mathrm{i}\,cz + k = 0 \quad \Rightarrow \quad z_{1,2} = \pm\omega_d + \mathrm{i}\,\xi\omega_n \tag{2.306}$$

with ω_n, ω_d being the natural undamped and damped frequencies, respectively. These two poles are shown as dots in Figure 2.37. Observe that the imaginary parts of these poles are both nonnegative. From the previous section, the impulse response function is then obtained from the inverse Fourier transformation

$$h(t) = \frac{1}{2\pi}\int_{-\infty}^{+\infty} H(\omega)\mathrm{e}^{\mathrm{i}\omega t}\,d\omega = -\frac{1}{2\pi}\int_{-\infty}^{+\infty} \frac{\mathrm{e}^{\mathrm{i}\omega t}}{m(\omega - z_1)(\omega - z_2)}\,d\omega \tag{2.307}$$

A direct evaluation of this integral is rather cumbersome. However, using the methods of contour integration for functions of complex variables, we can carry out the inversion with little effort. To this effect, we change the frequency ω into a complex variable z, and replace the improper Fourier integral by the equivalent contour integral

$$h(t) = \frac{-1}{2\pi m}\oint \frac{\mathrm{e}^{\mathrm{i}zt}}{(z - z_1)(z - z_2)}\,dz \tag{2.308}$$

which involves the complex frequency $z = x + \mathrm{i}\,y$ ($x \equiv \omega$). We must now distinguish three cases, depending on whether $t > 0, t < 0,$ or $t = 0$.

When $t > 0$, the exponential term e^{-yt} is bounded in the upper half-plane and unbounded in the lower one, so the integration contour *must* be closed in the *upper* half-plane. The residues of the integrand in this case are

$$R_1 = \lim_{z \to z_1} (z - z_1) \frac{e^{izt}}{(z - z_1)(z - z_2)} = \frac{e^{iz_1t}}{z_1 - z_2} = +\frac{e^{-\xi\omega_n t} e^{i\omega_d t}}{2\omega_d} \qquad (2.309)$$

$$R_1 = \lim_{z \to z_1} (z - z_2) \frac{e^{izt}}{(z - z_1)(z - z_2)} = \frac{e^{iz_1t}}{z_2 - z_1} = -\frac{e^{-\xi\omega_n t} e^{-i\omega_d t}}{2\omega_d} \qquad (2.310)$$

The integral is then

$$h(t) = \frac{-1}{2\pi m} 2\pi i (R_1 + R_2) = \frac{e^{-\xi\omega_n t}(e^{i\omega_d t} - e^{-i\omega_d t})}{2 i m \omega_d} = \frac{1}{m\omega_d} e^{-\xi\omega_n t} \sin \omega_d t \qquad (2.311)$$

which agrees with the known expression for the impulse response function.

When $t < 0$, the exponential term e^{-yt} is bounded in the lower half-plane and unbounded in the upper one, so the integration contour *must* be closed in the *lower* half-plane. Since the lower half-plane contains no poles, there are no residues and the contour integral is zero. Hence, we conclude that for negative times, *the system is at rest*, so the system is causal.

When $t = 0$, the exponential term is bounded *both* in the upper *and* lower half-planes, which means that the contour could be closed in either complex half-plane. The physical interpretation is that there is a discontinuity at this instant in time. At $t = 0^-$, that is, immediately before applying the impulse $\delta(t)$, the system is still at rest, while at $t = 0^+$, that is, immediately after application of the impulse, the system has acquired a finite momentum.

Location of Poles, Fourier Transforms, and Causality

A time function $f(t)$ is said to be causal if it is zero when its argument is negative, that is, $f(t < 0) = 0$. A dynamic system is defined as being causal if its response is causal when subjected to a causal dynamic excitation. More generally, the *causality principle* is a fundamental postulate in Newtonian physics that states that no *effect* can ever be observed in a system ahead of the agent or *cause* giving rise to that phenomenon. For example, a mass at rest with respect to an inertial system cannot begin moving before some external force actually acts on that mass.

In the case of dynamic systems, the validity of the causality principle is closely related to the location of the *poles* of that system. Indeed, the poles of any real, stable, *causal* mechanical system containing only energy sinks and no sources, must necessarily lie in the upper half-plane; in other words, the lower half-plane cannot contain any poles. Thus, the poles have a nonnegative imaginary part. This is true not only for the SDOF considered here, but also for multiple degree of freedom systems and for continuous systems.

Important note: The fact that our causal systems have all poles located in the upper complex half-plane is intimately related to our choice of a complex exponential term $e^{+i\omega t}$ in the derivation of the harmonic response functions. This choice relates in turn to, and is consistent with, the negative and positive signs we have used for the exponential terms in the forward and inverse Fourier transform, respectively. You should take note that some authors – notably in works on wave propagation – use instead a negative exponential term $e^{-i\omega t}$, which moves the poles to the lower half-plane. This implies a reversal not only in the signs of the forward and inverse Fourier transforms, but also in the representation of damping through complex moduli (these will be seen in Sections 2.8, 2.9.4, 3.8.5, and especially 8.1.6 and 8.1.7). Thus, the usual complex modulus of the form $G[1+2i\xi \mathrm{sgn}(\omega)]$ would have to be changed into $G[1-2i\xi \mathrm{sgn}(\omega)]$. Needless to add, these variations in standard notation are the source of much frustration and/or confusion, not to mention erroneous results.

2.7.6 Response Computation in the Frequency Domain

As we have seen, the response of an SDOF system, starting from rest, to a causal excitation $p(t)$ can be obtained from the Fourier integral

$$u(t) = \frac{1}{2\pi} \int_{-\infty}^{\infty} H(\omega)\, \tilde{p}(\omega)\, e^{i\omega t}\, d\omega \tag{2.312}$$

In most cases, this integral cannot be evaluated by analytical means, but must instead be evaluated numerically in a digital computer. Such a numerical evaluation, however, has a number of consequences:

(a) The improper limits must be changed into finite limits by an appropriate choice of a cutoff frequency v, that is,

$$u(t) = \frac{1}{2\pi} \int_{-v}^{+v} H(\omega)\, \tilde{p}(\omega)\, e^{i\omega t}\, d\omega \tag{2.313}$$

because otherwise the computation would never stop. To avoid truncation errors, the excitation $\tilde{p}(\omega)$ must then contain little or no energy above the chosen limit. This means in turn that the forcing function $p(t)$ must be sampled at a sufficiently small time interval Δt such that no significant energy exists beyond the cutoff or *Nyquist* frequency $v = \pi / \Delta t$ rad/s (see also Chapter 6, Section 6.6 on Fourier Methods).

(b) More importantly, the integral cannot be carried out in terms of a continuous variable ω. Instead, the integrand (or kernel) must be sampled at finite intervals ω_j involving a certain sampling rate in the frequency domain $\Delta \omega$.

$$u(t) = \frac{\Delta \omega}{2\pi} \overline{\sum_{j=-N}^{N}} H_j\, \tilde{p}_j\, e^{i\omega_j t} \tag{2.314}$$

with the bar through the summation signaling the so-called *split summation*. This means that the first and last elements (i.e., for $j = -N, j = N$) must carry a weight of 1/2 so as to smooth out the discontinuity taking place at the cutoff frequency, see section on Fourier Methods. The above summation is usually accomplished by means of the superbly efficient Fast Fourier Transform (FFT) algorithm.

However, the very moment that we discretize the integral as in Eq. 2.314 we are automatically making the response periodic with period $t_p = 2\pi / \Delta \omega$ (compare with the expression for the response under periodic loads). Thus, the response computed by numerical means is inevitably periodic, since the continuous Fourier integral is replaced by a discrete Fourier series. Unless the fictitious period t_p is taken sufficiently large, the free vibration response after termination of the excitation, that is, the tail, will not have decayed sufficiently before the start of the next period, and a spillover or *wraparound* of the response will take place. In other words, the periodic response and the sought after transient response will differ significantly. This is particularly problematic when the system has light damping and/or the tail is short, because the tail then decays only slowly, and wraparound can hardly be avoided. There are at least two alternatives to deal with this problem:

(1) Trailing Zeros

Append a fictitious tail or *quiet zone* to the excitation, often referred to as the *trailing zeros*. This tail must be taken long enough to ensure an adequate decay of the free vibrations before the start of the next period. To determine the length of this tail, we can proceed as follows.

Let t_d be the duration of the load, and t_p the period of the excitation to be determined. Thus, the oscillator will experience a free vibration in the interval $t_d < t < t_p$, during which the amplitudes will have decayed by $\exp\left(-\xi\omega_n(t_p - t_d)\right)$. Equating this decay to some arbitrary tolerance, say 10^{-m} (typically m is something like 2 or 3), we obtain

$$\xi\omega_n(t_p - t_d) = 2\pi\xi(t_p - t_d)/T_n = m\log 10 \qquad (2.315)$$

that is,

$$\frac{t_p}{T_n} = \frac{t_d}{T_n} + \frac{m\log 10}{2\pi\xi} \qquad (2.316)$$

As can be seen, the length of the tail $t_p - t_d$ depends both on the period of the system being solved and the level of damping. The smaller the damping, the longer the quiet zone. For very light damping, this length can be intolerably long. In the case of multiple degree of freedom systems, the above estimation of the length of the tail is made with the fundamental (i.e., lowest) frequency of the system, that is, with the longest period, which must either be computed or estimated.

(2) Exponential Window Method: The Preferred Strategy

Use the *Exponential Window Method* (EWM) described in Chapter 6, Section 6.6.14, which in essence is a numerical implementation of the Laplace transform (Bromwich integral). When problems are formulated in the frequency domain and then inverted back into the time domain, the EWM not only provides very accurate response computations for discrete and continuous systems alike, but it also dispenses altogether with the inconveniences posed by trailing zeros while completely avoiding wraparound. And just as nicely, it even works with fully undamped systems. Thus, this is the method of choice!

2.8 Dynamic Stiffness or Impedance

An alternative interpretation of the concepts of viscous and hysteretic damping, and of damping in general, can be made with the notion of complex *dynamic stiffness*, which is

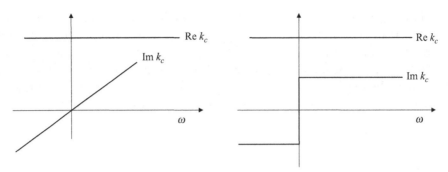

Figure 2.38. Complex stiffness or impedance. (*Left*) Spring + viscous damper. (*Right*) Hysteretic spring system.

also referred to as *impedance*. For this purpose, the equation of motion is written in the same way as for an undamped oscillator, but the stiffness is thought to be a complex function of frequency k_c:

$$m\ddot{u} + k_c(\omega)u = p_0 e^{i\omega t} \tag{2.317}$$

For this equation to represent a physically meaningful system, the dynamic stiffness k_c must exhibit complex-conjugate properties with respect to frequency, that is,

$$\mathrm{Re}\big[k_c(-\omega)\big] = \mathrm{Re}\big[k_c(\omega)\big] \tag{2.318}$$

$$\mathrm{Im}\big[k_c(-\omega)\big] = -\mathrm{Im}\big[k_c(\omega)\big] \tag{2.319}$$

The dynamic stiffness of the viscous and hysteretic oscillators are then

$$k_c = k + i\omega c \qquad k_c = k(1 + 2 i \xi_h \operatorname{sgn}\omega) \tag{2.320}$$

Choosing a hysteretic oscillator with stiffness $k_h = k(1 - 2\xi^2)$ and fraction of hysteretic damping $\xi_h = \xi\sqrt{1 - \xi^2}/(1 - 2\xi^2)$, this oscillator will have a dynamic stiffness

$$k_c = k\Big[1 - 2\xi^2 + 2 i \xi\sqrt{1 - \xi^2} \operatorname{sgn}(\omega)\Big] \tag{2.321}$$

which has the property that the complex number in square brackets has unit norm. Hence, this dynamic stiffness can be written in the compact exponential form

$$k_c = k e^{i\delta} \tag{2.322}$$

in which δ is called the *loss angle*. This angle gives the phase lag between the applied load and the response. The relationship between the fraction of damping and the loss angle is

$$\tan\delta = \frac{2\xi\sqrt{1 - \xi^2}}{1 - 2\xi^2} \tag{2.323}$$

so that for small damping $\delta \approx 2\xi$.

In the ensuing section we shall denote impedances by the generic symbol $Z \equiv k_c$. Their inverses Z^{-1} are the dynamic flexibilities, which are also sometimes referred to as

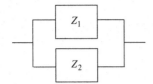

Figure 2.39. Impedances in parallel and in series.

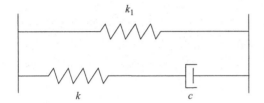

Figure 2.40. Standard linear solid, or Zener solid.

receptances; in the case of the reciprocal of velocity impedances (i.e., relating force to velocity), they may be called *mobilities*.

2.8.1 Connection of Impedances in Series and/or Parallel

More generally, the impedances of sets of spring and dashpots in any combination, but which do *not* contain any masses (inertial elements), can be obtained by the standard rules of connections in series and parallel. Let Z_1 and Z_2 be the impedances of two spring–dashpots systems. Their combined impedance Z is then

(a) Connection in parallel $Z = Z_1 + Z_2$ (2.324)

(b) Connection in series $\dfrac{1}{Z} = \dfrac{1}{Z_1} + \dfrac{1}{Z_2}$ (2.325)

That is, $Z = \dfrac{Z_1 Z_2}{Z_1 + Z_2}$ (2.326)

These expressions can be demonstrated by simple equilibrium considerations. Applying repeatedly these rules, we can obtain the impedances for complicated spring–dashpot arrangements. For example, the impedance for the *standard linear solid* (or *Zener solid*) shown in Figure 2.40 is

$$k_c = k_1 + \frac{i\omega ck}{k + i\omega c} = k_1 + \frac{\omega^2 c^2 k}{k^2 + \omega^2 c^2} + i\left(\frac{\omega ck^2}{k^2 + \omega^2 c^2}\right) \qquad (2.327)$$

In particular, if $k_1 = 0$, we obtain the so-called *Maxwell solid* (a single spring–single dashpot system in series), while if $k = \infty$, we obtain the *Voigt solid* (a spring–dashpot system in parallel).

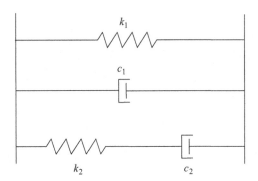

Figure 2.41. Combination of parallel and serial systems.

Standard Solid

We consider also a standard solid with an added dashpot in parallel, as shown in Figure 2.41. Also, we assume that both springs have hysteretic damping, that is, they have complex impedances of the form

$$z_j = k_j \left(1 + 2i\xi_j\right) \qquad j = 1, 2 \tag{2.328}$$

Hence, the impedance of the system is the sum of the systems in parallel and series

$$Z = k_1\left(1+2i\xi_1\right) + i\omega c_1 + \frac{i\omega c_2 k_2\left(1+2i\xi_2\right)}{k_2\left(1+2i\xi_2\right)+i\omega c_2} \tag{2.329}$$

Defining

$$\kappa = \frac{k_2}{k_1} \qquad \text{and} \qquad \frac{\omega c_j}{k_j} = 2\beta_j\frac{\omega}{\omega_0} = 2\beta_j\Omega \qquad \Omega = \frac{\omega}{\omega_0} \tag{2.330}$$

with ω_0 being an arbitrary reference frequency, then the system impedance can be written in dimensionless form involving the five independent parameters $\kappa,\beta_1,\beta_2,\xi_1,\xi_2$ as

$$\frac{Z}{k_1} = 1 + 2i\left(\xi_1 + \beta_1\Omega\right) + i\Omega\beta_2\kappa\frac{1+2i\xi_2}{1+2i\left(\xi_2+\Omega\beta_2\right)} \tag{2.331}$$

or in terms of the real and imaginary parts

$$\frac{Z}{k_1} = 1 + \frac{4\kappa\beta_2^2\,\Omega^2}{1+4\left(\xi_2+\Omega\beta_2\right)^2} + 2i\left\{\xi_1 + \Omega\left[\beta_1 + \beta_2\kappa\frac{1+4\xi_2\left(\xi_2+\Omega\beta_2\right)}{1+4\left(\xi_2+\Omega\beta_2\right)^2}\right]\right\} \tag{2.332}$$

2.9 Energy Dissipation through Damping

Damping forces in engineering materials and in soils result from a combination of several energy-dissipating mechanisms. It is generally accepted that most of this energy dissipation takes place through internal friction, nonlinear inelastic behavior, or some other irreversible material processes. In the case of multiphase materials, such as cohesive soils and saturated sands, energy dissipation may also occur as a result of sliding of the mineral particles together with viscous fluid flow through the pores in the skeleton. While an accurate description of these kinds of damping would require involved constitutive models, useful approximations can often be made in engineering applications, provided the inelastic processes are not unduly severe. The most common of these approximate models is that of *linear hysteretic damping*, in which the dissipation of energy in each cycle of deformation is assumed to be independent of the speed at which the material is being strained, that is, independent of the frequency of deformation. In contrast, viscous damping implies an energy dissipation that increases with the rate or frequency of deformation, as will be seen.

2.9.1 Viscous Damping

Instantaneous Power and Power Dissipation

Consider an SDOF system with mass, viscous dashpot, and stiffness constants m, c, and k that is subjected to a harmonic excitation with frequency ω. As we have already seen, the equation of motion for such a system is

$$m\ddot{u} + c\dot{u} + ku = F\,\mathrm{e}^{\mathrm{i}\omega t} \tag{2.333}$$

whose solution is of the form

$$u(t) = u_s A\,\mathrm{e}^{\mathrm{i}(\omega t - \varphi)} \qquad \text{and} \qquad \dot{u}(t) = \mathrm{i}\,\omega u_s A\,\mathrm{e}^{\mathrm{i}(\omega t - \varphi)} \tag{2.334}$$

where $u_s = F/k$ is the static deflection, A is the amplitude of motion and ϕ is the phase angle. The instantaneous power supplied by the external excitation is the product of the force times the velocity, that is,

$$
\begin{aligned}
\Pi(t) &= \mathrm{Re}[F\,\mathrm{e}^{\mathrm{i}\omega t}]\,\mathrm{Re}[\mathrm{i}\,\omega u_s A\,\mathrm{e}^{\mathrm{i}(\omega t - \varphi)}] \\
&= u_s FA\cos\omega t\left[-\omega\sin(\omega t - \varphi)\right] \\
&= \omega u_s FA\left[\cos^2\omega t\,\sin\varphi - \tfrac{1}{2}\sin 2\omega t\cos\varphi\right]
\end{aligned}
\tag{2.335}
$$

The average net power supplied during one cycle of motion is then

$$\left\langle\Pi_{supp}\right\rangle = \frac{1}{T}\int_0^T \Pi(t)\,dt = \tfrac{1}{2}\,\omega u_s FA\sin\varphi \tag{2.336}$$

On the other hand, the instantaneous power being dissipated by the dashpot is

$$
\begin{aligned}
\Pi_{diss}(t) &= \mathrm{Re}[c\dot{u}]\,\mathrm{Re}[\dot{u}] \\
&= c[-\omega u_s A\,\sin(\omega t - \varphi)]^2 \\
&= c\,u_s^2\,\omega^2 A^2\sin^2(\omega t - \varphi)
\end{aligned}
\tag{2.337}
$$

The average power dissipated in the dashpot is then

$$
\begin{aligned}
\left\langle\Pi_{diss}\right\rangle &= \frac{1}{T}\int_0^T \Pi_{diss}(t)\,dt \\
&= \tfrac{1}{2}c\,u_s^2\,\omega^2 A^2
\end{aligned}
\tag{2.338}
$$

Clearly, the average power supplied by the external force must equal the average power dissipated by the dashpot, so that

$$\tfrac{1}{2}\omega u_s A F \sin\varphi = \tfrac{1}{2} c u_s^2 \omega^2 A^2 \tag{2.339}$$

Hence

$$c\,\omega u_s A = F \sin\varphi \tag{2.340}$$

Therefore, the average power dissipated over one cycle of motion can be written as

$$\langle \Pi_{diss} \rangle = \frac{1}{2c} F^2 \sin^2\varphi = \frac{1}{2c} F^2 \frac{2\xi r}{\sqrt{\left(1 - r^2\right)^2 + 4\xi^2 r^2}} \tag{2.341}$$

which in addition to the physical parameters depends only on the phase angle of the response. This quantity attains its maximum value at resonance ($\omega = \omega_n$), when $\varphi = \tfrac{1}{2}\pi$ and $\sin\varphi = 1$, in which case

$$\boxed{\langle \Pi_{diss} \rangle = \frac{F^2}{2c}} \tag{2.342}$$

This quantity may be used to estimate the energy being converted into heat at resonance. If excessive heat were dissipated, it could literally lead to a meltdown of the damper.

Human Power

During physical exercise, the average human adult is able to deliver a relatively low level of sustained power, typically on the order of some 75 to 100 watts. However, an Olympic class athlete can deliver, even if only for very brief moments, well in excess of 1 HP, or 0.75 kW.

Average Power Dissipated in Harmonic Support Motion

We consider here the average power dissipated by an SDOF system when subjected to a harmonic support motion with amplitude u_{g0} and frequency ω. Clearly, all of the energy is being dissipated by the dashpot whose deformation is the relative velocity with respect to the support. Hence, we must consider the frequency response for relative displacement, which can be expressed as

$$v = |H_v| e^{i(\omega t - \phi)} u_{g0} \tag{2.343}$$

where $|H_v|$ and ϕ are, respectively, the absolute value and phase angle of the transfer function for relative motion

$$|H_v| = \frac{r^2}{\sqrt{\left(1 - r^2\right)^2 + 4\xi^2 r^2}}, \qquad \tan\phi = \frac{2\xi r}{1 - r^2} \tag{2.344}$$

and $r = \omega / \omega_n$ is the *tuning ratio*. This transfer function approaches unity as the frequency grows without bound, which means that the oscillator remains stationary while the support moves. From the preceding, we infer that the relative velocity is

$$\dot{v} = i\omega |H_v| e^{i(\omega t - \phi)} u_{g0} \tag{2.345}$$

whose real part is

$$\mathrm{Re}(\dot{v}) = -\omega |H_v| \sin(\omega t - \phi) u_{g0} \tag{2.346}$$

Now, the instantaneous power being dissipated by the dashpot is the product of the force in the dashpot and the velocity of deformation, so the average power dissipated in one cycle of motion is given by the integral

$$\begin{aligned}
\langle \Pi \rangle_{diss} &= \tfrac{1}{T} \int_0^T \mathrm{Re}(c\dot{v})\,\mathrm{Re}(\dot{v})\,dt \\
&= c\omega^2 |H_v|^2 u_{g0}^2 \tfrac{1}{T} \int_0^T \sin^2(\omega t - \phi)\,dt \\
&= \tfrac{1}{2} c\omega^2 |H_v|^2 u_{g0}^2 = \tfrac{1}{2} c |H_v|^2 \dot{u}_{g0}^2 \\
&= \tfrac{1}{2} c \dot{u}_{g0}^2 \frac{r^4}{(1-r^2)^2 + 4\xi^2 r^2}
\end{aligned} \tag{2.347}$$

in which $\dot{u}_{g0}$ is the peak ground velocity. Observe that $c\dot{u}_{g0}^2 \equiv 2\xi\omega_n m\dot{u}_{g0}^2$ has dimensions of inverse time (ω_n) multiplied by kinetic energy ($m\dot{u}_{g0}^2$), so this expression possesses indeed the requisite dimensions of power. At resonance and in terms of peak ground velocity, the power dissipated is

$$< \Pi >_{diss} = \frac{c \dot{u}_{g0}^2}{8\xi^2} = \frac{km}{2c} \dot{u}_{g0}^2 = \frac{\omega_n}{4\xi} m \dot{u}_{g0}^2 \tag{2.348}$$

Ratio of Energy Dissipated to Energy Stored

As we have just seen, the force in the viscous damper is

$$F_v = \mathrm{Re}[c\dot{u}] = -\omega c A \sin(\omega t - \varphi) \tag{2.349}$$

which can also be expressed as

$$F_v = -2\xi_v k \frac{\omega}{\omega_n} A \cos(\omega t - \varphi) \tag{2.350}$$

with ξ_v being the fraction of viscous critical damping. The total energy dissipated in the dashpot in one cycle of motion equals the average power times the period. From the preceding section, this energy is

$$E_d = T \langle \Pi_{diss} \rangle = \pi c \omega A^2 \tag{2.351}$$

On the other hand, the average elastic energy stored in the spring element during one cycle of motion is

$$E_s = \frac{1}{T}\int_0^T ku^2 dt = \frac{\omega}{2\pi}k\,A^2 \int_0^{2\pi/\omega}\sin^2(\omega t - \varphi)\,dt = \tfrac{1}{2}kA^2 \tag{2.352}$$

which in this case is equal to the maximum strain energy stored in the spring at maximum elongation. We now define the energy ratio as

$$\varepsilon = \frac{E_d}{E_s} = \frac{\pi c\omega A^2}{\tfrac{1}{2}kA^2} = \frac{2\pi c}{k}\frac{m}{m}\omega = 2\pi\frac{2\xi_v \omega_n}{\omega_n^2}\omega \tag{2.353}$$

That is,

$$\boxed{\varepsilon = 4\pi\xi_v \frac{\omega}{\omega_n}}, \qquad \rightarrow \qquad \boxed{\xi_v = \frac{1}{4\pi}\frac{E_d}{E_s}} \tag{2.354}$$

This energy ratio varies directly with the driving frequency, indicating that the energy dissipated is proportional to the speed with which the oscillations are being performed. A corollary is that the energy ratio at resonance is a direct measure of the fraction of critical damping.

Hysteresis Loop for Spring–Dashpot System

Consider a spring–dashpot system, without any mass, subjected to a harmonic force. The maximum force is F_0, while the maximum displacement is u_0. Our aim is to find a relationship between the instantaneous force F and the instantaneous displacement u. The dynamic equilibrium equation in this case is

$$F(t) = F_0 \cos\omega t = ku + c\dot{u} = ku_0 \cos(\omega t - \varphi) - \omega c u_0 \sin(\omega t - \varphi) \tag{2.355}$$

in which $F_0 = u_0\sqrt{k^2 + \omega^2 c^2}$ and $\tan\varphi = \omega c/k$. From this expression, we obtain

$$\begin{aligned}
F(u) &= u_0\left[k\cos(\omega t - \varphi)\mp\omega c\sqrt{1-\cos^2(\omega t - \varphi)}\right]\\
&= F_0\left[\cos\varphi\frac{u}{u_0}\mp\sin\varphi\sqrt{1-\left(\frac{u}{u_0}\right)^2}\right]
\end{aligned} \tag{2.356}$$

We can express this result as

$$\left(\frac{u}{u_0}\right)^2 + \left(\frac{F}{F_0}\right)^2 - 2\left(\frac{u}{u_0}\right)\left(\frac{F}{F_0}\right)\cos\varphi - \sin^2\varphi = 0 \tag{2.357}$$

This is the equation of an ellipse, whose area equals the energy dissipated in one cycle of motion, as shown in Figure 2.42.

When the frequency of the excitation is increased while the maximum displacement is maintained, the size of the ellipse increases, and so does also the inclination of the principal axis of the ellipse as well as the energy dissipated in each cycle. This ellipse defines the *hysteresis loop* for the system at hand.

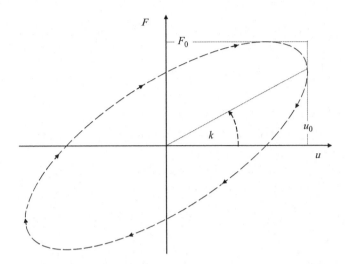

F

F_0

k

u_0

u

Figure 2.42. Hysteresis cycle for linear viscous oscillator.

2.9.2 Hysteretic Damping

Ratio of Energy Dissipated to Energy Stored

Experiments show that for many engineering materials, including soils, the amount of energy dissipated when strained cyclically is, to a large extent, *independent* of frequency, and that it is a function of the maximum deformation instead. This observation has motivated the creation of the concept of *linear hysteretic damping* (or material damping, or structural damping), which is properly formulated only in the frequency domain. By definition, in an oscillation having linear hysteretic damping, the damping force is assumed to be in phase with the rate of strain (i.e., the velocity), and proportional to the maximum deformation (displacement). Hence, the hysteretic force is

$$F_h = 2\xi_h k A \left[\text{sgn}(\omega) \cos(\omega t - \varphi) \right] \tag{2.358}$$

in which ξ_h is the hysteretic damping ratio, and A is the displacement amplitude. Other related constants sometimes used are $\delta = \arctan(2\xi_h)$ = the *loss angle*, and $Q = 1/(2\xi_h)$ = the *quality factor* (chiefly used by seismologists).

The term $2\xi_h k A$ – which does not contain the driving frequency – shows that the hysteretic force is proportional to the maximum displacement, while the factor within the square brackets specifies that the forces are in phase with the velocity. The sign function in the latter term, which is defined as

$$\text{sgn}(\omega) = \begin{cases} 1 & \text{if } \omega > 0 \\ 0 & \text{if } \omega = 0 \\ -1 & \text{if } \omega < 0 \end{cases} \tag{2.359}$$

is introduced here to account for the fact that the velocity reverses phase when the frequencies are negative, and to ensure the damping model to be physically meaningful. The energy dissipated by this system in each cycle of motion is

$$E_d = 2\pi \xi_h k A^2 \text{sgn}(\omega) \tag{2.360}$$

so that the energy ratio is now

$$\boxed{\varepsilon = 4\pi\,\xi_h\,\text{sgn}(\omega)}, \qquad \boxed{\xi_h = \frac{1}{4\pi}\frac{E_d}{E_s}} \tag{2.361}$$

Instantaneous Power and Power Dissipation via the Hilbert Transform

The instantaneous power delivered by an arbitrary external source to a hysteretic spring-damper system can be obtained by the inverse Fourier transform

$$\Pi(t) = F(t)\dot{u}(t) = \left\{\frac{1}{2\pi}\int_{-\infty}^{\infty} k\left[1+2\,i\,\xi_h\,\text{sgn}(\omega)\right]\tilde{u}(\omega)e^{i\omega t}\,d\omega\right\}\dot{u}(t)$$
$$= k\left\{u(t)+2\xi_h\,\hat{u}(t)\right\}\dot{u}(t) \tag{2.362}$$

in which $\hat{u}(t)$ is the *Hilbert transform* of the instantaneous displacement (see Chapter 8, Section 8.1 for further details). The first term represents the elastic power stored in, or released by, the spring, while the second term is the power dissipated by hysteresis. While the Hilbert transform $\hat{u}(t)$ is not strictly a causal function, the instantaneous power still has this property, because the velocity factor is causal. In the case of a harmonic excitation, the Hilbert transforms of $\sin\omega t$ and $\cos\omega t$ are, respectively $\cos\omega t$ and $-\sin\omega t$. The accumulation in time of the energy dissipated is then

$$E_{diss}(t) = 2k\xi_h\int_0^t \hat{u}(\tau)\dot{u}(\tau)\,d\tau \tag{2.363}$$

which in the case of harmonic motion reduces to

$$E_{diss}(t) = 2k\xi_h\int_0^t \left[-A\sin\omega\tau\right]\left[-\omega A\sin\omega\tau\right]d\tau$$
$$= \tfrac{1}{2}k\xi_h A^2\left[2\omega t-\sin 2\omega t\right] \tag{2.364}$$

The energy dissipated grows monotonically in a nearly linear fashion. In particular, at the end of the first cycle $t=T=2\pi/\omega$, so the released energy is $E_{diss}=2\pi k\xi_h A^2$, as we had before.

2.9.3 Power Dissipation during Broadband Base Excitation

Consider an SDOF system with mass m, viscous dashpot c, and stiffness k that is subjected to a support acceleration $\ddot{u}_g(t)$. The equation of motion is

$$m\ddot{v}+c\dot{v}+kv = -m\ddot{u}_g = f(t) \tag{2.365}$$

in which $v=u-u_g$ is the relative displacement, and $f(t)=-m\ddot{u}_g$ is the equivalent external forcing function. Formally, the velocity response function is given by the convolution

$$\dot{v}(t) = f*\dot{h} = \int_0^t f(\tau)\dot{h}(t-\tau)d\tau = \frac{1}{2\pi}\int_{-\infty}^{+\infty} i\omega F(\omega)H(\omega)e^{i\omega t}d\omega$$
$$= -\frac{1}{2\pi}\int_{-\infty}^{+\infty} i\omega F^*(\omega)H^*(\omega)e^{-i\omega t}d\omega \tag{2.366}$$

where $\dot{h} = \frac{d}{dt}h$ is the time derivative of the impulse response function, $i\omega H(\omega)$ is the Fourier transform of $\frac{d}{dt}h$, $F(\omega)$ is the Fourier transform of $f(t)$, and F^*, H^* are the complex conjugates of these functions. Formally, these functions are defined by

$$f(\tau) = \frac{1}{2\pi}\int_{-\infty}^{+\infty} F(\omega)e^{i\omega\tau}d\omega \qquad H(\omega) = \frac{1}{k(1-r^2+2i\,\xi r)}, \qquad r = \frac{\omega}{\omega_n} \qquad (2.367)$$

The *instantaneous* power being dissipated is (after removing an $-i^2 = 1$ factor)

$$\Pi(t) = c\dot{v}^2 = c\left\{\int_0^t f(\tau)\dot{h}(t-\tau)d\tau\right\}^2$$

$$= c\left\{\frac{1}{2\pi}\int_{-\infty}^{+\infty}\Omega F(\Omega)H(\Omega)e^{i\Omega t}d\Omega\right\}\left\{\frac{1}{2\pi}\int_{-\infty}^{+\infty}\omega F^*(\omega)H^*(\omega)e^{-i\omega t}d\omega\right\} \qquad (2.368)$$

The average power dissipated over some time T is then

$$\langle\Pi(t)\rangle_T = \frac{c}{T}\frac{1}{2\pi}\int_{-\infty}^{+\infty}\int_{-\infty}^{+\infty}\omega\Omega\ F(\Omega)\ F^*(\omega)H(\Omega)\ H^*(\omega)\left[\frac{1}{2\pi}\int_0^T e^{i(\Omega-\omega)t}dt\right]d\Omega\,d\omega \qquad (2.369)$$

Since the excitation is causal, we can replace the lower limit in the last integral to negative infinity without affecting results. Also,

$$\frac{1}{2\pi}\int_{-\infty}^{T}e^{i(\Omega-\omega)t}dt \to \frac{1}{2\pi}\int_{-\infty}^{+\infty}e^{i(\Omega-\omega)t}dt = \delta(\Omega-\omega) \qquad (2.370)$$

so for sufficiently large T, the average power dissipated tends to

$$\langle\Pi(t)\rangle_T = \frac{c}{T}\left[\frac{1}{2\pi}\int_{-\infty}^{+\infty}\omega^2\ F(\omega)\ F^*(\omega)H(\omega)H^*(\omega)\ d\omega\right]$$

$$= \frac{c}{T}\left[\frac{1}{2\pi}\int_{-\infty}^{+\infty}\omega^2\ |F(\omega)|^2\ |H(\omega)|^2\ d\omega\right] \qquad (2.371)$$

2.9.4 Comparing the Transfer Functions for Viscous and Hysteretic Damping

If we compared the steady-state harmonic responses of two SDOF systems with the same mass m and stiffness k, but the first with viscous damping ξ_v and the second with hysteretic damping ξ_h, we would observe equal motions at only one frequency. This frequency is the one that provides the same energy ratio for both systems. For lightly damped oscillators, the best agreement for all frequencies is found when the two systems have the same energy ratio at resonance. This implies choosing $\xi_v = \xi_h$. To illustrate this point, let us consider the transfer functions for these two systems:

$$H_v = \frac{1/k}{1-\left(\frac{\omega}{\omega_n}\right)^2 + 2i\,\xi_v\,\frac{\omega}{\omega_n}} \qquad (2.372)$$

and

$$H_h = \frac{1/k}{1-\left(\frac{\omega}{\omega_n}\right)^2 + 2i\,\xi_h\,\mathrm{sgn}(\omega)} \qquad (2.373)$$

If we compare the absolute values and phase angles for these two expressions, say with $\xi_v = \xi_h = 0.1$ (i.e., 10% of critical damping), we observe that both functions are equal only at resonance. Although they differ at other points, these discrepancies are noticeable only at low frequencies, and are particularly evident in the phase angle. In most cases, however, differences in the response to transient excitations are much smaller than the differences in the transfer functions, and deviations are small enough that they can be neglected in engineering applications. Hence, the hysteretic oscillator is essentially equivalent to the viscous oscillator. We should add, however, that the hysteretic oscillator does not rigorously satisfy *causality*, so it does not truly constitute a physically realizable system. Indeed, careful analyses by Crandall[2] have shown that when motions in hysteretic systems are transformed from the frequency domain into the time domain, small non-causal response precursors may precede the excitation.

Best Match between Viscous and Hysteretic Oscillator

It is interesting to note that it is possible to design a hysteretic oscillator perfectly matching at *all* frequencies the amplitude, but not the phase, of a viscous oscillator. This merely requires changing the stiffness and damping constants slightly. Let k_v, ξ_v and k_h, ξ_h be the stiffness and fractions of damping of the viscous and hysteretic oscillators, respectively. Choosing the equivalent values

$$k_h = k_v(1 - 2\xi_v^2), \qquad \xi_h = \xi_v \frac{\sqrt{1 - \xi_v^2}}{1 - 2\xi_v^2} \qquad \text{or} \qquad k_v = \frac{k_h}{2 - \sqrt{1 + 4\xi_h^2}}, \qquad \xi_v = \sqrt{\frac{\sqrt{1 + 4\xi_h^2} - 1}{2}}$$

$$(2.374)$$

It follows that the natural frequencies of these two systems are slightly different, namely $\omega_h = \omega_v \sqrt{1 - 2\xi_v^2}$. With these choices, the absolute value of the transfer function for the hysteretic oscillator is

$$|H_h| = \left| \frac{1}{k(1 - 2\xi_v^2)\left[1 - \left(\dfrac{\omega}{\omega_n\sqrt{1 - 2\xi_v^2}}\right)^2 + 2\,i\,\xi_v \dfrac{\sqrt{1 - \xi_v^2}}{1 - 2\xi_v^2}\,\mathrm{sgn}(\omega)\right]} \right| = \frac{1}{k}\frac{1}{\left|1 - 2\xi_v^2 - \left(\dfrac{\omega}{\omega_n}\right)^2 + 2\,i\,\xi_v\sqrt{1 - \xi_v^2}\,\mathrm{sgn}(\omega)\right|}$$

$$= \frac{1}{k}\frac{1}{\sqrt{\left[1 - \left(\dfrac{\omega}{\omega_n}\right)^2\right]^2 + 4\xi_v^2\left(\dfrac{\omega}{\omega_n}\right)^2}}$$

$$(2.375)$$

which agrees with the amplitude of the transfer function of the viscous oscillator. However, the phase angles of the viscous and hysteretic oscillators are

[2] S. H. Crandall, "Dynamic response of systems with structural damping," in *Air, Space and Instruments*, Draper anniversary volume (New York: McGraw-Hill, 1963), 183–193.

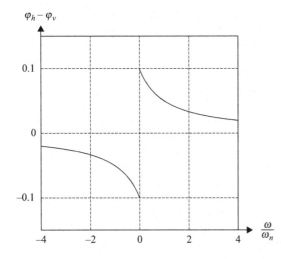

$\varphi_h - \varphi_v$

Figure 2.43. Phase angle difference between viscous and hysteretic systems.

$$\tan \varphi_v = \frac{2\xi_v \frac{\omega}{\omega_n}}{1 - \left(\frac{\omega}{\omega_n}\right)^2} \quad \text{and} \quad \tan \varphi_h = \frac{2\xi_v \sqrt{1 - \xi_v^2}}{1 - 2\xi_v^2 - \left(\frac{\omega}{\omega_n}\right)^2} \, \text{sgn}(\omega) \tag{2.376}$$

Nonetheless, the phase angle differences are not large if damping is not large. If we substitute the phase angles of the viscous and hysteretic oscillators into the trigonometric identity

$$\tan(\varphi_h - \varphi_v) = \frac{\tan \varphi_h - \tan \varphi_v}{1 + \tan \varphi_h \tan \varphi_v} \tag{2.377}$$

and neglecting terms in powers of the damping higher than one, we find

$$\tan(\varphi_h - \varphi_v) = \frac{2\xi_v \, \text{sgn}(\omega)}{1 + \left|\frac{\omega}{\omega_n}\right|} < 2\xi_v \tag{2.378}$$

At resonance, the difference in phase angle is essentially equal to the fraction of damping, which in most cases is small (because the tangent of a small angle can be approximated by the angle itself). This shows again that a hysteretic SDOF system is essentially equivalent to a viscous SDOF system.

2.9.5 Locus of Viscous and Hysteretic Transfer Function

We shall show first that the locus (geometric figure) for the real and imaginary parts of the transfer function of a hysteretic oscillator is a circle, albeit an incomplete one. By contrast, the locus of a viscous oscillator exhibits greater complexity: for low damping values, it approximates a circle, while large damping values are associated with oblong figures.

Let x, y be the real and (negative) imaginary parts times the stiffness of the transfer function for a hysteretic oscillator. If $r = \omega/\omega_n$ is the tuning ratio, then

$$kH(\omega) = x - \mathrm{i}\, y = \frac{1}{1 - r^2 + 2\mathrm{i}\,\xi} \tag{2.379}$$

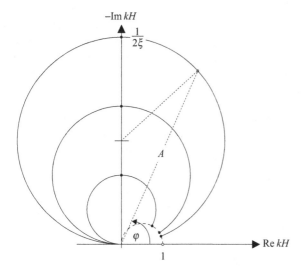

Figure 2.44. Locus of transfer function for hysteretic oscillator.

From the imaginary part of the inverse to this expression, we obtain

$$2\xi = \frac{y}{x^2 + y^2} \tag{2.380}$$

which can be expressed as

$$y^2 - 2\left(\frac{y}{4\xi}\right) + x^2 = 0 \tag{2.381}$$

Finally,

$$\left(y - \frac{1}{4\xi}\right)^2 + x^2 = \left(\frac{1}{4\xi}\right)^2 \tag{2.382}$$

This is the equation of a circle of radius $1/4\xi$ with center on the vertical axis at a height equal to the radius. The abscissa and ordinate are the real and (negative) imaginary parts of the transfer function, while the linear distance from the origin and the angle it forms with the abscissa are the absolute value and phase angle, respectively. Figure 2.44 shows three of these circles for damping values of 0.10, 0.15, and 0.30. As can be seen, the circles are not complete: the $\omega = 0^+$ value start at an offset point that itself lies on a semicircle (dashed lines). The intersection of the circles and the vertical axis represent the resonant frequency, while the origin corresponds to an infinite frequency.

We consider next the more complicated case of a viscous oscillator. The inverse transfer function is now

$$\frac{1}{kH(\omega)} = \frac{1}{x - iy} = \frac{x + iy}{x^2 + y^2} = 1 - r^2 + 2i\xi r \tag{2.383}$$

From the real and imaginary parts, we obtain

$$1 - r^2 = \frac{x}{x^2 + y^2} \tag{2.384}$$

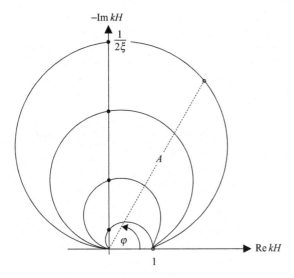

Figure 2.45. Locus of transfer function for viscous oscillator.

$$r = \frac{y}{2\xi(x^2 + y^2)} \tag{2.385}$$

Substituting the second expression into the first, we obtain

$$1 - \left(\frac{y}{2\xi(x^2 + y^2)}\right)^2 = \frac{x}{x^2 + y^2} \tag{2.386}$$

which can be changed into the biquadratic equation

$$y^4 - \left(\frac{1}{4\xi^2} + x - 2x^2\right)y^2 - x^3 + x^4 = 0 \tag{2.387}$$

This equation is shown in Figure 2.45 for light and heavy damping (0.1, 0.15, 0.30, 1.0). As can be seen, light damping produces near circles, while heavy damping is associated with oblong figures. At Re[kH]=1, the curves have a vertical tangent, which at the scale of the drawings is not readily apparent.

Finally, it is of interest to consider also the locus of the *velocity transfer function* of the viscous oscillator. Defining now y as the *positive* imaginary part, this function is

$$\sqrt{km}\, H_{v|p}(\omega) = x + \mathrm{i}\, y = \frac{\mathrm{i}r}{1 - r^2 + 2\,\mathrm{i}\,\xi r} \tag{2.388}$$

$$2\xi = \frac{x}{x^2 + y^2} \tag{2.389}$$

from which we obtain

$$\left(x - \frac{1}{4\xi}\right)^2 + y^2 = \left(\frac{1}{4\xi}\right)^2 \tag{2.390}$$

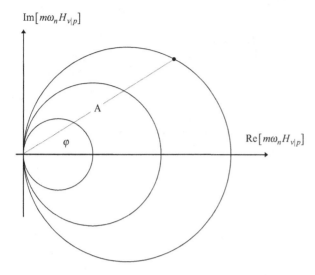

Figure 2.46. Locus of velocity transfer function for viscous oscillator.

This is once more the equation of a circle, but now the center is on the x-axis at a distance from the origin equal to the radius, namely $1/4\xi$; see Figure 2.46. Unlike the circles of the hysteretic oscillator, these circles are complete, because the transfer function is zero for both zero and infinite frequencies. As the frequency increases continuously, a point on the circle travels in clockwise direction. The half-power frequencies correspond to the points above and below the center of the circle.

3 Multiple Degree of Freedom Systems

3.1 Multidegree of Freedom Systems

A viscously damped, linear multidegree of freedom system (MDOF) is characterized by the matrix equation

$$\mathbf{M\ddot{u}} + \mathbf{C\dot{u}} + \mathbf{Ku} = \mathbf{p}(t) \tag{3.1}$$

in which $\mathbf{M}$, $\mathbf{C}$, and $\mathbf{K}$ are the mass, damping, and stiffness matrices, respectively, and $\mathbf{u}(t)$, $\mathbf{p}(t)$ are the displacement and load vectors. These matrices are symmetric and either positive semidefinite or positive definite. The latter condition expresses the fact that the energies involved during motion and deformation cannot be negative. The kinetic energy cannot be zero unless the system does not move, or the mass associated with the DOF in motion is zero. In addition, neither the dissipated energy nor the strain energy can be negative, because the system is not a source of energy, and motions generally involve structural deformations. If the support constraints allow it, however, the latter two energies can be zero, such as during the rigid body motion of an aircraft. In most practical cases, $\mathbf{M}$ is assumed to be positive definite.

The solution of the previous equation is more difficult than that of SDOF systems, not only because it is a matrix equation, but also because the damping term introduces important complications, as will be seen. Hence, we shall first consider undamped vibrations, and then generalize the results for damped ones.

3.1.1 Free Vibration Modes of Undamped MDOF Systems

The free vibration equation of an undamped system is

$$\mathbf{M\ddot{u}} + \mathbf{Ku} = \mathbf{0} \tag{3.2}$$

In analogy to the undamped single-DOF (SDOF) case, we try a sinusoidal solution of the form

$$\mathbf{u}(t) = \boldsymbol{\phi}\left(a\cos\omega t + b\sin\omega t\right) \tag{3.3}$$

$$\mathbf{\ddot{u}}(t) = -\omega^2\,\boldsymbol{\phi}\left(a\cos\omega t + b\sin\omega t\right) \tag{3.4}$$

in which ϕ is a shape vector that does not depend on time, and a, b are any constants. When we substitute this trial solution into the differential equation and cancel the common time-varying term, we obtain

$$\boxed{\mathbf{K}\phi = \omega^2\,\mathbf{M}\phi} \tag{3.5}$$

This a linear eigenvalue problem in ω^2. In a system with n degrees of freedom (the maximum rank of the matrices), this equation has n solutions (ordered from smallest to largest eigenvalue)

$$
\begin{aligned}
&\omega_1^2 &&\text{and} &&\phi_1 = \textit{the fundamental mode}\\
&\omega_2^2 &&\text{and} &&\phi_2\\
&\;\vdots\\
&\omega_n^2 &&\text{and} &&\phi_n
\end{aligned}
$$

in which ω_j^2 are the eigenvalues, and ϕ_j are the eigenvectors. Since $\mathbf{K}$, $\mathbf{M}$ are symmetric and positive semidefinite, all eigenvalues ω_j^2 are *real* and *nonnegative*. Hence, the square roots ω_j of the eigenvalues are real, and constitute the natural frequencies of the discrete system, while the eigenvectors are the associated modes of vibration at those frequencies. On the other hand, if λ is any arbitrary scalar, it is clear that it can be applied to the eigenvalue equation

$$\lambda\mathbf{K}\phi = \lambda\omega^2\mathbf{M}\phi \tag{3.6}$$

or equivalently

$$\mathbf{K}(\lambda\phi) = \omega^2\,\mathbf{M}(\lambda\phi) \tag{3.7}$$

which in turn implies that the eigenvectors are defined only up to a scalar factor, that is, they only define a direction in the n-dimensional space, and can be scaled to any arbitrary length.

Let $i = 1, \cdots, n$ be a generic index identifying a specific DOF, and $j = 1, \cdots, n$ the modal index identifying an individual mode. We then define the $n \times n$ matrices

$$\mathbf{\Omega} = \begin{Bmatrix} \omega_1 & & \\ & \ddots & \\ & & \omega_n \end{Bmatrix} = \mathrm{diag}\{\omega_j\} \qquad = \textit{the spectral matrix} \tag{3.8}$$

$$\mathbf{\Phi} = \{\phi_j\} \equiv \{\phi_1 \quad \cdots \quad \phi_n\} \equiv \{\phi_{ij}\} \qquad = \textit{the modal matrix} \tag{3.9}$$

in which ϕ_{ij} is the ith component of the jth mode. In terms of these matrices, the eigenvalue problem can be written as

$$\mathbf{K}\mathbf{\Phi} = \mathbf{M}\mathbf{\Phi}\mathbf{\Omega}^2 \tag{3.10}$$

Orthogonality Conditions

Consider the eigenvalue equations for two distinct modes i, j:

$$\mathbf{K}\phi_i = \omega_i^2\,\mathbf{M}\phi_i \tag{3.11}$$

$$\mathbf{K}\boldsymbol{\phi}_j = \omega_j^2 \, \mathbf{M}\boldsymbol{\phi}_j \tag{3.12}$$

Multiplying each of these by the transposed vector of the other equation, we obtain the *scalars*

$$\boldsymbol{\phi}_j^T \mathbf{K}\boldsymbol{\phi}_i = \omega_i^2 \, \boldsymbol{\phi}_j^T \, \mathbf{M}\boldsymbol{\phi}_i \tag{3.13}$$

$$\boldsymbol{\phi}_i^T \mathbf{K}\boldsymbol{\phi}_j = \omega_j^2 \, \boldsymbol{\phi}_i^T \mathbf{M}\boldsymbol{\phi}_j \tag{3.14}$$

Since $\mathbf{M}, \mathbf{K}$ are symmetric, when we transpose Eq. 3.14 we obtain

$$\boldsymbol{\phi}_j^T \mathbf{K} \, \boldsymbol{\phi}_i = \omega_j^2 \, \boldsymbol{\phi}_j^T \mathbf{M} \, \boldsymbol{\phi}_i \tag{3.15}$$

Subtracting this from Eq. 3.13 yields

$$0 = (\omega_i^2 - \omega_j^2) \, \boldsymbol{\phi}_j^T \mathbf{M} \, \boldsymbol{\phi}_i \tag{3.16}$$

If the two modes have distinct eigenvalues $\omega_i^2 \neq \omega_j^2$, their difference cannot be zero, so it must be that $\boldsymbol{\phi}_j^T \mathbf{M} \, \boldsymbol{\phi}_i = 0$, which in turn implies $\boldsymbol{\phi}_j^T \mathbf{K} \, \boldsymbol{\phi}_i = 0$. These two results indicate that all distinct modes (with nonrepeated eigenvalues) are *orthogonal* ("perpendicular") with respect to the mass and stiffness matrices, as illustrated schematically in Figure 3.1. Observe that $\mathbf{K}\boldsymbol{\phi}_1, \mathbf{M}\boldsymbol{\phi}_1$ are collinear vectors which merely differ in length by a factor ω_1^2, as indicated by the eigenvalue equation 3.11. On the other hand, when $i = j$, these products are generally nonzero, because the matrices are both positive semidefinite or positive definite. (In the case of repeated eigenvalues, it is always possible to find corresponding eigenvectors that will satisfy the orthogonality condition; see books on linear algebra.) In general, the orthogonality conditions are

$$\boldsymbol{\phi}_i^T \mathbf{M} \, \boldsymbol{\phi}_j = \begin{cases} \mu_j & \text{if } i = j \\ 0 & \text{if } i \neq j \end{cases} \tag{3.17}$$

$$\boldsymbol{\phi}_i^T \mathbf{K} \, \boldsymbol{\phi}_j = \begin{cases} \kappa_j & \text{if } i = j \\ 0 & \text{if } i \neq j \end{cases} \tag{3.18}$$

in which μ_i is the *modal mass*, and $\kappa_j = \omega_j^2 \mu_j$ is the *modal stiffness*. In matrix form

$$\boldsymbol{\Phi}^T \mathbf{M} \, \boldsymbol{\Phi} = \mathbb{M} = \text{diag}\{\mu_j\} \tag{3.19}$$

$$\boldsymbol{\Phi}^T \mathbf{K} \, \boldsymbol{\Phi} = \mathbb{K} = \text{diag}\{\kappa_j\} = \text{diag}\{\mu_j \, \omega_j^2\} \tag{3.20}$$

Often, we encounter systems in which two or more modes are *different*, yet have the *same* eigenvalue. For example, the fundamental torsional and translational modes of a building could have exactly the same frequencies. In that case, any linear combination of the repeated modes is also a mode; that is, any direction in the subspace defined by the repeated modes is an eigen-direction. A common strategy is then to choose the eigenvectors within this subspace in such way that they satisfy the orthogonality conditions.

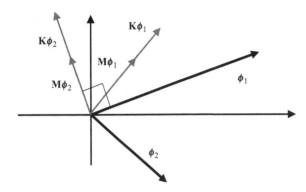

Figure 3.1. Orthogonality of eigenvectors.

Normalized Eigenvectors

Since the eigenvectors can be scaled to any arbitrary length, a convenient scaling factor is the square root of the modal mass, the application of which yields the *normalized* modes

$$\bar{\phi}_j = \frac{1}{\sqrt{\mu_j}} \phi_j \tag{3.21}$$

It is easy to see that the normalized modes have unit modal mass and satisfy the simpler orthogonality conditions

$$\bar{\phi}_i^T \mathbf{M} \bar{\phi}_j = \begin{cases} 1 & \text{if } i = j \\ 0 & \text{if } i \neq j \end{cases}, \qquad \bar{\phi}_i^T \mathbf{K} \bar{\phi}_j = \begin{cases} \omega_j^2 & \text{if } i = j \\ 0 & \text{if } i \neq j \end{cases} \tag{3.22}$$

which in matrix form reads

$$\boxed{\bar{\Phi}^T \mathbf{M} \bar{\Phi} = \mathbf{I}} \qquad \boxed{\bar{\Phi}^T \mathbf{M} \bar{\Phi} = \Omega^2} \tag{3.23}$$

with **I** being the identity matrix.

3.1.2 Expansion Theorem

If the eigenvectors span the full n-dimensional space (i.e., they are not degenerate), then they can be used as base to express any other vector:

$$\mathbf{u} = \sum_{j=1}^{n} \phi_j x_j \equiv \Phi \mathbf{x} \tag{3.24}$$

To determine the modal coordinates x_j, we use the orthogonality property:

$$\phi_i^T \mathbf{M} \mathbf{u} = \sum_{j=1}^{n} \phi_i^T \mathbf{M} \phi_j x_j = \phi_i^T \mathbf{M} \phi_i x_i \tag{3.25}$$

which implies modal coordinates

$$\boxed{x_j = \frac{\phi_j^T \mathbf{M} \mathbf{u}}{\phi_j^T \mathbf{M} \phi_j}} \tag{3.26}$$

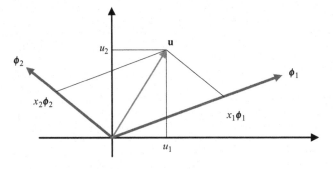

Figure 3.2. Eigenvectors as coordinate basis.

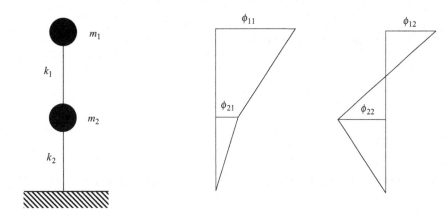

Figure 3.3. 2-DOF system and its two eigenvectors.

This is illustrated schematically in Figure 3.2 for the simple case of a 2-DOF system and an arbitrary vector $\mathbf{u}$.

Example Solve eigenvalue problem for 2-DOF system

Consider the close-coupled 2-DOF system shown in Figure 3.3, which has mass and stiffness matrices

$$\mathbf{M} = \begin{Bmatrix} m_1 & \\ & m_2 \end{Bmatrix} \qquad \mathbf{K} = \begin{Bmatrix} k_1 & -k_1 \\ -k_1 & k_1 + k_2 \end{Bmatrix} \tag{3.27}$$

These can conveniently be expressed as

$$\mathbf{M} = m_1 \begin{Bmatrix} 1 & \\ & \alpha \end{Bmatrix} \qquad \mathbf{K} = k_1 \begin{Bmatrix} 1 & -1 \\ -1 & \beta \end{Bmatrix} \tag{3.28}$$

with dimensionless parameters $\alpha = m_2 / m_1$ and $\beta = (k_1 + k_2) / k_1$. Defining also the dimensionless eigenvalue $\lambda = \omega^2 m_1 / k_1$, the eigenvalue problem can then be written as

$$\left| \mathbf{K} - \omega^2 \mathbf{M} \right| = k_1 \left| \begin{Bmatrix} 1-\lambda & -1 \\ -1 & \beta - \alpha\lambda \end{Bmatrix} \right| = k_1 \left[(1-\lambda)(\beta - \alpha\lambda) - 1 \right] = 0 \tag{3.29}$$

which leads to the quadratic equation

$$\alpha\lambda^2 - (\alpha+\beta)\lambda + (\beta-1) = 0 \tag{3.30}$$

with two solutions

$$\lambda_j = \frac{\alpha+\beta \mp \sqrt{(\alpha+\beta)^2 - 4\alpha(\beta-1)}}{2\alpha} \qquad j = 1,2 \tag{3.31}$$

which can also be expressed as

$$\lambda_j = \frac{\alpha+\beta \mp \sqrt{(\alpha-\beta)^2 + 4\alpha}}{2\alpha} \qquad j = 1,2 \tag{3.32}$$

For example, if $\alpha = 4$ and $\beta = 7$, we obtain

$$\lambda_1 = 0.75 \qquad \rightarrow \qquad \omega_1 = 0.866\sqrt{\frac{k_1}{m_1}} \tag{3.33}$$

$$\lambda_2 = 2 \qquad \rightarrow \qquad \omega_2 = 1.414\sqrt{\frac{k_1}{m_1}} \tag{3.34}$$

To obtain the eigenvectors, we substitute the eigenvalues found into the eigenvalue equation

$$k_1 \begin{Bmatrix} 1-\lambda_j & -1 \\ -1 & \beta - \alpha\lambda_j \end{Bmatrix} \begin{Bmatrix} \phi_{1j} \\ \phi_{2j} \end{Bmatrix} = \begin{Bmatrix} 0 \\ 0 \end{Bmatrix} \tag{3.35}$$

Next, we can set $\phi_{1j} = 1$, since the eigenvectors are defined only up to a multiplicative constant, so we can choose one component arbitrarily. We can then solve for ϕ_{2j} from the first equation $\phi_{2j} = 1 - \lambda_j$. The modal matrix is then

$$\Phi = \begin{Bmatrix} \phi_{11} & \phi_{12} \\ \phi_{21} & \phi_{22} \end{Bmatrix} = \begin{Bmatrix} 1 & 1 \\ 1-\lambda_1 & 1-\lambda_2 \end{Bmatrix} = \begin{Bmatrix} 1 & 1 \\ 0.25 & -1 \end{Bmatrix} \tag{3.36}$$

To find the normalized modes, we begin by computing the modal masses:

$$\mu_j = \phi_j^T \mathbf{M} \phi_j = m_1 \{1 \quad 1-\lambda_j\} \begin{Bmatrix} 1 & \\ & \alpha \end{Bmatrix} \begin{Bmatrix} 1 \\ 1-\lambda_j \end{Bmatrix} = m_1 \left[1 + \alpha(1-\lambda_j)^2\right] \tag{3.37}$$

which in our case yields $\mu_1 = \frac{5}{4}m_1$ and $\mu_2 = 5m_1$. The normalized modal matrix is then

$$\bar{\Phi} = \frac{1}{\sqrt{m_1}} \begin{Bmatrix} 0.894 & 0.447 \\ 0.224 & -0.447 \end{Bmatrix} \tag{3.38}$$

To avoid proliferation of symbols, we shall write in the ensuing the normalized eigenvectors without the superscript bar, indicating only where necessary if they are, or are not, normalized.

3.1.3 Free Vibration of Undamped System Subjected to Initial Conditions

Consider an undamped MDOF system that is subjected to the initial conditions in displacement and velocity

$$\mathbf{u}\big|_{t=0} = \mathbf{u}_0 \qquad \dot{\mathbf{u}}\big|_{t=0} = \dot{\mathbf{u}}_0 \tag{3.39}$$

Since for $t > 0$ the system is executing free vibrations, the response will consist in a linear, but as yet unknown, combination of all the normal modes. Thus, the response can be expressed as

$$\mathbf{u}(t) = \sum_{j=1}^{n} \left[A_j \cos \omega_j t + B_j \sin \omega_j t \right] \phi_j \tag{3.40}$$

in which the A_j, B_j are constants to be determined. Specializing this expression for $t = 0$, we obtain

$$\mathbf{u}_0 = \sum_{j=1}^{n} A_j \, \phi_j \qquad \dot{\mathbf{u}}_0 = \sum_{j=1}^{n} \omega_j B_j \, \phi_j \tag{3.41}$$

Making use of the expansion theorem, we obtain

$$A_j = \frac{\phi_j^T \mathbf{M} \mathbf{u}_0}{\phi_j^T \mathbf{M} \phi_j} \qquad \text{and} \qquad B_j = \frac{\phi_j^T \mathbf{M} \dot{\mathbf{u}}_0}{\omega_j \, \phi_j^T \mathbf{M} \phi_j} \tag{3.42}$$

which when substituted in the modal summation can be used to obtain the complete response in time.

3.1.4 Modal Partition of Energy in an Undamped MDOF System

We consider next the question of how the vibrational energy is divided among the various modes. For this purpose, consider once more the free vibration equation of an undamped system

$$\mathbf{M}\ddot{\mathbf{u}} + \mathbf{K}\mathbf{u} = \mathbf{0} \tag{3.43}$$

subjected to arbitrary initial conditions $\mathbf{u}_0$, $\dot{\mathbf{u}}_0$. Expressing the motion in terms of the modes, we obtain

$$\mathbf{u} = \sum_{j=1}^{n} \phi_j \, q_j \equiv \mathbf{\Phi}\mathbf{q} \tag{3.44}$$

The instantaneous *elastic* and *kinetic* energies for this system are

$$V = \tfrac{1}{2}\mathbf{u}^T \mathbf{K}\mathbf{u} \qquad \text{and} \qquad K = \tfrac{1}{2}\dot{\mathbf{u}}^T \mathbf{M}\dot{\mathbf{u}} \tag{3.45}$$

Expressing the displacement and velocity vectors in terms of the modes, these energies are

$$V = \tfrac{1}{2}\mathbf{q}^T \mathbf{\Phi}^T \mathbf{K} \mathbf{\Phi}\mathbf{q} = \tfrac{1}{2}\mathbf{q}^T \mathbb{K}\mathbf{q} = \sum_{j=1}^{n} \tfrac{1}{2}\kappa_j \, q_j^2 > 0 \tag{3.46}$$

and

$$K = \tfrac{1}{2}\dot{\mathbf{q}}^T \boldsymbol{\Phi}^T \mathbf{M} \boldsymbol{\Phi}\dot{\mathbf{q}} = \tfrac{1}{2}\dot{\mathbf{q}}^T \mathbb{M}\dot{\mathbf{q}} = \sum_{j=1}^{n}\tfrac{1}{2}\mu_j\,\dot{q}_j^2 \geq 0 \tag{3.47}$$

The inequalities result from the fact that the mass and stiffness matrices are positive definite and semidefinite, respectively. As can be seen, the potential and kinetic energies separate cleanly into modal contributions that have the same form as the energies in an SDOF system. It suffices to replace the modal stiffness and mass in place of the spring stiffness and lumped mass of the SDOF system, and add together the modal contributions.

3.1.5 What If the Stiffness and Mass Matrices Are Not Symmetric?

In some cases, and depending in how one chooses the independent DOF, one may end up with stiffness and mass matrices that are not symmetric. For example, this can happen when the active DOF are not at the center of mass of a rigid body. In principle, a simple combination of equations is all that it would take to get them into a symmetric form.

But what happens if one sticks to the nonsymmetric form? The consequence is that orthogonality and modal superposition will break down, at least partially. To illustrate matters, say we start from the ideal, fully *symmetric* form:

$$\mathbf{M\ddot{u} + Ku = 0} \tag{3.48}$$

which satisfies the standard eigenvalue problem and orthogonality conditions

$$\mathbf{K}\boldsymbol{\psi}_n = \omega_n^2 \mathbf{M}\boldsymbol{\psi}_n \qquad \boldsymbol{\psi}_i^T \mathbf{M}\boldsymbol{\psi}_j = \mu_j\,\delta_{ij} \qquad \boldsymbol{\psi}_i^T \mathbf{K}\boldsymbol{\psi}_j = \kappa_j\,\delta_{ij} \tag{3.49}$$

Next, we carry out the transformation

$$\mathbf{u = Tv} \tag{3.50}$$

where $\mathbf{T}$ is an arbitrary, nonsingular, linear transformation matrix. Then

$$\mathbf{MT\ddot{v} + KTv = 0} \tag{3.51}$$

or

$$\tilde{\mathbf{M}}\ddot{\mathbf{v}} + \tilde{\mathbf{K}}\mathbf{v} = 0 \tag{3.52}$$

where

$$\tilde{\mathbf{M}} = \mathbf{MT}, \qquad \tilde{\mathbf{K}} = \mathbf{KT} \tag{3.53}$$

are now *nonsymmetric* mass and stiffness matrices. If we then set up the eigenvalue problem

$$\tilde{\mathbf{K}}\boldsymbol{\phi}_n = \omega_n^2 \tilde{\mathbf{M}}\boldsymbol{\phi}_n \tag{3.54}$$

it is easy for us to see that the eigenvalues are exactly as above, yet the eigenvectors are related as $\boldsymbol{\phi}_n = \mathbf{T}\boldsymbol{\psi}_n$. But these new eigenvectors do *not* satisfy the standard orthogonality condition. Instead, these so-called "right" eigenvectors must be supplemented with the "left" eigenvectors

$$\boldsymbol{\varphi}_n^T \tilde{\mathbf{K}} = \omega_n^2 \boldsymbol{\varphi}_n^T \tilde{\mathbf{M}}, \quad \text{or} \quad \mathbf{K}^T \boldsymbol{\varphi}_n = \omega_n^2 \tilde{\mathbf{M}}^T \boldsymbol{\varphi}_n \tag{3.55}$$

obtained with the transposed matrices. Hence, orthogonality is now given by

$$\boldsymbol{\varphi}_i^T \tilde{\mathbf{K}} \boldsymbol{\phi}_j = \kappa_j \delta_{ij}, \quad \boldsymbol{\varphi}_i^T \tilde{\mathbf{M}} \boldsymbol{\phi}_j = \mu_j \delta_{ij} \tag{3.56}$$

This affects modal superposition. For example,

$$q_{oj} = \frac{\boldsymbol{\varphi}_j^T \tilde{\mathbf{M}} \mathbf{u}_0}{\boldsymbol{\varphi}_j^T \tilde{\mathbf{M}} \boldsymbol{\phi}_j} \quad \text{instead of simply} \quad q_{oj} = \frac{\boldsymbol{\psi}_j^T \tilde{\mathbf{M}} \mathbf{u}_0}{\boldsymbol{\psi}_j^T \tilde{\mathbf{M}} \boldsymbol{\psi}_j} \tag{3.57}$$

which is formed with both left and right eigenvectors.

The bottom line is that if one chooses to work with nonsymmetric matrices, then it behooves to compute both left and right eigenvectors and apply a generalization of the modal superposition method to that case. But in general, it is a bad idea to choose to work with large nonsymmetric matrices.

3.1.6 Physically Homogeneous Variables and Dimensionless Coordinates

It is sometimes advantageous to modify some (or all) of the variables involved in a vibration problem so that that all degrees of freedom have either the same physical dimension, or they are dimensionless. This is especially true when solving a small system by hand, say one that has only 2 or 3 DOF, but is also true when solving larger systems, say via MATLAB®, without having to specify the actual physical dimensions. This has the very significant advantage that all matrices and the eigenvalues can then be written in dimensionless form, that is, all matrices, vectors, and eigenvalues are just numbers, which greatly simplifies the algebra. We illustrate this concept with two examples.

Example 1: Cantilever Bending Beam with Mass Lumped at Top

Consider a massless, homogeneous cantilever beam as shown in Figure 3.4 that has a concentrated mass with translational and rotational inertias m, J lumped at its free end. The beam has bending rigidity EI and length L, and is subjected to a pair of time-varying force and moment F, M. The 2 active DOF are the translation and rotation at the free end, namely u, θ (the latter taken positive counterclockwise). The equation of motion is then

$$\begin{Bmatrix} m & 0 \\ 0 & J \end{Bmatrix} \begin{Bmatrix} \ddot{u} \\ \ddot{\theta} \end{Bmatrix} + \begin{bmatrix} 12\frac{EI}{L^3} & 6\frac{EI}{L^2} \\ 6\frac{EI}{L^2} & 4\frac{EI}{L} \end{bmatrix} \begin{Bmatrix} u \\ \theta \end{Bmatrix} = \begin{Bmatrix} F \\ M \end{Bmatrix} \tag{3.58}$$

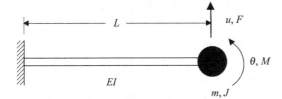

Figure 3.4. Lumped mass with translational and rotational inertia attached to massless beam. Active DOF are u, θ.

Dividing the second row in Eq. 3.58 by L, then extracting from the second column a factor L so that it forms part of the second variable, we obtain

$$\begin{Bmatrix} m & 0 \\ 0 & J/L^2 \end{Bmatrix} \begin{Bmatrix} \ddot{u} \\ L\ddot{\theta} \end{Bmatrix} + \frac{EI}{L^3} \begin{Bmatrix} 12 & 6 \\ 6 & 4 \end{Bmatrix} \begin{Bmatrix} u \\ L\theta \end{Bmatrix} = \begin{Bmatrix} F \\ M/L \end{Bmatrix} \tag{3.59}$$

We observe that $L\theta$ has the same physical dimension as u, M/L has the same dimension as the force F, and J/L^2 has dimensions of mass. Hence, this equation can be written in dimensionless form as

$$m\begin{Bmatrix} 1 & 0 \\ 0 & \mu \end{Bmatrix} \begin{Bmatrix} \ddot{u} \\ \ddot{u}_\theta \end{Bmatrix} + \frac{EI}{L^3} \begin{Bmatrix} 12 & 6 \\ 6 & 4 \end{Bmatrix} \begin{Bmatrix} u \\ u_\theta \end{Bmatrix} = \begin{Bmatrix} F \\ F_\theta \end{Bmatrix} \tag{3.60}$$

where $u_\theta = \theta L$, $F_\theta = M/L$, and $\mu = J/(mL^2)$ is just a dimensionless number. Hence, dividing the entire equation by the flexural rigidity EI/L^3 and in the absence of external loads, the eigenvalue problem for free vibration simplifies to

$$\begin{Bmatrix} 12-\lambda_j & 6 \\ 6 & 4-\mu\lambda_j \end{Bmatrix} \begin{Bmatrix} \psi_{1j} \\ \psi_{2j} \end{Bmatrix} = \begin{Bmatrix} 0 \\ 0 \end{Bmatrix}, \qquad \lambda_j = \frac{mL^3}{EI}\omega_j^2 \;\rightarrow\; \omega_j = \frac{EI}{L^3}\sqrt{\lambda_j} \tag{3.61}$$

which leads to the characteristic equation

$$(12-\lambda_j)(4-\mu\lambda_j)-36 = 0 \qquad \text{i.e.} \qquad \mu\lambda_j^2 - 4(1+3\mu)\lambda_j + 12 = 0 \tag{3.62}$$

$$\lambda_j = \frac{2}{\mu}\left[(1+3\mu)\lambda_j \mp \sqrt{9\mu^2+3\mu+1}\right] \tag{3.63}$$

Having found the eigenvalues, we proceed to find the eigenvectors, assuming for this purpose that $\psi_{1j} = 1$, in which case from the first eigenvalue equation we find $(12-\lambda_j)(1)+6\psi_{2j} = 0$, that is, $\psi_{2j} = \frac{1}{6}\lambda_j - 2$, so

$$\Psi = \{\psi_1 \quad \psi_2\} = \begin{Bmatrix} 1 & 1 \\ \frac{1}{6}\lambda_1-2 & \frac{1}{6}\lambda_2-2 \end{Bmatrix} \rightarrow \begin{Bmatrix} 6 & 6 \\ \lambda_1-12 & \lambda_2-12 \end{Bmatrix} \tag{3.64}$$

where in the last step we arbitrarily rescaled the dimensionless eigenvectors by a factor 6. The actual, physical eigenvectors are then

$$\Phi = \{\phi_1 \quad \phi_2\} = \begin{Bmatrix} 6 & 6 \\ \frac{1}{L}(\lambda_1-12) & \frac{1}{L}(\lambda_2-12) \end{Bmatrix} \tag{3.65}$$

Thus, finding the frequencies and modal shapes now merely requires knowledge of the dimensionless ratio μ. Had we tried to solve this problem in its original form, we would have been encumbered by a lengthy and error-prone algebra involving products and ratios of the physical variables involved.

Example 2: Cantilever Beam Modeled with Finite Elements
Consider again a homogeneous cantilever bending beam, but this time modeled by means of an arbitrary number N of finite elements, each with the stiffness and consistent mass matrices as given in Chapter 1, Section 1.3.8 while disregarding self-rotational

inertia $(j_z = 0)$ and shear deformation $(\phi_z = 0)$. Hence, this problem has a total of $2N$ DOF, namely N translations and N rotations. Making use of homogeneous dimensions as in the previous example, we obtain the beam element stiffness and mass matrices

$$\bar{\mathbf{K}} = \frac{EI}{L^3} \begin{bmatrix} 12 & 6 & -12 & 6 \\ 6 & 4 & -6 & 2 \\ -12 & -6 & 12 & -6 \\ 6 & 2 & -6 & 4 \end{bmatrix}, \quad \bar{\mathbf{M}} = \frac{m}{420} \begin{bmatrix} 156 & 22 & 54 & -13 \\ 22 & 4 & 13 & -3 \\ 54 & 13 & 156 & -22 \\ -13 & -3 & -22 & 4 \end{bmatrix} \tag{3.66}$$

Temporarily ignoring the leading factors EI/L^3 and $m/420$, we then proceed to overlap the element matrices as is usually done in the finite element method, and thus construct the dimensionless system matrices $\mathbf{K}, \mathbf{M}$ for all N elements and $2N$ DOF. This is a very simple operation that can be accomplished with a brief MATLAB® function script that takes N as an input argument. Defining the vector of dimensionless eigenvalues as $x = \left\{ \frac{mL^3}{420EI} \omega_j^2 \right\}$, we could then solve for all eigenvalues via the MATLAB® command $x = \text{eig}(\mathbf{K}, \mathbf{M})$ and thereafter be able to tell how good those eigenvalues are when compared to the actual eigenvalues of a continuous cantilever bending beam, the so-called Euler beam considered in Chapter 4. We could then decide how good and accurate the discrete finite element representation actually is. Moreover, we could just as easily proceed to do various other numerical experiments, such as what happens if one neglects the rotational inertias. In MATLAB®, that would easily be done as follows.

Say $\mathbf{K}$, $\mathbf{M}$ are the assembled dimensionless stiffness and mass matrices and $\mathbf{N}$ is the number of DOF. We then proceed to reorder the dimensionless DOF so that all displacements come first and then all of the rotations. We can accomplish this by

```
b = [2:2:2*N]; a = b-1;              % indices, even and odd DOF
K = [K(a,a),K(a,b); K(b,a), K(b,b)]; % reorder dof
O = zeros(N,N);                      % null matrix
M = [M(a,a); O; O, O];               % neglect rotational mass
```

Execution of the command $x = \text{eig}(K,M)$ will now yield N finite eigenvalues with rotational inertias neglected, and N infinite eigenvalues (inf), which shows that it is not necessary to carry out any static condensation to dispose of the now static rotational DOF. Observe also that all of this can be accomplished without the need to specify the actual physical dimensions. Thus, the dimensionless formulation allows for simple numerical experiments.

3.2 Effect of Static Loads on Structural Frequencies: Π-Δ Effects

3.2.1 Effective Lateral Stiffness

When we say that the forces associated with dead loads are static in nature, we are stating the obvious. Less evident is the fact that these static forces may have the capacity to alter the natural frequencies and mode shapes of a building. The reason is that gravity loads do affect the elastic stability of tall structural systems by reducing their capacity to carry lateral loads, which lengthens the resonant periods. The loss of apparent lateral stiffness may be especially important in the case of high-rise buildings, since the columns in the lower elevations must then carry large axial forces. While the most common designation for this phenomenon is Π–Δ *effects*, it is also sometimes referred to as *inverted pendulum effects*. We focus here on the particular case of buildings in which the structural system consists of plane frames.

Imagine a simple, plane rectangular frame, which, at least initially, carries only loads caused by gravity. Thus, in the absence of lateral vibrations, these loads cause compression forces in the columns, which can be computed (or estimated) by standard methods in structural analysis; thus, these loads are known a priori. Of course, because the system is elastic, these loads will also slightly shorten the columns, but this is of no concern to us here. If the building is then affected by wind or earthquake forces, it will respond to these new forces by oscillating laterally, which will also cause the columns to deform in bending and rotate with respect to the vertical. These deformations produce, in turn, an oscillatory vertical motion of the structural nodes and, therefore, of the gravity loads themselves, which causes their potential energy to be released and transformed into flexural deformation energy in the structure. Indeed, if a greater amount of flexural energy were required to accomplish a lateral perturbation than what the gravity loads could release in potential energy, then the system would be stable; contrariwise, the structure is unstable. Clearly, stability is related to the tendency of gravity loads to descend in response to lateral deformation, and this is what we shall examine in the following.

Consider an individual column in its deflected position, displaced $y(x)$ from the vertical. By elementary calculus, the difference in vertical length Δ between the originally straight column of length L and the rotated and flexed column is

$$\Delta = \tfrac{1}{2}\int_0^L \left(\frac{dy}{dx}\right)^2 dx \equiv \tfrac{1}{2}\int_0^L y'^2 dx \tag{3.67}$$

which, of course, is the amount by which A approaches B. Since there are no external lateral forces acting on the body of the column other than those at its two ends, we know that the *elastica* $y(x)$ (the shape of the elastic deformation) is a cubic parabola of the form

$$y = (1-3\xi^2+2\xi^3)u_A + \xi^2(3-2\xi)u_B + \xi(1-\xi)^2\,\theta_A L - \xi^2(1-\xi)\theta_B L \tag{3.68}$$

in which $\xi = 1 - x/L$, as shown in Figure 3.5. The slope is thus

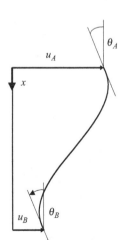

Figure 3.5. Column in deformed configuration (elastica).

$$y' = 6\xi(\xi-1)(u_A - u_B)/L + (1 - 4\xi + 3\xi^2)\theta_A + \xi(3\xi-2)\theta_B \tag{3.69}$$

in which $u_A, u_B, \theta_A, \theta_B$ are the lateral displacements and counterclockwise rotations observed at the two ends of the column, and $\xi = x/L$ (with the x-axis pointing from A to B). Substituting Eq. 3.69 into Eq. 3.67 and evaluating the straightforward even if somewhat tedious integral, we obtain

$$\Delta/L = \frac{1}{30}\left[18\left(\frac{u_A - u_B}{L}\right)^2 + 3\left(\frac{u_A - u_B}{L}\right)(\theta_A + \theta_B) + 2\theta_A^2 + 2\theta_B^2 - \theta_A\theta_B\right] \tag{3.70}$$

It is convenient to express this result in matrix form as

$$\Delta = \tfrac{1}{2}\mathbf{u}^T \mathbf{A}\,\mathbf{u} \qquad \text{with} \qquad \mathbf{u}^T = \{u_A \quad \theta_A \quad u_B \quad \theta_B\} \tag{3.71}$$

and

$$\mathbf{A} = \frac{1}{30L}\begin{Bmatrix} 36 & 3L & -36 & 3L \\ 3L & 4L^2 & -3L & -L^2 \\ -36 & -3L & 36 & -3L \\ 3L & -L^2 & -3L & 4L^2 \end{Bmatrix} \tag{3.72}$$

Let P be the axial force in the column, which we assume to be constant and independent of the lateral motion. We then perturb the system by applying a lateral deformation, in which case the strain energy increases by

$$E = \tfrac{1}{2}\mathbf{u}^T \mathbf{K}\,\mathbf{u} \tag{3.73}$$

with $\mathbf{K}$ being the lateral stiffness matrix of the column. In our case, this matrix is given by

$$\mathbf{K} = \frac{EI}{(1+\phi)L^3}\begin{Bmatrix} 12 & 6L & -12 & 6L \\ 6L & (4+\phi)L^2 & -6L & (2-\phi)L^2 \\ -12 & -6L & 12 & -6L \\ 6L & (2-\phi)L^2 & -6L & (4+\phi)L^2 \end{Bmatrix} \tag{3.74}$$

with

$$\phi = \frac{12EI}{GA_s L^2} = 24(1+v)\frac{I}{A_s L^2} \tag{3.75}$$

in which E, G are the Young and shear moduli; v is the Poisson's ratio; I, A_s are the column's moment of inertia and shear area, respectively; and L is the length of the column.

The lateral perturbation of our column causes the gravity load P applied at the upper end to lose potential energy in the amount $P\Delta = \tfrac{1}{2}P\mathbf{u}^T\mathbf{A}\mathbf{u}$. Hence, the net amount of *external* energy required to produce the deflection $\mathbf{u}$ for this system is

$$\Pi = \tfrac{1}{2}\mathbf{u}^T\left[\mathbf{K} - P\mathbf{A}\right]\mathbf{u} \tag{3.76}$$

Clearly, it would be possible for us to achieve this very same energy change by considering a virtual column lacking the axial force, but possessing instead the stiffness matrix

$$\boxed{\tilde{\mathbf{K}} = \mathbf{K} - P\mathbf{A}} \tag{3.77}$$

Thus, all we need do to include the effects of gravity on the vibration of a structural frame is to modify the stiffness matrices of each member according to the previous formula, using for P in each case the known axial forces elicited by the gravity loads. More generally, if we regard the modified member stiffness matrix $\tilde{\mathbf{K}}$ as being given in local member coordinates, then we can obviously extend the formulation to systems with inclined columns by simply adding an appropriate rotation of that member into global coordinates. We could even account for members in tension by reversing the sign of P for those members. At the system level, the equation of motion for lateral vibration would then be of the form

$$\mathbf{M}\ddot{\mathbf{u}} + \tilde{\mathbf{K}}\mathbf{u} = \mathbf{p} \tag{3.78}$$

whose natural frequencies of vibration are obtained from the eigenvalue problem

$$\tilde{\mathbf{K}}\boldsymbol{\phi} = \omega^2 \mathbf{M}\boldsymbol{\phi} \tag{3.79}$$

Notice that when $\tilde{\mathbf{K}}$ is singular, the natural frequencies drop to zero, and the system no longer can resist lateral loads. This situation arises when gravity loads are large enough to elicit *system buckling* (as opposed to member buckling, or local buckling), at which point the structure has failed.

3.2.2 Vibration of Cantilever Column under Gravity Loads

To show that buckling loads can indeed be obtained from the present formulation, we begin by evaluating the elastic stability of a cantilever column with a lumped mass m at the top, which transfers in turn a weight $P = mg$ onto the column; thereafter, we examine the natural frequency of that system. Expressing the matrices in dimensionless form and imposing the boundary conditions at the bottom of the column, we obtain

$$\tilde{\mathbf{K}}\mathbf{u} = \left\{ \frac{EI}{L^3} \begin{bmatrix} 12 & 6 \\ 6 & 4 \end{bmatrix} - \frac{P}{30L} \begin{bmatrix} 36 & 3 \\ 3 & 4 \end{bmatrix} \right\} \left\{ \begin{matrix} u_A \\ \theta_A L \end{matrix} \right\} = \left\{ \begin{matrix} 0 \\ 0 \end{matrix} \right\} \tag{3.80}$$

Defining the eigenvalue $\lambda = PL^2/EI$ and setting the determinant to zero, we obtain the characteristic equation as

$$15\lambda^2 - 520\lambda + 1200 = 0, \qquad \lambda = \frac{52}{3} \mp \frac{8}{3}\sqrt{31} \tag{3.81}$$

whose smallest root is

$$P_{crit} = 2.486 \frac{EI}{L^2} = 1.008 \frac{\pi^2 EI}{4L^2} \tag{3.82}$$

This is virtually the exact result for the buckling load of a cantilever column. Thus, the column is stable only as long as $P = mg < P_{crit}$. The effective lateral stiffness matrix is now

$$\tilde{\mathbf{K}} = \frac{EI}{L^3}\begin{bmatrix} 12(1-3\alpha) & 3(2-\alpha) \\ 3(2-\alpha) & 4(1-\alpha) \end{bmatrix}, \qquad \alpha = \frac{2.486}{30}\frac{P}{P_{crit}} \tag{3.83}$$

Condensing out the rotational DOF (there is no rotational mass attached), we obtain the effective lateral stiffness as seen from the free end as

$$k_{eff} = \frac{EI}{L^3}\left[12(1-3\alpha)-\frac{9(2-\alpha)^2}{4(1-\alpha)}\right] = \frac{3EI}{4(1-\alpha)L^3}(4-52\alpha+45\alpha^2) \tag{3.84}$$

The natural frequency of this system is then

$$\omega_n = \sqrt{\frac{k_{eff}}{m}} = \sqrt{\frac{3EI}{mL^3}}\sqrt{\frac{1-13\alpha+\frac{45}{4}\alpha^2}{1-\alpha}} \tag{3.85}$$

When $\alpha = 0$, then $k_{eff} = 3EI/L^3 \equiv k$, and we recover the classical solution with no Π–Δ effects included. At the opposite extreme, when $P = P_{crit}$, that is, $\alpha = 2.486/30$, the effective lateral stiffness drops to zero, and the natural frequency vanishes. Finally, since $\alpha \le \frac{2.486}{30} = 0.0829 \ll 1$, we can simplify the square root term as follows:

$$\sqrt{\frac{(1-13\alpha+\frac{45}{4}\alpha^2)(1+\alpha)}{1-\alpha^2}} \approx \sqrt{1-12\alpha} \tag{3.86}$$

Clearly, at the buckling load $12\alpha = 12\frac{2.486}{30} = 0.9944 \approx 1$, which leads us to the important result

$$\omega_n = \sqrt{\frac{k}{m}}\sqrt{1-\frac{P}{P_{crit}}} \tag{3.87}$$

which is a widely used approximation to estimate the effects of axial loads on the frequencies of a structure. That is, if $1/\lambda > 1$ were the factor by which either the earth's gravity or the magnitude of all of the masses would have to be increased so as to cause system buckling, then at the current gravity loads all natural frequencies could be estimated to decrease in proportion to $\sqrt{1-\lambda}$.

3.2.3 Buckling of Column with Rotations Prevented

To further illustrate the accuracy of the Π–Δ formulation, we consider next a column attached to a frame with rigid girders that prevent the end rotations of the columns. It suffices to examine only the stability problem, inasmuch as the dynamic problem is similar to that of the previous example. The boundary conditions are now $u_B = 0$, $\theta_A = 0$, $\theta_B = 0$, which leads us to the scalar equation

$$\left(12\frac{EI}{L^3}-\frac{36}{30}\frac{P}{L}\right)u_A = 0 \tag{3.88}$$

and so

$$P_{crit} = \frac{10EI}{L^2} = 1.013\frac{\pi^2 EI}{L^2} \tag{3.89}$$

which is again very nearly the exact result for the buckling load.

Now, you may have perhaps wondered why we did *not* obtain the absolutely exact answer in each case. The reason has to do with the *trial function* we used to compute the energies of deformation, namely a cubic parabola, which does *not* happen to be the exact buckling shape. However, the accuracy of the formulation does increase when we add more and more members to our system. In addition, the cubic parabola has the added benefit of better modeling the effects of lateral loads, and is indeed exact when no gravity loads are present.

By the way, if you try to apply this formulation to a frame that is laterally braced and, in addition, the girders are infinitely rigid so that the column ends cannot rotate, you will discover that the present formulation fails. This is so because this formulation cannot model the stability problem for a doubly clamped column, inasmuch as the *elastica* will be zero everywhere if the end motions are zero. A possible solution: in each column, add an intermediate (moment connecting) node at mid-height.

3.2.4 Vibration of Cantilever Shear Beam

Finally, we consider the case of a cantilevering shear beam with shear area A_s and a lumped mass m at its end, a structure that is not usually associated with buckling. This can be regarded as an idealization of a multistory frame with rigid girders. The deflected shape under lateral load is a straight line, so the vertical descent of the free end is

$$\Delta = L(1 - \cos\theta) = L\left[1 - (1 - \tfrac{1}{2}\theta^2 + \cdots)\right] = \tfrac{1}{2}\theta^2 = \tfrac{1}{2}u^2 / L \tag{3.90}$$

Hence, the effective lateral stiffness under an axial load $P = mg$ is

$$\tilde{k} = k - \frac{P}{L} = \frac{GA_s}{L} - \frac{P}{L} = \frac{GA_s}{L}\left(1 - \frac{P}{GA_s}\right) \tag{3.91}$$

which is zero at the buckling load $P_{crit} = GA_s$. In terms of the original frequency $\omega_0 = \sqrt{k/m}$, the Π–Δ effected frequency is then

$$\omega = \omega_0 \sqrt{1 - \frac{P}{GA_s}} = \omega_0 \sqrt{1 - \frac{P}{P_{crit}}} \tag{3.92}$$

You should notice that the equations we present in this section do not include any changes in axial forces caused by the vibrations themselves.

3.3 Estimation of Frequencies

Although there are well-established numerical algorithms for finding some or all of the frequencies in the eigenvalue problem $|\mathbf{K} - \omega^2\mathbf{M}|$, in many engineering applications one may need only an approximate value for the fundamental frequency, or for the highest frequency, in the discrete MDOF system being represented by this equation. In such cases, it is not necessary to solve directly for the eigenvalues, but to estimate those frequencies instead. There exist several methods to accomplish this goal, and some of them are simple enough that computations can often be carried out by hand, at least if the system has a manageable number of DOF. We present two such methods, namely the *Rayleigh quotient* and the *Dunkerley–Mikhlin* method.

3.3.1 Rayleigh Quotient

Let $\mathbf{v}$ be any arbitrary vector in the nth dimensional space. The *Rayleigh quotient* is then defined by the ratio

$$R = \frac{\mathbf{v}^T \mathbf{K} \mathbf{v}}{\mathbf{v}^T \mathbf{M} \mathbf{v}} \tag{3.93}$$

If $\mathbf{v}$ has approximately the shape of the fundamental mode, then R will provide an *upper bound* to the fundamental eigenvalue, that is, to the square of the fundamental frequency. Of course, if $\mathbf{v}$ is exactly in the shape of that mode, then the ratio will coincide exactly with the eigenvalue. As it turns out, however, the Rayleigh quotient is not too sensitive to imperfections in estimations of the fundamental mode, which means that even coarse shapes $\mathbf{v}$ used in this ratio will lead to close estimates for the fundamental eigenvalue. More generally, if $\mathbf{v}$ is in the approximate shape of a higher mode, then R will be close to the eigenvalue for that mode. Indeed, the Rayleigh quotient changes only slowly in the *vicinity* of a mode, because it is *stationary* in that vicinity. In other words, if R can be imagined as a surface in the $(n + 1)^{\text{th}}$ dimensional space (n for the vectors and one for the quotient itself), then this surface will have saddle points (i.e., points with horizontal tangents) along the directions of the eigenvectors. To see why this is so, consider the modal expansion of the arbitrary vector $\mathbf{v}$:

$$\mathbf{v} = \mathbf{\Phi} \mathbf{c} = \sum_{j=1}^{n} \phi_j c_j \tag{3.94}$$

in which the c_j are the modal coordinates of $\mathbf{v}$. Of course, the right-hand side in this equation is not known, as the eigenvectors themselves are not known. Introducing this expansion into the Rayleigh quotient (and assuming without loss of generality that the modes are normalized), we obtain

$$R = \frac{\mathbf{c}^T \mathbf{\Phi}^T \mathbf{K} \mathbf{\Phi} \mathbf{c}}{\mathbf{c}^T \mathbf{\Phi}^T \mathbf{M} \mathbf{\Phi} \mathbf{c}} = \frac{\mathbf{c}^T \mathbf{\Omega}^2 \mathbf{c}}{\mathbf{c}^T \mathbf{c}} = \frac{c_1^2 \omega_1^2 + c_2^2 \omega_2^2 + \cdots c_n^2 \omega_n^2}{c_1^2 + c_2^2 + \cdots + c_n^2} \tag{3.95}$$

It is clear from this expression that R must lie in the interval $\omega_1^2 \leq R \leq \omega_n^2$, a result that is known as the *Enclosure Theorem*. (Proof: set all eigenvalues equal to either the first or to the last eigenvalue, which makes the ratio smallest or largest; the unknown coefficients will then cancel).

To show that the Rayleigh quotient is stationary in the vicinity of an eigenvalue, we compute the partial derivatives of this ratio with respect to the modal coefficients. The easiest way to accomplish this is by considering small perturbations or *variations* in these coefficients:

$$\delta(\mathbf{c}^T \mathbf{c} R) = \delta(\mathbf{c}^T \mathbf{\Omega}^2 \mathbf{c}) \tag{3.96}$$

that is,

$$(\delta\mathbf{c}^T \mathbf{c} + \mathbf{c}^T \delta\mathbf{c})R + \mathbf{c}^T \mathbf{c} \, \delta R = \delta\mathbf{c}^T \mathbf{\Omega}^2 \mathbf{c} + \mathbf{c}^T \mathbf{\Omega}^2 \, \delta\mathbf{c} \tag{3.97}$$

Since the transpose of a scalar is the scalar itself,

$$\delta R = 2 \frac{\delta \mathbf{c}^T \mathbf{\Omega}^2 \mathbf{c} - \delta \mathbf{c}^T \mathbf{c}\, R}{\mathbf{c}^T \mathbf{c}} \tag{3.98}$$

A saddle point (i.e., horizontal tangent) occurs when $\delta R = 0$, which requires

$$\delta \mathbf{c}^T (\mathbf{\Omega}^2 - R\,\mathbf{I})\mathbf{c} = 0 \tag{3.99}$$

For arbitrary variations $\delta \mathbf{c}$, this can occur only when R is one of the eigenvalues (say, $R = \omega_j^2$). In such case, all but one of the elements of $\mathbf{c}$ will be zero, namely c_j, the coefficient for the corresponding eigenvector. Hence, arbitrary perturbations $\delta \mathbf{c}$ will not greatly affect the value of the Rayleigh quotient in the neighborhood of that eigenvalue, since $\delta R = 0$ at that point.

Example: Fundamental frequency of n-story building

Consider an nth-story steel-frame building with regular geometry, equal floor heights and stiffnesses, and uniform mass distribution. If the axial deformation in the columns as well as the flexibility of girder-floor structural elements are neglected, then the building can be idealized as a homogeneous, discrete shear beam with lumped masses at each floor. In other words, it can be modeled as a close-coupled assembly of n equal springs k and masses m, except that the nth mass at the elevation of the roof is only half as large. The system is supported at the base. Numbering the masses from the roof down, this system has mass and stiffness matrices of the form

$$\mathbf{M} = m \begin{Bmatrix} 0.5 & & & \\ & 1 & & \\ & & \ddots & \\ & & & 1 \end{Bmatrix} \qquad \mathbf{K} = k \begin{Bmatrix} 1 & -1 & & \\ -1 & 2 & -1 & \\ & & \ddots & \\ & & -1 & 2 \end{Bmatrix} \tag{3.100}$$

Guessing that the fundamental mode is not far off from a straight line, we can take $\mathbf{v}$ as

$$\mathbf{v}^T = \{ n \quad n-1 \quad \cdots \quad 1 \} \tag{3.101}$$

This shape implies a Rayleigh quotient

$$R = \frac{\mathbf{v}^T \mathbf{K} \mathbf{v}}{\mathbf{v}^T \mathbf{M} \mathbf{v}} = \frac{nk}{[1^2 + 2^2 + \cdots (n-1)^2 + 0.5n^2]\,m} = \frac{6k}{(2n^2+1)m} \tag{3.102}$$

On the other hand, the exact fundamental eigenvalue for this case happens to be

$$\omega_1^2 = 4 \frac{k}{m} \sin^2 \frac{\pi}{4n} \tag{3.103}$$

which we can use as yardstick for comparisons. For example, if $n = 10$, this implies $R = 0.02985 \frac{k}{m}$ whereas the exact value is $\omega_1^2 = 0.02462 \frac{k}{m}$. Thus, the Rayleigh quotient is about 21% *larger* than the exact eigenvalue. However, the error committed in the actual frequency is less, because frequencies vary with the square root of the eigenvalues. In this case, $\sqrt{R / \omega_1} = 1.101$, which represents only a 10% error. For very large n, the frequency ratio tends to $\sqrt{12} / \pi = 1.102$, while for $n = 1$, we get the exact result with both formulas.

Of course, real buildings will not be exactly uniform, and simple formulas to assess the accuracy of R or ω_1, such as Eq. 3.103, will not exist. However, the computation of the Rayleigh quotient will continue to be a simple task. The frequency thus estimated should not be too far off from the exact value, particularly if a better (albeit less simple) choice is made for the fundamental mode. This example also illustrates how the Rayleigh quotient is always larger than the fundamental eigenvalue.

Rayleigh–Schwarz Quotients

A method often used to improve estimations for the fundamental eigenvector is by means of the so-called *inverse iteration*

$$\mathbf{K}\mathbf{v}_{k+1} = \mathbf{M}\mathbf{v}_k \qquad k = 0,1,2,\ldots \tag{3.104}$$

which can be shown to converge to the first mode. The iteration is started with an arbitrary approximation. The Rayleigh–Schwarz quotients are then defined as follows[1]:

$$R_{00} = \frac{\mathbf{v}_0^T \mathbf{K}\mathbf{v}_0}{\mathbf{v}_0^T \mathbf{M}\mathbf{v}_0} \tag{3.105}$$

$$R_{01} = \frac{\mathbf{v}_0^T \mathbf{K}\mathbf{v}_1}{\mathbf{v}_0^T \mathbf{M}\mathbf{v}_1} = \frac{\mathbf{v}_0^T \mathbf{M}\mathbf{v}_0}{\mathbf{v}_1^T \mathbf{K}\mathbf{v}_1} = \frac{\mathbf{v}_0^T \mathbf{K}\mathbf{v}_1}{\mathbf{v}_1^T \mathbf{K}\mathbf{v}_1} = \frac{\mathbf{v}_0^T \mathbf{M}\mathbf{v}_0}{\mathbf{v}_1^T \mathbf{M}\mathbf{v}_0} \tag{3.106}$$

$$R_{11} = \frac{\mathbf{v}_1^T \mathbf{K}\mathbf{v}_1}{\mathbf{v}_1^T \mathbf{M}\mathbf{v}_1} = \frac{\mathbf{v}_1^T \mathbf{M}\mathbf{v}_0}{\mathbf{v}_1^T \mathbf{M}\mathbf{v}_1} \qquad \text{etc.} \tag{3.107}$$

that is,

$$\boxed{R_{k-1,k} = \frac{\mathbf{v}_{k-1}^T \mathbf{M}\mathbf{v}_{k-1}}{\mathbf{v}_k^T \mathbf{M}\mathbf{v}_{k-1}}} \qquad \text{and} \qquad \boxed{R_{kk} = \frac{\mathbf{v}_k^T \mathbf{M}\mathbf{v}_{k-1}}{\mathbf{v}_k^T \mathbf{M}\mathbf{v}_k}} \tag{3.108}$$

These mixed quotients are particularly useful and easy to compute, because $\mathbf{M}$ is often diagonal.

3.3.2 Dunkerley–Mikhlin Method

As we saw earlier, the Rayleigh quotient always gives an *upper bound* to the fundamental eigenvalue, that is, it is always larger than the true eigenvalue. In this section, we present Dunkerley's method[2], which provides a *lower bound* to the fundamental eigenvalue. Thus, when used together, these two methods can be used to bracket a root.

In its simplest incarnation, Dunkerley's method is based on two fundamental theorems found in textbooks on linear algebra that state that for the *special* eigenvalue problem $\mathbf{A}\mathbf{x} = \lambda \mathbf{x}$, the determinant equals the product of the eigenvalues $A = |\mathbf{A}| = \lambda_1 \lambda_2 \ldots \lambda_n$ and the *trace* of the matrix (= the sum of its diagonal values) equals the sum of its eigenvalues,

[1] S. Crandall, *Engineering Analysis* (New York: McGraw-Hill, 1956), 96.
[2] S. Dunkerley, "On the whirling and vibration of shafts," *Philos. Trans. R. Soc.*, V. 185, 1895, 269–360.

that is, $\mathrm{tr}(\mathbf{A}) = \sum a_{ii} = \sum \lambda_i$. In our case, we can write the eigenvalue in special form by writing

$$\mathbf{K}^{-1}\mathbf{M}\phi_j = \frac{1}{\omega_j^2}\phi_j \tag{3.109}$$

which implies

$$\begin{aligned}
\mathrm{tr}(\mathbf{K}^{-1}\mathbf{M}) &= \frac{1}{\omega_1^2} + \frac{1}{\omega_2^2} + \cdots \frac{1}{\omega_n^2} \\
&= \frac{1}{\omega_1^2}\left[1 + \left(\frac{\omega_1}{\omega_2}\right)^2 + \cdots \left(\frac{\omega_1}{\omega_n}\right)^2\right]
\end{aligned} \tag{3.110}$$

Since the fundamental eigenvalue is the smallest, all squared ratios in Eq. 3.110 are less than 1 in value, indeed much less than 1 for the highest eigenvalues. Hence, if we neglect their contribution to the summation, we obtain

$$\boxed{\omega_1 \approx \frac{1}{\sqrt{\mathrm{tr}(\mathbf{K}^{-1}\mathbf{M})}}} \tag{3.111}$$

Clearly, since all neglected terms in the summation were positive, the trace must be an upper bound estimate for the reciprocal of the eigenvalue, which implies that its inverse must be a *lower bound* for the eigenvalue itself. This implies that the related formula for the fundamental period

$$T_1 \approx 2\pi\sqrt{\mathrm{tr}(\mathbf{K}^{-1}\mathbf{M})} \tag{3.112}$$

always gives values that are larger than the actual period. A corollary of these formulas is

$$\omega_n \approx \sqrt{\mathrm{tr}(\mathbf{M}^{-1}\mathbf{K})} \tag{3.113}$$

which provides a convenient estimate for the highest eigenvalue.

The disadvantage of this method is that it requires the availability of the flexibility matrix $\mathbf{F} = \mathbf{K}^{-1}$, or at the very minimum, of its diagonal values. In some cases, these may not be difficult to obtain. For example, the diagonal flexibility elements for a close-coupled, fixed-free spring assembly (i.e., a chain of springs) are

$$f_{ii} = \frac{1}{k_1} + \frac{1}{k_2} + \cdots \frac{1}{k_i} \tag{3.114}$$

where we have assumed that the springs are numbered from the bottom up (i.e., from the support to the free end). If we return to the example we presented earlier in the section on the Rayleigh quotient, we see that it is a special case in which all springs are equal, so $f_{ii} = i/k$. In that case

$$\frac{1}{\omega_1^2} \approx \sum_{i=1}^{n} f_{ii}m_i = \frac{m}{k}[1 + 2 + \cdots + 0.5n] = \frac{n^2 m}{2k} \tag{3.115}$$

which gives $\omega_1 \approx \sqrt{\frac{k}{m}}\sqrt{2}/n$. When $n = 10$, we obtain an estimate $\omega_1 \approx 0.1414$, which is 89% of the exact value (a lower bound!). For large n, the ratio between the Dunkerly estimate and the exact frequency converges to $\sqrt{8}/\pi = 0.9003$. Thus, the exact frequency is in this case halfway between the results obtained with the methods by Dunkerley and Rayleigh. In general, however, the Rayleigh quotient is more robust and tends to give closer results.

A generalization of Dunkerley's rule can be obtained by observing that if a matrix $\mathbf{A}$ satisfies the simple eigenvalue problem $\mathbf{A}\mathbf{x} = \lambda\mathbf{x}$, then by raising the matrix to the pth power, so will also its eigenvalues, that is, $\mathbf{A}^p\mathbf{x} = \lambda^p\mathbf{x}$, while the eigenvectors remain the same. Hence, from the trace theorem, we obtain $\mathrm{tr}(\mathbf{A}^p) = \sum \lambda_i^p$. For our problem, this implies

$$\mathrm{tr}(\mathbf{K}^{-1}\mathbf{M})^p = \frac{1}{\omega_1^{2p}} + \frac{1}{\omega_2^{2p}} + \cdots \frac{1}{\omega_n^{2p}}$$

$$= \frac{1}{\omega_1^{2p}}\left[1 + \left(\frac{\omega_1}{\omega_2}\right)^{2p} + \cdots \left(\frac{\omega_1}{\omega_n}\right)^{2p}\right] \tag{3.116}$$

which constitutes the pth order Dunkerley–Mikhlin estimation of the fundamental eigenvalue. In principle, we obtain substantially more accurate results using powers $p > 1$ (usually p is at most 2), because the eigenvalues are better separated. In most cases, however, this refinement may not be warranted, because it does require considerable effort, even if only the diagonal elements are required.

Example Rocking-swaying frequeny of rigid cylinder on soil springs
Consider a rigid structure with total mass m and mass moment of inertia J_0 with respect to the center of mass, which is at a height h above the ground. The structure rests on a soft soil, the stiffness of which can be idealized by means of two springs, a translational spring k_x and a rotational spring k_θ that are attached at the base, as shown in Figure 3.6. If u, θ are the translation and rotation of the base (assumed positive when counterclockwise), and if $J = J_0 + mh^2$ is the mass moment of inertia with respect to the base, then the free vibration frequencies can be shown to follow from the eigenvalue problem

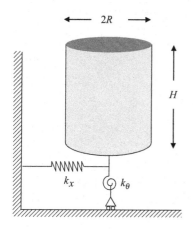

Figure 3.6. Rigid cylinder on lateral and rotational spring

$$\begin{Bmatrix} k_x & 0 \\ 0 & k_\theta \end{Bmatrix} \begin{Bmatrix} u \\ \theta \end{Bmatrix} = \omega^2 \begin{Bmatrix} m & -mh \\ -mh & J \end{Bmatrix} \begin{Bmatrix} u \\ \theta \end{Bmatrix} \tag{3.117}$$

or in compact matrix form,

$$\mathbf{K}\phi = \omega^2 \, \mathbf{M}\phi. \tag{3.118}$$

In terms of the (uncoupled) reference frequencies

$$\omega_x = \sqrt{\frac{k_x}{m}} \qquad \text{swaying frequency} \tag{3.119}$$

$$\omega_\theta = \sqrt{\frac{k_\theta}{J}} \qquad \text{rocking frequency}$$

the *exact* coupled frequencies of this system are

$$\omega^2 = \frac{1}{\alpha}\left(\omega_x^2 + \omega_\theta^2 \mp \sqrt{\left(\omega_x^2 + \omega_\theta^2\right)^2 - 2\,\alpha\,\omega_x^2\,\omega_\theta^2} \right) \tag{3.120}$$

in which

$$\alpha = \frac{2\left(J - mh^2\right)}{J} = \frac{2J_0}{J_0 + mh^2} = \frac{2J_0}{J} \le 2 \tag{3.121}$$

Alternatively, the exact coupled periods are

$$T^2 = \frac{1}{2}\left(T_x^2 + T_\theta^2\right)\left(1 \pm \sqrt{1 - 2\,\alpha\frac{T_x^2\,T_\theta^2}{\left(T_x^2 + T_\theta^2\right)^2}} \right) \tag{3.122}$$

On the other hand, the product $\mathbf{K}^{-1}\mathbf{M} = \mathbf{F}\mathbf{M}$ is

$$\begin{Bmatrix} k_x^2 & 0 \\ 0 & k_\theta \end{Bmatrix}^{-1} \begin{Bmatrix} m & -mh \\ -mh & J \end{Bmatrix} = \begin{Bmatrix} \dfrac{1}{\omega_x^2} & -\dfrac{h}{\omega_x^2} \\ -\dfrac{h}{\omega_x^2}\dfrac{k_x}{k_\theta} & \dfrac{1}{\omega_\theta^2} \end{Bmatrix} \tag{3.123}$$

The first-order Dunkerley approximation for the fundamental coupled frequency is then

$$\frac{1}{\omega^2} = \mathrm{tr}\,(\mathbf{F}\mathbf{M}) = \frac{1}{\omega_x^2} + \frac{1}{\omega_\theta^2} \tag{3.124}$$

or in terms of the periods,

$$T^2 = T_x^2 + T_\theta^2 \tag{3.125}$$

This expression is particularly simple, and it is quite useful in practice. Indeed, it can be shown that if the structure were not rigid, but had periods T_1, T_2, ... when supported by a rigid base, then a Dunkerley-based approximation for the coupled period of the soil-structure system would be given by

$$T^2 = T_1^2 + T_x^2 + T_\theta^2 \tag{3.126}$$

For a second-order approximation, we need only the diagonal elements of $(\mathbf{FM})^2$, which are

$$(\mathbf{FM})^2 = \begin{Bmatrix} \dfrac{1}{\omega_x^4} + \dfrac{h^2}{\omega_x^4}\dfrac{k_x}{k_\theta} & * \\ * & \dfrac{1}{\omega_\theta^4} + \dfrac{h^2}{\omega_x^4}\dfrac{k_x}{k_\theta} \end{Bmatrix} \tag{3.127}$$

After some simple transformations of the trace of this matrix, the second-order approximation can be shown to be

$$\frac{1}{\omega^4} = \frac{1}{\omega_x^4} + \frac{1}{\omega_\theta^4} + \frac{2}{\omega_x^2\,\omega_\theta^2}\left(1 - \frac{J_0}{J}\right) \tag{3.128}$$

or

$$T^4 = \left(T_x^2 + T_\theta^2\right)^2 - \alpha\, T_x^2\, T_\theta^2 \tag{3.129}$$

To assess the accuracy of these approximations, let's assume that the structure is a homogeneous cylinder of radius R and height $H = 2h$, and the foundation is a rigid circular plate resting on a homogeneous elastic half-space with shear modulus G and Poisson's ratio ν. The stiffness ratio is then

$$\frac{k_x}{k_\theta} = \frac{\dfrac{8GR}{2-\nu}}{\dfrac{8GR^3}{3(1-\nu)}} = \frac{3(1-\nu)}{2\left(1-\frac{1}{2}\nu\right)R^2} \approx \frac{3}{2\,R^2} \tag{3.130}$$

If we also define the aspect ratio as $\lambda = R/H$, we can write

$$\alpha = \frac{2J_0}{J} = \frac{2m\left(\frac{1}{4}R^2 + \frac{1}{12}H^2\right)}{m\left(\frac{1}{4}R^2 + \frac{1}{3}H^2\right)} = 2\frac{1+\frac{1}{3}\lambda^2}{1+\frac{4}{3}\lambda^2} \tag{3.131}$$

$$\frac{T_\theta^2}{T_x^2} = \frac{\omega_x^2}{\omega_\theta^2} = \frac{k_x R^2}{k_\theta}\left(\frac{1}{4} + \frac{1}{3}\lambda^2\right) = \frac{3}{8}\left(1 + \frac{4}{3}\lambda^2\right) \tag{3.132}$$

$$2\alpha\frac{T_\theta^2}{T_x^2} = \frac{3}{2}\left(1 + \frac{1}{3}\lambda^2\right) \tag{3.133}$$

In terms of these quantities, the exact fundamental period as well as the first- and second-order Dunkerley approximations can be found to be given by

$$\frac{T}{T_x} = \sqrt{\frac{1}{16}\left[11 + 4\lambda^2 + \sqrt{(7+4\lambda^2) - 24}\right]} \qquad \text{Exact} \tag{3.134}$$

$$\frac{T}{T_x} = \sqrt{\frac{1}{8}\left(11 + 4\lambda^2\right)} \qquad\qquad\qquad \text{First-order} \tag{3.135}$$

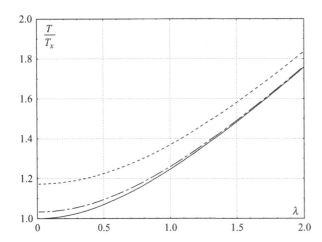

Figure 3.7. Frequencies of rigid cylinder on soil springs, exact vs. Dunkerley
approximation.

$$\frac{T}{T_x} = \sqrt{\frac{1}{8}\sqrt{\left(9+4\lambda^2\right)^2 - 8}} \qquad \text{Second-order} \tag{3.136}$$

A graphical comparison of these three expressions is shown in Figure 3.7. As can be seen, the first-order approximation has an error of about 18% for small λ (i.e., a tall cylinder), but this error diminishes as the aspect ratio increases. The second order approximation improves the solution considerably, but at an increase in the computational effort and complexity. In most cases of systems with many DOF, this additional effort may not be warranted, but it suffices to use instead the first-order approximation.

Dunkerely's Method for Systems with Rigid-Body Modes

Unconstrained systems, such as aircraft, contain a certain number r of rigid body modes whose frequencies are all zero. In addition to these rigid-body modes (which are known by inspection), the system also contains n other dynamic degrees of freedom. Thus, the system has a total of $n + r$ DOFs. If $r > 0$, the stiffness matrix is singular and has no inverse. There exists then a nonsingular modal matrix $\mathbf{\Phi}$, not necessarily unique, and a diagonal spectral matrix $\mathbf{\Omega}$ in which the frequencies are arranged in nondescending order such that[3]

$$\mathbf{K\Phi} = \mathbf{M\Phi\Omega}^2 \qquad \text{and} \qquad \mathbf{\Phi}^T\mathbf{M\Omega} = \mathbf{I} \tag{3.137}$$

Clearly, it is always possible to partition the matrices into submatrices of sizes related to the number of rigid body modes so that

$$\mathbf{K} = \begin{Bmatrix} \mathbf{K}_{rr} & \mathbf{K}_{rn} \\ \mathbf{K}_{nr} & \mathbf{K}_{nn} \end{Bmatrix} \qquad \mathbf{\Phi} = \left\{ \mathbf{\Phi}_r \quad \mathbf{\Phi}_n \right\} = \begin{Bmatrix} \mathbf{\Phi}_{rr} & \mathbf{\Phi}_{rn} \\ \mathbf{\Phi}_{nr} & \mathbf{\Phi}_{nn} \end{Bmatrix} \qquad \mathbf{\Omega}^2 = \begin{Bmatrix} \mathbf{O} & \\ & \mathbf{\Omega}_n^2 \end{Bmatrix} \tag{3.138}$$

[3] J. E. Brock, "Dunkerley–Mikhlin estimates of gravest frequency of vibrating system," *J. Appl. Mech.*, June 1976, 345–348.

We also define the matrices

$$\mathbf{F} = \begin{Bmatrix} \mathbf{O} & \mathbf{O} \\ \mathbf{O} & \mathbf{K}_{nn}^{-1} \end{Bmatrix} \qquad \mathbf{J} = \begin{Bmatrix} \mathbf{O} & \\ & \mathbf{I}_n \end{Bmatrix} \qquad \mathbf{\Lambda} = \begin{Bmatrix} \mathbf{O} & \\ & \mathbf{\Omega}_n^{-2} \end{Bmatrix} \tag{3.139}$$

that obviously satisfy

$$\mathbf{FJ} = \mathbf{F} \qquad \mathbf{\Omega}^2 \mathbf{\Lambda} = \mathbf{J} \qquad \mathbf{J\Lambda} = \mathbf{\Lambda} \tag{3.140}$$

On the other hand, the rigid body modes involve no internal or external forces, that is,

$$\mathbf{K}\mathbf{\Phi} = \left\{ \mathbf{K}\mathbf{\Phi}_r \quad \mathbf{K}\mathbf{\Phi}_n \right\} = \left\{ \mathbf{O} \quad \mathbf{K}\mathbf{\Phi}_n \right\} = \mathbf{K}\mathbf{\Phi}\mathbf{J} \tag{3.141}$$

It follows that the eigenvalue problem can also be written as

$$\mathbf{K}\mathbf{\Phi}\mathbf{\Lambda} = \mathbf{M}\mathbf{\Phi}\mathbf{J} \tag{3.142}$$

Multiplying this expression by the flexibility matrix $\mathbf{F}$ we obtain

$$\mathbf{FK}\mathbf{\Phi}\mathbf{\Lambda} = \mathbf{FM}\mathbf{\Phi}\mathbf{J} \tag{3.143}$$

We need now two matrices $\mathbf{R}$ and $\mathbf{S}$ such that $\mathbf{FK}\mathbf{\Phi} = \mathbf{\Phi}\mathbf{R}$ and $\mathbf{\Phi}\mathbf{J} = \mathbf{S}\mathbf{\Phi}$, which we shall obtain in the following. From the rigid body condition $\mathbf{K}\mathbf{\Phi}_r = \mathbf{O}$, we know that

$$\mathbf{K}_{nn}^{-1}\mathbf{K}_{nr} = -\mathbf{\Phi}_{nr}\mathbf{\Phi}_{rr}^{-1} \tag{3.144}$$

which in turn implies

$$\mathbf{K}_{nn}^{-1}\mathbf{K}_{nr}\mathbf{\Phi}_{rn} = -\mathbf{\Phi}_{nr}\mathbf{\Phi}_{rr}^{-1}\mathbf{\Phi}_{rn} \tag{3.145}$$

Hence

$$\mathbf{FK}\mathbf{\Phi} = \begin{Bmatrix} \mathbf{O} & \mathbf{O} \\ \mathbf{O} & \mathbf{K}_{nn}^{-1}\mathbf{K}_{nr}\mathbf{\Phi}_{rn} + \mathbf{\Phi}_{nn} \end{Bmatrix} \equiv \begin{Bmatrix} \mathbf{O} & \mathbf{O} \\ \mathbf{O} & -\mathbf{\Phi}_{nr}\mathbf{\Phi}_{rr}^{-1}\mathbf{\Phi}_{rn} + \mathbf{\Phi}_{nn} \end{Bmatrix} = \mathbf{\Phi}\mathbf{R} \tag{3.146}$$

with

$$\boxed{\mathbf{R} = \begin{Bmatrix} \mathbf{O} & -\mathbf{\Phi}_{rr}^{-1}\mathbf{\Phi}_{rn} \\ \mathbf{O} & \mathbf{I}_n \end{Bmatrix}} \tag{3.147}$$

Also

$$\mathbf{\Phi}\mathbf{J} = \mathbf{S}\mathbf{\Phi} \tag{3.148}$$

which implies

$$\mathbf{S} = \mathbf{\Phi}\mathbf{J}\mathbf{\Phi}^{-1} = \mathbf{\Phi}\left[\mathbf{I} - (\mathbf{I} - \mathbf{J})\right]\mathbf{\Phi}^T\mathbf{M} = \mathbf{\Phi}\mathbf{\Phi}^T\mathbf{M} - \mathbf{\Phi}_r\mathbf{\Phi}_r^T\mathbf{M} \tag{3.149}$$

That is,

$$\boxed{\mathbf{S} = \mathbf{I} - \mathbf{\Phi}_r\mathbf{\Phi}_r^T\mathbf{M}} \tag{3.150}$$

Hence

$$\mathbf{\Phi R \Lambda} = \mathbf{F M S \Phi} \tag{3.151}$$

so that

$$\mathbf{R \Lambda} = \mathbf{\Phi}^{-1}(\mathbf{F M S})\mathbf{\Phi} \tag{3.152}$$

Now, if p is any positive integer, then the pth power of this equation is

$$(\mathbf{R \Lambda})^{p} = \left(\mathbf{\Phi}^{-1}(\mathbf{F M S})\mathbf{\Phi}\right)^{p} = \mathbf{\Phi}^{-1}(\mathbf{F M S})^{p}\,\mathbf{\Phi} \tag{3.153}$$

Hence

$$\mathrm{tr}\,(\mathbf{R \Lambda})^{p} = \mathrm{tr}\left[\mathbf{\Phi}^{-1}(\mathbf{F M S})^{p}\,\mathbf{\Phi}\right] = \mathrm{tr}\,(\mathbf{F M S})^{p} \tag{3.154}$$

The last equality on the right follows from a theorem in linear algebra that states that *similar transformations* do not change the eigenvalues of a matrix, so they do not change the trace. Because of the special structure of $\mathbf{R}$, it follows that

$$\mathrm{tr}\left[(\mathbf{R \Lambda})^{p}\right] = \mathrm{tr}\left[\mathbf{\Lambda}^{p}\right] \tag{3.155}$$

Finally

$$\boxed{\frac{1}{\omega_{1}^{2p}} \approx \mathrm{tr}\left[\mathbf{\Lambda}^{p}\right] = \mathbf{tr}\left[(\mathbf{FMS})^{p}\right]} \tag{3.156}$$

which constitutes the pth order Dunkerley–Mikhlin estimation of the fundamental eigenvalue for an unconstrained system. If this system has no rigid body modes, then $\mathbf{S} = \mathbf{I}$, and the formula reduces then to the one presented earlier.

Example Fundamental frequency of unrestrained 4-DOF system

Consider a chain of four equal springs and masses that are free to slide along the chain's axial coordinate direction. Thus, this system has one rigid body mode. The stiffness and mass matrices are

$$\mathbf{K} = k\begin{Bmatrix} +1 & -1 & & \\ -1 & +2 & -1 & \\ & -1 & +2 & -1 \\ \hline & & -1 & +1 \end{Bmatrix} \qquad \mathbf{M} = m\begin{Bmatrix} 1 & & & \\ & 1 & & \\ & & 1 & \\ \hline & & & 1 \end{Bmatrix} \tag{3.157}$$

We remove the rigid body mode by constraining the bottom end (as suggested by the lines in the matrices); this allows us to define the flexibility matrix

$$\mathbf{F} = \frac{1}{k}\begin{Bmatrix} 3 & 2 & 1 & 0 \\ 2 & 2 & 1 & 0 \\ 1 & 1 & 1 & 0 \\ \hline 0 & 0 & 0 & 0 \end{Bmatrix} \tag{3.158}$$

By inspection, the rigid body mode is

$$\phi_r^T = \alpha\{1 \quad 1 \quad 1 \quad 1\} \tag{3.159}$$

in which α is a constant that we can determine from the normality condition $\phi_r^T \mathbf{M} \phi_r = 1 = 4\alpha^2 m$, which gives $\alpha = 0.5/\sqrt{m}$. The filter matrix $\mathbf{S}$ is then

$$\mathbf{S} = \mathbf{I} - \phi_r \phi_r^T \mathbf{M} = \frac{1}{4}\begin{bmatrix} 3 & -1 & -1 & -1 \\ -1 & 3 & -1 & -1 \\ -1 & -1 & 3 & -1 \\ -1 & -1 & -1 & 3 \end{bmatrix} \tag{3.160}$$

Hence, the product **FMS** is

$$\mathbf{FMS} = \frac{m}{4k}\begin{bmatrix} 6 & \cdot & \cdot & \cdot \\ \cdot & 3 & \cdot & \cdot \\ \cdot & \cdot & 1 & \cdot \\ \cdot & \cdot & \cdot & 0 \end{bmatrix} \tag{3.161}$$

where the dots represent irrelevant elements, because only the diagonal elements are of interest. The fundamental (nonzero) frequency predicted by Dunkerley's formula is then the square root of the inverse of the trace, that is,

$$\omega = \sqrt{0.4}\sqrt{\frac{k}{m}} = 0.6324\sqrt{\frac{k}{m}} \tag{3.162}$$

By comparison, the exact value for this problem is

$$\omega = \sqrt{2 - \sqrt{2}}\sqrt{\frac{k}{m}} = 0.7654\sqrt{\frac{k}{m}} \tag{3.163}$$

A second-order approximation ($p = 2$) would improve the result to $\omega = 0.72\sqrt{\frac{k}{m}}$, while a Rayleigh quotient with a trial mode $\mathbf{v}^T = \{3 \quad 1 \quad -1 \quad -3\}$ gives $\omega = 0.775\sqrt{\frac{k}{m}}$, which is much closer to the exact root.

3.3.3 Effect on Frequencies of a Perturbation in the Structural Properties

An interesting question with practical applications is: What happens to the frequencies of a system if small masses or stiffness elements are either added or subtracted from the structure? To answer this question, we begin by defining perturbation matrices $\Delta\mathbf{K}$ and $\Delta\mathbf{M}$, which we assume to be positive semidefinite. Addition of the small masses or stiffness leads then to the perturbed stiffness and mass matrices

$$\mathbf{M} = \mathbf{M}_0 + \varepsilon\Delta\mathbf{M} \tag{3.164}$$

$$\mathbf{K} = \mathbf{K}_0 + \lambda\Delta\mathbf{K} \tag{3.165}$$

in which $\mathbf{M}_0$, $\mathbf{K}_0$ are the initial, unperturbed matrices, and ε, λ are small perturbation parameters that define the intensity of the perturbation. Without loss of generality, we can assume also that the unperturbed modes, which are known, have been normalized with respect to the mass matrix.

Perturbation of Mass Matrix

Let's consider first a perturbation of the mass matrix only. Taking derivatives of orthogonality conditions with respect to ε, we obtain

$$\frac{\partial}{\partial \varepsilon}\left(\mathbf{\Phi}^T \mathbf{M} \mathbf{\Phi}\right) = \frac{\partial}{\partial \varepsilon}\mathbf{I} = \mathbf{0} \tag{3.166}$$

$$\frac{\partial}{\partial \varepsilon}\left(\mathbf{\Phi}^T \mathbf{K} \mathbf{\Phi}\right) = \frac{\partial}{\partial \varepsilon}\mathbf{\Omega}^2 = \mathrm{diag}\left\{\frac{\partial}{\partial \varepsilon}\,\omega_j^2\right\} \tag{3.167}$$

Since the derivatives of the eigenvectors with respect to the perturbation parameter are in turn vectors in the n-dimensional space, they can be expressed in terms of the eigenvectors themselves:

$$\frac{\partial}{\partial \varepsilon}\mathbf{\Phi} = \mathbf{\Phi}\,\mathbf{\Gamma} \qquad \text{or} \qquad \frac{\partial \phi_j}{\partial \varepsilon} = \sum_{i=1}^{N} \phi_i\, \gamma_{ij} \tag{3.168}$$

in which $\mathbf{\Gamma}$ is a matrix containing as yet unknown coefficients. The derivatives in the previous two equations can then be expanded and written as

$$\mathbf{\Gamma}^T\left(\mathbf{\Phi}^T \mathbf{M} \mathbf{\Phi}\right) + \left(\mathbf{\Phi}^T \mathbf{M} \mathbf{\Phi}\right)\mathbf{\Gamma} + \mathbf{\Phi}^T \Delta\mathbf{M}\,\mathbf{\Phi} = \mathbf{0} \tag{3.169}$$

$$\mathbf{\Gamma}^T\left(\mathbf{\Phi}^T \mathbf{K} \mathbf{\Phi}\right) + \left(\mathbf{\Phi}^T \mathbf{K} \mathbf{\Phi}\right)\mathbf{\Gamma} = \frac{\partial}{\partial \varepsilon}\mathbf{\Omega}^2 \tag{3.170}$$

which in view of the orthogonality conditions reduce to

$$\mathbf{\Gamma}^T + \mathbf{\Gamma} = -\mathbf{\Phi}^T \Delta\mathbf{M}\,\mathbf{\Phi} \tag{3.171}$$

$$\mathbf{\Gamma}^T\mathbf{\Omega}^2 + \mathbf{\Omega}^2\,\mathbf{\Gamma} = \frac{\partial}{\partial \varepsilon}\mathbf{\Omega}^2 \tag{3.172}$$

Defining the symmetric matrix (which can be computed, since the modes are known),

$$\left\{a_{ij}\right\} = \mathbf{\Phi}^T \Delta\mathbf{M}\,\mathbf{\Phi} \tag{3.173}$$

then Eqs. 3.171 and 3.172 can be written in scalar form as

$$\gamma_{ij} + \gamma_{ji} = -a_{ij} \tag{3.174}$$

$$\omega_j^2\,\gamma_{ji} + \omega_i^2\,\gamma_{ij} = \delta_{ij}\frac{\partial}{\partial \varepsilon}\,\omega_i^2 \tag{3.175}$$

in which the Kronecker delta is $\delta_{ij} = 1$ if $i = j$, and 0 otherwise. From these two equations, we find that $\gamma_{ii} = -\tfrac{1}{2}a_{ii}$, so that

$$\boxed{\frac{\partial}{\partial \varepsilon}\,\omega_i^2 = -a_{ii}\,\omega_i^2} \qquad \text{or} \qquad \frac{\partial \omega_i^2}{\partial \varepsilon} = -\omega_i^2\,\frac{\phi_i^T \Delta\mathbf{M}\,\phi_i}{\phi_i^T \mathbf{M}\,\phi_i} \tag{3.176}$$

The off-diagonal terms γ_{ij} can also be obtained from Eqs. 3.174 and 3.175, which gives

$$\gamma_{ij} = \frac{\omega_j^2 \, a_{ij}}{\omega_i^2 - \omega_j^2} \tag{3.177}$$

In principle, this equation could be used to evaluate the effect of a mass perturbation on the modes. However, inasmuch as the computation for all modes would involve considerable effort, it is not practical to use the formula for that purpose. However, this result is theoretically interesting because the γ_{ij} express the intensity with which mode i affects mode j as a result of the perturbation in mass properties. If $\omega_j \ll \omega_i$, the impact is virtually nil, because the second term in the denominator is negligible, so $\gamma_{ij} \approx a_{ij} \, \omega_j^2 / \omega_i^2 \approx 0$. This means that the high modes have almost no effect on the low modes. The reverse, however, is not true. When $\omega_j \gg \omega_i$, then $\gamma_{ij} \approx -a_{ij}$, which means that the low modes do significantly affect the high modes. This has consequences as far as the reliability with which high modes (or even transfer functions in the high-frequency range) can be computed. Imagine, for example, an insect landing on a light, homogeneous, simply supported beam. This insect will have virtually no effect on the low modes of the beam, but potentially a large one on the very high modes.

Perturbation of Stiffness Matrix

We turn next to perturbations in the stiffness matrix. Following an entirely analogous analysis as for the mass perturbation, we obtain

$$\Gamma^T + \Gamma = \mathbf{O} \tag{3.178}$$

$$\Gamma^T \Omega^2 + \Omega^2 \, \Gamma = \frac{\partial}{\partial \lambda} \Omega^2 - \Phi^T \, \Delta \mathbf{K} \, \Phi \tag{3.179}$$

with Γ referring now to perturbations of the eigenvectors with respect to λ. In scalar form

$$\gamma_{ij} + \gamma_{ji} = 0 \tag{3.180}$$

$$\omega_j^2 \gamma_{ji} + \omega_i^2 \gamma_{ij} = \delta_{ij} \frac{\partial}{\partial \lambda} \omega_i^2 - b_{ij} \tag{3.181}$$

with coefficients defined by the symmetric (and computable) matrix

$$\{b_{ij}\} = \Phi^T \, \Delta \mathbf{K} \, \Phi \tag{3.182}$$

These equations imply in turn $\gamma_{ii} = 0$ and

$$\boxed{\frac{\partial}{\partial \lambda} \omega_i^2 = b_{ii}} \quad \text{or} \quad \frac{\partial \omega_i^2}{\partial \lambda} = \omega_i^2 \frac{\phi_i^T \, \Delta \mathbf{K} \, \phi_i}{\phi_i^T \, \mathbf{K} \, \phi_i} \tag{3.183}$$

Also,

$$\gamma_{ij} = \frac{b_{ij}}{\omega_j^2 - \omega_i^2} \tag{3.184}$$

which again is theoretically interesting, but computationally not attractive.

Qualitative Implications of Perturbation Formulas

Since both $\Delta \mathbf{M}$ and $\Delta \mathbf{K}$ are positive semidefinite, it follows that $a_{ii} \geq 0$ and $b_{ii} \geq 0$. Hence, we arrive at the following two important lemmas.

Lemma 1

> *If a mass element is added somewhere to a structure, then some or all of its natural frequencies will be decreased* (i.e., no frequency can increase in value).

Lemma 2

> *If a stiffness element or constraint is added somewhere to a structure, then some or all of its natural frequencies will be increased* (i.e., no frequency can decrease in value).

Example 1: Addition of an Upper Floor to a Building

Consider an n-story building with rigid floors and flexible columns and walls to which we wish to add one more story. Clearly, the addition of that extra story can be imagined as being accomplished in two steps: first a floor is added that has only stiffness but no mass, and thereafter the mass is "switched on." Since the floors are rigid, the massless floor added at the top in the first step does not materially change the frequencies of the existing building, except that it adds one or more modes with an infinite frequency. Next, we switch on the mass of the added floor. By the first lemma, some or all frequencies of that system will then be lowered, including those that were infinitely large. Hence, all frequencies prior to the floor addition will either be lowered or stay the same, and none will be increased.

Example 2: Coupled Pendulums

Consider two pendulums hanging side by side, with equal masses but slightly different lengths, as shown in Figure 3.8. At first, these two pendulums are not coupled, but can oscillate independently of one another. We then couple the pendulums via a weak spring k attached at a distance a from the supports, and wish to estimate the effect of this stiffness perturbation on the frequencies and modal shapes. The free vibration equations for the fully coupled system can be shown to be given by

$$\begin{Bmatrix} ml_1^2 & \\ & ml_2^2 \end{Bmatrix} \begin{Bmatrix} \ddot{\theta}_1 \\ \ddot{\theta}_2 \end{Bmatrix} + \begin{Bmatrix} mgl_1 + ka^2 & -ka^2 \\ -ka^2 & mgl_2 + ka^2 \end{Bmatrix} \begin{Bmatrix} \theta_1 \\ \theta_2 \end{Bmatrix} = \begin{Bmatrix} 0 \\ 0 \end{Bmatrix} \tag{3.185}$$

in which l_1, l_2 are the two lengths of the pendulums. Since these are similar, we can write them in terms of an intermediate reference length l and a dimensionless, small perturbation parameter $\lambda \ll 1$ defined as follows:

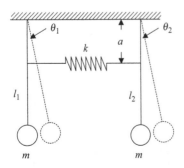

Figure 3.8. Weakly coupled pendulums.

$$\lambda = \frac{l_1 - l_2}{l_1 + l_2}, \qquad l = \frac{2 l_1 l_2}{l_1 + l_2}, \qquad \frac{l}{l_1} = 1 - \lambda, \qquad \frac{l}{l_2} = 1 + \lambda \tag{3.186}$$

Defining also $u_1 = l_1 \theta_1$ and $u_2 = l_2 \theta_2$, we can write the system equation in the fully symmetric form

$$\begin{Bmatrix} \ddot{u}_1 \\ \ddot{u}_2 \end{Bmatrix} + \frac{g}{l} \begin{Bmatrix} \frac{l}{l_1} + \frac{l_2}{l_1}\kappa & -\kappa \\ -\kappa & \frac{l}{l_1} + \frac{l_1}{l_2}\kappa \end{Bmatrix} \begin{Bmatrix} u_1 \\ u_2 \end{Bmatrix} = \begin{Bmatrix} 0 \\ 0 \end{Bmatrix} \tag{3.187}$$

in which κ is the dimensionless stiffness of the spring,

$$\kappa = \frac{k\, a^2 l}{mg\, l_1 l_2} = \frac{\left(\frac{k}{m}\right) a^2}{\left(\frac{g}{l}\right) l_1 l_2} \ll 1 \tag{3.188}$$

On the other hand,

$$\frac{l_2}{l_1} = \frac{1-\lambda}{1+\lambda} \approx 1 - 2\lambda \qquad \frac{l_1}{l_2} = \frac{1+\lambda}{1-\lambda} \approx 1 + 2\lambda \tag{3.189}$$

$$\frac{l_2}{l_1}\kappa \approx (1-2\lambda)\kappa \approx \kappa \qquad \frac{l_1}{l_2}\kappa \approx (1+2\lambda)\kappa \approx \kappa \tag{3.190}$$

Hence, we can approximate the system equations and express them in the dimensionless form:

$$\begin{Bmatrix} \ddot{u}_1 \\ \ddot{u}_2 \end{Bmatrix} + \frac{g}{l} \begin{Bmatrix} 1 - \lambda + \kappa & -\kappa \\ -\kappa & 1 + \lambda + \kappa \end{Bmatrix} \begin{Bmatrix} u_1 \\ u_2 \end{Bmatrix} = \begin{Bmatrix} 0 \\ 0 \end{Bmatrix} \tag{3.191}$$

If $\omega_0^2 = g/l$ is the frequency of a pendulum whose length equals the reference length, then the perturbed stiffness matrix can be written as

$$\mathbf{K} = \mathbf{K}_0 + \kappa \Delta \mathbf{K} \tag{3.192}$$

in which the unperturbed stiffness matrix together with the perturbation matrix are

$$\mathbf{K}_0 = \omega_0^2 \begin{Bmatrix} 1 - \lambda & 0 \\ 0 & 1 + \lambda \end{Bmatrix} \quad \text{and} \quad \Delta \mathbf{K} = \omega_0^2 \begin{Bmatrix} 1 & -1 \\ -1 & 1 \end{Bmatrix} \tag{3.193}$$

The eigenvalue problem for the unperturbed (i.e., uncoupled) system ($\kappa = 0$) possess the frequencies and modal shapes

$$\omega_1 = \omega_0 \sqrt{1-\lambda}, \qquad \omega_2 = \omega_0 \sqrt{1+\lambda}, \qquad \Phi = \begin{Bmatrix} 1 & 0 \\ 0 & 1 \end{Bmatrix} \tag{3.194}$$

From here, we obtain

$$\mathbf{B} = \{b_{ij}\} = \Phi^T \Delta \mathbf{K}\, \Phi \equiv \Delta \mathbf{K}, \qquad b_{11} = \frac{\partial \omega_1^2}{\partial \kappa} = \omega_0^2, \qquad b_{22} = \frac{\partial \omega_2^2}{\partial \kappa} = \omega_0^2 \tag{3.195}$$

Hence, we find the perturbed eigenvalues to be

$$\Omega_1^2 = \omega_1^2 + \kappa\omega_0^2 = \omega_0^2(1+\kappa-\lambda), \qquad \Omega_2^2 = \omega_2^2 + \kappa\omega_0^2 = \omega_0^2(1+\kappa+\lambda) \tag{3.196}$$

By comparison, the actual eigenvalues of the coupled system are

$$\Omega_1^2 = \omega_0^2\left(1+\kappa-\lambda\sqrt{1+(\kappa/\lambda)^2}\right), \qquad \Omega_2^2 = \omega_0^2\left(1+\kappa+\lambda\sqrt{1+(\kappa/\lambda)^2}\right) \tag{3.197}$$

which agree with the above if $\kappa \ll \lambda$. On the other hand, the two nonzero modal perturbation coefficients are found to be $\gamma_{12} = -1/2\lambda$ and $\gamma_{21} = 1/2\lambda$, so the modal perturbation matrix and the perturbed eigenvectors are

$$\tilde{\mathbf{A}} = \frac{1}{2\lambda}\begin{Bmatrix} 0 & -1 \\ 1 & 0 \end{Bmatrix}, \quad \boldsymbol{\Phi}_p = \boldsymbol{\Phi} + \kappa\boldsymbol{\Phi}\boldsymbol{\Gamma} = \begin{Bmatrix} 1 & 0 \\ 0 & 1 \end{Bmatrix} + \frac{\kappa}{2\lambda}\begin{Bmatrix} 0 & -1 \\ 1 & 0 \end{Bmatrix} = \frac{1}{2\lambda}\begin{Bmatrix} 2\lambda & -\kappa \\ \kappa & 2\lambda \end{Bmatrix} \tag{3.198}$$

The term $1/2\lambda$ in front of the matrix of eigenvectors is a nonsignificant scaling factor, and can safely be ignored. By comparison, the true eigenvectors are

$$\boldsymbol{\Phi}_p = \begin{Bmatrix} \lambda\left(1+\sqrt{1+(\kappa/\lambda)^2}\right) & -\kappa \\ \kappa & \lambda\left(1+\sqrt{1+(\kappa/\lambda)^2}\right) \end{Bmatrix} \tag{3.199}$$

As in the case of the eigenvalues, approximate and exact eigenvectors agree again when $\kappa \ll \lambda$. However, since the pendulums have nearly equal length, then λ will be a small number, so even a weak spring may lead to values of κ that fail the smallness test. Hence, this is a case where the perturbation method may fail to give accurate predictions. The reason for this situation is that the initially uncoupled pendulums have close eigenvalues, and after addition of the spring are only weakly coupled. Hence, small perturbations can lead to large differences in the modes.

3.4 Spacing Properties of Natural Frequencies

We present in this section some fundamental theorems concerning the range of values that the natural frequencies of a dynamic system can attain. These theorems not only are of great theoretical and practical interest, but they also play an important role in the numerical estimation of the eigenvalues, and how they are changed when external constraints are added to (or removed from) the system.

3.4.1 The *Minimax* Property of Rayleigh's Quotient

Courant's *minimax* characterization of the eigenvalues of a dynamic system is a fundamental theorem that addresses how Rayleigh's quotient changes in the presence of kinematic constraints.

Let $\mathbf{K}, \mathbf{M}$ be the real, symmetric stiffness and mass matrices of a dynamic system with n degrees of freedom, of which $\mathbf{K}$ may be positive semidefinite, and $\mathbf{M}$ is positive definite. These matrices satisfy the eigenvalue problem $\mathbf{K}\boldsymbol{\Phi} = \mathbf{M}\boldsymbol{\Phi}\boldsymbol{\Lambda}$, in which $\boldsymbol{\Phi}$ is the modal matrix and $\boldsymbol{\Lambda} = \text{diag}\{\lambda_j\}$ is the diagonal matrix of eigenvalues (where for notational brevity we have denoted the eigenvalues simply as $\omega_j^2 \equiv \lambda_j$). Without loss of generality, we may also assume the modes to be normalized with respect to the mass matrix, that is, $\mathbf{\Phi}^T \mathbf{M} \mathbf{\Phi} = \mathbf{I}$.

As we have seen previously, the enclosure theorem states that in a system with a finite number of degrees of freedom, the Rayleigh quotient is bounded by the system's smallest and largest eigenvalues, that is,

$$\lambda_1 \leq R(\mathbf{v}) = \frac{\mathbf{v}^T \mathbf{K} \mathbf{v}}{\mathbf{v}^T \mathbf{M} \mathbf{v}} = \frac{c_1^2 \lambda_1 + c_2^2 \lambda_2 + \cdots + c_n^2 \lambda_n}{c_1^2 + c_2^2 + \cdots + c_n^2} \leq \lambda_n \tag{3.200}$$

in which

$$\mathbf{v} = \sum_{j=1}^{n} \phi_j c_j = \mathbf{\Phi} \mathbf{c} \tag{3.201}$$

is an arbitrary vector, and $R(\mathbf{v})$ can be understood as a function of $\mathbf{v}$. Notice that only the direction and not the magnitude of $\mathbf{v}$ has an effect on Rayleigh's quotient. Geometrically, R can be visualized as the square of the distance to an ellipsoid in the n-dimensional space whose smallest and largest half-axes are $\sqrt{\lambda_1} = \omega_1$ and $\sqrt{\lambda_n} = \omega_n$ respectively. Indeed, if x_j are the coordinates of a point on the ellipsoid, then its parametric representation is

$$x_j = \omega_j \cos \theta_j \qquad \text{with} \qquad \cos \theta_j = \frac{c_j}{\sqrt{c_1^2 + c_2^2 + \cdots c_n^2}} \tag{3.202}$$

which implies

$$\left(\frac{x_1}{\omega_1} \right)^2 + \left(\frac{x_2}{\omega_2} \right)^2 + \cdots \left(\frac{x_n}{\omega_n} \right)^2 = 1 \qquad \text{and} \qquad R = r^2 = \sum_{j=1}^{n} \omega_j^2 \cos^2 \theta_j \tag{3.203}$$

We next modify this problem and ask, What is the range $[a, b]$ of values that the Rayleigh quotient can take if we add the constraint that $\mathbf{v}$ be mass-perpendicular to some arbitrary vector $\mathbf{w}_1$? We can write this problem symbolically as $a \leq R(\mathbf{w}_1, \mathbf{v}) \leq b$, by which we mean

$$R(\mathbf{v}) = \frac{\mathbf{v}^T \mathbf{K} \mathbf{v}}{\mathbf{v}^T \mathbf{M} \mathbf{v}} \qquad \text{subjected to } \mathbf{w}_1^T \mathbf{M} \mathbf{v} = 0 \tag{3.204}$$

Clearly, this range will depend on the choice of $\mathbf{w}_1$, but in any case it will generally be smaller than the range of the unconstrained quotient. To determine this range, we begin by expressing $\mathbf{w}_1$ in terms of modal coordinates as $\mathbf{w}_1 = \mathbf{\Phi} \mathbf{d}_1$ so that $\mathbf{d}_1 = \mathbf{\Phi}^T \mathbf{M} \mathbf{w}_1$. The constraint equation is then

$$\mathbf{w}_1^T \mathbf{M} \mathbf{v} = \mathbf{d}^T \mathbf{\Phi}^T \mathbf{M} \mathbf{\Phi} \mathbf{c} = \mathbf{d}^T \mathbf{c} = \sum_{j=1}^{n} d_j c_j = 0 \tag{3.205}$$

The above condition states that we can choose $n-1$ coefficients c_j arbitrarily and obtain the remaining coefficient from the constraint equation. By a straightforward modification with the definition $x_j = \omega_j c_j$, this equation can also be written as

$$\sum_{j=1}^{n} \left(\frac{d_j}{\omega_j} \right) x_j = 0 \tag{3.206}$$

which is the equation of a plane in the n-space that passes through the origin. Its intersection with the ellipsoid produces another ellipsoid of dimension $n-1$ whose points define the constrained Rayleigh quotient. For example, in the 3-D space ($n = 3$), the intersection with a plane gives rise to an ellipse, while in the 2-D case, the plane degenerates into a straight line that intersects the ellipse at two diametrically opposite points.

Points on the reduced ellipsoid of dimension $n-1$ will generally have components x_j along all coordinate directions (i.e., all $d_j \neq 0$). In particular, there will always be directions $\mathbf{w}_1$ for which at least $d_1 \neq 0, d_2 \neq 0$. Selecting also $c_3 = \cdots c_n = 0$ (a choice that lowers the value of the quotient), the constraint equation is $c_1 d_1 + c_2 d_2 = 0$, which implies

$$\lambda_1 \leq a = \frac{c_1^2 \lambda_1 + c_2^2 \lambda_2}{c_1^2 + c_2^2} = \frac{d_2^2 \lambda_1 + d_1^2 \lambda_2}{d_1^2 + d_2^2} \leq \lambda_2 \tag{3.207}$$

Similarly, there will be directions $\mathbf{w}_1$ for which at least $d_{n-1} \neq 0, d_n \neq 0$. When combined with coefficients $c_1 = \cdots c_{n-2} = 0$ and a constraint equation $c_{n-1} d_{n-1} + c_n d_n = 0$, we obtain

$$\lambda_{n-1} \leq b = \frac{c_{n-1}^2 \lambda_{n-1} + c_n^2 \lambda_n}{c_{n-1}^2 + c_n^2} = \frac{d_n^2 \lambda_{n-1} + d_{n-1}^2 \lambda_n}{d_{n-1}^2 + d_n^2} \leq \lambda_n \tag{3.208}$$

This means that the extremal values of the constrained Rayleigh quotient, that is, smallest and largest values for $R(\mathbf{w}_1, \mathbf{v})$ obtained by varying both $\mathbf{v}$ and $\mathbf{w}_1$, will satisfy the inequalities $\lambda_1 \leq a \leq \lambda_2$ and $\lambda_{n-1} \leq b \leq \lambda_n$, that is,

$$\lambda_1 \leq \min_{\mathbf{v}} R(\mathbf{w}_1, \mathbf{v}) \leq \lambda_2 \qquad \lambda_{n-1} \leq \max_{\mathbf{v}} R(\mathbf{w}_1, \mathbf{v}) \leq \lambda_n \tag{3.209}$$

in which the min and max operations are attained by varying $\mathbf{v}$. It remains to establish if the previous result is generally true. For this purpose, consider first the special case in which $\mathbf{w}_1$ coincides in direction with the fundamental eigenvector, that is, $\mathbf{w}_1 = \phi_1$, which implies $d_1 \neq 0$ and $d_j = 0$ for $j > 1$. In this case the constraint plane has no components along x_1 and the reduced ellipsoid is fully contained in (and spans) the $n-1$ space defined by the remaining eigenvectors. The constraint equation demands $c_1 = 0$, but all other c_j can be chosen arbitrarily, implying that the minimum and maximum values for Rayleigh's quotient are $a = \lambda_2, b = \lambda_n$. Similarly, if $\mathbf{w}_1$ coincides in direction with the last eigenvector, that is, $\mathbf{w}_1 = \phi_n$, then $a = \lambda_1, b = \lambda_{n-1}$. Finally, if $\mathbf{w}_1$ coincides with any eigenvector ϕ_i other than the first and last, then $c_i = 0$, and all other coefficients can be chosen arbitrarily. If so, the smallest and largest values of the Rayleigh quotient are again those of the unconstrained system, that is, $a = \lambda_1$ and $b = \lambda_n$. We conclude that the largest value that a can ever attain is $a = \lambda_2$ (which happens when $\mathbf{w}_1 = \phi_1$), and the smallest value that b can attain is $b = \lambda_{n-1}$ (which happens when $\mathbf{w}_1 = \phi_n$). This leads us to the complementary *maximin* and *minimax* theorems

$$\lambda_2 = \min_{\mathbf{v}} R(\phi_1, \mathbf{v}) = \max_{\mathbf{w}_1} \min_{\mathbf{v}} R(\mathbf{w}_1, \mathbf{v}) \tag{3.210}$$

and

$$\lambda_{n-1} = \max_{\mathbf{v}} R(\phi_n, \mathbf{v}) = \min_{\mathbf{w}_1} \max_{\mathbf{v}} R(\mathbf{w}_1, \mathbf{v}) \tag{3.211}$$

More generally, if we introduce additional constraints of the form $\mathbf{w}_i^T \mathbf{M} \mathbf{v} = 0$ using linearly independent vectors $\mathbf{w}_i$, we find that

$$\boxed{\min_{\mathbf{v}} R(\mathbf{w}_1, \cdots \mathbf{w}_{j-1}, \mathbf{v}) \leq \lambda_j} \tag{3.212}$$

and

$$\max_{\mathbf{v}} R(\mathbf{w}_1, \cdots \mathbf{w}_j, \mathbf{v}) \geq \lambda_{n-j} \tag{3.213}$$

which leads us to the extended *maximin* and *minimax* theorems

$$\boxed{\lambda_j = \min_{\mathbf{v}} R(\boldsymbol{\phi}_1, \cdots \boldsymbol{\phi}_{j-1}, \mathbf{v}) = \max_{\mathbf{w}_1, \cdots \mathbf{w}_{j-1}} \min_{\mathbf{v}} R(\mathbf{w}_1, \cdots \mathbf{w}_{j-1}, \mathbf{v})} \tag{3.214}$$

and

$$\lambda_{n-j} = \max_{\mathbf{v}} R(\boldsymbol{\phi}_{n-j+1}, \cdots \boldsymbol{\phi}_n, \mathbf{v}) = \min_{\mathbf{w}_1, \cdots \mathbf{w}_j} \max_{\mathbf{v}} R(\mathbf{w}_1, \cdots \mathbf{w}_j, \mathbf{v}) \tag{3.215}$$

In words, the maximin theorem states that the largest value that the lower bound to Rayleigh's quotient, $a = \min R(\mathbf{v})$, can ever attain in the presence of $j-1$ linearly independent constraints of the form $\mathbf{w}_i^T \mathbf{M} \mathbf{v} = 0$ is the jth eigenvalue. This value is attained when the constraint vectors $\mathbf{w}_i$ equal the first $j-1$ eigenvectors. The minimax theorem is basically the same theorem in reverse order.

3.4.2 Interlacing of Eigenvalues for Systems with Single External Constraint

Armed as we are now with the maximin theorem, we proceed to ask a question with far-reaching implications. How do the eigenvalues of an arbitrary system with stiffness and mass matrices $\mathbf{K}, \mathbf{M}$ relate to those of the same system when any one of its DOF is removed by means of an external constraint? In other words, what are the eigenvalues of the new system that is obtained by deleting an arbitrary row and corresponding column in the matrices? Clearly, we can always shuffle rows and columns in such way that the DOF being removed is the last one, an action that does not change the eigenvalues. Thus, we can assume, without loss of generality, that the deleted row and column is indeed the last one.

Denote the eigenvalues for the constrained system by λ'_j. The Rayleigh quotient obtained with the reduced matrices can be related to that of the original system by introducing the constraint that $\mathbf{v}$ be orthogonal to $\mathbf{e}_n$, the unit vector for the nth dimension (i.e., the DOF removed). This is equivalent to requiring that $\mathbf{v}$ be mass-orthogonal with respect to $\bar{\mathbf{e}}_n = \mathbf{M}^{-1} \mathbf{e}_n$. From the maximin theorem, we have then

$$\lambda'_j = \max_{\mathbf{w}_1, \cdots \mathbf{w}_{j-1}} \min_{\mathbf{v}} R(\mathbf{w}_1, \ldots, \mathbf{w}_{j-1}, \bar{\mathbf{e}}_n, \mathbf{v}) \leq \max_{\mathbf{w}_1, \cdots \mathbf{w}_j} \min_{\mathbf{v}} R(\mathbf{w}_1, \ldots, \mathbf{w}_j, \mathbf{v}) = \lambda_{j+1} \tag{3.216}$$

The inequality sign stems from the fact that $\bar{\mathbf{e}}_n = \mathbf{w}_j$ on the left is not optimized to achieve the maximum possible value for the min R, an extreme that defines the eigenvalue of the original system. On the other hand, we also have

$$\lambda_j = \max_{\mathbf{w}_1, \cdots \mathbf{w}_{j-1}} \min_{\mathbf{v}} R(\mathbf{w}_1, \ldots, \mathbf{w}_{j-1}, \mathbf{v}) \leq \max_{\mathbf{w}_1, \cdots \mathbf{w}_{j-1}} \min_{\mathbf{v}} R(\mathbf{w}_1, \ldots, \mathbf{w}_{j-1}, \bar{\mathbf{e}}_n, \mathbf{v}) = \lambda'_j \tag{3.217}$$

$$\lambda_1^{(n)}$$

$$\cdots$$

$$\lambda_1'' \quad \cdots \quad \lambda_{n-2}''$$

$$\lambda_1' \quad \lambda_2' \quad \cdots \quad \lambda_{n-2}' \quad \lambda_{n-1}'$$

$$\lambda_1 \quad \lambda_2 \quad \cdots \quad \lambda_{n-1} \quad \lambda_n$$

Figure 3.9. Interlacing of eigenvalues with the addition of constraints.

Here, the inequality sign stems from the fact that the expression on the right has one more constraint, which generally raises min R. We conclude that

$$\boxed{\lambda_j \le \lambda_j' \le \lambda_{j+1}} \tag{3.218}$$

which is known as *Rayleigh's*[4] *eigenvalue separation property*, or *eigenvalue interlacing theorem*.

If the original system has no coincident frequencies, and none of its natural modes has a node (i.e., stationary point) coinciding with the constraint, then

$$\lambda_j < \lambda_j' < \lambda_{j+1} \tag{3.219}$$

Also, if λ_j is a root of multiplicity m, then the constrained structure must have either $m-1$, m or $m+1$ eigenvalues λ_j' coinciding with this root. The converse is also true, that is, if λ_j' has multiplicity m, then λ_j must have multiplicity $m-1, m$ or $m+1$ at this same value.

Clearly, if this result holds true for the eigenvalues of the matrices of size n and $n-1$, it must also hold for the eigenvalues of the matrices of size $n-1, n-2$ (i.e., if two columns and rows of the stiffness and mass matrices are deleted). If the eigenvalues of the latter are denoted as λ_j'', it follows that

$$\lambda_j' \le \lambda_j'' \le \lambda_{j+1}' \tag{3.220}$$

and so on for smaller matrices. The eigenvalues of successively constrained systems then follow the pyramidal pattern shown in Figure 3.9.

Notice, however, that the pyramid need not be symmetric, but may be skewed to either the right or the left. Thus, one cannot decide if, say, λ_1'' is greater or smaller than λ_2, and so forth. Nonetheless, it can be guaranteed that the smallest and largest eigenvalues of the constrained systems can't be smaller or larger than those of the full system, that is, $\lambda_1 \le \lambda_1' \le \cdots \le \lambda_1^{(n)}$, and $\lambda_1^{(n)} \le \cdots \le \lambda_{n-1}' \le \lambda_n$.

Single Elastic External Support

The interlacing property previously described also applies to systems to which a single external *elastic* constraint is added. This arises when a spring element of stiffness k is added to any arbitrary diagonal element of the stiffness matrix **K**, or equivalently, when an elastic support is added to some DOF. The proof is simple: as we already know, adding a stiffness element generally raises some or all frequencies of the system, and none is decreased. The increase is largest when the added stiffness is infinitely large, that is, $k = \infty$,

---

[4] John William Strutt (Lord Rayleigh), *The Theory of Sound*, Vol. I, Section 92a (New York: Dover Publications, 1894), 119.

which is exactly the case of an external constraint just considered. Hence, the interlacing property applies. In addition, the nth frequency of the elastically restrained system is interlaced between the infinitely large nth frequency of the fully restrained system and the finite nth frequency of the unrestrained system.

3.4.3 Interlacing of Eigenvalues for Systems with Single Internal Constraint

The interlacing property previously demonstrated for systems with a single *external* constraint also applies to systems with a single *internal* constraint, that is, for systems in which any 2 DOF are coupled internally via a single kinematic constraint. The typical case is when the ith and jth DOF are forced to move in synchrony by means of an infinitely rigid element connecting the 2 DOF. If $\mathbf{v}$ is an arbitrary vector in the n-dimensional space, then its ith and jth components are simply $v_i = \mathbf{v}^T \mathbf{e}_i$ and $v_j = \mathbf{v}^T \mathbf{e}_j$, with $\mathbf{e}_i, \mathbf{e}_j$ being unit vectors in the two chosen directions. Hence, the kinematic requirement that $v_i = v_j$ can be expressed as $\mathbf{v}^T \left(\mathbf{e}_i - \mathbf{e}_j \right) = 0$, which is of the form $\mathbf{v}^T \mathbf{M} \mathbf{w}_1 = 0$, with $\mathbf{w}_1 = \mathbf{M}^{-1} (\mathbf{e}_i - \mathbf{e}_j)$. Hence, this is again a system with a single constraint $\mathbf{w}_1$, for which the interlacing property of the previous section applies.

Single Elastic Internal Constraint

Finally, the interlacing property also holds if the 2 DOF are internally coupled via an elastic spring, that is, if the stiffness matrix is modified by an increment of the form

$$\Delta \mathbf{K} = \begin{Bmatrix} 0 & & & \\ & k & -k & \\ & & 0 & \\ & -k & k & \\ & & & 0 \end{Bmatrix} \tag{3.221}$$

in which the k elements appear in the ith and jth columns and rows, respectively. The proof relies again on the fact that the elastic connection represents an intermediate stage between no connection whatsoever and the infinitely rigid connection just described. In addition, the nth frequency of the elastically restrained system either equals or lies above the nth frequency of the unrestrained system.

3.4.4 Number of Eigenvalues in Some Frequency Interval

We consider next the problem of determining the number of eigenvalues for the matrix pair $\mathbf{K}, \mathbf{M}$ that lie in some preestablished frequency interval, that is, the number of roots that the eigenvalue problem has in that interval. This problem has significant interest in the numerical determination of the natural frequencies of both discrete and continuous systems.

Sturm Sequence Property

Denote by $\mathbf{K}_k, \mathbf{M}_k$ the leading minors of $\mathbf{K}, \mathbf{M}$ of order k, that is, the submatrices obtained by ignoring the last $j = n - k$ rows and columns. The eigenvalues for each of the successively constrained systems can then be thought as the roots of the sequence of characteristic polynomials

$$p_n(\lambda) = \det(\mathbf{K}_n - \lambda \mathbf{M}_n) = M_n \prod_{i=1}^{n} (\lambda_i - \lambda) \tag{3.222}$$

$$p_{n-1}(\lambda') = \det(\mathbf{K}_{n-1} - \lambda' \mathbf{M}_{n-1}) = M_{n-1} \prod_{i=1}^{n-1} (\lambda_i' - \lambda) \tag{3.223}$$

$$p_{n-2}(\lambda'') = \det(\mathbf{K}_{n-2} - \lambda'' \mathbf{M}_{n-2}) = M_{n-2} \prod_{i=1}^{n-2} (\lambda_i'' - \lambda) \tag{3.224}$$

and so forth, where the number j of primes identifies the eigenvalues for the systems with j constrains. Also, $M_k = \det \mathbf{M}_k > 0$ is the determinant of the leading minor of $\mathbf{M}$ of order $k = n - j$. Since $\mathbf{M}$ is positive definite, all its minor determinants on the diagonal, including the leading ones, must be positive. The *separation property* then states that the roots $\lambda_i^{(j)}$ of p_k interlace the roots $\lambda_i^{(j+1)}$ of p_{k-1}, for $k = n, n-1, \dots 2$. A set of polynomials that exhibit this interlacing property are said to form a *Sturm sequence*. We shall use this property in the next section.

The Sign Count of the Shifted Stiffness Matrix

Consider now the shifted stiffness matrix $\mathbf{A}(\lambda) = \mathbf{K} - \lambda \mathbf{M}$, in which λ is an arbitrary positive parameter (but typically chosen to lie in the range of eigenvalues of $\mathbf{K}, \mathbf{M}$, a goal that can be accomplished by means of the Rayleigh quotient). Carrying out a Cholesky decomposition (i.e., Gaussian reduction) on this matrix, we can express $\mathbf{A}$ as

$$\mathbf{A}(\lambda) = \mathbf{L} \mathbf{D} \mathbf{L}^T \tag{3.225}$$

in which $\mathbf{L}$ is a unit, lower triangular matrix (i.e., a matrix whose diagonal elements are all equal to 1, while all elements above the diagonal are zero), and $\mathbf{D}$ is a diagonal matrix. Thus, the determinant of $\mathbf{A}$ is

$$p_n(\lambda) = \det \mathbf{A}(\lambda) = \det \mathbf{L} \det \mathbf{D} \det \mathbf{L}^T = \det \mathbf{D} = d_{11} \times d_{22} \times \cdots d_{nn} \tag{3.226}$$

that is, the value of the characteristic polynomial p_n equals the product of the diagonal elements of the matrix obtained after a Gaussian reduction of $\mathbf{A}$. On the other hand, if $\mathbf{A}_k(\lambda)$ is the kth leading minor of $\mathbf{A}$, which is obtained by considering only the first k rows and columns of $\mathbf{A}$, it follows that

$$\mathbf{A}_k(\lambda) = \mathbf{L}_k \mathbf{D}_k \mathbf{L}_k^T \tag{3.227}$$

in which $\mathbf{L}_k$ and $\mathbf{D}_k$ are the kth leading minors of $\mathbf{L}$ and $\mathbf{D}$. Hence,

$$p_k(\lambda) = \det \mathbf{A}_k(\lambda) = \det \mathbf{L}_k \det \mathbf{D}_k \det \mathbf{L}_k^T = \det \mathbf{D}_k = d_{11} \times d_{22} \times \cdots d_{kk} \tag{3.228}$$

which implies the sequence

$$[d_{11}, d_{22}, \dots, d_{nn}] = \left[\frac{p_1}{p_0}, \frac{p_2}{p_1}, \dots \frac{p_n}{p_{n-1}} \right] \tag{3.229}$$

with $p_0 = +1$. In the light of this result, it follows that each diagonal element d_{kk} satisfies

$$d_{kk} = \frac{p_k(\lambda)}{p_{k-1}(\lambda)} = \frac{M_k \prod_{i=1}^{k}(\lambda_i^{n-k} - \lambda)}{M_{k-1} \prod_{i=1}^{k-1}(\lambda_i^{n-k-1} - \lambda)} \tag{3.230}$$

Let's assume that the parameter λ is such that there are s roots λ_i of p_n that are lower in value than λ, and that this parameter does not coincide with any root. It follows that the expansion of p_n in terms of products of the roots will contain s terms that are negative, and $n - s$ that are positive. On the other hand, because of the interlacing theorem, we are guaranteed that there will be at least $s - 1$ roots of p_{n-1} that will also lie to the left of λ, and certainly no more than s such roots. In the former case, the negative factors in p_n, p_{n-1} will differ by one in number, so their ratio will be negative; contrariwise, it will be positive. A similar argument can be made for any other element d_{kk}. Hence, we conclude that if $d_{kk} < 0$, the count of roots in p_k, p_{k-1} that lie to the left of λ decreases by one in number either because the root of p_k closest to λ moved to the right in p_{k-1} (arrow in Figure 3.10), or because the number of roots on the left equals the order k of p_k (so there are no more roots on the right, and none can move there), in which case all elements d_{kk} from this point up (i.e., for descending k) will be negative. Either way, a negative element d_{kk} will signal each and every time a single root (and not more than one) is lost in the count on the left when going from p_k to p_{k-1}. By the time the zero-order polynomial p_0 is reached, all roots will have been lost, as illustrated at the top of Figure 3.10. Notice that there are three negative signs in d_{kk}, a number that equals the count of roots to the left of λ in p_5.

In the light of the previous reasoning, we conclude that the count of all negative elements d_{kk} equals the number s of roots λ_i of p_n that lie to the left of λ. This is the *sign count* of the shifted stiffness matrix $\mathbf{A}(\lambda) = \mathbf{K} - \lambda \mathbf{M}$, which is defined as

$$s(\mathbf{A}) = \text{number of } d_{kk} < 0 = \text{number of } \lambda_i < \lambda \tag{3.231}$$

This result holds with the proviso that no pivoting is used in the Gaussian reduction, because pivoting is an operation that introduces additional sign changes. A corollary of the above sign count is that if m constraints are *removed* from a dynamic system, then the number s of eigenvalues that lie below a chosen parameter λ increases by r, where $0 \leq r \leq m$.

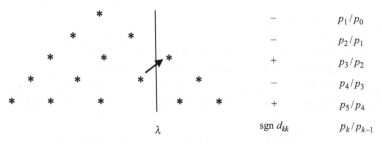

Figure 3.10. Sign count in the course of Gaussian elimination.

In general, the sign count can be used to determine how many eigenvalues lie in a given range $[a \le \lambda \le b]$, namely

$$N_\lambda = s[\mathbf{A}(b)] - s[\mathbf{A}(a)] \tag{3.232}$$

This quantity is extremely useful in the iterative solution of eigenvalue problems, such as inverse iteration with shift by Rayleigh quotient, or in determinant search techniques. It provides a reliable check on missed roots, proximal roots, or even on multiple roots.

Root Count for Dynamically Condensed Systems

We consider next the problem of assessing the number of roots lying in some frequency interval when the shifted stiffness matrix has been subjected to *dynamic condensation*. The motivation for this lies in the development of eigenvalue search techniques for systems composed of substructures with known eigenvalues (e.g., spatially periodic structures), and more importantly, in the application of such techniques to continuous systems. We shall basically follow here the method described by Wittrick and Williams.[5]

Consider once more the shifted stiffness matrix of the previous sections, and assume that it has been partitioned into two parts, the first containing the *external* degrees of freedom $\mathbf{u}_1$ being retained (say, at the interface of the substructures), and the second part, the *internal* degrees of freedom $\mathbf{u}_2$, that is, those that are being condensed out. Also, if $\mathbf{u}$ denotes the displacement vector for the complete system, we obtain a free vibration problem of the form

$$(\mathbf{K} - \lambda \mathbf{M})\mathbf{u} = \mathbf{A}(\lambda)\mathbf{u} = \begin{Bmatrix} \mathbf{A}_{11} & \mathbf{A}_{12} \\ \mathbf{A}_{21} & \mathbf{A}_{22} \end{Bmatrix} \begin{Bmatrix} \mathbf{u}_1 \\ \mathbf{u}_2 \end{Bmatrix} = \begin{Bmatrix} \mathbf{0} \\ \mathbf{0} \end{Bmatrix}$$
$$= \begin{Bmatrix} \mathbf{I} & \mathbf{A}_{12} \\ \mathbf{0} & \mathbf{A}_{22} \end{Bmatrix} \begin{Bmatrix} \mathbf{A}_{11} - \mathbf{A}_{12}\mathbf{A}_{22}^{-1}\mathbf{A}_{21} & \mathbf{0} \\ \mathbf{A}_{22}^{-1}\mathbf{A}_{21} & \mathbf{I} \end{Bmatrix} \begin{Bmatrix} \mathbf{u}_1 \\ \mathbf{u}_2 \end{Bmatrix} = \begin{Bmatrix} \mathbf{0} \\ \mathbf{0} \end{Bmatrix} \tag{3.233}$$

which can be satisfied for $\mathbf{u} \ne \mathbf{0}$ only if $\det \mathbf{A} = 0$. Alternatively, we conclude from the structure of the two matrices in the second row that

$$\det \mathbf{A} = \det \mathbf{A}_1 \det \mathbf{A}_{22}, \quad \text{in which} \quad \mathbf{A}_1(\lambda) = \mathbf{A}_{11} - \mathbf{A}_{12}\mathbf{A}_{22}^{-1}\mathbf{A}_{21} \tag{3.234}$$

After condensation of $\mathbf{u}_2$, the characteristic system equations are

$$\mathbf{A}_1(\lambda)\mathbf{u}_1 = \left(\mathbf{A}_{11} - \mathbf{A}_{12}\mathbf{A}_{22}^{-1}\mathbf{A}_{21}\right)\mathbf{u}_1 = \mathbf{0} \tag{3.235}$$

and

$$\mathbf{u}_2 = -\mathbf{A}_{22}^{-1}\mathbf{A}_{21}\mathbf{u}_1 \tag{3.236}$$

In principle, the free vibration condition $\mathbf{A}_1\mathbf{u}_1 = \mathbf{0}$ should lead to the same natural frequencies as the uncondensed system in both value and number, but it no longer constitutes a linear eigenvalue problem. Indeed, whereas in the original, uncondensed matrix, the elements of $\mathbf{A}$ are linear functions of the parameter λ, in the condensed matrix $\mathbf{A}_1$, the

---

[5] W. H. Wittrick and F. W. Williams, "A general algorithm for computing natural frequencies of elastic structures," *Q. J. Mech. Appl.Math.*, XXIV, Pt. 3, 1971, 263–284.

elements are ratios of polynomials in λ. In addition, any values of λ that make $\mathbf{A}_{22}$ singular will produce infinitely large elements in $\mathbf{A}_1$. This will take place at the natural frequencies of the internal system with external constraints $\mathbf{u}_1 = \mathbf{0}$, or exceptionally, at frequencies for which these external DOF happen to be stationary points of a mode that involves only the internal DOF, in which case λ is also an eigenvalue of the complete, unrestrained system, and $\mathbf{A}_{22}$ is singular. Thus, the free vibration condition is

$$\det \mathbf{A}_1 = 0 \quad \text{and exceptionally also} \quad \mathbf{u}_1 = \mathbf{0}, \quad \det \mathbf{A}_1 \neq 0, \quad \det \mathbf{A}_{22} = 0 \quad (3.237)$$

Let's proceed now to evaluate the root count for the condensed discrete system, which we choose to characterize solely by a finite number of external DOF $\mathbf{u}_1$. For this purpose, let's consider once more the characteristic equation in partitioned form:

$$\begin{bmatrix} \mathbf{A}_{11} & \mathbf{A}_{12} \\ \mathbf{A}_{21} & \mathbf{A}_{22} \end{bmatrix} \begin{Bmatrix} \mathbf{u}_1 \\ \mathbf{u}_2 \end{Bmatrix} = \begin{Bmatrix} \mathbf{0} \\ \mathbf{0} \end{Bmatrix} \quad (3.238)$$

and assume that the number of internal DOF (i.e., size of $\mathbf{A}_{22}$) is large but finite. We proceed to carry out a Gaussian reduction of the above matrix, starting from the lower right-hand corner of $\mathbf{A}_{22}$ and moving up from this point, and we stop this process at the row immediately below the submatrix $\mathbf{A}_{11}$. As a result of this manipulation, $\mathbf{A}_{11}$ changes into the condensed matrix $\mathbf{A}_1$; $\mathbf{A}_{12}$ changes into a null matrix; and $\mathbf{A}_{22}$ changes into a lower triangular matrix, the signs of whose diagonal elements d_{kk} provide us with the first partial root count for the complete system. Thereafter, we obtain the second part of the root count by carrying out the Gaussian reduction all the way up to the first row of $\mathbf{A}_{11}$, but the count of negative elements there happens to equal the sign count of the condensed matrix $\mathbf{A}_1$. In addition, we can just as well interpret the sign count of $\mathbf{A}_{22}$ as that of an internal system so restrained externally that $\mathbf{u}_1 = \mathbf{0}$, which leads us to conclude that the sign count $N(\lambda)$ of the full system must be given by

$$\begin{aligned} N(\lambda) &= s\big[\mathbf{A}_1(\lambda)\big] + s\big[\mathbf{A}_{22}(\lambda)\big] \\ &= s\big[\mathbf{A}_1(\lambda)\big] + N_0(\lambda) \end{aligned} \quad (3.239)$$

in which $N_0(\lambda)$ is the number of eigenvalues to the left of λ (i.e., smaller than λ) when the external degrees of freedom are fully constrained and only the internal nodes can vibrate. Also, $s\big[\mathbf{A}_1(\lambda)\big]$ is the sign count of $\mathbf{A}_1$ as defined in the previous section. If $N_0(\lambda) > 0$, this means that $\mathbf{A}_1$ must have singularities to the left of λ at which the elements of $\mathbf{A}_1$ are infinite.

Example Sign count for system with 2 internal and 2 external DOF
Consider the 4-DOF system shown in Figure 3.11. We choose to define as "external" the DOF associated with the shaded masses 1 and 4, and as "internal" the remaining two masses at the center. Numbering the DOF according to their characteristic as external or internal, the stiffness and mass matrices in partitioned form are

$$\mathbf{K} = k \left\{ \begin{array}{cc|cc} 1 & 0 & -1 & 0 \\ 0 & 3 & 0 & -3 \\ \hline -1 & 0 & 3 & -2 \\ 0 & -3 & -2 & 5 \end{array} \right\}, \quad \mathbf{M} = m \left\{ \begin{array}{cc|cc} 3 & 0 & 0 & 0 \\ 0 & 1 & 0 & 0 \\ \hline 0 & 0 & 1 & 0 \\ 0 & 0 & 0 & 2 \end{array} \right\} \quad (3.240)$$

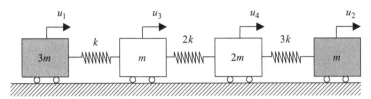

Figure 3.11. Demonstration of sign count theorem with simple 4-DOF system. The shaded masses are defined as external, the other two as internal.

Defining $\lambda = \omega^2 m / k$, then

$$\mathbf{K} - \omega^2 \mathbf{M} = k \left\{ \begin{array}{cc|cc} 1-3\lambda & 0 & -1 & 0 \\ 0 & 3-\lambda & 0 & -3 \\ \hline -1 & 0 & 3-\lambda & -2 \\ 0 & -3 & -2 & 5-2\lambda \end{array} \right\} = \left\{ \begin{array}{cc} \mathbf{A}_{11} & \mathbf{A}_{12} \\ \mathbf{A}_{21} & \mathbf{A}_{22} \end{array} \right\} \tag{3.241}$$

so

$$\begin{aligned}
\mathbf{A}_1(\lambda) &= \mathbf{A}_{11} - \mathbf{A}_{12} \mathbf{A}_{22}^{-1} \mathbf{A}_{21} \\
&= k \left[\left\{ \begin{array}{cc} 1-3\lambda & 0 \\ 0 & 3-\lambda \end{array} \right\} - \left\{ \begin{array}{cc} -1 & 0 \\ 0 & -3 \end{array} \right\} \left\{ \begin{array}{cc} 3-\lambda & -2 \\ -2 & 5-2\lambda \end{array} \right\}^{-1} \left\{ \begin{array}{cc} -1 & 0 \\ 0 & -3 \end{array} \right\} \right] \\
&= \frac{k}{2\lambda^2 - 11\lambda + 11} \left\{ \begin{array}{cc} 6 - 42\lambda + 35\lambda^2 - 6\lambda^3 & -6 \\ -6 & 6 - 35\lambda + 17\lambda^2 - 2\lambda^3 \end{array} \right\}
\end{aligned} \tag{3.242}$$

and

$$\det \mathbf{A}_1 = k^2 \frac{(6\lambda^3 - 53\lambda^2 + 120\lambda - 42)\lambda}{2\lambda^2 - 11\lambda + 8} \tag{3.243}$$

whose four roots are

$$\lambda_1 = 0., \quad \lambda_2 = 0.4264 \quad \lambda_3 = 3.0836 \quad \lambda_4 = 5.3233 \tag{3.244}$$

These agree with the roots of the original eigenvalue problem. Also, there are two singularities of $\mathbf{A}_1$ at the roots of the denominator $2\lambda^2 - 11\lambda + 11 = 0$ that correspond to the eigenvalues of the internal substructure with fixed external supports $u_1 = u_2 = 0$, namely 1.3139 and 4.1861. Also, if we consider the arbitrary value $\lambda = 1$, then $N_0(\lambda = 1) = 0$ (there are no roots of the internal structure less than or equal to $\lambda = 1$), and $s(\mathbf{A}_1(1)) = 2$, because the Choleski decomposition of $\mathbf{A}_1(1)$ is

$$\mathbf{A}_1(1) = \tfrac{1}{2} k \left\{ \begin{array}{cc} -7 & -6 \\ -6 & -14 \end{array} \right\} = \tfrac{1}{2} k \left\{ \begin{array}{cc} 1 & 0 \\ \frac{6}{7} & 1 \end{array} \right\} \left\{ \begin{array}{cc} -7 & 0 \\ 0 & -\frac{62}{7} \end{array} \right\} \left\{ \begin{array}{cc} 1 & \frac{6}{7} \\ 0 & 1 \end{array} \right\} \tag{3.245}$$

which has two negative diagonal elements. Thus, $N(\lambda = 1) = 2 + 0 = 2$, so there are two eigenvalues in the condensed system to the left of $\lambda = 1$, which we know are $\lambda_1 = 0$ and $\lambda_2 = 0.4264$.

Generalization to Continuous Systems

At this stage, we can generalize this result by proceeding to increase the number of *internal* nodes without limit, that is, by making the internal structure a *continuous* system characterized solely by a finite number of external degrees of freedom. For instance, in the ideal limit of infinitely many internal degrees of freedom, a finite element model of a beam could approach a continuous beam for which the external nodes are the displacements and rotations at its two ends. Imagining further that the continuum is approached by steadily increasing the number of internal DOF, we see that the elements of the dynamic stiffness matrix $\mathbf{A}_1(\lambda)$ become ratios of ever higher order polynomials in λ. In the limit of the continuum, these elements become transcendental functions of λ. Of course, such a dynamic stiffness (or spectral) matrix would not be obtained by dynamic condensation, but by some other direct means, the details of which need not concern us here (e.g., see Chapter 4, Section 4.3 and Chapter 5, Section 5.2). However, in this case $\mathbf{A}_1(\lambda)$ will generally have an infinite number of frequencies at which its elements become infinitely large. Conceptually, this will occur when the now infinitely large matrix $\mathbf{A}_{22}$ becomes "singular," that is, when λ is an eigenvalue of the system so constrained that $\mathbf{u}_1 = \mathbf{0}$. Exceptionally, the system may also possess eigenvalues for which $\det \mathbf{A}_1 \neq 0$ and $\mathbf{u}_1 = \mathbf{0}$. This again occurs when the external nodes happen to be stationary points in a normal mode, and λ is indeed an eigenvalue of the complete system. Now, the previous logic and formula concerning the root count of the now continuous system remain valid, except that because the size of $\mathbf{A}_{22}$ increased without limit, we must now have other means at our disposition for obtaining the root count $N_0(\lambda)$ for the continuous system with fixed supports $\mathbf{u}_1 = \mathbf{0}$. In other words, for the modal count equation to be useful for a continuous system, it is necessary that the constrained structure's modal count $N_0(\lambda)$ be independently available. In general, this will be possible if the solution for the constrained system is known ahead of time, which will be the case for a structure composed of ideal members. We illustrate this concept by means of three examples.

Example 1: Sign count in a rod with fixed and free boundary conditions
Consider a uniform, continuous rod of length L, Young's modulus E, mass density ρ, and cross section A. When this rod is simply supported (i.e., restrained) at both ends, it is known to have natural frequencies $\omega_j = j\pi C/L$, for $j = 1, 2, 3, \ldots$, in which $C = \sqrt{E/\rho}$ is the rod wave velocity. Expressed in dimensionless form, these resonant frequencies can be written as $\theta_j = \omega_j L/C = j\pi$.

Consider next a *free* rod. Its dynamic stiffness matrix can be shown to be given by (Chapter 4, Section 4.3)

$$\mathbf{K}(\omega) = k \frac{\theta}{\sin\theta} \begin{Bmatrix} \cos\theta & -1 \\ -1 & \cos\theta \end{Bmatrix} \qquad \theta = \frac{\omega L}{C} \qquad k = \frac{EA}{L} \qquad (3.246)$$

Notice that the elements of this matrix become infinitely large when $\sin\theta = 0$, that is, $\theta = j\pi$, which are the frequencies of the end-restrained rod. Also, the determinant of this matrix is

$$\det \mathbf{K} = (k\theta)^2 \frac{\cos^2\theta - 1}{\sin^2\theta} = -(k\theta)^2 \neq 0 \qquad (3.247)$$

which is nonzero for all positive frequencies! Still, when $\sin\theta = 0$, then $\cos\theta = \pm1$, and the two rows are identical, except for a sign change, so the matrix is singular. That the matrix can be singular and yet its determinant nonzero results from the fact that the elements themselves are infinitely large.

Let's now determine the sign count of **K**. For this purpose, we carry out a Gaussian reduction, the result of which is

$$\hat{\mathbf{K}}(\theta) = k\theta \begin{Bmatrix} \cot\theta & -\sec\theta \\ 0 & -\tan\theta \end{Bmatrix} \tag{3.248}$$

Thus, the sign count depends on the signs of the two diagonal terms above. Since both $\cot\theta$ and $\tan\theta$ are positive in the first and third quadrants while they are negative in the second and fourth, it follows that on account of the negative sign in front of $\tan\theta$, they will always have opposite signs. Hence, the sign count for any of the four quadrants (and integer multiples thereof) is always 1. Hence, the modal counts for the free rod is

$$N(\theta) = N_0(\theta) + 1 \tag{3.249}$$

which is the correct result, because the natural frequencies of the free rod are the same as those of the constrained rod, except that the former possesses also the zero-frequency rigid-body mode.

Example 2: Sign count in rod with one end fixed and the other free

Let's examine now the case of a rod that is free at one end and fixed at the other. Eliminating the fixed DOF by suppressing the second column and row in the above dynamic stiffness matrix, we obtain the dynamic stiffness for the remaining DOF as $K(\theta) = k\theta\cot\theta$. This expression is negative when θ lies in the second or fourth quadrant. On the other hand, the fully restrained system has normal modes at each transition from the second to the third quadrant, and from the fourth to the first. Thus, the total modal count follows the following pattern:

Quadrant	I	II	III	IV	I (etc.)
Angle θ	$0 \to \pi/2$	$\pi/2 \to \pi$	$\pi \to 3\pi/2$	$3\pi/2 \to 2\pi$	$2\pi \to 5\pi/2$
N_0	0	0	1	1	2
$s(K)$	0	1	0	1	0
N (total)	0	1	1	2	2

This simple system has, of course, an exact solution, which is $\theta_j = \omega_j L / C = (2j-1)/2$. Hence, it has natural frequencies at $\pi/2$, $3\pi/2$, $5\pi/2$, and so forth. The accumulated modal count in the table above is in agreement with this exact result.

Example 3: Sign count for two dissimilar rods that have been joined

Finally, consider two equally long and materially identical rods with distinct cross sections A_1, A_2 that are connected together in series. Thus, this system has 3 *external* DOF, 2 at the ends, and 1 at the junction of the rods. Its 3×3 tridiagonal dynamic stiffness (or impedance) matrix **K** is assembled from the dynamic stiffness matrices of each of the rods by overlapping appropriately the elements at the middle joint; see Chapter 4, Section 4.3. If $\alpha = A_2/A_1$, then the ratio of axial stiffnesses is $k_2/k_1 = \alpha$, and the dynamic stiffness matrix is

$$\mathbf{K}(\omega) = k_1 \frac{\theta}{\sin \theta} \begin{Bmatrix} \cos \theta & -1 & 0 \\ -1 & (1+\alpha)\cos \theta & -\alpha \\ 0 & -\alpha & \alpha \cos \theta \end{Bmatrix}, \qquad k_1 = \frac{EA_1}{L}, \qquad \theta = \frac{\omega L}{C} \tag{3.250}$$

whose diagonal elements in an LU (lower–upper triangular) decomposition can be shown to be

$$\mathbf{D} = \{d_{\ell\ell}\} = k_1 \frac{\theta}{\sin \theta} \operatorname{diag} \left\{ \cos \theta, \ \frac{(1+\alpha)\cos^2 \theta - 1}{\cos \theta}, \ \frac{\alpha \cos \theta (1+\alpha)(\cos^2 \theta - 1)}{(1+\alpha)\cos^2 \theta - 1} \right\} \tag{3.251}$$

whose determinant is

$$|\mathbf{K}| = |\mathbf{D}| = -\alpha(1+\alpha)k_1^3 \theta^3 \cot \theta \tag{3.252}$$

The determinant is zero when $\cot \theta = 0$, that is, $\theta_k = \frac{1}{2}\pi(2k-1)$, which lies at the midpoint between the resonant frequencies of the fully restrained rods. The modal count is then

$$\begin{aligned} N(\omega) &= N_{01}(\omega) + N_{02}(\omega) + s[\mathbf{K}(\omega)] \\ &= 2N_0(\omega) + s[\mathbf{D}(\omega)] \end{aligned} \tag{3.253}$$

in which N_{01} and N_{02} are the modal counts for each of the two fully restrained rods. In this particular case, $N_{01} = N_{02}$ because the natural frequency of a fully restrained rod, $\omega_j = j\pi C/L$, $j = 1,2,\ldots$ does not depend on the cross section. In particular, at a frequency slightly higher than the jth frequency of each of the two rods, $\operatorname{sgn}(\sin \theta_j^+) = (-1)^j$, $\cos \theta_j^+ \approx (-1)^j$, $\cos^2 \theta - 1 < 0$ $N_{01}(\theta_j^+) = N_{02}(\theta_j^+) = j$, then $d_{11} > 0, d_{22} > 0$ but $d_{33} < 0$, so

$$s[\mathbf{D}(\theta_j^+)] = 1, \qquad \text{and} \qquad N(\omega) = 2j+1 \tag{3.254}$$

that is, there is one additional frequency. On the other hand, at a frequency slightly higher than the midpoint between the jth and the $(j+1)$th frequency, then $\operatorname{sgn}(\sin \theta_{j+0.5}^+) = (-1)^j$ but $\operatorname{sgn}(\cos \theta_{j+0.5}^+) = (-1)^{j+1}$, and now $d_{11} < 0, d_{22} < 0$ but $d_{33} > 0$, so

$$s[\mathbf{D}(\theta_{j+0.5}^+)] = 2 \qquad \text{and} \qquad N(\omega) = 2(j+1)$$

In summary, the sign count for this problem is

$$\left. \begin{array}{ll} N(\omega) = 2j+1 & \omega_j < \omega < \omega_{j+0.5} \\ N(\omega) = 2j+2 & \omega_{j+0.5} < \omega < \omega_{j+1} \end{array} \right|, \qquad \omega_j = j\pi C/L \tag{3.255}$$

That is, the composite structure consisting of two rods has twice as many natural frequencies as the individual rods, and pairs of these interlace the frequencies of any one of the two rods in series.

3.5 Vibrations of Damped MDOF Systems

As we stated earlier, damped vibrations in MDOF systems are more complicated than those in SDOF systems, and this is true for a number of reasons. In general, damping forces in structural systems are the result of many complex energy-dissipating mechanisms related to internal friction and inelastic effects that are difficult to describe mathematically. It is then not surprising that in most applications, the damping is assumed to be of a viscous nature (i.e., proportional to the rate of deformation and/or velocities), since this type of damping leads to linear equations for which a vast arsenal of solution techniques is available.

Even the choice of viscous damping for a dynamic model, however, does not remove all of the difficulties in the analysis of MDOF systems. The main problem relates to the evaluation of the member viscosities, as most structural systems (such as beams, frames, etc.) do *not* have dashpots associated with, or built into, them. For this reason, in the majority of cases, it is easier to estimate experimentally the overall damping characteristics of a structure than to compute the individual viscous components. In certain types of analyses, this leads then to the question as to how to set up a viscous damping matrix that reasonably represents the energy-dissipating characteristics of the structure as a whole.

Alternatively, when the damping matrix *is* available (or can be constructed) and the solution is to be obtained by *modal superposition* (as will be considered next), an issue is whether or not the system has *normal modes* of vibration. Such modes will exist only if the linear transformation of the damping matrix with the *undamped* modes of the system leads to a diagonal matrix, that is, if

$$\mathbb{C} = \mathbf{\Phi}^T \mathbf{C} \mathbf{\Phi} \qquad = \text{a diagonal matrix?} \tag{3.256}$$

Most often, this transformation does *not* produce a diagonal matrix, in which case normal modes do not exist. If, however, this matrix *happens* to be diagonal, then the damping matrix is said to be *proportional*, and the system has *classical normal modes*. Later, we shall examine the conditions and develop a test for proportionality that does not require computing any of the modes. First, however, we consider the special case of *diagonalizable damping matrices* (i.e., proportional damping), and present later on strategies for dealing with nonproportional damping.

3.5.1 Vibrations of Proportionally Damped MDOF Systems

The equation of motion of a viscously damped system is

$$\mathbf{M}\ddot{\mathbf{u}} + \mathbf{C}\dot{\mathbf{u}} + \mathbf{K}\mathbf{u} = \mathbf{p}(t) \tag{3.257}$$

If the damping matrix $\mathbf{C}$ is proportional, we can solve this equation by *modal superposition*, that is, we can express the displacement vector in terms of the *undamped* natural modes of the system (which we assume to be known):

$$\mathbf{u}(t) = \sum_{j=1}^{n} \phi_j \, q_j(t) = \mathbf{\Phi}\mathbf{q}(t) \tag{3.258}$$

When we substitute this expression into the differential equation of motion and multiply the result by the transposed modal matrix, we obtain

$$(\mathbf{\Phi}^T \mathbf{M} \mathbf{\Phi})\ddot{\mathbf{q}} + (\mathbf{\Phi}^T \mathbf{C} \mathbf{\Phi})\dot{\mathbf{q}} + (\mathbf{\Phi}^T \mathbf{K} \mathbf{\Phi})\mathbf{q} = \mathbf{\Phi}^T \mathbf{p} \qquad (3.259)$$

But from the orthogonality condition *and* the assumption of proportional damping, we have

$$\mathbf{\Phi}^T \mathbf{M} \mathbf{\Phi} = \mathbb{M} = \text{diag}\{\mu_j\} = \text{diag}\{\phi_j^T \mathbf{M} \phi_j\} \qquad (3.260)$$

$$\mathbf{\Phi}^T \mathbf{C} \mathbf{\Phi} = \mathbb{C} = \text{diag}\{\eta_j\} = \text{diag}\{\phi_j^T \mathbf{C} \phi_j\} \qquad (3.261)$$

$$\mathbf{\Phi}^T \mathbf{K} \mathbf{\Phi} = \mathbb{K} = \text{diag}\{\kappa_j\} = \text{diag}\{\mu_j \, \omega_j^2\} = \text{diag}\{\phi_j^T \mathbf{K} \phi_j\} \qquad (3.262)$$

Also, we define the modal load vector

$$\mathbf{\Phi}^T \mathbf{p} = \mathbb{P} = \{\pi_j(t)\} = \left\{\sum_{i=1}^{n} \phi_{ij} \, p_i(t)\right\} \qquad (3.263)$$

Hence, the equation of motion transforms into

$$\mathbb{M}\ddot{\mathbf{q}} + \mathbb{C}\dot{\mathbf{q}} + \mathbb{K}\mathbf{q} = \mathbb{P}(t) \qquad (3.264)$$

Since the matrices $\mathbb{M}, \mathbb{C}, \mathbb{K}$ are *diagonal*, this is equivalent to a system of *uncoupled* differential equations

$$\mu_j \, \ddot{q}_j + \eta_j \dot{q}_j + \kappa_j \, q_j = \pi_j(t) \qquad j = 1,\dots,n \qquad (3.265)$$

Each of these equations is analogous to that for an SDOF system with mass, damping, and stiffness μ_j, η_j, κ_j. Its solution must then be of the same form as the response of a damped SDOF system subjected to both dynamic forces and prescribed initial conditions:

$$q_j(t) = e^{-\xi_j \omega_j t}\left[q_{0j} \cos \omega_{dj} t + \frac{\dot{q}_{0j} + \xi_j \, q_{0j}}{\omega_{dj}} \sin \omega_{dj} t\right] + h_j * \pi_j \qquad (3.266)$$

in which q_{0j}, $\dot{q}_{0j}$ are the (as yet unknown) initial values of q_j, and

$$\mu_j = \phi_j^T \mathbf{M} \phi_j \qquad = \qquad \text{modal mass} \qquad (3.267)$$

$$\eta_j = \phi_j^T \mathbf{C} \phi_j \qquad = \qquad \text{modal dashpot} \qquad (3.268)$$

$$\kappa_j = \phi_j^T \mathbf{K} \phi_j \qquad = \qquad \text{modal stiffness} \qquad (3.269)$$

$$\omega_j = \sqrt{\frac{\kappa_j}{\mu_j}} \qquad = \qquad \text{undamped modal frequency} \qquad (3.270)$$

$$\xi_j = \frac{\eta_j}{2\sqrt{\kappa_j \mu_j}} \qquad = \qquad \text{modal damping ratio} \qquad (3.271)$$

$$\omega_{dj} = \omega_j \sqrt{1 - \xi_j^2} \qquad = \qquad \text{damped modal frequency} \qquad (3.272)$$

$$h_j(t) = \frac{1}{\mu_j \omega_{dj}} e^{-\xi_j \omega_j t} \sin \omega_{dj} t \qquad = \qquad \text{modal impulse response function} \qquad (3.273)$$

The actual displacement is then

$$\mathbf{u}(t) = \sum_{j=1}^{n} \left(e^{-\xi_j \omega_j t} \left[q_{0j} \cos \omega_{dj} t + \frac{\dot{q}_{0j} + \xi_j \omega_j q_{0j}}{\omega_{dj}} \sin \omega_{dj} t \right] + h_j * \pi_j \right) \boldsymbol{\phi}_j \qquad (3.274)$$

Taking the derivative of this expression, we obtain the velocity vector

$$\dot{\mathbf{u}}(t) = \sum_{j=1}^{n} \left(e^{-\xi_j \omega_j t} \left[\dot{q}_{0j} \cos \omega_{dj} t + \frac{\omega_j}{\omega_{dj}} (\xi_j \dot{q}_{0j} + \omega_j q_{0j}) \sin \omega_{dj} t \right] + \dot{h}_j * \pi_j \right) \boldsymbol{\phi}_j \qquad (3.275)$$

In particular, at $t = 0$, the displacement and velocity vectors are

$$\mathbf{u}_0 = \sum_{j=1}^{n} q_{0j} \boldsymbol{\phi}_j \qquad \dot{\mathbf{u}}_0 = \sum_{j=1}^{n} \dot{q}_{0j} \boldsymbol{\phi}_j \qquad (3.276)$$

Finally, we apply the *expansion theorem*, which gives us the missing initial conditions q_{0j}, $\dot{q}_{0j}$

$$q_{0j} = \frac{\boldsymbol{\phi}_j^T \mathbf{M} \mathbf{u}_0}{\boldsymbol{\phi}_j^T \mathbf{M} \boldsymbol{\phi}_j} \qquad \text{and} \qquad \dot{q}_{0j} = \frac{\boldsymbol{\phi}_j^T \mathbf{M} \dot{\mathbf{u}}_0}{\boldsymbol{\phi}_j^T \mathbf{M} \boldsymbol{\phi}_j} \qquad (3.277)$$

Hence, to determine the *damped forced vibration* in a system with proportional damping and given initial conditions $\mathbf{u}_0, \dot{\mathbf{u}}_0$, it suffices to compute the natural modes of vibration, determine the modal coefficients q_{0j}, $\dot{q}_{0j}$, compute the modal convolution integrals, and apply modal superposition.

Example of Modal Superposition

Consider the undamped 2-DOF system shown in Figure 3.12, which is subjected to an arbitrary dynamic force $p_2(t)$ applied on mass 2, and assume it to be at rest at time $t = 0$. Find the response.

The mass and stiffness matrices together with the load vector are

$$\mathbf{M} = m \begin{Bmatrix} 2 & 0 \\ 0 & 3 \end{Bmatrix} \qquad \mathbf{K} = 2k \begin{Bmatrix} 1 & -1 \\ -1 & 2 \end{Bmatrix} \qquad \mathbf{p} = \begin{Bmatrix} 0 \\ p_2 \end{Bmatrix} = \begin{Bmatrix} 0 \\ 1 \end{Bmatrix} f(t) \qquad (3.278)$$

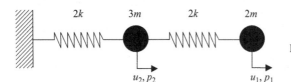

Figure 3.12. Closely coupled 2-DOF system.

Normal modes:

With the definition $\lambda = \omega^2 m / 2k$ (i.e., $\omega = \sqrt{2k/m}\sqrt{\lambda}$) the eigenvalue problem is

$$\begin{vmatrix} 1-2\lambda & -1 \\ -1 & 2-3\lambda \end{vmatrix} = 0 \quad \rightarrow \quad (1-2\lambda)(2-3\lambda)-1 = 0 \tag{3.279}$$

that is,

$$6\lambda^2 - 7\lambda + 1 = 0 \quad \rightarrow \quad \lambda = \frac{7 \mp \sqrt{7^2 - 4 \times 6 \times 1}}{2 \times 6} = \frac{7 \mp 5}{12} = \begin{cases} \lambda_1 = \frac{1}{6} \\ \lambda_2 = 1 \end{cases} \tag{3.280}$$

so

$$\omega_1 = \sqrt{\frac{2k}{6m}} = \sqrt{\frac{k}{3m}}, \qquad \omega_2 = \sqrt{\frac{2k}{m}} \tag{3.281}$$

To determine the modes, we consider once more the eigenvalue equation with the dimensionless eigenvalues:

$$\begin{Bmatrix} 1-2\lambda_j & -1 \\ -1 & 2-3\lambda_j \end{Bmatrix} \begin{Bmatrix} \phi_{1j} \\ \phi_{2j} \end{Bmatrix} = \begin{Bmatrix} 0 \\ 0 \end{Bmatrix} \tag{3.282}$$

Choosing arbitrarily $\phi_{1j} = 1$, we obtain from the first row in Eq. 3.282 the second components as

$$\phi_{2j} = 1-2\lambda_j \quad \rightarrow \quad \phi_{21} = 1-2\tfrac{1}{6} = \tfrac{2}{3}, \ \phi_{22} = 1-2\times1 = -1 \tag{3.283}$$

The modal matrix is then

$$\Phi = \begin{Bmatrix} 1 & 1 \\ \frac{2}{3} & -1 \end{Bmatrix} \tag{3.284}$$

To avoid working with fractions, we choose to rescale the first eigenvector by a factor 3, which changes the modal matrix into

$$\Phi = \begin{Bmatrix} 3 & 1 \\ 2 & -1 \end{Bmatrix} \quad \rightarrow \quad \phi_1 = \begin{Bmatrix} 3 \\ 2 \end{Bmatrix}, \qquad \phi_2 = \begin{Bmatrix} 1 \\ -1 \end{Bmatrix} \tag{3.285}$$

We now proceed to determine the modal parameters.
Modal mass:

$$\mu_1 = \phi_1^T \mathbf{M} \phi_1 = m\begin{bmatrix} 3 & 2 \end{bmatrix} \begin{bmatrix} 2 & 0 \\ 0 & 3 \end{bmatrix} \begin{bmatrix} 3 \\ 2 \end{bmatrix} = 30\,m \tag{3.286}$$

$$\mu_2 = \phi_2^T \mathbf{M} \phi_2 = m\begin{bmatrix} 1 & -1 \end{bmatrix} \begin{bmatrix} 2 & 0 \\ 0 & 3 \end{bmatrix} \begin{bmatrix} 1 \\ -1 \end{bmatrix} = 5\,m \tag{3.287}$$

Modal stiffness:

$$\kappa_1 = \phi_1^T \mathbf{K} \phi_1 = 2k\begin{bmatrix} 3 & 2 \end{bmatrix}\begin{bmatrix} 1 & -1 \\ -1 & 2 \end{bmatrix}\begin{bmatrix} 3 \\ 2 \end{bmatrix} = 10k \qquad (3.288)$$

$$\kappa_2 = \phi_2^T \mathbf{K} \phi_2 = 2k\begin{bmatrix} 1 & -1 \end{bmatrix}\begin{bmatrix} 1 & -1 \\ -1 & 2 \end{bmatrix}\begin{bmatrix} 1 \\ -1 \end{bmatrix} = 10k \qquad (3.289)$$

Observe that $\omega_j = \sqrt{\kappa_j / \mu_j}$ agrees with the frequencies determined previously.

Modal load:

$$\pi_1 = \phi_1^T \mathbf{p} = \begin{bmatrix} 3 & 2 \end{bmatrix}\begin{bmatrix} 0 \\ 1 \end{bmatrix} f(t) = 2f(t) \qquad (3.290)$$

$$\pi_2 = \phi_2^T \mathbf{p} = \begin{bmatrix} 1 & -1 \end{bmatrix}\begin{bmatrix} 0 \\ 1 \end{bmatrix} f(t) = -f(t) \qquad (3.291)$$

Modal impulse response functions (no damping!)

$$h_1 = \frac{1}{\mu_1 \omega_1}\sin\omega_1 t \qquad h_2 = \frac{1}{\mu_2 \omega_2}\sin\omega_2 t \qquad (3.292)$$

Hence, the modal equation is

$$\mu_j \ddot{q}_j + \kappa_j q_j = \pi_j(t) \qquad (3.293)$$

which for zero initial conditions $q_{0j} = 0$, $\dot{q}_{0j} = 0$ has a solution

$$q_j(t) = h_j * \pi_j \qquad (3.294)$$

Finally, the displacements are

$$\begin{aligned} \mathbf{u} &= \phi_1\, q_1(t) + \phi_2\, q_2(t) \\ &= \phi_1\, h_1 * \pi_1(t) + \phi_2\, h_2 * \pi_2(t) \\ &= 2\begin{bmatrix} 3 \\ 2 \end{bmatrix} h_1 * f(t) - \begin{bmatrix} 1 \\ -1 \end{bmatrix} h_2 * f(t) \end{aligned} \qquad (3.295)$$

which must be evaluated numerically using a computer. In the case of a simple forcing function $f(t)$, it may be possible to provide closed-form solutions. For example, for a step load of infinite duration $f(t) = p_0\, \mathcal{H}(t)$ with amplitude p_0, the convolution is

$$h_j * p_0 \mathcal{H}(t) = \frac{p_0}{\kappa_j}\left(1 - \cos\omega_j t\right) \qquad (3.296)$$

in which case

$$\begin{aligned} \mathbf{u} &= p_0\left\{ \frac{2}{\kappa_1}\begin{bmatrix} 3 \\ 2 \end{bmatrix}(1 - \cos\omega_1 t) - \frac{1}{\kappa_2}\begin{bmatrix} 1 \\ -1 \end{bmatrix}(1 - \cos\omega_2 t) \right\} \\ &= \frac{p_0}{10k}\begin{Bmatrix} 5 - 6\cos\omega_1 t + \cos\omega_2 t \\ 5 - 4\cos\omega_1 t - \cos\omega_2 t \end{Bmatrix} \end{aligned} \qquad (3.297)$$

3.5.2 Proportional versus Nonproportional Damping Matrices

When a structural or mechanical system includes viscous dampers (e.g., the passive vibration control devices in a tall building, or the shock absorbers in an automobile), the assembly of the damping matrix is readily carried out by appropriate superposition of the damping elements, which in most cases will lead to a damping matrix that is not proportional. Then again, most structural systems do not have dampers anywhere, yet they still exhibit damping due to material hysteresis and internal friction. In those cases, the damping is typically prescribed directly at the level of the modes, and is usually taken to be uniform, that is, the fraction of critical damping is chosen to be the same in each and every mode. However, when modal analysis is *not* used but alternative solution methods are used, such as time-step integration, it is sometimes necessary to add a damping matrix to simulate the energy losses at low deformations and avoid an undamped condition. This has led to ways for constructing damping matrices with physically reasonable damping characteristics, and such matrices are almost invariably of the proportional kind.

Still, it may seem paradoxical that the construction of proportional damping matrices is rarely useful in the context of a solution via modal superposition. Indeed, if classical modes and frequencies are already available to the analyst and damping is prescribed at the level of the modes, then there is absolutely no need to construct a diagonalizable matrix to begin with, since after a modal transformation of that matrix, one would simply recover the very damping values from which one started. Instead, proportional damping matrices are most useful when both the existence of normal modes and the orthogonality of such matrices with respect to the modal transformation are completely irrelevant, for example, in the context of a nonlinear problem solved directly by time step integration. The motivation for their use may be to provide realistic levels of damping at small vibration amplitudes, or to guarantee numerical stability, or for other similar reasons. If so, the computation of modal frequencies is preferably avoided – indeed they might not even be computable if a nonlinear problem is being dealt with. Instead, the natural frequencies are *estimated*, and this fact has important theoretical and physical implications for the damping matrices thus constructed, as will be seen later on in the context of so-called Caughey damping matrices.

3.5.3 Conditions under Which a Damping Matrix Is Proportional

As we stated previously, a damping matrix is said to be proportional if it can be diagonalized by the modal transformation, that is, if $\mathbb{C} = \mathbf{\Phi}^T \mathbf{C} \mathbf{\Phi}$ is a diagonal matrix. What intrinsic properties must $\mathbf{C}$ possess in order for this condition to be satisfied? Of course, if the modes are computed and the product is carried out, it will be readily apparent whether or not the result is diagonal. Surprisingly, however, it is possible to predict ahead of time whether or not this will turn out to be the case, even if the modes are not computed. We shall show here that a *necessary and sufficient* condition for a damping matrix be diagonalizable under the modal transformation is that it satisfy the expression[6]

$$\boxed{\mathbf{K}\mathbf{M}^{-1}\mathbf{C} = (\mathbf{K}\mathbf{M}^{-1}\mathbf{C})^T}$$

(3.298)

---

[6] T. K. Caughey and M. E. J. O'Kelly, "Classical normal modes in damped linear dynamic systems," *J. Appl. Mech. ASME*, 32, 1965, 583–588.

that is, the product $\mathbf{KM^{-1}C}$ must be symmetric. The proof is as brief as this condition is simple. Assume without loss of generality that the undamped modes of the system have been normalized. In other words, that $\mathbf{\Phi}^T\mathbf{M}\mathbf{\Phi} = \mathbf{I}$ and $\mathbf{\Phi}^T\mathbf{K}\mathbf{\Phi} = \mathbf{\Omega}^2$. Define also the product (not necessarily diagonal) $\mathbb{C} = \mathbf{\Phi}^T\mathbf{C}\mathbf{\Phi}$.

For a well-posed problem, $\mathbf{M}$ is a positive definite matrix, and $\mathbf{\Phi}$ is nonsingular. It follows that

$$\mathbf{M}^{-1} = \mathbf{\Phi}\mathbf{\Phi}^T \qquad \mathbf{K} = \mathbf{\Phi}^{-T}\mathbf{\Omega}^2\,\mathbf{\Phi}^{-1} \qquad \mathbf{C} = \mathbf{\Phi}^{-T}\mathbb{C}\mathbf{\Phi}^{-1} \tag{3.299}$$

Hence,

$$\mathbf{KM^{-1}C} = \mathbf{\Phi}^{-T}\mathbf{\Omega}^2\,\mathbf{\Phi}^{-1}\mathbf{\Phi}\,\mathbf{\Phi}^T\mathbf{\Phi}^{-T}\mathbb{C}\mathbf{\Phi}^{-1} = \mathbf{\Phi}^{-T}\mathbf{\Omega}^2\mathbb{C}\mathbf{\Phi}^{-1} \tag{3.300}$$

and

$$\mathbf{CM^{-1}K} = (\mathbf{KM^{-1}C})^{-T} = \mathbf{\Phi}^{-T}\mathbb{C}^T\mathbf{\Omega}^2\,\mathbf{\Phi}^{-1} \tag{3.301}$$

Since $\mathbf{\Omega}^2$ is a diagonal matrix, the condition $\mathbf{\Omega}^2\mathbb{C} = \mathbb{C}^T\mathbf{\Omega}^2$ can be satisfied only if $\mathbb{C}$ is a diagonal matrix. From Eqs. 3.300 and 3.301, it follows that the product $\mathbf{KM^{-1}C}$ is symmetric if and only if $\mathbb{C}$ is diagonal. Hence, if the triple product is symmetric, then $\mathbb{C}$ is diagonal.

It is remarkable that this condition can be established a priori without having to compute the modes. Also, in most cases, $\mathbf{K}$ and $\mathbf{C}$ are banded matrices, and $\mathbf{M}$ is diagonal, so testing isolated off-diagonal values of this product for symmetry does not require much effort. It must be admitted, however, that this condition is more of theoretical than of practical interest, as a complete proof would require testing all values, which is computationally impractical (what if the test fails near the end?). In addition, a failure of the test, and thus the knowledge that normal modes do not exist, does not advance us much in the solution of the problem.

Example of Nonproportional Damping

Consider the damped 2-DOF system shown in Figure 3.13. Does this system have proportional damping?

The exact modes for this system are

$$\mathbf{\Phi} = \begin{Bmatrix} 3 & 1 \\ 2 & -1 \end{Bmatrix} \quad \rightarrow \quad \phi_1 = \begin{Bmatrix} 3 \\ 2 \end{Bmatrix}, \quad \phi_2 = \begin{Bmatrix} 1 \\ -1 \end{Bmatrix} \tag{3.302}$$

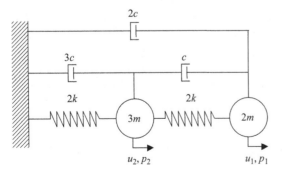

Figure 3.13. 2-DOF system with nonproportional damping.

We could now answer the question of proportionality in two ways, both of which require that we first assemble the damping matrix:

$$\mathbf{C} = c \begin{Bmatrix} 2+1 & -1 \\ -1 & 3+1 \end{Bmatrix} = c \begin{Bmatrix} 3 & -1 \\ -1 & 4 \end{Bmatrix} \tag{3.303}$$

1. We first try out the modal transformation, and because we only need to verify whether or not it has off-diagonal terms, we compute just the coupling term:

$$\phi_1^T \mathbf{C} \phi_2 = c \{3 \quad 2\} \begin{Bmatrix} 3 & -1 \\ -1 & 4 \end{Bmatrix} \begin{Bmatrix} 1 \\ -1 \end{Bmatrix} = c \{3 \quad 2\} \begin{Bmatrix} 4 \\ -5 \end{Bmatrix} = 2c \tag{3.304}$$

Since this number is not zero, we already know that the matrix is not proportional.

2. We next verify the symmetry, or lack thereof, of the product $\mathbf{CM^{-1}K}$

$$\begin{aligned} \mathbf{CM^{-1}K} &= \frac{c(2k)}{(6m)} \begin{Bmatrix} 3 & -1 \\ -1 & 4 \end{Bmatrix} \begin{Bmatrix} 3 & 0 \\ 0 & 2 \end{Bmatrix} \begin{Bmatrix} 1 & -1 \\ -1 & 2 \end{Bmatrix} \\ &= \frac{ck}{3m} \begin{Bmatrix} 3 & -1 \\ -1 & 4 \end{Bmatrix} \begin{Bmatrix} 3 & -3 \\ -2 & 4 \end{Bmatrix} \\ &= \frac{ck}{3m} \begin{Bmatrix} 11 & -13 \\ -11 & 19 \end{Bmatrix} \end{aligned} \tag{3.305}$$

Since this matrix is not symmetric, we confirm once more that the damping matrix is not of the *proportional* type.

3.5.4 Bounds to Coupling Terms in Modal Transformation

We now show that when a modal transformation is carried out on a nonproportional damping matrix, the off-diagonal terms that arise in that transformation cannot attain arbitrarily large values, but are subjected to specific bounds that depend on the diagonal terms. For this purpose, let $\mathbf{C}$ be an arbitrary viscous damping matrix, to which we apply the modal transformation. We can write the result as

$$(\mathbf{\Phi}^T \mathbf{M} \mathbf{\Phi})^{-1/2} (\mathbf{\Phi}^T \mathbf{C} \mathbf{\Phi}) (\mathbf{\Phi}^T \mathbf{M} \mathbf{\Phi})^{-1/2} = \left\{ \frac{\phi_i^T \mathbf{C} \phi_j}{\sqrt{\phi_i^T \mathbf{M} \phi_i} \sqrt{\phi_j^T \mathbf{M} \phi_j}} \right\} = 2 (\Xi \Omega)^{1/2} \mathbf{A} (\Omega \Xi)^{1/2} \tag{3.306}$$

in which $\mathbf{A} = \{a_{ij}\}$ is an as yet unknown, symmetric, fully populated matrix, Ω is the diagonal matrix of undamped frequencies ω_j, and Ξ is the diagonal matrix with classical modal damping ratios ξ_j, that is,

$$\Xi = \text{diag} \left\{ \frac{1}{2\omega_j} \frac{\phi_j^T \mathbf{C} \phi_j}{\phi_j^T \mathbf{M} \phi_j} \right\} = \text{diag} \{\xi_j\} \tag{3.307}$$

Clearly, if the damping matrix $\mathbf{C}$ is of the proportional type, then $\mathbf{A}$ reduces to the identity matrix. From the preceding, it can be seen that the elements of $\mathbf{A}$ are of the form

$$a_{ij} = \frac{\phi_i^T \mathbf{C} \phi_j}{\sqrt{2\mu_i \xi_i \omega_i} \sqrt{2\mu_j \xi_j \omega_j}} = \frac{\phi_i^T \mathbf{C} \phi_j}{\sqrt{\phi_i^T \mathbf{C} \phi_i} \sqrt{\phi_j^T \mathbf{C} \phi_j}} \tag{3.308}$$

in which the μ_j are the modal masses. Since $\mathbf{C}$ is a real, symmetric, positive semidefinite matrix, it follows that all of its eigenvalues λ_i in the eigenvalue problem $\mathbf{C}\boldsymbol{\psi}_i = \lambda_i \boldsymbol{\psi}_i$ are real and nonnegative. Hence, if we express $\mathbf{C}$ in terms of its own modes (which, without loss of generality, we may assume to be normalized), then we can write the damping matrix as $\mathbf{C} = \boldsymbol{\Psi}^T \boldsymbol{\Lambda} \boldsymbol{\Psi}$ in which $\boldsymbol{\Lambda} = \mathrm{diag}\{\lambda_i\}$, and $\boldsymbol{\Psi} = \{\boldsymbol{\psi}_j\}$. Defining $\mathbf{z}_j = \boldsymbol{\Lambda}^{1/2} \boldsymbol{\Psi} \boldsymbol{\phi}_j$, the squares of the elements of $\mathbf{A}$ are then

$$a_{ij}^2 = \frac{(\mathbf{z}_i^T \mathbf{z}_j)(\mathbf{z}_i^T \mathbf{z}_j)}{(\mathbf{z}_i^T \mathbf{z}_i)(\mathbf{z}_j^T \mathbf{z}_j)} = \frac{\sum_k z_{ik} z_{jk} \sum_k z_{ik} z_{jk}}{\sum_k z_{ik}^2 \sum_k z_{jk}^2} \leq 1 \tag{3.309}$$

Taking the square root of this result, we conclude that $-1 \leq a_{ij} \leq 1$, and $a_{ii} = 1$. Hence, we can write the off-diagonal elements of $\mathbf{A}$ in the convenient form $a_{ij} = \cos\theta_{ij}$, with $\theta_{ii} = 0$, that is, $\mathbf{A} = \{\cos\theta_{ij}\}$. It follows that

$$\boxed{\boldsymbol{\phi}_i^T \mathbf{C} \boldsymbol{\phi}_j = 2\sqrt{\mu_i \mu_j} \sqrt{\xi_i \xi_j} \sqrt{\omega_i \omega_j} \cos\theta_{ij}} \,, \qquad \text{with} \qquad \cos\theta_{jj} = 1 \tag{3.310}$$

We conclude that the nonzero off-diagonal elements are proportional to the geometric mean of the diagonal elements, with the proportionality constant being a number less than one in absolute value. A corollary is that the off-diagonal terms never exceed the geometric mean of the two corresponding diagonal terms. This establishes a bound on how large the off-diagonal elements can ever get to be.

3.5.5 Rayleigh Damping

Perhaps the most widely used damping matrix of the proportional type – indeed the one that gave rise to the "proportional" adjective – is the so-called Rayleigh damping

$$\mathbf{C} = a_0 \mathbf{M} + a_1 \mathbf{K} \tag{3.311}$$

in which a_0, a_1 are arbitrary coefficients. Thus, this damping matrix is a linear combination of the mass and stiffness matrix, and is one of the simplest matrices of the proportional type. Carrying out the modal transformation, we obtain

$$\eta_j = \boldsymbol{\phi}_j^T \mathbf{C} \boldsymbol{\phi}_j = a_0 \mu_j + a_1 \kappa_j = 2\xi_j \omega_j \mu_j \tag{3.312}$$

which after division by $\omega_j \mu_j$ leads immediately to

$$\xi_j = \frac{1}{2}\left[\frac{a_0}{\omega_j} + a_1 \omega_j \right] \tag{3.313}$$

For given coefficients a_0, a_1, the implied modal damping ratios are points on a hyperbola with two asymptotes, as shown in Figure 3.14. In most engineering applications, these constants are determined by specifying damping ratios ξ_1, ξ_2 at two arbitrary frequencies thought to be representative for the problem at hand. Let ω_1, ω_2 be these frequencies, which may or not coincide with two actual modal frequencies. We have then

$$\begin{Bmatrix} \omega_1^{-1} & \omega_1 \\ \omega_2^{-1} & \omega_2 \end{Bmatrix} \begin{Bmatrix} a_0 \\ a_1 \end{Bmatrix} = \begin{Bmatrix} 2\xi_1 \\ 2\xi_2 \end{Bmatrix} \tag{3.314}$$

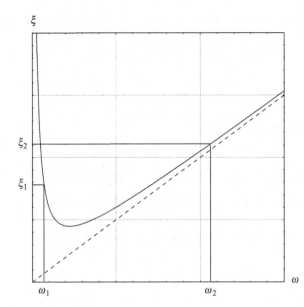

Figure 3.14. Rayleigh damping.

which is a system of two equations in two unknowns. The minimum damping predicted by these coefficients will occur at the geometric mean $\omega_{\min} = \sqrt{\omega_1\omega_2}$. In particular, if $\xi_1 = \xi_2 \equiv \xi$, then

$$a_0 = \frac{2\xi\omega_1\omega_2}{\omega_1 + \omega_2} \quad \text{and} \quad a_1 = \frac{2\xi}{\omega_1 + \omega_2} \tag{3.315}$$

and the lowest value is $\xi_{\min} = 2\xi\sqrt{\omega_1\omega_2}/(\omega_1 + \omega_2)$ (i.e., the damping times the ratio between the geometric and arithmetic means). In general, actual modes whose natural frequencies fall in between the frequency pivots will have less critical damping than the pivots, and those outside of those bounds will have more (indeed, in the case of the highest modes, perhaps excessive damping).

Rayleigh damping matrices are easy to set up, and do not require computation of any of the modes. Thus, they are most often used in the context of time domain solutions via numerical integration of the equations of motion (i.e., without using modal superposition), particularly for systems that exhibit nonlinear behavior. Considering, however, that they will make some modes to have little damping and others too much (even if not explicitly computed), these matrices may not be satisfactory.

3.5.6 Caughey Damping

The damping matrices proposed by Caughey[7] are also of the proportional type, and can be regarded as a generalization of the Rayleigh damping matrices. They are of the form

$$\boxed{\mathbf{C} = \mathbf{M}^{1/2}\left\{\sum_{r=0}^{n-1} a_{rs}\left[\mathbf{M}^{-1/2}\mathbf{K}\,\mathbf{M}^{-1/2}\right]^{r/s}\right\}\mathbf{M}^{1/2}} \tag{3.316}$$

[7] T. K. Caughey, "Classical normal modes in damped linear system," *J. Appl. Mech.* 27, 1960.

in which s is an integer (which can be negative, if $\mathbf{K}$ is not singular), the a_{rs} are arbitrary numerical coefficients, and $n \leq N$ is the number of terms in the summation, with N being the number of DOF. Higher powers in r do not enter here on account of the Cayley–Hamilton theorem of linear algebra. In particular, if $s = 1$, this expression is the equivalent to

$$\mathbf{C} = \mathbf{M} \sum_{r=0}^{n-1} a_r \left[\mathbf{M}^{-1} \mathbf{K} \right]^r = a_0 \mathbf{M} + a_1 \mathbf{K} + a_2 \mathbf{K} \mathbf{M}^{-1} \mathbf{K} + \cdots \tag{3.317}$$

where the a_r are again arbitrary coefficients. As can be seen, Rayleigh damping is obtained by setting all but the first two coefficients to zero, that is, $n = 2$. The fractions of critical damping implied by this matrix are obtained by applying the modal transformation

$$\mathbb{C} = \mathbf{\Phi}^T \mathbf{C} \mathbf{\Phi} = \mathbf{\Phi}^T \left\{ \mathbf{M} \sum_{r=0}^{n-1} a_r \left[\mathbf{M}^{-1} \mathbf{K} \right]^r \right\} \mathbf{\Phi} = \mathbf{\Phi}^T \left\{ a_0 \mathbf{M} + a_1 \mathbf{K} + a_2 \mathbf{K} \mathbf{M}^{-1} \mathbf{K} + \cdots \right\} \mathbf{\Phi} \tag{3.318}$$

But from the eigenvalue equation

$$\mathbf{K} \mathbf{\Phi} = \mathbf{M} \mathbf{\Phi} \mathbf{\Omega}^2 \quad \rightarrow \quad \mathbf{M}^{-1} \mathbf{K} = \mathbf{\Phi} \, \mathbf{\Omega}^2 \, \mathbf{\Phi}^{-1} \tag{3.319}$$

$$(\mathbf{M}^{-1} \mathbf{K})^r = (\mathbf{\Phi} \, \mathbf{\Omega}^2 \, \mathbf{\Phi}^{-1})^r = \mathbf{\Phi} \, \mathbf{\Omega}^2 \, \mathbf{\Phi}^{-1} \mathbf{\Phi} \, \mathbf{\Omega}^2 \, \mathbf{\Phi}^{-1} \cdots = \mathbf{\Phi} \, \mathbf{\Omega}^{2r} \, \mathbf{\Phi}^{-1} \tag{3.320}$$

so that

$$\mathbf{\Phi}^T \mathbf{M} (\mathbf{M}^{-1} \mathbf{K})^r \mathbf{\Phi} = (\mathbf{\Phi}^T \mathbf{M} \mathbf{\Phi}) \mathbf{\Omega}^{2r} \mathbf{\Phi}^{-1} \mathbf{\Phi} = \mathbb{M} \mathbf{\Omega}^{2r} = \operatorname{diag} \left\{ \mu_j \omega_j^{2r} \right\} \tag{3.321}$$

It follows that

$$\mathbb{C} = \operatorname{diag} \left\{ \eta_j \right\} = \operatorname{diag} \left\{ \mu_j \sum_{r=0}^{n-1} a_r \omega_j^{2r} \right\} \tag{3.322}$$

which shows that the Caughey damping matrices are indeed diagonalizable by the modal transformation and are thus proportional. If the a_r coefficients are known, the fractions of damping in each mode are

$$\xi_j = \frac{1}{2} \sum_{r=0}^{n-1} a_r \omega_j^{2r-1} \tag{3.323}$$

Alternatively, if the fractions of damping ξ_j are prescribed at n arbitrary frequencies or *pivot* $\varpi_0 \, \varpi_j \cdots \varpi_{n-1}$ (not necessarily the actual modal frequencies!), the coefficients can then be obtained from the system of equations

$$\begin{bmatrix} 1 & \varpi_0^2 & \cdots & \varpi_0^{2n-2} \\ \cdots & \cdots & \cdots & \cdots \\ 1 & \varpi_{n-1}^2 & \cdots & \varpi_{n-1}^{2n-2} \end{bmatrix} \left\{ \begin{array}{c} a_0 \\ \vdots \\ a_{n-1} \end{array} \right\} = \left\{ \begin{array}{c} 2\xi_0 \varpi_0 \\ \vdots \\ 2\xi_{n-1} \varpi_{n-1} \end{array} \right\} \tag{3.324}$$

The direct solution of this system of equations is, however, not attractive, for at least two reasons. On the one hand, the matrix of coefficients has the form of a Vandermonde matrix, which is notoriously ill conditioned, so special care is needed in the solution. More importantly, the fractions of damping implied by a damping matrix so constructed at frequencies other than the pivots used to define the coefficients a_r is controlled by the high-order polynomial given in Eq. 3.323, which can oscillate substantially between the frequency pivots, even to the point of turning negative. Worse still, if the nth term is a negative number (i.e., $a_{n-1} < 0$), then the polynomial at some point will attain negative values, implying that some modes could be negatively damped! This will indeed happen whenever $n > 1$ is an odd number. Consequently, the fractions of damping at the *actual* modal frequencies are highly uncertain or even unacceptable. If, on the other hand, the actual frequencies (or subset of frequencies) were used to determine the coefficients, then the modal damping ratios would be as prescribed. Alas, in such a case the proportional damping matrix would no longer be attractive, because with the availability of the modes, a direct modal superposition could entirely bypass the assembly of a proportional damping matrix.

To demonstrate these assertions, consider the implied damping values at various frequencies when the frequency pivots are taken in the ratios 1:2:3: …:8 and $1^2:2^2:3^2…:8^2$ and the damping at the pivots is uniform. These two ratios are meant to simulate the natural frequencies in a rod and in a bending beam. Figure 3.15 shows the Caughey damping as a function of frequency normalized by the pivot damping. Seven cases are considered, namely $n = 2, 3, …8$, the pivots for which are shown as marks. Observe the following:

1. Whenever n is odd, the curve drops to negative values after the last pivot.
2. While the damping is close to the prescribed damping when the pivot separation is linear, this is not the case when the separation is quadratic. Indeed, there is a range of frequencies between the fifth and sixth pivots, and between the seventh and eight where all curves drop precipitously to negative values. Then again, at other frequencies damping rises steeply and is thus excessive. It follows that a quadratic separation of the pivots, and very probably also any other nonuniform spacing of pivots (or when damping at the pivots is not uniform), is most certainly *not* acceptable.

We conclude that in most practical situations in which the damping matrix is constructed by using pivots and not the actual modal frequencies, it behooves to choose an *even* number of such pivots that are *uniformly* spaced – or nearly so – lest damping be

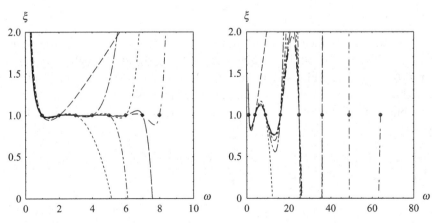

Figure 3.15. Pitfalls of Caughey damping.

grossly falsified. This would be true even when considering a structure whose normal frequencies are known to be spaced quadratically. For example, a simply supported bending beam, when modeled with finite elements, will exhibit some unavoidable dispersion, so its natural frequencies will not satisfy exactly the ratios $1^2 : 2^2 : 3^2 \cdots$, but will lie close to these ratios instead. However, as can be seen from Figure 3.15 on the right, even small deviations of the actual frequencies from the pivots will lead to unacceptable values of damping at the actual frequencies of the discretized bending beam, and the problem will become worse as more coefficients M are added. Conceivably, gross oscillations might also arise, even when using uniform spacing of the pivots, if the damping at those pivots is not uniform. This strongly suggests that Caughey damping matrices constructed with a large number M of terms are eminently suspect, unless, of course, the actual normal frequencies are used, which in most practical cases is not of interest.

Should you still wish to make use of Caughey damping matrices with a relatively small number of terms n, the following simple *recursive* procedure might be of help in constructing such a matrix. Begin by expressing the polynomial for the damping in terms of Newton interpolation coefficients

$$
\begin{aligned}
F(\omega) = 2\xi\omega &= a_0 + a_1\omega^2 + a_2\omega^4 + \cdots \\
&= c_0 + c_1(\omega^2 - \varpi_0^2) + c_2(\omega^2 - \varpi_0^2)(\omega^2 - \varpi_1^2) + \cdots
\end{aligned}
\tag{3.325}
$$

so that

$$
2\xi_0\varpi_0 = c_0
\tag{3.326}
$$

$$
2\xi_1\varpi_1 = c_0 + c_1(\varpi_1^2 - \varpi_0^2)
\tag{3.327}
$$

$$
2\xi_2\varpi_2 = c_0 + c_1(\varpi_2^2 - \varpi_0^2) + c_2(\varpi_2^2 - \varpi_0^2)(\varpi_2^2 - \varpi_1^2)
\tag{3.328}
$$

etc.

Define the recursive sequence

$$
S_\alpha = 2\xi_\alpha\varpi_\alpha, \qquad S_{\alpha\beta} = S_{\beta\alpha} = \frac{S_\alpha - S_\beta}{\varpi_\alpha^2 - \varpi_\beta^2}, \qquad S_{\alpha\beta\gamma} = S_{\gamma\beta\alpha} = \frac{S_{\alpha\beta} - S_{\beta\gamma}}{\varpi_\alpha^2 - \varpi_\gamma^2} \qquad \text{etc.}
\tag{3.329}
$$

In terms of this sequence, the coefficients c_j are then

$$
c_0 = S_0, \qquad c_1 = S_{10}, \qquad c_2 = S_{210}, \qquad c_3 = S_{3210} \qquad \text{etc.}
\tag{3.330}
$$

In particular, the last coefficient in the summation satisfies $a_{n-1} = c_{n-1}$. This implies that when $n = 3$, the last coefficient is negative (i.e., $a_2 = c_2 < 0$), so a Caughey sequence with three terms (or more generally with an odd number $n > 1$) is a priori not acceptable!

Once the coefficients c_r have been found, they can be used to obtain the a_r by appropriate combinations, or better still, the proportional damping matrix can be constructed directly as

$$
\begin{aligned}
\mathbf{C} &= \mathbf{M}\left[c_0\mathbf{I} + c_1\left(\mathbf{M}^{-1}\mathbf{K} - \varpi_0^2\mathbf{I}\right) + c_2\left(\mathbf{M}^{-1}\mathbf{K} - \varpi_0^2\mathbf{I}\right)\left(\mathbf{M}^{-1}\mathbf{K} - \varpi_1^2\mathbf{I}\right) + \cdots\right] \\
&= c_0\mathbf{M} + c_1\left(\mathbf{K} - \varpi_0^2\mathbf{M}\right) + c_2\left(\mathbf{K} - \varpi_0^2\mathbf{M}\right)\mathbf{M}^{-1}\left(\mathbf{K} - \varpi_1^2\mathbf{M}\right) + \cdots
\end{aligned}
\tag{3.331}
$$

The advantage of this formula vis-à-vis the Caughey series formula is that the coefficients c_j do *not* depend on the total number n of terms in the Caughey series. Hence, the number of terms included could be adjusted on the fly, inasmuch as the product of matrices in Eq. 3.331 can be obtained recursively as more terms are being added. While the matrix factors $\mathbf{K} - \omega_j^2\mathbf{M}$ will be singular whenever the pivots coincide with the actual normal frequencies, that fact is inconsequential here.

In the special case of uniform damping $\xi_0 = \xi_1 = \xi_2 \ldots$ at the pivots, application of this method yields for the first four coefficients

$$c_0 = 2\xi_0 \varpi_0, \qquad c_1 = \frac{2\xi_0}{\varpi_0 + \varpi_1}, \qquad c_2 = \frac{-2\xi_0}{(\varpi_0 + \varpi_1)(\varpi_0 + \varpi_2)(\varpi_1 + \varpi_2)} \tag{3.332}$$

$$c_3 = \frac{2\xi_0(\varpi_0 + \varpi_1 + \varpi_2 + \varpi_3)}{(\varpi_0 + \varpi_1)(\varpi_0 + \varpi_2)(\varpi_0 + \varpi_3)(\varpi_1 + \varpi_2)(\varpi_1 + \varpi_3)(\varpi_2 + \varpi_3)} \tag{3.333}$$

Of interest is mainly the case of uniform damping $\xi_0 = \xi_1 = \cdots \xi_{n-1} \equiv \xi$ and pivots spaced uniformly at frequency steps $\Delta\varpi = (\varpi_{M-1} - \varpi_0)/(n-1)$. In this case, the pivots can be expressed as

$$\begin{aligned} \varpi_j &= \varpi_0 + j\Delta\varpi \\ &= \varpi_0(1+jh) \end{aligned}, \qquad h = \frac{\Delta\varpi}{\varpi_0} \tag{3.334}$$

The explicit solution for this case can be shown to be given by

$$c_0 = 2\xi\varpi_0 \qquad \text{and} \qquad c_k = \frac{(-1)^{k-1}}{\varpi_0^{2k}} \frac{\alpha_k}{\displaystyle\prod_{j=1}^{2k-1}(2+jh)} c_0, \qquad k = 1,2,3,\ldots \tag{3.335}$$

with $\alpha_1 = 1$, $\alpha_2 = 1$, $\alpha_3 = 2$, $\alpha_4 = 5$, $\alpha_5 = 14$, $\alpha_7 = 42$, etc. These are known as the *Catalan numbers*, which are encountered in many different types of counting problems. An explicit formula is

$$\alpha_n = \frac{(2n-2)!}{(n-1)!n!} \tag{3.336}$$

which can be determined recursively as well.

This procedure is not effective for large n, in which case the method for Vandermonde matrices described in *Numerical Recipes*[8] should be preferred.

3.5.7 Damping Matrix Satisfying Prescribed Modal Damping Ratios

In some cases, it may be desirable to construct a damping matrix $\mathbf{C}$ that will ensure reasonable (or measured) values of damping in the structure throughout the range of interest. This may be particularly relevant when using solution procedures like time step integration. Since modal superposition is not used in those cases, the central issue is not the proportionality of $\mathbf{C}$ but its physical adequacy. It is then ironic that the construction of such a matrix requires the availability of some or all of the undamped modes, even if they

---

[8] W. H. Press, S. A. Teukolsky, W. T. Vetterling, and B. P. Flannery, *Numerical Recipes in C*, 2nd ed. (Cambridge: Cambridge University Press, 1992).

will not be used later to determine the response. For if they were indeed used, it would not be necessary to determine a damping matrix in the first place. Clearly, this task could be accomplished with the Caughey matrices just presented. We describe here an alternative method often used in practice.

To obtain a damping matrix satisfying prescribed fractions of damping ξ_j in each mode ω_j, assume that the undamped modes are available. We then impose the condition

$$\mathbf{\Phi}^T \mathbf{C} \mathbf{\Phi} = 2 \mathbb{M} \, \mathbf{\Xi} \, \mathbf{\Omega} = \text{diag} \left\{ 2 \mu_j \xi_j \omega_j \right\} \tag{3.337}$$

Hence

$$\mathbf{C} = 2 \, \mathbf{\Phi}^{-T} \, \mathbb{M} \, \mathbf{\Xi} \, \mathbf{\Omega} \, \mathbf{\Phi}^{-1} \tag{3.338}$$

But from the orthogonality condition

$$\mathbf{\Phi}^{-T} = \mathbf{M} \, \mathbf{\Phi} \, \mathbb{M}^{-1} \qquad \text{and} \qquad \mathbf{\Phi}^{-1} = \mathbb{M}^{-1} \mathbf{\Phi}^T \, \mathbf{M} \tag{3.339}$$

It follows finally

$$\mathbf{C} = 2 \, \mathbf{M} \, \mathbf{\Phi} \, \mathbf{\Xi} \, \mathbf{\Omega} \, \mathbb{M}^{-1} \mathbf{\Phi}^T \, \mathbf{M} = \mathbf{M} \, \mathbf{\Phi} \, \text{diag} \left\{ 2 \mu_j^{-1} \xi_j \omega_j \right\} \mathbf{\Phi}^T \, \mathbf{M} \tag{3.340}$$

which is equivalent to

$$\mathbf{C} = \mathbf{M} \left\{ \sum_{j=1}^{n} \frac{2 \xi_j \omega_j}{\mu_j} \boldsymbol{\phi}_j \, \boldsymbol{\phi}_j^T \right\} \mathbf{M} \tag{3.341}$$

Notice that if the damping of the highest modes is set to zero, the summation involves only a subset of the modes. This is equivalent to neglecting damping in the absent modes.

Although this damping matrix may indeed satisfy the prescribed damping ratios, it should be noted that it suffers from the following pitfalls:

1. It is generally fully populated (i.e., does not have a banded structure).
2. It may imply the presence of isolated dashpots in the structure with negative constants (i.e., local sources rather than sinks of energy cannot be ruled out a priori).
3. It requires a large computational effort.
4. Modes not included in the summation have zero damping.

It is interesting to observe that if $\xi_j = \xi$ is constant in all modes, then

$$\mathbf{C} = 2 \xi \mathbf{M} \left\{ \sum_{j=1}^{n} \frac{\omega_j}{\mu_j} \boldsymbol{\phi}_j \, \boldsymbol{\phi}_j^T \right\} \mathbf{M} = 2 \xi \mathbf{M} \left(\mathbf{M}^{-1} \mathbf{K} \right)^{1/2} = 2 \xi \mathbf{M}^{1/2} \left(\mathbf{M}^{-1/2} \mathbf{K} \mathbf{M}^{-1/2} \right)^{1/2} \mathbf{M}^{1/2} \tag{3.342}$$

that is, the resulting damping matrix is the same as a Caughey matrix with just one term, $r = 1$ and $s = 2$. This can be shown to be true by means of the spectral representation $\left(\mathbf{M}^{-1} \mathbf{K} \right)^{1/2} = \mathbf{\Phi} \mathbf{\Omega} \mathbf{\Phi}^{-1}$ together with the orthogonality condition. This is the matrix generalization of the scalar expression for an SDOF system $c = 2 \xi m \sqrt{k / m} = 2 \xi \sqrt{k m}$. Although interesting, the above is not practical for engineering purposes in that the computation of the square root of a matrix is numerically intensive (see *Wikipedia* under "square root of matrix").

3.5.8 Construction of Nonproportional Damping Matrices

As mentioned earlier, the realm of application of proportional matrices lies in problems in which the existence of normal modes and the proportionally property are not relevant. In fact, the only reason for using these is because they are relatively simple to construct, and especially so when the Rayleigh damping variety is chosen. But if the Caughey variety is employed, and the number of terms is high, and the number of DOF is large, chances are good that some modes will either be damped in excess, or worse still, be negatively damped. A question is then, Why not something else? As will be seen, it is easy to construct damping matrices that guarantee reasonable values of structural damping, provided that we abandon the (irrelevant) requirement for these matrices to be proportional.[9]

As is well known (see Chapter 2), any proportional damping matrix $\mathbf{C}$ can be written in the form

$$\mathbf{C} = \mathbf{M}\,\mathbf{\Phi}\,\mathbf{\Xi}\,\mathbf{\Phi}^T\mathbf{M} = \mathbf{M}\left\{\sum_{j=1}^{N}\frac{2\xi_j\omega_j}{\mu_j}\phi_j\phi_j^T\right\}\mathbf{M} \tag{3.343}$$

in which $\mathbf{\Xi} = \mathrm{diag}\left(2\xi_j\omega_j/\mu_j\right)$ is a diagonal matrix, $\mu_j = \phi_j^T\mathbf{M}\phi_j$ is the modal mass, and each of the dyadic modal products $\phi_j\phi_j^T$ is a square matrix. Any term missing in the summation above implies that the corresponding mode is undamped. Similarly, if the sum is truncated at $M < N$, the modes $j > M$ will remain undamped.

Taking inspiration in the *Weighted Residuals Method*, Eq. 3.343 suggests choosing an arbitrary number M of trial vectors ψ_r – our a priori estimates of some of the relevant modes – which we choose to normalize so that $\bar{\mu}_r = \psi_r^T\mathbf{M}\psi_r = 1$. For each of these trial vectors there is also an associated frequency pivot that follows from the Rayleigh quotient:

$$\bar{\omega}_r^2 = \frac{\psi_r^T\mathbf{K}\psi_r}{\psi_r^T\mathbf{M}\psi_r} = \psi_r^T\mathbf{K}\psi_r, \qquad r = 1,2,\dots M \tag{3.344}$$

and from the enclosure theorem, we know that this quotient satisfies the inequality $\omega_1 \le \bar{\omega}_r \le \omega_N$. Thus equipped, we proceed to construct a damping matrix according to the rule

$$\mathbf{C} = \sum_{r=1}^{M}\frac{2\xi_r\bar{\omega}_r}{\bar{\mu}_r}\mathbf{M}\psi_r\psi_r^T\mathbf{M} = 2\sum_{r=1}^{M}\xi_r\bar{\omega}_r\mathbf{M}\psi_r\psi_r^T\mathbf{M} \tag{3.345}$$

Of course, in most cases such a matrix will not be proportional, but that is not a hindrance to our purposes. Instead, the only important issue is that it must provide appropriate levels of damping. This can be verified by assessing the ratio of energy dissipated to energy stored in one cycle of harmonic motion of frequency ω and duration $T = 2\pi/\omega$. For this purpose, we impose onto the system a forced vibration of the general form $\mathbf{u}(\mathbf{x},t) = \mathbf{v}(\mathbf{x})\sin\omega t$, $\dot{\mathbf{u}}(\mathbf{x},t) = \omega\mathbf{v}(\mathbf{x})\cos\omega t$ and without loss of generality we proceed to scale this vector so that $\mathbf{v}^T\mathbf{M}\mathbf{v} = 1$ (it is the ratios that matter, and not the actual amplitudes).

---

[9] E. Kausel, "Damping matrices revisited," *J. Eng. Mech. ASCE*, 140 (8), 2014.

To assess the energy-dissipating characteristics, we start by expressing the arbitrary shape vector $\mathbf{v}$ as well as the trial functions in terms of the normal modes, and for the sake of simplicity, we also assume that the modes are so normalized that $\mathbf{\Phi}^T \mathbf{M} \mathbf{\Phi} = \mathbf{I}$. Hence,

$$\mathbf{v} = \mathbf{\Phi} \mathbf{c} = \sum_{j=1}^{N} \phi_j c_j \tag{3.346}$$

$$\mathbf{\psi}_r = \mathbf{\Phi} \mathbf{d}_r = \sum_{j=1}^{N} \phi_j d_{rj} \tag{3.347}$$

in which the c_j, d_{rj} are unknown coefficients. This implies in turn

$$\mathbf{v}^T \mathbf{M} \mathbf{v} = \mathbf{c}^T \mathbf{\Phi} \mathbf{M} \mathbf{\Phi} \mathbf{c} = \mathbf{c}^T \mathbf{c} = |\mathbf{c}|^2 = 1 \tag{3.348}$$

$$\mathbf{v}^T \mathbf{K} \mathbf{v} = \mathbf{c}^T \mathbf{\Phi} \mathbf{K} \mathbf{\Phi} \mathbf{c} = \mathbf{c}^T \mathbf{\Omega}^2 \mathbf{c} = \sum_{j=1}^{N} c_j^2 \, \omega_j^2 \tag{3.349}$$

$$\mathbf{\psi}_r^T \mathbf{M} \mathbf{\psi}_r = \mathbf{d}_r^T \mathbf{\Phi} \mathbf{M} \mathbf{\Phi} \mathbf{d}_r = \mathbf{d}_r^T \mathbf{d}_r = |\mathbf{d}_r|^2 = 1 \tag{3.350}$$

Also,

$$
\begin{aligned}
\mathbf{v}^T \mathbf{C} \mathbf{v} &= \sum_{r=1}^{M} \left(2\xi_r \varpi_r \right) \left(\mathbf{v}^T \mathbf{M} \mathbf{\psi}_r \right) \left(\mathbf{\psi}_r^T \mathbf{M} \mathbf{v} \right) = 2 \sum_{r=1}^{M} \xi_r \varpi_r \left(\mathbf{c}^T \mathbf{d}_r \right) \left(\mathbf{d}_r^T \mathbf{c} \right) \\
&= 2 \sum_{r=1}^{M} \xi_r \varpi_r |\mathbf{c}|^2 |\mathbf{d}_r|^2 \cos^2 \theta_r \\
&= 2 \sum_{r=1}^{M} \xi_r \, \varpi_r \cos^2 \theta_r
\end{aligned}
\tag{3.351}
$$

where θ_r is the angle between the vectors $\mathbf{c}$ and $\mathbf{d}_r$, which in most cases will neither be parallel nor orthogonal to each other. The effective damping in one cycle of motion of frequency ω and amplitude $\mathbf{v}$ is then assessed from the ratio of energy dissipated to energy stored during that cycle, that is,

$$\xi_{\text{eff}} \frac{\omega}{\omega_{ref}} = \frac{1}{4\pi} \frac{E_d}{E_s} = \frac{1}{4\pi} \frac{\int_0^T \dot{\mathbf{u}}^T \mathbf{C} \dot{\mathbf{u}} \, dt}{\frac{1}{2} \mathbf{v}^T \mathbf{K} \mathbf{v}} = \frac{1}{4\pi} \frac{\mathbf{v}^T \mathbf{C} \mathbf{v}}{\frac{1}{2} \mathbf{v}^T \mathbf{K} \mathbf{v}} \int_0^{2\pi/\omega} \omega^2 \cos^2 \omega t \, dt \tag{3.352}$$

where ω_{ref} is an arbitrary reference frequency used to define the effective viscous damping ξ_{eff}. Carrying out the substitutions and integrating, we obtain

$$\xi_{\text{eff}} = \frac{\sum\limits_{r=1}^{M} \xi_r \omega_{ref} \varpi_r \cos^2 \theta_r}{\sum\limits_{j=1}^{N} c_j^2 \, \omega_j^2} < \frac{\sum\limits_{r=1}^{M} \xi_r \, \omega_{ref} \, \varpi_r}{\sum\limits_{j=1}^{N} c_j^2 \, \omega_j^2} \tag{3.353}$$

whose denominator satisfies the bounds $\omega_1^2 \leq \sum\limits_{j=1}^{N} c_j^2 \, \omega_j^2 \leq \omega_N^2$, so

$$0 < \xi_{eff} < \frac{\displaystyle\sum_{r=1}^{M} \xi_r\, \omega_{ref}\, \varpi_r}{\displaystyle\sum_{j=1}^{N} c_j^2\, \omega_j^2} \tag{3.354}$$

We conclude that $\xi_{eff} > 0$, that is, *the effective damping is never negative*, no matter the frequency or shape in which we should drive the system. Also, from the enclosure theorem, we know that the pivots ϖ_r obtained with the trial functions are bracketed by the fundamental and highest frequency, $\omega_1 \le \varpi_r \le \omega_N$.

Example Nonproportional Damping Matrix for 5-DOF System

Consider an upright, discrete cantilever shear beam composed of five equal springs k and four equal masses m which are topped by a fifth mass that is half as large. We choose to number the masses from the top down, in which case the downward order of the masses matches the order of the DOF in the matrices and vectors. We proceed to construct a damping matrix by choosing the three arbitrary trial functions

$$\psi_1^T = \{5 \quad 4 \quad 3 \quad 2 \quad 1\} \tag{3.355}$$

$$\psi_2^T = \{3 \quad 1 \quad -3 \quad -2 \quad -1\} \tag{3.356}$$

$$\psi_3^T = \{1 \quad -1 \quad 1 \quad -1 \quad 1\} \tag{3.357}$$

and we assign the same fraction of damping ξ to each. Hence

$$\mathbf{C} = 2\xi\mathbf{M}\left\{\sum_{r=1}^{3} \sqrt{\frac{\psi_r \mathbf{K} \psi_r^T}{\psi_r^T \mathbf{M} \psi_r}}\; \frac{\psi_r \psi_r^T}{\psi_r^T \mathbf{M} \psi_r}\right\}\mathbf{M} \tag{3.358}$$

In this expression, the square root is the Rayleigh quotient for the frequency ϖ_r of the trial functions and the denominator of the dyadic product is the normalization. The results are as follows.

Damping Matrix

$$\mathbf{C} = \xi\sqrt{km}\left\{\begin{array}{ccccc} 0.5675 & -0.1034 & 0.0517 & -0.6854 & 0.3052 \\ -0.1034 & 1.2335 & -1.0043 & 0.7702 & -0.9107 \\ 0.0517 & -1.0043 & 2.0116 & -0.0987 & 1.2464 \\ -0.6854 & 0.7702 & -0.0987 & 1.3740 & -0.6088 \\ 0.3052 & -0.9107 & 1.2464 & -0.6088 & 0.9914 \end{array}\right\} \tag{3.359}$$

Frequencies of Normal Modes (Exact): (rad/s)

$$\omega_n = \sqrt{k/m}\,[0.3129,\ 0.9080,\ 1.4142,\ 1.7820,\ 1.9754] \tag{3.360}$$

Frequencies Associated with Trial Functions (Pivots)

$$\varpi_r = \sqrt{k/m}\,[0.3430,\ 1.0860,\ 1.9437] \tag{3.361}$$

Test of Orthogonality

(Using normalized modes, which have dimension $1/\sqrt{m}$)

$$\boldsymbol{\Phi}^{T}\mathbf{C}\boldsymbol{\Phi} = \xi\sqrt{\frac{k}{m}}\begin{Bmatrix} 0.7710 & 0.2996 & -0.1811 & 0.0546 & 0.1691 \\ 0.2996 & 1.4928 & -0.8528 & 0.2792 & 0.5340 \\ -0.1811 & -0.8528 & 0.6338 & 0.0522 & 0.3885 \\ 0.0546 & 0.2792 & 0.0522 & 0.3582 & 1.1035 \\ 0.1691 & 0.5340 & 0.3885 & 1.1035 & 3.4896 \end{Bmatrix} \tag{3.362}$$

This matrix is not diagonal; hence the damping matrix is nonproportional. The damping in each normal mode is obtained dividing the diagonal elements by twice the modal frequencies, which gives

$$\xi_{eff} = \xi\left\{1.2322 \quad 0.8220 \quad 0.2241 \quad 0.1005 \quad 0.8833\right\} \tag{3.363}$$

Observe that the fractions of damping in the first, second, and last modes are close to the damping prescribed via the trial functions. This is because the chosen trial functions are rough estimates for each of these modes, as can be seen by comparing the respective frequencies given earlier.

3.5.9 Weighted Modal Damping: The Biggs–Roësset Equation

Certain classes of problems in structural dynamics can involve damping characteristics that not only exhibit large variations between parts of the structure, but also may differ in their nature. For example, in dynamic models for soil–structure interaction problems, the damping in the structure may be low to moderate in value and of a hysteretic type (i.e., independent of the rate of deformation), while the radiation damping in the subgrade – the energy lost to the system through waves transmitted to the soil – is better modeled with viscous elements, and can attain relatively large values. Hence, a typical soil–structure interaction system does not have normal modes, a situation that has motivated the development of simple procedures to circumvent this difficulty. In essence, these procedures establish rules for assigning "equivalent" modal damping ratios to the modes of the undamped system so as to preserve the advantages of classical modal analysis.

In 1973, J. M. Roësset presented a generalization of a procedure originally devised by J. M. Biggs to assign damping values to a system that does not possess modes in the classical sense. In essence, this procedure uses weighted averages of the damping values in the components. The weighting factors used are based on the ratios between the energies dissipated in each component to the strain energy stored in the system when it is forced to vibrate in a given normal mode:

$$\xi_j^{eq} = \frac{1}{4\pi}\frac{\sum\limits_{i}\varepsilon_{ij}E_{sij}}{\sum\limits_{i}E_{sij}} \tag{3.364}$$

in which i is the index for each structural component, j is the modal index, and

ξ_j^{eq} = equivalent damping to be used for the jth mode

E_{sij} = strain energy in the ith component of the system when the system vibrates in the jth mode

ε_{ij} = ratio of energy dissipated to energy stored in ith component when vibrating in the jth mode

In an earlier section, we saw that the energy ratios for SDOF systems with viscous and hysteretic damping were

$\varepsilon_v = 4\pi \xi_v \dfrac{\omega}{\omega_n}$ for viscous damping, and

$\varepsilon_h = 4\pi \xi_h$ for hysteretic damping (assuming positive driving frequency)

Clearly, if the SDOF system has both viscous and hysteretic damping, then the combined energy ratio is

$$\varepsilon_v = 4\pi \left[\xi_v \frac{\omega}{\omega_n} + \xi_h \right] \tag{3.365}$$

In the case of a structure with many DOF, the system can be imagined as consisting of many structural components (members, elements, or springs) that have both stiffness and dampers associated with them. The energy ratio in a given ith component when the system is forced to vibrate in its jth classical mode with frequency ω_j is then

$$\varepsilon_{ij} = 4\pi \left[\xi_{vi} \frac{\omega_j}{\omega_i} + \xi_{hi} \right] \tag{3.366}$$

in which

$$\frac{\xi_{vi}}{\omega_i} \equiv \frac{c_i}{2k_i} \qquad \left(= \frac{\xi_{vi}\,\omega_i}{\omega_i^2} = \frac{c_i}{2m_i}\frac{m_i}{k_i} \right) \tag{3.367}$$

with k_i, c_i being the stiffness and damping constants for the component. Notice that in this expression, ω_i is just an arbitrary (auxiliary) frequency that we use to *define* (or interpret physically) the fraction of viscous damping ξ_{vi} in the ith component. The ratio actually used depends solely on the stiffness and damper constants of the component. The *Biggs-Roësset equation* is then

$$\boxed{\xi_j^{eq} = \frac{\displaystyle\sum_i \left[\xi_{vi} \frac{\omega_j}{\omega_i} + \xi_{hi} \right] E_{sij}}{\displaystyle\sum_i E_{sij}}} \tag{3.368}$$

For example, in the case of an MDOF consisting of hysteretically damped springs, viscous dampers, and masses, the strain energies are $E_{sij} = \frac{1}{2} k_i \Delta_{ij}^2$, in which Δ_{ij} is the (local) modal distortion of the ith spring when the structure vibrates in the jth mode (i.e., the difference in modal values for the two nodes being connected by the ith spring, projected along its direction of deformation). In this case, the weighted modal damping is

$$\xi_j^{eq} = \frac{\sum_i \left[\frac{1}{2} c_i \omega_j + \xi_{hi} k_i\right] \Delta_{ij}^2}{\sum_i k_i \Delta_{ij}^2} \tag{3.369}$$

We will show next that in a structure that has viscous damping only, the Biggs–Roësset criterion is equivalent to keeping the diagonal terms in the modal transformation $\mathbf{\Phi}^T \mathbf{C} \mathbf{\Phi}$ and discarding the off-diagonal terms. Assume that we have assembled the damping matrix $\mathbf{C}$, and that the system is forced to vibrate in the jth mode. Then

$$\mathbf{u} = \boldsymbol{\phi}_j \sin \omega_j t \qquad \dot{\mathbf{u}} = \boldsymbol{\phi}_j \, \omega_j \cos \omega_j t \tag{3.370}$$

The energies stored and dissipated in one cycle of motion are then

$$E_s = \tfrac{1}{2} \boldsymbol{\phi}_j^T \mathbf{K} \boldsymbol{\phi}_j = \tfrac{1}{2} \omega_j^2 \boldsymbol{\phi}_j^T \mathbf{M} \boldsymbol{\phi}_j \qquad E_d = \pi \omega_j \, \boldsymbol{\phi}_j^T \mathbf{C} \boldsymbol{\phi}_j \tag{3.371}$$

which implies

$$\xi_j^{eq} = \frac{1}{4\pi} \frac{E_d}{E_s} = \frac{1}{4\pi} \frac{\pi \omega_j \, \boldsymbol{\phi}_j^T \mathbf{C} \boldsymbol{\phi}_j}{\tfrac{1}{2} \omega_j^2 \, \boldsymbol{\phi}_j^T \mathbf{M} \boldsymbol{\phi}_j} = \frac{1}{2\omega_j} \frac{\boldsymbol{\phi}_j^T \mathbf{C} \boldsymbol{\phi}_j}{\boldsymbol{\phi}_j^T \mathbf{M} \boldsymbol{\phi}_j} \tag{3.372}$$

That is,

$$2\omega_j \, \xi_j^{eq} = \frac{\boldsymbol{\phi}_j^T \mathbf{C} \boldsymbol{\phi}_j}{\boldsymbol{\phi}_j^T \mathbf{M} \boldsymbol{\phi}_j} \tag{3.373}$$

which corresponds exactly to the definition of modal damping in terms of the diagonal terms in the modal transformation of the equations of motion. Hence, discarding the off-diagonal terms is the optimal strategy from the point of view of average energy dissipation, and it works surprisingly well, even when these terms are not small.

3.6 Support Motions in MDOF Systems

In general, support motions may consist of translations and rotations in (or about) all three coordinate directions (i.e., multiple seismic components). For simplicity, however, we shall begin by considering the much simpler case of a structure with only one translational DOF at each mass point that is contained in a single plane. This structure responds dynamically to a single component support motion at the base that acts in the same direction as the motion of the mass points (usually the lateral motion). In a later section, we will extend these results to the general case of more than 1 DOF for each mass, and multiple components of support motion. Also, while the support motion could be the product of various physical causes, we shall refer to it simply as the earthquake or seismic motion.

3.6.1 Structure with Single Translational DOF at Each Mass Point

Consider a lumped mass idealization of a structure with one translational DOF at each mass point, say a building that deforms laterally in response to an earthquake. We choose to number the masses *from the top down*, because in that case the order of the masses (i.e., the floor "numbers") agree with the indices that we use for the components of vectors and matrices for the structure. We define the following vectors:

$$\mathbf{u} = \begin{Bmatrix} u_1 \\ \vdots \\ u_n \end{Bmatrix} \quad = absolute \text{ displacements} \tag{3.374}$$

$$\mathbf{v} = \begin{Bmatrix} v_1 \\ \vdots \\ v_n \end{Bmatrix} \quad = relative \text{ displacements} \tag{3.375}$$

Clearly, the relationship between these two is

$$\mathbf{v} = \begin{Bmatrix} u_1 - u_g \\ \vdots \\ u_n - u_g \end{Bmatrix} = \begin{Bmatrix} u_1 \\ \vdots \\ u_n \end{Bmatrix} - \begin{Bmatrix} 1 \\ \vdots \\ 1 \end{Bmatrix} u_g = \mathbf{u} - \mathbf{e}\,u_g \tag{3.376}$$

where $\mathbf{e}$ is the unit *rigid body* displacement vector, which in this case is a column of ones. On the other hand, the dynamic equilibrium equation is

$$\mathbf{M}\ddot{\mathbf{u}} + \mathbf{C}\dot{\mathbf{v}} + \mathbf{K}\mathbf{v} = \mathbf{0} \tag{3.377}$$

which states that the net forces on the structure, which is the sum of the inertia forces and the forces associated with the deformation (which depends on the relative motions), must be zero (no external forces acting on the masses). From here, we obtain the two equations

$$\boxed{\mathbf{M}\ddot{\mathbf{v}} + \mathbf{C}\dot{\mathbf{v}} + \mathbf{K}\mathbf{v} = -\mathbf{M}\mathbf{e}\,\ddot{u}_g} \qquad \text{in terms of relative motions} \tag{3.378}$$

$$\boxed{\mathbf{M}\ddot{\mathbf{u}} + \mathbf{C}\dot{\mathbf{u}} + \mathbf{K}\mathbf{u} = \mathbf{C}\mathbf{e}\,\dot{u}_g + \mathbf{K}\mathbf{e}\,u_g} \qquad \text{in terms of absolute motions} \tag{3.379}$$

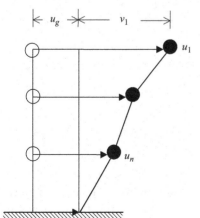

Figure 3.16. Absolute versus relative displacements.

Solution by Modal Superposition (Proportional Damping)

Assuming that the damping matrix is *proportional*, normal modes exist, so we can obtain the solution by means of modal synthesis:

(a) Relative motions

$$\mathbf{v} = \mathbf{\Phi}\mathbf{q} = \sum_{j=1}^{n} \phi_j\, q_j(t) \tag{3.380}$$

Substituting into the equation of motion and multiplying by the transpose of the modal matrix, we obtain

$$(\mathbf{\Phi}^T\mathbf{M}\mathbf{\Phi})\ddot{\mathbf{q}} + (\mathbf{\Phi}^T\mathbf{C}\mathbf{\Phi})\dot{\mathbf{q}} + (\mathbf{\Phi}^T\mathbf{K}\mathbf{\Phi})\mathbf{q} = -(\mathbf{\Phi}^T\mathbf{M}\mathbf{e})\ddot{u}_g \tag{3.381}$$

Since normal modes exist, each of the expressions in parentheses on the left-hand side are diagonal matrices. It follows that the modal equations decouple, and the previous equation is equivalent to the scalar equations

$$\mu_j\,\ddot{q}_j + \eta_j\,\dot{q}_j + \kappa_j\,q_j = -\mu_j\,\gamma_j\,\ddot{u}_g \qquad j = 1,\dots n \tag{3.382}$$

or

$$\ddot{q}_j + 2\xi_j\omega_j\,\dot{q}_j + \omega_j^2\,q_j = -\gamma_j\,\ddot{u}_g \tag{3.383}$$

in which

$$\mu_j = \phi_j^T\mathbf{M}\phi_j \qquad = \text{modal mass} \tag{3.384}$$

$$\eta_j = \phi_j^T\mathbf{C}\phi_j \qquad = \text{modal dashpot} \tag{3.385}$$

$$\kappa_j = \phi_j^T\mathbf{K}\phi_j \qquad = \text{modal stiffness} \tag{3.386}$$

$$\gamma_j = \frac{\phi_j^T\mathbf{M}\mathbf{e}}{\phi_j^T\mathbf{M}\phi_j} \qquad = \text{modal participation factor} \tag{3.387}$$

Except for the participation factor, these equations are the same as those for an SDOF system. Indeed, if we interpret the product $\gamma_j\,\ddot{u}_g$ as the modal ground excitation, then the analogy is complete. Hence, the modal response for quiescent initial conditions (i.e., system starting from rest) is simply

$$\boxed{q_j(t) = -\mu_j\,\gamma_j\,h_j * \ddot{u}_g} \tag{3.388}$$

with

$$h_j(t) = \frac{1}{\mu_j\,\omega_{dj}}\,e^{-\xi_j\omega_j t}\sin\omega_{dj}t \qquad = \textit{modal impulse response function} \tag{3.389}$$

Now that we have found the relative displacements, we may also wish to compute the absolute displacements and velocities. These response functions, however, require a numerical integration of the ground motion record, an operation that is fraught with potential problems. While the true absolute accelerations could be obtained by manipulating the relative displacements, we will defer their evaluation until the next section. On the other hand, the *pseudo-accelerations*, which involve neglecting the damping terms in the modal solution and are a close approximation to the true absolute accelerations, can easily be determined as a byproduct of the relative displacements, as we shall see next.

From the equation of motion, we obtain directly

$$\ddot{\mathbf{u}} = -\mathbf{M}^{-1}(\mathbf{C}\dot{\mathbf{v}} + \mathbf{K}\mathbf{v}) \tag{3.390}$$

Neglecting damping and expressing the relative displacement in terms of the model expansion, we have

$$\ddot{\mathbf{u}} \approx -\mathbf{M}^{-1}\mathbf{K}\mathbf{v} = -\mathbf{M}^{-1}\mathbf{K}\mathbf{\Phi}\mathbf{q} \tag{3.391}$$

But from the eigenvalue problem, $\mathbf{M}^{-1}\mathbf{K}\mathbf{\Phi} = \mathbf{\Phi}\mathbf{\Omega}^2$, so

$$\ddot{\mathbf{u}} \approx -\mathbf{\Phi}\mathbf{\Omega}^2\mathbf{q} = -\sum_{j=1}^{n} \phi_j\, \omega_j^2\, q_j \qquad = pseudo-acceleration \tag{3.392}$$

These accelerations can readily be computed once the modal responses q_j are known.

(b) Absolute motions

The modal expansion is now

$$\mathbf{u} = \mathbf{\Phi}\mathbf{q} = \sum_{j=1}^{n} \phi_j\, q_j(t) \tag{3.393}$$

Please observe that the meaning of $\mathbf{q}$ in this expression is *not* the same as in the previous section. If we use here the same symbol, it is merely to avoid notational proliferation; and since we shall not mix the two formulations, there won't be any danger of confusion.

Substituting the modal expansion into the differential equation for absolute motions, and multiplying by the transposed modal matrix, we obtain

$$\mathbf{\Phi}^T\mathbf{M}\mathbf{\Phi}\ddot{\mathbf{q}} + \mathbf{\Phi}^T\mathbf{C}\mathbf{\Phi}\dot{\mathbf{q}} + \mathbf{\Phi}^T\mathbf{K}\mathbf{\Phi}\mathbf{q} = \mathbf{\Phi}^T\mathbf{C}\mathbf{e}\dot{u}_g + \mathbf{\Phi}^T\mathbf{K}\mathbf{e}u_g \tag{3.394}$$

Now, the rigid body vector can be expanded again in terms of the normal modes by means of the modal participation factors, that is,

$$\mathbf{e} = \mathbf{\Phi}\tilde{\mathbf{a}} \tag{3.395}$$

Hence

$$(\mathbf{\Phi}^T\mathbf{M}\mathbf{\Phi})\ddot{\mathbf{q}} + (\mathbf{\Phi}^T\mathbf{C}\mathbf{\Phi})\dot{\mathbf{q}} + (\mathbf{\Phi}^T\mathbf{K}\mathbf{\Phi})\mathbf{q} = (\mathbf{\Phi}^T\mathbf{C}\mathbf{\Phi})\tilde{\mathbf{a}}\dot{u}_g + (\mathbf{\Phi}^T\mathbf{K}\mathbf{\Phi})\tilde{\mathbf{a}}u_g \tag{3.396}$$

Once more, if normal modes exist, all matrix products in parentheses are diagonal matrices. Hence, this is equivalent to the scalar modal equations

$$\mu_j\,\ddot{q}_j + \eta_j\,\dot{q}_j + \kappa_j\,q_j = \gamma_j(\eta_j\,\dot{u}_g + \kappa_j\,u_g) \qquad j = 1,\ldots n \tag{3.397}$$

or

$$\ddot{q}_j + 2\xi_j\omega_j\,\dot{q}_j + \omega_j^2\,q_j = \gamma_j(2\xi_j\omega_j\,\dot{u}_g + \omega_j^2\,u_g) \tag{3.398}$$

These equations are the same as the corresponding ones for an SDOF system, the only difference being that the ground motion is scaled up (or down) by the participation factors. It follows that

$$q_j(t) = \gamma_j \, h_{aj} * u_g \tag{3.399}$$

in which

$$h_{aj}(t) = \omega_j e^{-\xi_j \omega_j t} \left\{ 2\xi_j \cos \omega_{dj} t + \frac{\omega_j}{\omega_{dj}} (1 - 2\xi_j^2) \sin \omega_{dj} t \right\} \tag{3.400}$$

is the impulse response function for absolute displacement due to absolute ground displacement. This function has the important property that it is invariant under a change of input–output type, so it is identical to the impulse response function for absolute acceleration due to ground acceleration. Hence,

$$\boxed{\ddot{q}_j(t) = \gamma_j \, h_{aj} * \ddot{u}_g} \tag{3.401}$$

and

$$\boxed{\ddot{\mathbf{u}} = \mathbf{\Phi} \ddot{\mathbf{q}} = \sum_{j=1}^{n} \boldsymbol{\phi}_j \, \ddot{q}_j(t)} \tag{3.402}$$

Hence, we can obtain the absolute acceleration response directly from the ground acceleration, without any intermediate steps. Of course, if we also wished to compute the relative displacements from this expression without repeating convolutions with appropriate impulse response functions, we would have to neglect damping, which would lead us to

$$\mathbf{v} \approx -\mathbf{\Phi}^{-2} \mathbf{\Omega}^{-2} \ddot{\mathbf{q}} = -\sum_{j=1}^{n} \boldsymbol{\phi}_j \, \omega_j^{-2} \ddot{q}_j(t) \tag{3.403}$$

3.6.2 MDOF System Subjected to Multicomponent Support Motion

We consider now the general case of a structure that rests on a *single* rigid support system, say a foundation, that can both translate and rotate, as illustrated in Figure 3.17. Each mass point or *joint* (we avoid another synonym, *node*, which could be confused with *mode*) has up to *six* DOF, namely three translations and three rotations. We assume, as usual, that the rotations are sufficiently small that the tangents of the angles of rotation can be replaced by the angles themselves, and that these can be treated as vectors. In addition, we consider a right-handed system of coordinates in which rotations follow the usual right-hand rule.

Let x_i, y_i, z_i be the coordinates of the ith joint in the structure, and x_0, y_0, z_0 those of the support point (i.e., the foundation). Omitting for simplicity the index i in each of the component of motion, the absolute displacement vector for the ith joint has the six components

$$\mathbf{u}_i^T = \left\{ u_x \quad u_y \quad u_z \quad \theta_x \quad \theta_y \quad \theta_z \right\} \tag{3.404}$$

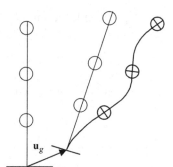

Figure 3.17. Structure undergoing multicomponent support motion.

On the other hand, the support motion vector is of the form

$$\mathbf{u}_g^T = \left\{ u_{g1} \quad \cdots \quad u_{g6} \right\} \tag{3.405}$$

Here, we use for the ground motion components numerical subindices instead of the usual x, y, and so forth, to allow for simple expressions involving summations over these indices. The first three are translations, while the remaining three are rotations.

If the structure had *no* masses whatsoever, then no inertia forces would develop anywhere, and the structure would simply follow the ground motion at all times and would execute a *rigid body motion*. If so, the rigid body translations and rotations would be given by

$$\mathbf{u}_{iR} = \mathbf{E}_i \mathbf{u}_g \tag{3.406}$$

in which

$$\mathbf{E}_i = \left\{ \begin{array}{cccccc} 1 & 0 & 0 & 0 & +(z_i - z_0) & -(y_i - y_0) \\ 0 & 1 & 0 & -(z_i - z_0) & 0 & +(x_i - x_0) \\ 0 & 0 & 1 & +(y_i - y_0) & -(x_i - x_0) & 0 \\ 0 & 0 & 0 & 1 & 0 & 0 \\ 0 & 0 & 0 & 0 & 1 & 0 \\ 0 & 0 & 0 & 0 & 0 & 1 \end{array} \right\} \tag{3.407}$$

is the rigid-body matrix for the ith joint. This rigid body motion does not deform the structure in any way, so it produces no internal elastic or damping forces. Of course, the actual structure does have mass, so it will deform in the course of the support excitation and the motion will deviate from that predicted by the rigid body formula. Clearly, the total motion must be the sum of the rigid body motions and the displacements relative to the support, that is,

$$\mathbf{u}_i = \mathbf{v}_i + \mathbf{E}_i \mathbf{u}_g \tag{3.408}$$

The system's absolute displacement vector is then

$$\left\{ \begin{array}{c} \mathbf{u}_1 \\ \vdots \\ \mathbf{u}_N \end{array} \right\} = \left\{ \begin{array}{c} \mathbf{v}_1 \\ \vdots \\ \mathbf{v}_N \end{array} \right\} + \left\{ \begin{array}{c} \mathbf{E}_1 \\ \vdots \\ \mathbf{E}_N \end{array} \right\} \mathbf{u}_g \tag{3.409}$$

with N being the total number of joints. Hence, there are $n = 6N$ degrees of freedom. We write this more briefly as

$$\mathbf{u} = \mathbf{v} + \mathbf{E}\mathbf{u}_g \tag{3.410}$$

$\mathbf{E}$ is the rigid-body matrix for the structure as a whole, which is of size $6N \times 6$; it consists of the six rigid body vectors:

$$\mathbf{E} = \{\mathbf{e}_1 \quad \cdots \quad \mathbf{e}_6\} \tag{3.411}$$

The equation of motion is again the sum of the inertia forces, which are proportional to the absolute accelerations, and the internal elastic and damping forces, which depend on the relative motions:

$$\mathbf{M}\ddot{\mathbf{u}} + \mathbf{C}\dot{\mathbf{v}} + \mathbf{K}\mathbf{v} = \mathbf{0} \tag{3.412}$$

Using, as before, the relationship between relative and absolute motions, we obtain

$$\boxed{\mathbf{M}\ddot{\mathbf{v}} + \mathbf{C}\dot{\mathbf{v}} + \mathbf{K}\mathbf{v} = -\mathbf{M}\mathbf{E}\,\ddot{\mathbf{u}}_g} \qquad \text{in terms of } \textit{relative} \text{ motions} \tag{3.413}$$

$$\boxed{\mathbf{M}\ddot{\mathbf{u}} + \mathbf{C}\dot{\mathbf{u}} + \mathbf{K}\mathbf{u} = \mathbf{C}\mathbf{E}\,\dot{\mathbf{u}}_g + \mathbf{K}\mathbf{E}\mathbf{u}_g} \qquad \text{in terms of } \textit{absolute} \text{ motions} \tag{3.414}$$

We apply once more modal superposition, which leads us to modal equations of motion that are entirely similar to those we saw before, so they need not be repeated. The only difference is that now the right-hand sides will contain summations over the six components of ground motion, and that each mode and each ground motion component will have different participation factors, which together will form a matrix of participation factors. The latter are in turn the modal coordinates of the rigid body matrix:

$$\mathbf{E} = \mathbf{\Phi}\,\mathbf{\Gamma} \qquad\qquad\qquad \text{Modal expansion of rigid body matrix} \tag{3.415}$$

$$\mathbf{\Gamma} = \mathbf{\Phi}^{-1}\mathbf{E} = \mathbb{M}^{-1}\mathbf{\Phi}^T\mathbf{M}\mathbf{E} \qquad \text{Matrix of participation factors} \tag{3.416}$$

The matrix of participation factors consists of six vectors

$$\mathbf{\Gamma} = \{\boldsymbol{\gamma}_1 \quad \cdots \quad \boldsymbol{\gamma}_6\} \tag{3.417}$$

For example, the participation factors for ground motion in direction x are now

$$\boldsymbol{\gamma}_1 = \{\gamma_{j1}\} = \left\{\frac{\boldsymbol{\phi}_j^T\mathbf{M}\mathbf{e}_1}{\boldsymbol{\phi}_j^T\mathbf{M}\boldsymbol{\phi}_j}\right\} \tag{3.418}$$

In general, since $\mathbf{E} = \{\mathbf{e}_k\} = \mathbf{\Phi}\mathbf{\Gamma} = \{\mathbf{\Phi}\boldsymbol{\gamma}_k\}$, and recalling also the modal expansion theorem, we see that the participation factors can be interpreted as the *modal coordinates of the rigid body vectors*. The modal equation for *relative motions* and its solution are now

$$\ddot{q}_j + 2\xi_j\omega_j\,\dot{q}_j + \omega_j^2\,q_j = -\sum_{k=1}^{6}\gamma_{jk}\,\ddot{u}_{gk} \tag{3.419}$$

$$q_j(t) = -\mu_j\,h_j * \sum_{k=1}^{6}\gamma_{jk}\,\ddot{u}_{gk} \tag{3.420}$$

As before, the modal summation is then

$$\mathbf{v} = \boldsymbol{\Phi}\mathbf{q} = \sum_{j=1}^{n} \boldsymbol{\phi}_j \, q_j(t) \qquad (3.421)$$

3.6.3 Number of Modes in Modal Summation

As we saw in the previous sections, the evaluation of the response of an n-DOF system requires, at least in principle, determining all modes and frequencies, computing n convolutions, and carrying out an equal number of modal summations for each response quantity of interest. As it turns out, however, it is often not necessary to include *all* of the modes to obtain acceptable answers. In fact, a few modes may well be sufficient for this purpose. How many modes are necessary? The answer depends on a number of considerations:

1. *Modal frequencies versus frequency spectrum of excitation.* Generally, only the modes whose frequencies fall in the range where the excitation has any significant power are important. In the case of ground motions, earthquakes at short epicentral distances have most of their energy in the range from 1 Hz to 5 Hz, and, virtually all of it below 30 Hz. Hence, modes whose frequencies are above 10 or 15 Hz are comparatively unimportant. Recalling also the rule of thumb that states that the fundamental frequency of a steel-frame building is approximately 10 divided by the number of stories (in Hz), and that the frequencies for the higher modes follow roughly in the ratios 1:3:5..., it can be seen that in many cases, the first few modes are the most important. On the other hand, some of the higher modes may play a more dominant role in very tall buildings.

2. *Geometry of the excitation.* The spatial distribution of the structural loads is important in the sense that loads whose spatial variation is in the shape of a higher mode will excite preferentially (or exclusively) that mode. For example, a load near the top of the structure will excite mostly the lower modes, while a load near the base may also significantly excite the higher modes. In the case of seismic excitations, the distribution of the fictitious inertia loads over the height of the structure is controlled by the rigid body vector. Its modal expansion is affected mostly by the first few modes, inasmuch as the relative importance of each mode is determined by the product of the participation factor and the largest modal amplitude. For this reason, the fundamental mode plays the most important role in seismic response analyses.

3. *Type and location of response.* Depending on what physical quantity is being computed and where in the structure the response takes place (absolute accelerations at the top, relative displacements at the center, shear and overturning moment at the base, etc.), more or less modes may be necessary.

In the case of seismic loads, it is often possible to measure the adequacy of the modal summation by keeping tab on the total modal mass. Indeed, since participation factors are the modal coordinates of the rigid body vector, it follows that the total mass m of the building is the summation of the structural masses, that is

$$m = \mathbf{e}^T \mathbf{M}\mathbf{e} = \boldsymbol{\gamma}^T \boldsymbol{\Phi}^T \mathbf{M} \boldsymbol{\Phi} \boldsymbol{\gamma} = \boldsymbol{\gamma}^T \mathbb{M} \boldsymbol{\gamma} = \sum_{j=1}^{n} \gamma_j^2 \, \mu_j \qquad (3.422)$$

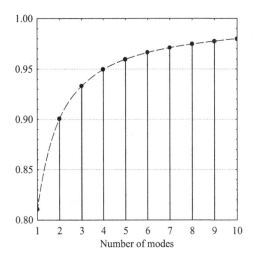

Figure 3.18. Modal mass in shear beam.

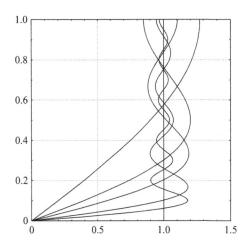

Figure 3.19. Rigid body vector modeled with 1, 2, 3, 6, and 10 shear beam modes.

Hence, by comparing the total mass m of the building against the running summation of the product of modal masses and the squares of the participation factors, one can decide whether or not a sufficient number of modes has already been added. An example of eq. 3.422 applied to a cantilever shear beam evaluated with a variable number n of terms is shown in Figure 3.18. As can be seen, with $n=10$ modes the total modal mass is already 98% of the structural mass, which means that 10 modes suffice to evaluate the seismic response of the shear beam considered herein. On the other hand, Figure 3.19 shows also how well the rigid-body vector $\mathbf{e}$ is modeled by means of a partial superposition with n modes. The closeness of this approximation too is an indication of how well the partial modal summation can be expected to perform. The modal mass check is often made in conjunction with a simple remedial procedure to compensate for the modes neglected, the *static correction*, which will be seen next.

   In the more general case of multiple degrees of freedom in each joint and/or multiple seismic components, the total structural mass properties are described by the rigid body mass matrix

$$\mathbf{M}_0 = \mathbf{E}^T \mathbf{M} \mathbf{E} = \sum_{i=1}^{N} \mathbf{E}_i^T \mathbf{M}_i \, \mathbf{E}_i \tag{3.423}$$

in which the summation is carried out over all mass points. The $\mathbf{M}_i$ are the (usually diagonal) 6×6 mass matrices for each joint, whose first three (identical) elements are the joint mass, while the remaining three are the mass moments of inertia. On the other hand, $\mathbf{M}_0$ is the 6×6 mass matrix of the structure as a whole, as if it were rigid, relative to the base. If we partition $\mathbf{M}_0$ into four 3×3 sub-matrices of the form

$$\mathbf{M}_0 = \begin{Bmatrix} \mathbf{M}_{uu} & \mathbf{M}_{u\theta} \\ \mathbf{M}_{\theta u} & \mathbf{M}_{\theta\theta} \end{Bmatrix} \tag{3.424}$$

then $\mathbf{M}_{uu}$ will contain the total mass of the structure (three identical diagonal elements), $\mathbf{M}_{\theta\theta}$ has the mass moments and mass products of inertia relative to x_0, y_0, z_0, and $\mathbf{M}_{u\theta}$ will have the *static moments* of the masses relative to the base, with which we could compute the location of the center of mass relative to the support.

On the other hand, from the modal expansion of the rigid body matrix, we have

$$\mathbf{M}_0 = \mathbf{E}^T \mathbf{M} \mathbf{E} = \mathbf{\Gamma}^T \mathbf{\Phi}^T \mathbf{M} \mathbf{\Phi} \mathbf{\Gamma} = \mathbf{\Gamma}^T \mathbb{M} \mathbf{\Gamma} \tag{3.425}$$

that is,

$$\mathbf{M}_0 = \sum_{j=1}^{n} \mu_j \begin{Bmatrix} \gamma_{1j}^2 & \cdots & \gamma_{1j}\gamma_{6j} \\ \vdots & \ddots & \vdots \\ \gamma_{6j}\gamma_{1j} & \cdots & \gamma_{6j}^2 \end{Bmatrix} \tag{3.426}$$

As can be seen, we have again an expression with which we could assess if the modal defect is, or is not, acceptable, that is, if we have included sufficient number of modes in our modal summation. Notice, however, that participation factors can be both positive and negative. Hence, only the diagonal elements are always nonnegative, and only they will converge monotonically to the corresponding elements of the mass matrix of the rigid structure (the total mass, and the total mass moments of inertia).

3.6.4 Static Correction

The static correction is a simple procedure that is often used to compensate, albeit in an indirect and approximate fashion, for modes left out in a modal summation. In essence, the static correction is based on the assumption that the neglected modes have an infinite modal frequency. This is equivalent to saying that their contribution to the response is the same as if the loads were quasistatic (i.e., loads varying slowly in comparison to the modal response time).

Assume we decided to include up to s modes in the modal summation, and neglected the remaining r modes, so $n = r + s$ is the total number of DOFs. In accordance with these numbers, we partition the modal matrices and vectors as

$$\mathbf{\Phi} = \{ \mathbf{\Phi}_s \quad \mathbf{\Phi}_r \}, \qquad \mathbf{q} = \begin{Bmatrix} \mathbf{q}_s \\ \mathbf{q}_r \end{Bmatrix} \tag{3.427}$$

The total modal summation in a formulation in terms of *absolute displacements* is then

$$\mathbf{u} = \boldsymbol{\Phi}_s \mathbf{q}_s + \boldsymbol{\Phi}_r \mathbf{q}_r \tag{3.428}$$

Case 1: Seismic Loads

We substitute this expression into the differential equation, multiply by $\boldsymbol{\Phi}_r^T$ (the transpose of the neglected modes), take orthogonality into consideration, and obtain the modal equation

$$\boldsymbol{\Phi}_r^T \mathbf{M} \boldsymbol{\Phi}_r \ddot{\mathbf{q}}_r + \boldsymbol{\Phi}_r^T \mathbf{C} \boldsymbol{\Phi}_r \dot{\mathbf{q}}_r + \boldsymbol{\Phi}_r^T \mathbf{K} \boldsymbol{\Phi}_r \mathbf{q}_r = \boldsymbol{\Phi}_r^T \mathbf{C} \boldsymbol{\Phi}_r \boldsymbol{\gamma}_r \dot{u}_g + \boldsymbol{\Phi}_r^T \mathbf{K} \boldsymbol{\Phi}_r \boldsymbol{\gamma}_r u_g \tag{3.429}$$

If we assume that the response of these modes is quasistatic, this allows us to neglect the inertia terms, in which case the previous modal equation is obviously satisfied if

$$\mathbf{q}_r = \boldsymbol{\gamma}_r u_g \tag{3.430}$$

But

$$\mathbf{e} = \boldsymbol{\Phi}_s \boldsymbol{\gamma}_s + \boldsymbol{\Phi}_r \boldsymbol{\gamma}_r \tag{3.431}$$

so

$$\boldsymbol{\Phi}_r \mathbf{q}_r = \boldsymbol{\Phi}_r \boldsymbol{\gamma}_r u_g = (\mathbf{e} - \boldsymbol{\Phi}_s \boldsymbol{\gamma}_s) u_g \tag{3.432}$$

The term in parentheses is the *modal defect*. It follows that

$$\mathbf{u} = \boldsymbol{\Phi}_s \mathbf{q}_s + (\mathbf{e} - \boldsymbol{\Phi} \boldsymbol{\gamma}_s) u_g = \boldsymbol{\Phi}_s (\mathbf{q}_s - \boldsymbol{\gamma}_s u_g) + \mathbf{e} u_g \tag{3.433}$$

Finally

$$\begin{aligned}
\ddot{\mathbf{u}} &= \mathbf{e} \ddot{u}_g + \boldsymbol{\Phi}_s (\ddot{\mathbf{q}}_s - \boldsymbol{\gamma}_s \ddot{u}_g) \\
&= \mathbf{e} \ddot{u}_g + \sum_{j=1}^{s} \boldsymbol{\phi}_j \, \gamma_j (h_{aj} * \ddot{u}_g - \ddot{u}_g)
\end{aligned} \tag{3.434}$$

Notice that the upper limit in the summation on the right is s, the number of included modes.

Case 2: Structural Loads

Consider a time-varying structural load that changes in amplitude but not in shape, that is, it is of the form $\mathbf{p} = \mathbf{p}_0 f(t)$ where $\mathbf{p}_0$ is a constant vector (more general loads can readily be obtained by superposition of these shape-invariant loads). If $f(t)$ varies slowly in time in comparison to the response time for the structure (i.e., all natural frequencies are much higher than the characteristic frequencies of the load) then the response will be quasistatic and given by

$$\mathbf{u}_{QS} = \mathbf{K}^{-1} \mathbf{p}_0 f(t) = f(t) \sum_{j=1}^{N} \boldsymbol{\phi}_j \frac{\pi_j}{\kappa_j}, \qquad \pi_j = \boldsymbol{\phi}_j^T \mathbf{p}_0, \qquad \kappa_j = \boldsymbol{\phi}_j^T \mathbf{K} \boldsymbol{\phi}_j = \mu_j \omega_j^2 \tag{3.435}$$

where $\mu_j = \boldsymbol{\phi}_j^T \mathbf{M} \boldsymbol{\phi}_j$ is the modal mass. It follows that the quasistatic contribution to the response of the higher modes to be neglected is the difference of the total quasistatic response and the partial summation up to the sth mode to be included in the modal super-position, that is,

$$\Delta u_{QS} = \mathbf{\Phi}_r \mathbf{q}_r = \left[\mathbf{K}^{-1} \mathbf{p}_0 - \sum_{j=1}^{s} \phi_j \frac{\pi_j}{\kappa_j} \right] f(t) \tag{3.436}$$

Hence, the total response is

$$\mathbf{u} = \sum_{j=1}^{s} \phi_j \, \pi_j \, h_j * f + \left[\mathbf{K}^{-1} \mathbf{p}_0 - \sum_{j=1}^{s} \phi_j \frac{\pi_j}{\mu_j \omega_j^2} \right] f(t), \qquad h_j(t) = \frac{e^{-\xi_j \omega_j t}}{\mu_j \omega_{dj}} \sin(\omega_{dj} t) \tag{3.437}$$

Observe that the term in square brackets is computed just once, and that only s modes are needed to evaluate the response.

3.6.5 Structures Subjected to Spatially Varying Support Motion

Consider a *free* body in space that has a total of $N = n + m$ degrees of freedom. This body is subjected to dynamic motions that are prescribed at support nodes encompassing a total of m degrees of freedom $\mathbf{u}_g$ that describe the ground motion. It will thus obey an equation of motion of the form

$$\begin{bmatrix} \mathbf{M}_{11} & \mathbf{M}_{12} \\ \mathbf{M}_{21} & \mathbf{M}_{22} \end{bmatrix} \begin{Bmatrix} \ddot{\mathbf{u}}_1 \\ \ddot{\mathbf{u}}_g \end{Bmatrix} + \begin{bmatrix} \mathbf{C}_{11} & \mathbf{C}_{12} \\ \mathbf{C}_{21} & \mathbf{C}_{22} \end{bmatrix} \begin{Bmatrix} \dot{\mathbf{u}}_1 \\ \dot{\mathbf{u}}_g \end{Bmatrix} + \begin{bmatrix} \mathbf{K}_{11} & \mathbf{K}_{12} \\ \mathbf{K}_{21} & \mathbf{K}_{22} \end{bmatrix} \begin{Bmatrix} \mathbf{u}_1 \\ \mathbf{u}_g \end{Bmatrix} = \begin{Bmatrix} \mathbf{0} \\ \mathbf{p}_g \end{Bmatrix} \tag{3.438}$$

where $\mathbf{u}_1$ is the absolute displacement vector for the n free (unrestrained) degrees of freedom in the superstructure; $\mathbf{u}_g = \mathbf{u}_g(t)$ is the ground motion prescribed at the m degrees of freedom of the support nodes; and $\mathbf{p}_g = \mathbf{p}_g(t)$ is the vector of external reaction forces which are necessary to drive those support points. Hence, from the first equation we infer that

$$\mathbf{M}_{11} \ddot{\mathbf{u}}_1 + \mathbf{C}_{11} \dot{\mathbf{u}}_1 + \mathbf{K}_{11} \mathbf{u}_1 = -\left(\mathbf{M}_{12} \ddot{\mathbf{u}}_g + \mathbf{C}_{12} \dot{\mathbf{u}}_g + \mathbf{K}_{12} \mathbf{u}_g \right) \tag{3.439}$$

The classical approach to solve for spatially varying support motions is based on decomposing the response into two parts, namely

$$\mathbf{u}_1 = \mathbf{u}^s + \mathbf{u}^d \tag{3.440}$$

where

$$\mathbf{u}^s = -\mathbf{K}_{11}^{-1} \mathbf{K}_{12} \mathbf{u}_g \equiv \mathbf{Q} \mathbf{u}_g, \qquad = \text{quasi-static part} \tag{3.441}$$

$$\mathbf{u}^d = \mathbf{u}_1 - \mathbf{u}^s, \qquad = \text{dynamic part} \tag{3.442}$$

The dynamic part satisfies the differential equation of motion

$$\mathbf{M}_{11} \ddot{\mathbf{u}}^d + \mathbf{C}_{11} \dot{\mathbf{u}}^d + \mathbf{K}_{11} \mathbf{u}^d = -\left[\left(\mathbf{M}_{11} \mathbf{Q} + \mathbf{M}_{12} \right) \ddot{\mathbf{u}}_g + \left(\mathbf{C}_{11} \mathbf{Q} + \mathbf{C}_{12} \right) \dot{\mathbf{u}}_g \right] \tag{3.443}$$

On the other hand, the complete system with *all* DOF must be able to undergo (slow) rigid body motions $\mathbf{R}$ such that these cause no deformations, that is, motions that in the absence of inertial forces do not elicit any elastic or damping forces. This implies that $\mathbf{KR} = \mathbf{O}, \mathbf{CR} = \mathbf{O}$, where $\mathbf{R}$ has the structure

$$\mathbf{R} = \left\{ \begin{matrix} \mathbf{R}_1 \\ \mathbf{R}_2 \end{matrix} \right\}, \qquad \mathbf{R}_1 = \left\{ \begin{matrix} \mathbf{T}_1 \\ \vdots \\ \mathbf{T}_n \end{matrix} \right\}, \qquad \mathbf{R}_2 = \left\{ \begin{matrix} \mathbf{T}_{n+1} \\ \vdots \\ \mathbf{T}_{n+m} \end{matrix} \right\} \tag{3.444}$$

where

$$\mathbf{T}_j = \left\{ \begin{matrix} 1 & 0 & 0 & \vdots & 0 & +(z_j - z_0) & -(y_j - y_0) \\ 0 & 1 & 0 & \vdots & -(z_j - z_0) & 0 & -(x_j - x_0) \\ 0 & 0 & 1 & \vdots & +(y_j - y_0) & +(x_j - x_0) & 0 \\ \hdashline 0 & 0 & 0 & \vdots & 1 & 0 & 0 \\ 0 & 0 & 0 & \vdots & 0 & 1 & 0 \\ 0 & 0 & 0 & \vdots & 0 & 0 & 1 \end{matrix} \right\}, \qquad j = 1, \dots N \tag{3.445}$$

in which j is the nodal index, $\mathbf{x}_j = (x_j, y_j, z_j)$ are the nodal coordinates, and $\mathbf{x}_0 = (x_0, y_0, z_0)$ are the coordinates of the arbitrary reference point at which the six rigid-body ground motions are defined. This is typically the geometric center of the ground nodes. This in turn implies that

$$\mathbf{K}_{12}\mathbf{R}_2 = -\mathbf{K}_{11}\mathbf{R}_1, \quad \rightarrow \quad \mathbf{R}_1 = -\mathbf{K}_{11}^{-1}\mathbf{K}_{12}\mathbf{R}_2, \quad \mathbf{R}_1 = \mathbf{Q}\mathbf{R}_2, \quad \mathbf{Q} = -\mathbf{K}_{11}^{-1}\mathbf{K}_{12} \tag{3.446}$$

$$\mathbf{C}_{12}\mathbf{R}_2 = -\mathbf{C}_{11}\mathbf{R}_1 \quad \rightarrow \quad \mathbf{R}_1 = -\mathbf{C}_{11}^{-1}\mathbf{C}_{12}\mathbf{R}_2 \quad \mathbf{R}_1 = \mathbf{Q}\mathbf{R}_2, \quad \mathbf{Q} = -\mathbf{C}_{11}^{-1}\mathbf{C}_{12} \tag{3.447}$$

so

$$\boxed{\mathbf{Q} = -\mathbf{K}_{11}^{-1}\mathbf{K}_{12}} = -\mathbf{C}_{11}^{-1}\mathbf{C}_{12} \tag{3.448}$$

and hence $\mathbf{C}_{11}\mathbf{Q} + \mathbf{C}_{12} = -\mathbf{C}_{12} + \mathbf{C}_{12} = \mathbf{O}$ (here we are assuming that $\mathbf{C}_{11}$ is not singular). The equation of motion then simplifies to

$$\mathbf{M}_{11}\ddot{\mathbf{u}}^d + \mathbf{C}_{11}\dot{\mathbf{u}}^d + \mathbf{K}_{11}\mathbf{u}^d = -\left(\mathbf{M}_{11}\mathbf{Q} + \mathbf{M}_{12}\right)\ddot{\mathbf{u}}_g \tag{3.449}$$

We next proceed to decompose the support motion into two parts, $\mathbf{u}_g = \mathbf{u}_g^{RMB} + \mathbf{u}_g^d$, that is, first a set of six rigid body translation and rotations $\mathbf{u}_g^{RMB} = \mathbf{R}_2\,\mathbf{r}_g$, and then a pure deformational motion $\mathbf{u}_g^d = \mathbf{u}_g - \mathbf{R}_2\mathbf{r}_g$. The two components $\mathbf{u}_g^d, \mathbf{r}_g$ are defined so that $\mathbf{R}_2\mathbf{u}_g^d = \mathbf{0}$, that is, taking the average translation and rotation of the deformational component of the ground motion to be zero. If the nodal points are regarded as unit mass points, this implies in turn that $\mathbf{R}_2^T\mathbf{u}_g = \mathbf{R}_2^T\mathbf{R}_2\,\mathbf{r}_g = \mathbf{S}\,\mathbf{r}_g$, in which $\mathbf{S} = \mathbf{R}_2^T\mathbf{R}_2$ is a 6×6 nonsingular square matrix whose elements represent the total nodal mass, the static moments of inertia and the products of inertia defined by the support nodes relative to the arbitrary (but fixed) rotation center $\mathbf{x}_0$. It follows that

$$\boxed{\mathbf{S} = \mathbf{R}_2^T\mathbf{R}_2}, \qquad \boxed{\mathbf{r}_g = \mathbf{S}^{-1}\mathbf{R}_2^T\mathbf{u}_g}, \qquad \boxed{\mathbf{u}_g^d = \mathbf{u}_g - \mathbf{R}_2\,\mathbf{r}_g} \tag{3.450}$$

Observe that $\mathbf{r}_g$ has 6×1 elements (three translations and three rotations) while $\mathbf{u}_g^d, \mathbf{u}_g$ have $m \times 1$ elements. The differential equations of motion for $\mathbf{u}^s$ and $\mathbf{u}^d$ are then as follows:

$$\mathbf{u}^s = -\mathbf{K}_{11}^{-1}\mathbf{K}_{12}\left(\mathbf{u}_g^d + \mathbf{R}_2\,\mathbf{r}_g\right) = \mathbf{Q}\mathbf{u}_g^d + \mathbf{Q}\mathbf{R}_2\mathbf{r}_g \tag{3.451}$$

and since $\mathbf{Q}\mathbf{R}_2 = \mathbf{R}_1$, then

$$\boxed{\mathbf{u}^s = \mathbf{u}^{sd} + \mathbf{u}^{sRBM}}, \qquad \boxed{\mathbf{u}^{sd} = \mathbf{Q}\mathbf{u}_g^d}, \qquad \boxed{\mathbf{u}^{sRBM} = \mathbf{R}_1\mathbf{r}_g} \tag{3.452}$$

We see that the quasistatic motion consists of two parts: the deformation and the rigid-body motion. The latter causes no deformations in the structure, and thus elicits no stresses. Also,

$$\begin{aligned}\mathbf{M}_{11}\ddot{\mathbf{u}}^d + \mathbf{C}_{11}\dot{\mathbf{u}}^d + \mathbf{K}_{11}\mathbf{u}^d &= -\left(\mathbf{M}_{11}\mathbf{Q} + \mathbf{M}_{12}\right)\left(\mathbf{R}_2\ddot{\mathbf{r}}_g + \ddot{\mathbf{u}}_g^d\right) \\ &= -\left[\left(\mathbf{M}_{11}\mathbf{R}_1 + \mathbf{M}_{12}\mathbf{R}_2\right)\ddot{\mathbf{r}}_g + \left(\mathbf{M}_{11}\mathbf{Q} + \mathbf{M}_{12}\right)\mathbf{u}_g^d\right]\end{aligned} \tag{3.453}$$

So

$$\mathbf{M}_{11}\ddot{\mathbf{u}}^d + \mathbf{C}_{11}\dot{\mathbf{u}}^d + \mathbf{K}_{11}\mathbf{u}^d = -\left[\left(\mathbf{M}_{11}\mathbf{R}_1 + \mathbf{M}_{12}\mathbf{R}_2\right)\ddot{\mathbf{r}}_g + \left(\mathbf{M}_{11}\mathbf{Q} + \mathbf{M}_{12}\right)\ddot{\mathbf{u}}_g^d\right] \tag{3.454}$$

Finally, if the mass matrix is lumped (or the coupling mass can be neglected), then $\mathbf{M}_{12} = \mathbf{O}$ and

$$\mathbf{M}_{11}\ddot{\mathbf{u}}^d + \mathbf{C}_{11}\dot{\mathbf{u}}^d + \mathbf{K}_{11}\mathbf{u}^d = -\mathbf{M}_{11}\left(\mathbf{R}_1\ddot{\mathbf{r}}_g + \mathbf{Q}\ddot{\mathbf{u}}_g^d\right) \tag{3.455}$$

The first loading term on the right-hand side describes the classical equation of motion for a structure supported on a rigid foundation, and the second accounts for the variability of the ground motion across the support points. At this point, we can drop the subindices in the matrices and simply write

$$\boxed{\mathbf{M}\ddot{\mathbf{u}}^d + \mathbf{C}\dot{\mathbf{u}}^d + \mathbf{K}\mathbf{u}^d = -\mathbf{M}\left(\mathbf{R}\,\ddot{\mathbf{r}}_g + \mathbf{Q}\ddot{\mathbf{u}}_g^d\right) \equiv -\mathbf{M}\left(\mathbf{R}\,\ddot{\mathbf{r}}_g + \ddot{\mathbf{u}}^{sd}\right)} \tag{3.456}$$

where $\ddot{\mathbf{u}}^{sd} = \mathbf{Q}\ddot{\mathbf{u}}_g^d$ is the acceleration time history that corresponds to the deformational component $\mathbf{u}^{sd}$ of the quasistatic response obtained earlier, and the matrices are those of the superstructure. This linear system of equations could be solved either by modal superposition, discrete time integration, or using a formulation in the frequency domain.

In summary:

1. Define the ground motion vector $\mathbf{u}_g\left(t\right)$ at the m support nodes.
2. Choose a convenient reference point $\mathbf{x}_0$ relative to which the rigid body components of motion are defined, then form $\mathbf{R}_2$ and find the 6×6 geometric matrix $\mathbf{S} = \mathbf{R}_2^T\mathbf{R}_2$.
3. Find the rigid-body ground motion component $\mathbf{r}_g = \mathbf{S}^{-1}\mathbf{R}_2^T\mathbf{u}_g$, and from there, the deformational ground motion component $\mathbf{u}_g^d = \mathbf{u}_g - \mathbf{R}_2\,\mathbf{r}_g$.
4. Find the quasistatic response function $\mathbf{u}^s = \mathbf{Q}\mathbf{u}_g^d + \mathbf{R}_1\mathbf{r}_g$, with $\mathbf{Q} = -\mathbf{K}_{11}^{-1}\mathbf{K}_{12}$.
5. Find the deformational response function $\mathbf{u}^d$ from Eq. 3.456.
6. Find the total structural response $\mathbf{u}_1 = \mathbf{u}^s + \mathbf{u}^d$

3.7 Nonclassical, Complex Modes

The equation of motion for structural systems having nonproportional damping matrices can still be decoupled provided the modal superposition technique is generalized to include complex modes and frequencies. However, the application of such a generalization removes many of the advantages of classical modal analysis, since it is necessary to use complex algebra and a special quadratic eigenvalue problem algorithm. Also, it makes the use of response and shock spectra much more difficult. Nonetheless, this alternative has great theoretical interest, being particularly useful in investigations on the dynamic characteristics of relatively simple mechanical systems.

3.7.1 Quadratic Eigenvalue Problem

Consider once more the dynamic equilibrium equation

$$\mathbf{M\ddot{u}} + \mathbf{C\dot{u}} + \mathbf{Ku} = \mathbf{p}(t) \tag{3.457}$$

which in the absence of loads turns into the free vibration equation

$$\mathbf{M\ddot{u}} + \mathbf{C\dot{u}} + \mathbf{Ku} = \mathbf{0} \tag{3.458}$$

This equation admits solutions of the form

$$\mathbf{u}(t) = \mathbf{z}\,e^{-\lambda t} \tag{3.459}$$

with an as yet undetermined parameter λ and shape vector $\mathbf{z}$. When we substitute this trial solution into the homogeneous equations above, we obtain

$$(\lambda^2 \mathbf{M} - \lambda \mathbf{C} + \mathbf{K})\,\mathbf{z} = \mathbf{0} \tag{3.460}$$

This equation is an nth order *quadratic eigenvalue problem*, which generally has $2n$ complex eigenvalues λ and complex eigenvectors $\mathbf{z}$. Clearly, if some eigenvalue is indeed complex, then its complex conjugate λ^c together with the complex conjugate eigenvector $\mathbf{z}^c$ is also a solution. This can be verified by simply conjugating the eigenvalue equation, and taking into account that the matrices are real. Hence, *complex eigenvalues occur in complex conjugate pairs*. On the other hand, if the eigenvalue is real, then the associated eigenvector must be real, since all terms within the parenthesis in the eigenvalue equation are then real.

The eigenvalue equation can also be written in matrix form as

$$\mathbf{M}\,\mathbf{Z}\Lambda^2 - \mathbf{C}\,\mathbf{Z}\Lambda + \mathbf{K}\,\mathbf{Z} = \mathbf{O} \tag{3.461}$$

in which

$$\Lambda = \operatorname{diag}\left\{\lambda_j\right\} \qquad = \text{spectral matrix} \tag{3.462}$$

$$\mathbf{Z} = \left\{\mathbf{z}_j\right\} \qquad = \text{modal matrix (a } n \times 2n \text{ rectangular matrix!).} \tag{3.463}$$

$$\mathbf{O} = \left\{0\right\} \qquad = \text{null matrix} \tag{3.464}$$

3.7.2 Poles or Complex Frequencies

Some of the properties of the complex modes can be assessed even before solving the eigenvalue problem. To demonstrate this assertion, we first redefine the eigenvalue as $\lambda = -\mathrm{i}\,\omega_c$, the complex frequency, which changes the eigenvalue equation into

$$(-\omega_c^2 \mathbf{M} + i\,\omega_c \mathbf{C} + \mathbf{K})\mathbf{z} = \mathbf{0} \tag{3.465}$$

Let $\mathbf{z}$ be an arbitrary eigenvector (real or complex), and consider the hermitian forms

$$\mathbf{z}^* \mathbf{M} \mathbf{z} = \mu > 0 \qquad (\text{since } \mathbf{M} \text{ is positive definite}) \tag{3.466}$$

$$\mathbf{z}^* \mathbf{C} \mathbf{z} = \eta \geq 0 \tag{3.467}$$

$$\mathbf{z}^* \mathbf{K} \mathbf{z} = \kappa \geq 0 \tag{3.468}$$

We notice that these computations always result in *real, nonnegative* numbers μ, η, κ. Premultiplying the eigenvalue equation by the *transposed conjugate* eigenvector $\mathbf{z}^* \equiv (\mathbf{z}^c)^T$, we obtain

$$\mathbf{z}^*(-\omega_c^2 \mathbf{M} + i\,\omega_c \mathbf{C} + \mathbf{K})\mathbf{z} = \mathbf{0} \tag{3.469}$$

That is,

$$-\omega_c^2 \mu + i\,\omega_c \eta + \kappa = 0 \tag{3.470}$$

which is a quadratic equation in ω_c. Dividing this equation by μ and defining the *Rayleigh quotients*

$$b = \frac{1}{2}\frac{\eta}{\mu} = \frac{1}{2}\frac{\mathbf{z}^* \mathbf{C} \mathbf{z}}{\mathbf{z}^* \mathbf{M} \mathbf{z}} \geq 0 \tag{3.471}$$

$$r^2 = \frac{\kappa}{\mu} = \frac{\mathbf{z}^* \mathbf{K} \mathbf{z}}{\mathbf{z}^* \mathbf{M} \mathbf{z}} \geq 0 \tag{3.472}$$

we obtain

$$\omega_c^2 - 2\,i\,b\,\omega_c + r^2 = 0 \tag{3.473}$$

whose solution is

$$\omega_c = \pm\sqrt{r^2 - b^2} + i\,b = \pm a + i\,b \qquad \text{if} \qquad r \geq b \tag{3.474}$$

$$\omega_c = i\left[b \mp \sqrt{b^2 - r^2}\right] = i(b \mp a) \qquad \text{if} \qquad r < b \tag{3.475}$$

In either case, the imaginary part is nonnegative, and a is nonnegative.

To focus ideas, let us return to the particular case of a *proportional* damping matrix. Since in that case the quadratic eigenvalue equation can be diagonalized by the classical modes, it follows that $\mathbf{z}_j \equiv \phi_j$. The Rayleigh quotients then reduce to

$$b \equiv \xi_j \omega_j \qquad r \equiv \omega_j \qquad \text{and} \qquad a = \sqrt{r^2 - b^2} \equiv \omega_{dj} \tag{3.476}$$

in which the ω_j are the *classical* undamped natural frequencies, and the modal damping is assumed to be subcritical. The two complex frequencies or *poles* that correspond to each of these modes are then

$$\boxed{\omega_{cj} = \pm\omega_{dj} + i\,\xi_j \omega_j = \omega_j\left(\pm\sqrt{1 - \xi_j^2} + i\,\xi_j\right)} \tag{3.477}$$

Extending this analogy to the nonproportional damping case, we can always interpret r, b/r and a as the *nonclassical* modal frequencies, modal damping, and damped modal frequency, respectively. In that case, we will have the following relationships with the original eigenvalues:

$$\omega_c = \pm a + i b \qquad \lambda = b \mp i a \qquad (b \geq 0) \tag{3.478}$$

$$\omega_c = i(b \mp a) \qquad \lambda = b \mp a \geq 0 \tag{3.479}$$

In the case of complex frequencies $\omega_c = \pm a + i b$ and associated modes $\mathbf{z} = \mathbf{x} \pm i\mathbf{y}$, they must be combined together so that they produce a real displacement vector. Hence, if C is any arbitrary complex constant, the pair of complex conjugate modes must be combined as

$$\mathbf{u} = \left[C\,\mathbf{z}\,e^{iat} + C^c\,\mathbf{z}^c\,e^{-iat} \right] e^{-bt} \tag{3.480}$$

Expressing the complex constant in polar form $C = A e^{-i\vartheta}$, the free vibration can also be written as

$$\mathbf{u} = 2A\left[\mathbf{x}\cos(at - \vartheta) - \mathbf{y}\sin(at - \vartheta) \right] e^{-bt} \tag{3.481}$$

which represents a harmonic vibration with damped frequency a and an amplitude decaying exponentially with time. Since $\mathbf{x}$ and $\mathbf{y}$ are not proportional, the phase angles of the components are different from each other, and structural masses do *not* move in phase. By contrast, in the proportionally damped case, all masses move simultaneously when the system vibrates in a single mode. Thus, initial conditions consisting of *both* displacements *and* velocities must be imposed on the system in order to achieve a free vibration in a single damped mode (actually, a pair of such modes).

On the other hand, if the complex frequency is purely imaginary $\omega_c = i(b \mp a)$, this represents an exponentially decaying solution, and the mode is overdamped. In this case, however, the mode itself is real, so that all masses will move in phase.

In summary:

1. If $b < r$, the damped modal frequency is real, the eigenvalue is complex, the mode is complex, and the oscillation in that mode is *underdamped*. Masses do *not oscillate in phase*.
2. If $b > r$, the damped modal frequency is purely imaginary, the eigenvalue is real and positive, the mode is real, and the oscillation in that mode is *overdamped*. All masses *oscillate in phase*.

Although the nonclassical frequencies play a role similar to those in the undamped case, it should be noted that they do not necessarily coincide with them. However, they are enclosed by these frequencies, as will be seen next.

We can get an idea of the nature of the solutions by attempting to interpret the Rayleigh quotients previously defined. For this purpose, consider the two real, linear eigenvalue problems

$$\mathbf{K}\boldsymbol{\phi} = \omega^2 \mathbf{M}\boldsymbol{\phi} \qquad \text{(the undamped eigenvalue problem)} \tag{3.482}$$

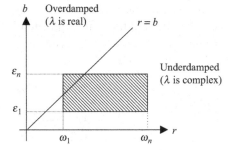

Underdamped
(λ is complex)

Figure 3.20. Bounded domain enclosing points for complex eigenvalues.

$$\mathbf{C}\boldsymbol{\psi} = 2\varepsilon \mathbf{M}\boldsymbol{\psi} \qquad \text{(an auxiliary eigenvalue problem)} \qquad (3.483)$$

Since $\mathbf{M}$ is real, symmetric and positive definite, these two eigenvalue problems have *real* nonnegative eigenvalues, which can be ordered by increasing values as $\omega_1, \cdots \omega_n$ and $\varepsilon_1, \cdots \varepsilon_n$. From the enclosure theorem, it then follows that the Rayleigh quotients r, b, will be bounded by

$$\omega_1^2 \le r^2 \le \omega_n^2 \qquad \text{and} \qquad \varepsilon_1 \le b \le \varepsilon_n \qquad (3.484)$$

It follows that

$$\frac{\boldsymbol{\phi}_1^T \mathbf{K}\boldsymbol{\phi}_1}{\boldsymbol{\phi}_1^T \mathbf{M}\boldsymbol{\phi}_1} \le \frac{\mathbf{z}^* \mathbf{K}\mathbf{z}}{\mathbf{z}^* \mathbf{M}\mathbf{z}} \le \frac{\boldsymbol{\phi}_n^T \mathbf{K}\boldsymbol{\phi}_n}{\boldsymbol{\phi}_n^T \mathbf{M}\boldsymbol{\phi}_n} \qquad (3.485)$$

$$\frac{\boldsymbol{\psi}_1^T \mathbf{C}\boldsymbol{\psi}_1}{\boldsymbol{\psi}_1^T \mathbf{M}\boldsymbol{\psi}_1} \le \frac{\mathbf{z}^* \mathbf{C}\mathbf{z}}{\mathbf{z}^* \mathbf{M}\mathbf{z}} \le \frac{\boldsymbol{\psi}_n^T \mathbf{C}\boldsymbol{\psi}_n}{\boldsymbol{\psi}_n^T \mathbf{M}\boldsymbol{\psi}_n} \qquad (3.486)$$

In general, $\mathbf{z}_j \ne \boldsymbol{\phi}_j \ne \boldsymbol{\psi}_j$, so that a priori predictions on the relative values of the Rayleigh quotients cannot be made; however, the maximum and minimum values (the bounds) could certainly be estimated. Indeed, we notice that if the system has proportional damping, then $\mathbf{z}_j \equiv \boldsymbol{\phi}_j \equiv \boldsymbol{\psi}_j$. Thus, it would be reasonable to apply the undamped modes to estimate the Rayleigh quotients b and r^2, and in particular, their highest and lowest possible values. On the basis of these bounds, we could then decide on the nature of the possible solutions, as shown in Figure 3.20. In this figure, the range of values of r, b is defined by the rectangular area. Depending on whether the point (r,b) is below, on, or above the 45° line, we will have underdamped, critically damped or overdamped oscillations, respectively.

3.7.3 Doubled-Up Form of Differential Equation

By adding the trivial equation $\dot{\mathbf{u}} - \dot{\mathbf{u}} = \mathbf{0}$, the dynamic equilibrium equation can be written as a symmetric system of first-order equations of double dimension, namely

$$\begin{bmatrix} \mathbf{K} & \mathbf{O} \\ \mathbf{O} & -\mathbf{M} \end{bmatrix} \frac{d}{dt} \begin{Bmatrix} \mathbf{u} \\ -\dot{\mathbf{u}} \end{Bmatrix} + \begin{bmatrix} \mathbf{O} & \mathbf{K} \\ \mathbf{K} & -\mathbf{C} \end{bmatrix} \begin{Bmatrix} \mathbf{u} \\ -\dot{\mathbf{u}} \end{Bmatrix} = \begin{Bmatrix} \mathbf{0} \\ \mathbf{p} \end{Bmatrix} \qquad (3.487)$$

or

$$\begin{bmatrix} \mathbf{C} & -\mathbf{M} \\ -\mathbf{M} & \mathbf{O} \end{bmatrix} \frac{d}{dt} \begin{Bmatrix} \mathbf{u} \\ -\dot{\mathbf{u}} \end{Bmatrix} + \begin{bmatrix} \mathbf{K} & \mathbf{O} \\ \mathbf{O} & -\mathbf{M} \end{bmatrix} \begin{Bmatrix} \mathbf{u} \\ -\dot{\mathbf{u}} \end{Bmatrix} = \begin{Bmatrix} \mathbf{p} \\ \mathbf{0} \end{Bmatrix} \qquad (3.488)$$

in which $\mathbf{O}$ is again the null matrix and $\mathbf{0}$ is a null vector. We included a negative sign in front of the velocity term $\dot{\mathbf{u}}$ for reasons of convenience. While both of these are equivalent equations, we prefer the second form, because it has a simpler structure (the mass matrix $\mathbf{M}$ is most often diagonal). This doubled-up differential equation is of the form

$$\mathbf{A}\dot{\mathbf{v}} + \mathbf{B}\mathbf{v} = \hat{\mathbf{p}}(t) \tag{3.489}$$

in which $\hat{\mathbf{p}}$ is the load vector augmented with zeros, and the combined displacement-velocity vector $\mathbf{v}$ is the *state vector*. If $\hat{\mathbf{p}}(t) = \mathbf{0}$, we can once more obtain the free vibration problem by setting $\mathbf{u} = \mathbf{z}e^{-\lambda t}$ $\dot{\mathbf{u}} = -\lambda \mathbf{z}e^{-\lambda t}$, so that

$$\mathbf{v} = \begin{Bmatrix} \mathbf{z} \\ \lambda\mathbf{z} \end{Bmatrix} e^{-\lambda t} = \mathbf{w}\,e^{-\lambda t} \tag{3.490}$$

which leads us to the eigenvalue equation

$$\lambda\mathbf{A}\mathbf{w} = \mathbf{B}\mathbf{w} \tag{3.491}$$

or in full

$$\lambda \begin{Bmatrix} \mathbf{C} & -\mathbf{M} \\ -\mathbf{M} & \mathbf{O} \end{Bmatrix} \begin{Bmatrix} \mathbf{z} \\ \lambda\mathbf{z} \end{Bmatrix} = \begin{Bmatrix} \mathbf{K} & \mathbf{O} \\ \mathbf{O} & -\mathbf{M} \end{Bmatrix} \begin{Bmatrix} \mathbf{z} \\ \lambda\mathbf{z} \end{Bmatrix} \tag{3.492}$$

Clearly, this is a *linear* eigenvalue problem in λ involving real and symmetric matrices $\mathbf{A}, \mathbf{B}$ and eigenvectors $\mathbf{w}$. It follows that many of the techniques and properties of linear eigenvalue problems apply also to this case. It should be noted, however, that the two matrices are *not* positive definite, but in fact they are *indefinite*, so the eigenvalues are not necessarily real, as we already know.

Suppose, for example, that $\mathbf{w}$ is a complex eigenvector with corresponding complex eigenvalue λ. Hence, when we multiply the eigenvalue equation by the transposed complex conjugate eigenvector $\mathbf{w}^* = (\mathbf{w}^c)^T$ we obtain

$$\lambda\mathbf{w}^*\mathbf{A}\mathbf{w} = \mathbf{w}^*\mathbf{B}\mathbf{w} \tag{3.493}$$

Now, both $\mathbf{A}$ and $\mathbf{B}$ are real and symmetric matrices, so the two hermitian forms $\mathbf{w}^*\mathbf{A}\mathbf{w}$ and $\mathbf{w}^*\mathbf{B}\mathbf{w}$ must be real. Since λ is the only complex quantity left in this expression, the only way that the two sides can then be equal is if they are zero, that is, if both of the hermitian forms vanish identically. Considering that both $\mathbf{A}$ and $\mathbf{B}$ are *indefinite*, these hermitian forms can indeed be zero. This result is not as strange as it seems; it is simply an expression of the orthogonality condition of two complex conjugate modes. For if λ and $\mathbf{w}$ are one solution to the eigenvalue equation, then the conjugates λ^c and $\mathbf{w}^c$ are also a distinct solution, and are orthogonal to each other with respect to $\mathbf{A}$ and $\mathbf{B}$, as will be seen. More generally, the doubled-up eigenvalue problem can be written in matrix form as

$$\mathbf{A}\mathbf{W}\mathbf{\Lambda} = \mathbf{B}\mathbf{W} \tag{3.494}$$

with

$$\mathbf{W} = \{\mathbf{w}_j\} = \begin{Bmatrix} \mathbf{Z} \\ \mathbf{Z}\mathbf{\Lambda} \end{Bmatrix} \qquad = \text{expanded modal matrix} \tag{3.495}$$

It should be noted that this expanded modal matrix has an internal structure

$$\mathbf{W} = \{ \mathbf{W}_1 \quad \mathbf{W}_2 \quad \mathbf{W}_2^c \} \tag{3.496}$$

That is, it consists of an *even* number of real eigenvectors $\mathbf{W}_1$ that correspond to real eigenvalues $\mathbf{\Lambda}_1$, and a set of complex-conjugate pairs $\mathbf{W}_2$ and $\mathbf{W}_2^c$ with complex conjugate eigenvalues $\mathbf{\Lambda}_2$ and $\mathbf{\Lambda}_2^c$.

3.7.4 Orthogonality Conditions

Like linear eigenvalue problems, the quadratic eigenvalue problem satisfies also an orthogonality condition, which for the doubled up form can be obtained in the same way as for the classical modes. This results in

$$\boxed{\mathbf{W}^T \mathbf{A} \mathbf{W} = \mathbb{D}} \qquad (\mathbb{D} = \mathbf{Z}^T \mathbf{C} \mathbf{Z} - \mathbf{Z}^T \mathbf{M} \mathbf{Z} - \mathbf{Z}^T \mathbf{M} \mathbf{Z} \mathbf{\Lambda}) \tag{3.497}$$

$$\boxed{\mathbf{W}^T \mathbf{B} \mathbf{W} = \mathbb{D} \mathbf{\Lambda}} \qquad (\mathbb{D} \mathbf{\Lambda} = \mathbf{Z}^T \mathbf{K} \mathbf{Z} - \mathbf{\Lambda} \mathbf{Z}^T \mathbf{M} \mathbf{Z} \mathbf{\Lambda}) \tag{3.498}$$

in which $\mathbb{D} = \mathbf{diag}\{D_j\}$ is a diagonal matrix. The equivalent expressions on the right are obtained by carrying out the implied operations with the submatrices of $\mathbf{A}, \mathbf{B},$ and $\mathbf{W}$. The resulting values in $\mathbb{D}$ can then be used to normalize the eigenvectors in some appropriate way, if so desired.

For example, we could normalize the eigenvectors so that

$$\mathbb{D} \mathbf{\Lambda} = \mathbf{\Omega}^2 + \mathbf{\Omega}_c^2 = \mathbf{\Omega}^2 - \mathbf{\Lambda}^2 \tag{3.499}$$

in which

$$\mathbf{\Omega}_c = i \mathbf{\Lambda} \qquad\qquad = \text{the complex frequencies or poles} \tag{3.500}$$

$$\mathbf{\Omega}^2 = \mathbf{\Omega}_c^c \, \mathbf{\Omega}_c \equiv \mathbf{\Lambda}_c^c \, \mathbf{\Lambda}_c \qquad = \text{the absolute values} \tag{3.501}$$

implying

$$\mathbb{D} = \mathbf{\Lambda}^c - \mathbf{\Lambda} = i \left(\mathbf{\Omega}_c^c + \mathbf{\Omega}_c \right) \tag{3.502}$$

The reason for choosing this particular normalization is to make it consistent with the one for proportional subcritical damping, for which $\mathbf{Z} \equiv \mathbf{\Phi}, \mathbf{\Phi}^T \mathbf{M} \mathbf{\Phi} = \mathbf{I}$, and $\mathbf{\Phi}^T \mathbf{K} \mathbf{\Phi} = \mathbf{\Omega}^2$. It should be noted, however, that an alternative normalization is necessary for modes with hypercritical damping (i.e., modes with real eigenvalues, or purely imaginary complex frequencies), because otherwise the diagonal elements of $\mathbf{D}$ for those modes would be zero. In addition, critically damped modes are degenerate in that they constitute double roots, so they require special treatment. Notice also that if damping is not proportional, then the individual quadratic forms $\mathbf{Z}^T \mathbf{M} \mathbf{Z}$ and $\mathbf{Z}^T \mathbf{K} \mathbf{Z}$ are *not* diagonal.

The orthogonality conditions just presented can also be derived directly from the quadratic eigenvalue problem. Indeed, consider the eigenvalue equations for two distinct modes $\lambda_i, \mathbf{z}_i$ and $\lambda_j, \mathbf{z}_j$, and multiply each equation by the eigenvalue for the other:

$$(\lambda_i^2 \lambda_j \mathbf{M} - \lambda_i \lambda_j \mathbf{C} + \lambda_j \mathbf{K}) \mathbf{z}_i = \mathbf{0} \tag{3.503}$$

$$(\lambda_i \lambda_j^2 \, \mathbf{M} - \lambda_i \lambda_j \, \mathbf{C} + \lambda_i \, \mathbf{K}) \mathbf{z}_j = \mathbf{0} \tag{3.504}$$

Multiply next each of these equations by the transposed eigenvector for the dual equation, and also transpose the result of the first of these transformations while taking into account the symmetry of the matrices. This produces

$$\mathbf{z}_i^T (\lambda_i^2 \lambda_j \, \mathbf{M} - \lambda_i \lambda_j \mathbf{C} + \lambda_j \, \mathbf{K}) \mathbf{z}_j = 0 \tag{3.505}$$

$$\mathbf{z}_i^T (\lambda_i \lambda_j^2 \, \mathbf{M} - \lambda_i \lambda_j \mathbf{C} + \lambda_i \, \mathbf{K}) \mathbf{z}_j = 0 \tag{3.506}$$

Subtracting the second equation from the first produces

$$(\lambda_j - \lambda_i)(\mathbf{z}_i^T \, \mathbf{K} \mathbf{z}_j - \lambda_i \lambda_j \, \mathbf{z}_i^T \, \mathbf{M} \mathbf{z}_j) = 0 \tag{3.507}$$

If the modes are distinct, the first factor cannot be zero. On the other hand, if the eigenvalues are the same, then the second factor can have any arbitrary value, which can be used to scale (normalize) the eigenvectors at will. Hence, we conclude that

$$\mathbf{z}_i^T \, \mathbf{K} \mathbf{z}_j - \lambda_i \lambda_j \, \mathbf{z}_i^T \, \mathbf{M} \mathbf{z}_j = 0 \qquad \text{if} \quad i \neq j \quad (\lambda_i \neq \lambda_j) \tag{3.508}$$

$$\mathbf{z}_j^T \, \mathbf{K} \mathbf{z}_j - \lambda_j^2 \, \mathbf{z}_j^T \, \mathbf{M} \mathbf{z}_j = D_j \, \lambda_j \neq 0 \qquad \text{if} \quad i = j \tag{3.509}$$

which agrees with the second orthogonality condition. To obtain the dual condition, we add the eigenvalue equations for each of the two modes multiplied by the eigenvector for the other, that is,

$$(\lambda_i^2 + \lambda_j^2) \mathbf{z}_i^T \, \mathbf{M} \mathbf{z}_j - (\lambda_i + \lambda_j) \mathbf{z}_i^T \, \mathbf{C} \mathbf{z}_j + 2 \mathbf{z}_i^T \, \mathbf{K} \mathbf{z}_j = 0 \tag{3.510}$$

or

$$(\lambda_i + \lambda_j)\left[\mathbf{z}_i^T \, \mathbf{C} \mathbf{z}_j - (\lambda_i + \lambda_j) \mathbf{z}_i^T \, \mathbf{M} \mathbf{z}_j \right] - 2\left[\mathbf{z}_i^T \, \mathbf{K} \mathbf{z}_j - \lambda_i \lambda_j \mathbf{z}_i^T \, \mathbf{M} \mathbf{z}_j \right] = 0 \tag{3.511}$$

Taking into account the orthogonality condition for the second term and the fact that the eigenvalues cannot be negative, it finally follows that

$$\mathbf{z}_i^T \, \mathbf{C} \mathbf{z}_j - (\lambda_i + \lambda_j) \mathbf{z}_i^T \, \mathbf{M} \mathbf{z}_j = 0 \qquad \text{if} \quad i \neq j \quad (\lambda_i \neq \lambda_j) \tag{3.512}$$

$$\mathbf{z}_j^T \, \mathbf{C} \mathbf{z}_j - 2\lambda_j \, \mathbf{z}_j^T \, \mathbf{M} \mathbf{z}_j = D_j \qquad \text{if} \quad i = j \quad (\lambda_j \neq 0) \tag{3.513}$$

3.7.5 Modal Superposition with Complex Modes

To solve the doubled-up differential equation, we apply a complex-modes superposition

$$\mathbf{v} = \sum_{j=1}^{2n} \mathbf{w}_j \, q_j(t) = \mathbf{W} \mathbf{q} \tag{3.514}$$

which at time $t = 0$ evaluates to

$$\mathbf{v}_0 = \sum_{j=1}^{2n} \mathbf{w}_j \, q_{0j} = \mathbf{W} \mathbf{q}_0 \tag{3.515}$$

The initial values of the modal coordinates $\mathbf{q}_0$ can be found by means of the orthogonality conditions

$$\mathbf{W}^T \mathbf{B} \mathbf{v}_0 = (\mathbf{W}^T \mathbf{B} \mathbf{W}) \mathbf{q}_0 = \mathbb{D} \mathbf{\Lambda} \mathbf{q}_0 \qquad (3.516)$$

from which we can solve for $\mathbf{q}_0$

$$\mathbf{q}_0 = \mathbb{D}^{-1} \mathbf{\Lambda}^{-1} \mathbf{W}^T \mathbf{B} \mathbf{v}_0 = \mathbb{D}^{-1} \left(\mathbf{\Lambda}^{-1} \mathbf{Z}^T \mathbf{K} \mathbf{u}_0 + \mathbf{Z}^T \mathbf{M} \dot{\mathbf{u}}_0 \right) \qquad (3.517)$$

That is,

$$q_{0j} = D_j^{-1} \mathbf{z}_j^T \left(\lambda_j^{-1} \mathbf{K} \mathbf{u}_0 + \mathbf{M} \dot{\mathbf{u}}_0 \right) \qquad (3.518)$$

Next, we substitute the modal expansion into the differential equation

$$\mathbf{A} \mathbf{W} \dot{\mathbf{q}} + \mathbf{B} \mathbf{W} \mathbf{q} = \hat{\mathbf{p}}(t) \qquad (3.519)$$

and multiplying by the transposed expanded matrix, we obtain

$$(\mathbf{W}^T \mathbf{A} \mathbf{W}) \dot{\mathbf{q}} + (\mathbf{W}^T \mathbf{B} \mathbf{W}) \mathbf{q} = \mathbf{W}^T \hat{\mathbf{p}}(t) \qquad (3.520)$$

Because of orthogonality, the two products in parentheses are diagonal matrices. Hence

$$\mathbb{D} (\dot{\mathbf{q}} + \mathbf{\Lambda} \mathbf{q}) = \boldsymbol{\pi}(t) \qquad (3.521)$$

with

$$\boldsymbol{\pi}(t) = \mathbf{W}^T \hat{\mathbf{p}}(t) \equiv \mathbf{Z}^T \mathbf{p}(t) = \left\{ \mathbf{z}_j^T \mathbf{p} \right\} = \left\{ \pi_j(t) \right\} \qquad (3.522)$$

being the modal load. Hence

$$\dot{\mathbf{q}} + \mathbf{\Lambda} \mathbf{q} = \mathbb{D}^{-1} \boldsymbol{\pi}(t) \qquad (3.523)$$

which is equivalent to the scalar equations

$$\dot{q}_j + \lambda_j q_j = D_j^{-1} \pi_j(t) \qquad j = 1, 2, \ldots 2n \qquad (3.524)$$

This is a linear first-order differential equation. Its solution for $t \geq 0$ can be obtained as the sum of the homogeneous part, and a particular solution in the form of a convolution of the right-hand side with the impulse response function (or Green's function) for this equation. This Green's function is the solution to $\dot{q}_j + \lambda_j q_j = \delta(t)$, which in this case is simply $q_j = H(t) e^{-\lambda t}$ (the unit step function times the decaying exponential). Hence

$$q_j(t) = q_{j0} e^{-\lambda_j t} + D_j^{-1} \int_0^t \pi_j(t - \tau) e^{-\lambda_j \tau} d\tau \qquad (3.525)$$

For each real mode, $\lambda_j > 0$ (that is, the complex frequency ω_{cj} is a positive, purely imaginary number), all terms are real, the exponential terms decay with t, and the solution is bounded.

In the case of complex modes, the eigenvalues and eigenvectors occur in complex conjugate pairs and have the form

$$\omega_{cj} = \pm a_j + i b_j \qquad \lambda_j = b_j \mp i a_j \qquad (3.526)$$

Hence, the modal summation for two complex conjugate modes is

$$
\begin{aligned}
\mathbf{u}_j = \Big[& \mathbf{z}_j \, q_{0j} \, e^{i a_j t} + \mathbf{z}_j^c \, q_{0j}^c \, e^{-i a_j t} \Big] e^{-b_j t} \\
& + D_j^{-1} \mathbf{z}_j \int_0^t \mathbf{z}_j^T \mathbf{p}(t - \tau) e^{(i a_j - b_j)\tau} d\tau \\
& + D_j^{-c} \mathbf{z}_j^c \int_0^t \mathbf{z}_j^* \, \mathbf{p}(t - \tau) e^{-(i a_j + b_j)\tau} d\tau
\end{aligned}
\tag{3.527}
$$

Without loss of generality, we can assume $D_j = 2 i a_j$, that is, we normalize the complex eigenvectors as suggested earlier. Hence, we can write the initial modal values as

$$
\begin{aligned}
q_{0j} &= \frac{1}{2 i a_j} (\mathbf{x}_j^T + i \, \mathbf{y}_j^T)\Big(\frac{1}{b_j - i a_j} \mathbf{K} \mathbf{u}_0 + \mathbf{M} \dot{\mathbf{u}}_0 \Big) \\
&= \frac{1}{2 a_j} \left\{ \left[\frac{(a_j \mathbf{x}_j^T + b_j \mathbf{y}_j^T)\mathbf{K} \mathbf{u}_0}{a_j^2 + b_j^2} + \mathbf{y}_j^T \mathbf{M} \dot{\mathbf{u}}_0 \right] + i \left[\frac{(a_j \mathbf{y}_j^T - b_j \mathbf{x}_j^T)\mathbf{K} \mathbf{u}_0}{a_j^2 + b_j^2} - \mathbf{x}_j^T \mathbf{M} \dot{\mathbf{u}}_0 \right] \right\} \\
&= \frac{1}{2 a_j} Q_j \, e^{-i \vartheta_j}
\end{aligned}
\tag{3.528}
$$

The term $Q_j \, e^{-i \vartheta_j}$ represents the polar representation of the expression in braces. Next, we assume that $\mathbf{p}$ can be written as

$$
\mathbf{p}(t) = \mathbf{p}_0 \, p(t)
\tag{3.529}
$$

that is, that the spatial and temporal dependencies of the load can be separated. (More general loads can still be written as summations over separated functions.) With these definitions, we can write the partial modal summation as

$$
\begin{aligned}
\mathbf{u}_j = \frac{1}{a_j} \Big\{ & \Big[Q_j \, e^{-b_j t} \cos(a_j t - \vartheta_j) + (\mathbf{x}_j^T \mathbf{p}_0) S_j + (\mathbf{y}_j^T \mathbf{p}_0) C_j \Big] \mathbf{x}_j \\
& - \Big[Q_j \, e^{-b_j t} \sin(a_j t - \vartheta_j) + (\mathbf{y}_j^T \mathbf{p}_0) S_j - (\mathbf{x}_j^T \mathbf{p}_0) C_j \Big] \mathbf{y}_j \Big\}
\end{aligned}
\tag{3.530}
$$

in which

$$
S_j = \int_0^t p(t - \tau) \sin(a_j \tau) e^{-b_j \tau} d\tau \qquad C_j = \int_0^t p(t - \tau) \cos(a_j \tau) e^{-b_j \tau} d\tau
\tag{3.531}
$$

The partial modal summation involves only real numbers.

In the case of proportional damping, this expression reduces to that of conventional modal analysis:

$$
\mathbf{x}_j \equiv \boldsymbol{\phi}_j \qquad \mathbf{y}_j = \mathbf{0}
\tag{3.532}
$$

$$
\boldsymbol{\phi}_j^T \mathbf{K} \mathbf{u}_0 = \omega_j^2 \, \boldsymbol{\phi}_j^T \mathbf{M} \mathbf{u}_0 \qquad \boldsymbol{\phi}_j^T \mathbf{M} \boldsymbol{\phi}_j = 1
\tag{3.533}
$$

$$
a_j \equiv \omega_{dj} \qquad b_j \equiv \xi_j \omega_j \qquad a_j^2 + b_j^2 = \omega_j^2
\tag{3.534}
$$

$$
q_{0j} = \frac{1}{2} \left[\boldsymbol{\phi}_j^T \mathbf{M} \dot{\mathbf{u}}_0 - i \frac{\boldsymbol{\phi}_j^T \mathbf{M} \dot{\mathbf{u}}_0 + \xi_j \omega_j \boldsymbol{\phi}_j^T \mathbf{M} \mathbf{u}_0}{\omega_{dj}} \right]
\tag{3.535}
$$

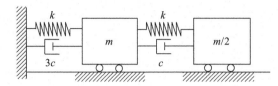

Figure 3.21. System with nonclassical modes.

$$\mathbf{u}_j = \left\{ e^{-\xi_j \omega_j t} \left[\boldsymbol{\phi}_j^T \mathbf{M} \mathbf{u}_0 \cos \omega_{dj} t + \frac{\boldsymbol{\phi}_j^T \mathbf{M} \dot{\mathbf{u}}_0 + \xi_j \omega_j \boldsymbol{\phi}_j^T \mathbf{M} \mathbf{u}_0}{\omega_{dj}} \sin \omega_{dj} t \right] \right.$$
$$\left. + \frac{\boldsymbol{\phi}_j^T \mathbf{p}_0}{\omega_{dj}} \int_0^t p(t-\tau) \sin(\omega_{dj}\tau) e^{-\xi_j \omega_j \tau} d\tau \right\} \boldsymbol{\phi}_j \qquad (3.536)$$

Example 2-DOF system with non-classical modes

Consider a two simply connected spring-mass system, shown in Figure 3.21. The mass, damping and stiffness matrices are

$$\mathbf{M} = m \begin{Bmatrix} \frac{1}{2} & 0 \\ 0 & 1 \end{Bmatrix} \qquad \mathbf{C} = c \begin{Bmatrix} 1 & -1 \\ -1 & 4 \end{Bmatrix} \qquad \mathbf{K} = k \begin{Bmatrix} 1 & -1 \\ -1 & 2 \end{Bmatrix} \qquad (3.537)$$

We define also

$$\omega = \sqrt{\frac{k}{m}} \qquad \xi = \frac{c}{2\sqrt{km}} = 0.05 \qquad (3.538)$$

The symmetry test for normal modes yields

$$\mathbf{C} \mathbf{M}^{-1} \mathbf{K} = \frac{ck}{m} \begin{Bmatrix} 3 & -4 \\ -6 & 10 \end{Bmatrix} \qquad (3.539)$$

which is not symmetric. Hence, normal modes do not exist. However, to show the parallel with the case where normal modes do exist, we start by presenting the classical, normal modes for the undamped system, together with the modal transformation on the damping matrix. These are

$$\omega_1 = \omega\sqrt{2-\sqrt{2}} = 0.7654\,\omega \qquad \omega_2 = \omega\sqrt{2+\sqrt{2}} = 1.8478\,\omega \qquad (3.540)$$

$$\boldsymbol{\Phi} = \begin{Bmatrix} \sqrt{2} & -\sqrt{2} \\ 1 & 1 \end{Bmatrix} \qquad \boldsymbol{\Phi}^T \mathbf{C} \boldsymbol{\Phi} = 2c \begin{Bmatrix} 3-\sqrt{2} & -1 \\ -1 & 3+\sqrt{2} \end{Bmatrix} \qquad (3.541)$$

Since the system does not have normal modes, the modal transformation on the damping matrix is not diagonal, that is, $\mathbf{C}$ is not proportional. The weighted modal damping values that would be obtained by neglecting the off-diagonal terms would be

$$\xi_1 = \frac{1}{2\omega_1} \frac{\boldsymbol{\phi}_1^T \mathbf{C}\, \boldsymbol{\phi}_1}{\boldsymbol{\phi}_1^T \mathbf{M}\, \boldsymbol{\phi}_1} = \frac{3-\sqrt{2}}{\sqrt{2-\sqrt{2}}} \xi = 2.0719\,\xi = 0.1036 \qquad (3.542)$$

$$\xi_2 = \frac{1}{2\omega_2} \frac{\boldsymbol{\phi}_2^T \mathbf{C}\, \boldsymbol{\phi}_2}{\boldsymbol{\phi}_2^T \mathbf{M}\, \boldsymbol{\phi}_2} = \frac{3+\sqrt{2}}{\sqrt{2+\sqrt{2}}} \xi = 2.3890\,\xi = 0.1194 \qquad (3.543)$$

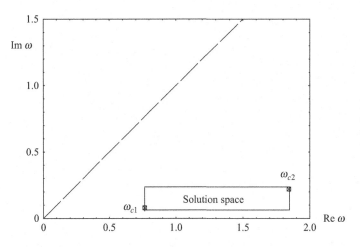

Figure 3.22. Bounded domain of eigenvalues for the 2-DOF system with nonclassical modes used as illustration.

To assess the possible types of damped modes, we solve the auxiliary linear eigenvalue problem $\mathbf{C}\boldsymbol{\psi}_j = 2\,\varepsilon_j \mathbf{M}\boldsymbol{\psi}_j$. Its solution is

$$\varepsilon_1 = \left(3 - \sqrt{3}\right)\xi\,\omega = 1.2679\,\xi\,\omega = 0.0634\,\omega \tag{3.544}$$

$$\varepsilon_2 = \left(3 + \sqrt{3}\right)\xi\,\omega = 4.7321\,\xi\,\omega = 0.2366\,\omega \tag{3.545}$$

The rectangle defined by $\varepsilon_1, \omega_1, \varepsilon_2, \omega_2$ is shown in Figure 3.22. Also shown in this figure are the actual values for the parameters b_j, r_j in the two complex frequencies, which we defined earlier as

$$\omega_c^2 = \pm\sqrt{r^2 - b^2} + \mathrm{i}\,b \tag{3.546}$$

Hence, $r_j \leftrightarrow \omega_j$ plays the role of "undamped" frequency, the square root terms is the "damped" frequency, and $b_j \leftrightarrow \xi_j\omega_j$ is the imaginary part of the complex frequency. The fact that the rectangle is entirely below the 45° line indicates that both complex modes must have less than critical damping, and that they represent stable, attenuated vibrations. If the viscosity c of the dashpots were increased by a factor 2.32, the rectangle would intercept the diagonal, in which case the system could also admit overdamped solutions.

The solution of the quadratic eigenvalue problem $\left(\mathbf{K} + \mathrm{i}\,\omega_c\mathbf{C} - \omega_c^2\mathbf{M}\right)\mathbf{z} = \mathbf{0}$ is simple, but numerically tedious if done by hand. Its four complex roots and complex modes are found to be

$$\omega_{c1} = \left(\pm 0.7626 + 0.0795\,\mathrm{i}\right)\omega \qquad \omega_{c2} = \left(\pm 1.8312 + 0.2205\,\mathrm{i}\right)\omega \tag{3.547}$$

$$\mathbf{Z} = \begin{Bmatrix} 1.4100 \mp 0.0769\,\mathrm{i} & -1.3900 \mp 0.1834\,\mathrm{i} \\ 1 & 1 \end{Bmatrix} \tag{3.548}$$

In general, all four modes are necessary to describe the response of the system to dynamic loads. Extracting r_j and b_j from the real and imaginary parts of ω_{cj} given above, we obtain the associated "undamped" frequencies and damping ratios

$$\omega_1 = 0.7667\omega \qquad \xi_1 = 0.1037 \tag{3.549}$$

$$\omega_2 = 1.844\omega \qquad \xi_2 = 0.1196 \tag{3.550}$$

These frequencies and damping ratios are almost the same as the true undamped frequencies and weighted modal damping ratios computed earlier for the normal modes. Thus, the attenuated vibration can be simulated almost exactly with the classical modal solution. One important difference remains, however: the modal shapes of the nonclassical modes are complex. This implies that the two masses will not vibrate in phase in any given mode. For instance, a free vibration in the first mode with amplitude A will be described by

$$u_1 = A \ \mathrm{Re}\left[(1.4100 - 0.0769\,\mathrm{i})\,e^{\mathrm{i}0.7626\,\omega t}\right]e^{-0.0795\,\omega t}$$
$$= 1.412\,A\,e^{-0.0795\,\omega t}\cos(0.762\,\omega t - 0.0545) \tag{3.551}$$

$$u_2 = A \ \mathrm{Re}\left[1.0\ e^{\mathrm{i}0.7626\,\omega t}\right]e^{-0.0795\,\omega t}$$
$$= A\,e^{-0.0795\,\omega t}\cos(0.7626\,\omega t) \tag{3.552}$$

Thus, the motion of the first mass lags behind that of the second mass by an angle $\theta = 0.0545$ rad $= 3.12°$. Although this difference appears to be modest in this case, it could have important consequences in more general systems with heavy damping. For example, in an MDOF structure having normal modes, there are certain locations – which change from mode to mode – that do not vibrate at all; these correspond to stationary points on the system. If the structure were subjected to dynamic loads with dominant frequency in the neighborhood of a given natural frequency, the motion at the location of a stationary point would exhibit little motion. This absence would be particularly noticeable in the response spectrum. On the other hand, if the system did not have normal modes, no such stationary points could develop. In such case, less filtering and selectivity would be evidenced in the response of the system.

3.7.6 Computation of Complex Modes

Before proceeding with the computation of the complex modes, the classical modes and complex poles obtained by using only the diagonal terms of the modal transformation on the damping matrix should be obtained first, to be used as starting values for the computational scheme described in the ensuing. Thereafter, a convenient and relatively simple way of computing the complex eigenvalues and modal shapes is by means of inverse iteration with shift by the (mixed) Rayleigh quotient. If an eigenvalue is shifted by a known approximation λ_k so that $\lambda = \lambda_k + \varepsilon$, then the two shifted eigenvalue forms are

First Doubled-Up Form

$$\begin{Bmatrix} -\lambda_k \mathbf{K} & \mathbf{K} \\ \mathbf{K} & \lambda_k \mathbf{M} - \mathbf{C} \end{Bmatrix} \begin{Bmatrix} \mathbf{x} \\ \mathbf{y} \end{Bmatrix} = \varepsilon \begin{Bmatrix} \mathbf{K} & \mathbf{O} \\ \mathbf{O} & -\mathbf{M} \end{Bmatrix} \begin{Bmatrix} \mathbf{x} \\ \mathbf{y} \end{Bmatrix} \tag{3.553}$$

which after some simple transformations leads to the inverse iteration scheme

$$\begin{Bmatrix} \mathbf{K} - \lambda_k \mathbf{C} + \lambda_k^2 \mathbf{M} & \mathbf{O} \\ \lambda_k \mathbf{I} & -\mathbf{I} \end{Bmatrix} \begin{Bmatrix} \mathbf{x}'_{k+1} \\ \mathbf{y}'_{k+1} \end{Bmatrix} = \begin{Bmatrix} \mathbf{C} & -\mathbf{M} \\ -\mathbf{I} & \mathbf{O} \end{Bmatrix} \begin{Bmatrix} \mathbf{x}_k \\ \mathbf{y}_k \end{Bmatrix} \tag{3.554}$$

If m is the number of modes already found, and $\mathbf{x}_0$, $\mathbf{y}_0$ are starting vectors that together span the full $2N$-dimensional space, then this eigenvalue problem can be solved in the following steps:

Rinse starting vector
$$\begin{Bmatrix} \mathbf{x}_0 \\ \mathbf{y}_0 \end{Bmatrix} \Leftarrow \begin{Bmatrix} \mathbf{x}_0 \\ \mathbf{y}_0 \end{Bmatrix} - \sum_{j=1}^{m} \alpha_j \begin{Bmatrix} \mathbf{z}_j \\ \lambda_j \mathbf{z}_j \end{Bmatrix}$$

$$\alpha_j = \frac{\mathbf{z}_j^T \mathbf{K} \mathbf{x}_0 - \lambda_j \mathbf{z}_j^T \mathbf{M} \mathbf{y}_0}{\mathbf{z}_j^T \mathbf{K} \mathbf{z}_j - \lambda_j \mathbf{z}_j^T \mathbf{M} \mathbf{z}_j} \tag{3.555}$$

Inverse iteration
$$\left(\mathbf{K} - \lambda_k \mathbf{C} + \lambda_k^2 \mathbf{M} \right) \mathbf{y}_{k+1}' = \mathbf{K} \mathbf{x}_k - \lambda_k \mathbf{M} \mathbf{y}_k \tag{3.556}$$

Back – substitution
$$\mathbf{x}_{k+1}' = \left(\mathbf{y}_{k+1}' - \mathbf{x}_k \right) / \lambda_k \tag{3.557}$$

Rayleigh quotient
$$\lambda_{k+1} = \frac{\mathbf{x}_k^T \mathbf{K} \mathbf{x}_k - \mathbf{y}_k^T \mathbf{M} \mathbf{y}_k}{\mathbf{x}_k^T \mathbf{K} \mathbf{x}_{k+1}' - \mathbf{y}_k^T \mathbf{M} \mathbf{y}_{k+1}'} \tag{3.558}$$

Scaling
$$\sigma = \sqrt{\mathbf{x}_{k+1}'^T \mathbf{K} \mathbf{x}_{k+1}' - \mathbf{y}_{k+1}'^T \mathbf{M} \mathbf{y}_{k+1}'} \qquad \begin{Bmatrix} \mathbf{x}_{k+1} \\ \mathbf{y}_{k+1} \end{Bmatrix} = \frac{1}{\sigma} \begin{Bmatrix} \mathbf{x}_{k+1}' \\ \mathbf{y}_{k+1}' \end{Bmatrix} \tag{3.559}$$

With the scaling factor used here, the numerator in the mixed Rayleigh quotient above is unity in each iteration. Other scaling factors are, of course, also possible. Notice that for each complex eigenvalue found (but not for real ones), the complex conjugate is also a solution, so that one must be rinsed out too.

Second Doubled-Up Form

$$\begin{Bmatrix} \mathbf{K} - \lambda_k \mathbf{C} & \lambda_k \mathbf{M} \\ \lambda_k \mathbf{M} & -\mathbf{M} \end{Bmatrix} \begin{Bmatrix} \mathbf{x} \\ \mathbf{y} \end{Bmatrix} = \varepsilon \begin{Bmatrix} \mathbf{C} & -\mathbf{M} \\ -\mathbf{M} & \mathbf{O} \end{Bmatrix} \begin{Bmatrix} \mathbf{x} \\ \mathbf{y} \end{Bmatrix} \tag{3.560}$$

which after some simple transformations leads to the inverse iteration scheme

$$\begin{Bmatrix} \mathbf{K} - \lambda_k \mathbf{C} + \lambda_k^2 \mathbf{M} & \mathbf{O} \\ \lambda_k \mathbf{I} & -\mathbf{I} \end{Bmatrix} \begin{Bmatrix} \mathbf{x}_{k+1}' \\ \mathbf{y}_{k+1}' \end{Bmatrix} = \begin{Bmatrix} \mathbf{C} & -\mathbf{M} \\ -\mathbf{I} & \mathbf{O} \end{Bmatrix} \begin{Bmatrix} \mathbf{x}_k \\ \mathbf{y}_k \end{Bmatrix} \tag{3.561}$$

Rinse starting vector
$$\begin{Bmatrix} \mathbf{x}_0 \\ \mathbf{y}_0 \end{Bmatrix} \Leftarrow \begin{Bmatrix} \mathbf{x}_0 \\ \mathbf{y}_0 \end{Bmatrix} - \sum_{j=1}^{m} \alpha_j \begin{Bmatrix} \mathbf{z}_j \\ \lambda_j \mathbf{z}_j \end{Bmatrix}$$

$$\alpha_j = \frac{\mathbf{z}_j^T \mathbf{C} \mathbf{x}_0 - \lambda_j \mathbf{z}_j^T \mathbf{M} \mathbf{x}_0 - \mathbf{x}_j^T \mathbf{M} \mathbf{y}_0}{\mathbf{z}_j^T \mathbf{C} \mathbf{z}_j - 2\lambda_j \mathbf{z}_j^T \mathbf{M} \mathbf{z}_j} \tag{3.562}$$

Inverse iteration
$$\left(\mathbf{K} - \lambda_k \mathbf{C} + \lambda_k^2 \mathbf{M} \right) \mathbf{x}_{k+1}' = \mathbf{C} \mathbf{x}_k - \mathbf{M} \left(\lambda_k \mathbf{x}_k + \mathbf{y}_k \right) \tag{3.563}$$

Back-substitution
$$\mathbf{y}_{k+1}' = \mathbf{x}_k + \lambda_k \mathbf{x}_{k+1}' \tag{3.564}$$

Rayleigh quotient
$$\lambda_{k+1} = \frac{\mathbf{x}_k^T \mathbf{C} \mathbf{x}_k - 2\mathbf{x}_k^T \mathbf{M} \mathbf{y}_k}{\mathbf{x}_k^T \mathbf{C} \mathbf{x}_{k+1}' - \mathbf{x}_k^T \mathbf{M} \mathbf{y}_{k+1}' - \mathbf{y}_k^T \mathbf{M} \mathbf{x}_{k+1}'} \tag{3.565}$$

Scaling
$$\sigma = \sqrt{\mathbf{x}_{k+1}'^T \mathbf{C} \mathbf{x}_{k+1}' - 2\mathbf{x}_{k+1}'^T \mathbf{M} \mathbf{y}_{k+1}'} \qquad \begin{Bmatrix} \mathbf{x}_{k+1} \\ \mathbf{y}_{k+1} \end{Bmatrix} = \frac{1}{\sigma} \begin{Bmatrix} \mathbf{x}_{k+1}' \\ \mathbf{y}_{k+1}' \end{Bmatrix} \tag{3.566}$$

3.8 Frequency Domain Analysis of MDOF Systems

The analysis of MDOF systems in the frequency domain is largely analogous to that of SDOF systems, and similar numerical techniques are used. However, because they involve a larger number of DOF, MDOF systems require the use of vectors and matrices, which make the theory of such systems more complicated. For example, the task required in some analyses of finding the complex *poles* and *zeros* of the transfer functions entails far more effort.

3.8.1 Steady-State Response of MDOF Systems to Structural Loads

Consider an MDOF system subjected to harmonic loads. This system has stiffness, damping and mass matrices $\mathbf{K}$, $\mathbf{C}$, and $\mathbf{M}$, respectively, and the load vector is $\mathbf{p}(t)$. The dynamic equilibrium equation is once more

$$\mathbf{M}\ddot{\mathbf{u}} + \mathbf{C}\dot{\mathbf{u}} + \mathbf{K}\mathbf{u} = \mathbf{p}(t) \tag{3.567}$$

Unlike modal analysis, we make no a priori assumptions about the structure of the damping matrix $\mathbf{C}$, so we need not be concerned as to whether or not it is of the proportional type. For harmonic excitation with driving frequency ω, the load vector is of the form

$$\mathbf{p}(t) = \tilde{\mathbf{p}}\, e^{i\omega t} \tag{3.568}$$

in which $\tilde{\mathbf{p}}$ indicates the spatial variation of the harmonic load. In the interest of greater generality, we allow this vector to depend on the driving frequency, that is, $\tilde{\mathbf{p}} = \tilde{\mathbf{p}}(\omega)$. We assume then a harmonic solution

$$\mathbf{u}(t) = \tilde{\mathbf{u}}(\omega)\, e^{i\omega t} \tag{3.569}$$

whose derivatives are

$$\dot{\mathbf{u}}(t) = i\,\omega\,\tilde{\mathbf{u}}(\omega)\, e^{i\omega t} \tag{3.570}$$

$$\ddot{\mathbf{u}}(t) = -\omega^2\, \tilde{\mathbf{u}}(\omega)\, e^{i\omega t} \tag{3.571}$$

After we substitute these expressions into the equilibrium equation 3.567, factor out common terms, and cancel the nonvanishing exponential term appearing on both sides of the equation, we obtain

$$\left[-\omega^2 \mathbf{M} + i\,\omega \mathbf{C} + \mathbf{K}\right]\tilde{\mathbf{u}} = \tilde{\mathbf{p}} \tag{3.572}$$

Next, we define the term in square brackets as the dynamic stiffness matrix or *impedance matrix*:

$$\boxed{\mathbf{Z}(\omega) = \mathbf{K} + i\,\omega \mathbf{C} - \omega^2 \mathbf{M}} \tag{3.573}$$

With this definition, we can formally write the steady-state solution as

$$\boxed{\tilde{\mathbf{u}} = \mathbf{Z}^{-1}\,\tilde{\mathbf{p}}} \tag{3.574}$$

We emphasize that this is a formal result only. In practical situations, we determine $\tilde{\mathbf{u}}$ by solving the system of equations $\mathbf{Z}\tilde{\mathbf{u}} = \tilde{\mathbf{p}}$. This leads us in turn to the first complication: unlike SDOF systems – and excepting very simple systems with at most 2 or 3 DOF – the frequency response functions for MDOF systems are generally known only in numerical form.

If the βth element of $\tilde{\mathbf{p}}$ has unit value while all of its other elements are zero, it is clear that the frequency response functions will be numerically equal to the elements in the βth column of $\mathbf{Z}^{-1}$. Thus, this column contains the *transfer functions* for a unit load in that position. We conclude that

$$\mathbf{Z}^{-1} = \mathbf{H}(\omega) = \left\{ H_{\alpha\beta}(\omega) \right\} \tag{3.575}$$

in which $H_{\alpha\beta}(\omega)$ is the transfer function at DOF α due to a unit load at DOF β.

3.8.2 Steady-State Response of MDOF System Due to Support Motion

In our earlier analysis of MDOF systems subjected to seismic excitations, we arrived at the equation

$$\mathbf{M}\ddot{\mathbf{u}} + \mathbf{C}\dot{\mathbf{u}} + \mathbf{K}\mathbf{u} = \mathbf{C}\mathbf{E}\dot{\mathbf{u}}_g + \mathbf{K}\mathbf{E}\mathbf{u}_g \qquad \text{in terms of } absolute \text{ motions} \tag{3.576}$$

in which $\mathbf{u}_g = \mathbf{u}_g(t)$ is the support motion vector, with up to 6 DOF, and $\mathbf{E}$ is the rigid-body matrix for the structure (see *MDOF system subjected to multicomponent support motion*).

In particular, if we consider a harmonic support motion of the form

$$\mathbf{u}_g(t) = \tilde{\mathbf{u}}_g(\omega) e^{i\omega t} \tag{3.577}$$

$$\dot{\mathbf{u}}_g(t) = i\omega \tilde{\mathbf{u}}_g(\omega) e^{i\omega t} \tag{3.578}$$

$$\ddot{\mathbf{u}}_g(t) = -\omega^2 \tilde{\mathbf{u}}_g(\omega) e^{i\omega t} \tag{3.579}$$

we obtain the seismic equation in the frequency domain as

$$\left[-\omega^2 \mathbf{M} + i\omega \mathbf{C} + \mathbf{K} \right] \tilde{\mathbf{u}} = \left[i\omega \mathbf{C} + \mathbf{K} \right] \mathbf{E} \tilde{\mathbf{u}}_g \tag{3.580}$$

or more compactly

$$\mathbf{Z}\tilde{\mathbf{u}} = \left[i\omega \mathbf{C} + \mathbf{K} \right] \mathbf{E} \tilde{\mathbf{u}}_g \tag{3.581}$$

in which $\mathbf{Z}$ is again the dynamic stiffness (or impedance) matrix. Solving for $\tilde{\mathbf{u}}$, the steady-state harmonic response is

$$\boxed{\tilde{\mathbf{u}} = \mathbf{Z}^{-1} \left[i\omega \mathbf{C} + \mathbf{K} \right] \mathbf{E} \tilde{\mathbf{u}}_g} \tag{3.582}$$

The rectangular matrix

$$\mathbf{H}(\omega) = \mathbf{H}_{\mathbf{u}|\mathbf{u}_g}(\omega) = \mathbf{Z}^{-1} \left[i\omega \mathbf{C} + \mathbf{K} \right] \mathbf{E} \tag{3.583}$$

contains the transfer functions for absolute displacements due to absolute ground displacements. As in the case of SDOF systems, these functions are invariant under a change of input–output type. In other words, if we consider either absolute velocities or accelerations, then

$$\mathbf{H}_{u|u_g} \equiv \mathbf{H}_{\dot{u}|\dot{u}_g} \equiv \mathbf{H}_{\ddot{u}|\ddot{u}_g} \tag{3.584}$$

Clearly, many more transfer functions could be obtained by considering different types of input–output, including relative motions, internal forces such as bending moments, or even support reactions.

Example: 2-DOF System

Consider first a general lumped-mass 2-DOF system:

$$\mathbf{K} = \begin{Bmatrix} k_{11} & k_{12} \\ k_{21} & k_{22} \end{Bmatrix} \qquad \mathbf{C} = \begin{Bmatrix} c_{11} & c_{12} \\ c_{21} & c_{22} \end{Bmatrix} \qquad \mathbf{M} = \begin{Bmatrix} m_1 & \\ & m_2 \end{Bmatrix} \tag{3.585}$$

The load vector and frequency response vectors are

$$\tilde{\mathbf{u}} = \begin{Bmatrix} \tilde{u}_1 \\ \tilde{u}_2 \end{Bmatrix} \qquad \tilde{\mathbf{p}} = \begin{Bmatrix} \tilde{p}_1 \\ \tilde{p}_2 \end{Bmatrix} \tag{3.586}$$

Hence, the dynamic stiffness matrix is

$$\mathbf{Z} = \begin{Bmatrix} k_{11} + i\,\omega c_{11} - \omega^2 m_1 & k_{12} + i\,\omega c_{12} \\ k_{21} + i\,\omega c_{21} & k_{22} + i\,\omega c_{22} - \omega^2 m_2 \end{Bmatrix} \tag{3.587}$$

whose inverse is

$$\mathbf{Z}^{-1} = \frac{1}{\Delta} \begin{Bmatrix} k_{22} + i\,\omega c_{22} - \omega^2 m_2 & -(k_{12} + i\,\omega c_{12}) \\ -(k_{21} + i\,\omega c_{21}) & k_{11} + i\,\omega c_{11} - \omega^2 m_1 \end{Bmatrix} = \begin{bmatrix} H_{11} & H_{12} \\ H_{21} & H_{22} \end{bmatrix} \tag{3.588}$$

$$\Delta = (k_{11} + i\,\omega c_{11} - \omega^2 m_1)(k_{22} + i\,\omega c_{22} - \omega^2 m_2) - (k_{12} + i\,\omega c_{12})(k_{21} + i\,\omega c_{21}) \tag{3.589}$$

To focus ideas, let's consider the 2-DOF system depicted in Figure 3.23. In this case, $k_{11} = k$, $k_{12} = k_{21} = -k, k_{22} = k + 6k = 7k, m_1 = m$, and $m_2 = 4m$. The undamped natural frequencies of this system are

$$\omega_1 = \sqrt{\frac{3k}{4m}} \qquad \omega_2 = \sqrt{\frac{2k}{m}} \qquad \Rightarrow \qquad \frac{\omega_2}{\omega_1} = \sqrt{\frac{8}{3}} = 1.633 \tag{3.590}$$

If the system were indeed undamped, these natural frequencies would coincide with the roots of the determinant $\Delta = 0$. If so, the response would be infinitely large at those frequencies. However, since our system has damping, the response is not infinite, although it may still be large.

For the sake of simplicity and facilitate the interpretation of the results, we shall choose a proportional damping matrix producing equal modal damping ξ in both modes. As we saw earlier in the section on Rayleigh damping, this can be achieved with a damping matrix $\mathbf{C} = a_0\mathbf{M} + a_1\mathbf{K}$ with coefficients

$$a_0 = \frac{2\xi\,\omega_1\,\omega_2}{\omega_1 + \omega_2} \qquad \text{and} \qquad a_1 = \frac{2\xi}{\omega_1 + \omega_2} \tag{3.591}$$

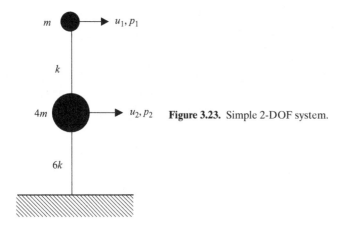

Figure 3.23. Simple 2-DOF system.

so

$$\mathbf{C} = \frac{2\xi}{1 + \omega_2 / \omega_1} \left[\omega_2 \, \mathbf{M} + \frac{1}{\omega_1} \, \mathbf{K} \right]$$

$$= \frac{4\xi\sqrt{km}}{\sqrt{3} + \sqrt{8}} \begin{Bmatrix} 1 + \sqrt{\tfrac{3}{2}} & -1 \\ -1 & 7 + 4\sqrt{\tfrac{3}{2}} \end{Bmatrix} = \xi\sqrt{km} \begin{Bmatrix} 1.951 & -0.877 \\ -0.877 & 10.437 \end{Bmatrix} \tag{3.592}$$

We choose k and m so as to give $\omega_1 = 1$ (or equivalently, we normalize the driving frequency by the fundamental frequency), and we evaluate the Rayleigh damping matrix with fractions of modal damping $\xi = 0.01, 0.02$, and 0.05. We then compute the transfer functions and display them in Figures 3.24 (top, middle, and bottom) in terms of their absolute values and phase angles (because of the symmetry of the dynamic stiffness matrix, $H_{21} = H_{12}$, so there are only three distinct transfer functions for this system shown). We define the sign of the phase angles so that

$$H_{\alpha\beta} = \left| H_{\alpha\beta} \right| e^{-i\varphi} \tag{3.593}$$

As you may have expected, the transfer functions are peaked at the two resonant frequencies $\omega_1 / \omega_1 = 1$ and $\omega_2 / \omega_1 = 1.633$. Also, the less the damping, the greater the maximum amplification. Notice that the two peaks are not of the same height, despite the fact that the fractions of modal damping are identical. The reason for this is that the two modes do not contribute in equal amounts to the total response. Indeed, the degree of modal contribution changes with both the location of output and the point of application of the source. If we were to express the transfer function in terms of the modes, we would find that $\mathbf{H} = \mathbf{\Phi} \, \mathbb{A} \, \mathbf{\Phi}^T$, where $\mathbb{A}$ is the diagonal matrix of modal amplification functions $\mathbb{A} = diag \left\{ \mu_j \left(\omega_j^2 - \omega^2 + 2i \, \xi_j \omega \omega_j \right) \right\}^{-1}$. This shows that the resonant response at $\omega = \omega_n$ is controlled not just by the mode with the current frequency, but also by the other modes with modal amplitude $A_j = \left[\mu_j \left(\omega_j^2 - \omega_n^2 + 2i \, \xi_j \omega_n \omega_j \right) \right]^{-1}$.

You should also notice that the phase angle in the vicinity of each resonant frequency is $\pm 90°$. This means that the real part of the transfer functions changes sign in the neighborhood of these frequencies, the exact location of which depends on the damping; for zero damping, they equal the resonant frequencies.

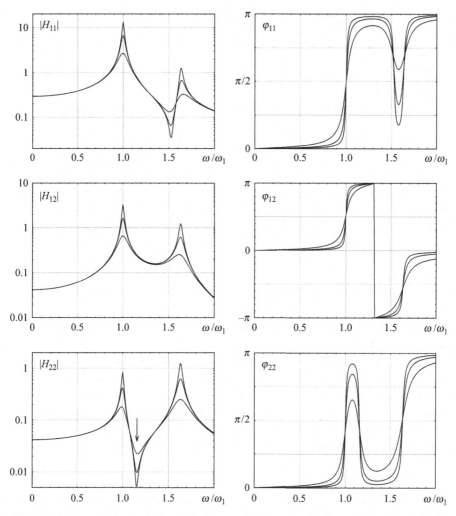

Figure 3.24. 2-DOF system, absolute value and phase angle of transfer functions for 1%, 2%, and 5% modal damping. Top = H_{11}, middle = H_{12}, bottom = H_{22}.

Observe an interesting feature in the H_{22} transfer function: the response is very small at the frequency pointed at by the arrow, namely $\omega/\omega_1 = \sqrt{4/3} = 1.155$. This happens to be the natural frequency of the superstructure, which in this case is just a single spring-mass system. If the modal damping had been zero, the response would have been exactly zero at this frequency. A similar result also holds true for more complicated structures: the transfer function for the loaded mass is zero at each and every natural frequency of the substructure obtained by removing (i.e., constraining) the degree of freedom at the driven point, as will be shown in a subsequent section.

Example 2: Submersible Launching System

We consider the problem of a mother ship in the shape of a catamaran that is used as launching pad for (or the recovery of) a submersible, as shown schematically in Figure 3.25. The submersible is lowered into the water by means of a cradle suspended

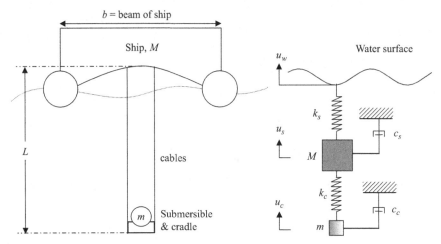

Figure 3.25. Submersible launching system (left) and its 2-DOF idealization (right).

by plastic cables made of polypropylene, which are somewhat less dense than water and thus float. As the cradle with the submersible is gradually lowered into the water, the cables gain length, which causes their effective axial stiffness to decrease. As a result, the vibration characteristics in heaving motion of the system consisting of the catamaran and the submersible change with the depth of the submersible. To a good approximation, this problem can be analyzed as a 2 DOF system "attached" to the ocean surface that consists of the masses M, m of the ship and the submersible, the buoyancy stiffness of the ship k_s, and the stiffness of the cables k_c. In addition, there exist two viscous dampers c_s, c_c that model the wave radiation damping and viscous damping on the ship as well as the interaction of the submersible with the water around it. The dampers themselves are not attached to the surface, but to some fixed reference frame denoting the standing water. In addition, the waves on the surface of the water can be modeled as a support motion $u_w(t)$ underneath the mother ship. This last approximation is valid only for waves with wavelength much greater than the dimensions of the ship. This turns out to be acceptable at the natural frequency that leads to significant submersible motions, but not at higher frequencies. Hence, the transfer functions shown here are only approximately valid at wave periods longer than about 8 seconds.

If u_c, u_s are the absolute displacements of the ship and cradle, respectively, then the dynamic idealization is governed by the equation

$$\begin{Bmatrix} M & 0 \\ 0 & m \end{Bmatrix} \begin{Bmatrix} \ddot{u}_s \\ \ddot{u}_c \end{Bmatrix} + \begin{Bmatrix} c_s & 0 \\ 0 & c_c \end{Bmatrix} \begin{Bmatrix} \dot{u}_s \\ \dot{u}_c \end{Bmatrix} + \begin{Bmatrix} k_c + k_s & -k_c \\ -k_c & k_c \end{Bmatrix} \begin{Bmatrix} u_s \\ u_c \end{Bmatrix} = \begin{Bmatrix} k_c + k_s & -k_c \\ -k_c & k_c \end{Bmatrix} \begin{Bmatrix} 1 \\ 1 \end{Bmatrix} u_w \qquad (3.594)$$

in which

$$k_s = \rho_w A_w \, g, \qquad k_c = n \frac{EA}{L} \qquad (3.595)$$

with A_w = the cross section (area) of the catamaran at the water line, n is the number of cables supporting the cradle, and A, L, E are the cross section, length, and Young's modulus of each of the cables. Observe that the damping matrix does not contribute to the fictitious load on the right-hand side due to support motion. Hence, the characteristic equation for free vibration is

$$\begin{bmatrix} k_c + k_s & -k_c \\ -k_c & k_c \end{bmatrix} \begin{Bmatrix} \phi_{cj} \\ \phi_{sj} \end{Bmatrix} = \omega_j^2 \begin{bmatrix} M & 0 \\ 0 & m \end{bmatrix} \begin{Bmatrix} \phi_{cj} \\ \phi_{sj} \end{Bmatrix} \tag{3.596}$$

Concerning the wave motion, we can assume that they are harmonic waves whose frequency depends on sea conditions and are of the form

$$u_w(x,t) = a_w \exp\left[2\pi i(t/T - x/\lambda)\right] \tag{3.597}$$

in which a, T, λ are respectively, the amplitude, the dominant period, and the wavelength of the waves. In deep water, the period and wavelength are related as

$$\lambda = \frac{g}{2\pi} T^2 \tag{3.598}$$

with g being the acceleration of gravity. In addition, the wave amplitude is generally less than one seventh of the wavelength, for otherwise the wave would break. Hence,

$$a_w = \alpha\lambda < \tfrac{1}{7}\lambda, \qquad \alpha < \tfrac{1}{7} \tag{3.599}$$

In particular, let's consider the following data:

$$\left.\begin{aligned} n &= 4 \\ E &= 1.5\times10^9 \text{ N/m}^2 \text{ (polypropylene)} \\ \sigma_Y &= 3\times10^7 \text{ N/m}^2 \text{ (tensile strength)} \\ A &= 0.5\times10^{-4}\text{m} \quad (\phi = 8 \text{ mm}) \\ L &= 10\,\text{m} \end{aligned}\right\} \quad k_c = \frac{4\times1.5\times10^9\times0.5\times10^{-4}}{10} = 3\times10^4 \text{ N/m}$$

$$A_w = 200 \text{ m}^2 \qquad \rightarrow \qquad k_s = 10^3\times200\times9.8 = 2\times10^6 \text{ N/m}$$

$$M = 8\times10^5 \text{ kg}$$
$$m = 5\times10^4 \text{ kg}$$

After simplification by the common factor 10^4, the eigenvalue problem is then

$$\begin{bmatrix} 203 & -3 \\ -3 & 3 \end{bmatrix} \begin{Bmatrix} \phi_{cj} \\ \phi_{sj} \end{Bmatrix} = \omega_j^2 \begin{bmatrix} 80 & 0 \\ 0 & 5 \end{bmatrix} \begin{Bmatrix} \phi_{cj} \\ \phi_{sj} \end{Bmatrix} \tag{3.600}$$

whose solution is

$$\omega_1 = 0.7671 \text{ rad/sec}, \qquad T_1 = 8.19\,\text{sec}, \qquad \phi_1^T = \{1 \quad 52\} \tag{3.601}$$

$$\omega_2 = 1.5966 \text{ rad/sec} \qquad T_2 = 3.94\,\text{sec} \qquad \phi_2^T = \{-3.25 \quad 1\} \tag{3.602}$$

We observe that in the first mode, the mother ship hardly moves while the submersible executes large-amplitude vibrations. This is because this 2-DOF system is weakly coupled, and the submersible acts as an SDOF system with frequency and period

$$\omega_1 \sim \sqrt{\frac{k}{m}} = \sqrt{\frac{3\times10^4}{5\times10^4}} = 0.7745 \qquad T_1 \sim 8.11\,\text{sec} \tag{3.603}$$

When the submersible is at depths greater than $L = 10$ m, the modal decoupling only increases, which causes the above SDOF approximation to improve, in which case the ship can be taken as a fixed support whose heaving motion (if any) is not affected by the presence of the submersible. If so, the fundamental period of the cradle with the submersible as a function of cable length L can be simply expressed in terms of the period for $L = 10$ m as

$$T_1 = 8.11\sqrt{\frac{L}{10}} \tag{3.604}$$

This gives the following results:

L [m]	T_1 [sec]
5	5.7
10	8.1
30	14.0
60	19.9
100	25.6

Now, ocean waves of significant size have periods in the range from 3 to 18 seconds, which means that resonance will always be observed for some cable length L and given ocean conditions.

We next go on to determine the dynamic response elicited by the waves, but first we construct an appropriate damping matrix for this problem. Assuming mass-proportional damping, the $\mathbf{C}$ matrix is of the form $\mathbf{C} = a_0\mathbf{M}$, in which case the modal damping ratios are

$$\xi_j = \frac{\phi_j^T \mathbf{C}\phi_j}{2\omega_j\,\phi_j^T \mathbf{M}\phi_j} = \frac{a_0\,\phi_j^T \mathbf{M}\phi_j}{2\omega_j\,\phi_j^T \mathbf{M}\phi_j} = \frac{a_0}{2\omega_j} \tag{3.605}$$

Let's assume that we either know, or know how to estimate, the fraction of damping in the first mode, the most dangerous one. Let's assume further that this damping is 10% of critical, in which case $a_0 = 2\xi_1\omega_1 = 2\times 0.1\times 0.7671 = 0.1534$.

We now are ready to consider the response to surface waves. For this purpose, we formulate the equations of motion in the frequency domain:

$$\left[\mathbf{K} + i\omega\mathbf{C} - \omega^2\mathbf{M}\right]\mathbf{u} = a_w\mathbf{K}\,\mathbf{e} \tag{3.606}$$

That is,

$$\begin{aligned}\mathbf{u} &= a_w\left[\mathbf{K} + i\omega\mathbf{C} - \omega^2\mathbf{M}\right]^{-1}\mathbf{K}\,\mathbf{e}, \\ &= a_w\,\mathbf{H}\end{aligned} \tag{3.607}$$

$$\mathbf{H} = \left[\mathbf{K} + \left(i\omega a_0 - \omega^2\right)\mathbf{M}\right]^{-1}\mathbf{K}\,\mathbf{e} \tag{3.608}$$

in which

$$\mathbf{H} = \begin{Bmatrix} H_s \\ H_c \end{Bmatrix} \equiv \begin{Bmatrix} H_{u_s/u_w} \\ H_{u_c/u_w} \end{Bmatrix} \tag{3.609}$$

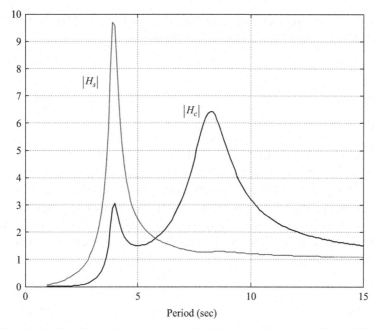

Figure 3.26. Transfer functions for surface wave motion. In this example, the submersible exhibits large motions at periods at which the mother ship barely moves.

are the two transfer functions for unit harmonic motion of the ocean surface, and a_w is the amplitude of the surface waves. These are depicted in Figure 3.26, obtained using a simple MATLAB® program. Do notice that these are plotted in terms of period and not frequency as usual. This is to facilitate interpretation of the results in terms of the dominant period of the ocean waves. Observe the strong dynamic response of the submersible at the fundamental period of about 8 seconds, while the ship remains virtually quiescent.

In conclusion, this example demonstrates how a submersible can potentially experience very large motions even while the launching mother ship remains relatively quiescent, and thus, while the crew remains largely unaware of the dangers to the cradle-submersible system.

3.8.3 In-Phase, Antiphase, and Opposite-Phase Motions

We examine herein somewhat more carefully the concept of phase angle of a frequency response function, and elaborate on the meaning of the relative phase when two such functions are compared. For this purpose, consider the transfer response function $H_{ij}(\omega)$ for the response at DOF i due to a unit harmonic load at DOF j. The response is then

$$u_{ij} = H_{ij}e^{i\omega t} = A_{ij}e^{i(\omega t - \phi_{ij})} \tag{3.610}$$

where

$$A_{ij} = |H_{ij}|, \qquad \tan\phi_{ij} = \frac{\operatorname{Im}H_{ij}}{\operatorname{Re}H_{ij}} \tag{3.611}$$

A more careful analysis reveals, however, that if $H_{ij} = \text{Re}\, H_{ij} + \text{i}\, \text{Im}\, H_{ij} \equiv z = a + \text{i}\,b$ and we *define* the phase angle as $\phi = \arctan\left(|b|/|a|\right)$, then

$$\phi_{ij} = \begin{cases} \phi & a > 0 \quad b > 0 \\ \pi - \phi & a < 0 \quad b > 0 \\ \pi + \phi & a < 0 \quad b < 0 \\ -\phi & a > 0 \quad b < 0 \end{cases} \tag{3.612}$$

that is, the angle depends on $\text{sgn}\,(\text{Re}\,z)$ and $\text{sgn}\,(\text{Im}\,z)$. In MATLAB®, this is accomplished with the function call `phi=atan2(imag(z)),real(z))`, where `z = Hij`.

Similarly, $u_{kj} = H_{kj}\mathrm{e}^{\mathrm{i}\omega t} = A_{kj}\mathrm{e}^{\mathrm{i}\left(\omega t - \phi_{kj}\right)}$, which defines the motion at some other point for the response of DOF $k \neq i$ due to a unit load at j. If we were to compare the phase angles of these two motions, we would generally find that they are different, but in some cases these motions could have the same phase, opposite phase, or even be in antiphase, as will be seen.

In-Phase Motions

If both have the *same* phase, then $\phi_{ij} = \phi_{kj}$, and the situation is clear. This requires both the real and imaginary parts to have the same sign, and for these parts to be proportional.

Opposite-Phase Motions

On the other hand, u_{ij} and u_{kj} will have *opposite* phase if $\phi_{ij} = -\phi_{kj}$, that is, if one motion lags behind the excitation and the other is ahead of the excitation by that same angle. This requires that the real parts agree in sign and the imaginary parts have opposite sign. Physically, one motion is then delayed by $\tau = \phi_{kj}/\omega$ while the other is advanced by that same time (or the other way around). The time delay between the two components is then 2τ.

Antiphase Motions

The preceding should not be confused with the concept of "180 degrees out-of-phase" or "antiphase," which would require $\phi_{ij} = \phi_{kj} + \pi$. This can be satisfied only if the real and imaginary parts of H_{ij} *simultaneously* have signs that are the opposite of those of H_{kj}. For example, consider the 2-DOF system

$$\mathbf{K} = \begin{Bmatrix} k_{11} & k_{12} \\ k_{12} & k_{22} \end{Bmatrix}, \qquad \mathbf{C} = \begin{Bmatrix} c_{11} & c_{12} \\ c_{12} & c_{22} \end{Bmatrix} \qquad \mathbf{M} = \begin{Bmatrix} m_1 & 0 \\ 0 & m_2 \end{Bmatrix} \tag{3.613}$$

with positive semidefinite matrices $\mathbf{K}, \mathbf{C}$. This requires

$$k_{11} > 0, \qquad k_{22} > 0, \qquad k_{11}k_{22} - k_{12}^2 > 0 \qquad \text{(i.e. } |k_{12}| < \sqrt{k_{11}k_{22}}) \tag{3.614}$$

$$c_{11} > 0, \qquad c_{22} > 0, \qquad c_{11}c_{22} - c_{12}^2 > 0 \qquad \text{(i.e. } |c_{12}| < \sqrt{c_{11}c_{22}}) \tag{3.615}$$

The impedance matrix is then

$$
\mathbf{Z} = \begin{Bmatrix} k_{11} - \omega^2 m_1 + i\omega c_{11} & k_{12} + i\omega c_{12} \\ k_{12} + i\omega c_{12} & k_{22} - \omega^2 m_1 + i\omega c_{22} \end{Bmatrix} \tag{3.616}
$$

so

$$
\mathbf{H} = \mathbf{Z}^{-1} = \frac{1}{\Delta} \begin{Bmatrix} k_{22} - \omega^2 m_1 + i\omega c_{22} & -\left(k_{12} + i\omega c_{12}\right) \\ -\left(k_{12} + i\omega c_{12}\right) & k_{11} - \omega^2 m_1 + i\omega c_{11} \end{Bmatrix} \tag{3.617}
$$

where

$$
\Delta = \left(k_{11} - \omega^2 m_1 + i\omega c_{11}\right)\left(k_{22} - \omega^2 m_1 + i\omega c_{22}\right) - \left(k_{12} + i\omega c_{12}\right)^2 \tag{3.618}
$$

Clearly, if we choose $0 < c_{12} < \sqrt{c_{11}c_{22}}$ (i.e., a positive coupling term), then undoubtedly the imaginary parts of the off-diagonal terms of $\mathbf{H}$ will be opposite to that of the diagonal terms, that is, $\mathrm{Im}\left(H_{11}\right) = -\mathrm{Im}\left(H_{21}\right)$ and $\mathrm{Im}\left(H_{12}\right) = -\mathrm{Im}\left(H_{22}\right)$ for *any* frequency ω. We can then always find frequencies such that $\mathrm{Re}\left(H_{11}\right) = -\mathrm{Re}\left(H_{21}\right)$ (or if alternatively, $\mathrm{Re}\left(H_{12}\right) - \mathrm{Re}\left(H_{22}\right)$ for a source applied at DOF 2), that is, both the real and imaginary parts are opposite to each other, in which case the motions at the 2 DOF will be in antiphase.

In summary, whenever one talks of "opposite phase," one should be crystal clear as to what that means, to avoid confusion with the concept of "antiphase."

3.8.4 Zeros of Transfer Functions at Point of Application of Load

We now proceed to prove that the transfer functions for an undamped N DOF system, when observed at the point of application of the load, exhibit zeros (i.e., are quiescent) at frequencies that correspond to the natural frequencies of the smaller $(N-1)$ DOF subsystem obtained by fully constraining the driven point, that is, making it fixed.

Consider an arbitrary *undamped* vibrating system with N degrees of freedom that is subjected to a single harmonic load p_β applied at some mass point in some specific direction associated with DOF β. This elicits a response elsewhere u_α in direction α, the transfer function $H_{u_\alpha|p_\beta}(\omega)$ of which will exhibit infinite peaks at each of the N resonant frequencies $\omega_1, \omega_2, \ldots \omega_N$ of the complete system. Then from the linearity of this problem and the absence of sources elsewhere, we can express the response at points other than the driven point as

$$
H_{u_\alpha|p_\beta} = H_{u_\beta|p_\beta} \times H_{u_\alpha|u_\beta=1} \tag{3.619}
$$

where $H_{u_\beta|p_\beta}$ is the magnitude of the response observed at the driven point in the direction of the load, and $H_{u_\alpha|u_\beta=1}$ is the response anywhere else due to a prescribed *unit* displacement at the driven point, that is, after applying a single constraint to the system at that point and forcing it to execute a harmonic vibration of unit amplitude, a situation that reduces the total number of DOF by one. Now, from the interlacing theorem, we know that the frequencies $\omega'_j, j = 1, 2, \ldots N-1$ of the constrained structure do interlace with those of the unconstrained structure, that is,

$$
\omega_1 \leq \omega'_1 \leq \omega_2 \leq \omega'_2 \leq \cdots \leq \omega'_{N-1} \leq \omega_N \tag{3.620}
$$

where the primed quantities are the frequencies of the constrained structure, and the unprimed are those for the whole structure. Clearly, $H_{u_\alpha | P_\beta}\left(\omega_j'\right) \neq \infty$ because the frequencies of the constrained structure are *not* resonant frequencies of the whole structure. However, $H_{u_\alpha | u_\beta = 1}\left(\omega_j'\right) = \infty$, because the amplification function of the constrained structure must have resonant peaks at its own natural frequencies. This means that the first factor must be zero so that the product $0 \times \infty$ may remain finite. This is the same as saying that

$$H_{u_\beta | P_\beta}\left(\omega_j'\right) = \frac{H_{u_\alpha | P_\beta}\left(\omega_j'\right)}{H_{u_\alpha | u_\beta = 1}\left(\omega_j'\right)} = \frac{finite}{\infty} = 0, \qquad j = 1, 2, \ldots N - 1 \tag{3.621}$$

We conclude that when a system of N DOF is driven by a harmonic load acting in some fixed direction at a given location, the transfer function at the driven point in the direction of the load will exhibit zeros at each and every one of the frequencies of the system with $N - 1$ DOF obtained by constraining the DOF acted upon by the load. Hence, these zeros interlace with the singularities at the resonant frequencies of the system. In other words, poles and zeros interlace along the frequency axis, which means in turn that the transfer functions exhibit phase reversals at each of the zeros and at each of the poles. Inasmuch as N is arbitrary, this means that this theorem continues to be valid in the limit when $N \to \infty$, that is, for continuous systems.

It must be added that if the system were to have light damping, then both the zeros as well as the poles would become complex numbers, in which case the response would not be zero, but remain small. We also rush to add that the properties described herein for the zeros do not hold for points other than the driven point. Indeed, transfer functions elsewhere may have fewer than $N - 1$ zeros or even none, and the imaginary parts of those zeros could well have of any sign, either positive or negative.

3.8.5 Steady-State Response of Structures with Hysteretic Damping

In analogy to SDOF systems, the complex impedance of an assembly of elements with frequency-independent structural damping can be written as

$$\mathbf{K}_c = \mathbf{K} + i\,\mathbf{D}\,\mathrm{sgn}(\omega) \tag{3.622}$$

In particular, if all elements have the same fraction of hysteretic damping ξ_h, then $\mathbf{D}$ is of the form

$$\mathbf{D} = 2\xi_h\,\mathrm{sgn}(\omega)\,\mathbf{K} \tag{3.623}$$

Hence, the dynamic stiffness matrix of an MDOF system with hysteretic damping is either

$$\mathbf{Z} = \mathbf{K} + i\,\mathbf{D}\,\mathrm{sgn}(\omega) - \omega^2\mathbf{M} \tag{3.624}$$

or more simply

$$\mathbf{Z} = \left[1 + 2\,i\,\xi\,\mathrm{sgn}(\omega)\right]\mathbf{K} - \omega^2\mathbf{M} \tag{3.625}$$

Using a modal representation of the transfer functions, it is easy to show that the latter equation would lead to uncoupled *modal* equations with uniform hysteretic damping. Now, each modal equation is analogous to the equation for a single DOF system. Since for such systems there is only a negligible difference between viscous and hysteretic damping, it follows that uniform modal damping can readily be simulated in the frequency domain by simply multiplying the stiffness matrix by the complex scalar factor $1 + 2\,i\,\xi\,\mathrm{sgn}(\omega)$, as in Eq. 3.625.

3.8.6 Transient Response of MDOF Systems via Fourier Synthesis

If $\mathbf{p}(t)$ is an arbitrary time-varying load vector whose Fourier transform is $\tilde{\mathbf{p}}(\omega)$, then the response to this load is simply the inverse Fourier transform

$$\mathbf{u}(t) = \frac{1}{2\pi}\int_{-\infty}^{\infty} \tilde{\mathbf{u}}(\omega)\,e^{i\,\omega t}\,d\omega = \frac{1}{2\pi}\int_{-\infty}^{\infty} \mathbf{Z}^{-1}(\omega)\,\tilde{\mathbf{p}}(\omega)\,e^{i\,\omega t}\,d\omega \tag{3.626}$$

In most practical cases, the Fourier transform of the loads is computed with the Fourier Fast Transform (FFT) algorithm and could exhibit an erratic variation with frequency. By contrast, the transfer functions are generally smooth between each of the natural frequencies. It is then customary to compute the transfer function at a lesser number of frequencies that those in the FFT, and to interpolate as needed to obtain intermediate values. The motivation lies in the fact that from a computational point of view, the most demanding part lies in the determination of the transfer functions.

Hermite Interpolation

In some cases, it may be advantageous to interpolate the transfer functions using both the functions themselves as well as their slopes, that is, using *Hermite interpolation*. The reason for this is that the derivatives of the transfer functions with respect to frequency can be obtained with very little extra effort, once the dynamic stiffness matrix has been reduced (i.e., brought to triangular form in the course of a Gaussian decomposition). Indeed, from the equations of motion in the frequency domain with unit load vectors, we have

$$\left[-\omega^2\mathbf{M} + i\,\omega\mathbf{C} + \mathbf{K}\right]\mathbf{H} = \mathbf{I} \qquad \text{or briefly} \qquad \mathbf{Z}\mathbf{H} = \mathbf{I} \tag{3.627}$$

Taking derivatives with respect to frequency, we obtain

$$\left[-2\omega\mathbf{M} + i\,\mathbf{C}\right]\mathbf{H} + \left[-\omega^2\mathbf{M} + i\,\omega\mathbf{C} + \mathbf{K}\right]\frac{d}{d\omega}\mathbf{H} = \mathbf{O} \tag{3.628}$$

With the shorthand $\mathbf{H}' \equiv \frac{d}{d\omega}\mathbf{H}$, we can finally write the derivatives as

$$\boxed{\mathbf{H}' = \mathbf{Z}^{-1}\left[2\omega\mathbf{M} - i\,\mathbf{C}\right]\mathbf{H}} \tag{3.629}$$

Clearly, since the triangular form of $\mathbf{Z}^{-1}$ is already available, solving for $\mathbf{H}'$ requires only modest extra effort.

In the case of support motions, the equation of motion for transfer functions is

$$\left[-\omega^2\mathbf{M} + i\,\omega\mathbf{C} + \mathbf{K}\right]\mathbf{H} = \left[i\,\omega\mathbf{C} + \mathbf{K}\right]\mathbf{E} \tag{3.630}$$

Taking once more derivatives with respect to frequency, we obtain

$$\left[-2\omega\mathbf{M} + i\,\mathbf{C}\right]\mathbf{H} + \left[-\omega^2\mathbf{M} + i\,\omega\mathbf{C} + \mathbf{K}\right]\mathbf{H}' = i\,\mathbf{C}\mathbf{E} \tag{3.631}$$

Hence

$$\boxed{\mathbf{H}' = \mathbf{Z}^{-1}\left[(2\omega\mathbf{M}\,\mathbf{H} - \mathrm{i}\,\mathbf{C}(\mathbf{H} - \mathbf{E})\right]}$$

(3.632)

In the case of lightly damped systems, however, the transfer functions are very sharply peaked, and it may be difficult or impractical, even if not impossible, to sample at a sufficient number of frequencies to appropriately resolve those peaks. In such cases, it is preferable to use the (highly recommended) *exponential window method* described in Chapter 6, Section 6.6.14 of this book.

3.8.7 Decibel Scale

Amplification functions and other frequency response functions such as *transfer functions* are often measured in a logarithmic scale known as the decibel scale. This scale is a measure of the power in a signal at a given frequency ω, which in turn is a function of the square of the amplitude $A(\omega)$ of the signal at that frequency. Its practical measuring unit is the decibel (dB), a designation that derives from the name of Alexander Graham Bell. It is defined as

$$I = 10 \, \log \frac{\text{Power}}{\text{Ref. power level}} = 10 \, \log \left(\frac{A}{A_{ref}}\right)^2 = 20 \, \log \frac{A}{A_{ref}} \quad [\text{dB}]$$

(3.633)

In acoustics, the reference power level is 10^{-12} Watt ($= 1$ dB) so 1 Watt corresponds to a sound intensity of 120 dB.

Often, the decibel scale is used to measure relative changes in power, so

$$\Delta I = 10 \, \log \frac{\text{power}_2}{\text{power}_0} - 10 \, \log \frac{\text{power}_1}{\text{power}_0} = 10 \, \log \frac{\text{power}_2}{\text{power}_1}$$

(3.634)

Thus, a doubling of power corresponds to an increase of about 3 dB ($10 \log 2 = 3.01$). An often-used relative scale is also the so-called signal-to-noise ratio,

$$R = \frac{I_{\text{signal}}}{I_{\text{noise}}}$$

(3.635)

which measures how much a coherent signal raises above the background noise. (More precise definitions of the signal-to-noise ratio are given in Chapter 7, Section 7.1.7 and in books on signal processing.)

3.8.8 Reciprocity Principle

As we have seen in an earlier section, the elements of the dynamic flexibility matrix $\mathbf{H} = \mathbf{K}_d^{-1}$ are the transfer functions for the response observed at one location in a dynamic system due to a unit load applied at another. Inasmuch as the dynamic stiffness matrix is symmetric, so must also be the inverse. This implies that

$$H_{\alpha\beta}(\omega) = H_{\beta\alpha}(\omega)$$

(3.636)

that is, the response at DOF α due to a unit load at DOF β is identical to the response at DOF β due to a unit load at DOF α. Let's now carry out two separate loading experiments. First, we apply an impulsive load at β and observe the response time history at α. After the system comes to rest, we apply the impulsive load at α, and observe the new response at β. The time histories in these two experiments are given by the inverse Fourier transforms

$$h_{\alpha\beta}(t) = \frac{1}{2\pi}\int_{-\infty}^{+\infty} H_{\alpha\beta}(\omega)e^{i\omega t}d\omega \tag{3.637}$$

$$h_{\beta\alpha}(t) = \frac{1}{2\pi}\int_{-\infty}^{+\infty} H_{\beta\alpha}(\omega)e^{i\omega t}d\omega \tag{3.638}$$

Since the two transfer functions above are identical, it follows that the two impulse response functions are also identical,

$$h_{\alpha\beta}(t) = h_{\beta\alpha}(t) \tag{3.639}$$

Finally, if we apply an arbitrary load $p(t)$ at one location, then at the other, the observed time histories are then

$$u_{\alpha\beta}(t) = u_{\beta\alpha}(t) = h_{\alpha\beta} * p = h_{\beta\alpha} * p \tag{3.640}$$

which are once more identical to each other. This is the fundamental *reciprocity principle*, which constitutes the dynamic counterpart of the well-known Maxwell–Betti law in elasticity.

Example 1: Reciprocity Principle for an Acoustic Problem
As an illustration of the applicability of this principle to dynamic systems, consider a person giving a lecture in a classroom from a fixed location, such as the lectern. Depending on the location of each of the listeners in the audience, the intensity of the lecturer's voice will be perceived differently across the room. Suppose now that at some point in time, the lecturer swaps location with one listener in the audience, so this person goes to the lectern while the lecturer sits down in that person's seat and then repeats there his narration. The listener standing at the lectern should then hear the lecturer's voice exactly the same as he did before from his seat, provided that there is no additional noise in the room. It should be clear, however, that with the lecturer in the new location, the other listeners elsewhere will not hear the lecture as they did before; the reciprocity principle applies only to the two exchanged locations, that is, the two swapped persons.

Example 2: Measure Unbalanced Load without Dismounting Motor
A ship's engine is slightly unbalanced and produces excessive vibrations when rotating at its fixed operating frequency ω. As a first step toward resolving this problem, it has been decided to measure the intensity of the unbalanced load. This can be accomplished as follows: Let A be the location of the engine, and B an observation point nearby on the floor. The running engine exerts on the floor a harmonic load with an unknown amplitude P_A = the unbalanced load.

1. While the engine is running, measure the amplitude U_{BA} of the response at B.
2. Turn off the engine, and apply at B a harmonic load with a known intensity P_B having the same frequency as the operating frequency of the engine. Measure the response U_{AB} elicited at A.
3. From the reciprocity principle, we have that $\dfrac{U_{BA}}{P_A} = \dfrac{U_{AB}}{P_B}$, that is, $P_A = \dfrac{U_{BA}}{U_{AB}} P_B$.

Effect of Noise on Reciprocity Principle

In general, in the presence of noise, signals do not obey the reciprocity principle implied by the symmetry of the *transfer functions*:

$$H_{ij}(\omega) = H(\mathbf{x}_i, \mathbf{x}_j, \omega) = H(\mathbf{x}_j, \mathbf{x}_i, \omega) = H_{ji}(\omega) \tag{3.641}$$

that is, the input $\mathbf{x}_j$ and output $\mathbf{x}_i$ (i.e., source and receiver) locations and directions cannot be traded when the environment is affected by noise. To see why this is so, consider the following scenario in a college dormitory:

While a student is taking a shower, the dorm telephone rings, which is then answered by her roommate. She then calls from the bathroom door with "telephone for you," to which the student in the shower replies "What? I can't hear you," a reply that is, however, perfectly understandable to the student at the door. This lack of symmetry in intelligibility is due to the following. Let's say the first student calls with a voice intensity of 80 dB, which by the time it reaches the student in the shower has attenuated to 60 dB. She in turn answers with 80 dB, which also attenuates to 60 dB at the door. Now, if the shower makes a noise of 80 dB, this noise will also attenuate to 60 dB by the time it reaches the student at the door. The signal-to-noise ratios (i.e., voice to shower noise) are then 60:80 for the student in the shower, and 60:60 for the student at the door. Clearly, the latter ratio is larger, which explains why the student outside can understand, while the other cannot.

3.9 Harmonic Vibrations Due to Vortex Shedding

When a slender, flexible, cylindrical body capable of sustaining vibrations is embedded in a fluid that flows transversely to the body with some constant velocity, the fluid develop vortices at the interface with the body that promptly detach from the contact surface and are carried downstream. As the vortices are shed alternatingly from side to side, they elicit alternating pressures that can excite a structure to nearly resonant conditions in a direction perpendicular to the flow, if the frequency of vortex shedding happens to be comparable to one of the natural frequency of the structure. In fact, vibrations transverse to the flow due to vortex shedding are often more damaging to structures than vibrations and stresses in the direction of the flow, either through excessive oscillations or because of fatigue effects, so this is a phenomenon that must be accounted for in design. Examples where vortex shedding is a concern is in tall buildings and slender chimneys subjected to winds, or ocean currents causing oscillations in marine risers and offshore platforms, and so on. Vortex shedding is also the reason for the whooshing sound heard when a slender rod or a rope is rapidly moved by hand in circular motion. In some cases, corkscrew devices on the surface of the cylinder known as *strakes* may be used to disturb the fluid flow and disrupt vortex shedding.

The frequency (or rate) at which the vortices are shed depends on the speed of flow and the characteristic width of the body that impedes the flow. In the case of a cylindrical body (a pipe, an offshore platform leg, a marine riser, or a tall chimney), the frequency can be estimated as

$$f = S\frac{V}{D}[\text{Hz}]$$ (3.642)

where D is the diameter of the vibrating body in the path of the fluid moving with velocity V, f is the frequency of vortex shedding in cycles per second, and S is a dimensionless parameter known as the *Strouhal number*. This number is fairly constant, namely $S \approx 0.21$ for fluid flows whose Reynolds numbers are anywhere in the range $R = 10^2$ to $R = 10^6$.

Example: Vortex Shedding in Tall Chimney

A chimney (tall stack) of $D = 2$ m is known to have a fundamental lateral frequency of 1.5 Hz. What wind velocity would elicit a resonant condition due to vortex shedding?

$$V = \frac{fD}{S} = \frac{1.5 \times 2}{0.21} = 14.3 [\text{m}/\text{s}]$$ (3.643)

or what is the same, $V = 51$ km/h (i.e., 32 mph). This is a moderate wind speed likely to arise, and thus must be accounted for in design. Then again, the next higher mode of the chimney is probably at least three times higher (assuming it to be a shear beam), which would require a wind speed of some 100 mph, which is already a hurricane condition at which other design considerations, and not vortex shedding, are more important.

3.10 Vibration Absorbers

A vibration absorber is a passive mechanism or device that can be installed at strategic locations in a structure so as to prevent excessive vibrations caused by dynamic loads, that is, to avoid or ameliorate resonant effects. Typical examples are: the tuned mass dampers placed in tall buildings so as to limit the response of a building to wind loads; vibration isolation devices underneath buildings or instruments (e.g., a telescope) to insulate that system from detrimental ground-borne vibrations in its surroundings; or vibration-suppressing devices attached to reciprocating machinery, engines or motors with rotating imbalance. Their main purpose may be to improve comfort to building occupants, increase the service life of some expensive or motion-sensitive piece of equipment, or even prevent damage and improve serviceability. We consider briefly some of these devices in turn, and provide a few examples of their use.

3.10.1 Tuned Mass Damper

In a nutshell, a tuned mass damper is an appropriately sized device in the form of an SDOF system that is installed at some optimal location and whose frequency is adjusted so as to be nearly coincident with the vibration mode one wishes to suppress. The design assumes that the structure to be controlled can be modeled in turn as an SDOF system with a frequency equal to that of the mode to be suppressed.

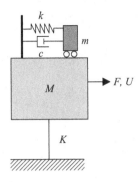

Figure 3.27. TMD, simple model as a 2-DOF system.

With reference to Figure 3.27, let K, M denote the stiffness and mass of a structure, modeled as an SDOF system, and let k, m be the stiffness and mass of a vibration absorber or tuned-mass damper (TMD). The TMD is used to ameliorate the intensity of motion in the structure when the latter is subjected to dynamic forces F. We define

$$\Omega = \sqrt{\frac{K}{M}} \qquad \text{Frequency of structure alone (without the TMD)}$$

$$\omega_o = \sqrt{\frac{k}{m}} \qquad \text{Frequency of TMD alone (without the struct}$$

$$\omega_r = \sqrt{\frac{K}{m+M}} \qquad \text{Frequency of structure with infinitely rigid oscillator}$$

$$\omega_1, \omega_2 \qquad \text{Frequencies of coupled system}$$

$$\mu = \frac{m}{M} \qquad \text{Mass ratio}$$

The stiffness and mass matrices of the coupled system as well as the displacement and load vectors are

$$\mathbf{K} = \begin{Bmatrix} k & -k \\ -k & k+K \end{Bmatrix} \quad \mathbf{M} = \begin{Bmatrix} m & \\ & M \end{Bmatrix} \quad \mathbf{u} = \begin{Bmatrix} u \\ U \end{Bmatrix} \quad \mathbf{p} = \begin{Bmatrix} 0 \\ F \end{Bmatrix} \qquad (3.644)$$

Solving the eigenvalue problem for this 2-DOF system, we obtain the coupled frequencies

$$\left(\frac{\omega_j}{\Omega}\right)^2 = \frac{1}{2}\left\{(1+\mu)\left(\frac{\omega_o}{\Omega}\right)^2 + 1 \mp \sqrt{\left[(1+\mu)\left(\frac{\omega_o}{\Omega}\right)^2 - 1\right]^2 + 4\mu\left(\frac{\omega_o}{\Omega}\right)^2}\right\} \qquad j=1,2 \quad (3.645)$$

Next, we evaluate the dynamic response in the structure elicited by a harmonic force with amplitude F acting on the structural mass M. Neglecting damping in the structure (but not in the oscillator), the dynamic equilibrium equation is then

$$\begin{Bmatrix} k+i\omega c - \omega^2 m & -(k+i\omega c) \\ -(k+i\omega c) & K+k+i\omega c - \omega^2 M \end{Bmatrix} \begin{Bmatrix} u \\ U \end{Bmatrix} = \begin{Bmatrix} 0 \\ F \end{Bmatrix} \qquad (3.646)$$

Solving for the response in the structure U, we obtain after brief algebra

$$U = \frac{F\left[k - \omega^2 m + i\omega c\right]}{\left[K - \omega^2(m+M)\right]\left[k - \dfrac{(K - \omega^2 M)\omega^2 m}{K - \omega^2(m+M)} + i\omega c\right]} \tag{3.647}$$

We shall show now that there exist two frequencies for which the response is independent of the damping constant c. Hence, all amplification functions have these points in common, no matter what the damping. These two points occur when the complex terms in the numerator and denominator cancel identically. This is satisfied if, and only if, the two real parts are equal, that is, if

$$\pm\left[k - \omega^2 m\right] = k - \frac{(K - \omega^2 M)\omega^2 m}{K - \omega^2(m+M)} \tag{3.648}$$

The plus/minus sign on the left-hand side is to allow for frequencies greater than that of the tuned mass damper (a condition that would make the left-hand side term negative). If we consider first the positive sign, we find that it is satisfied only if $\omega = 0$. This solution is not interesting, because it represents a static problem. On the other hand, if we consider the negative sign, we obtain the biquadratic equation

$$m(m+2M)\omega^4 - 2\left[k(m+M)+Km\right]\omega^2 + 2Kk = 0 \tag{3.649}$$

whose solution is

$$\left(\frac{\omega_{P,Q}}{\Omega}\right)^2 = \frac{1}{2+\mu}\left\{(1+\mu)\left(\frac{\omega_o}{\Omega}\right)^2 + 1 \mp \sqrt{\left[\left(\frac{\omega_o}{\Omega}\right)^2 - 1\right]^2 + \mu(2+\mu)\left(\frac{\omega_o}{\Omega}\right)^2}\right\} \tag{3.650}$$

The two frequencies given by this equation represent two points P, Q through which all transfer functions must pass, no matter what their damping should be (Figure 2.28 on the left). In particular, these two points must also be traversed by the two transfer functions that correspond to zero damping and to infinite damping, that is, $c = 0$ and $c = \infty$. In the latter case, the system behaves the same as if the TMD was perfectly rigid. This in turn represents an undamped SDOF system with stiffness K, total mass $m + M$, and frequency ω_r as defined earlier. This frequency, and that of the oscillator alone (ω_o), are bracketed by the two natural frequencies of the coupled system, as can be shown by considering Rayleigh's quotient with an arbitrary vector $\mathbf{v}^T = \{a \ \ b\}$. From the enclosure theorem, we have

$$\omega_1^2 \le R = \frac{\mathbf{v}^T \mathbf{K} \mathbf{v}}{\mathbf{v}^T \mathbf{M} \mathbf{v}} = \frac{(a-b)^2 k + b^2 K}{a^2 m + b^2 M} \le \omega_2^2 \tag{3.651}$$

in which a and b can be chosen arbitrarily; if we consider in turn the two choices $a = b$ as well as $b = 0$, we obtain the two inequalities

$$\omega_1^2 \le \frac{K}{m+M} \le \omega_2^2 \qquad \text{and} \qquad \omega_1^2 \le \frac{k}{m} \le \omega_2^2 \tag{3.652}$$

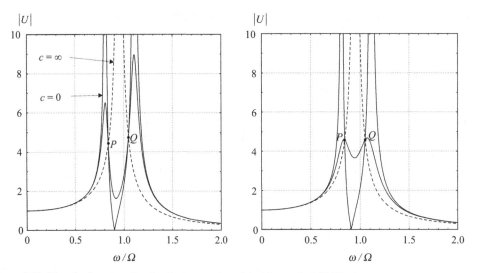

Figure 3.28. Transfer functions for simple structural model with attached TMD.

that is, $\omega_1 \le \omega_r \le \omega_2$ and $\omega_1 \le \omega_o \le \omega_2$. It follows that the two points P, Q lie somewhere in between the two natural frequencies at the intersection of the amplification function of the undamped coupled system and the undamped SDOF system with augmented mass $m + M$. The response amplitudes at these two frequencies are obtained by substituting the left-hand side of Eq. 3.648 (negative sign case) into the denominator of Eq. 3.647, and setting the dashpot constant to zero. The result is

$$U_{P,Q} = -\frac{F}{K - \omega_{P,Q}^2 (m + M)} \tag{3.653}$$

In general, the response amplitudes at the two frequencies for P, Q will not be equal. Optimal tuning of the mass damper can be achieved by enforcing these two amplitudes to be the same, as illustrated in Figure 3.28 on the right:

$$K - \omega_P^2 (m + M) = \pm \left[K - \omega_Q^2 (m + M) \right] \tag{3.654}$$

The case where both amplitudes are equal and have the same phase cannot be satisfied, since it implies equal frequencies for P and Q. Alternatively, if we consider equal amplitudes and opposite phase, we obtain

$$\omega_P^2 + \omega_Q^2 = \frac{2K}{m + M} = \frac{2\Omega^2}{1+\mu} \tag{3.655}$$

Equating this to the sum of the two roots in Eq. 3.647, we obtain

$$\frac{\omega_P^2 + \omega_Q^2}{2\Omega^2} = \frac{1}{1+\mu} = \frac{(1+\mu)\left(\frac{\omega_o}{\Omega}\right)^2 + 1}{2+\mu} \tag{3.656}$$

From here, we obtain the optimal tuning condition

$$\boxed{\frac{\omega_o}{\Omega} = \frac{1}{1+\mu} = \sqrt{\frac{k\,M}{K\,m}}}$$

which relates the optimal frequency of the oscillator to the design mass ratio. This ratio ensures that the two points P, Q have the same height. The coupled frequencies observed with optimal tuning are

$$\boxed{\frac{\omega_j}{\Omega} = \sqrt{\frac{1+\frac{1}{2}\mu \mp \sqrt{\mu + \frac{1}{4}\mu^2}}{1+\mu}}} \tag{3.657}$$

The optimal damping constant that should be assigned to an optimally tuned mass damper is the one that would cause the transfer function at the two points P, Q to have a horizontal slope. However, the analysis for this condition is rather cumbersome, and exact expressions are not available. A reasonably close approximation is given by the expression[10]

$$\boxed{\frac{c}{2\,m\Omega} \equiv \frac{\xi\omega_o}{\Omega} = \sqrt{\frac{3\mu}{8(1+\mu)^3}}} \tag{3.658}$$

The transfer function for an optimally tuned mass damper is shown in Figure 3.28b. The maximum amplification for this case is $A_{\max} = \sqrt{1+2/\mu}$.

In summary, to design a TMD, we proceed along the following lines:

1. Choose appropriately the mass of the TMD, neither too small not too large. This choice depends on the mass of the structure to be controlled. This defines the mass ratio μ.
2. Impose optimal tuning, which yields the stiffness of the TMD and at the same time also guarantees that the maximum amplification at the two points P, Q will be nearly the same.
3. Choose optimal damping, that is, the dashpot constant c.

3.10.2 Lanchester Mass Damper

A Lanchester tuned mass damper is one in which the stiffness of the damper is zero (or nearly zero). The optimal parameters for this case are

$$A_{\max} = 1 + 2/\mu \tag{3.659}$$

$$\xi = \frac{c}{2\,m\Omega} = \sqrt{\frac{1}{2(2+\mu)(1+\mu)}} \tag{3.660}$$

$$\frac{\omega_Q}{\Omega} = \sqrt{1 - \frac{\mu}{2(1+\mu)}} \approx \sqrt{\frac{1}{1+\frac{1}{2}\mu}} \tag{3.661}$$

---

[10] J. P. Den Hartog, *Mechanical Vibration*, 4th ed. (New York: McGraw-Hill, 1956).

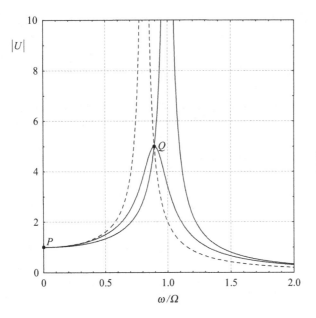

Figure 3.29. Transfer function for Lanchester mass damper.

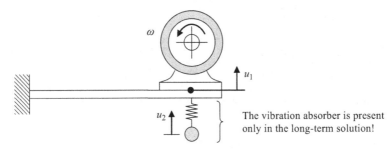

The vibration absorber is present only in the long-term solution!

Figure 3.30. TMD application to a slab that supports an unbalanced motor.

This case is illustrated in Figure 3.29.

3.10.3 Examples of Application of Vibration Absorbers

Example 1: Structural Vibration Caused by Unbalanced Motor
Figure 3.30 shows a lightweight, *flexible* cantilever beam supporting a heavy electric motor, which has an imbalance at the 60 Hz rotation rate of the rotor (or what is the same, 3,600 rpm). Assume that the mass of the motor, which is 100 kg, is large compared to the mass of the beam, and that the vertical imbalance force has a magnitude of 50 N.

The natural frequency of the beam with the motor is also 60 Hz, that is, it coincides with the operating speed of the rotor. In a transient decay test, you also determine that damping is about 1% of critical. Your job is to present the owner with possible solutions for reducing the vibration. You may not change the speed of the motor or move the rotor to another location on the beam.

1. Propose a solution that could be implemented easily and inexpensively the same day with materials commonly found around a shop. Your objective is to reduce the

vibration by 50% immediately, to give you time to implement a dynamic absorber later. Explain how you would accomplish that quick fix solution, and how it works.

2. Design a long-term solution using a simple dynamic absorber that is required to work well only at the problem frequency. Assume that a mass ratio $\mu = 0.1$ (i.e., 10%) is specified.
  I. What are the mass and spring constants of the absorber?
  II. What is the theoretical or ideal, steady state, dynamic response amplitude of the machine with the dynamic absorber attached? Sketch the H_{11} transfer function for this system and indicate the point on the curve corresponding to the operating frequency of the motor with the absorber attached and working properly.

(a) Simple, quick fix: Add mass

An on-the-spot reduction of the peak can easily be accomplished by adding mass Δm right next to the motor. The amplification function for this system is now (see also Section 2.6.3)

$$A(\omega_n, \omega) = \frac{\hat{r}^2}{\sqrt{(1 - \hat{r}^2)^2 + 4\xi^2 \hat{r}^2}} \tag{3.662}$$

$$\xi = \tfrac{1}{2} c / \sqrt{k(m + \Delta m)} = \tfrac{1}{2} c \sqrt{km(1 + \Delta m / m)} = \xi_0 / \hat{r} \tag{3.663}$$

$$A_{\text{resonance}} = \frac{1}{2\xi_0} = \frac{1}{2 \times 0.01} = 50 \, (\text{before adding mass}) \tag{3.664}$$

where $\hat{r} = \omega / \omega_n = f / f_n$ is the reciprocal of the tuning ratio (see Eq. 2.243). Addition of the mass reduces the natural frequency from $f_n = 60$ Hz to $f_{new} = 60\sqrt{m / (m + \Delta m)}$, so the new reciprocal of the tuning ratio squared is $r^2 = (f_{new} / f_n)^2 = m / (m + \Delta m) = 1 / (1 + \Delta m / m)$. Since damping is light, we can ignore damping in the new condition away from resonance and estimate the new amplification simply as

$$A_{50\%} = \frac{\hat{r}^2}{1 - \hat{r}^2} = \tfrac{1}{2} \frac{1}{2\xi_0} = \frac{1}{4\xi_0}, \qquad \hat{r}^2 = \frac{1}{1 + 4\xi_0} = \frac{1}{1 + \Delta m / m} \tag{3.665}$$

so

$\Delta m = 4\xi m = 4 \times 0.01 \times 100 = 4 \, \text{kg} \ \text{kg}$

Hence, it suffices to add 4 kg (about 10 lbs) to reduce the amplification by half. It should be straightforward to find enough scrap metal, which is quite heavy, to do the job.

(b) Tuned mass damper

1. Optimal tuning with $\mu = \tfrac{1}{10}$ implies

$$m = \frac{1}{10} M = \frac{1}{10} 100 = 10 \, \text{kg} \tag{3.666}$$

Also

$$\frac{f_0}{f_n} = \frac{1}{1 + \mu} = \sqrt{\frac{k}{m} \frac{M}{K}} = \sqrt{\frac{k}{\mu M} \frac{M}{K}} \tag{3.667}$$

That is,

$$k = \frac{K}{M}\frac{\mu M}{(1+\mu)^2} = 4\pi^2 f_n^2 \frac{\mu M}{(1+\mu)^2} = 4\pi^2 60^2 \frac{\frac{1}{10}100}{(1+\frac{1}{10})^2} = 1,174,564 [\text{N}/\text{m}] \qquad (3.668)$$

2. The theoretical or ideal dynamic amplitude with the TMD is

$$A_{\max} = \sqrt{1+\frac{2}{\mu}} = \sqrt{1+\frac{2}{0.1}} = 4.58 \qquad (3.669)$$

which is much better than the reduction accomplished by simply adding mass. Now, the damper of the motor is

$$C = 2\xi\Omega M = 2\times 0.01\times(2\pi\times 60)\times 100 = 754 [\text{N}-\text{s}/\text{m}] \qquad (3.670)$$

Also, the ideal dashpot constant for the tuned mass damper is

$$c = 2m\Omega\sqrt{\frac{3\mu}{8(1+\mu)^3}} = 2\times 10\times(2\pi\times 60)\sqrt{\frac{3\frac{1}{10}}{8(1+\frac{1}{10})^3}} = 1,266 [\text{N}-\text{s}/\text{m}] \qquad (3.671)$$

which can be used to obtain the amplification function. If, as shown in Figure 3.30, the motor is the first degree of freedom, then

$$\begin{Bmatrix} K+k+i\omega(c+C)-\omega^2 M & -(k+i\omega c) \\ (k+i\omega c) & k+i\omega c-\omega^2 m \end{Bmatrix}\begin{Bmatrix} u_1 \\ u_2 \end{Bmatrix} = F\frac{\omega^2}{\Omega^2}\begin{Bmatrix} 1 \\ 0 \end{Bmatrix} \qquad (3.672)$$

A sketch of $|u_1|$ is shown in Figure 3.31.

Example 2: Isolation System to Suppress Floor Vibrations

A machine with mass of 20 kg exerts a vertical force on the floor of a laboratory building, given by $F(t) = F_0 \cos\omega_0 t$, and illustrated in Figure 3.32a. This force causes a vertical

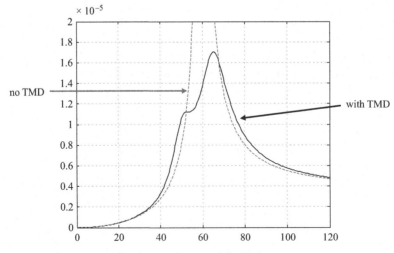

Figure 3.31. Comparison demonstrating the effectiveness of the TMD.

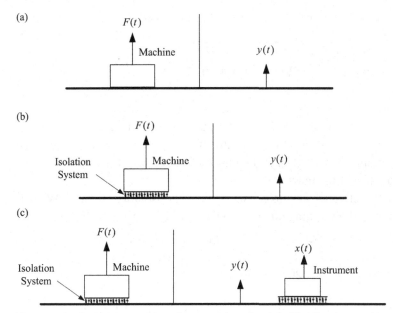

Figure 3.32. (a)Schematic view of floor subjected to excessive vibration. (b) Vibrating machine mounted on isolation pad. (c) Both the machine and the instrument mounted on pads.

vibration $y(t)$ on the floor of the next room with an amplitude $y_0 = 10^{-4}$ m at the machine's operating frequency of $f_0 = 30$ Hz.

A commercially available vibration isolation support system for the machine is purchased. It consists of a lightweight rigid plate resting on an *elastic* pad, which has some internal damping, as depicted in Figure 3.32b.

The machine on the isolation pad has a natural frequency f_m such that $f_0 / f_M = 3$, where f_0 is the problem-frequency observed in the next room. The damping ratio of the system is estimated to be $\xi_m = 7\%$ of critical damping.

  a. At the operating frequency of 30 Hz, what is the magnitude of the ratio of the vertical force transmitted to the floor through the isolation pad to the force produced by the machine, $|F_T/F_0|$? What is the ratio of the floor vibration amplitude next door after the isolation pad has been installed compared to before?

Ans.: The natural frequency of the machine on the pad is

$$f_M = \tfrac{1}{3}f_0 = \tfrac{1}{3}30 = 10 \text{ Hz}, \tag{3.673}$$

while the force transmitted to the floor is

$$F_T = ku + c\dot{u}, \text{with } u = F_0 A e^{i(\omega t - \varphi)}/k \tag{3.674}$$

So

$$F_T = F_0 A_M \left[1 + 2i\xi r\right]e^{i(\omega t - \varphi)}, \qquad A_M = \frac{1}{\sqrt{\left(1 - r^2\right)^2 + 4\xi^2 r^2}}, \qquad r = \frac{f_0}{f_M} = \frac{\omega_0}{\omega_M} = 3 \tag{3.675}$$

The ratio of transmitted to applied force is then reduced to

$$\left|\frac{F_T}{F_0}\right| = \sqrt{\frac{1+4\xi^2 r^2}{\left(1-r^2\right)^2+4\xi^2 r^2}} = \sqrt{\frac{1+4\times0.07^2\times3^2}{\left(1-3^2\right)^2+4\times0.07^2\times3^2}} = 0.1354 \tag{3.676}$$

Note: It might have seemed logical to have used instead the amplification function for the eccentric mass vibrator, but that would have been wrong, because the operating frequency f_0 is fixed at 30 Hz, and thus the magnitude of the force F_0 is also fixed. What is being changed here is the natural frequency of the oscillator, not the operating frequency of the machine.

b. What is the effective spring constant k for the isolation system? What is the static deflection of the machine due to its own weight as it rests on the support?

Ans.: Since the natural frequency and the mass are known, then

$$k = \omega_m^2 m_M = 4\pi^2 \times 10^2 \times 20 = 78,957 \ [\text{N/m}] \tag{3.677}$$

$$\Delta = \frac{m_M g}{k} = \frac{g}{\omega_n^2} = \frac{9.81}{(2\pi10)^2} = 0.00248 \ [\text{m}] \quad \text{or} \quad \Delta = 2.48 \ [\text{mm}] \tag{3.678}$$

The engineers are so happy with the result that they decide to do the same with the instrument in the room next door, see Figure 3.32c. They put it on an identical support – they got a great deal from the supplier by buying two at the same time. Due to its own weight, the instrument has a static deflection $\frac{1}{4}$ of that of the machine when placed on the pad.

c. What is the natural frequency f_1 of the instrument on its isolation pad and what is the frequency ratio f_0 / f_1 for the instrument at the problem frequency?

Ans.: Since the mats are identical, they have the same stiffness. Hence, if the static deflection due to weight is $\frac{1}{4}$ of that of the machine, the instrument must weigh $m_I = \frac{1}{4}m_M = \frac{1}{4}\times20 = 5$ [kg]. The natural frequency of the instrument is then

$$f_1 = f_M\sqrt{\frac{m_M}{m_I}} = f_M\sqrt{4} = 2f_M = 20 \ [\text{Hz}], \quad \frac{f_0}{f_1} = \frac{30}{20} = 1.5 \tag{3.679}$$

d. What damping ratio ξ_1 would you expect to have for the instrument on its vibration isolation pad?

Ans.: Again, since the pad is the same as for the machine, the dashpot constant must be the same, so the new damping is

$$\xi_1 = \frac{c}{2\sqrt{km_I}} = \frac{c}{2\sqrt{k\,m_M}}\sqrt{\frac{m_M}{m_I}} = \xi_M\sqrt{\frac{m_M}{m_I}} = 2\xi_M = 0.14 \ \rightarrow \ 14\% \tag{3.680}$$

e. What is your prediction for the response amplitude of the instrument compared to the floor motion at 30 Hz?

Ans.: The amplification for support motion is

$$A_I = \sqrt{\frac{1 + 4\xi_I^2 r_I^2}{\left(1 - r_I^2\right)^2 + 4\xi_I^2 r_I^2}} = \sqrt{\frac{1 + 4 \times 0.14^2 \times 1.5^2}{\left(1 - 1.5^2\right)^2 + 4 \times 0.14^2 \times 1.5^2}} = 0.8225 \tag{3.681}$$

So addition of the second pad ameliorated the response only by 18%.

  f. The engineers are not thrilled with the result. What is the problem? Without buying another system – that would be embarrassing to explain to the boss – what simple inexpensive fix would you suggest they try to improve the performance of the system, so as to make the natural frequency such that $f_0 / f_1 = 3$?

Ans.: The problem is that the natural frequency of the instrument is much closer to the operating frequency than the machine is, so it is closer to resonance. An inexpensive fix would be to add mass. For example, by adding 15 kg, the instrument will weigh as much as the machine, and thus have the same natural frequency, in which case $f_0 / f_I = 3$.

  g. Assume both isolation systems are fixed properly such that each one has $f_0/f_m = f_0/f_1 = 3$ and the damping $\xi_m = \xi_1 = 0.07$; what is the total reduction in the vibration of the instrument due to the machine compared to the vibration with no isolation pads at all?

Ans.: The net amplification or reduction is obtained from the product $A = A_M A_I$, where $A_M = |F_T / F|$ and A_I is as above, but with the damping and tuning ratio of the adjusted instrument. Since both systems have now the same frequency and damping, then $A_I = A_M$, in which case the net reduction is the square of the amplification function,

$$A = A_M^2 = 0.1354^2 = 1.83 \times 10^{-2} \approx 0$$

So the instrument is now virtually quiescent.

3.10.4 Torsional Vibration Absorber

A torsional mass damper is a device used to attenuate vibrations in a mechanical system.[11] In a typical application, it may consist of pendulums that automatically adjust their resonant frequencies to the rotational speed of wheels and shafts. It is used to ameliorate vibrations in such systems and typically dissipates energy through friction instead of viscous dampers.

Consider a flywheel to which two simple pendulums of length L are attached, as shown in Figure 3.33. The pendulums pivot about diametrically opposite points that are distant a from the axis, and are maintained in place by small springs. As the wheel turns with rotational speed ω, the pivoting points experience a centripetal acceleration that elicits a fictitious centrifugal gravity field $g' = \omega^2 (a + L) \gg g$, which in turn imparts on the pendulums a resonant frequency

$$\omega_o = \sqrt{\frac{g'}{L}} = \omega\sqrt{1 + \frac{a}{L}} \tag{3.682}$$

---

[11] A. H. Hsieh, "Optimum performance on pendulum-type torsional vibration absorber," *J. Aeronaut. Sci.*, July 1942, 337–340.

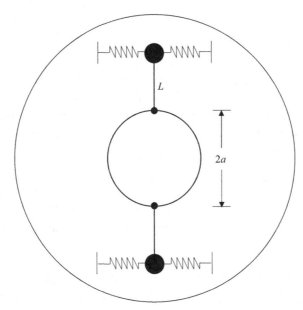

Figure 3.33. Flywheel on Shaft with Mounted Torsional TMDs.

If $a \ll L$, then $\omega_o \approx \omega$, implying a vibration absorber that is tuned to the rotational speed.

Example Machine shaft with torsional vibration absorber

As an example of application of a torsional vibration absorber, consider a machine shaft whose flexural vibration mode is being excited by unavoidable eccentricities, and assume that the frequency of this mode is twice the operational speed of the shaft. To suppress this vibration, we must design a tuned mass damper that is tuned to that frequency, that is, having a natural frequency $\omega_o = 2\omega$. This can be accomplished by setting $4\omega^2 = \omega^2(1 + a/L)$, which yields $a = 3L$.

4 Continuous Systems

4.1 Mathematical Characteristics of Continuous Systems

Although continuous systems constitute – at least in principle – the logical extension of discrete structural systems to mechanical objects with an infinitely large number of degrees of freedom (DOF), the fact is that the vast majority of continuous systems are governed by intractable equations that only rarely can be solved in closed form. Hence, exact solutions for continuous systems are scant and few in number. However, this does not mean that they cannot be solved and analyzed, but simply that in most cases they need to be approached with the aid of numerical tools, of which an excellent variety is available. Examples are the method of weighted residuals, the assumed modes method (based on Lagrange's equations), or the powerful set of discrete tools embodied in the finite element and finite differences methods.

Most numerical tools used rely on certain mathematical properties exhibited by the differential equations that characterize the *bodies* of vibrating continua in combination with their *boundary conditions*. An important such property is the generalized symmetry that guarantees that the elastic and kinetic energies will never be negative. But before analyzing in detail some of these properties in the context of the most often encountered continuous systems, we shall examine first the mathematical underpinning of continuous systems in general. We start this by examining the governing linear differential equations and their ancillary boundary conditions for some typical, simple systems, and then comment on their similarities and differences.

4.1.1 Taut String

Here, the *body* is the string itself, and the *boundaries* are the two anchoring points. Each point in the body has 1 DOF, which is the transverse displacement. The string is subjected to an initial tension T, which is uniform throughout the length. With reference to Figure 4.1 and from the dynamic equilibrium of a differential element, we then obtain:

$$\rho A \ddot{u} - T \frac{\partial^2 u}{\partial x^2} = b(x,t) \tag{4.1}$$

with boundary conditions

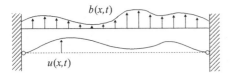

Figure 4.1. Taut string.

$$u\big|_{x=0} = 0 \qquad u\big|_{x=L} = 0 \tag{4.2}$$

This is a *second*-order differential equation, which has *one* boundary condition at *each* boundary point.

4.1.2 Rods and Bars

A rod is an ideal one-dimensional continuous system that is subjected solely to axial loads and axial stresses, both of which are uniformly distributed in cross section. Thus, vibrations in a rod result from longitudinal waves only, that is, from compressive-dilatational waves involving particle motions that coincide in direction with the axis of the rod. The rod model is only an approximation in that it neglects stresses and vibrations in planes perpendicular to the axis, which must surely arise as a result of Poisson's effect. Indeed, longitudinal extensions of the rod must cause lateral contractions, which in turn elicit transverse inertia forces and guided waves that travel near the surface of the rod. However, if the characteristic wavelengths of the motion are much greater than the width of the rod, the plane stress approximation is very good indeed.

The equations of the rod with variable axial stiffness are similar to those of the string, that is,

$$\rho A \ddot{u} - \frac{\partial}{\partial x}\left(EA \frac{\partial u}{\partial x} \right) = b(x,t) \tag{4.3}$$

This is again a differential equation of order $2\kappa = 2$ that has $\kappa = 1$ boundary conditions at *each* boundary point. If x_b is the coordinate of one of the two boundary points, then the boundary conditions are either one of

$$u\big|_{x_b} = 0 \qquad \text{Fixed end where the displacement is prescribed} \tag{4.4}$$

or

$$\frac{\partial u}{\partial x}\bigg|_{x_b} = 0 \qquad \text{Free end where the axial stress is prescribed } (\sigma = E\frac{\partial u}{\partial x} = 0) \tag{4.5}$$

Clearly, the highest derivative in any of the boundary conditions is less than the order of the differential equation. This system is formally equivalent to the string, so similar solutions apply.

4.1.3 Bending Beam, Rotational Inertia Neglected

The simplest model for a bending beam (with either constant or variable cross section) is that where only flexural deformations and transverse inertia forces are taken

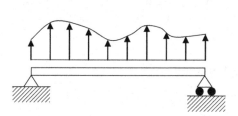
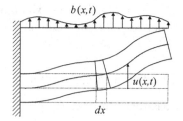

Figure 4.2. Simply supported beam.

into account, while plane sections are assumed to remain plane (Figure 4.2). Thus, shear deformations and rotational inertia effects are neglected. This model is referred to as the *Euler–Bernoulli beam.*

If loads and displacements are defined to be positive when upwards and the rotations are positive when counterclockwise, then the differential equation for a Euler–Bernoulli beam with variable cross section can readily be shown to be given by

$$\rho A \frac{\partial^2 u}{\partial t^2} + \frac{\partial^2}{\partial x^2}\left(EI(x)\frac{\partial^2 u}{\partial x^2}\right) = b(x,t) \tag{4.6}$$

which is a differential equation of order $2\kappa = 4$, so each of its two boundary points has $\kappa = 2$ boundary conditions, each of order less than 4. If x_b is the coordinate of a boundary point, then the admissible boundary conditions are consistent pairs of any of the following:

$$u\big|_{x_b} = 0 \qquad \text{Transverse displacement prevented} \tag{4.7}$$

$$\frac{\partial u}{\partial x}\bigg|_{x_b} = 0 \qquad \text{Rotation prevented } (\theta = \tfrac{\partial u}{\partial x} = 0) \tag{4.8}$$

$$\frac{\partial^2 u}{\partial x^2}\bigg|_{x_b} = 0 \qquad \text{Free rotation (moment vanishes, } M = EI\tfrac{\partial^2 u}{\partial x^2} = 0) \tag{4.9}$$

$$\frac{\partial}{\partial x}\left(EI \frac{\partial^2 u}{\partial x^2}\right)\bigg|_{x_b} = 0 \qquad \text{Free translation (shear vanishes, } S = \tfrac{\partial}{\partial x}\left(EI \tfrac{\partial^2 u}{\partial x^2}\right) = 0) \tag{4.10}$$

If the bending stiffness is constant in the neighborhood of the boundary point, then the vanishing shear condition is equivalent to $\frac{\partial^3 u}{\partial x^3} = 0$. Comparing all of the above, we see that the highest order of the boundary condition is again less than that of the differential equation, as was also the case for the rod.

Trick Question on Boundary Conditions

Observe that if we decide not to neglect axial deformations in a beam, that is, if we model it as a beam column, then the system acts simultaneously as a beam and as a rod. But be careful: the boundary conditions can be deceitful. For example, Figure 4.2 shows a simply supported beam with transverse displacements prevented – and of course also vanishing end moments – on both sides. But a rod with those boundary conditions would constitute a cantilever rod in which the axial motion is prevented on the left support yet the horizontal roller on the right would allow axial motions and thus cause the axial force there to be zero.

4.1.4 Bending Beam, Rotational Inertia Included

This case is very similar to the previous one:

$$\left(\rho A - \frac{\partial}{\partial x}\left(\rho I \frac{\partial}{\partial x}\right)\right)\ddot{u} + \frac{\partial^2}{\partial x^2}\left(EI \frac{\partial^2 u}{\partial x^2}\right) = b(x,t) \tag{4.11}$$

The boundary conditions are as before. Notice the differential operator acting on the acceleration term. It is also interesting to observe that, despite the fact that this problem includes distributed rotational inertia, it involves only 1 DOF at every point. The reason is that the translation and rotation remain functionally related to each other, because of kinematic constraints: the latter is simply the spatial derivative of the former. In contrast, when considering a lumped mass beam with discrete lumped masses that include both translational and rotational inertias, those mass points have 2 DOF each, because the translations at the finite number of mass points are not sufficient to fully describe the deformation of the beam between the mass points.

4.1.5 Timoshenko Beam

A bending beam that includes the effects of shear deformations and rotational inertia is commonly referred to as a *Timoshenko beam*. Let

b_y = distributed lateral (transverse) load
b_θ = distributed rotational load (positive counterclockwise)
M = bending moment (positive when compressing the top fiber)
S = shear (positive when acting upward from right, downward from left)
θ = rotation of cross section due to bending (positive counterclockwise)
γ = shear distortion or strain (positive counterclockwise)
$u' = \theta + \gamma$ = slope of neutral axis (slope of beam)

Because of the shear distortion, the neutral axis in a Timoshenko beam does *not* remain perpendicular to the cross section. With reference to Figure 4.3, the governing equations are as follows:

Equilibrium Equations

$$\frac{\partial M}{\partial x} + S + b_\theta - \rho I \ddot{\theta} = 0 \tag{4.12}$$

$$\frac{\partial S}{\partial x} + b_y - \rho A \ddot{u} = 0 \tag{4.13}$$

Force-Deformation Equations

$$\frac{\partial \theta}{\partial x} = \frac{M}{EI} \tag{4.14}$$

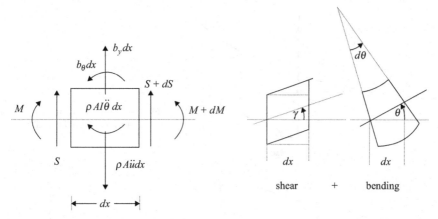

Figure 4.3. Equilibrium and deformation of Timoshenko beam.

$$\gamma = \frac{S}{GA_s} \tag{4.15}$$

Strain-Displacement Equation

$$\frac{\partial u}{\partial x} = \theta + \gamma \tag{4.16}$$

Combining the previous equations, we obtain the *system* of two *second*-order differential equations

$$\left\{ \begin{matrix} \rho I & \\ & \rho A \end{matrix} \right\} \left\{ \begin{matrix} \ddot{\theta} \\ \ddot{u} \end{matrix} \right\} - \left\{ \begin{matrix} \dfrac{\partial}{\partial x}\left(EI \dfrac{\partial}{\partial x} \right) - GA_s & GA_s \dfrac{\partial}{\partial x} \\[2mm] -\dfrac{\partial}{\partial x}(GA_s) & \dfrac{\partial}{\partial x}\left(GA_s \dfrac{\partial}{\partial x} \right) \end{matrix} \right\} \left\{ \begin{matrix} \theta \\ u \end{matrix} \right\} = \left\{ \begin{matrix} b_\theta \\ b_y \end{matrix} \right\} \tag{4.17}$$

which is a second-order differential equation ($2\kappa = 2$) in $n = 2$ field variables. Thus there are $\kappa = 1$ boundary conditions at each boundary point, as described next.

Boundary Conditions
Fixed end: $u = 0$, $\theta = 0$ (and not $u' = 0$!!!). In matrix form

$$\left\{ \begin{matrix} 1 & \\ & 1 \end{matrix} \right\} \left\{ \begin{matrix} \theta \\ u \end{matrix} \right\} = \left\{ \begin{matrix} 0 \\ 0 \end{matrix} \right\} \tag{4.18}$$

Free end: $M = 0$, $S = 0$, that is, $\theta' = 0$, $\gamma = 0$. In matrix form:

$$\left\{ \begin{matrix} \dfrac{\partial}{\partial x} & \\ -1 & \dfrac{\partial}{\partial x} \end{matrix} \right\} \left\{ \begin{matrix} \theta \\ u \end{matrix} \right\} = \left\{ \begin{matrix} 0 \\ 0 \end{matrix} \right\} \tag{4.19}$$

Roller end: $u = 0, M = 0$ ($\theta' = 0$). In matrix form:

$$\left\{ \begin{matrix} \dfrac{\partial}{\partial x} & \\ & 1 \end{matrix} \right\} \left\{ \begin{matrix} \theta \\ u \end{matrix} \right\} = \left\{ \begin{matrix} 0 \\ 0 \end{matrix} \right\} \tag{4.20}$$

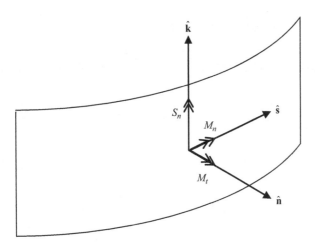

Figure 4.4. Forces acting at an edge of plate along a tangential coordinate s.

4.1.6 Plate Bending

For simplicity, we consider herein only a plate of constant thickness and material properties. Let

$D = \frac{1}{12(1-v^2)} Eh^3$ = plate rigidity parameter (equivalent to the EI in beams)

x, y = coordinates in the plane of the plate

n, s = normal and tangential directions at boundary point ($\perp$ and $\parallel$ to edge)

M_n = bending moment at edge, in normal direction (i.e., *about* an axis along s)

M_{ns} = twisting moment at edge (i.e., *about* a local axis along n)

S_n = shearing force/length at edge (parallel to the z-direction)

u = transverse displacement (positive upward)

b = transverse loading (positive upward)

The equation of motion is then[1]

$$\rho h \ddot{u} + D \nabla^4 u = b(x, y, t), \qquad \nabla^4 = \frac{\partial^4}{\partial x^4} + \frac{2\partial^4}{\partial x^2 \partial y^2} + \frac{\partial^4}{\partial y^4} \tag{4.21}$$

This is a *fourth*-order differential equation that has *two* boundary conditions at *each* boundary point. The natural (force boundary) conditions are controlled by the differential expressions for the edge shear S_n, flexural moment M_n, and torsional moment M_t acting along a free edge with unit normal edge vector $\hat{\mathbf{n}}$, as shown in Figure 4.4. These are

$$S_n = -D \frac{\partial}{\partial n} \nabla^2 u = -D \left(\frac{\partial^3 u}{\partial n^3} + \frac{\partial^3 u}{\partial n \partial s^2} \right) \tag{4.22}$$

$$M_n = -D \left(\frac{\partial^2 u}{\partial n^2} + v \frac{\partial^2 u}{\partial s^2} \right) \tag{4.23}$$

$$M_t = D(1-v) \frac{\partial^2 u}{\partial n \partial s} \tag{4.24}$$

---

[1] See S. Timoshenko and S. Woinowsky-Krieger, *Theory of Plates and Shells* (New York: McGraw-Hill, 1959).

Observe that if $u = 0$ along an edge for any s, then automatically $\frac{\partial}{\partial s} u = 0$.

The various possible boundary conditions are then as follows.

Fixed Edge

$$u = 0 \quad \text{and} \quad \frac{\partial u}{\partial n} = 0 \tag{4.25}$$

Simply Supported Edge

$$u = 0 \quad \text{and} \quad \frac{\partial^2 u}{\partial n^2} = 0 \tag{4.26}$$

Free Edge

There is a long an interesting story on the historical difficulties that existed with the formulation of this boundary condition. A detailed account is given in Timoshenko's *Theory of Plates and Shells*. The final expressions are

$$S_n - \frac{\partial M_{ns}}{\partial s} = 0 \quad \Rightarrow \quad \frac{\partial^3 u}{\partial n^3} + (2 - v)\frac{\partial^3 u}{\partial n\, \partial s^2} = 0 \tag{4.27}$$

$$M_n = 0 \quad \Rightarrow \quad \frac{\partial^2 u}{\partial n^2} + v\frac{\partial^2 u}{\partial s^2} = 0 \tag{4.28}$$

Strain Energy

It can be shown that the strain energy in a plate is

$$U = \tfrac{1}{2} \iint D\left[w_{xx}^2 + 2v\, w_{xx} w_{yy} + w_{yy}^2 + 2(1 - v)w_{xy}^2 \right] dA \tag{4.29}$$

with the subindices indicating partial derivatives.

4.1.7 Vibrations in Solids

We can readily write down the equations of motion of solid continua by merely adding the inertia (or D'Alembert) forces $-\rho\ddot{u}$ to the equilibrium equations shown in standard books on theory of elasticity. The final (formal) result is the *vector wave equation*

$$\rho\ddot{\mathbf{u}} + G\,\nabla \times \nabla \times \mathbf{u} - (\lambda + 2G)\nabla\nabla \cdot \mathbf{u} = \mathbf{b}(\mathbf{x}, t) \tag{4.30}$$

in which

$$\mathbf{u} = \begin{Bmatrix} u_x \\ u_y \\ u_z \end{Bmatrix} \text{ is the displacement vector,} \quad \text{and} \quad \mathbf{b} = \begin{Bmatrix} b_x \\ b_y \\ b_z \end{Bmatrix} \text{ is the load vector} \tag{4.31}$$

(For detailed expressions, see textbooks on the theory of elasticity and wave propagation.) There are now $n = 3$ DOF at each of the infinitely many points within the body, which means that we must deal with a system of three partial differential equations of order $2\kappa = 2$ involving $s = 3$ spatial coordinates. There are also $\kappa = 1$ boundary conditions involving $n = 3$ DOF at each boundary point. This is equivalent to *one* 3×1 matrix boundary condition at *each* boundary point. For example, in the case of a boundary that is parallel to the $y-z$ plane, and for which the normal displacement and the tangential shearing stresses are zero, the boundary condition would be of the form

$$
\begin{Bmatrix}
1 & & \\
\dfrac{\partial}{\partial y} & \dfrac{\partial}{\partial x} \\
\dfrac{\partial}{\partial z} & & \dfrac{\partial}{\partial x}
\end{Bmatrix}
\begin{Bmatrix} u_x \\ u_y \\ u_z \end{Bmatrix}
=
\begin{Bmatrix} 0 \\ 0 \\ 0 \end{Bmatrix}
\tag{4.32}
$$

at each point on that boundary.

4.1.8 General Mathematical Form of Continuous Systems

As can be discerned from the previous sections, continuous systems are characterized by a *body* throughout which the elastic and inertia properties as well as the *body forces* are continuously distributed – except, perhaps, at some discrete locations exhibiting material discontinuities – and a *boundary* at which either displacements, stresses, or are a combination of both are prescribed. The dynamic loads acting at any point in the system are referred to as the *body loads*. The material and load parameters are all functions of position. All continuous systems have infinitely many DOF as a whole, but only a finite number of DOF at each *point* – of which there are, of course, infinitely many. These systems are characterized mathematically by *systems of partial differential equations* in the spatial and temporal variables, and have the following defining qualities:

- Number s of spatial coordinates.
- Number n of DOF at *each point in the body*. This equals the number of partial differential equations.
- Order of the partial differential equations. This is normally an even number, 2κ.
- Number of boundary conditions *at each boundary point*. There are κ such conditions (i.e., half the order of the partial differential equations).
- Highest order of the differential operators in the boundary conditions. It is usually less than the order of the partial differential equations.

For example, both a simple bending beam and a thin plate have only 1 DOF at each body point, namely the transverse displacement, and both involve fourth-order differential equations (i.e., $2\kappa = 4$). However, the beam has only one spatial coordinate (x), while the plate has two (x, y). Each of these has $\kappa = 2$ boundary conditions at each boundary point. Yet the beam has only two boundary points (i.e., the two supports), while the plate has infinitely many such points (i.e., the edges). Thus, the two boundary conditions for a plate are defined in terms of two differential equations that are functions of the boundary coordinates, as we shall see later.

Comparing the equations for a string, rod, beam, plate, or solid, it can be seen that the governing linear differential equations and boundary conditions for a continuous system can be written in the general symbolic form

$$
\mathfrak{M}_{2\mu}\ddot{\mathbf{u}} + \mathfrak{K}_{2\kappa}\mathbf{u} = \mathbf{b}(\mathbf{x}, t) \qquad \text{in } V \text{ (in the body)} \tag{4.33}
$$

$$
\mathfrak{B}_{\ell}\mathbf{u} = \mathbf{0}, \qquad \ell = 1, 2 \ldots \kappa \qquad \text{on } S \text{ (on the boundaries)} \tag{4.34}
$$

in which $\mathfrak{M}, \mathfrak{K}$ are mass and stiffness *differential operator matrices* of orders $2\mu, 2\kappa$ respectively (with $\mu < \kappa$), and $\mathfrak{B}_i$ is a matrix differential operator of order smaller than 2κ. Furthermore, there are κ such boundary conditions at *each boundary point*, although they need not be the same at each such boundary point (i.e., $\mathfrak{B}_\ell(\mathbf{x}_b)$ may be a function of $\mathbf{x}_b$, namely the coordinates of the boundary points).

The boundary conditions can be classified into the following two fundamental groups:

- *Essential*, geometric or Dirichlet boundaries: These are boundaries at which geometric conditions, such as displacements or rotations, are prescribed. They involve differential operators whose orders are in the range $[0, \kappa - 1]$.
- *Natural*, additional or Neumann boundaries: These are boundaries at which stresses, forces, or moments are prescribed. They involve differential operators whose orders are in the range $[\kappa, 2\kappa - 1]$.

In conjunction with the boundary conditions, the differential operators $\mathfrak{M}, \mathfrak{K}$ satisfy two important mathematical relations, namely they are *self-adjoint* and are also either *positive semidefinite* or *positive definite*. In a nutshell, these properties are defined as follows. Let $\mathbf{v}(\mathbf{x}), \mathbf{w}(\mathbf{x})$ be two distinct, *arbitrary* test functions satisfying the boundary conditions, but *not* necessarily the differential equation. The operators satisfy then the following conditions (their proof can be obtained via integration by parts):

Self-adjoint property:

$$\int_V \mathbf{v}^T \mathfrak{M} \mathbf{w} \, dV = \int_V \mathbf{w}^T \mathfrak{M} \mathbf{v} \, dV \qquad \text{and} \qquad \int_V \mathbf{v}^T \mathfrak{K} \mathbf{w} \, dV = \int_V \mathbf{w}^T \mathfrak{K} \mathbf{v} \, dV \qquad (4.35)$$

which is analogous to the symmetry property of matrices: a matrix $\mathbf{A}$ is symmetric if and only if the equality $\mathbf{x}^T \mathbf{A} \mathbf{y} = \mathbf{y}^T \mathbf{A} \mathbf{x}$ holds for arbitrary nonzero vectors $\mathbf{x}, \mathbf{y}$.

Positive semidefinite property

$$\int_V \mathbf{v}^T \mathfrak{M} \mathbf{v} \, dV \geq 0 \qquad \text{and} \qquad \int_V \mathbf{v}^T \mathfrak{K} \mathbf{v} \, dV \geq 0 \qquad (4.36)$$

If the equal sign is satisfied only when $\mathbf{v} = \mathbf{0}$, then the operator is said to be *positive definite* (instead of semidefinite). This property relates to the nonnegativity of the kinetic and strain energies.

Important: A differential operator such as $\mathfrak{K}$ is not *eo ipso* self-adjoint and positive definite, that is, not just by itself. It achieves these properties only in combination with appropriate boundary conditions. Indeed, unphysical boundary conditions could potentially lead to a failure of self-adjointness or positive definiteness.

4.1.9 Orthogonality of Modes in Continuous Systems

Let $\phi_i = \phi_i(\mathbf{x}), \phi_j = \phi_j(\mathbf{x})$ constitute two distinct modes. By definition, they satisfy automatically the differential equation of free vibration together with its boundary conditions, namely

$$\mathfrak{K}_{2\kappa} \phi_i = \omega_i^2 \, \mathfrak{M}_{2\mu} \phi_j, \qquad \mathfrak{B}_\ell \, \phi_i = \mathbf{0}, \qquad \ell = 1, \cdots \kappa \qquad (4.37)$$

$$\mathfrak{K}_{2\kappa} \phi_j = \omega_j^2 \, \mathfrak{M}_{2\mu} \phi_j, \qquad \mathfrak{B}_\ell \, \phi_j = \mathbf{0}, \qquad \ell = 1, \cdots \kappa \qquad (4.38)$$

Multiplying each equation by the mode of the other, integrating over the volume of the body, we obtain

$$\int_V \phi_j^T \, \mathfrak{K} \, \phi_i^T \, dV = \omega_i^2 \int_V \phi_j^T \, \mathfrak{M} \, \phi_i^T \, dV, \qquad \int_V \phi_i^T \, \mathfrak{K} \, \phi_j^T \, dV = \omega_j^2 \int_V \phi_i^T \, \mathfrak{M} \, \phi_j^T \, dV \tag{4.39}$$

Also, from the self-adjoint property of the two operators $\mathfrak{K}, \mathfrak{M}$ together with their boundary conditions, it follows that the second of these two integrals could just as well be written as

$$\int_V \phi_j^T \, \mathfrak{K} \, \phi_i^T \, dV = \omega_j^2 \int_V \phi_j^T \, \mathfrak{M} \, \phi_i^T \, dV \tag{4.40}$$

Subtracting this expression from the integral in ω_i^2 and collecting terms, we obtain

$$0 = \left(\omega_i^2 - \omega_j^2 \right) \int_V \phi_j^T \, \mathfrak{M} \, \phi_i^T \, dV \tag{4.41}$$

which for $i \neq j$ (or more precisely, $\omega_i^2 \neq \omega_j^2$) requires the integral to be zero. That in turn also requires the integral in the stiffness operator to be zero. Finally, from the positive definiteness we also know that when $i = j$, that is, $\omega_i^2 = \omega_j^2$, the integrals are nonnegative numbers, namely the modal stiffness κ_i and modal mass μ_i. We conclude finally that

$$\int_V \phi_j^T \, \mathfrak{K} \, \phi_i^T \, dV = \begin{cases} \kappa_i & i = j \\ 0 & i \neq j \end{cases}, \qquad \int_V \phi_j^T \, \mathfrak{M} \, \phi_i^T \, dV = \begin{cases} \mu_i & i = j \\ 0 & i \neq j \end{cases} \tag{4.42}$$

We shall return to these properties later on in Chapter 6, Sections 6.2–6.4 when we discuss approximate solution methods, and where these properties play an essential role.

4.2 Exact Solutions for Simple Continuous Systems

We consider in the ensuing a number of simple mechanical system whose equations can be solved either in a closed form, or nearly so, namely miscellaneous homogeneous rods, shear beams, and bending beams. We begin with the simplest of these, which is the rod with constant axial stiffness, and then move on to other, more complicated systems. We mention in passing that the solution for homogeneous shear beams with appropriate boundary conditions as well as cylindrical shafts subjected to torsional waves are identical to those of a rod subjected to axial motions, inasmuch as all of these are governed by the same classical 1-D wave equation.

4.2.1 Homogeneous Rod

We begin by deriving the governing equation for wave propagation in a homogeneous rod of infinite length subjected to longitudinal (or axial) wave motion, and thereafter consider the boundary conditions for a rod of finite length.

With reference to the differential rod element shown in Figure 4.5, and assuming small deformations and homogeneous properties, the equations of equilibrium, force deformation (stress–strain), and strain–displacement are, respectively

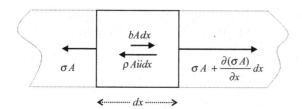

Figure 4.5. Equilibrium of a differential rod element.

$$\frac{\partial N}{\partial x} + bA - \rho A \ddot{u} = 0, \qquad \varepsilon = \frac{\sigma}{E} = \frac{N}{EA}, \qquad \varepsilon = \frac{\partial u}{\partial x} \tag{4.43}$$

where $N = \sigma A$ is the net axial force, σ is the axial stress, ε is the axial strain, ρ is the mass density, E is Young's modulus, and A is the cross section. Also, $u = u(x,t)$ is the axial displacement (assumed uniform across the section), and $b(x,t)$ is the body load per unit volume.

Substituting the last two equations into the first and dividing by ρA, we obtain

$$\frac{1}{C_r^2} \frac{\partial^2 u}{\partial t^2} - \frac{\partial^2 u}{\partial x^2} = \frac{b}{E} \tag{4.44}$$

in which

$$\boxed{C_r = \sqrt{\frac{E}{\rho}}} = \text{rod wave velocity } (\text{or } celerity) \tag{4.45}$$

In particular, if there are no external (body) forces acting on the system (i.e., free vibrations), then

$$\frac{\partial^2 u}{\partial x^2} = \frac{1}{C_r^2} \frac{\partial^2 u}{\partial t^2} \tag{4.46}$$

which is the classical *wave equation*. Waves in the rod then consist of compressional waves that propagate with speed C_r, which are of the form

$$u = F(t - x/C_r) + G(t + x/C_r) \tag{4.47}$$

where F,G are any arbitrary functions (or "pulses"). The first propagates from left to right, and the second from right to left, as will be seen later on in Chapter 5, Section 5.1 dealing with wave propagation. In particular, harmonic waves can be found in the form

$$\begin{aligned} u &= C_1 \sin(\omega t - kx) + C_2 \sin(\omega t + kx) \\ &= C_1 \sin[\omega(t - x/V_{ph})] + C_2 \sin[\omega(t + x/V_{ph})] \end{aligned} \tag{4.48}$$

where $\omega = 2\pi f$ is the frequency in rad/s, f is the frequency in Hz, $k = 2\pi / \lambda$ is the wavenumber, λ is the wavelength, and $V_{ph} = \omega / k = f\lambda$ is the phase velocity. Also, C_1, C_2 are arbitrary constants (wave amplitudes). The first term models once more waves propagating in the positive x- direction, while the second term represents a harmonic wave moving in the opposite direction. Clearly, to satisfy the wave equation, we must have $V_{ph} = C_r$, which means that whatever their frequency and wavelength, rod waves propagate with constant speed equal to the rod wave velocity. Thus, they are said to be nondispersive.

In particular, choosing $C_1 = -\frac{1}{2}, C_2 = \frac{1}{2}$, we obtain immediately

$$u = \sin kx \, \cos \omega t$$

which constitutes a *standing wave*, that is, one that oscillates harmonically but does not have any net propagation. This is because $\sin kx = 0$ when $kx = j\pi$, that is, when $x = j\frac{1}{2}\lambda$. These are fixed position, stationary nodes separated by a distance of half wavelength each. We shall use this property in the ensuing section for an alternative derivation of the modes and frequencies, and will return to the subject of rod waves and examine these in more depth in Chapter 5, Section 5.1.

Normal Modes of a Finite Rod

The formal method to find the normal modes of the rod is to assume a harmonic solution in time, that is, $u(x,t) = \phi(x) \sin \omega t$ and solve the resulting differential equation, that is, to change

$$\frac{1}{C_r^2} \frac{\partial^2 u}{\partial t^2} \rightarrow (-) \frac{\omega^2}{C_r^2} \phi, \quad \text{in which case} \tag{4.49}$$

$$\frac{\partial^2 \phi}{\partial x^2} + k^2 \phi = 0, \quad k = \frac{\omega}{C_r} \tag{4.50}$$

which is commonly referred to as the *Helmholtz equation*, subjected to appropriate boundary conditions. Its general solution for the displacement together with the axial stress at a point is

$$\phi = C_1 \cos kx + C_2 \sin kx, \quad \sigma = E\frac{\partial \phi}{\partial x} = Ek(-C_1 \sin kx + C_2 \cos kx)$$

where C_1, C_2 are constants of integration that are defined by the actual boundary conditions.

Note: In the ensuing, we shall visualize the modes by showing a deflected shape that is transverse to the rod, but in reality the motion is aligned with the rod, even if it cannot be drawn that way.

Fixed–Fixed Rod

For a doubly fixed rod (Figure 4.6a), the boundary conditions at the two supports are

$$\phi\big|_{x=0} = 0, \quad \phi\big|_{x=L} = 0 \tag{4.51}$$

which implies

$$C_1 \times 1 + C_2 \times 0 = 0, \quad \text{i.e.,} \quad C_1 = 0 \tag{4.52}$$

$$C_2 \sin kL = 0, \quad \text{i.e.,} \quad \sin kL = 0, \ kL = j\pi, \ j = 1,2,3\cdots \tag{4.53}$$

(a)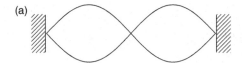

Figure 4.6a. Fixed–fixed rod.

(b)

Figure 4.6b. Free–free rod.

so from the definition of k and changing $\omega \to \omega_j$ we infer that

$$\omega_j = \frac{j\pi C_r}{L} [\text{rad}/\text{s}] \quad \text{or} \quad \boxed{f_j = \frac{jC_r}{2L}} [\text{Hz}] \tag{4.54}$$

This equation gives the natural frequencies for a homogeneous rod of finite length L that is fixed at the supports. Observe that the fundamental period $(j=1)$ is $T_1 = 2L/C_r$, which is the time it takes for a disturbance to travel back and forth between the two supports. The modal shape follows simply by taking $C_2 = 1$, in which case

$$\phi_j(x) = \sin kx = \sin \frac{j\pi x}{L} \quad (x = 0 \text{ at left end}) \tag{4.55}$$

Free–Free Rod

The boundary conditions for a free-free rod, which is shown in Figure 4.6b, are now given by the two stress-free conditions

$$\left.\frac{\partial \phi}{\partial x}\right|_{x=0} = 0, \quad \left.\frac{\partial u}{\partial x}\right|_{x=L} = 0 \tag{4.56}$$

which leads to

$$k(-C_1 \times 0 + C_2 \times 1) = 0 \quad \to \quad C_2 = 0 \tag{4.57}$$

$$-kC_1 \sin kL = 0, \quad\quad \to \quad \sin kL = 0 \tag{4.58}$$

which is the same condition as for the previous case. Hence, the frequencies are now

$$\boxed{f_j = \frac{jC_r}{2L}} [\text{Hz}] \quad j = 0,1,2,3, \dots \tag{4.59}$$

These are identical to those of the fixed-fixed case, except that the rod now accepts also a zero frequency $(j=0)$. This null frequency is associated with the *rigid body mode*. Because now it is the second integration constant that is zero, after setting $C_1 = 1$ the mode changes into (c)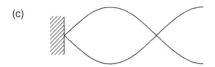

Figure 4.6c. Fixed–free rod.

$$\phi_j(x) = \cos\frac{j\pi x}{L} \tag{4.60}$$

The rigid body mode ($j = 0$) is then $\phi_0 = 1$.

Fixed–Free Rod

Finally, we consider a fixed condition at the left end, and a free condition at the right end, as shown in Figure 4.6c. The boundary conditions are now

$$u\big|_{x=0} = 0, \qquad \frac{\partial u}{\partial x}\bigg|_{x=L} = 0 \tag{4.61}$$

which translates into the conditions

$$C_1 \times 1 + C_2 \times 0 = 0 \qquad \text{i.e.,} \qquad C_1 = 0 \tag{4.62}$$

$$kC_2 \cos kL = 0, \qquad \text{i.e.,} \qquad \cos kL = 0 \tag{4.63}$$

The solution is now

$$kL = \left(j - \tfrac{1}{2}\right)\pi, \qquad j = 1, 2, \dots \tag{4.64}$$

$$f_j = \frac{(2j-1)C_r}{4L} \qquad j = 1, 2, 3, \dots \tag{4.65}$$

The fundamental period ($j = 1$) is now $T_1 = 4L/C_r$, which is the time it takes for a disturbance to travel *twice* back and forth between the two extremes. Considering that $k = k_j = (j - \tfrac{1}{2})\pi/L$, the modal shape is now

$$\phi_j(x) = \sin\frac{(2j-1)\pi x}{2L} \tag{4.66}$$

Normal Modes of a Rod without Solving a Differential Equation

As it turns out, one can also derive the normal modes of the homogeneous rod by considering a standing wave consisting of two identical compressional waves that travel in opposite directions.

As we have already seen, one special kind of standing rod wave of unit amplitude is given by the equation $u = \sin kx \cos \omega t$. Now, if in the as yet infinite rod we focus attention onto a finite segment of length $L = j\tfrac{1}{2}\lambda$ such that both $x = 0$ and $x = L$ coincide with two nodes (i.e., stationary points) in the standing wave, it is clear that we could cut out that segment and replace the nodes with simple supports. Thus, adding for clarity a subindex j to the frequency, wavenumber, and wavelength, we obtain

$$L = \tfrac{1}{2} j \lambda_j = \tfrac{1}{2} j C_r / f_j \quad \text{so} \quad f_j = j \frac{C_r}{2L}, \qquad j = 1, 2, \cdots \tag{4.67}$$

$$k_j = 2\pi / \lambda_j = j \tfrac{\pi}{L} \tag{4.68}$$

Furthermore, the deflected configuration between those nodes is simply $\sin kx = \sin j\pi \tfrac{x}{L}$, which agrees with what we had before with the analytical solution for a fixed–fixed rod.

The free–free rod is obtained similarly, except that instead of considering the distance L between two nodes, the segment of length L is made to coincide with the crests and troughs of the standing wave, which is where the slope is always zero, so the stress is zero there at all times. If so, then we can cut the rod segment at those two locations without altering equilibrium and thus it becomes a free–free rod.

Finally, the solution for the fixed–free rod is obtained by considering the distance L between a node and a crest, that is, an odd multiple of a quarter wavelength,

$$L = (2j - 1)\tfrac{1}{2}\lambda \equiv \left(j - \tfrac{1}{2}\right)\lambda, \qquad j = 1, 2, \ldots \tag{4.69}$$

The results for the frequency and mode follow then immediately and agree with the analytical solution.

Orthogonality of Rod Modes

Although we have already formally proved the orthogonality of the modes for any mechanical system in Section 4.1.9, for purposes of illustration we repeat the proof for the special case of the rod. As we have seen, the free vibration problem is characterized by the partial differential equation

$$\rho A \frac{\partial^2 u}{\partial t^2} - \frac{\partial}{\partial x}\left(EA \frac{\partial u}{\partial x} \right) = 0 \tag{4.70}$$

together with appropriate boundary conditions. If the system vibrates in one of its modes, the motion and acceleration attain the form

$$u(x,t) = \phi_j(x) \sin \omega_j t \tag{4.71}$$

$$\ddot{u}(x,t) = -\omega_j^2 \phi_j(x) \sin \omega_j t \tag{4.72}$$

Hence

$$-\left[\omega_j^2 \rho A \phi_j + \frac{\partial}{\partial x}\left(EA \frac{\partial \phi_j}{\partial x} \right) \right] \sin \omega_j t = 0 \tag{4.73}$$

and since this must be valid at all times, then

$$\omega_j^2 \rho A \phi_j + \frac{\partial}{\partial x}\left(EA \frac{\partial \phi_j}{\partial x} \right) = 0 \tag{4.74}$$

Next, we premultiply this expression by some other distinct mode $\phi_i(x)$, and integrate over the length of the rod. The result is

$$\omega_j^2 \int_0^L \rho A\, \phi_i \phi_j\, dx + \int_0^L \phi_i \frac{\partial^2}{\partial x^2}\left(EA\frac{\partial^2 \phi_j}{\partial x^2}\right) dx = 0 \tag{4.75}$$

Integrating the second term by parts, we obtain

$$\omega_j^2 \int_0^L \rho A\, \phi_i\, \phi_j\, dx + \phi_i\, EA\frac{\partial \phi_j}{\partial x}\bigg|_0^L - \int_0^L \frac{\partial \phi_i}{\partial x} EA \frac{\partial \phi_j}{\partial x}\, dx = 0 \tag{4.76}$$

Now, the term in the middle is zero because of boundary conditions. Indeed, at a fixed end, $\phi_i = 0$, while at a free end, $\partial \phi_j / \partial x = 0$. Hence

$$\omega_j^2 \int_0^L \phi_i\, \phi_j\, \rho A\, dx = \int_0^L \frac{\partial \phi_i}{\partial x}\frac{\partial \phi_j}{\partial x} EA\, dx \tag{4.77}$$

If we next exchange the roles of modes ϕ_i and ϕ_j, we can also write

$$\omega_i^2 \int_0^L \phi_j\, \phi_i\, \rho A\, dx = \int_0^L \frac{\partial \phi_j}{\partial x}\frac{\partial \phi_i}{\partial x} EA\, dx \tag{4.78}$$

Subtracting these two equations while observing that the right-hand sides are identical, we obtain

$$\left(\omega_i^2 - \omega_j^2\right)\int_0^L \phi_i\, \phi_j\, \rho A\, dx = 0 \tag{4.79}$$

Finally, since the modes are distinct, that is, $\omega_i \neq \omega_j$, we conclude that

$$\int_0^L \phi_i\, \phi_j\, \rho A\, dx = 0 \qquad i \neq j \tag{4.80}$$

which in turn implies

$$\int_0^L \frac{\partial \phi_i}{\partial x}\frac{\partial \phi_j}{\partial x} EA\, dx = 0 \qquad i \neq j \tag{4.81}$$

These are the two orthogonality conditions for the modes of the rod, whatever the boundary conditions. On the other hand, when $i = j$, then $\omega_j^2 - \omega_j^2 = 0$ so the product is zero and the integral above need not be zero. Hence, we define the modal mass and modal stiffness as

$$\boxed{\mu_j = \int_0^L \phi_j^2\, \rho A\, dx} > 0 \qquad \boxed{\kappa_j = \int_0^L \left(\frac{\partial \phi_j}{\partial x}\right)^2 EA\, dx} \geq 0 \tag{4.82}$$

Clearly, the modal mass must be a positive number, because all terms in the integral are positive. By contrast, the modal stiffness can be both a positive number or zero (i.e., a nonnegative number), inasmuch as for the rigid body mode of a free rod the derivative term is zero.

4.2.2 Euler–Bernoulli Beam (Bending Beam)

The simplest model for a bending beam is the so-called *Euler–Bernoulli beam*, which disregards both rotational inertia as well as shear deformations. Its characteristic equation is most easily obtained by adding the inertia (i.e., D'Alembert) forces as fictitious external loads to the classical, static equilibrium equation, namely

$$\frac{\partial^2}{\partial x^2}\left(EI(x)\frac{\partial^2 u}{\partial x^2}\right) = b(x,t) - \rho A \frac{\partial^2 u}{\partial t^2} \tag{4.83}$$

where EI is the flexural rigidity and ρA is the mass per unit length. This leads immediately to

$$\boxed{\rho A \frac{\partial^2 u}{\partial t^2} + \frac{\partial^2}{\partial x^2}\left(EI(x)\frac{\partial^2 u}{\partial x^2}\right) = b(x,t)} \tag{4.84}$$

subjected to appropriate boundary conditions. If the beam has constant cross section and no body forces are applied, this equation simplifies to

$$\rho A \ddot{u} + EI \frac{\partial^4 u}{\partial x^4} = 0 \tag{4.85}$$

We now try a harmonic solution for this equation in the form of a wave propagating from left to right, that is,

$$u = C \sin(\omega t - kx) = C \sin\left[\omega\left(t - x/V_{ph}\right)\right] \tag{4.86}$$

in which C is a constant, ω is the frequency, k is the wavenumber, and $V_{ph} = \omega/k$ is the *phase velocity*. Substituting this trial solution into the differential equation, we obtain

$$\left(-\rho A \omega^2 + EI\, k^4\right) C \sin(\omega t - kx) = 0 \tag{4.87}$$

which implies

$$k = \sqrt[4]{\frac{\rho A}{EI}\,\omega^2} = \sqrt{\frac{\omega}{C_r R}}\sqrt[4]{1} = \begin{cases} \pm\sqrt{\dfrac{\omega}{C_r R}} \\[2ex] \pm i\sqrt{\dfrac{\omega}{C_r R}} \end{cases} \tag{4.88}$$

in which $R = \sqrt{I/A}$ is the radius of gyration, and $C_r = \sqrt{E/\rho}$ is the rod-wave velocity. For the real, positive root of k, the phase velocity is

$$V_{ph} = \frac{\omega}{k} = \frac{\omega}{\sqrt{\dfrac{\omega}{C_r R}}} = \sqrt{R C_r\, \omega} \tag{4.89}$$

which gives the phase velocity (i.e., propagation velocity) for *flexural waves*. As can be seen, this velocity increases with the square root of the frequency, which means that the flexural waves of different frequencies travel at different speeds. Hence, flexural waves are said to be *dispersive*. The implication is that when pulses of arbitrary shape propagate in a bending beam, these pulses change shape and stretch out in space as they propagate, because the higher frequency components of the pulse outrun the lower frequency components.

You should already notice that the previous expression for the speed of flexural waves cannot remain valid for arbitrarily large frequencies. This is because V_{ph} can never exceed the rod wave velocity C_r, inasmuch as that happens to be the largest speed with which *physical* waves can propagate under plane stress conditions. Hence, this expression begins to fail as we approach the limit $V_{ph} \sim C_r$, that is, as the frequency approaches $\omega = C_r/R$. It follows that the wavelength must obey the restriction

$$\frac{R}{\lambda} = \frac{kR}{2\pi} \quad \frac{\omega R}{C_r} = 1 \tag{4.90}$$

That is, the wavelengths must be much longer than the radius of gyration, which is on the order of the depth of the beam. The reason why the phase velocity for flexural waves in an Euler beam fails to remain bounded is that this beam model neglects rotational inertia, shear deformations, and "skin" effects related to the plane stress assumption (i.e., neglecting stresses in the transverse direction, and assuming plane sections to remain plane).

Normal Modes of a Finite-Length Euler–Bernoulli Beam

The superposition of two flexural waves of equal frequency, equal amplitude, and opposite direction of travel does produce, as expected, a standing wave with stationary nodes and crests, as was the case for the rod. However, we shall not be able to use this artifact to derive the modal solution for a beam as it could be done for a rod, and this is because a beam must satisfy *two* boundary conditions at each boundary point, while the standing wave allows choosing only *one* such condition. Hence, we must resort to the direct, conventional method of solution to find the normal modes and frequencies.

Consider again the free vibration equation for a beam with constant properties, namely

$$\rho A \ddot{u} + EI \frac{\partial^4 u}{\partial x^4} = 0 \tag{4.91}$$

Assuming a modal solution of the form $u(x,t) = \phi(x) \sin \omega t$, we obtain

$$EI \frac{\partial^4 \phi}{\partial x^4} - \rho A \omega^2 \phi = 0 \tag{4.92}$$

Defining

$$k^4 = \frac{\rho A}{EI} \omega^2 = \left(\frac{\omega}{RC_r} \right)^2 \tag{4.93}$$

the general solution is then

$$\phi(x) = C_1 \cos kx + C_2 \sin kx + C_3 \cosh kx + C_4 \sinh kx \tag{4.94}$$

in which the C_i are constants of integration. They are obtained by imposing an appropriate set of boundary conditions, namely two conditions at each boundary point. There are four types of such conditions available, namely prescribed displacement ϕ, rotation θ, moment M, and shear S. Omitting the harmonic factor $\sin \omega t$, the last three involve the expressions

$$\theta = \phi' = k\left(-C_1 \sin kx + C_2 \cos kx + C_3 \sinh kx + C_4 \cosh kx\right) \tag{4.95}$$

$$M = EI\,\phi'' = EI\,k^2\left(-C_1 \cos kx - C_2 \sin kx + C_3 \cosh kx + C_4 \sinh kx\right) \tag{4.96}$$

$$S = EI\,\phi''' = EI\,k^3\left(C_1 \sin kx - C_2 \cos kx + C_3 \sinh kx + C_4 \cosh kx\right) \tag{4.97}$$

Simply Supported Beam

For example, a simply supported beam must satisfy boundary conditions of zero displacement and zero bending moment at both ends, that is, $\phi(0) = 0$, $\phi''(0) = 0$, $\phi(L) = 0$, $\phi''(L) = 0$. Hence, we are led to the system

$$\begin{Bmatrix} \begin{bmatrix} 1 & 0 & 1 & 0 \\ -1 & 0 & 1 & 0 \\ \cos kL & \sin kL & \cosh kL & \sinh kL \\ -\cos kL & -\sin kL & \cosh kL & \sinh kL \end{bmatrix} \begin{bmatrix} C_1 \\ C_2 \\ C_3 \\ C_4 \end{bmatrix} = \begin{bmatrix} 0 \\ 0 \\ 0 \\ 0 \end{bmatrix} \end{Bmatrix} \tag{4.98}$$

which constitutes a transcendental eigenvalue problem in the wavenumber k, which generally remains "hidden" as an argument of trigonometric functions. A nontrivial solution requires the vanishing of the determinant, which after brief algebra leads to the conditions $C_1 = C_3 = C_4 = 0$ and $C_2 \sin kL = 0$. The latter condition is satisfied if $kL \equiv k_j L = j\pi$, that is, if

$$k_j^2 = \left(\frac{j\pi}{L}\right)^2 = \frac{\omega_j}{RC_r} \tag{4.99}$$

It follows that the natural frequencies of vibration for a simply supported beam are

$$\boxed{\omega_j = \left(j\pi\right)^2 \frac{RC_r}{L^2}} \qquad \text{Frequencies of simply supported beam} \tag{4.100}$$

Choosing $C_2 = 1$, the modal shape is found to be $\phi(x) = \sin k_j x$, that is,

$$\boxed{\phi_j = \sin \frac{j\pi x}{L}} \tag{4.101}$$

Other Boundary Conditions

Finding the solution for other boundary conditions requires considerably more effort, because the determinant equations lead to transcendental equations with no exact solution. Table 4.1 summarizes close approximations (exact for the simply supported beam) to the natural frequencies and modal shapes of a Euler–Bernoulli beam presented by Dugundji.[2] For $j = 1$ the table gives modes with error smaller than 1%, except for the

---

[2] J. Dugundji, "Simple expressions for higher vibration modes of uniform Euler beams," *AIAA J.*, 26(8), 1988, 1013–1014.

Table 4.1

Boundary condition	β_j	λ_j	ϑ	A	B
SS-SS	$j\pi$	$2L/j$	0	0	0
CL-FR	$(j-\frac{1}{2})\pi^a$	$4L/(2j-1)$	$-\pi/4$	1	1
CL-CL	$(j+\frac{1}{2})\pi$	$4L/(2j+1)$	$-\pi/4$	1	1
FR-FR	$(j+\frac{1}{2})\pi$	$4L/(2j+1)$	$+3\pi/4$	1	1
SS-CL	$(j+\frac{1}{4})\pi$	$8L/(4j+1)$	0	0	1
SS-FR	$(j+\frac{1}{4})\pi$	$8L/(4j+1)$	0	0	-1

^{a} $\beta_1 = 0.597\pi$.

clamped-free case, whose exact first coefficient is $\beta_1 = 0.597\pi$ and the approximate expression for ϕ_1 fails. For $j > 2$, these expressions give frequencies to better than 0.01% accuracy. The various boundary conditions of interest are abbreviated as FR, SS, CL to denote a free end, a simply supported end, and a clamped end, respectively. Also,

$$\beta_j = k_j L, \qquad \xi = x/L, \qquad \lambda_j = \frac{2\pi L}{\beta_j} \qquad C_r = \sqrt{\frac{E}{\rho}} \qquad R = \sqrt{\frac{I}{A}} \tag{4.102}$$

$$\boxed{\omega_j = \frac{\beta_j^2}{L^2}\sqrt{\frac{EI}{\rho A}} = \frac{4\pi^2 C_r R}{\lambda_j^2}} \qquad \boxed{\phi_j(x) \approx \sqrt{2}\sin(\beta_j\xi + \vartheta) + A\,e^{-\beta_j\xi} - (-1)^j\,B\,e^{-\beta_j(1-\xi)}} \tag{4.103}$$

(for CL-FR, this fails for ϕ_1)

The modes in Table 4.1 are normalized so that $\int_0^L \phi_j^2(x)\,dx = 1$. Observe that the two exponential terms in the modal shape are important only in the immediate vicinity of the supports, while the sine term dominates in the interior and can be interpreted as a standing wave of wavelength λ_j and apparent propagation velocity $c_j = \beta_j\,C_r R/L$.

Note: These approximate modes do *not* satisfy the geometric boundary conditions exactly, so they cannot be used as trial functions in the context of a weighted residual formulation!

Normal Modes of a Free–Free Beam

The eigenvalue problem for a free bending beam leads to the characteristic equation

$$\begin{Bmatrix}\cosh kL - \cos kL & \sinh kL - \sin kL \\ \sinh kL + \sin kL & \cosh kL - \cos kL\end{Bmatrix}\begin{Bmatrix}C_1 \\ C_2\end{Bmatrix} = \begin{Bmatrix}0 \\ 0\end{Bmatrix} \tag{4.104}$$

which reduces to the transcendental equation

$$\cos kL = \frac{1}{\cosh kL} \tag{4.105}$$

If $kL \neq 0$, then we can choose $C_1 = 1$, and solve for C_2 from either equation in the characteristic system, say the first. This results in a modal shape of the form

$$\phi(x) = \cos kx + \cosh kx - \frac{\cosh kL - \cos kL}{\sinh kL - \sin kL}(\sin kx + \sinh kx) \tag{4.106}$$

However, Eq. 4.106 cannot be used in practice to compute the modes, because the hyperbolic terms cause severe cancellation errors. We could rewrite this expression using trigonometric identities and the definition of the hyperbolic functions, but we prefer an alternative solution in which we set the origin of coordinates at the *center* of the beam. The eigenvalue problem in this case is

$$\begin{Bmatrix}\begin{bmatrix} -\cos\theta & \sin\theta & \cosh\theta & -\sinh\theta \\ -\cos\theta & -\sin\theta & \cosh\theta & \sinh\theta \\ -\sin\theta & -\cos\theta & -\sinh\theta & \cosh\theta \\ \sin\theta & -\cos\theta & \sinh\theta & \cosh\theta \end{bmatrix}\begin{bmatrix} C_1 \\ C_2 \\ C_3 \\ C_4 \end{bmatrix}\end{Bmatrix} = \begin{Bmatrix} 0 \\ 0 \\ 0 \\ 0 \end{Bmatrix}, \qquad \theta = \tfrac{1}{2}kL \qquad (4.107)$$

Adding and subtracting the first row to the second and likewise for the third and fourth, exchanging appropriately rows and columns and finally dividing by 2, we eventually obtain

$$\begin{Bmatrix}\begin{bmatrix} \cos\theta & -\cosh\theta & 0 & 0 \\ \sin\theta & \sinh\theta & 0 & 0 \\ 0 & 0 & \sin\theta & -\sinh\theta \\ 0 & 0 & -\cos\theta & \cosh\theta \end{bmatrix}\begin{bmatrix} C_1 \\ C_3 \\ C_2 \\ C_4 \end{bmatrix}\end{Bmatrix} = \begin{Bmatrix} 0 \\ 0 \\ 0 \\ 0 \end{Bmatrix} \qquad (4.108)$$

This equation has two independent solutions, the symmetric and anti-symmetric modes. Setting the first minor determinant to zero, we obtain the transcendental eigenvalue problem for the symmetric modes:

$$\boxed{\tan\theta = -\tanh\theta}, \qquad C_1 = \frac{1}{\cos\theta}, \qquad C_3 = \frac{1}{\cosh\theta}, \qquad C_2 = C_4 = 0 \qquad (4.109)$$

$$\boxed{\phi(x) = \frac{\cos\theta\xi}{\cos\theta} + \frac{\cosh\theta\xi}{\cosh\theta}} \qquad \xi = 2x/L, \qquad -1 \le \xi \le 1 \qquad (4.110)$$

Next, we set to zero the second minor determinant and obtain the eigenvalue problem for the antisymmetric modes:

$$\boxed{\tan\theta = \tanh\theta}, \qquad C_2 = \frac{1}{\sin\theta}, \qquad C_4 = \frac{1}{\sinh\theta}, \qquad C_1 = C_3 = 0 \qquad (4.111)$$

$$\boxed{\phi(x) = \frac{\sin\theta\xi}{\sin\theta} + \frac{\sinh\theta\xi}{\sinh\theta}} \qquad \xi = 2x/L, \qquad -1 \le \xi \le 1 \qquad (4.112)$$

These revised expressions for the modes are well behaved numerically and can safely be used for direct computation. A graphic solution to the symmetric and antisymmetric eigenvalue problems is shown in Figure 4.7; the solutions are defined by the intersections of the curves for $\tan\theta$ and $\mp\tanh\theta$.

In addition to the symmetric and antisymmetric modes, a free beam admits also two rigid body modes. These follow simply from the previous expressions by setting $\theta = 0$, which gives the two rigid-body modes as $\phi_0' = 2$ and $\phi_0'' = 2\xi$. Alternatively, these could also have been obtained by setting $\omega = 0$ in the differential equation, and solving the resulting expression, namely $EI\,\phi^{IV} = 0$.

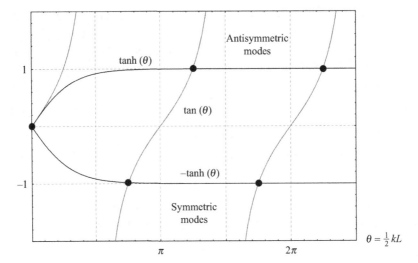

Figure 4.7. Graphic solution of transcendental eigenvalue problem for a free–free beam.

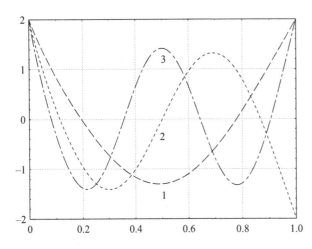

Figure 4.8. First three modal shapes of a free–free bending beam, plotted vs. $\xi = x/L$.

As we will be seen shortly, all modes must satisfy certain orthogonality conditions involving pairs of modes. In particular, all higher modes are orthogonal with respect to the two rigid body modes. Hence, these higher modes cannot contain any net translation or net rotation about of the center of mass. A consequence of that is that all higher modes satisfy simultaneously the principles of conservation of linear and angular momenta. Indeed, in the illustration of the first three modal shapes shown in Figure 4.8, none of the modes exhibits any net translation or rotation. Observe also that the antisymmetric mode has two large peaks at about $x/_L = \pm 0.21$, which compensate for the peaks in the opposite directions at the two ends. The modal shape is such that its first moment with respect to the center vanishes, that is, $\int_{-\frac{1}{2}L}^{-\frac{1}{2}L} x\,\phi_j\,dx = 0$. This is necessary to satisfy the principle of angular momentum. Another way of looking at this is to recognize that all nonzero

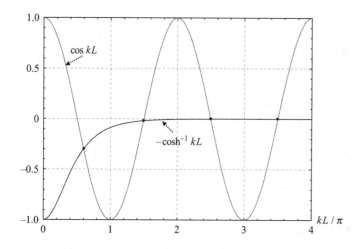

Figure 4.9. Graphic solution of transcendental eigenvalue problem for a cantilever beam.

frequency modes must remain orthogonal to the two rigid body modes of pure translation and pure rotation

The zero-crossing points $\xi_j = 2x/L$ (with $x = 0$ at the center) of the first three nonzero modes are

First mode: $\xi_j = -0.552, +0.552$

Second mode: $\xi_j = -0.736, \ 0.0, \ +0.736$

Third mode : $\xi_j = -0.811, \ -0.288, \ +0.288, +0.811$

Normal Modes of a Cantilever Beam

The eigenvalue problem for a cantilever bending beam leads to the characteristic equation

$$\begin{Bmatrix} \cosh kL + \cos kL & \sinh kL + \sin kL \\ \sinh kL - \sin kL & \cosh kL + \cos kL \end{Bmatrix} \begin{Bmatrix} C_1 \\ C_2 \end{Bmatrix} = \begin{Bmatrix} 0 \\ 0 \end{Bmatrix} \tag{4.113}$$

the determinant of which is

$$\left(\cosh kL + \cos kL\right)^2 - \sinh^2 kL + \sin^2 kL = 0 \tag{4.114}$$

After straightforward trigonometric transformations, this leads to the characteristic equation

$$\cos k_j L = -\frac{1}{\cosh k_j L}, \qquad j = 1, 2, 3, \ldots \tag{4.115}$$

the solution of which can be found graphically, as shown in Figure 4.9. Also, we define the ratio of modal constants as

$$B_j = -\frac{C_2}{C_1} = \frac{\sinh k_j L - \sin k_j L}{\cosh k_j L + \cos k_j L} \equiv \frac{\cosh k_j L + \cos k_j L}{\sinh k_j L + \sin k_j L} \tag{4.116}$$

yielding a modal shape

$$\begin{aligned}\phi_j &= \cos k_j x - \cosh k_j x + B_j \left(\sinh k_j x - \sin k_j x\right) \\ &= \cos k_j x - B_j \sin k_j x + \tfrac{1}{2}\left[\left(B_j - 1\right)e^{k_j L \xi} - \left(B_j + 1\right)e^{-k_j L \xi}\right]\end{aligned} \tag{4.117}$$

This shape is normalized in the sense that $\int_0^1 \phi_j^2\left(\xi\right)d\xi = 1$, with $\xi = x / L$ The modal amplitudes B_j and characteristic wavenumbers $\beta_j = k_j L$ are

$$B_1 = 0.7341, \qquad B_2 = 1.0185, \qquad B_3 = 0.9992, \qquad B_j = 1.000 \quad \text{for } j \geq 4$$

$$b_1 = 0.597\pi, \qquad \beta_j = \left(j - \tfrac{1}{2}\right)\pi \quad \text{for } j \geq 2$$

Orthogonality Conditions of a Bending Beam

By a development that entirely parallels that of the rod (which also requires integrating twice by parts the integral in EI and discarding the terms associated with boundary conditions), we are ultimately led to the orthogonality conditions

$$\int_0^L \phi_i(x)\phi_j(x)\rho A\, dx = \mu_j \delta_{ij}, \qquad \int_0^L \phi_i''(x)\, \phi_j''(x)EI\, dx = \kappa_j \delta_{ij} \tag{4.118}$$

in which μ_j, κ_j are the modal mass and modal stiffness for the jth mode, and δ_{ij} is the Kronecker delta.

Strain and Kinetic Energies of a Beam

The expression for the strain and kinetic energies of a beam involve integrals that are similar to those of the orthogonality conditions, that is,

$$V = \int_0^L u\,\frac{\partial}{\partial x^2}\left(EI\,\frac{\partial u}{\partial x^2}\right)dx = \int_0^L \left(\frac{\partial^2 u}{\partial x^2}\right)^2 EI\, dx, \qquad K = \int_0^L \left(\frac{\partial u}{\partial t}\right)^2 \rho A\, dx \tag{4.119}$$

(The equivalence of the two terms in V is obtained via integration by parts.) In particular, with the substitutions $u \leftrightarrow \phi_j \sin \omega_j t$, $\dot{u} \leftrightarrow \omega_j \phi_j \cos \omega_j t$, then $V = \kappa_j \sin^2\left(\omega_j t\right)$, $K = \omega_j^2 \mu_j \cos^2\left(\omega_j t\right)$. We see that when the beam oscillates in a given mode, the kinetic energy is maximum when the elastic (deformation) energy is zero, and vice versa. Hence, the Rayleigh quotient $\omega_j^2 = \kappa/\mu$ can be interpreted as a statement on the conservation of energy.

4.2.3 **Bending Beam Subjected to Moving Harmonic Load**

Consider an undamped bending beam of length L with appropriate boundary conditions that is being acted upon by a uniformly distributed and moving wavelike load that oscillates harmonically with frequency ω and travels with speed $V = \omega/k$ in the positive x direction, where k is the load's wavenumber, or alternatively, the load has wavelength $2\pi/k$. As parts of the wave exit the beam on the right, a steady wave source replenishes the load on the left, so there is always a full load acting on the beam. We are interested in finding the *steady-state solution*, that is, the long-term solution after any initial conditions have died out. (**Note:** although we assume no damping, a minute amount does actually exist so that initial conditions will in due time have been erased.) The relevant equation of motion can be written as

$$\rho A \frac{\partial^2 u}{\partial t^2} + EI \frac{\partial^4 u}{\partial x^4} = p_0 \, e^{i(\omega t - kx)} \tag{4.120}$$

subjected to some boundary conditions (free, fixed, simply supported, etc.). The complete solution is obtained by superposition of the homogeneous solution and the particular solution, that is,

$$u(x,t) = u_h + u_p. \tag{4.121}$$

Homogeneous Solution

$$\rho A \frac{\partial^2 u_h}{\partial t^2} + EI \frac{\partial^4 u_h}{\partial x^4} = 0 \qquad \text{or} \qquad \left(\frac{\rho A}{EI}\right) \frac{\partial^2 u_h}{\partial t^2} + \frac{\partial^4 u_h}{\partial x^4} = 0 \tag{4.122}$$

the solution to which is

$$u_h = e^{i\omega t} \left(C_1 \, e^{-i \, k_0 x} + C_2 e^{+i \, k_0 x} + C_3 e^{-k_0 x} + C_4 e^{+k_0 x} \right) \tag{4.123}$$

where the C_j are constants of integration and the wavenumber k_0 satisfies the equation

$$-\omega^2 \frac{\rho A}{EI} + k_0^4 = 0, \qquad \text{so} \qquad k_0 = \sqrt[4]{\frac{\omega^2 \rho A}{EI}} = \sqrt[4]{\frac{\omega^2}{C_r^2 R^2}} = \sqrt{\frac{\omega}{C_r R}} \tag{4.124}$$

Observe that $\sqrt[4]{1} = \pm 1, \pm i$, which is the reason for the four solutions above. Also,

$$C_r = \sqrt{\frac{E}{\rho}} \qquad \text{rod wave velocity}, \qquad R = \sqrt{\frac{I}{A}} \qquad \text{radius of gyration} \tag{4.125}$$

Particular Solution

By inspection, a particular solution is

$$u_p = D \, e^{i(\omega t - kx)}, \qquad \left(-\rho A \, \omega^2 + EI \, k^4\right) D = p_0 \tag{4.126}$$

So

$$D = \frac{p_0}{EI\left(k^4 - k_0^4\right)} = \frac{p_0}{\omega^2 \rho A \left(\kappa^4 - 1\right)} \tag{4.127}$$

where $\kappa = k / k_0$. Hence, the complete complex solution is

$$u = e^{i\omega t} \left(C_1 e^{-i \, k_0 x} + C_2 e^{+i \, k_0 x} + C_3 e^{-k_0 x} + C_4 e^{+k_0 x} + \frac{p_0}{\omega^2 \rho A \left(\kappa^4 - 1\right)} e^{-ikx} \right) \tag{4.128}$$

This solution is subjected to four boundary conditions. For example, for a *simply supported beam*

$$u\big|_{x=0} = 0, \qquad u''\big|_{x=0} = 0 \tag{4.129}$$

$$u\big|_{x=L} = 0, \qquad u''\big|_{x=L} = 0 \tag{4.130}$$

that is,

$$C_1 + C_2 + C_3 + C_4 + \frac{p_0}{EI\left(k^4 - k_0^4\right)} = 0 \tag{4.131}$$

$$k_0^2\left(-C_1 - C_2 + C_3 + C_4\right) - k^2 \frac{p_0}{EI\left(k^4 - k_0^4\right)} = 0 \tag{4.132}$$

$$C_1\,\mathrm{e}^{-\mathrm{i}\,k_0 L} + C_2 \mathrm{e}^{\mathrm{i}\,k_0 L} + C_3 \mathrm{e}^{-k_0 L} + C_4 \mathrm{e}^{k_0 L} + \frac{p_0}{EI\left(k^4 - k_0^4\right)}\mathrm{e}^{-\mathrm{i}kL} = 0 \tag{4.133}$$

$$k_0^2\left(-C_1\,\mathrm{e}^{-\mathrm{i}\,k_0 L} - C_2 \mathrm{e}^{\mathrm{i}\,k_0 L} + C_3 \mathrm{e}^{-k_0 L} + C_4 \mathrm{e}^{k_0 L}\right) - k^2 \frac{p_0}{EI\left(k^4 - k_0^4\right)}\mathrm{e}^{-\mathrm{i}kL} = 0 \tag{4.134}$$

Defining $\xi = x/L$, $\kappa = k/k_0$, $\lambda = k_0 L$, the above system of equations can be written compactly as

$$\begin{Bmatrix} \begin{bmatrix} 1 & 1 & 1 & 1 \\ -1 & -1 & 1 & 1 \\ \mathrm{e}^{-\mathrm{i}\,\lambda} & \mathrm{e}^{\mathrm{i}\,\lambda} & \mathrm{e}^{-\lambda} & \mathrm{e}^{\lambda} \\ -\mathrm{e}^{-\mathrm{i}\,\lambda} & -\mathrm{e}^{\mathrm{i}\,\lambda} & \mathrm{e}^{-\lambda} & \mathrm{e}^{\lambda} \end{bmatrix} \end{Bmatrix} \begin{bmatrix} C_1 \\ C_2 \\ C_3 \\ C_4 \end{bmatrix} = \left(\frac{p_0}{\omega^2 \rho A}\right)\frac{1}{\kappa^4 - 1} \begin{Bmatrix} -1 \\ \kappa^2 \\ -\mathrm{e}^{-\mathrm{i}\kappa\lambda} \\ \kappa^2 \mathrm{e}^{-\mathrm{i}\kappa\lambda} \end{Bmatrix} \tag{4.135}$$

the solution to which can be shown to be

$$\begin{Bmatrix} C_1 \\ C_2 \\ C_3 \\ C_4 \end{Bmatrix} = \frac{p_0}{4\,\omega^2 \rho A} \begin{Bmatrix} \dfrac{\mathrm{i}\left(\mathrm{e}^{\mathrm{i}\lambda} - \mathrm{e}^{-\mathrm{i}\kappa\lambda}\right)}{\left(\kappa^2 - 1\right)\sin\lambda} \\[2mm] \dfrac{-\mathrm{i}\left(\mathrm{e}^{-\mathrm{i}\lambda} - \mathrm{e}^{-\mathrm{i}\kappa\lambda}\right)}{\left(\kappa^2 - 1\right)\sin\lambda} \\[2mm] \dfrac{\mathrm{e}^{\lambda} - \mathrm{e}^{-\mathrm{i}\kappa\lambda}}{\left(\kappa^2 + 1\right)\sinh\lambda} \\[2mm] \dfrac{-\left(\mathrm{e}^{-\lambda} - \mathrm{e}^{-\mathrm{i}\kappa\lambda}\right)}{\left(\kappa^2 + 1\right)\sinh\lambda} \end{Bmatrix} \tag{4.136}$$

in which case the steady-state response is

$$u = \frac{p_0}{4\omega^2 \rho A}\mathrm{e}^{\mathrm{i}\omega t}\left\{ \frac{\mathrm{i}}{\left(\kappa^2 - 1\right)\sin\lambda}\left[\left(\mathrm{e}^{\mathrm{i}\lambda} - \mathrm{e}^{-\mathrm{i}\kappa\lambda}\right)\mathrm{e}^{-\mathrm{i}\,\xi\lambda} - \left(\mathrm{e}^{-\mathrm{i}\lambda} - \mathrm{e}^{-\mathrm{i}\kappa\lambda}\right)\mathrm{e}^{\mathrm{i}\,\xi\lambda}\right] \right.$$
$$\left. + \frac{1}{\left(\kappa^2 + 1\right)\sinh\lambda}\left[\left(\mathrm{e}^{\lambda} - \mathrm{e}^{-\mathrm{i}\kappa\lambda}\right)\mathrm{e}^{-\xi\lambda} - \left(\mathrm{e}^{-\lambda} - \mathrm{e}^{-\mathrm{i}\kappa\lambda}\right)\mathrm{e}^{\xi\lambda}\right] + \frac{4}{\kappa^4 - 1}\mathrm{e}^{-\mathrm{i}\xi\kappa\lambda} \right\} \tag{4.137}$$

In combination with the harmonic factor $e^{i\omega t}$, the various terms can be interpreted as propagating or evanescent waves that move (and perhaps decay) from left to right, or the other way around. Observe that when $\kappa = 1$ (i.e., $k = k_0$) we have a resonant condition at which the last term that divides by $(\kappa^4 - 1)$ blows up while the remaining terms converge to the following finite limits:

$$\lim_{\kappa \to 1} C_1 = \frac{-p_0}{8\,\omega^2 \rho A} \frac{\lambda\, e^{-i\lambda}}{\sin \lambda}, \qquad \lim_{\kappa \to 1} C_2 = \frac{p_0}{8\,\omega^2 \rho A} \frac{\lambda\, e^{-i\lambda}}{\sin \lambda}$$

$$\lim_{\kappa \to 1} C_3 = \frac{p_0}{8\,\omega^2 \rho A} \frac{e^{\lambda} - e^{-i\lambda}}{\sinh \lambda}, \qquad \lim_{\kappa \to 1} C_4 = \frac{-p_0}{8\,\omega^2 \rho A} \frac{e^{-\lambda} - e^{-i\lambda}}{\sinh \lambda} \tag{4.138}$$

In general, if $V = \omega/k < V_{ph} = \omega/k_0 = \sqrt{\omega RC_r}$, that is, $k > k_0$ or $\kappa > 1$, the load is said to be subsonic or *subcritical*, that is, it propagates at a speed lower than that of flexural waves. The resonant condition develops when $V = V_{ph}$, which defines the *critical speed*. Also, if $V > V_{ph}$ ($k < k_0$ or $\kappa < 1$) the load is supersonic or *supercritical*, and it outruns the flexural waves that develop in response to that load.

If the beam has damping, say of the Rayleigh type, then the differential equation changes into

$$\rho A \frac{\partial^2 u}{\partial t^2} + EI \frac{\partial^4 u}{\partial x^4} + \frac{\partial}{\partial t}\left(\alpha\, \rho A\, u + \beta EI \frac{\partial^4 u}{\partial x^4}\right) = p_0\, e^{i(\omega t - kx)} \tag{4.139}$$

the solution to which is substantially more complicated than for the undamped case.

4.2.4 Nonuniform Bending Beam

Although the vast majority of inhomogeneous beams cannot be solved by purely analytical means, a few tapered types can indeed be treated that way, as will be seen herein.[3] Consider for now a beam of length L with variable material properties and arbitrary boundary conditions, which is characterized by the differential equation

$$\frac{\partial^2}{\partial x^2}\left(EI \frac{\partial u}{\partial x^2}\right) + \rho A \frac{\partial u}{\partial t^2} = 0 \tag{4.140}$$

Define

$$EI(x) = E_0 I_0\, e(x) \qquad \rho A(x) = \rho_0 A_0\, r(x) \tag{4.141}$$

where $E_0 I_0, \rho_0 A_0$ are the flexural rigidity and the mass per unit length of the beam at $x = 0$, while $e(x), r(x)$ give the variation of these quantities with x. Specializing the differential equation for harmonic vibration of the form $\ddot{u} = -\omega^2 u$, and making use of the above definitions we obtain

$$\frac{\partial^2}{\partial x^2}\left(e(x) \frac{\partial u}{\partial x^2}\right) - \omega^2 \left(\frac{\rho_0 A_0}{E_0 I_0}\right) r(x) u = 0 \tag{4.142}$$

---

[3] C. Y. Wang and C. M. Wang, "Exact vibration solution for a class of non-uniform beams," *J. Eng. Mech. ASCE*, 139 (7), 2013, 928–931.

Let the taper of the beam and variable properties be defined by

$$z = 1 - c\tfrac{x}{L}, \qquad e(x) = z^{n+4}, \qquad r(x) = z^n \tag{4.143}$$

where $0 \le c < 1$ and z – both dimensionless – are the taper parameter and the new spatial variable. If $c = 0$, the beam is uniform. It follows that $x = \tfrac{L}{c}(1-z)$ and

$$\frac{\partial}{\partial x} = -\frac{c}{L}\frac{\partial}{\partial z} \tag{4.144}$$

Hence

$$\left(\frac{C}{L}\right)^4 \frac{\partial}{\partial z^2}\left(z^{n+4}\frac{\partial u}{\partial z^2}\right) - \omega^2\left(\frac{\rho_0 A_0}{E_0 I_0}\right)z^n u = 0 \tag{4.145}$$

Making use of the ansatz and the substitution

$$u = L z^\lambda \tag{4.146}$$

$$k^4 = \omega^2 \left(\frac{L}{c}\right)^4\left(\frac{\rho A_0}{EI_0}\right) = \frac{1}{c^4}\left(\frac{\omega L}{C_{r0}}\right)^2\left(\frac{L}{R_0}\right)^2, \qquad C_{r0} = \sqrt{\frac{E_0}{\rho_0}}, \qquad R_0 = \sqrt{\frac{I_0}{A_0}} \tag{4.147}$$

we are led to the characteristic equation

$$\lambda(\lambda-1)(n+\lambda+2)(n+\lambda+1) - k^4 = 0 \tag{4.148}$$

which leads to a fourth-order equation in λ that ultimately can be factorized as

$$\tfrac{1}{16}\left[(2\lambda+n+1)^2 - (n^2+4n+5)\right]^2 - 2\left(\tfrac{1}{2}n+1\right) - k^4 = 0 \tag{4.149}$$

whose solution is

$$\boxed{\lambda = \tfrac{1}{2}\left(\pm\sqrt{n^2+4n+5\pm4\sqrt{\left(\tfrac{1}{2}n+1\right)^2+k^4}} - (n+1)\right)} \tag{4.150}$$

of which at least the first two roots are real. If all four roots are real, then the general solution is

$$\boxed{u = L\left(C_1 z^{\lambda_1} + C_2 z^{\lambda_2} + C_3 z^{\lambda_3} + C_4 z^{\lambda_4}\right)} \tag{4.151}$$

where the C_j are constants of integration. On the other hand, if the subradical is negative when using the negative sign for the inner square root, then

$$\boxed{u = L\left\{C_1 z^{\lambda_1} + C_2 z^{\lambda_2} + z^{-a}\left[C_3 \cos(b \ln z) + C_4 \sin(b \ln z)\right]\right\}} \tag{4.152}$$

where

$$a = \tfrac{1}{2}(n+1), \qquad b = \sqrt{\sqrt{\left(\tfrac{1}{2}n+1\right)^2+k^4} - \tfrac{1}{4}(n^2+4n+5)} \tag{4.153}$$

The boundary conditions are

$$u = 0, \qquad \frac{\partial u}{\partial z} = 0 \qquad \text{at a clamped end} \tag{4.154}$$

$$u = 0, \qquad \frac{\partial^2 u}{\partial z^2} = 0 \qquad \text{at a pinned end} \tag{4.155}$$

$$\frac{\partial^2 u}{\partial z^2} = 0, \qquad \frac{\partial}{\partial z}\left(e \frac{\partial^2 u}{\partial z^2} \right) = 0 \qquad \text{at a free end} \tag{4.156}$$

which can be used to find the vibration modes.

4.2.5 Nonclassical Modes of Uniform Shear Beam

Arguably, the simplest of structures with continuous mass and stiffness are the string, the rod, and the shear beam. Indeed, if we repeated the derivation of the dynamic equilibrium equation for the shear beam along the lines of that for the rod, considering a homogeneous beam without body forces applied, we would find the equation

$$\frac{\partial^2 u}{\partial x^2} = \frac{1}{C_s^2}\frac{\partial^2 u}{\partial t^2} \qquad C_s = \sqrt{\frac{GA_s}{\rho A}} = \text{shear wave velocity} \tag{4.157}$$

in which G is the shear modulus, ρ the mass density, A_s the shear area, and A the cross-sectional area. Hence, we arrive once more at the classical wave equation. The implication is that the normal modes of the shear beam must be identical to those of the homogeneous rod, except for the fact that the shear wave velocity must be used in place of the rod wave velocity. Also, particle motions take place in a transverse instead of a longitudinal direction. In the light of this finding, it would appear at first that the shear beam would not need to be considered separately. As we shall see, however, there is an interesting aspect concerning the normal modes of *partially restrained* shear beams (i.e., beams that are free to rotate) whose behavior differs from that of the rod, namely the effect that a rotational motion has on some (but not all) of the normal modes. We shall explore this issue in this section.[4]

As we saw in the case of the rod, an easy way of deriving the classical frequencies and normal modes for a shear beam is by considering a standing wave and observing that the displacements and stresses are at all times zero at the nodes and crests, respectively. By taking a beam with length equal to the distance between any pair of such nodes or crests, any set of boundary conditions can be modeled. The advantage of this approach is that one need not solve a differential equation. In particular, in the case of a free-free *shear beam* of length L and shear wave velocity C_s, the standing wave model would lead us to believe that the frequencies and modal shapes (with $x = 0$ at left end) are given by

$$f_j = \frac{jC_s}{2L}[\text{Hz}] \qquad \phi_j(x) = \cos\frac{j\pi x}{L} \qquad j = 0,1,2,3,\ldots \tag{4.158}$$

---

[4] E. Kausel, "Non-classical modes of unrestrained shear beams," *J. Eng. Mech. ASCE*, 128, (6), 2002, 663–667.

These results do indeed satisfy the wave equation, and lead to shearing stresses that vanish at the two ends, which would appear to validate the solution. However, in the case of a free–free shear beam (but not a rod), this solution does have a serious problem: the *odd* modes ($j = 1, 3, 5 \ldots$) violate the *principle of conservation of angular momentum*. This is because the shear beam, unlike the rod, has two rigid body modes, namely a translational mode and a rotational mode. Thus, the simplistic standing-wave approach suggested earlier is incorrect, at least as far as the free–free shear beam is concerned. Whichever the true odd modes are, they must not produce net angular momentum in the beam. Another way to look at this problem is to see that there is a silent equation not normally considered in the analysis of shear beams, namely the *moment equilibrium equation*. In general, this moment equation does not enter the formulation of the solution, because the sections of a shear beam cannot rotate if at least one of the ends is fixed. However, this is not true if both ends are free, or if one is free and the other end is a pivot that allows rotation. Thus, we reconsider this problem in more careful detail, and obtain the correct frequencies and modal shapes for the modes for these two systems.

Dynamic Equations of Shear Beam

By definition, an ideal shear beam exhibits no flexural deformations, but deforms in shear only. Nonetheless, equilibrium considerations dictate that the beam must also be subjected to bending moments, even if these moments do not contribute to the deformation. If the beam is fully restrained at one of its ends, these bending moments are absorbed by the support and thus play no role in the vibration problem. However, if the shear beam happens to be able to rotate freely, then the rotational component of motion can, and does, play a role in at least some of the normal modes of the beam, as will be seen. For this reason, in the ensuing we shall take into consideration the effect of the bending moments and rotational inertia on the dynamic equilibrium of the shear beam.

Consider a shear beam of length L, cross section A, shear area A_s, area-moment of inertia I, mass density ρ, and shear modulus G. Also, let x be the axial coordinate with origin at left end, such that $\xi = x/L$ is the dimensionless abscissa. We define also the following field variables:

$R = \sqrt{I/A}$ is the radius of gyration; $\lambda = R/L$ is the thickness ratio; $C_s = \sqrt{G/\rho}\,\sqrt{A_s/A}$ is the shear wave velocity; $b = b(x,t)$ is the lateral load density; M is the bending moment; S is the shear; u is the transverse displacement; θ is the rigid-body rotation of the beam (positive counterclockwise); and γ is the shear deformation. The latter three quantities define the slope of the axis as

$$\frac{\partial u}{\partial x} = \theta + \gamma \tag{4.159}$$

If the beam is fully clamped at one of its ends, then the beam cannot rotate as a rigid body, $\theta = 0$, and the behavior is the classical one. However, this will not be true if the rotation is not restrained, because the net angular momentum about the rotation point must be conserved. In the case of a free beam, the rotation point lies at the center of mass, whereas for the pivoted beam it lies at support. Considering the equilibrium and

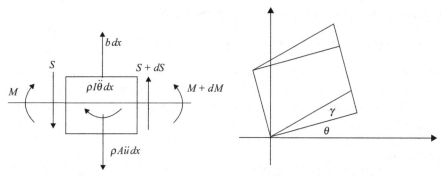

Figure 4.10. Equilibrium (left) and deformation (right) of differential shear beam element.

deformation of the beam element of length dx as shown in Figure 4.10, the relevant equations are as follows.

Equilibrium Equations

$$\frac{\partial S}{\partial x} + b - \rho A \ddot{u} = 0 \tag{4.160}$$

$$\frac{\partial M}{\partial x} + S - \rho I \ddot{\theta} = 0 \tag{4.161}$$

Force-Deformation Equations

$$\frac{\partial \theta}{\partial x} = 0 \tag{4.162}$$

$$\gamma = \frac{S}{GA_s} \tag{4.163}$$

Strain-Displacement Equation

$$\frac{\partial u}{\partial x} = \theta + \gamma \tag{4.164}$$

Combination of the previous expressions leads to the dynamic equilibrium equations

$$\rho A \ddot{u} - \frac{\partial}{\partial x}\left(GA_s \frac{\partial u}{\partial x}\right) = b(x,t) \tag{4.165}$$

$$M(x) = M_0 + \int_0^x \left(\rho I \ddot{\theta} - S\right) d\xi \tag{4.166}$$

Modes of Rotationally Unrestrained Shear Beam

In the absence of body forces and for uniform properties, the first of the two equations above reduces to the classical wave equation

$$\frac{\partial^2 u}{\partial t^2} = \frac{1}{C_s^2} \frac{\partial^2 u}{\partial x^2} \tag{4.167}$$

For harmonic motion, its general solution is of the form

$$u(x,t) = \left(C_1 \sin kx + C_2 \cos kx\right) \sin \omega t = \phi(x) \sin \omega t \tag{4.168}$$

in which $k = \omega / C_s$ is the wavenumber whose value depends on the boundary conditions, and C_1, C_2 are constants of integration. We proceed to examine the solution for the two cases of interest here, namely the free–free and pinned–free beams, whose modes differ from those of the classical solution.

Free–Free Beam

This case is complicated by the fact that both the shear and the bending moments must vanish at the two ends. These conditions are

$$\gamma_0 = 0, \quad \text{i.e.,} \quad \left.\left(\frac{\partial u}{\partial x} - \theta\right)\right|_{x=0} = 0 \tag{4.169}$$

$$\gamma_L = 0, \quad \text{i.e.,} \quad \left.\left(\frac{\partial u}{\partial x} - \theta\right)\right|_{x=L} = 0 \tag{4.170}$$

$$\int_0^L \left(\rho I \ddot{\theta} - S\right) dx = 0 \tag{4.171}$$

The first two conditions are satisfied if

$$k \begin{Bmatrix} 1 & 0 \\ \cos kL & -\sin kL \end{Bmatrix} \begin{Bmatrix} C_1 \\ C_2 \end{Bmatrix} = \theta \begin{Bmatrix} 1 \\ 1 \end{Bmatrix} \tag{4.172}$$

which implies

$$C_1 = \frac{\theta}{k} \qquad C_2 = -\frac{\theta}{k}\left(\frac{1-\cos kL}{\sin kL}\right) = -\left(\frac{\theta}{k}\right)\tan\frac{kL}{2} \tag{4.173}$$

The solution in this case is then

$$u(x) = \frac{\theta}{k}\left(\sin kx - \tan\frac{kL}{2}\cos kx\right) = \frac{\theta \sin k\left(x - \frac{L}{2}\right)}{k \cos \frac{kL}{2}} \tag{4.174}$$

On the other hand, the third condition can be written as

$$\omega^2 \rho I \theta L + GA_s \int_0^L (u' - \theta)\, dx = 0 \tag{4.175}$$

which after integration can be changed into

$$\theta\left(1 - k^2 R^2\right) = \frac{u_L - u_0}{L} = \frac{\theta}{\frac{kL}{2}}\tan\frac{kL}{2} \tag{4.176}$$

with R being the radius of gyration of the cross section. If $\theta \neq 0$ (a condition that excludes the even modes), then it can be canceled out, which leads to the transcendental equation

Table 4.2. Frequencies of free-free shear beam (no distributed rotational inertia, i.e., R=0)

J	$\frac{1}{\pi}\alpha_j$	$\theta_j L$
0	0.000	0.000
1	2.861	−0.976
2	2.000	0.000
3	4.918	+0.992
4	4.000	0.000
5	6.942	−0.996

$$\frac{\tan\frac{1}{2}\alpha_j}{\frac{1}{2}\alpha_j} = 1 - \lambda^2\alpha_j^2 \tag{4.177}$$

with $\lambda = R/L$ = the thickness ratio, and $\alpha_j = kL = \omega_j L / C_s$ = the dimensionless natural frequencies. It is convenient to normalize the modes so that

$$\theta L = kL\cos\frac{kL}{2} = \alpha_j\cos\frac{1}{2}\alpha_j \tag{4.178}$$

$$\phi_j(x) = \sin k\left(x - \frac{L}{2}\right) = \sin\alpha_j\left(\xi - \frac{1}{2}\right) \tag{4.179}$$

To assess the effect of the rotational inertia on the frequencies of the odd modes, consider first the extreme case where $R = \infty$. This means that $\tan\frac{1}{2}\alpha_j = -\infty$, which is satisfied by $\alpha_j = j\pi$, with odd $j = 1, 3, 5\ldots$ (the even roots j = 0, 2, 4…are not valid, because they imply $\theta = 0$ for which the transcendental equation is not valid). Thus, we recover the classical odd modes. At the opposite extreme is the case where the rotational inertia is zero, that is, $R = 0$. This leads in turn to the transcendental equation

$$\tan\frac{1}{2}\alpha_j = \frac{1}{2}\alpha_j \tag{4.180}$$

whose first three roots and rigid body rotation angles are listed in Table 4.2. These are the roots of the antisymmetric modes, for which j is odd (1, 3, 5). Also included in this table are the first three roots of the symmetric (classical) even modes (0, 2, 4) which cause no rotation.

As can be seen, the odd roots in this case are close to the classical odd roots ($\frac{1}{\pi}\alpha_j = 3, 5, 7\ldots$), except that the first, third, and fifth odd root moved into the vicinity of the classical third, fifth and seventh root, respectively. Thus, the first even root ($j = 2$) is now the fundamental mode of vibration of the free–free shear beam. Observe also that the motion of the beam consists of both rigid body rotations and shear deformations. Rather counterintuitive is the fact that the rotational effect on the frequencies should be largest when the rotational inertia of the beam vanishes.

The case in which R is finite yields roots α_j that lie somewhere in between the roots for two cases considered previously. They can easily be found by trial and error for any arbitrary value of R. A graphical solution can be obtained by considering the intersection of $\tan\frac{1}{2}\alpha/\frac{1}{2}\alpha$ with the parabola $1 - (\alpha\lambda)^2$.

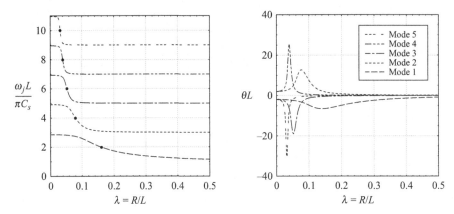

Figure 4.11. Nonclassical roots of free–free shear beam, and rigid body associated rotation.

Figure 4.11 illustrates on the left the first 10 natural frequencies of the free shear beam. The curved segmented lines depict the first five *odd* roots as a function of the thickness ratio, while the horizontal grid lines at the vertical positions 0, 2, 4, 6, and 8 depict the classical *even* roots, which are independent of λ. Notice that each odd branch intersects an even branch at a point that constitutes a double root (heavy dots in Figure 4.11 indicate coincident frequencies). These points occur at thickness ratios $\lambda = 1/(j\pi)$, $j = 2, 4, 6,\ldots$ For thickness ratios to the left of these points, the classical even mode moves down in the modal count with respect to the nonclassical odd mode, and becomes a lower (i.e., more fundamental) mode. The figure on the right, on the other hand, shows the rigid body rotation angle as a function of the thickness. The maximum rotation occurs near the double root, and is quite large in comparison to the total modal displacement. This rather counterintuitive result will be taken up again later and explained in the context of the pinned-free case.

Finally, it can be shown that the nonclassical modes thus found satisfy the principle of conservation of angular momentum with respect to the center of mass, that is,

$$
\int_L \dot\theta \rho I\, dx + \int_L \left(x - \tfrac{1}{2}L\right)\ddot u \rho A\, dx = \omega\cos\omega t\left[\theta\rho I L + \frac{\rho A\,\theta}{k\,\cos\frac{1}{2}kL}\int_{-L/2}^{+L/2} x'\sin kx'\, dx'\right]
$$
$$
= \frac{\theta\rho A L\cos\omega t}{k^2}\left[(kR)^2 - 1 + \frac{\tan\frac{1}{2}kL}{\frac{1}{2}kL}\right] = 0
\tag{4.181}
$$

The term in square bracket is zero, because it is the same as the equation for the roots.

Pinned–Free Shear Beam

This case differs from the previous in that the pin restrains the translation while allowing the rotation. Thus, the beam has only one rigid body mode. The boundary conditions are now

$$
u\big|_{x=0} = 0
\tag{4.182}
$$

$$
\gamma_L = 0, \text{ i.e., } \left(\frac{\partial u}{\partial x} - \theta\right)\bigg|_{x=L} = 0
\tag{4.183}
$$

$$\int_0^L \left(\rho I \ddot{\theta} - S \right) dx = 0 \tag{4.184}$$

The first boundary condition implies $C_2 = 0$, so the displacement, its spatial derivative, and the shear strain are

$$u = C_1 \sin kx \tag{4.185}$$

$$\frac{\partial u}{\partial x} = k C_1 \cos kx \tag{4.186}$$

$$\gamma = k C_1 \cos kx - \theta \tag{4.187}$$

On the other hand, from the second boundary condition, we obtain

$$C_1 = \frac{\theta}{k \cos kL} \tag{4.188}$$

Finally, the third condition can again be written as

$$\omega^2 \rho I \theta L + GA_s \int_0^L (u' - \theta)\, dx = 0 \tag{4.189}$$

which in turn yields

$$\omega^2 \rho I + GA_s \left(\frac{\tan kL}{kL} - 1 \right) = 0 \tag{4.190}$$

That is,

$$\left(\frac{\omega R}{C_s} \right)^2 + \frac{\tan kL}{kL} - 1 = 0 \tag{4.191}$$

or more compactly

$$\frac{\tan \alpha_j}{\alpha_j} = 1 - \lambda^2 \alpha_j^2 \tag{4.192}$$

which is similar to the characteristic equation for a free–free beam. Again, it is convenient to normalize the mode of vibration so that

$$\theta L = \alpha_j \cos \alpha_j \tag{4.193}$$

$$\phi_j = \sin kx = \sin \alpha_j \xi \tag{4.194}$$

Once more we begin by considering the extreme case where $R = \infty$. This means that $\tan \alpha_j = -\infty$, which is satisfied by $\alpha_j = (2j-1)\frac{\pi}{2}$ ($j = 1, 2, 3 \ldots$), implying frequencies

$$\omega_j = \frac{(2j-1)C_s \pi}{2L} \tag{4.195}$$

Thus, we recover the classical modes of a base-supported shear beam. At the opposite extreme is the case where the rotational inertia is zero, that is, $R = 0$. This leads to the

Table 4.3   Frequencies of pinned-free shear beam (no rotational inertia)

J	$\frac{2}{\pi}\alpha_j$	$\theta_j L$
1	2.861	−1.952
2	4.918	+1.983
3	6.942	−1.992
4	8.955	+1.995
5	10.963	−1.997

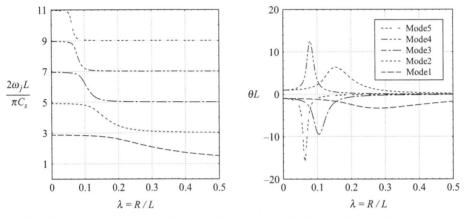

Figure 4.12. Nonclassical roots of pinned-free shear beam, and rigid body associated rotation.

transcendental equation $\tan\alpha_j = \alpha_j$ whose first five roots are listed in Table 4.3, all of which are nonclassical.

Figure 4.12 shows on the left the first five roots (do notice the factor 2 on the vertical axis), and on the right, the rigid-body rotation angles as a function of the thickness ratio. Again, we observe that the rotation angles exhibit peaks that are numerically large in comparison to the displacements (although there is no double root in this case). The reason for this relates to the fact that the shear distortions are also large and compensate the rotation in such a way that the displacements remain small. This is illustrated in Figure 4.13, which shows a schematic of the first mode of a pinned-free shear beam when $kL = \pi$, which occurs for a thickness ratio $\lambda = 1/\pi$.

Finally, the modes thus found satisfy the principle of conservation of angular momentum. Taking moment of the momentum with respect to the pinned support, we obtain

$$\int_L \dot\theta \rho I\, dx + \int_L x\,\ddot u\,\rho A\, dx = \omega\cos\omega t \left[\theta\rho I L + \frac{\rho A\theta}{k\,\cos kL}\int_0^L x\,\sin kx\, dx\right]$$

$$= \frac{\theta\rho A L\cos\omega t}{k^2}\left[(kR)^2 - 1 + \frac{\tan kL}{kL}\right] = 0 \qquad (4.196)$$

The term in square brackets is once more the transcendental equation for the roots, so it is zero.

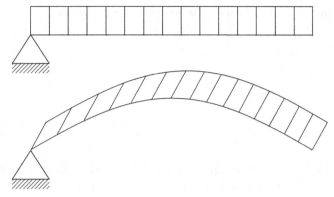

Figure 4.13. Modal form showing displacement + rotation.

Concluding Observations

Shear beams that are either free in space, or are simply pivoted at one end, allow rigid-body rotations that invalidate the classical solution for the natural frequencies of such systems. The reason is that the standard method of derivation neglects the effect that the rotational inertia must have on the modes and frequencies, and thus violates the principle of conservation of angular momentum. Indeed, this effect takes place even when the rotational inertia of the beam vanishes or is neglected, because the translational mass continues to make an important contribution to the angular momentum. It is found that the frequencies are significantly affected by this phenomenon at small thickness ratios, while it vanishes when the rotational inertia is infinitely large, because it prevents the rotation. In addition, it is found that the free–free shear beam has double frequencies at certain values of the thickness ratios below which a mode conversion takes place, the higher mode crossing over into a lower mode. These phenomena have implications for shear-beam-like structures that allow rigid-body rotations, such as space trusses or tall buildings on elastic foundation.

4.2.6 Inhomogeneous Shear Beam

When sedimentary soil deposits are subjected to seismic excitations, the motions within the soil, and particularly those at the surface, are subjected to dynamic magnification referred to as *soil amplification*. One of the principal components of seismic motion at or near the surface consists of shear waves that propagate vertically up and down, or nearly so. These are referred to as SH waves, and they can be studied by modeling the soil as a shear beam of unit cross section. As is well known, the stiffness of the soil generally increases with depth as a result of the increase in confining stress with depth caused by the weight of the soil. Often, the increase of the shear modulus is gradual, so that a simple power law may suffice to represent the change in stiffness with depth, at least near the surface. This idealization is considered herein.

Solution for Shear Modulus Growing Unboundedly with Depth

Consider a shear beam with constant material density ρ, but whose shear modulus increases with depth according to the power law

$$G = G_0 \left(\frac{z}{z_0}\right)^m = G_0 \lambda^m \tag{4.197}$$

in which G_0 is the shear modulus at an arbitrary reference elevation z_0 (usually the free surface), m is a nonnegative exponent, and the z-axis is positive in the downward direction. Clearly, a large range of soil models can be represented by appropriate choices of these parameters. It should be noted, however, that for any $m > 0$, the soil stiffness increases without bound as z increases, so the modulus attains nonphysical values at large depths. The dynamic equilibrium equation for this system is

$$\rho \ddot{u} - \left(Gu'\right)' = 0 \tag{4.198}$$

in which the primes denote differentiation with respect to z and the dots differentiation with respect to t. Carrying out the differentiation, we obtain

$$\rho \ddot{u} - G_0 \left[\left(\frac{z}{z_0}\right)^m u'' + \frac{m}{z_0}\left(\frac{z}{z_0}\right)^{m-1} u'\right] = 0 \tag{4.199}$$

For harmonic motion, $\ddot{u} = -\omega^2 u$, this equation can be written as

$$z^m u'' + m z^{m-1} u' + r^2 u = 0 \tag{4.200}$$

in which $r = \dfrac{\omega z_0^{m/2}}{C_s}$ and $C_s = \sqrt{\dfrac{G_0}{\rho}}$ is the shear wave velocity at the reference surface. By an appropriate change of variables and function of the form

$$z = x^n, \qquad n = 1 - \tfrac{1}{2}m \tag{4.201}$$

and

$$u = x^p w = z^{np} w, \qquad p = \frac{1-m}{2-m} \tag{4.202}$$

the differential equation reduces to

$$x^2 \frac{\partial^2 w}{\partial x^2} + x \frac{\partial w}{\partial x} + \left[\left(\frac{r}{n}\right)^2 x^2 - p^2\right] w = 0 \tag{4.203}$$

which has the form of a Bessel differential equation of pth order. The final solution is then

$$\boxed{u = \zeta^{np}\left[C_1 J_{\pm p}(a_0 \zeta^n) + C_2 Y_{\pm p}(a_0 \zeta^n)\right]} \tag{4.204}$$

in which

$$\zeta = \frac{z}{z_0}, \qquad a_0 = \frac{\omega z_0}{|n| C_s}, \qquad p = \frac{1-m}{2-m}, \qquad n = 1 - \tfrac{1}{2}m, \qquad np = \tfrac{1}{2}(1-m) \tag{4.205}$$

and the C_1, C_2 are constants of integration. If the problem under consideration does not include the origin of coordinates (i.e., if the shear modulus at the soil surface is finite), then either the positive or negative sign for p can be used. As we shall see later, however, when the modulus is zero at the surface, then only the negative sign before p leads to meaningful solutions.

The shearing stresses at any point are simply $\tau = Gu'$. Carrying out the required differentiation, and taking into account the recursion properties of the Bessel functions, we obtain

$$\boxed{\tau = \pm\rho\omega C_s \, \mathrm{sgn}(n)\sqrt{\zeta}\left[C_1 J_{\pm(p-1)}(a_0\zeta^n) + C_2 Y_{\pm(p-1)}(a_0\zeta^n)\right]} \tag{4.206}$$

For $m = 0$ (homogeneous soil), $n = 1$, $p = \frac{1}{2}$, the half-order Bessel functions reduce to ordinary sine and cosine functions, and we recover the well-known solution of the homogeneous shear beam.

For $m = \frac{1}{2}$ ("sand"), $n = \frac{3}{4}$, $p = \frac{1}{3}$, the solution involves Bessel functions of order $\frac{1}{3}$, which can be represented in terms of Airy functions.[5]

Finally, for $m = 1$, $n = \frac{1}{2}$, $p = 0$, we obtain

$$u = C_1 J_0(a_0\sqrt{\zeta}) + C_2 Y_0(a_0\sqrt{\zeta}) \tag{4.207}$$

Finite Layer of Inhomogeneous Soil

Combining the displacement together with the shear stress, we can write

$$\begin{Bmatrix} u \\ \tau \end{Bmatrix} = \sqrt{\zeta}\begin{Bmatrix} \zeta^{-m/2}J_{\pm p}(a_0\zeta^n) & \zeta^{-m/2}Y_{\pm p}(a_0\zeta^n) \\ \pm\alpha J_{\pm(p-1)}(a_0\zeta^n) & \pm\alpha Y_{\pm(p-1)}(a_0\zeta^n) \end{Bmatrix}\begin{Bmatrix} C_1 \\ C_2 \end{Bmatrix} \tag{4.208}$$

in which $\alpha = \rho\omega C_s \, \mathrm{sgn}(n)$. Choosing the reference surface at elevation z_0 to be at free surface, $\zeta = 1$, $u = u_0$, $\tau = 0$, then

$$\begin{Bmatrix} u_0 \\ 0 \end{Bmatrix} = \begin{Bmatrix} J_{\pm p} & Y_{\pm p} \\ \pm\alpha J_{\pm(p-1)} & \pm\alpha Y_{\pm(p-1)} \end{Bmatrix}\begin{Bmatrix} C_1 \\ C_2 \end{Bmatrix} \tag{4.209}$$

in which $J_{\pm p} \equiv J_{\pm p}(a_0)$, and so forth. This implies

$$\begin{aligned}
\begin{Bmatrix} C_1 \\ C_2 \end{Bmatrix} &= \frac{1}{\pm\alpha\left(J_{\pm p}Y_{\pm(p-1)} - J_{\pm(p-1)}Y_{\pm p}\right)}\begin{Bmatrix} \pm\alpha Y_{\pm(p-1)} & -Y_{\pm p} \\ \mp\alpha J_{\pm(p-1)} & J_{\pm p} \end{Bmatrix}\begin{Bmatrix} u_0 \\ 0 \end{Bmatrix} \\
&= \frac{u_0}{J_{\pm p}Y_{\pm(p-1)} - J_{\pm(p-1)}Y_{\pm p}}\begin{Bmatrix} Y_{\pm(p-1)} \\ -J_{\pm(p-1)} \end{Bmatrix}
\end{aligned} \tag{4.210}$$

In terms of the displacements at the surface, the solution at depth z (i.e., $\zeta > 1$) is then

$$\begin{Bmatrix} u \\ \tau \end{Bmatrix} = \frac{\sqrt{\zeta}}{J_{\pm p}Y_{\pm(p-1)} - J_{\pm(p-1)}Y_{\pm p}}\begin{Bmatrix} \zeta^{-m/2}\left(J_{\pm p}(a_0\zeta^n)Y_{\pm(p-1)} - Y_{\pm p}(a_0\zeta^n)J_{\pm(p-1)}\right) \\ \pm\alpha\left(J_{\pm(p-1)}(a_0\zeta^n)Y_{\pm(p-1)} - Y_{\pm(p-1)}(a_0\zeta^n)J_{\pm(p-1)}\right) \end{Bmatrix}u_0 \tag{4.211}$$

---

[5] M. Abramovitz and I.A. Stegun, *Handbook of Mathematical Functions* (National Bureau of Standards, 1970), p. 447.

In particular, the amplification function from depth z to the surface $z = z_0$ is

$$\frac{u_0}{u} = \frac{J_{\pm p} Y_{\pm(p-1)} - J_{\pm(p-1)} Y_{\pm p}}{\zeta^{np} \left(J_{\pm p}(a_0 \zeta^n) Y_{\pm(p-1)} - Y_{\pm p}(a_0 \zeta^n) J_{\pm(p-1)} \right)} \tag{4.212}$$

When the soil column of depth h is driven harmonically at $z = z_0 + h$, (i.e., for $\zeta = 1 + h / z_0$), resonance will take place when the denominator of the amplification function takes on zero values, that is, when

$$J_{\pm p}(a_0 \zeta^n) Y_{\pm(p-1)}(a_0) - Y_{\pm p}(a_0 \zeta^n) J_{\pm(p-1)}(a_0) = 0 \tag{4.213}$$

This is a transcendental equation whose roots can be found by means of search techniques. These roots provide in turn the dimensionless resonant frequencies a_0 of the inhomogeneous, finite soil column.

Special Case: Shear Modulus Zero at Free Surface

This situation arises when the origin of coordinates coincides with the free surface, a situation that requires defining the reference surface at some arbitrary depth below, usually at the bottom of the soil column (if the soil has finite depth h). To obtain the displacements and stresses at the free surface, we must now consider the limiting expressions of the Bessel functions for zero arguments. From the ascending series of the Bessel functions, their values for small argument are

$$\lim_{z \to 0} J_{\pm v}(z) \quad \frac{1}{\Gamma(v+1)} \left(\tfrac{1}{2} z\right)^v \qquad v \neq -1, -2, -3 \ldots \tag{4.214}$$

$$\lim_{z \to 0} Y_v(z) \quad -\frac{\Gamma(v)}{\pi} \left(\tfrac{1}{2} z\right)^{-v} \qquad v > 0 \tag{4.215}$$

$$\lim_{z \to 0} Y_{-v}(z) \quad -\frac{\cos v\pi\, \Gamma(v)}{\pi} \left(\tfrac{1}{2} z\right)^{-v} - \frac{\sin v\pi}{\Gamma(v+1)} \left(\tfrac{1}{2} z\right)^v \qquad v > 0 \tag{4.216}$$

$$\lim_{z \to 0} Y_0(z) \quad \frac{2}{\pi} \ln z \tag{4.217}$$

Using these expressions to evaluate the limits $\lambda^{np} J_{\pm p}$, and so forth in the solution for displacements and stresses, it can be shown that a finite displacement and a vanishing stress at the free surface can be satisfied simultaneously only by taking the solution for $-p$. Furthermore, solutions exist only in the exponent range $0 \leq m < 2$. The displacements and stresses are

$$u = C_1 \zeta^{(1-m)/2} J_{-p}(a_0 \zeta^n) \tag{4.218}$$

$$\tau = -C_1 \rho \omega C_s \sqrt{\zeta}\, J_{1-p}(a_0 \zeta^n) \tag{4.219}$$

The displacement at the surface is then

$$u_0 = \frac{C_1}{\Gamma(p+1)} \left(\tfrac{1}{2} a_0\right)^{np} \tag{4.220}$$

and the amplification function from depth z to the free surface is

$$\frac{u_0}{u} = \frac{\zeta^{-(1-m)/2}}{\Gamma(p+1) J_{-p}(a_0 \zeta^n)} \left(\tfrac{1}{2} a_0\right)^{np} \tag{4.221}$$

Resonance takes place when the Bessel function in the denominator attains zero values, that is, when $J_{-p}(a_0 \zeta^n) = 0$. In particular, if $h = z_0$, that is, the reference surface used to define G_0 is taken at the base of the soil column, then $\zeta = 1$ there, in which case the resonance condition is $J_{-p}(a_0) = 0$. The resonance frequencies are then

$$\omega_j = \frac{|n| C_s}{h} z_{jp} \qquad j = 1, 2, 3, \cdots \tag{4.222}$$

where z_{jp} are the zeros of the Bessel function of order $-p$, and C_s is the shear wave velocity at the base of the soil.

Special Case: Linearly Increasing Shear Wave Velocity

The case $m = 2$, $n = 0$, $p = \infty$ requires special consideration. The differential equation in this case is

$$z^2 u'' + 2 z u' + a_0^2 u = 0 \tag{4.223}$$

which admits solutions of the form

$$u = C_1 \zeta^{\alpha_1} + C_2 \zeta^{\alpha_2} \tag{4.224}$$

$$\tau = \frac{G_0}{z_0} \left(\alpha_1 C_1 \zeta^{\alpha_1 + 1} + \alpha_1 C_2 \zeta^{\alpha_2 + 1} \right) \tag{4.225}$$

with C_1, C_2 being constants of integration, and α_1, α_2 are the two roots of the quadratic equation $\alpha^2 + \alpha + r^2 = 0$. These roots have the form

$$\alpha_j = \tfrac{1}{2} \left[-1 \pm i \sqrt{(2a_0)^2 - 1} \right] = \tfrac{1}{2} (-1 \pm i b) \tag{4.226}$$

If we wrote these roots in complex form it is because no real roots can exist in a stratum of finite depth h. The reason is that they violate the boundary conditions of zero stresses at the surface $z = z_1$ (i.e., $\zeta = \zeta_1$) and zero displacement at the base $z_2 = z_1 + h$ (i.e., $\zeta = \zeta_2 > \zeta_1$). In matrix form, these conditions are

$$\begin{Bmatrix} \zeta_2^{\alpha_1} & \zeta_2^{\alpha_2} \\ \alpha_1 \zeta_1^{\alpha_1} & \alpha_1 \zeta_1^{\alpha_2} \end{Bmatrix} \begin{Bmatrix} C_1 \\ C_2 \end{Bmatrix} = \begin{Bmatrix} 0 \\ 0 \end{Bmatrix} \tag{4.227}$$

which admits nontrivial solutions only if the determinant is zero, that is, $\alpha_2 \lambda_1^{\alpha_2} \lambda_2^{\alpha_1} - \alpha_1 \lambda_1^{\alpha_2} \lambda_2^{\alpha_1} = 0$. This equation cannot be satisfied for real values of α. For complex values, on the other hand, we can write

$$\left(\frac{\zeta_2}{\zeta_1} \right)^{\alpha_2 - \alpha_1} = \frac{\alpha_2}{\alpha_1} \qquad \text{or} \qquad \left(\frac{\zeta_2}{\zeta_1} \right)^{ib} = \frac{1 - ib}{1 + ib} \tag{4.228}$$

With the definition $\theta = \ln\dfrac{\zeta_2}{\zeta_1}$, the determinant equation can be reduced to the transcendental equation

$$\tan\frac{\theta b}{2} + b = 0 \qquad\qquad (4.229)$$

which can be used to find the resonant frequencies. If $z_1 = 0$, that is, if the shear modulus at the surface vanishes, then $\theta = \infty$, and no solution exists for the normal modes.

4.2.7 Rectangular Prism Subjected to SH Waves

Consider a rectangular prism of lateral dimensions a, b and infinite length *into* the plane of the figure as shown in Figure 4.14, with fixed boundaries on two contiguous edges. We are interested in finding the normal modes for antiplane shear waves or SH waves that propagate in the plane x, z and whose particle motion is in direction y perpendicular to the plane of the body. This problem is identical to that of acoustic waves in a 2-D rectangular room, and of stretch waves in an elastic membrane subjected to uniform tension. We shall refer to this body as an SH plate.

Normal Modes

The plate satisfies the Helmholtz equation

$$\nabla^2 u + \left(\frac{\omega}{\beta}\right)^2 u = 0 \qquad\qquad (4.230)$$

where β is the speed of shear waves in the plate. We start by assuming a modal solution $u = \phi(x, z)$ of the form

$$\phi = A\sin m\xi \sin n\zeta, \qquad \xi = \tfrac{x}{a}, \qquad \zeta = \tfrac{z}{b} \qquad\qquad (4.231)$$

Substituting this ansatz into the differential equation, we obtain

$$\left(\frac{m}{a}\right)^2 + \left(\frac{n}{b}\right)^2 = \left(\frac{\omega}{\beta}\right)^2 \qquad\qquad (4.232)$$

Also, from the boundary conditions

$$\left.\frac{\partial\phi}{\partial x}\right|_{x=a} = 0 \qquad \cos m\xi\big|_{\xi=1} = 0, \qquad m = \tfrac{1}{2}\pi(2j-1) \qquad\qquad (4.233)$$

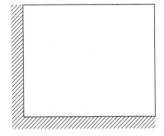

Figure 4.14. SH plate clamped at left and bottom.

$$\left.\frac{\partial \phi}{\partial z}\right|_{z=b} = 0 \qquad \cos n\zeta\big|_{\zeta=1} = 0, \qquad n = \tfrac{1}{2}\pi(2k-1) \tag{4.234}$$

Hence

$$\boxed{\phi_{jk} = A\,\sin\tfrac{1}{2}\pi(2j-1)\xi\,\sin\tfrac{1}{2}\pi(2k-1)\zeta,} \qquad j,k = 1,2,\cdots \tag{4.235}$$

$$\boxed{\omega_{jk} = \tfrac{1}{2}\pi\beta\sqrt{\left(\frac{2j-1}{a}\right)^2 + \left(\frac{2k-1}{b}\right)^2}} \tag{4.236}$$

which provide the normal modes and the natural frequencies for the plate.

Forced Vibration

We consider next a loading case in which the plate is subjected to a harmonic antiplane source or traction with distribution $p = p(a,z,\omega)$ applied along the right vertical boundary at $x = a$. This source can be expressed in terms of modal terms as follows:

$$p(a,z,\omega) = \sum_{n=1}^{\infty} p_n \sin\left[\tfrac{1}{2}\pi(2n-1)\zeta\right], \qquad \zeta = z/b \tag{4.237}$$

To obtain the modal components p_n, we multiply both sides by $\sin\left[\tfrac{1}{2}\pi(2m-1)\tfrac{z}{b}\right]$, with m being an arbitrary integer index, and integrate with respect to z in the interval $0:b$. From the orthogonality of the resulting expression in which only the mth term in the series survives, we ultimately infer that

$$\boxed{p_n = \tfrac{2}{b}\int_0^b p\,\sin\left[\tfrac{1}{2}\pi(2n-1)\tfrac{z}{b}\right]dz} \tag{4.238}$$

In particular, if the source is in the form of a unit line (or "point") load applied at the relative height $\zeta_0 = z_0/b$, then $p = \delta(z - z_0)$ and

$$\begin{aligned} p_n &= \tfrac{2}{b}\int_0^b \delta(z - z_0)\,\sin\left[\tfrac{1}{2}\pi(2n-1)\tfrac{z}{b}\right]dz \\ &= \tfrac{2}{b}\sin\left[\tfrac{1}{2}\pi(2n-1)\tfrac{z_0}{b}\right] \end{aligned} \quad = \text{unit point load} \tag{4.239}$$

Proceeding further, we shall next assume a modal solution for the forced vibration of the form

$$u(x,z,\omega) = \sum_{n=1}^{\infty} f_n(x,\omega)\sin\left[\tfrac{1}{2}\pi(2n-1)\zeta\right] \tag{4.240}$$

which automatically satisfies the free boundary conditions at the top and the bottom. From here, we infer the second derivatives

$$\frac{\partial^2 u}{\partial x^2}(x,z,\omega) = \sum_{n=1}^{\infty} \frac{d^2 f_n}{dx^2}\sin\left[\tfrac{1}{2}\pi(2n-1)\zeta\right] \tag{4.241}$$

$$\frac{\partial^2 u}{\partial z^2}(x,z,\omega) = -\left(\frac{\pi}{2b}\right)^2 \sum_{n=1}^{\infty} (2n-1)^2 \, f_n \sin\left[\tfrac{1}{2}\pi(2n-1)\zeta\right]$$ (4.242)

Substituting these expressions into the differential equation, we obtain

$$\sum_{n=1}^{\infty} \frac{d^2 f_n}{dx^2} \sin\left[\tfrac{1}{2}\pi(2n-1)\zeta\right] - \left(\frac{\pi}{2b}\right)^2 \sum_{n=1}^{\infty} (2n-1)^2 \, f_n \sin\left[\tfrac{1}{2}\pi(2n-1)\zeta\right] =$$
$$-\left(\frac{\omega}{\beta}\right)^2 \sum_{n=1}^{\infty} f_n \sin\left[\tfrac{1}{2}\pi(2n-1)\zeta\right]$$ (4.243)

which in turn leads us to the modal differential equation and boundary conditions

$$\frac{d^2 f_n}{dx^2} + \left[\left(\frac{\omega}{\beta}\right)^2 - \left(\frac{\pi}{2b}\right)^2 (2n-1)^2\right] f_n = 0, \qquad f_n\big|_{x=0} = 0, \qquad \frac{df_n}{dx}\bigg|_{x=a} = \frac{p_n}{\mu}$$ (4.244)

where p_n is the nth component of a source p applied on the right boundary. The differential Eq. 4.244 admits simple solutions of the form

$$f_n(x) = A_n e^{ik_n x} + B_n e^{-ik_n x}$$ (4.245)

involving the modal wavenumbers

$$k_n = \sqrt{\left(\frac{\omega}{\beta}\right)^2 - \left(\frac{\pi}{2b}\right)^2 (2n-1)^2}$$ (4.246)

From the first boundary condition at $x = 0$, it follows that $B_n = -A_n$. Also, from the second boundary condition, we infer

$$\frac{df_n}{dx}\bigg|_{x=a} = ik_n \left[A_n e^{ik_n x} - B_n e^{-ik_n x}\right]_{x=a} = \frac{p_n}{\mu}$$ (4.247)

$$A_n = \frac{p_n}{2ik_n \mu \cos k_n a}$$ (4.248)

and

$$f_n(x) = p_n \frac{\sin k_n x}{k_n \mu \cos k_n a}$$ (4.249)

so

$$\boxed{u(x,z,\omega) = \frac{a}{\mu} \sum_{n=1}^{\infty} \tau_n \frac{\sin(k_n a \xi)}{(k_n a)\cos(k_n a)} \sin\left[\tfrac{1}{2}\pi(2n-1)\zeta\right]} \qquad \xi = \frac{x}{a}, \qquad \zeta = \frac{z}{b}$$ (4.250)

with $\qquad \boxed{k_n a = \frac{\pi}{2}\sqrt{\left(\frac{\omega}{\omega_1}\right)^2 - \left(\frac{a}{b}\right)^2 (2n-1)^2}}, \qquad n = 1,2,3,\ldots$ (4.251) where $\omega_1 = \frac{1}{2}\pi\beta/a$ is the fundamental cutoff frequency in the width direction. In particular, for a unit line load

$$u(x,z,z_0,\omega) = \frac{2}{\mu}\frac{a}{b}\sum_{n=1}^{\infty}\frac{\sin(k_n a\xi)}{(k_n a)\cos(k_n a)}\sin\left[\tfrac{1}{2}\pi(2n-1)\zeta_0\right]\sin\left[\tfrac{1}{2}\pi(2n-1)\zeta\right] \qquad (4.252)$$

For example, the displacement along the right edge due to a point load applied at the upper right vertex $(z_0 = b)$ is

$$u(a,z,b,\omega) = \frac{2}{\mu}\frac{a}{b}\sum_{n=1}^{\infty}(-1)^{(n-1)}\frac{\tan k_n a}{k_n a}\sin\left[\tfrac{1}{2}\pi(2n-1)\zeta\right] \qquad (4.253)$$

4.2.8 Cones, Frustums, and Horns

Consider a cylindrical, solid rod of variable cross section $A(x)$ in the shape of a cone, frustum or horn, which is subjected to compressional waves in the axial direction $x \geq 0$. If the vibrational frequencies are sufficiently low that the wavelengths are long compared to the lateral dimensions of the rod, then guided waves on the rod's outer boundary (i.e., skin) can be neglected. This implies that the axial stresses are uniformly distributed across the rod section, and the problem is one-dimensional. With appropriate changes in physical parameters, the material presented in this section applies also to torsional waves in cones, and to acoustic waves in horns and pipes.

Consider the equilibrium of an elementary volume of length dx, cross section $A(x)$, and mass density ρ. If σ is the axial stress, then the axial force and the dynamic equilibrium equations are

$$N = A\sigma = EA(x)\varepsilon = EA(x)\frac{\partial u}{\partial x}, \qquad \frac{\partial N}{\partial x} = \rho A(x)\frac{\partial^2 u}{\partial t^2} \qquad (4.254)$$

That is, the change in axial force equals the D'Alembert force. The differential equation of motion is then

$$E\left[A\frac{\partial^2 u}{\partial x^2} + \frac{\partial A}{\partial x}\frac{\partial u}{\partial x}\right] = \rho A\frac{\partial^2 u}{\partial t^2} \qquad (4.255)$$

We shall consider four different types of cross section, namely

(a) $A(x) = A_0\,e^{2ax}$
(b) $A(x) = A_1\left(\frac{x}{x_1}\right)^m$
(c) $A(x) = A_\infty - (A_\infty - A_0)(1+ax)^{-1}$
(d) $A(x) = A_0 + (A_\infty - A_0)(1-e^{-ax})$

in which A_0, A_1, and A_∞ are the cross sections at $x = 0$, $x = x_1$, and $x = \infty$, respectively, and a, m are arbitrary, nonnegative parameters. In the first two cases, the cross section grows without bound as $x\to\infty$, while in the last two cases, it approaches asymptotically a finite value.

(a) Exponential Horn

We consider first the case, $A(x) = A_0\, e^{2ax}$, which in the case of acoustic waves represents a horn. The factor 2 is for convenience only, and reflects the fact that areas grow with the square of linear dimensions. Substituting into the differential equation, we obtain

$$\frac{\partial}{\partial x}\left[A_0\, e^{2ax}\, \frac{\partial u}{\partial x} \right] = \rho A_0\, e^{2ax}\, \frac{\partial^2 u}{\partial t^2} \tag{4.256}$$

That is,

$$u'' + 2au' = \frac{1}{C_r^2}\ddot{u} \tag{4.257}$$

where $C_r = \sqrt{E/\rho}$ is the rod wave velocity (or celerity), and primes and double dots denote differentiation with respect to x and t, respectively. For harmonic waves of the form $u = e^{i(\omega t - kx)}$, we obtain by substitution into the differential equation,

$$k^2 + 2i\,ak - k_0^2 = 0 \tag{4.258}$$

in which $k_0 = \omega/C_r$. This is a quadratic equation in k, with solutions

$$k = \pm\sqrt{k_0^2 - a^2} - ia \tag{4.259}$$

The general solution to the differential equation is

$$u(x,t) = e^{-ax}\left[C_1 e^{i\left(\omega t + x\sqrt{k_0^2 - a^2}\right)} + C_2 e^{i\left(\omega t - x\sqrt{k_0^2 - a^2}\right)} \right] \tag{4.260}$$

provided that $k_0^2 > a^2$. This solution represents a combination of evanescent waves that propagate in the negative and positive x-directions with amplitudes C_1 and C_2, respectively, and with a phase velocity

$$v = \frac{\omega}{k} = \frac{\omega}{\sqrt{k_0^2 - a^2}} \tag{4.261}$$

Since this velocity is a function of the frequency, the waves are *dispersive*. On the other hand, when $k_0^2 < a^2$, the square root returns an imaginary number, and the solution becomes a purely evanescent harmonic motion of the form

$$u(x,t) = e^{i\omega t}\left[C_1 e^{-\left(a + \sqrt{a^2 - k_0^2}\right)x} + C_2 e^{-\left(a - \sqrt{a^2 - k_0^2}\right)x} \right] \tag{4.262}$$

The limit $k_0 = a$, that is, $\omega = ca$ is referred to as the *cutoff frequency*, below which no waves propagate. Thus, this frequency defines the separation between the so-called *stopping* and *starting* bands, namely $[0, \omega c]$ and $[\omega c, \infty]$. In the former, waves are evanescent and do not propagate, while in the latter, the waves propagate with finite speed while decaying in amplitude in the positive x-direction. This decay is not the result of energy losses, but instead because the vibrational energy spreads out as the cross section grows.

Exponential Horn of Finite Length with Fixed Boundaries

Let's consider next an exponential horn of finite length defined in the interval $[0, L]$. If both ends are fixed, the boundary conditions are

$$u\big|_{x=0} = 0 \qquad u\big|_{x=L} = 0 \tag{4.263}$$

So

$$u(0,t) = e^{i\omega t}\left(C_1 + C_2\right) = 0 \qquad \text{i.e.,} \qquad C_2 = -C_1 \tag{4.264}$$

$$u(L,t) = e^{i\omega t}e^{-aL}\left[C_1 e^{iL\sqrt{k_0^2 - a^2}} + C_2 e^{-iL\sqrt{k_0^2 - a^2}}\right] = 0 \tag{4.265}$$

This last equation can be satisfied only if the term in square brackets is zero. Taking into account the relationship between the two constants of integration, this requires

$$e^{iL\sqrt{k_0^2 - a^2}} - e^{-iL\sqrt{k_0^2 - a^2}} = 2\sinh iL\sqrt{k_0^2 - a^2} = i\sin L\sqrt{k_0^2 - a^2} = 0 \tag{4.266}$$

That is,

$$\sin L\sqrt{k_0^2 - a^2} = 0 \tag{4.267}$$

which has solutions

$$L\sqrt{k_0^2 - a^2} = \pi j \tag{4.268}$$

with j being an integer. Solving for the frequency from k_0,

$$\boxed{\omega_j = C_r\sqrt{\left(\frac{\pi j}{L}\right)^2 + a^2}} \qquad j = 1,2,3,\ldots \tag{4.269}$$

The modal shape follows from the displacement solution together with the value found for k_0, taking $C_1 = 1$, and omitting the time-dependent exponential term. This gives

$$\boxed{\phi(x) = e^{-ax}\sin\frac{j\pi x}{L}} \tag{4.270}$$

Setting $a = 0$, we recover the solution for the homogeneous rod with fixed boundary conditions.

Exponential Horn of Finite Length with Fixed–Free Boundaries

A free boundary condition at $x = L$ requires setting the derivative of the displacement at that location to zero, that is,

$$\frac{\partial u}{\partial x} = e^{i\omega t}e^{-aL}\left[C_1\left(iL\sqrt{k_0^2 - a^2} - a\right)e^{iL\sqrt{k_0^2 - a^2}} - C_2\left(iL\sqrt{k_0^2 - a^2} + a\right)e^{-iL\sqrt{k_0^2 - a^2}}\right] = 0 \tag{4.271}$$

which together with the fixed condition at $x = 0$ implies

$$\left(iL\sqrt{k_0^2 - a^2} - a\right)e^{iL\sqrt{k_0^2 - a^2}} + \left(iL\sqrt{k_0^2 - a^2} + a\right)e^{-iL\sqrt{k_0^2 - a^2}} = 0 \tag{4.272}$$

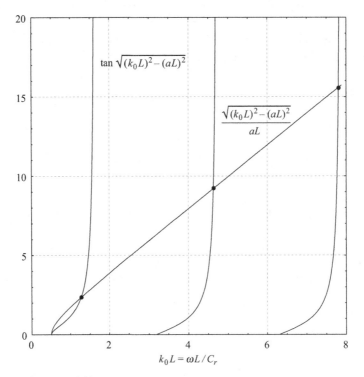

Figure 4.15. Roots of exponential horn.

That is,

$$\sqrt{k_0^2 - a^2}\,\cos L\sqrt{k_0^2 - a^2} - a\sin L\sqrt{k_0^2 - a^2} = 0 \tag{4.273}$$

or

$$\boxed{\tan\sqrt{\left(k_0L\right)^2 - \left(aL\right)^2} = \frac{\sqrt{\left(k_0L\right)^2 - \left(aL\right)^2}}{aL}} \tag{4.274}$$

This transcendental equation can be solved numerically by root-finding methods, or graphically, by searching for the intersection of the curves defined by the left- and right-hand sides, as shown in Figure 4.15 for the case $aL = 0.5$. The modes, on the other hand, are obtained by substitution of the obtained roots k_0 into the expression of the displacement, but without the time-dependent term:

$$\boxed{\phi(x) = e^{-ax}\sin x\sqrt{k_0^2 - a^2}} \tag{4.275}$$

In particular, we see that $k_0 = a$ satisfies the transcendental equation identically, namely $0 = 0$, which occurs at the cutoff frequency. However, this is not a normal mode of the horn, for it would imply $\phi(x) = 0$, which is a trivial solution (i.e., no vibration at all).

On the other hand, in the limit $a \to 0$ (i.e., rod of uniform cross section with fixed–free boundary conditions), the frequency equation reduces to $\tan k_0 L = \infty$, which has solutions

$$k_0 L = \frac{\omega_j L}{C_r} = \frac{\pi}{2}(2j - 1) \tag{4.276}$$

That is,

$$f_j = \frac{C_r}{4L}(2j - 1) \ \ [\text{Hz}] \tag{4.277}$$

which agrees with the classical solution for a uniform rod.

(b) Frustum Growing as a Power of the Axial Distance

In this case

$$A(x) = A_1 \left(\frac{x}{x_1}\right)^m \tag{4.278}$$

which we restrict to $x > 0, m \geq 0$, to avoid a vanishing (or singularity) of the cross section at $x = 0$. Substituting this expression into the differential equation and after straightforward manipulations, we obtain

$$\frac{\partial^2 u}{\partial x^2} + \frac{m}{x}\frac{\partial u}{\partial x} = \frac{1}{C_r^2}\frac{\partial^2 u}{\partial t^2} \tag{4.279}$$

where again $C_r = \sqrt{E/\rho}$ is the rod wave velocity. For harmonic motion, this reduces to

$$\frac{\partial^2 u}{\partial x^2} + \frac{m}{x}\frac{\partial u}{\partial x} + k_0^2 u = 0 \tag{4.280}$$

in which once more $k_0 = \omega/C_r$. Substituting the trial solution

$$u(x,\omega) = x^n \, y(x,\omega) \tag{4.281}$$

with an as yet undetermined parameter n, we obtain after simplifications

$$y'' + \frac{2n + m}{x}y' + \left[k_0^2 + \frac{n(n - 1 + m)}{x^2}\right]y = 0 \tag{4.282}$$

where primes denote differentiation with respect to x. Choosing $2n + m = 1$, that is,

$$\boxed{n = -\tfrac{1}{2}(m - 1)} \tag{4.283}$$

the differential equation for y reduces to the Bessel equation

$$y'' + \frac{1}{x}y' + \left[k_0^2 - \frac{n^2}{x^2}\right]y = 0 \tag{4.284}$$

whose solutions are Bessel functions of order n. In the case of *infinite domains*, we express the solution for y in terms of Bessel functions of the third kind, that is, the complex-valued first and second Hankel functions, so as to represent propagating waves. In that case, the complete solution for u is

$$u = x^n \left[C_1 \, H_n^{(1)}(k_0 x) + C_2 \, H_n^{(2)}(k_0 x)\right]e^{i\omega t} \tag{4.285}$$

in which C_1 and C_2 are once again integration constants. In combination with the factor $\exp(i\omega t)$, the first term represents waves that propagate in the negative direction, while the second term corresponds to waves traveling in the positive x-direction.

On the other hand, in the case of *finite domains*, it is more convenient to express the solution in terms of Bessel functions of the first and second kind, that is, the real-valued Bessel and Neumann functions:

$$u = x^n \left[C_1 J_n(k_0 x) + C_2 Y_n(k_0 x) \right] e^{i\omega t} \tag{4.286}$$

For example, in the case of a horn defined in the interval $[x_1, x_2]$ with fixed boundary conditions, this implies

$$C_1 J_n(k_0 x_1) + C_2 Y_n(k_0 x_1) = 0 \quad \text{and} \quad C_1 J_n(k_0 x_2) + C_2 Y_n(k_0 x_2) = 0 \tag{4.287}$$

which in matrix form is

$$\begin{Bmatrix} J_n(k_0 x_1) & Y_n(k_0 x_1) \\ J_n(k_0 x_2) & Y_n(k_0 x_2) \end{Bmatrix} \begin{Bmatrix} C_1 \\ C_2 \end{Bmatrix} = \begin{Bmatrix} 0 \\ 0 \end{Bmatrix} \tag{4.288}$$

This eigenvalue problem has nontrivial solution if the determinant is zero, that is, if

$$J_n(k_0 x_1) Y_n(k_0 x_2) - J_n(k_0 x_2) Y_n(k_0 x_1) = 0 \tag{4.289}$$

This is a transcendental equation in k_0, the roots of which can be found by numerical methods. After finding this root, we can obtain the eigenvectors for C_1, C_2, and from here, the modal shapes for the cone or frustum. The result is of the form

$$\phi(x) = (k_0 x)^n \left[J_n(k_0 x) Y_n(k_0 x_1) - J_n(k_0 x_1) Y_n(k_0 x) \right] \tag{4.290}$$

*Special Case: Even **m***

When m is an even number, then n is a nonpositive half-integer, and the ordinary Bessel functions of fractional order reduce to spherical Bessel functions of nonpositive integer order $p = n - \frac{1}{2} = -\frac{1}{2} m$. In this case,

$$z^n J_n(z) = z^{p+1} \sqrt{\frac{2}{\pi}} \, j_p(z) = -(-1)^{\frac{1}{2} m - 1} \sqrt{\frac{2}{\pi}} \frac{y_{\frac{1}{2} m}(z)}{z^{\frac{1}{2} m - 1}} \tag{4.291}$$

and

$$z^n Y_n(z) = z^{p+1} \sqrt{\frac{2}{\pi}} \, y_p(z) = (-1)^{\frac{1}{2} m - 1} \sqrt{\frac{2}{\pi}} \frac{j_{\frac{1}{2} m}(z)}{z^{\frac{1}{2} m - 1}} \tag{4.292}$$

with $z = k_0 x$. Hence, absorbing the constants in the equivalences above within the integration constants, the solution can be written as

$$u(x,t) = \frac{C_1 j_{\frac{1}{2} m}(k_0 x) + C_2 y_{\frac{1}{2} m}(k_0 x)}{x^{\frac{1}{2} m - 1}} e^{i\omega t} \tag{4.293}$$

In particular, when $m = 0$ (i.e., homogeneous rod), then $p = 0$, and the spherical Bessel functions are

$$j_0(z) = \frac{\sin z}{z} \quad \text{and} \quad y_0(z) = -\frac{\cos z}{z} \tag{4.294}$$

in which case the solution is of the form

$$u = e^{i\omega t}\left[C_1 \sin k_0 x + C_2 \cos k_0 x \right] \tag{4.295}$$

which is the classical solution for the homogeneous rod.

(c) Cones of Infinite Depth with Bounded Growth of Cross Section

We consider now a cone with bounded growth of cross section of the form

$$A(x) = A_0 + (A_\infty - A_0)(1 - e^{-\alpha x}) \tag{4.296}$$

That is,

$$\frac{A}{A_0} = 1 + \left(\frac{A_\infty}{A_0} - 1\right)(1 - e^{-\alpha x}) \tag{4.297}$$

so that

$$A' = \frac{d}{dx}A = \alpha(A_\infty - A_0)e^{-\alpha x} \tag{4.298}$$

and

$$\frac{A'}{A_0} = \alpha\left(\frac{A_\infty}{A_0} - 1\right)e^{-\alpha x} \tag{4.299}$$

Using MATLAB®'s symbolic tool (see listing that follows later), it can be show that the solution is

$$\begin{aligned}
u = {}& e^{\frac{1}{2}\alpha x(1+\beta)}\left\{ C_1 \, {}_2F_1\left(\left[\tfrac{1}{2}(1+\beta+i\kappa), \tfrac{1}{2}(1+\beta-i\kappa)\right], [1+\beta], \, e^{\alpha x}\gamma/(\gamma-1)\right)\right.\\
& + e^{\frac{1}{2}\alpha x(1-\beta)}\left\{ C_2 \, {}_2F_1\left(\left[\tfrac{1}{2}(1-\beta+i\kappa), \tfrac{1}{2}(1-\beta-i\kappa)\right], [1-\beta], \, e^{\alpha x}\gamma/(\gamma-1)\right)\right.
\end{aligned} \tag{4.300}$$

with

$$\kappa = 2k_0/\alpha, \qquad k_0 = \frac{\omega}{C_r}, \qquad \beta = (1-\kappa^2)^{\frac{1}{2}}, \qquad \gamma = \frac{A_\infty}{A_0} \tag{4.301}$$

in which $_2F_1$ is the hypergeometric function.

Interestingly, if we change $\alpha \to -\alpha$ so that

$$\frac{A}{A_0} = 1 + \left(\frac{A_\infty}{A_0} - 1\right)(1 - e^{\alpha x}), \qquad \frac{A'}{A_0} = -\alpha\left(\frac{A_\infty}{A_0} - 1\right)e^{\alpha x} \tag{4.302}$$

the differential equation changes into

$$\frac{A}{A_0}\left(u'' + k^2 u\right) - \alpha \frac{A'}{A_0} u' = 0 \tag{4.303}$$

in which case MALAB®'s solution is

$$u = e^{-ikx}\left\{C_1 \,_2F_1\left(\left[\tfrac{1}{2}(1+\beta-i\kappa), \tfrac{1}{2}(1-\beta-i\kappa)\right],[1-\kappa], e^{\alpha x}(\gamma-1)/\gamma\right)\right.$$
$$\left. + e^{+ikx}\left\{C_2 \,_2F_1\left(\left[\tfrac{1}{2}(1+\beta+i\kappa), \tfrac{1}{2}(1-\beta+i\kappa)\right],[1+\kappa], e^{\alpha x}(\gamma-1)/\gamma\right)\right. \tag{4.304}$$

with the parameters defined as earlier. Thus, it would seem that by taking α as a negative number, which makes κ a negative number, we can obtain a "nicer" solution.

```
Matlab script for cone with bounded growth
% A = A_inf - A-0
y = dsolve('(1+A*(1-exp(-a*z)))*(D2y+K^2*y)+
a*A*exp(-a*z)*Dy=0','z')
u =
C1*exp((1+(1-(2*K/a)^2)^(1/2))*a*z/2)
*hypergeom([1/2*(1+2*i*K/a + (1-(2*K/a)^2)^(1/2)),
            1/2*(1-2*i*K/a + (1-(2*K/a)^2)^(1/2))],
            [(1+(1-(2*K/a)^2)^(1/2))], exp(a*z)*(1+A)/A)
+C2*exp((1-(1-(2*K/a)^2)^(1/2))*a*z/2)
*hypergeom([1/2*(1+2*i*K/a - (1-(2*K/a)^2)^(1/2)),
            1/2*(1-2*i*K/a - (1-(2*K/a)^2)^(1/2))],
            [(1-(1-(2*K/a)^2)^(1/2))], exp(a*z)*(1+A)/A)
```

4.2.9 Simply Supported, Homogeneous, Rectangular Plate

Inasmuch as plates constitute an extension of beams into a two-dimensional space, their vibrations can be expected to exhibit some similarities to those of beams. However, they also entail further complications and complexity that makes their analysis substantially more difficult.

We begin by examining the orthogonality conditions and then move on to analyze a homogeneous and simply supported rectangular plate.

Orthogonality Conditions of General Plate

Consider a homogeneous plate of thickness h and bending stiffness $D = \frac{1}{12} Eh^3 / (1-v^2)$. As we have seen earlier, the differential equation of motion for free vibration (i.e., when the body forces are zero) is

$$\rho h \frac{\partial^2 w}{\partial t^2} + D\left(\frac{\partial^4 w}{\partial x^4} + 2\frac{\partial^4 w}{\partial x^2 \partial y^2} + \frac{\partial^4 w}{\partial y^4}\right) = 0 \qquad \text{or simply} \qquad \rho h \ddot{w} + D\nabla^4 w = 0 \tag{4.305}$$

Thus, for a vibration in any arbitrary mode of the form $w(\mathbf{x}, t) = \phi_j(x, y)\sin(\omega_j t)$, then after division by $\sin \omega_j t \neq 0$ we obtain the free vibration problem as

$$D \nabla^4 \phi_j = \rho h \, \omega_j^2 \, \phi_j \qquad (4.306)$$

subjected to appropriate boundary conditions. We recognize $\mathfrak{K} \rightarrow D\nabla^4$ and $\mathfrak{M} \rightarrow \rho h$ as the operators for a plate. We now multiply both sides by some distinct eigenmode ϕ_i and

$$D \iint\limits_A \phi_i \, \nabla^4 \phi_j \, dx \, dy = \rho h \, \omega_j^2 \iint\limits_A \phi_i \phi_j \, dx \, dy \qquad (4.307)$$

Integrating the left-hand side twice by parts and imposing whichever boundary conditions the plate may have, all boundary terms disappear and we are led to the equivalent expression

$$D \iint\limits_A \phi_i \, \nabla^4 \phi_j \, dx \, dy$$

$$= \iint \left[\frac{\partial^2 \phi_i}{\partial x^2} \frac{\partial^2 \phi_j}{\partial x^2} + \frac{\partial^2 \phi_i}{\partial y^2} \frac{\partial^2 \phi_j}{\partial y^2} + v \left(\frac{\partial^2 \phi_i}{\partial x^2} \frac{\partial^2 \phi_j}{\partial y^2} + \frac{\partial^2 \phi_j}{\partial x^2} \frac{\partial^2 \phi_i}{\partial y^2} \right) + 2(1-v) \frac{\partial^2 \phi_i}{\partial x \partial y} \frac{\partial^2 \phi_j}{\partial x \partial y} \right] D \, dx \, dy$$

$$(4.308)$$

Actually, the equivalent expression on the right-hand side is more general in that D could be a function of position. Still, this would have corresponded to a more complicated form of the differential equation on the left-hand side, the details of which are best omitted herein. From the self-adjoint property of the plate operator in combination with the boundary conditions, and following a process entirely analogous to that of the rod, we obtain the orthogonality conditions for the modes of a plate as follows:

$$\mu_j \, \delta_{ij} = \iint\limits_A \phi_i \, \phi_j \, \rho h \, dA \qquad (4.309)$$

$$\kappa_j \delta_{ij} = \iint\limits_A \left[\frac{\partial^2 \phi_i}{\partial x^2} \frac{\partial^2 \phi_j}{\partial x^2} + \frac{\partial^2 \phi_i}{\partial y^2} \frac{\partial^2 \phi_j}{\partial y^2} + v \left(\frac{\partial^2 \phi_i}{\partial x^2} \frac{\partial^2 \phi_j}{\partial y^2} + \frac{\partial^2 \phi_j}{\partial x^2} \frac{\partial^2 \phi_i}{\partial y^2} \right) + 2(1-v) \frac{\partial^2 \phi_i}{\partial x \partial y} \frac{\partial^2 \phi_j}{\partial x \partial y} \right] D \, dA$$

$$(4.310)$$

where δ_{ij} is again the Kronecker delta. Although we started from the differential equation where both D, ρh were assumed to be constant, these orthogonality conditions remain also valid when these parameters depend on the position, and whatever the boundary conditions may actually be. But be careful: the boundary conditions given in Section 4.1.6 assume homogeneous properties, and would not be appropriate if the plate parameters were a function of position. Still, for any other variation of these properties, one could, in principle, figure out what the correct expressions for the boundary conditions are by simply integrating twice by parts and examining the boundary terms that one would need to set to zero.

Simply Supported, Homogeneous Rectangular Plate

In analogy to the simply supported beam, we begin by making the simple ansatz with the substitution $\phi_j \rightarrow \phi_{mn}$:

$$\boxed{\phi_{mn}\left(x,y\right) = \sin\frac{m\pi x}{a}\sin\frac{n\pi y}{b}}, \qquad m,n = 1,2,3,\ldots \tag{4.311}$$

whose partial derivatives are

$$\frac{\partial \phi_{mn}}{\partial x} = \frac{m\pi}{a}\cos\frac{m\pi x}{a}\sin\frac{n\pi y}{b}, \qquad \frac{\partial \phi_{mn}}{\partial y} = \frac{n\pi}{b}\sin\frac{m\pi x}{a}\cos\frac{n\pi y}{b} \tag{4.312}$$

$$\frac{\partial^2 \phi_{mn}}{\partial x^2} = -\left(\frac{m\pi}{a}\right)^2 \sin\frac{m\pi x}{a}\sin\frac{n\pi y}{b}, \qquad \frac{\partial^2 \phi_{mn}}{\partial y^2} = -\left(\frac{n\pi}{b}\right)^2 \sin\frac{m\pi x}{a}\sin\frac{n\pi y}{b} \tag{4.313}$$

$$\frac{\partial^2 \phi_{mn}}{\partial x \partial y} = \left(\frac{m\pi}{a}\right)\left(\frac{n\pi}{b}\right)\cos\frac{m\pi x}{a}\cos\frac{n\pi y}{b} \tag{4.314}$$

$$\frac{\partial^4 \phi_{mn}}{\partial x^4} = \left(\frac{m\pi}{a}\right)^4 \sin\frac{m\pi x}{a}\sin\frac{n\pi y}{b}, \qquad \frac{\partial^4 \phi_{mn}}{\partial y^4} = \left(\frac{n\pi}{b}\right)^4 \sin\frac{m\pi x}{a}\sin\frac{n\pi y}{b} \tag{4.315}$$

$$\frac{\partial^4 \phi_{mn}}{\partial x^2 \partial y^2} = \left(\frac{m\pi}{a}\right)^2\left(\frac{n\pi}{b}\right)^2 \sin\frac{m\pi x}{a}\sin\frac{n\pi y}{b} \tag{4.316}$$

Now, all of ϕ_{mn}, $\partial^2 \phi_{mn}/\partial x^2$ and $\partial^2 \phi_{mn}/\partial y^2$ vanish at all four edges, so the boundary conditions of the simply supported plate are satisfied. Also, substituting the expressions into the differential equation, we find that it is satisfied at very point x,y provided that

$$D\left[\left(\frac{m\pi}{a}\right)^4 + 2\left(\frac{m\pi}{a}\right)^2\left(\frac{m\pi}{b}\right)^2 + \left(\frac{m\pi}{b}\right)^4\right] = \rho h\,\omega_{mn}^2 \tag{4.317}$$

from where we obtain immediately

$$\boxed{\omega_{mn} = \sqrt{\frac{D}{\rho h}}\left[\left(\pi\frac{m}{a}\right)^2 + \left(\pi\frac{n}{b}\right)^2\right]} = \frac{\pi^2\,C_r\,h}{\sqrt{12\left(1-v^2\right)}}\left[\left(\frac{m}{a}\right)^2 + \left(\frac{n}{b}\right)^2\right] \tag{4.318}$$

which shows that the ansatz was correct, that is, those are indeed the actual modes of the plate.

Let's now verify the orthogonality conditions for this system, replacing the indices i,j by the pair of indices m',n' and m'',n''. For two distinct modes, only one of the two indices in these pairs need be different, that is, $m' \neq m''$ or $n' \neq n''$, or both can differ. Hence

$$\iint \phi_i\,\phi_j\,\rho h\,dx\,dy \to \iint \sin\frac{m'\pi x}{a}\sin\frac{n'\pi y}{b}\sin\frac{m''\pi x}{a}\sin\frac{n''\pi y}{b}\,dx\,dy$$
$$= \left[\int \sin\frac{m'\pi x}{a}\sin\frac{m''\pi x}{a}\,dx\right]\left[\int \sin\frac{n'\pi y}{b}\sin\frac{n''\pi y}{b}\,dy\right] = 0 \tag{4.319}$$

This is so because either the left or the right term in square brackets will be zero (and perhaps even both) if at least one pair of indices does not match. Thus, the first orthogonality condition is satisfied. As for the second orthogonality condition, we observe that for each of the terms,

$$\iint \frac{\partial^2 \phi_i}{\partial x^2} \frac{\partial^2 \phi_j}{\partial x^2} \, dx \, dy = \left(\frac{m'\pi}{a}\right)^2 \left(\frac{m''\pi}{a}\right)^2 \iint \sin\frac{m'\pi x}{a} \sin\frac{n'\pi y}{b} \sin\frac{m''\pi x}{a} \sin\frac{n''\pi y}{b} \, dx \, dy = 0 \tag{4.320}$$

$$\iint \frac{\partial^2 \phi_i}{\partial y^2} \frac{\partial^2 \phi_j}{\partial y^2} \, dx \, dy = \left(\frac{n'\pi}{b}\right)^2 \left(\frac{n''\pi}{b}\right)^2 \iint \sin\frac{m'\pi x}{a} \sin\frac{n'\pi y}{b} \sin\frac{m''\pi x}{a} \sin\frac{n''\pi y}{b} \, dx \, dy = 0 \tag{4.321}$$

$$\iint \frac{\partial^2 \phi_i}{\partial x^2} \frac{\partial^2 \phi_j}{\partial y^2} \, dx \, dy = \left(\frac{m'\pi}{a}\right)^2 \left(\frac{n''\pi}{b}\right)^2 \iint \sin\frac{m'\pi x}{a} \sin\frac{n'\pi y}{b} \sin\frac{m''\pi x}{a} \sin\frac{n''\pi y}{b} \, dx \, dy = 0 \tag{4.322}$$

$$\iint \frac{\partial^2 \phi_i}{\partial x \partial y} \frac{\partial^2 \phi_y}{\partial x \partial y} \, dx \, dy = \left(\frac{m'\pi}{a}\right)\left(\frac{n'\pi}{b}\right)\left(\frac{m''\pi}{a}\right)\left(\frac{n''\pi}{b}\right) \times$$

$$\iint \cos\frac{m'\pi x}{a} \cos\frac{n'\pi y}{b} \cos\frac{m''\pi x}{a} \cos\frac{n''\pi y}{b} \, dx \, dy = 0 \tag{4.323}$$

Hence, the second orthogonality condition is satisfied too. Finally, the modal mass and modal stiffness are

$$\mu_{mn} = \iint \phi_{mn}^2 \, \rho h \, dx \, dy \to \rho h \iint \left(\sin\frac{m\pi x}{a}\right)^2 \left(\sin\frac{n\pi y}{b}\right)^2 dx \, dy = \frac{\rho h a b}{4} = \frac{m}{4} \tag{4.324}$$

$$\kappa_{mn} = D\left[\left(\frac{m\pi}{a}\right)^4 + \left(\frac{n\pi}{b}\right)^4 + 2v\left(\frac{m\pi}{a}\right)^2\left(\frac{n\pi}{b}\right)^2 + 2(1-v)\left(\frac{m\pi}{a}\right)^2\left(\frac{n\pi}{b}\right)^2\right]\frac{ab}{4}$$

$$= D\frac{ab}{4}\left[\left(\frac{m\pi}{a}\right)^2 + \left(\frac{n\pi}{b}\right)^2\right]^2 \tag{4.325}$$

Thus, the natural frequencies of the simply supported rectangular plate are

$$\omega_{mn} = \sqrt{\frac{\kappa_{mn}}{\mu_{mn}}} = \pi^2 \left[\left(\frac{m}{a}\right)^2 + \left(\frac{n}{b}\right)^2\right]\sqrt{\frac{D}{\rho h}} \tag{4.326}$$

which agrees with the previous result.

4.3 Continuous, Wave-Based Elements (Spectral Elements)

Most continuous structures (or parts of structures) in structural dynamics are not amenable to closed-form solutions, but must be tackled instead by means of numerical methods, such as the finite-element method. Still, there exist a handful of continuous structures for which it is possible to obtain their *exact* dynamic impedances, which define what could be called wave-based (or continuous) elements. Some authors refer to these as *spectral elements*, but although pertinent, that designation conflicts with the name given to other mathematical artifacts, and especially *discrete* finite elements of very high interpolation order. While conceptually similar to those, the elements herein are *exact* and make no approximations.

These elements allow providing either closed-form solutions for simple configurations or very effective solutions for structures composed of several of these continuous elements. For example, in the case of 1-dimensional structures such as rods and pure bending beams, it is possible to define the active DOF, where loads are applied and displacements observed, at the two ends of the structure and relate these via an exact impedance matrix that is a function of the frequency of excitation. Clearly, this demands to formulate the problem in the frequency domain. We provide first the impedance matrices for beams and rods, and thereafter we also demonstrate their practical use.

4.3.1 Impedance of a Finite Rod

Consider the homogeneous equation for a rod of length L that has constant Young's modulus E, mass density ρ, and cross section A:

$$\rho A \frac{\partial^2 u}{\partial t^2} - EA \frac{\partial^2 u}{\partial x^2} = 0 \tag{4.327}$$

In the frequency domain, this partial differential equation simplifies into the Helmholtz equation

$$\frac{\partial^2 u}{\partial x^2} + k^2 u = 0, \quad k = \frac{\omega}{C_r}, \quad C_r = \sqrt{\frac{E}{\rho}} \tag{4.328}$$

If C_1, C_2 are arbitrary constants, then the solution to this equation is of the form

$$u = C_1 e^{ikx} + C_2 e^{-ikx} \tag{4.329}$$

$$\frac{\partial u}{\partial x} = i k \left[C_1 e^{ikx} - C_2 e^{-ikx} \right] \tag{4.330}$$

Hence, the displacements at the two ends are

$$u_A = C_1 + C_2 \tag{4.331}$$

$$u_B = C_1 e^{ikL} + C_2 e^{-ikL} \tag{4.332}$$

or

$$\begin{Bmatrix} u_A \\ u_B \end{Bmatrix} = \begin{Bmatrix} 1 & 1 \\ e^{ikL} & e^{-ikL} \end{Bmatrix} \begin{Bmatrix} C_1 \\ C_2 \end{Bmatrix} \tag{4.333}$$

So

$$\begin{Bmatrix} C_1 \\ C_2 \end{Bmatrix} = \frac{1}{e^{-ikL} - e^{ikL}} \begin{Bmatrix} e^{-ikL} & -1 \\ -e^{ikL} & 1 \end{Bmatrix} \begin{Bmatrix} u_A \\ u_B \end{Bmatrix} \tag{4.334}$$

Similarly, the axial (normal) forces at the two ends are

$$N_A = -EA \frac{\partial u}{\partial x}\bigg|_{x=0} = -EA\, ik \left[C_1 - C_2 \right] \tag{4.335}$$

$$N_B = EA \frac{\partial u}{\partial x}\bigg|_{x=L} = i\,k\,EA\left[C_1\,e^{ikL} - C_2\,e^{-ikL}\right] \tag{4.336}$$

or

$$
\begin{aligned}
\left\{\begin{matrix} N_A \\ N_B \end{matrix}\right\} &= i\,k\,EA \begin{Bmatrix} -1 & 1 \\ e^{ikL} & -e^{-ikL} \end{Bmatrix} \left\{\begin{matrix} C_1 \\ C_2 \end{matrix}\right\} \\
&= \frac{i\,k\,EA}{e^{-ikL} - e^{ikL}} \begin{Bmatrix} -1 & 1 \\ e^{ikL} & -e^{-ikL} \end{Bmatrix} \begin{Bmatrix} e^{-ikL} & -1 \\ -e^{ikL} & 1 \end{Bmatrix} \left\{\begin{matrix} u_A \\ u_B \end{matrix}\right\} \\
&= \frac{2\,i\,k\,EA}{e^{ikL} - e^{-ikL}} \begin{Bmatrix} \tfrac{1}{2}\left(e^{ikL} + e^{-ikL}\right) & -1 \\ -1 & \tfrac{1}{2}\left(e^{ikL} + e^{-ikL}\right) \end{Bmatrix} \left\{\begin{matrix} u_A \\ u_B \end{matrix}\right\}
\end{aligned} \tag{4.337}
$$

But

$$\tfrac{1}{2}\left(e^{ikL} + e^{-ikL}\right) = \cosh\left(ikL\right) = \cos kL \tag{4.338}$$

$$\tfrac{1}{2}\left(e^{ikL} - e^{-ikL}\right) = \sinh\left(ikL\right) = i\,\sin kL \tag{4.339}$$

So

$$\left\{\begin{matrix} N_A \\ N_B \end{matrix}\right\} = \frac{EA}{L}\frac{kL}{\sin kL} \begin{Bmatrix} \cos kL & -1 \\ -1 & \cos kL \end{Bmatrix} \left\{\begin{matrix} u_A \\ u_B \end{matrix}\right\} \tag{4.340}$$

and defining $\theta = kL = \omega L/C_r$, we obtain finally

$$\left\{\begin{matrix} N_A \\ N_B \end{matrix}\right\} = \left(\frac{EA}{L}\right)\frac{\theta}{\sin\theta} \begin{Bmatrix} \cos\theta & -1 \\ -1 & \cos\theta \end{Bmatrix} \left\{\begin{matrix} u_A \\ u_B \end{matrix}\right\} = \mathbf{Z}_{rod}\mathbf{u}, \qquad \theta = \frac{\omega L}{C_r}, \qquad \mathbf{u} = \left\{\begin{matrix} u_A \\ u_B \end{matrix}\right\} \tag{4.341}$$

$$\boxed{\mathbf{Z}_{rod} = \left(\frac{EA}{L}\right)\frac{\theta}{\sin\theta} \begin{Bmatrix} \cos\theta & -1 \\ -1 & \cos\theta \end{Bmatrix}} \tag{4.342}$$

The matrix $\mathbf{Z}_{rod}$ is the symmetric impedance matrix of the finite rod.

To obtain the displacement at any arbitrary point in the rod when forces and/or displacements are applied at the two ends, it suffices to combine the impedance matrices of two rods, one of length x, the other of length $L - x$, and assemble the 3×3 impedance matrix of that combination. Equation 4.340 – which is associated with no external force – can then be used to obtain the displacement at x in terms of the displacements at the two ends. This is referred to as *analytic continuation*.

Example 1: Response of Rod Subjected to Impulsive Load

A continuous, homogeneous rod is clamped at the left end and free on the right end, as shown in Figure 4.16. The material properties, namely the mass density ρ, Young's modulus E, length L, and cross section A are known. The rod supports axial vibrations only. Find the response anywhere due to an impulsive load applied at the free end.

From the material in the preceding section, if we set $u_A = 0$ and solve for u_B, which is the same as the inverse of the impedance of the rod as seen from the free end, we infer that the compliance u_B at the point of application of a load $N_B \equiv P$ is simply

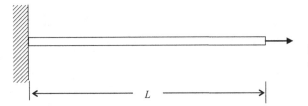

Figure 4.16. Rod subjected to unit impulse at the free end.

$$u_B = \frac{PL}{EA}\frac{\tan\theta}{\theta} = \frac{P}{k}\frac{\tan\theta}{\theta} = u_S\frac{\tan\theta}{\theta}, \qquad \theta = \frac{\omega L}{C}, \qquad C = \sqrt{\frac{E}{\rho}}, \qquad k = \frac{EA}{L}, \qquad u_s = \frac{P}{k} \tag{4.343}$$

So the frequency response function is

$$u(L,\omega) = u_s\frac{\tan\theta}{\theta}\exp(i\omega t) = u_s\, H(\omega)\exp(i\omega t) \tag{4.344}$$

with

$$H(\omega) = \frac{\tan\theta}{\theta} = \frac{1}{\cos\theta}\frac{\sin\theta}{\theta} \tag{4.345}$$

The resonant frequencies for the above system occur at the values at which $\tan\theta = \infty$, that is, at

$$\omega_j = \frac{\pi}{2}\frac{C}{L}j, \qquad j = -\infty\cdots-5,-3,-1,\,1,3,5,\cdots+\infty. \qquad \text{(odd integers!)} \tag{4.346}$$

The response at the free end of the rod due to an *impulsive* load $P\delta(t)$ applied there can then be obtained by contour integration as follows. Let, $\tau = tC/L$, then

$$\begin{aligned} u(L,t) &= \frac{u_s}{2\pi}\int_{-\infty}^{+\infty} H\exp(i\omega t)\,d\omega \\ &= \frac{u_s}{2\pi}\frac{C}{L}\int_{-\infty}^{+\infty} H\exp(i\theta\tau)\,d\theta = u_s\frac{C}{L}i\sum_{\text{odd }j=-\infty}^{\infty} R_j \end{aligned} \tag{4.347}$$

where the residues R_j are given by

$$R_j = \lim_{\theta\to\theta_j}\left[\frac{(\theta-\theta_j)}{\cos\theta}\frac{\sin(\theta)}{\theta}\exp(i\theta\tau)\right] = \lim_{\theta\to\theta_j}\left[-\frac{1}{\sin\theta_j}\frac{\sin(\theta_j)}{\theta_j}\exp(i\theta_j\,\tau)\right] = -\frac{\exp(i\theta_j\,\tau)}{\theta_j} \tag{4.348}$$

Hence

$$u(L,t) = -u_s\frac{C}{L}i\sum_{\text{odd }j=-\infty}^{\infty}\frac{\exp(i\theta_j\,\tau)}{\theta_j}, \qquad \theta_j = \frac{\omega_j L}{C} \tag{4.349}$$

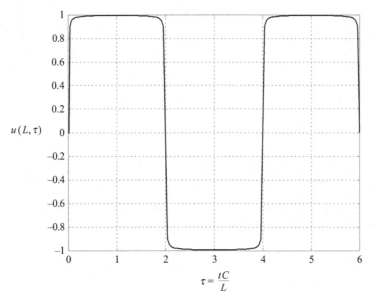

Figure 4.17. Response of rod to unit impulse.

and after trivial algebra

$$\boxed{u(L,t) = u_s \frac{C}{L} \sum_{\text{odd } j=1}^{\infty} \frac{\sin(\theta_j \tau)}{\theta_j}}$$ (4.350)

The series converges for sure because

$$\int_0^\infty \frac{\sin(\theta\tau)}{\theta} d\theta = \frac{\pi}{2}\text{sgn}(\tau) = \frac{\pi}{2}$$ (4.351)

Figure 4.17 shows the response in the interval from $0 \rightarrow \frac{3}{2}T$ (where T is the fundamental period) when using 50 integration terms $j = 1,3,5\ldots99$. The agreement with the theoretical solution, namely a square wave, is excellent.

Example 2: Natural Frequencies of Rod with Spring
A continuous, homogeneous rod is clamped at the left end and elastically supported at the right end, as shown in Figure 4.18. The material properties, namely the mass density ρ, Young's modulus E, length L, cross section A, and spring stiffness k, are known. In terms of these data, define the stiffness parameter $\kappa = kL/AE$. The rod supports axial vibrations only. In addition, the spring is massless, and has zero length.

We begin by considering the impedance of the cantilever rod, which requires setting $u_A = 0$. From the preceding expression for the impedance of the full rod, this yields immediately

$$Z_{22} = \frac{EA}{L}\frac{\theta}{\tan\theta}, \qquad \theta = \frac{\omega L}{C_r}$$ (4.352)

Hence, the total impedance at the point B where the spring is attached is simply

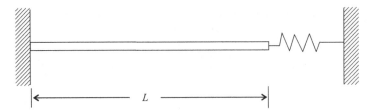

Figure 4.18. Rod coupled to an elastic spring at its free end.

$$Z = Z_{22} + k = \frac{EA}{L}\frac{\theta}{\tan\theta} + k = \frac{EA}{L}\left[\frac{\theta}{\tan\theta} + \kappa\right], \qquad \kappa = \frac{kL}{EA} \tag{4.353}$$

If we were to apply a harmonic force at this point, then resonance would be observed whenever the total impedance is zero, that is, when term in square brackets is zero. This happens when

$$\frac{\tan\theta}{\theta} = (-)\frac{1}{\kappa} \tag{4.354}$$

The preceding expression is a transcendental equation for the resonant frequencies, but it can easily be solved by graphical means, plotting the two curves represented by the left- and right-hand sides and determining their intersection points. When this is done in the particular case of $\kappa = 1$, one finds that the roots θ_j lie close to the value at which $\tan\theta = \pm\infty$, which are the points at which the rod without the spring has its own roots. Thus, except for the first two or three modes, which have a slightly higher frequency than the rod without the spring, all higher modes have natural frequencies that pretty much coincide with those of the rod.

The intersections of these two curves define the solutions. We can see that they are slightly to the right of the solution for no spring ($\kappa = 0$) at the transitions of the tangent function from plus to minus infinity (interlacing theorem!). More importantly, beyond the fourth or fifth root, there is virtually no difference of these roots with the frequencies for the free beam, that is, the added spring has almost no effect by then. Once we know where the roots lie, namely $\theta = \frac{\pi}{2}(2j-1) + \varepsilon$, where ε is a small angle such that $\tan\theta = -\cot\varepsilon$, then we can rewrite the transcendental equation as

$$\frac{\kappa}{\frac{\pi}{2}(2j-1)+\varepsilon} = \tan\varepsilon \tag{4.355}$$

for which it is almost trivial to write an algorithm to estimate its roots. For example, to a first approximation we could write

$$\frac{\kappa}{\frac{\pi}{2}(2j-1)+\varepsilon} \approx \frac{\kappa}{\frac{\pi}{2}(2j-1)} = \tan\varepsilon \approx \varepsilon + \tfrac{1}{3}\varepsilon^3 + \cdots \tag{4.356}$$

which can easily be solved for any given j. Clearly, once the modal index is sufficiently large, then

$$\varepsilon = \frac{\kappa}{\frac{\pi}{2}(2j-1)} \qquad \text{so} \qquad \boxed{\theta_j = \frac{\pi}{2}(2j-1) + \frac{\kappa}{\frac{\pi}{2}(2j-1)}} \tag{4.357}$$

For example, for $j = 2, \kappa = 1$ we get $\theta_2 = \frac{3\pi}{2} + \frac{2}{3\pi} = 4.924$, which is virtually exact.

To obtain the modes we would need to know how displacements vary within the rod. Making use of the analytic continuation technique alluded to in the previous section, it can be established that such displacements are given by

$$u(x,\omega) = u_B(\omega)\frac{\sin\left(\theta\frac{x}{L}\right)}{\sin(\theta)}, \qquad u_B = \frac{PL}{EA}\frac{\tan\theta}{\theta} \tag{4.358}$$

Evaluating this expression at the resonant frequencies θ_j of the combination with $u_B = 1$, we then obtain

$$\phi_j(x) = \frac{\sin\left(\theta_j\frac{x}{L}\right)}{\sin\theta_j} \tag{4.359}$$

with the θ_j being the roots obtained as earlier. However, since the modes are defined up to a multiplicative constant, we can simply scale them as

$$\boxed{\phi_j(x) = \sin\left(\theta_j\frac{x}{L}\right)} \tag{4.360}$$

which are identical to the modes of the rod without the spring, even if at a slightly different frequency such that θ_j is not an odd multiple of $\frac{1}{2}\pi$.

4.3.2 Impedance of a Semi-infinite Rod

Consider a semi-infinite rod that stretches from $x = -\infty$ to $x = 0$. This problem is characterized by the same equations as in the previous case, except that we must disallow waves emanating from infinity, that is, we must set $C_2 = 0$. This is because in combination with the implied time varying exponential, the term $\exp\left[i(\omega t - kx)\right]$ represents a wave propagating in the positive x-direction, which cannot exist because there are no sources at infinity. It follows that $u = C_1\,e^{ikx}$ and $N = EA\dfrac{\partial u}{\partial x} = i\,k\,EAC_1\,e^{ikx}$, so eliminating the integration constant, we obtain

$$N = i\,k\,EAu = i\,\omega\rho C_r Au \tag{4.361}$$

Thus, the impedance of a semi-infinite rod is

$$\boxed{Z(\omega) = i\,\omega\rho C_r A} \tag{4.362}$$

which is analogous to the impedance of a pure dashpot with viscosity $C = \rho C_r A$.

4.3.3 Viscoelastic Rod on a Viscous Foundation (Damped Rod)

Consider the dynamic equilibrium equation for a free rod of cross section A and rod wave velocity C_r. In addition, the rod is constituted of a viscoelastic material of viscosity α and rests on a uniformly distributed viscous foundation of viscosity β. This adds two more terms to the differential equation of an elastic rod, namely

$$\rho\frac{\partial^2 u}{\partial t^2} - E\frac{\partial^2 u}{\partial x^2} - \alpha\frac{\partial^3 u}{\partial x^2\partial t} + \beta\frac{\partial u}{\partial t} = 0 \tag{4.363}$$

To solve this equation, we make again an ansatz in the form of a propagating wave, i.e.,

$$u = U \exp\left[\,i(\omega t - kx)\right] \tag{4.364}$$

where $U = U(\omega)$ is an arbitrary constant of integration. After substitution this solution into the rod differential equation, we are led to the characteristic equation

$$-\rho\omega^2 + k^2\left(E + i\,\omega\alpha\right) + i\omega\beta = 0 \tag{4.365}$$

and solving for the axial wavenumber, we obtain

$$\frac{kC}{\omega} = \pm\sqrt{\dfrac{1 - i\dfrac{\beta}{\rho\omega}}{1 + i\dfrac{\omega\alpha}{E}}} = \pm\sqrt{\dfrac{\left(1 - i\dfrac{\beta}{\rho\omega}\right)\left(1 - i\dfrac{\omega\alpha}{E}\right)}{1 + \left(\dfrac{\omega\alpha}{E}\right)^2}} = \pm\sqrt{\dfrac{1 - \dfrac{\alpha\beta}{\rho E} - i\left(\dfrac{\omega\alpha}{E} + \dfrac{\beta}{\rho\omega}\right)}{1 + \left(\dfrac{\omega\alpha}{E}\right)^2}} \tag{4.366}$$

where

$$C = \sqrt{\frac{E}{\rho}}, \quad \frac{\omega\alpha}{E} = 2\xi\frac{\omega}{\omega_{\text{ref}}}, \quad \frac{\beta}{\rho\omega} = 2\zeta\frac{\omega_{\text{ref}}}{\omega} \tag{4.367}$$

and ω_{ref} is an arbitrary reference frequency. This can also be written as

$$\frac{kC}{\omega} = \pm\sqrt{\dfrac{1 - \dfrac{\alpha\beta}{\rho E}}{1 + \left(\dfrac{\omega\alpha}{E}\right)^2}}\sqrt{1 - i\left(\dfrac{\dfrac{\omega\alpha}{E} + \dfrac{\beta}{\rho\omega}}{1 - \dfrac{\alpha\beta}{\rho E}}\right)}, \quad \left(\text{assuming that } 1 - \frac{\alpha\beta}{\rho E} > 0!\right) \tag{4.368}$$

Define

$$Q(\omega) = \sqrt{\dfrac{1 - \dfrac{\alpha\beta}{\rho E}}{1 + \left(\dfrac{\omega\alpha}{E}\right)^2}} \quad R(\omega) = \dfrac{\dfrac{\omega\alpha}{E} + \dfrac{\beta}{\rho\omega}}{1 - \dfrac{\alpha\beta}{\rho E}} \tag{4.369}$$

with which the solution can be expressed in simple form with separated real and imaginary parts

$$\frac{kC}{\omega} = \pm Q\sqrt{1 - iR} = \pm\frac{\sqrt{2}}{2}Q\left\{\sqrt{\sqrt{1 + R^2} + 1} - i\sqrt{\sqrt{1 + R^2} - 1}\right\} \tag{4.370}$$

Choosing the root with positive sign, then the wave propagation solution in the rod is of the form

$$u = U \exp\left[i\omega\left(t - \frac{x}{C}\frac{\sqrt{2}}{2}Q\left\{\sqrt{\sqrt{1 + R^2} + 1} - i\sqrt{\sqrt{1 + R^2} - 1}\right\}\right)\right] \tag{4.371}$$

That is,

$$u = U \exp\left[i\omega\left(t - \frac{x}{C} \frac{\sqrt{2}}{2} Q\sqrt{\sqrt{1+R^2}+1} \right) \right] \exp\left[-\omega \frac{x}{C} \frac{\sqrt{2}}{2} Q\sqrt{\sqrt{1+R^2}-1} \right] \qquad (4.372)$$

which represents a wave that propagates from left to right (in the positive x direction) while it decays exponentially in that direction. The solution with the negative sign is then a wave that does the same, but in the negative x direction, i.e., from right to left. The first term gives the propagation phase, while the second factor provides the attenuation, both of which are frequency dependent.

Stress and Velocity

Consider now an infinite rod in which waves propagates from left to right. We now wish to find an expression for the mechanical impedance of that rod, that is, find the relationship between force and velocity when the rod is cut anywhere. By straightforward calculation, the stress and displacement components at an arbitrary point are

$$\sigma = E \frac{\partial u}{\partial x} = -i k\rho C^2 u, \quad \dot{u} = \frac{\partial u}{\partial t} = i\omega u \qquad (4.373)$$

so the *acoustic* impedance is $Z_V = -\sigma A / \dot{u}$ (the subscript V is for *velocity*). The negative sign arises because the internal stress acting on the left face of the rod must be balanced by an external traction acting in the negative direction, and thus must be reversed to conform to a positive external source:

$$\begin{aligned} Z_V &= (-)\frac{\sigma A}{\dot{u}} = \frac{k\rho}{\omega} AC^2 = \frac{kC}{\omega}\rho CA \\ &= \rho CA \frac{\sqrt{2}}{2} Q\left\{ \sqrt{\sqrt{1+R^2}+1} - i\sqrt{\sqrt{1+R^2}-1} \right\} \\ &= \rho CA Q\sqrt{1-iR} \end{aligned} \qquad (4.374)$$

which unlike the undamped rod is now both complex and frequency dependent. Its absolute value and phase angles are

$$|Z_V| = \rho CAQ\sqrt{1+R^2}, \quad \phi = \arg(Z_V) = -\tfrac{1}{2}\arctan R \qquad (4.375)$$

Mechanically, this corresponds to a dynamic stiffness (or displacement impedance)

$$K + i\omega D = i\omega Z_V = \rho CA \frac{\sqrt{2}}{2} Q\left\{ \omega\sqrt{\sqrt{1+R^2}-1} + i\,\omega\sqrt{\sqrt{1+R^2}+1} \right\} \qquad (4.376)$$

which can be interpreted as a frequency-dependent spring–damper system in parallel, with stiffness and dashpot

$$K = \tfrac{1}{2}\rho CA\omega Q\sqrt{2}\sqrt{\sqrt{1+R^2}-1}, \quad D = \tfrac{1}{2}\rho CAQ\sqrt{2}\sqrt{\sqrt{1+R^2}+1} \qquad (4.377)$$

That is, a combination of a spring–dashpot system K, D with the above values is fully equivalent to a semi-infinite rod, even if only at the current frequency. At high frequencies and when $\alpha = 0$, the stiffness and dashpot tend asymptotically to

$$\lim_{\omega \to \infty} K \to \tfrac{1}{2}\rho C A \sqrt{2}\; \omega \sqrt{1 + \tfrac{1}{2}\left(\frac{\beta}{\rho\omega}\right)^2 - 1} = \tfrac{1}{2}CA\beta \tag{4.378}$$

$$\lim_{\omega \to \infty} D = \tfrac{1}{2}\rho C A \sqrt{2}\sqrt{1 + \tfrac{1}{2}\left(\frac{\beta}{\rho\omega}\right)^2 + 1} = \rho C A \tag{4.379}$$

that is, to a constant spring–damper system.

Power Flow

In the preceding we have found that

$$\sigma = Z_V \dot{u} = \frac{K + i\omega D}{i\omega}\dot{u} = \left(D - i\frac{K}{\omega}\right)\dot{u} = |Z_V|\,e^{-i\phi}\dot{u} \tag{4.380}$$

where

$$\tan\phi = \frac{K}{\omega D} = \sqrt{\frac{\sqrt{1 + R^2} - 1}{\sqrt{1 + R^2} + 1}} \tag{4.381}$$

where Z_V is again the acoustic impedance, defined here as the relationship between velocity and stress. Assuming a harmonic velocity of the form $\dot{u} = V e^{i\omega t}$ and with $T = 2\pi / \omega$, the average power flow is then

$$\begin{aligned}
\langle \Pi \rangle &= \tfrac{1}{T}\int_0^T \mathrm{Re}(\sigma)\,\mathrm{Re}(\dot{u})\,dt = \tfrac{1}{T}\int_0^T \mathrm{Re}(Z_V\dot{u})\,\mathrm{Re}(\dot{u})\,dt = \tfrac{1}{T}V^2\int_0^T \mathrm{Re}(Z_V\,e^{i\omega t})\,\mathrm{Re}(e^{i\omega t})\,dt \\
&= \tfrac{1}{T}|Z_V|V^2\int_0^T \mathrm{Re}(e^{i(\omega t - \phi)})\,\mathrm{Re}(e^{i\omega t})\,dt = \tfrac{1}{T}|Z_V|V^2\int_0^T \cos(\omega t - \phi)\cos(\omega t)\,dt \\
&= \tfrac{1}{2}|Z_V|V^2 \cos\phi
\end{aligned} \tag{4.382}$$

But $|Z_V|\cos\phi = \mathrm{Re}\,Z_V = D$, so $\langle \Pi \rangle = \tfrac{1}{2}DV^2$, or in full

$$\boxed{\langle \Pi \rangle = \tfrac{1}{4}\rho C A V^2\, Q\sqrt{2}\sqrt{\sqrt{1 + R^2} + 1}} \tag{4.383}$$

When $\alpha = 0$ and in the limit of high frequencies, this tends to

$$\lim_{\omega \to \infty}\langle \Pi \rangle = \tfrac{1}{2}\rho C\,A V^2 \tag{4.384}$$

which agrees with the classical result for un undamped rod.

Example: Uniform Cantilever Rod Subjected to Harmonic Load
Consider a cantilever rod subjected to a harmonic load P applied at the free end. The full solution now consists of waves propagating in both directions; i.e., it is of the form

$$u = C_1 \exp\left[i x \tfrac{\omega}{C} Q\sqrt{1 - iR}\right] + C_2 \exp\left[-i x \tfrac{\omega}{C} Q\sqrt{1 - iR}\right] \tag{4.385}$$

where C_1, C_2 are constants of integration to be determined by imposing the two boundary conditions of this problem, namely the vanishing of the displacements at the support, and the total axial force at the free end equals the load. From the first boundary conditions, we obtain

$$u\big|_{x=0} = 0 \text{ so } C_1 + C_2 = 0 \text{ or } C_2 = -C_1 \tag{4.386}$$

Also, the second boundary condition is

$$EA \tfrac{\partial}{\partial x} u\big|_{x=L} = P \tag{4.387}$$

which in full reads as

$$\mathrm{i}\tfrac{\omega}{C} Q\sqrt{1-\mathrm{i}R} \Big[C_1 \exp\big[\mathrm{i}\tfrac{\omega}{C} L Q\sqrt{1-\mathrm{i}R}\big] - C_2 \exp\big[-\mathrm{i}\tfrac{\omega}{C} L Q\sqrt{1-\mathrm{i}R}\big]\Big] EA = P \tag{4.388}$$

Hence

$$C_1 = \frac{P}{EA} \frac{1}{\mathrm{i}\tfrac{\omega}{C} Q\sqrt{1-\mathrm{i}R}\Big[\exp\big[\mathrm{i}\tfrac{\omega}{C} L Q\sqrt{1-\mathrm{i}R}\big] + \exp\big[-\mathrm{i}\tfrac{\omega}{C} L Q\sqrt{1-\mathrm{i}R}\big]\Big]} \tag{4.389}$$

Thus, the solution anywhere is

$$u(x,\omega) = \frac{PL}{EA} \frac{1}{\mathrm{i}\tfrac{\omega L}{C} Q\sqrt{1-\mathrm{i}R}} \frac{\exp\big[\mathrm{i}\tfrac{\omega}{C} x Q\sqrt{1-\mathrm{i}R}\big] - \exp\big[-\mathrm{i}\tfrac{\omega}{C} x Q\sqrt{1-\mathrm{i}R}\big]}{\exp\big[\mathrm{i}\tfrac{\omega}{C} L Q\sqrt{1-\mathrm{i}R}\big] + \exp\big[-\mathrm{i}\tfrac{\omega}{C} L Q\sqrt{1-\mathrm{i}R}\big]} \tag{4.390}$$

In particular, at the free end where the load is applied

$$\begin{aligned} u_L = u(L,\omega) &= \frac{PL}{EA} \frac{1}{\mathrm{i}\tfrac{\omega L}{C} Q\sqrt{1-\mathrm{i}R}} \frac{\exp\big[\mathrm{i}\tfrac{\omega L}{C} Q\sqrt{1-\mathrm{i}R}\big] - \exp\big[-\mathrm{i}\tfrac{\omega L}{C} Q\sqrt{1-\mathrm{i}R}\big]}{\exp\big[\mathrm{i}\tfrac{\omega L}{C} Q\sqrt{1-\mathrm{i}R}\big] + \exp\big[-\mathrm{i}\tfrac{\omega L}{C} Q\sqrt{1-\mathrm{i}R}\big]} \\ &= \frac{PL}{EA} \frac{1}{\mathrm{i}\tfrac{\omega L}{C} Q\sqrt{1-\mathrm{i}R}} \frac{1 - \exp\big[-2\mathrm{i}\tfrac{\omega L}{C} Q\sqrt{1-\mathrm{i}R}\big]}{1 + \exp\big[-2\mathrm{i}\tfrac{\omega L}{C} Q\sqrt{1-\mathrm{i}R}\big]} \end{aligned} \tag{4.391}$$

But

$$2\mathrm{i} Q\sqrt{1-\mathrm{i}R} = \sqrt{2} Q\left\{\sqrt{\sqrt{1+R^2}-1} + \mathrm{i}\sqrt{\sqrt{1+R^2}+1}\right\} \tag{4.392}$$

So

$$\begin{aligned} \exp\big(-2\mathrm{i}\tfrac{\omega L}{C} Q\sqrt{1-\mathrm{i}R}\big) &= \exp\left[-\tfrac{\omega L}{C} Q\left(\sqrt{2}\sqrt{\sqrt{1+R^2}-1} + \mathrm{i}\sqrt{2}\sqrt{\sqrt{1+R^2}+1}\right)\right] \\ &= \exp\left[\left(-\tfrac{\omega L}{C} Q\sqrt{2\left(\sqrt{1+R^2}-1\right)}\right)\right]\exp\left[-\mathrm{i}\tfrac{\omega L}{C} Q\left(\sqrt{2\left(\sqrt{1+R^2}+1\right)}\right)\right] \end{aligned} \tag{4.393}$$

Hence

$$u_L = \frac{PL}{EA} \frac{1}{\mathrm{i}\tfrac{\omega L}{C} Q\sqrt{1-\mathrm{i}R}} \frac{1 - \exp\left(-\tfrac{\omega L}{C} Q\sqrt{2\left(\sqrt{1+R^2}-1\right)}\right)\exp\left(-\mathrm{i}\tfrac{\omega L}{C} Q\sqrt{2\left(\sqrt{1+R^2}+1\right)}\right)}{1 + \exp\left(-\tfrac{\omega L}{C} Q\sqrt{2\left(\sqrt{1+R^2}-1\right)}\right)\exp\left(-\mathrm{i}\tfrac{\omega L}{C} Q\sqrt{2\left(\sqrt{1+R^2}+1\right)}\right)} \tag{4.394}$$

The absolute value is then

$$
|u_L| = \frac{PL}{EA} \frac{1}{\frac{\omega L}{C} Q\sqrt{1+R^2}} \left| \frac{1 - \exp\left(-\frac{\omega L}{C} Q\sqrt{2\left(\sqrt{1+R^2}-1\right)}\right) \exp\left(-i\frac{\omega L}{C} Q\sqrt{2\left(\sqrt{1+R^2}+1\right)}\right)}{1 + \exp\left(-\frac{\omega L}{C} Q\sqrt{2\left(\sqrt{1+R^2}-1\right)}\right) \exp\left(-i\frac{\omega L}{C} Q\sqrt{2\left(\sqrt{1+R^2}+1\right)}\right)} \right| \quad (4.395)
$$

Here

$$
\frac{1}{Q\sqrt{1+R^2}} = \sqrt{\frac{1+\left(\frac{\omega\alpha}{E}\right)^2}{\left(1-\frac{\alpha\beta}{\rho E}\right)^2 + \left(\frac{\omega\alpha}{E}+\frac{\beta}{\rho\omega}\right)^2}} = \sqrt{\frac{1+\left(\frac{\omega\alpha}{E}\right)^2}{1+\left(\frac{\omega\alpha}{E}\right)^2 + \left(\frac{\alpha\beta}{\rho E}\right)^2 + \left(\frac{\beta}{\rho\omega}\right)^2}} < 1 \quad (4.396)
$$

Also, with the shorthand

$$
\gamma_1 = \tfrac{\omega L}{C} Q\sqrt{2\left(\sqrt{1+R^2}-1\right)}, \quad \gamma_2 = \tfrac{\omega L}{C} Q\sqrt{2\left(\sqrt{1+R^2}+1\right)} \quad (4.397)
$$

Then the second factor above is

$$
\left| \frac{1-\exp(-\gamma_1)\exp(-i\gamma_2)}{1+\exp(-\gamma_1)\exp(-i\gamma_2)} \right| = \left| \frac{1-\exp(-\gamma_1)(\cos\gamma_2 - i\sin\gamma_2)}{1+\exp(-\gamma_1)(\cos\gamma_2 - i\sin\gamma_2)} \right|
$$

$$
= \sqrt{\frac{(1-\exp(-\gamma_1)\cos\gamma_2)^2 + [\exp(-\gamma_1)\sin\gamma_2]^2}{(1+\exp(-\gamma_1)\cos\gamma_2)^2 + [\exp(-\gamma_1)\sin\gamma_2]^2}} = \sqrt{\frac{1+\exp(-2\gamma_1) - 2\exp(-\gamma_1)\cos\gamma_2}{1+\exp(-2\gamma_1) + 2\exp(-\gamma_1)\cos\gamma_2}}
$$

$$
= \sqrt{\frac{[1-\exp(-\gamma_1)]^2 + 2\exp(-\gamma_1)(1-\cos\gamma_2)}{[1-\exp(-\gamma_1)]^2 + 2\exp(-\gamma_1)(1+\cos\gamma_2)}} = \sqrt{\frac{[1-\exp(-\gamma_1)]^2 + 4\exp(-\gamma_1)\sin^2\tfrac{1}{2}\gamma_2}{[1-\exp(-\gamma_1)]^2 + 4\exp(-\gamma_1)\cos^2\tfrac{1}{2}\gamma_2}}
$$

$$(4.398)$$

Hence

$$
|u_L| = \frac{PL}{EA} \frac{1}{\left(\frac{\omega L}{C}\right)} \sqrt{\frac{1+\left(\frac{\omega\alpha}{E}\right)^2}{1+\left(\frac{\omega\alpha}{E}\right)^2 + \left(\frac{\alpha\beta}{\rho E}\right)^2 + \left(\frac{\beta}{\rho\omega}\right)^2}} \sqrt{\frac{[1-\exp(-\gamma_1)]^2 + 4\exp(-\gamma_1)\sin^2\tfrac{1}{2}\gamma_2}{[1-\exp(-\gamma_1)]^2 + 4\exp(-\gamma_1)\cos^2\tfrac{1}{2}\gamma_2}} \quad (4.399)
$$

For zero damping, $Q=1$, $R=0$, in which case $\gamma_1 = 0$, $\gamma_2 = 2\frac{\omega L}{C}$ and the elastic response is

$$
|u_L| = \frac{PL}{EA\left(\frac{\omega L}{C}\right)} \left| \tan\tfrac{1}{2}\gamma_2 \right| = \frac{PL}{EA} \frac{\tan\left(\frac{\omega L}{C}\right)}{\left(\frac{\omega L}{C}\right)} \quad (4.400)
$$

which agrees perfectly with the known solution for a cantilever rod.

Let's now compare the visco-elastic solution to the elastic one. For this purpose, we define

$$
\frac{\omega\alpha}{E} = 2\xi\frac{\omega}{\omega_{\text{ref}}} \qquad \frac{\beta}{\rho\omega} = 2\zeta\frac{\omega_{\text{ref}}}{\omega} \quad (4.401)
$$

in which ω_{ref} is an arbitrary reference frequency, and ξ, ζ are arbitrary fractions of critical damping defined in terms of the reference frequency. In this particular example, it is best to choose $\omega_{\text{ref}} = \omega_1 = \frac{\pi}{2} C / L$, namely the fundamental frequency of the cantilever rod. In that case

$$\frac{\omega\alpha}{E} = 2\xi\frac{\omega}{\omega_{\text{ref}}} = 2\xi\vartheta, \quad \frac{\beta}{\rho\omega} = 2\zeta\frac{\omega_{\text{ref}}}{\omega} = 2\zeta / \vartheta \tag{4.402}$$

$$\vartheta = \frac{\omega}{\omega_1}, \quad \frac{\omega L}{C} = \frac{\omega_1 L}{C}\frac{\omega}{\omega_1} = \frac{\pi}{2}\vartheta \tag{4.403}$$

Clearly, the term in ξ is analogous to stiffness-proportional damping while the term in ζ is analogous to mass-proportional damping. Then

$$\sqrt{\frac{1+\left(\frac{\omega\alpha}{E}\right)^2}{1+\left(\frac{\omega\alpha}{E}\right)^2+\left(\frac{\alpha\beta}{\rho E}\right)^2+\left(\frac{\beta}{\rho\omega}\right)^2}} = \sqrt{\frac{1+4\xi^2\vartheta^2}{1+4\xi^2\vartheta^2+4\zeta^2/\vartheta^2+16\xi^2\zeta^2}} \tag{4.404}$$

At zero frequency and for $\zeta \neq 0$, this expression is zero. For $\xi \neq 0, \zeta = 0$, this expression equals 1 at any frequency. At the opposite extreme, for very large frequency and whatever the values of ξ, ζ, it tends to 1. Thus, we see that the above expression lies in the bracket $0 \leq \sqrt{} \leq 1$. The auxiliary parameters, on the other hand, are

$$Q(\omega) = \sqrt{\frac{1-4\xi\zeta}{1+4\xi^2\vartheta^2}}, \quad R(\omega) = 2\frac{\xi\vartheta+\zeta/\vartheta}{1-4\xi\zeta} \tag{4.405}$$

$$R^2(\omega) = \left[\frac{2\xi\vartheta+2\zeta/\vartheta}{1-4\xi\zeta}\right]^2 = 4\frac{\xi^2\vartheta^2+\zeta^2/\vartheta^2+2\xi\zeta}{(1-4\xi\zeta)^2} \tag{4.406}$$

$$\sqrt{1+R^2} = \frac{1}{1-4\xi\zeta}\sqrt{1+16(\xi\zeta)^2+4\xi^2\vartheta^2+4\zeta^2/\vartheta^2} \tag{4.407}$$

$$\gamma_1 = \frac{\pi}{2}\vartheta\sqrt{2\frac{\sqrt{1+16(\xi\zeta)^2+4\xi^2\vartheta^2+4\zeta^2/\vartheta^2}-1+4\xi\zeta}{1+4\xi^2\vartheta^2}} \tag{4.408}$$

$$\gamma_2 = \frac{\pi}{2}\vartheta\sqrt{2\frac{\sqrt{1+16(\xi\zeta)^2+4\xi^2\vartheta^2+4\zeta^2/\vartheta^2}+1-4\xi\zeta}{1+4\xi^2\vartheta^2}} \tag{4.409}$$

So

$$|u_L| = \frac{PL}{EA}\frac{1}{\frac{\pi}{2}\vartheta}\sqrt{\frac{1+4\xi^2\vartheta^2}{1+4\xi^2\vartheta^2+4\zeta^2/\vartheta^2+16\xi^2\zeta^2}}\sqrt{\frac{[1-\exp(-\gamma_1)]^2+4\exp(-\gamma_1)\sin^2\frac{1}{2}\gamma_2}{[1-\exp(-\gamma_1)]^2+4\exp(-\gamma_1)\cos^2\frac{1}{2}\gamma_2}} \tag{4.410}$$

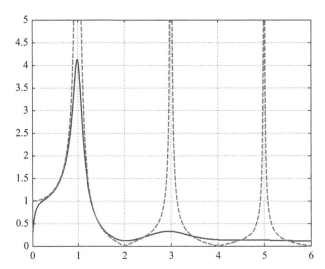

Figure 4.19. Frequency response function for cantilever rod, $\xi = \zeta = 0.05$ as function of frequency ratio ω / ω_1. Dashed line is the undamped solution.

Again, for $\xi = \zeta = 0$, $\gamma_1 = 0$, $\frac{1}{2}\gamma_2 = \frac{\pi}{2}\vartheta$ the above reduces to

$$|u_L| = \frac{PL}{EA}\frac{\tan\left(\frac{\pi}{2}\vartheta\right)}{\left(\frac{\pi}{2}\vartheta\right)}, \qquad \text{which is the correct solution.} \tag{4.411}$$

The damped amplification function is then of the form $|u_L| = u_0 |H(\vartheta, \xi, \zeta)|$, where $u_0 = PL / EA$.

We now consider the particular case $\xi = \zeta = 0.05$, and vary ϑ in the range $0 \le \vartheta \le 6$, which covers the range of three natural frequencies (1, 2, 5). The results are shown in Figure 4.19. Observe that in the region of amplification, the solid curve is underneath the dotted one, while in regions of deamplification, it is the other way around. This is the expected behavior.

4.3.4 Impedance of a Euler Beam

Consider the equation of motion of a Euler beam with constant properties:

$$EI\frac{\partial^4 v}{\partial x^4} + \rho A\frac{\partial^2 v}{\partial t^2} = 0, \qquad -\tfrac{1}{2}L < x < \tfrac{1}{2}L \tag{4.412}$$

Expressed in the frequency domain, this is

$$\frac{\partial^4 v}{\partial x^4} - \omega^2\frac{\rho}{E}\frac{A}{I}v = 0 \qquad \text{or} \qquad \boxed{\frac{\partial^4 v}{\partial x^4} - k^4 v = 0} \tag{4.413}$$

where

$$k^4 = \omega^2\frac{\rho}{E}\frac{A}{I} = \left(\frac{\omega}{C_r R}\right)^2, \qquad C_r = \sqrt{\frac{E}{\rho}}, \qquad R = \sqrt{\frac{I}{A}} \tag{4.414}$$

A general solution can be found of the form

$$v = C_1 \cos kx + C_2 \sin kx + C_3 \cosh kx + C_4 \sinh kx \tag{4.415}$$

$$\theta = \frac{\partial v}{\partial x} = k\left[-C_1 \sin kx + C_2 \cos kx + C_3 \sinh kx + C_4 \cosh kx\right] \tag{4.416}$$

$$M = EI \frac{\partial^2 v}{\partial x^2} = EI\,k^2\left[-C_1 \cos kx - C_2 \sin kx + C_3 \cosh kx + C_4 \sinh kx\right] \tag{4.417}$$

$$S = EI \frac{\partial^3 v}{\partial x^3} = EI\,k^3\left[C_1 \sin kx - C_2 \cos kx + C_3 \sinh kx + C_4 \cosh kx\right] \tag{4.418}$$

The bending moment given above is positive when the upper fibers are in compression, while the shear is positive up when acting from left to right and down when acting from right to left. Hence, at $x = 0, L$ and with

$$\eta = kL = L\sqrt{\frac{|\omega|}{C_r R}} = \sqrt{\frac{|\omega| L}{C_r}}\sqrt{\frac{L}{R}} \tag{4.419}$$

we obtain at the two ends A and B, with $M_A = -M\big|_{x=0}$ and $V_B = -S\big|_{x=L}$ ($\rightarrow$ reactions + up, moment + when counterclockwise)

$$V_A = EI\,k^3\left(-C_2 + C_4\right) \tag{4.420}$$

$$M_A = EI\,k^2\left(C_1 - C_3\right) \tag{4.421}$$

$$V_B = EI\,k^3\left(-C_1 \sin \eta + C_2 \cos \eta - C_3 \sinh \eta - C_4 \cosh \eta\right) \tag{4.422}$$

$$M_B = EI\,k^2\left(-C_1 \cos \eta - C_2 \sin \eta + C_3 \cosh \eta + C_4 \sinh \eta\right) \tag{4.423}$$

Also

$$u_A = C_1 + C_3 \tag{4.424}$$

$$\theta_A = k\left(C_2 + C_4\right) \tag{4.425}$$

$$u_B = C_1 \cos \eta + C_2 \sin \eta + C_3 \cosh \eta + C_4 \sinh \eta \tag{4.426}$$

$$\theta_B = k\left(-C_1 \sin \eta + C_2 \cos \eta + C_3 \sinh \eta + C_4 \cosh \eta\right) \tag{4.427}$$

It follows that

$$\begin{Bmatrix} v_A \\ \theta_A / k \\ v_B \\ \theta_B / k \end{Bmatrix} = \begin{Bmatrix} 1 & 0 & 1 & 0 \\ 0 & 1 & 0 & 1 \\ \cos \eta & \sin \eta & \cosh \eta & \sinh \eta \\ -\sin \eta & \cos \eta & \sinh \eta & \cosh \eta \end{Bmatrix} \begin{Bmatrix} C_1 \\ C_2 \\ C_3 \\ C_4 \end{Bmatrix} = \mathbf{A}\mathbf{c} \tag{4.428}$$

and

$$\begin{Bmatrix} V_A/k \\ M_A \\ V_B/k \\ M_B \end{Bmatrix} = k^2 EI \begin{bmatrix} 0 & -1 & 0 & 1 \\ 1 & 0 & -1 & 0 \\ -\sin\eta & \cos\eta & -\sinh\eta & -\cosh\eta \\ -\cos\eta & -\sin\eta & \cosh\eta & \sinh\eta \end{bmatrix} \begin{Bmatrix} C_1 \\ C_2 \\ C_3 \\ C_4 \end{Bmatrix} = \mathbf{Bc} \qquad (4.429)$$

Solving for the integration constants, we obtain

$$\mathbf{c} = \mathbf{A}^{-1}\tilde{\mathbf{u}} \qquad (4.430)$$

$$\tilde{\mathbf{p}} = \mathbf{BA}^{-1}\tilde{\mathbf{u}} \qquad (4.431)$$

Substituting the constants in the expression for the forces, we obtain (using MATLAB®)

$$\begin{Bmatrix} V_A/k \\ M_A \\ V_B/k \\ M_B \end{Bmatrix} = \frac{k^2 EI}{1-cC} \begin{bmatrix} sC+cS & sS & -(s+S) & C-c \\ sS & sC-cS & -(C-c) & S-s \\ -(s+S) & -(C-c) & sC+cS & -sS \\ C-c & S-s & -sS & sC-cS \end{bmatrix} \begin{Bmatrix} v_A \\ \theta_A/k \\ v_B \\ \theta_B/k \end{Bmatrix} = \tilde{\mathbf{Z}}\tilde{\mathbf{u}} \qquad (4.432)$$

with elements

$$s = \sin\eta = \sin kL \qquad c = \cos\eta = \cos kL \qquad (4.433)$$

$$S = \sinh\eta = \sinh kL \qquad C = \cosh\eta = \cosh kL \qquad (4.434)$$

As can be seen, the resulting matrix is symmetric. After brief transformations, we obtain

$$\mathbf{p} = \begin{Bmatrix} V_A \\ M_A \\ V_B \\ M_B \end{Bmatrix} = \frac{k\,EI}{1-cC} \begin{bmatrix} k^2(sC+cS) & ksS & -k^2(s+S) & k(C-c) \\ ksS & sC-cS & -k(C-c) & S-s \\ -k^2(s+S) & -k(C-c) & k^2(sC+cS) & -ksS \\ k(C-c) & S-s & -ksS & sC-cS \end{bmatrix} \begin{Bmatrix} v_A \\ \theta_A \\ v_B \\ \theta_B \end{Bmatrix} = \mathbf{Z}_{beam}\mathbf{u} \qquad (4.435)$$

where $\mathbf{Z}_{beam}$ is the symmetric impedance (or rigidity) matrix of the Euler beam. In combination with the results for the rod, we could also write down the impedance matrix for the beam column.

Example 1: Modes of Bending Beam with Arbitrary Boundary Conditions
A simply supported beam is obtained by setting $M_A = M_B = 0, u_A = u_B = 0$, which implies setting to zero the determinant of the submatrix $\Delta = \det[Z_{22}, Z_{24}; Z_{42}, Z_{44}] = 0$. It can be shown that this leads to the characteristic equation

$$\Delta = \frac{2\eta^2 \sinh\eta}{1-\cos\eta\cosh\eta}\sin\eta = 0 \qquad (4.436)$$

This equation admits nontrivial solutions only when $\sin\eta = 0$, that is, $\omega_j = j\pi C_r/L$, which is the correct solution. On the other hand, a cantilever beam with $u_A = \theta_A = 0$ leads to

$$\Delta = \det\left(Z_{33}, Z_{34}; Z_{43}, Z_{44}\right) = \eta^4 \frac{1 + \cos\eta\cosh\eta}{1 - \cos\eta\cosh\eta} = 0 \tag{4.437}$$

implying the nontrivial solution $1 + \cos\eta\cosh\eta = 0$, which is again the correct transcendental equation for the frequencies of a cantilever beam.

An apparent difficulty seems to appear in the case of a free–free beam, which requires the vanishing of the determinant of the complete rigidity matrix. This condition can be shown to be

$$\Delta = \eta^8 \left[\frac{1 - \cos\eta\cosh\eta}{1 - \cos\eta\cosh\eta}\right]^4 = \eta^8 [1] \neq 0 \tag{4.438}$$

which appears to contradict the true solution, but the ratio in square brackets is indeterminate when $1 - \cos\eta\cosh\eta = 0$, which happens to be the characteristic equation of the free–free beam. This anomaly occurs because when η satisfies one of these roots, the rigidity matrix is both singular *and* its elements are infinitely large, which allows the determinant to be nonzero (here $\Delta = \eta^8 \neq 0$). That a matrix can be singular without its determinant being zero is certainly a very surprising result.

Example 2: Beam with Single Pinned Support

Consider a beam whose left support is pinned, but is otherwise free. Clearly, this beam has a zero-frequency rigid body mode that consists in a rotation about the left support. It is of the form

$$\omega_0 = 0, \qquad \phi_0 = x/L \tag{4.439}$$

We now seek the remaining modes. This requires setting $v_A = 0$ in the impedance matrix, that is, the eigenvalue problem is $\Delta = \det\left[Z_{22}, Z_{23}, Z_{24}; Z_{32}, Z_{33}, Z_{34}; Z_{42}, Z_{43}, Z_{44}\right] = 0$, which after some algebra and simplifications works out to be

$$\tan kL = \tanh kL \tag{4.440}$$

Seeking the intersection of the left- and right-hand sides of this transcendental equation, we observe that – except for the rigid body mode at $kL = 0$ – the roots virtually coincide with the locations at which $\tanh kL \approx 1$, that is, $\tan k_j L \approx 1$ or

$$k_j L = \tfrac{\pi}{4}(4j+1), \qquad \text{i.e.,} \qquad \boxed{\omega_j = \left(\frac{\tfrac{\pi}{4}(4j+1)}{L}\right)^2 \sqrt{\frac{EI}{\rho A}}} \tag{4.441}$$

The modal shape is obtained by substituting the eigenvalue found into the characteristic equation while choosing arbitrarily one of the constants to be 1 (the modes are defined only up to a multiplicative constant). This requires solving a 2×2 system of equations in the two remaining unknowns, and using these in the expression for the displacement $v(x)$. The final result is

$$\phi_j(x) = \sin\left[\tfrac{\pi}{4}(4j+1)\tfrac{x}{L}\right] + \frac{\sin\left[\tfrac{\pi}{4}(4j+1)\right]}{\sinh\left[\tfrac{\pi}{4}(4j+1)\right]} \sinh\left[\tfrac{\pi}{4}(4j+1)\tfrac{x}{L}\right] \tag{4.442}$$

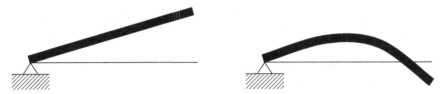

Figure 4.20. Rigid-body mode and first mode of pinned-free beam

and in particular

$$\omega_1 = \left(\frac{5\pi}{4L}\right)^2 \sqrt{\frac{EI}{\rho A}} \qquad \phi_1 = \sin\frac{5\pi x}{4L} - \frac{\sqrt{2}}{2}\frac{\sinh\frac{5\pi x}{4L}}{\sinh\frac{5\pi}{4}} \tag{4.443}$$

Figure 4.20 shows the rigid body mode and the first mode.

4.3.5 Impedance of a Semi-infinite Beam

Consider a semi-infinite Euler beam that stretches from $x = -\infty$ to $x = 0$, at which point external harmonic loads (i.e., a transverse force and a moment) are applied, and displacements are observed. We could start again with the solution for the finite beam in terms of trigonometric functions, but we prefer instead an equivalent solution with exponentials of the form

$$v = C_1 e^{ikx} + C_2 e^{-ikx} + C_3 e^{kx} + C_4 e^{-kx} \tag{4.444}$$

In combination with an implied harmonic exponential factor $e^{i\omega t}$, the first term represents waves that travel in the negative direction (from the origin towards negative infinity), while the second is one that travels from left to right and thus must be rejected because there are no sources to the left. Hence, $C_2 = 0$. Similarly, the third term decays toward the left (negative x), while the last one grows in that direction and must also be rejected, so $C_4 = 0$. Hence, this leads us to

$$v = C_1 e^{ikx} + C_3 e^{kx} \tag{4.445}$$

$$\theta = \frac{\partial v}{\partial x} = k\left[iC_1 e^{ikx} + C_3 e^{kx}\right] \tag{4.446}$$

$$M = EI\frac{\partial^2 v}{\partial x^2} = EI k^2\left[-C_1 e^{ikx} + C_3 e^{kx}\right] \tag{4.447}$$

$$V = -EI\frac{\partial^3 v}{\partial x^3} = EI k^3\left[iC_1 e^{ikx} - C_3 e^{kx}\right] \tag{4.448}$$

or in matrix form (observe that $V = -S$, that is, the negative of the shear)

$$\begin{Bmatrix} v \\ \theta/k \end{Bmatrix} = \begin{bmatrix} 1 & 1 \\ i & 1 \end{bmatrix}\begin{Bmatrix} C_1 e^{ikx} \\ C_3 e^{kx} \end{Bmatrix}, \qquad \begin{Bmatrix} V/k \\ M \end{Bmatrix} = k^2 EI\begin{bmatrix} i & -1 \\ -1 & 1 \end{bmatrix}\begin{Bmatrix} C_1 e^{ikx} \\ C_3 e^{kx} \end{Bmatrix} \tag{4.449}$$

Solving for the exponentials times the constants from the first set and substituting these into the second set, we obtain after brief algebra

$$\begin{Bmatrix} V \\ M \end{Bmatrix} = i\,k\,EI \begin{Bmatrix} k^2\,(1+i) & -k \\ -k & 1-i \end{Bmatrix} \begin{Bmatrix} v \\ \theta \end{Bmatrix} \tag{4.450}$$

The impedance of the left-semi-infinite beam, as perceived from $x = 0$, is then

$$\boxed{\mathbf{Z} = i\,k\,EI \begin{Bmatrix} k^2\,(1+i) & -k \\ -k & 1-i \end{Bmatrix}}, \qquad \boxed{k = \sqrt{\frac{|\omega|}{C_r R}}} \tag{4.451}$$

A right-semi-infinite beam would have reversed off-diagonal terms, and a fully infinite beam would be obtained by superposition of a left and right semi-infinite beam, which would give

$$\mathbf{Z} = 2\,i\,k\,EI \begin{Bmatrix} k^2\,(1+i) & 0 \\ 0 & 1-i \end{Bmatrix} \tag{4.452}$$

4.3.6 Infinite Euler Beam with Springs at Regular Intervals

Proceed next to add springs of rigidity $r = \tfrac{1}{2}\chi\,EI/L^3$ to each end of a finite beam and then connect a series of identical such beams, after which there will exist a spring of stiffness $r = \chi\,EI/L^3$ under each node. This will lead to an infinite system that accepts solutions of the form

$$\mathbf{u}_j = \boldsymbol{\phi}\,e^{i(\omega t - K x_j)} = \boldsymbol{\phi}\,e^{i(\omega t - jKL)} = \boldsymbol{\phi}\,e^{i\omega t}e^{-ij\xi} = \boldsymbol{\phi}\,e^{i\omega t}\left(e^{-i\xi}\right)^j, \qquad j = \cdots, -2, -1, 0, 1, 2, 3, \cdots \tag{4.453}$$

in which $\boldsymbol{\phi}$ is a shape vector, namely the mode of wave propagation, K is the effective wavenumber, the integer index j identifies the nodes underneath which the springs are attached, and

$$\boxed{\xi = KL} \tag{4.454}$$

is the dimensionless effective wavenumber. The characteristic equation for each node is then

$$\frac{EI}{L^3}\left(\frac{1}{1-\cos\eta\cosh\eta}\right)\left(\mathbf{K}_1\mathbf{u}_{j-1} + \mathbf{K}_2\mathbf{u}_j + \mathbf{K}_1^T\mathbf{u}_{j+1}\right) = 0 \tag{4.455}$$

in which the leading factor is never zero when $\eta > 0$, so it can be ignored. However, that factor could be singular, so one has to verify that the zeros of the denominator do not coincide with the roots of the last term in parentheses. Also,

$$\mathbf{K}_1 = \begin{Bmatrix} -\eta^3\,(\sin\eta+\sinh\eta) & -\eta^2\,(\cosh\eta-\cos\eta) \\ \eta^2\,(\cosh\eta-\cos\eta) & \eta(\sinh\eta-\sin\eta) \end{Bmatrix} \tag{4.456}$$

$$\mathbf{K}_2 = \begin{Bmatrix} 2\eta^3\,(\sin\eta\cosh\eta+\cos\eta\sinh\eta)+\chi(1-\cos\eta\cosh\eta) & 0 \\ 0 & 2\eta(\sin\eta\cosh\eta-\cos\eta\sinh\eta) \end{Bmatrix} \tag{4.457}$$

Defining

$$\boxed{\lambda = e^{i\xi}} \tag{4.458}$$

then the characteristic equation for free waves $\mathbf{u}_j = \boldsymbol{\phi}\, e^{i(\omega t - j\xi)}$ is $\left(\lambda \mathbf{K}_1 + \mathbf{K}_2 + \lambda^{-1}\mathbf{K}_1^T\right)\boldsymbol{\phi} = \mathbf{0}$, or

$$\left(\lambda^2\, \mathbf{K}_1 + \lambda\, \mathbf{K}_2 + \mathbf{K}_1^T\right)\boldsymbol{\phi} = \mathbf{0} \tag{4.459}$$

The nontrivial solutions of this equation are obtained by setting its determinant to zero, that is,

$$\Delta = \det \frac{1}{1-\cos\eta\cosh\eta}\left[\begin{cases} -\eta^3\left(\sin\eta+\sinh\eta\right)\left(\lambda^2+1\right) & -\eta^2\left(\cosh\eta-\cos\eta\right)\left(\lambda^2-1\right) \\ \eta^2\left(\cosh\eta-\cos\eta\right)\left(\lambda^2-1\right) & \eta(\sinh\eta-\sin\eta)\left(\lambda^2+1\right) \end{cases}\right.$$
$$\left. +\lambda\begin{cases} 2\eta^3\left(\sin\eta\cosh\eta+\cos\eta\sinh\eta\right)+\chi(1-\cos\eta\cosh\eta) & 0 \\ 0 & 2\eta(\sin\eta\cosh\eta-\cos\eta\sinh\eta) \end{cases}\right] \tag{4.460}$$

Using the symbolic tool in MATLAB®, the result of the above determinant can be shown to be given by

$$\Delta = \frac{\eta}{1-\cos\eta\cosh\eta}\left[a\left(1+\lambda^4\right)-b\lambda\left(1+\lambda^2\right)+c\lambda^2\right] \tag{4.461}$$

in which

$$a = 2\eta^3 \tag{4.462}$$

$$b = \chi\left(\sin\eta-\sinh\eta\right)+4\eta^3\left(\cos\eta+\cosh\eta\right) \tag{4.463}$$

$$c = 2\chi\left(\sin\eta\cosh\eta-\cos\eta\sinh\eta\right)+4\eta^3\left(1+2\cos\eta\cosh\eta\right) \tag{4.464}$$

For $\eta > 0$, the leading term of the determinant is not zero, in which case the solution for free waves is

$$a-b\lambda+c\lambda^2-b\lambda^3+a\lambda^4 = 0 \tag{4.465}$$

or

$$1-\frac{b}{a}\lambda+\frac{c}{a}\lambda^2-\frac{b}{a}\lambda^3+\lambda^4 = 0 \tag{4.466}$$

Because of the symmetry of the coefficients of this fourth-order equation, we observe that if λ is a solution, then so is also the reciprocal $1/\lambda$. In addition, if λ should be complex, then the complex conjugate solution λ^* must exist as well, because all coefficients of the equation are real.

Using MATLAB®'s symbolic tool, it can be shown that the four roots of this equation are

$$\lambda = \frac{b}{4a} \pm \sqrt{\frac{1}{2} + \left(\frac{b}{4a}\right)^2 - \frac{c}{4a} \pm \sqrt{\left(\frac{b}{4a} \pm \sqrt{\frac{1}{2} + \left(\frac{b}{4a}\right)^2 - \frac{c}{4a}}\right)^2 - 1}} \tag{4.467}$$

Of the six possible combinations of the three $\pm$ signs in the above expression, the first and third signs must be equal. Hence, valid triplets of signs are $\lambda_1 (+++), \lambda_2 (-+-), \lambda_3 (+-+), \lambda_4 (---)$, in which case the above solution can be written in the compact form

$$\lambda = z_\alpha \pm \sqrt{z_\alpha^2 - 1}, \qquad z_1 = \frac{b}{4a} + \sqrt{\frac{1}{2} + \left(\frac{b}{4a}\right)^2 - \frac{c}{4a}},$$

$$z_2 = \frac{b}{4a} - \sqrt{\frac{1}{2} + \left(\frac{b}{4a}\right)^2 - \frac{c}{4a}} \tag{4.468}$$

so that

$$\lambda_1 = z_1 + \sqrt{z_1^2 - 1}, \quad \lambda_2 = z_2 + \sqrt{z_2^2 - 1}, \quad \lambda_3 = z_1 - \sqrt{z_1^2 - 1} = \lambda_1^{-1}, \quad \lambda_4 = z_2 - \sqrt{z_2^2 - 1} = \lambda_2^{-1} \tag{4.469}$$

Observe that if z_1 is real, then $z_2 \le 0 \le z_1$ no matter what the sign of b/a should be.

Case 1: If z_α is real and $|z_\alpha| \le 1$, then with $z_\alpha = \cos \phi_\alpha$, we obtain

$$\lambda_\alpha = e^{\pm i \phi_\alpha} \tag{4.470}$$

Case 2: If z_α is real and $|z_\alpha| > 1$, then with $|z_\alpha| = \cosh \phi_\alpha$, we obtain

$$\lambda_\alpha = \begin{cases} \cosh \phi_\alpha \pm \sinh \phi_\alpha = e^{\pm \phi_\alpha} & \text{if } z_\alpha > 1 \\ -\cosh \phi_\alpha \mp \sinh \phi_\alpha = -e^{\pm \phi_\alpha} & \text{if } z_\alpha < -1 \end{cases} \tag{4.471}$$

Case 3: If z_α is complex (i.e., the square root term is imaginary), then setting

$$z_\alpha = \frac{b}{4a} \pm i\sqrt{\frac{c}{4a} - \frac{1}{2} - \left(\frac{b}{4a}\right)^2} = \cos \alpha \cosh \beta \pm i \sin \alpha \sinh \beta = \cos(\alpha \pm i\beta) \tag{4.472}$$

we obtain

$$\lambda_\alpha = z_\alpha \pm i\sqrt{1 - z_\alpha^2} = \cos(\alpha \pm i\beta) \pm i \sin(\alpha \pm i\beta) = e^{\pm i(\alpha \pm i\beta)} = e^{\mp \beta} e^{\pm i\alpha} = |\lambda_\alpha| e^{i\xi} \tag{4.473}$$

Thus, case 1 corresponds to $\alpha = \phi_\alpha, \beta = 0$ while case 2 corresponds to $\beta = \phi_\alpha$ and either $\alpha = 0$ or $\alpha = \pi$. More generally, and because the trigonometric functions are periodic, we must write the solution as $\lambda = |\lambda| e^{i\phi} \equiv |\lambda| e^{i(\phi + 2j\pi)} = e^{i\xi}$, with j being any integer. Taking the logarithm $\ln \lambda = \ln|\lambda| + i(\phi + 2j\pi) = i\xi$, then the effective wavenumber is simply

$$\xi = KL = \phi + 2j\pi - i \ln|\lambda| \tag{4.474}$$

Since λ^{-1} is also a solution and

$$\ln \lambda^{-1} = \ln|\lambda|^{-1} e^{-i(\phi + 2j\pi)} = -\ln \lambda - i(\phi + 2j\pi) \tag{4.475}$$

then

$$\xi = -(\phi + 2j\pi) + i \ln|\lambda| \tag{4.476}$$

which is the negative of the previous one. In addition, if λ is complex, then λ^* is also a solution, in which case $\xi = \pm\phi + 2j\pi \mp i \ln|\lambda|$. Hence, we see that not only do solutions appear in positive–negative pairs $\xi = \xi(\omega, \pm K)$, but that if the wavenumber is complex, then $\xi = \xi(\omega, \pm K^*)$ are also valid points in the wavenumber spectrum. Hence, there exist symmetrical branches in all four quadrants of the wavenumber spectrum that correspond to waves that travel and/or evanesce in opposite directions.

Cutoff Frequencies

The cutoff frequencies for this system, if any, satisfy $K = 0$, that is, $\lambda = 1$. This requires satisfaction of the condition $2a - 2b + c = 0$, or in full

$$4\eta^3(1 - \cos\eta)(1 - \cosh\eta) - \chi(\sin\eta(1 - \cosh\eta) - \sinh\eta(1 - \cos\eta)) = 0 \tag{4.477}$$

This can be expressed as

$$\sinh\tfrac{1}{2}\eta \sin\tfrac{1}{2}\eta\left[-4\eta^3 \sin\tfrac{1}{2}\eta \tanh\tfrac{1}{2}\eta + \chi\left(\cos\tfrac{1}{2}\eta \tanh\tfrac{1}{2}\eta + \sin\tfrac{1}{2}\eta\right)\right] = 0 \tag{4.478}$$

which admits solutions

$$\sin\tfrac{1}{2}\eta = 0, \qquad \eta = kL = 2j\pi = L\sqrt{\frac{\omega_j}{C_r R}} \tag{4.479}$$

$$\omega_j = \left(\frac{2j\pi}{L}\right)^2 C_r R \tag{4.480}$$

These coincide with the even modes of a simply supported beam that cause no translation at the points of attachment of the springs. On the other hand, the term in square brackets implies

$$\left(\tfrac{1}{2}\eta\right)^3 = \tfrac{1}{32}\chi\left(\coth\tfrac{1}{2}\eta + \cot\tfrac{1}{2}\eta\right) \tag{4.481}$$

which yields the nontrivial modes. The first of these is one in which the springs elongate synchronously in one direction and the beam opposes this motion by deforming symmetrically about the spring supports. As will be seen, the higher modes of this type virtually coincide with the even modes of the simply supported beam, but not quite so, because the springs do elongate, so there is a net translation of the beam segments.

  The above transcendental equation can be solved graphically (i.e., iteratively), as shown in Figure 4.21 when $\chi = 32$. Whatever the stiffness of the springs χ may be, the first root always lies between 0 and the point at which $\coth\tfrac{1}{2}\eta + \cot\tfrac{1}{2}\eta \approx 1 + \cot\tfrac{1}{2}\eta = 0$, which is very nearly $\eta = \tfrac{3}{2}\pi$. As can be seen, when $\chi = 32$, the first root is approximately 2.3, while when $\chi = 1$, this root is very close to $\eta = 1$.

  Then again, the higher roots occur – to an excellent approximation – at the singularities of $\cot\tfrac{1}{2}\eta$, namely $\eta = 2j\pi$ which, as already mentioned, virtually coincide with the even

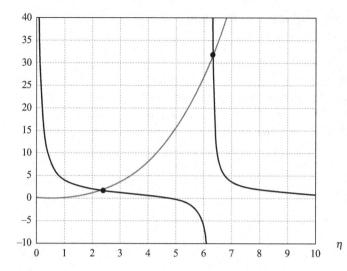

modes of a simply supported beam. Since in the preceding this was also found to be an exact root from the first factor of the characteristic equation, there exist two cutoff frequencies in very close proximity to each other, but one of these two cause no deformation of the springs, and no net translation of the beam.

Static Roots

For zero frequency (static case), $\eta = 0$, the solution to the characteristic equation gives the points of intersection of the complex branches with the complex, horizontal wavenumber plane. To obtain these, we must consider the limit of the determinant

$$\lim_{\eta \to 0}\left\{\frac{\eta}{1-\cos \eta \cosh \eta}\left[a\left(1+\lambda^4\right)-b\lambda\left(1+\lambda^2\right)+c\lambda^2\right]\right\}$$
$$= 12\left(1+\lambda^4\right)-\left(48-2\chi\right)\lambda\left(1+\lambda^2\right)+\left(72+8\chi\right)\lambda^2$$
$$12\left[\left(1+\lambda^4\right)-\left(4-\tfrac{1}{6}\chi\right)\lambda\left(1+\lambda^2\right)+\left(6+\tfrac{4}{6}\chi\right)\lambda^2\right] \tag{4.482}$$

Hence, the roots follow from

$$1-\left(4-\tfrac{1}{6}\chi\right)\lambda+2\left(3+2\tfrac{1}{6}\chi\right)\lambda^2-\left(4-\tfrac{1}{6}\chi\right)\lambda^3+\lambda^4 = 0 \tag{4.483}$$

the solution of which is

$$\lambda = \frac{1}{24}\left[24-\chi\pm\sqrt{\chi(\chi-144)}\mp\sqrt{2}\sqrt{\chi(\chi-96)\pm(24-\chi)\sqrt{\chi(\chi-144)}}\right] \tag{4.484}$$

Clearly, these roots are generally complex, depending on the value of the spring stiffness χ. The roots define displacements patterns that decay in one or the other direction, and generally require an external force at some fixed location or node, say at $x = 0$. When no springs are present (i.e., $\chi = 0$), the four roots equal 1, in which case $K = 0$, that is, the beam can only execute rigid-body static displacements.

4.3.7 Semi-infinite Euler Beam Subjected to Bending Combined with Tension

We consider next a uniform Euler beam subjected to constant tension T. In a differential beam element of length dx and curvature $\kappa \approx d\theta / dx = u''$, the tension contributes via an unbalanced force $T\,d\theta = T\,u''dx$ that is directed toward the center of curvature; in our sign convention this force is positive when the center of curvature is above the beam (i.e., increasing slope). Hence, the differential equation for lateral equilibrium of a beam element subjected to a distributed load per unit length $b(x,t)$ is

$$b\,dx + S - (S + dS) + T\,d\theta = 0 \tag{4.485}$$

which leads immediately to

$$\frac{dS}{dx} = b + T\frac{d\theta}{dx}, \qquad S = \frac{dM}{dx}, \qquad M = EI\frac{d\theta}{dx} = EI\,u'' \tag{4.486}$$

so for a dynamic problem

$$\boxed{\frac{\partial^2}{\partial x^2}\left(EI\frac{\partial^2 u}{\partial x^2}\right) - T\frac{\partial^2 u}{\partial x^2} + \rho A\frac{\partial^2 u}{\partial t^2} = b(x,t)} \tag{4.487}$$

We now examine the propagation of free waves of frequency ω in an unbounded, homogeneous beam. In that case, the differential equation reduces to

$$EI\frac{\partial^4 u}{\partial x^4} - T\frac{\partial^2 u}{\partial x^2} - \rho A\omega^2 u = 0 \qquad \text{or} \qquad \frac{\partial^4 u}{\partial x^4} - \frac{T}{\rho A}\frac{\rho A}{EI}\frac{\partial^2 u}{\partial x^2} - \omega^2\frac{\rho A}{EI}u = 0 \tag{4.488}$$

We define

$$C_r = \sqrt{\frac{E}{\rho}} \qquad \text{rod wave velocity} \tag{4.489}$$

$$C_s = \sqrt{\frac{T}{\rho A}} \qquad \text{string wave velocity} \tag{4.490}$$

$$R = \sqrt{\frac{I}{A}} \qquad \text{radius of gyration} \tag{4.491}$$

Hence

$$\frac{\partial^4 u}{\partial x^4} - \frac{C_s^2}{R^2 C_r^2}\frac{\partial^2 u}{\partial x^2} - \frac{\omega^2}{R^2 C_r^2}u = 0 \tag{4.492}$$

With the trial solution $u = \exp(-i\,k\,x)$, we obtain the characteristic equation

$$(kR)^4 + 2(kR)^2\frac{1}{2}\left(\frac{C_s}{C_r}\right)^2 - \left(\frac{\omega R}{C_r}\right)^2 = 0 \tag{4.493}$$

whose solution is

$$kR = \pm \frac{\sqrt{2}}{2} \frac{C_s}{C_r} \sqrt{\pm \sqrt{1 + 4\left(\frac{C_r}{C_s}\right)^4 \left(\frac{\omega R}{C_r}\right)^2} - 1} \qquad (4.494)$$

We are then led to the purely real and purely imaginary pairs

$$k = \pm k_1, \qquad \boxed{k_1 = \frac{\sqrt{2}}{2R} \frac{C_s}{C_r} \operatorname{sgn}(\omega) \sqrt{\sqrt{1 + 4\left(\frac{C_r}{C_s}\right)^4 \left(\frac{\omega R}{C_r}\right)^2} - 1}} \qquad (4.495)$$

$$k = \pm i k_2, \qquad \boxed{k_2 = \frac{\sqrt{2}}{2R} \frac{C_s}{C_r} \operatorname{sgn}(\omega) \sqrt{\sqrt{1 + 4\left(\frac{C_r}{C_s}\right)^4 \left(\frac{\omega R}{C_r}\right)^2} + 1}} \qquad (4.496)$$

where we have added the sign function to account for the case when the frequency is negative (the term $\exp\left[i(\omega t - kx)\right]$ should continue modeling waves propagating or decaying from left to right). It follows that

$$u = C_1 e^{-ik_1 x} + C_2 e^{-k_2 x} + C_3 e^{ik_1 x} + C_4 e^{k_2 x} \qquad (4.497)$$

If we consider a right semi-infinite beam that starts at $x = 0$ and extends to $x = +\infty$ where waves propagate and/or decay from left to right, this implies that $C_3 = C_4 = 0$, in which case

$$u = C_1 e^{-ik_1 x} + C_2 e^{-k_2 x} \qquad (4.498)$$

$$\theta = -\left(i k_1 C_1 e^{-ik_1 x} + k_2 C_2 e^{-k_2 x}\right) \qquad (4.499)$$

$$M = EI\, u'' = EI\left(-k_1^2 C_1 e^{-ik_1 x} + k_2^2 C_2 e^{-k_2 x}\right) \qquad (4.500)$$

$$S = EI\, u''' = EI\left(i k_1^3 C_1 e^{-ik_1 x} - k_2^3 C_2 e^{-k_2 x}\right) \qquad (4.501)$$

so at $x = 0$ (= point A) where the boundary conditions are $F_A + T\theta_A = S\big|_{x=0}$ and $M_A = -M\big|_{x=0}$, we have

$$\begin{Bmatrix} u_A \\ \theta_A \end{Bmatrix} = \begin{Bmatrix} 1 & 1 \\ -ik_1 & -k_2 \end{Bmatrix} \begin{Bmatrix} C_1 \\ C_2 \end{Bmatrix}, \qquad \begin{Bmatrix} C_1 \\ C_2 \end{Bmatrix} = \frac{1}{k_2 - ik_1} \begin{Bmatrix} k_2 & 1 \\ -ik_1 & -1 \end{Bmatrix} \begin{Bmatrix} u_A \\ \theta_A \end{Bmatrix} \qquad (4.502)$$

$$\begin{Bmatrix} F_A + T\theta_A \\ M_A \end{Bmatrix} = EI \begin{Bmatrix} ik_1^3 & -k_2^3 \\ k_1^2 & -k_2^2 \end{Bmatrix} \begin{Bmatrix} C_1 \\ C_2 \end{Bmatrix} = \frac{EI}{k_2 - ik_1} \begin{Bmatrix} ik_1^3 & -k_2^3 \\ k_1^2 & -k_2^2 \end{Bmatrix} \begin{Bmatrix} k_2 & 1 \\ -ik_1 & -1 \end{Bmatrix} \begin{Bmatrix} u_A \\ \theta_A \end{Bmatrix} \qquad (4.503)$$

That is,

$$\begin{Bmatrix} F_A + T\theta_A \\ M_A \end{Bmatrix} = \frac{EI}{k_2 - ik_1} \begin{Bmatrix} ik_1 k_2\left(k_1^2 + k_2^2\right) & ik_1^3 + k_2^3 \\ k_1 k_2\left(k_1 + ik_2\right) & k_1^2 + k_2^2 \end{Bmatrix} \begin{Bmatrix} u_A \\ \theta_A \end{Bmatrix} \qquad (4.504)$$

which after some algebra can be written as

$$\begin{Bmatrix} F_A \\ M_A \end{Bmatrix} = \begin{Bmatrix} i\,EI\,k_1 k_2\,(k_2 + i\,k_1) & i\,EI\left[k_1 k_2 + i\left(k_1^2 - k_2^2\right)\right] - T \\ i\,EI\,k_1 k_2 & i\,EI\,(k_1 - i\,k_2) \end{Bmatrix} \begin{Bmatrix} u_A \\ \theta_A \end{Bmatrix} \tag{4.505}$$

From the expressions for k_1, k_2, it can easily be shown that $(i\,EI)\,i\left(k_1^2 - k_2^2\right) = T$, so the last two terms in element 1, 2 drop out, and we obtain the simpler expression

$$\begin{Bmatrix} F_A \\ M_A \end{Bmatrix} = i\,EI \begin{Bmatrix} k_1 k_2\,(k_2 + i\,k_1) & k_1 k_2 \\ k_1 k_2 & k_1 - i\,k_2 \end{Bmatrix} \begin{Bmatrix} u_A \\ \theta_A \end{Bmatrix} \tag{4.506}$$

Hence, the symmetric impedance matrix is

$$\boxed{\mathbf{Z} = i\,EI \begin{Bmatrix} k_1 k_2\,(k_2 + i\,k_1) & k_1 k_2 \\ k_1 k_2 & k_1 - i\,k_2 \end{Bmatrix}} \qquad \text{Impedance matrix of beam in tension} \tag{4.507}$$

It also straightforward to show that $i\,EI\,k_1 k_2 = i\rho A\,|\omega|\,C_r R$.

Check 1: If $T = 0 \to C_s = 0$, we must recover the classical beam solution. In that case

$$k_1 = k_2 \equiv k = \frac{\omega}{\sqrt{|\omega|\,RC_r}} = \frac{\omega}{V_{ph}}, \qquad \mathbf{Z} = i k\,EI \begin{Bmatrix} k^2\,(1+i) & k \\ k & 1-i \end{Bmatrix} \tag{4.508}$$

which agrees with the solution obtained earlier for that special case. Recognizing $V_{ph} = \sqrt{\omega R C_r}$ as the flexural wave velocity, we can write the above coefficients as

$$i k^3\,EI = i\omega\rho A V_{ph}, \qquad i k^2\,EI = i\rho A\,V_{ph}^2, \qquad i k\,EI = i\rho A C_r R\,V_{ph} \tag{4.509}$$

so the impedance matrix is

$$\boxed{\mathbf{Z} = i\rho A V_{ph} \begin{Bmatrix} \omega(1+i) & V_{ph} \\ V_{ph} & C_r R\,(1-i) \end{Bmatrix}} \qquad \text{Bending beam with no tension} \tag{4.510}$$

Check 2: If $EI = 0$ we should recover the string solution. In that case,

$$k_1 \to \frac{|\omega|}{C_s}, \qquad EI\,k_2^2 = EI\,k_1^2 + T \to T, \qquad i\,EI\,k_1\left(k_2^2 + i k_1\right) \to i\frac{|\omega|}{C_s}T = i\rho A C_s\,|\omega| \tag{4.511}$$

$$EI\,k_1 k_2 \to 0, \qquad i\,EI\left[k_1 k_2 - i\left(k_1^2 - k_2^2\right)\right] = -T \tag{4.512}$$

So

$$\mathbf{Z} = \begin{Bmatrix} i\omega\rho A C_s & -T \\ 0 & 0 \end{Bmatrix} \tag{4.513}$$

and since M is zero everywhere, including $M_A = 0$, then together with the second equation this implies that θ is undefined everywhere and θ_A can be taken as zero. If so, we obtain

$$F_A = i\omega\rho A C_s u_A \tag{4.514}$$

which is again the correct solution.

Power Transmission

Let

$$\mathbf{u}(t) = \mathbf{U} e^{i\omega t}, \qquad \dot{\mathbf{u}}(t) = i\omega \mathbf{U} e^{i\omega t}, \qquad \mathbf{p}(t) = \mathbf{P} e^{i\omega t} = \mathbf{ZU} e^{i\omega t} \tag{4.515}$$

The average power transmitted in one cycle of motion of duration $T = 2\pi / \omega$ is then

$$
\begin{aligned}
<\Pi> &= \tfrac{1}{T}\int_0^T \text{Re}\left(\dot{\mathbf{u}}^T\right)\text{Re}\left(\mathbf{p}\right)dt = \tfrac{1}{T}\int_0^T \text{Re}\left(i\omega \mathbf{U}^T e^{i\omega t}\right)\text{Re}\left(\mathbf{ZU} e^{i\omega t}\right)dt \\
&= \tfrac{1}{4T}\int_0^T \left(i\omega \mathbf{U}^T e^{i\omega t} - i\omega \mathbf{U}^* e^{-i\omega t}\right)\left(\mathbf{ZU} e^{i\omega t} + \mathbf{Z}_c \mathbf{U}_c e^{-i\omega t}\right)dt \\
&= \tfrac{1}{4T}\int_0^T \left(i\omega\left[\mathbf{U}^T \mathbf{ZU} e^{2i\omega t} - \mathbf{U}^* \mathbf{Z}_c \mathbf{U}_c e^{-2i\omega t}\right] + i\omega\left[\mathbf{U}^T \mathbf{Z}_c \mathbf{U}_c - \mathbf{U}^* \mathbf{ZU}\right]\right)dt
\end{aligned} \tag{4.516}
$$

The first term drops out because the integrals of the exponential terms over one cycle yield zero. The second term, on the other hand, does not depend on time, so the integral is T. Hence

$$
\begin{aligned}
<\Pi> &= \tfrac{1}{4}\omega\left[i\mathbf{U}^T \mathbf{Z}_c \mathbf{U}_c - i\mathbf{U}^* \mathbf{ZU}\right] \\
&= \tfrac{1}{2}\omega \,\text{Im}\left(\mathbf{U}^* \mathbf{ZU}\right)
\end{aligned} \tag{4.517}
$$

Furthermore, if we split the displacement amplitudes into real and imaginary parts, that is, $\mathbf{U} = \mathbf{a} + i\mathbf{b}$, then a straightforward calculation will show that

$$<\Pi> = \tfrac{1}{2}\omega\left[\mathbf{a}^T \,\text{Im}\left(\mathbf{Z}\right)\mathbf{a} + \mathbf{b}^T \,\text{Im}\left(\mathbf{Z}\right)\mathbf{b}\right] \tag{4.518}$$

That is, the calculation depends solely on the imaginary part of the impedance matrix. If so, then

$$\boxed{<\Pi> = \tfrac{1}{2}\omega \mathbf{U}^* \,\text{Im}\left(\mathbf{Z}\right)\mathbf{U}} \tag{4.519}$$

where $\mathbf{U}$ is any arbitrary vector. In the case of the beam in tension, this yields

$$\text{Im}\left(\mathbf{Z}\right) = k_1 EI \begin{Bmatrix} k_2^2 & k_2 \\ k_2 & 1 \end{Bmatrix} \tag{4.520}$$

So

$$<\Pi> = \tfrac{1}{2}\omega k_1 EI\left[k_2^2 |u|^2 + 2k_2 \,\text{Re}\left(u^*\theta\right) + |\theta|^2\right] \tag{4.521}$$

with k_1, k_2 given by the expressions found earlier in this section. Also,

$$EI\, k_1 k_2 = \rho A |\omega|\, C_r R \tag{4.522}$$

Power Transmission after Evanescent Wave Has Decayed

Now, inasmuch as one of the two wave fields in the beam is evanescent, that is, k_2, after a short distance only the propagating phase will remain active. The displacement and rotation are then be characterized by

$$
\mathbf{U} = \left\{ \begin{matrix} u \\ \theta \end{matrix} \right\} = C_1 \left\{ \begin{matrix} 1 \\ -i\,k_1 \end{matrix} \right\} e^{-i\,k_1 x}
\tag{4.523}
$$

where C_1 is an arbitrary constant, in which case the transmitted power is

$$
\begin{aligned}
<\Pi> &= \tfrac{1}{2}\omega k_1 EI \left|C_1\right|^2 \{1 \quad i\,k_1\} \begin{bmatrix} k_2^2 & k_2 \\ k_2 & 1 \end{bmatrix} \left\{ \begin{matrix} 1 \\ -i\,k_1 \end{matrix} \right\} \\
&= \tfrac{1}{2}\omega k_1 EI \left|C_1\right|^2 \left(k_1^2 + k_2^2\right)
\end{aligned}
\tag{4.524}
$$

where

$$
k_1^2 + k_2^2 = \left(\frac{C_s}{RC_r}\right)^2 \sqrt{1 + 4\left(\frac{C_r}{C_s}\right)^4 \left(\frac{\omega R}{C_r}\right)^2}
\tag{4.525}
$$

Finally

$$
\boxed{<\Pi> = \tfrac{1}{2}\omega k_1 EI \left|C_1\right|^2 \left(\frac{C_s}{RC_r}\right)^2 \sqrt{1 + 4\left(\frac{C_r}{C_s}\right)^4 \left(\frac{\omega R}{C_r}\right)^2}}
\tag{4.526}
$$

where again

$$
C_r = \sqrt{\frac{E}{\rho}} \qquad \text{rod wave velocity}
\tag{4.527}
$$

$$
C_s = \sqrt{\frac{T}{\rho A}} \qquad \text{string wave velocity}
\tag{4.528}
$$

$$
R = \sqrt{\frac{I}{A}} \qquad \text{radius of gyration}
\tag{4.529}
$$

When the tension drops to zero, then $C_s = 0$ and

$$
\boxed{\begin{aligned}
<\Pi> &= \rho A V^2 \sqrt{C_r \left|\omega\right| R} \\
&= \rho A C_f V^2
\end{aligned}}
\tag{4.530}
$$

where $V = \omega\left|C_1\right|$ is the maximum transverse velocity in the beam after the evanescent wave has decayed, and $C_f = \sqrt{C_r \left|\omega\right| R}$ is the phase velocity of flexural waves. This is the classical result for the power transmission in an Euler beam.

5 Wave Propagation

5.1 Fundamentals of Wave Propagation

We present herein a brief review of the normal modes of wave propagation in simple systems such as rods and beams, and use these to illustrate the fundamental concepts of wave propagation, including complex wave spectra. Thereafter, we generalize these concepts to the much more complicated case of horizontally layered media. We present also a brief review of the Stiffness Matrix Method (SMM) for layered media and use it for the solution of wave propagation problems. Finally, we summarize the fundamental elements of the discrete counterpart to the SMM, the Thin-Layer Method (TLM), which constitutes a powerful tool to obtain the normal modes for both propagating and evanescent waves.

5.1.1 Waves in Elastic Bodies

As is well known, when an elastic medium is subjected to a local disturbance and then left on its own, the deformation and kinetic energies stored in the neighborhood of the disturbance give rise to waves that propagate away from this local region and carry elastic energy to other parts of the structure. Mathematically, this represents what is commonly referred to as an *initial value problem*, that is, a wave propagation problem. In essence, an initial value problem is one that involves the free vibration of an elastic body subjected to initial conditions in velocity and displacement, and can thus be interpreted in terms of a superposition of *normal modes* of vibration. Physically, such normal modes can be interpreted as *stationary waves* of appropriate fixed wavelength such that all boundary and material continuity conditions are satisfied simultaneously at all times. Alternatively, the modes may be visualized as resulting from constructive and destructive interference of waves elicited by reflections at material discontinuities and at the external boundaries.

In the case of *finite* elastic bodies with appropriate boundary conditions, there exist in general infinitely many, albeit distinct vibration modes, each of which is characterized by a real-valued, discrete natural frequency of vibration and an associated modal shape that describes the spatial variation of the motion for that mode. In such case, the free vibration for arbitrary initial conditions can be obtained by superposition over all modes of vibration, that is, by recourse to the classical *modal superposition* technique of mechanical vibration and structural dynamics. The weight assigned to each mode in this

333

summation – the participation factor– is obtained by expressing the initial conditions in terms of the modes and then applying certain orthogonality conditions. By appropriate generalization, this method can also be used to find the response to external sources with arbitrary spatial/temporal variation, that is, the so-called *forced vibration* problem.

By contrast, when a body is *unbounded* in some – or all – coordinate directions, say an elastic layer over an elastic half-space, it may admit a *continuous* (as opposed to discrete) spectrum of frequencies for which *guided waves* can propagate in the absence of external sources. Moreover, it is not always possible to express a source (i.e., forced vibration, or inhomogeneous wave propagation) problem solely in terms of normal modes, although very useful results can still be obtained using this method. This is so because the normal modes may not suffice to express all physical aspects of the source and response, and so additional terms (so-called branch integrals) are needed to capture the arrival and evolution of body waves, especially at points near the source. For example, an elastic half-space has only one mode – the Rayleigh wave – which is not enough to fully describe motions elicited by a source, even if these waves dominate the response in the far field, at least near the surface.

In general, there exist two kinds of normal modes, the real-valued *propagating modes* (which may decay, but only because of geometric spreading and/or material attenuation), and the complex-valued *evanescent modes*, which even in the absence of damping decay exponentially in the direction of apparent propagation, and thus exist only in the vicinity of the source. However, at large ranges (i.e., source-receiver or epicentral distances), a subset of the normal modes – the real-valued propagating modes – dominate the response, and they may indeed suffice to describe the wave motion.

5.1.2 Normal Modes and Dispersive Properties of Simple Systems

To demonstrate some of the fundamental concepts underlying the propagation of waves in unbounded systems in general – and in layered soils in particular – we begin with a brief discussion of the wave types and normal modes in some simple mechanical systems. This will allow us to review the concepts of wave modes, phase and group velocity, dispersion, and frequency-wavenumber spectra in the context of simple systems. For this purpose, we begin with waves in and infinite rod, and then move up gradually to other, more complicated systems.

An Infinite Rod

An infinite, homogeneous rod of constant cross section is the simplest of all mechanical systems that can sustain waves, namely compression-dilatation (or P) waves. In the absence of external sources, and neglecting "skin effects," the rod obeys the classical wave equation

$$C_r^2 \frac{\partial^2 u}{\partial x^2} = \frac{\partial^2 u}{\partial t^2} \qquad\qquad (5.1)$$

in which $u(x,t)$ is the axial displacement at some arbitrary point, x is the axial coordinate, $C_r = \sqrt{E/\rho}$ is the *rod-wave velocity*, and E, ρ are the Young's modulus and mass density, respectively. This equation admits solutions of the form

$$u(x,t) = F_1(x+C_r t) + F_2(x-C_r t) \tag{5.2}$$

in which F_1 and F_2 are arbitrary functions. This can be demonstrated by simple substitution into the wave equation. For example, if we consider the first term in F_1, and write the argument as $\xi = x + C_r t$, then by the chain rule

$$\frac{\partial u}{\partial x} = \frac{\partial F_1}{\partial \xi}\frac{\partial \xi}{\partial x} = \frac{\partial F_1}{\partial \xi} = F_1' \qquad \frac{\partial u}{\partial t} = \frac{\partial F_1}{\partial \xi}\frac{\partial \xi}{\partial t} = C_r \frac{\partial F_1}{\partial \xi} = C_r F_1' \tag{5.3}$$

$$\frac{\partial^2 u}{\partial x^2} = \frac{\partial F_1'}{\partial \xi}\frac{\partial \xi}{\partial x} = \frac{\partial F_1'}{\partial \xi} = F_1'' \qquad \frac{1}{C_r^2}\frac{\partial^2 u}{\partial t^2} = \frac{1}{C_r}\frac{\partial F_1'}{\partial \xi}\frac{\partial \tau}{\partial t} = \frac{\partial F_1'}{\partial \xi} = F_1'' \tag{5.4}$$

Thus, the wave equation is indeed satisfied. A similar proof applies to the second term in F_2, except for a slight modification in the sign of the first derivative.

Physically, F_1, F_2 can be interpreted as two *wavelets* or pulses that propagate with *celerity* (i.e., speed) C_r from right to left, and from left to right, respectively. Furthermore, if Δx and Δt are arbitrary space–time increments and we consider the case of a single wavelet F_1, then

$$u(x+\Delta x, t+\Delta t) = F_1(x+\Delta x - C_r(t+\Delta t))$$
$$= F_1(x-C_r t + (\Delta x - C_r \Delta t)) \tag{5.5}$$

Choosing $\Delta x - C_r \Delta t = 0$, i.e. $\Delta x/\Delta t = C_r$, we conclude that $u(x+\Delta x, t+\Delta t) = u(x,t)$, that is, the motion at $x+\Delta x$ is a delayed replica of the motion at x. Thus, the wavelet propagates in the rod with *phase velocity* C_r and suffers no distortions, that is, does not change shape as it propagates. Thus, rod waves are said to be *nondispersive*.

Consider next the special case of harmonic waves of the form

$$u(x,t) = A \sin(\omega t - kx) \tag{5.6}$$

where A is the amplitude, ω is the frequency, $k = 2\pi/\lambda$ is the *wavenumber* (i.e., spatial frequency), and λ is the *wavelength* (i.e., the spatial period) of the waves. Factoring out the frequency, this can be written as

$$u(x,t) = A\sin[\omega(t - xk/\omega)] = A\sin[\omega(t - x/V_{ph})] \tag{5.7}$$

in which

$$\boxed{V_{ph} = \omega/k} \qquad \text{slope of secant} = \textit{phase velocity} \tag{5.8}$$

Clearly, in this case $V_{ph} = C_r$, so the phase velocity is seen to be independent of the frequency of the waves; this is another way of saying that rod waves are nondispersive. It also follows that $V_{ph} = \omega\lambda/2\pi$, that is,

$$\boxed{V_{ph} = f\lambda} \tag{5.9}$$

with f being the frequency of the waves in cycles per second (Hz).

Gravity Waves in a Deep Ocean

Waves observed on the surface of the ocean as well as in other bodies of water are largely of the harmonic type whose characteristic period T usually changes with atmospheric and sea conditions. As weather conditions change from calm to very rough and stormy, their period increases from a few seconds, say 3 s to 5 s, to some 10 s to 15 s. Also, the height (amplitude) of open ocean waves will generally not exceed a threshold of about $\lambda/7$, because the waves would break if they were any higher. Waves on the surface of the ocean whose amplitude A is much smaller than the wavelength λ can be idealized as plane, linear harmonic waves of the form

$$y = A \sin(\omega t - kx) \tag{5.10}$$

where k is the wavenumber and ω is the frequency of the waves. For a body of water of depth d, the dispersion relationship for gravity waves is known to be given by

$$\omega^2 = g k \tanh kd \tag{5.11}$$

where g is the acceleration of gravity. This can be written in dimensionless form as

$$\Omega^2 = \kappa \tanh \kappa, \qquad \Omega = \omega \sqrt{\frac{d}{g}}, \qquad \kappa = kd = 2\pi\frac{d}{\lambda} \tag{5.12}$$

The phase velocity is then

$$\boxed{\frac{V_{ph}}{\sqrt{gd}} = \sqrt{\frac{\tanh \kappa}{\kappa}}} \tag{5.13}$$

which is plotted in Figure 5.1 in terms of the wavenumber.

(a) Deepwater Approximation

When $d > \frac{1}{4}\lambda$, then $\tanh kd \approx 1$ and $V_{ph}/\sqrt{g} = \sqrt{1/k}$, so

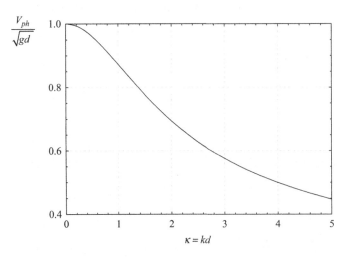

Figure 5.1. Phase velocity of ocean waves as function of depth.

$$V_{ph} = \frac{g}{\omega} = \frac{1}{2\pi} gT \qquad = \text{phase velocity} \qquad (5.14)$$

$$\omega = \sqrt{kg} \qquad = \text{wave spectrum} \qquad (5.15)$$

$$\lambda = 2\pi \frac{g}{\omega^2} = \frac{1}{2\pi} gT^2 \qquad = \text{wavelength} \qquad (5.16)$$

The limit of d in terms of λ guarantees that the phase velocity will be within 10% of the accurate value. It follows that the phase velocity of ocean waves in deepwater increases in inverse proportion to their frequency, and are thus dispersive.

(b) Shallow Water Approximation

When $d < \frac{1}{11}\lambda$, then $\tanh kd \approx kd$ and

$$\omega^2 = g k^2 d, \qquad V_{ph} = \sqrt{gd} \qquad (5.17)$$

where again the limit on depth guarantees the phase velocity to be within 10% of the accurate value. These waves are nondispersive, since the phase velocity is constant.

For example, assume that during a massive earthquake in the Aleutian Islands near Alaska, a large tsunami of wavelength $\lambda = 100$ km is generated that travels over the Pacific Ocean with a depth $d = 5$ km. Assuming this to be a wave over shallow water, then $V_{ph} = \sqrt{9.8 \times 5000} \approx 221$ m/s or $V_{ph} \approx 800$ km/h, which is the speed of a passenger jet. Thus, it reaches the California coast, some 4000 km away in about 5 hours or so.

An Infinite Bending Beam

As a third example, we consider an infinitely long Euler–Bernoulli bending beam, which takes only bending deformation and translational inertia into consideration. In the absence of body sources, a homogeneous beam is governed by the differential equation

$$\rho A \frac{\partial^2 u}{\partial t^2} + EI \frac{\partial^4 u}{\partial x^4} = 0 \qquad (5.18)$$

with u = transverse displacement, ρ = mass density, A = cross section, and I = moment of inertia. Assuming a harmonic wave of the form $u(x,t) = C \sin(\omega t - kx)$, we obtain

$$\left[-\rho A \omega^2 + EI k^4\right] u = 0 \qquad (5.19)$$

from which we deduce the phase velocity and frequency–wavenumber relationship as

$$V_{ph} = \sqrt{R C_r \omega} = k R C_r \qquad (5.20)$$

$$\omega = R C_r k^2 \qquad (5.21)$$

where $R = \sqrt{I/A}$ is the radius of gyration. Thus, the phase velocity is again a function of the frequency, yet unlike ocean waves, the speed of flexural waves does not fall, but rises with frequency. Hence, flexural waves are *dispersive*. The phase velocity can also be expressed in terms of the wavelength as

$$\frac{V_{ph}}{C_r} = 2\pi \frac{R}{\lambda} \tag{5.22}$$

Thus, low-frequency flexural waves whose wavelength is much greater than the beam's radius of gyration R (which is of the same order of magnitude as the beam's depth) move at substantially lower speeds than the speed of rod waves C_r. However, for frequencies high enough to elicit waves comparable in length to the thickness of the beam, the phase speed would appear to be able to exceed the rod wave velocity without bound. However, this is spurious effect of the theory, because no waves in a beam can exceed the largest physical speed, which is the speed of compressional waves. We conclude that at high frequencies, the simple theory of flexural waves in an Euler–Bernoulli beam breaks down.

A Bending Beam on an Elastic Foundation

The next level of complication arises when we consider a bending beam resting on an elastic (distributed Winkler) foundation of stiffness K. The dynamic equilibrium equation for this case is

$$\rho A \frac{\partial^2 u}{\partial t^2} + EI \frac{\partial^4 u}{\partial x^4} + Ku = 0 \tag{5.23}$$

For harmonic waves of the form $u(x,t) = C \exp i(\omega t - kx)$, we infer that

$$\left[-\rho A \omega^2 + EIk^4 + K \right] u = 0 \tag{5.24}$$

so that

$$k = \left[\frac{\rho A \omega^2 - K}{EI} \right]^{1/4} = \sqrt{\frac{\omega_0}{RC_r}} \left[\left(\frac{\omega}{\omega_0} \right)^2 - 1 \right]^{1/4} \tag{5.25}$$

with

$$\omega_0 = \sqrt{\frac{K}{\rho A}} \tag{5.26}$$

from which it becomes clear that all waves in a beam on elastic foundation are again frequency dependent and are thus *dispersive*. Observe that the frequency ω_0 is the frequency at which the beam moves up and down as a rigid body, which is analogous to the "breathing mode" of cylindrical shells. The fourth root term has four solutions that generally show up in complex conjugate pairs. Depending on the value of the frequency compared to the frequency of the rigid-body mode, the four roots for the wavenumber k are

$$k = \sqrt{\frac{\omega_0}{RC_r}} \left[\left| \left(\frac{\omega}{\omega_0} \right)^2 - 1 \right| \right]^{1/4} \times \left\{ \begin{array}{ll} \pm 1, \pm i & \omega \geq \omega_0 \\ \frac{1}{2}\sqrt{2}(\pm 1 \pm i) & \omega < \omega_0 \end{array} \right\} \tag{5.27}$$

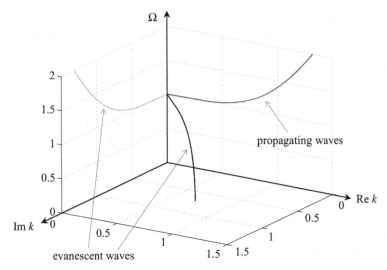

Figure 5.2. Wave spectrum for beam on elastic foundation, $\Omega = \omega / \omega_0$.

When the frequency ω is taken as a parameter and k is allowed to take on complex values, the above roots define four branches in the three-dimensional $\omega - \mathrm{Re}\,k - \mathrm{Im}\,k$ space. The combined set of branches in this space is referred to as the frequency–wavenumber spectrum, or simply the *wave spectrum*.

Figure 5.2 shows three such branches. The first one is complex and exists only below the *cutoff frequency* ω_0, after which another purely real branch begins. Also shown is a purely imaginary branch above ω_0, which is an evanescent wave. There exist also continuations of the branches along mirror image branches that lie in the graph's other octants. In addition, branches associated with the complex conjugate of these branches exist as well, so that the wave spectrum is symmetric with respect to the vertical axis. Real branches correspond to propagating modes, purely imaginary branches are exponentially decaying (or rising) modes, and complex branches are evanescent modes (i.e., modes that appear to decay or rise as they propagate). Specifically, a positive real part $\mathrm{Re}\,k > 0$ indicates propagation in the positive x-direction, and vice versa. Also, a negative imaginary part $\mathrm{Im}\,k < 0$ indicates exponential decay in the positive x-direction, while a positive imaginary part indicates exponential growth (i.e., decay in the negative x direction).

The four normal modes of the beam on elastic foundation exhibit the following characteristics:

(a) $\omega < \omega_0$

There exist four complex branches that appear in complex conjugate pairs. Two of these branches have identical *negative* imaginary parts and opposite real parts, while the other two are the complex conjugate counterparts. The first pair corresponds to waves that decay in the positive x-direction and appear to propagate in opposite coordinate directions, while the second pair represents waves that decay in the negative x-direction, and again appear to travel in opposite directions. It can be shown that in the presence of sources, each of the two branches in a pair combine with equal weights and form an evanescent standing wave. Thus, such modes do not carry any energy, and they may exist only if external sources are present in some neighborhood – otherwise, the wave would grow exponentially in one of the coordinate directions, and would violate the bounded-ness condition at infinity.

(b) $\omega = \omega_0$

Here, the four solutions for the wavenumber k degenerate to zero, which implies no spa-tial variation in x. Hence, the "wave" is simply an up and down motion of the beam exe-cuting a rigid body vibration.

(c) $\omega > \omega_0$

Two branches are real, and the other two are purely imaginary. Of the two real ones, one propagates in the positive x-direction, the other in the negative. These are the only modes that carry energy, and the only ones that can be observed at distances from a source. By contrast, the imaginary branches correspond to exponentially decaying motions that do not carry any energy and quickly die out.

We observe that waves do not necessarily propagate at all frequencies, but that they do so only in some specific frequency bands. Thus, it is said that wave spectra exhibit *starting* and *stopping bands*. In this particular case, the propagating mode exists only above the cutoff frequency. From the preceding it follows that when a low-frequency excitation is applied to a beam on elastic foundation, the beam responds only in the neighborhood of the source, and the disturbance does not propagate to any significant distance from this source. Hence, all waves are evanescent. However, as the frequency of the excitation is raised above the cutoff frequency, it begins to propagate waves that *radiate* energy away from the source.

A Bending Beam on an Elastic Half-Space

Next, we assume that the bending beam rests smoothly and without friction on an elastic half-space. This case is qualitatively similar to the preceding one, but differs from it that the half-space – unlike the elastic foundation – now allows guided waves and body waves on its own, which in turn causes a new phenomenon to arise, namely the so-called *radia-tion damping*. As a result, the cutoff frequency associated with the rigid-body mode is now highly damped, the wavenumber spectrum grows further in complexity, and additional propagation characteristics appear:

- At very low frequencies, all modes are evanescent, so no wave propagation takes place anywhere in either the beam or the half-space.
- At low frequencies above some threshold, very slow flexural waves may begin propa-gating in the beam, but in the half-space, the motion remains confined to the vicinity of the beam, and decays exponentially with depth.
- As the frequency rises further, the flexural wave speed eventually grows until it equals or exceeds the Rayleigh wave velocity of the half-space, at which point waves begin to be transmitted and radiated into the half-space. The frequency at which this transition

occurs is referred to as the *coincidence frequency*. Hence, energy begins leaking into the half-space, which means that the waves in the beam *must* decay as they propagate. Flexural waves whose speed exceeds any of the physical wave velocities in the half-space are said to be *supersonic*. To a first approximation, the coincidence frequency can be estimated by equating the speed of flexural waves in a free beam with the shear wave velocity of the half-space, that is, $\sqrt{RC_r}\,\omega = C_S$, but this is only an approximation, because the beam and the half-space will exhibit strong coupling well before reaching this frequency.

Closely related to this issue, consider the problem of why we can hear somebody knocking on a door: the brief (i.e., high-frequency) knocks elicit fast flexural waves in the door that are supersonic in comparison to the acoustic wave speed in air. Hence, these waves radiate strongly into the air and transmit to our ears the sound of the knocks.

Elastic Thick Plate (Mindlin Plate)

As a final example, we consider a homogeneous plate of thickness h subjected to plane strain. This system can sustain infinitely many wave modes – the so-called Lamb's waves – so the plate's spectrum has infinitely many branches. It can be shown that the characteristic wavenumbers for Lamb's modes can be obtained by solving the pair of equations

$$ps\tanh\tfrac{1}{2}kph - q^2\tanh\tfrac{1}{2}ksh = 0 \qquad \text{Symmetric modes} \tag{5.28}$$

$$ps\coth\tfrac{1}{2}kph - q^2\coth\tfrac{1}{2}ksh = 0 \qquad \text{Antisymmetric modes} \tag{5.29}$$

in which

$$p = \sqrt{1-(k_P/k)^2}, \qquad s = \sqrt{1-(k_S/k)^2}, \qquad q = \tfrac{1}{2}(1+s^2) \tag{5.30}$$

$$k_P = \omega/C_P, \qquad k_S = \omega/C_S \tag{5.31}$$

and C_P, C_S are the compressional and shear wave velocities of the plate. This deceptively simple looking transcendental eigenvalue problem happens to be extremely difficult to solve – at least by standard search techniques. The first investigator to fully unravel the intricacies of the complex wave spectrum of thick plates was Mindlin, so in his honor such plates are now called Mindlin plates. Today, effective solution techniques are available, some of which will be discussed briefly later on.

The wave spectrum for Lamb's waves in a thick plate is shown in Figure 5.3, which depicts the first few propagating and evanescent modes. More generally, at any given frequency there exists only a *finite* number of wave modes that propagate, that is, of real branches; their wavenumbers k lie between zero and the wavenumber of shear waves, that is, $\omega = C_S k$. As the frequency rises above each of the cutoff frequencies of the plate – which happen to coincide with the shear and dilatational resonant frequencies of the plate – one more complex branch turns real, and so an increasing number of modes participate in propagating energy.

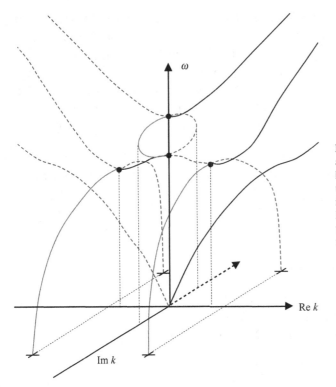

Figure 5.3. Wavenumber spectrum for a Mindlin plate. Solid lines: first few low-order branches that correspond to waves that carry energy or decay in the positive x-direction. Dashed lines: complex-conjugate branches for waves that propagate or decay in the negative x-direction.

5.1.3 Standing Waves, Wave Groups, Group Velocity, and Wave Dispersion

Standing Waves

Consider a 1-dimensional system such as a rod or a beam in which two harmonic waves of the same amplitude propagate in opposite directions. The combination is

$$
\begin{aligned}
u(x,t) &= C\sin(\omega t - kx) + C\sin(\omega t + kx) \\
&= 2C\sin\omega t\cos kx = 2C\sin\omega t\cos(2\pi x/\lambda)
\end{aligned}
\tag{5.32}
$$

which contains the product of a harmonic factor of time and a harmonic factor of space. Since $\sin kx = 0$ for $x = j\lambda/2$, $j = 0,1,2\ldots$, we see that the motion is now harmonic both in space and in time. More importantly, we observe that there exist points or *nodes* $x_j = 2j\pi\lambda$, $j = 0, \pm 1, \pm 2, \ldots$ at which the motion is zero at all times; that is, these points do not move but remain stationary. Thus, the motion is confined to the regions between the nodes, and the wave exhibits no net propagation. The maximum occurs at the midpoint between nodes – the crests and troughs – at which the axial strain (i.e., the stress) is zero. Such waves are referred to as *standing waves* (see Figure 5.4).

The axial stresses elicited by the standing wave are

$$
\sigma = Eu' = kE\,\cos\omega t\,\cos kx
\tag{5.33}
$$

which is also a standing waveform, albeit shifted in space by a quarter wavelength ($\frac{1}{4}\lambda$) with respect to the displacement waveform.

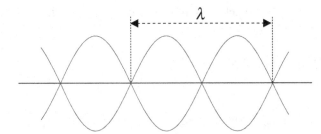

Figure 5.4. Standing wave in rod.

We can now use the results of a standing wave to obtain an alternative (and simpler) solution for the normal modes of a rod of *finite* length L and arbitrary boundary conditions – and do so without the need to solve a boundary value problem. It suffices for us to choose a rod length L equal to an even or odd multiple of a quarter wavelength $\frac{1}{4}\lambda$ – the distance between a node (where the displacement is zero) and/or a trough (where the stress is zero).

For a free–free, fixed–free and fixed–fixed rod, this gives

$$
\begin{aligned}
\text{FR} - \text{FR}: \quad & L = 2j\frac{\lambda_j}{4} = j\frac{C_r}{2f_j} && \rightarrow \quad f_j = j\frac{C_r}{2L}, && j = 0,1,2,3,\cdots \\[2mm]
\text{FX} - \text{FR}: \quad & L = (2j-1)\frac{\lambda_j}{4} = (j-\tfrac{1}{2})\frac{C_r}{2f_j} && \rightarrow \quad f_j = (j-\tfrac{1}{2})\frac{C_r}{2L}, && j = 1,2,3,\cdots \\[2mm]
\text{FX} - \text{FX}: \quad & L = 2j\frac{\lambda_j}{4} = j\frac{C_r}{2f_j} && \rightarrow \quad f_j = j\frac{C_r}{2L}, && j = 1,2,3,\cdots
\end{aligned}
\tag{5.34}
$$

In the first case, the 0th mode ($j = 0$) is the rigid-body mode, which has zero frequency, infinite period, and an infinite wavelength (i.e., constant displacement). Observe that the frequencies of the second case interlace those of the first and third. Also, although the frequencies of the first and third cases appear to be identical despite the different boundary conditions, this does not violate the fundamental principle that states that when kinematic constraints are added to a system, then some or all of its natural frequencies must rise. Indeed, the first frequency of the free–free case (i.e., the rigid-body mode zero) rises to the first frequency of the fixed–free case, and as we add a further constraint this one rises in turn to the first frequency of the fixed–fixed case, which happens to coincide with the *second* frequency of the free–free case.

The aforementioned observation explains why the pitch of an open organ pipe (or flute) seemingly *drops* when its free end is covered with a stopper (i.e., a kinematic constraint is added): the fundamental frequency of the open pipe is zero, and thus cannot be heard, so the audible tone that it emits is really the second mode, which is higher in pitch than the fundamental mode of the fixed-free pipe.

Groups and Group Velocity

Consider next a dispersive system, such as a bending beam or the deep ocean. Assume further that two harmonic waves of equal amplitude and nearly identical frequencies

$\omega \pm \Delta \omega$ and wavenumbers $k \pm \Delta k$ propagate in such a system. Hence, the wave field is given by

$$u = \tfrac{1}{2}C\left\{\sin\left[(\omega+\Delta\omega)t - (k+\Delta k)x)\right] + \sin\left[(\omega-\Delta\omega)t - (k-\Delta k)x)\right]\right\}$$
$$= C\sin(\omega t - kx)\cos(\Delta\omega t - \Delta k\, x) \qquad (5.35)$$

In the limit when $\Delta\omega \to 0$ and $\Delta k \to 0$, the ratio $\dfrac{\Delta\omega}{\Delta k} \to \dfrac{d\omega}{dk} \underset{def}{=} V_{gr}$. This limit is termed the *group velocity*. Hence, the combination of two waves can be written as

$$u(x,t) = C\sin\omega\left(t - x/V_{ph}\right)\cos\Delta\omega\left(t - x/V_{gr}\right) \qquad (5.36)$$

This equation can be interpreted as a *carrier wave* of low frequency $\Delta\omega$ that propagates at the group velocity V_{gr} which *modulates* the amplitude of another wave of frequency ω propagating with speed V_{ph} (see Figure 5.4). If the phase velocity is greater than the group velocity, as in the case of ocean waves, the waves appear to emanate from a node, move forward from there, grow within the group, and then vanish at the subsequent node. The opposite is true when the phase velocity is less than the group velocity, such as in the case of flexural waves: the waves appear to move backwards relative to the group (or more precisely, be left behind by the group), so they appear to emanate from a node and vanish at the preceding node.

It is easy to show that the group and phase velocities are related as

$$\frac{V_{gr}}{V_{ph}} = \frac{1}{1 - \dfrac{\omega}{V_{ph}}\dfrac{dV_{ph}}{d\omega}} = 1 - \frac{\lambda}{V_{ph}}\frac{dV_{ph}}{d\lambda} \qquad (5.37)$$

It can also be shown that energy flows at the speed of the group velocity. Table 5.1 summarizes the phase and group velocities in the simple systems considered earlier.

Wave Groups and the Beating Phenomenon

The groups are intimately related to the well-known beating phenomenon. Figure 5.5 depicts a group of waves in space, and as time elapses, the group with the contained wavelets moves steadily to the right. On the other hand, if we observe the motion at some arbitrary fixed location and focus on the variation in time instead, then at that location, the wave amplitude seems to pulse with time, and this constitutes the beating phenomenon, which is well known to people used to tuning musical instruments. Observe that the waves need not be dispersive for this phenomenon to occur, but it is enough for sources with slightly different frequencies to interfere with each other.

Summary of Concepts

- Wave spectra provide fundamental information on propagation modes in a mechanical system.
- Real branches represent propagating waves.

Table 5.1

	Wave spectrum $\omega = \omega(k)$	Phase velocity $V_{ph} = \dfrac{\omega}{k}$	Group velocity $V_{gr} = \dfrac{d\omega}{dk}$	Restriction	Reason
Rod	$\omega = C_r k$	$V_{ph} = C_r$	$V_{gr} = V_{ph}$	$\lambda \gg h$ (thickness)	Skin effects, Poisson effect, Lamb waves
Beam	$\omega = R C_r k^2$	$V_{ph} = \sqrt{\omega R C_r}$	$V_{gr} = 2V_{ph}$	$\lambda \gg h$ (thickness)	Rotational inertia, shear deformation, skin effects
Ocean, shallow water	$\omega = k\sqrt{gd}$	$V_{ph} = \sqrt{gd}$	$V_{gr} = V_{ph}$	$d < \frac{1}{11}\lambda$	Neither shallow nor deep if exceeded
Ocean, deep water	$\omega = \sqrt{kg}$	$V_{ph} = \dfrac{g}{\omega}$	$V_{gr} = \frac{1}{2}V_{ph}$	$d > \frac{1}{4}\lambda$	Ocean bottom effects if below

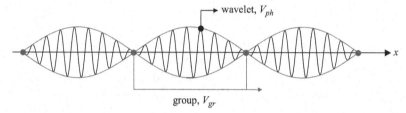

Figure 5.5. Group of two near-frequency waves.

- Imaginary branches do not propagate, but decay exponentially instead.
- Complex branches represent evanescent waves that both propagate and decay.
- Wave spectra are normally plotted in terms of frequency–wavenumber (ω–k), or alternatively in terms of either phase velocity–frequency (V_{ph}-f) or phase velocity–wavelength (V_{ph}–λ). However, the latter two are only used for real, propagating waves.
- The *phase velocity* V_{ph} equals the slope of the secant from the origin to a point on a given branch.
- The *group velocity* V_{gr} equals the slope of the tangent to a given branch. It can be shown that wave energy flows at the speed of the group velocity.
- If both the phase and group velocities have the same sign, the mode is normally dispersive. However, if they have opposite sign, the mode is said to exhibit *anomalous* dispersion. In general, to the right of a source in a Mindlin plate (and more generally, in a horizontally layered medium) only waves with positive group velocity can exist, because energy always radiates away from a source. Hence, anomalous modes, if any, will involve waves that while appearing to move toward the source, they still carry energy away from it.

5.1.4 Impedance of an Infinite Rod

We now go on to explore the "dynamic stiffness" of an infinite rod, that is, its impedance, and the relationship of that impedance with waves in the rod, a subject that we had already taken up in a different context in Chapter 4, Section 4.3.1. Consider then an infinitely long rod subjected to a wavelet F_1 that propagates from right to left, that is,

$$u = F_1(t + x / C_r) \tag{5.38}$$

Clearly, this wavelet elicits a *particle* velocity $\dot{u} = \dot{F}_1(t + x / C_r)$. On the other hand, the net axial forces produced by this wavelet are

$$N = EA\varepsilon = EA\dot{F}_1 / C_r = \rho C_r A\dot{F}_1 = \rho C_r A\dot{u} \tag{5.39}$$

As can be seen, the axial force at a given point is at all times proportional to the particle velocity there. Thus, this relationship is analogous to the force–deformation equation for a viscous dashpot whose damping constant is

$$D = \rho C_r A \tag{5.40}$$

Hence, if we were to cut the rod at any arbitrary section and replace the removed semi-infinite part extending to the left by a dashpot with a constant D as given previously, on impingement of the wavelet onto the dashpot, the wavelet will be completely absorbed, and no reflection will take place. Hence, the dashpot is an *exact* – and compact – representation of a semi-infinite rod; its value equals the *impedance* of the rod, which is the relationship between axial stresses and particle velocity.

From the foregoing it follows that if we apply a load $F(t)$ to the free end of a semi-infinite rod, this is the same as if we applied the load to a single dashpot. Thus, the point of application will respond with a velocity $\dot{u}(t) = F(t)/\rho C_r A$. Thereafter, a particle velocity wavelet $\dot{u}(t + x / C_r)$ will begin to propagate through the rod to the left of the point of application of the load, as shown in Figure 5.6 for the case of a triangular load of finite duration t_d. Notice that after passage of the triangular wavelet, the rod remains motionless and stress free in a displaced position.

Example 1: Infinite Rod with Lumped Mass Attached

Consider an infinite rod with a concentrated mass m lumped at $x = 0$. The rod is subjected to a wavelet $U(t + x/C_r)$ that approaches the mass from the right. Determine the motion of the mass and the scattered waves (i.e., the transmitted and reflected waves).

To solve this problem, we begin by observing that if the mass were not there, no interaction would take place, and the motion would simply be that caused by the incident wavelet. We refer to this motion as the *incident field*, which at $x = 0$ we denote as $u_0 = U(t)$. The deviation in motion $u - u_0$ of the mass from the incident field, which is the *scattered field*, must surely be caused by the reaction (inertia) forces transmitted by the mass to the rod. These forces are local to the neighborhood $x = 0$, so it suffices to consider that location only. Hence, if we isolate the mass and represent each of the two semi-infinite rods to the left and to the right by means of a dashpot, we see that the dynamic equilibrium equation for the scattered field must be that of a single degree of freedom (SDOF) system, namely

$$2D(\dot{u} - \dot{u}_0) = -m\ddot{u} \tag{5.41}$$

That is,

$$m\ddot{u} + 2D\dot{u} = 2D\dot{u}_0(t) \tag{5.42}$$

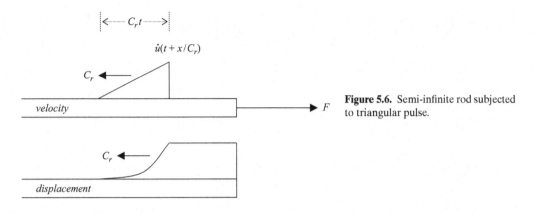

Figure 5.6. Semi-infinite rod subjected to triangular pulse.

Figure 5.7. Infinite rod with lumped mass.

Integrating once and disposing of the integration constant on account of the fact that there is no motion of the mass before the wavelet arrives, we obtain

$$m\ddot{u} + 2Du = 2Du_0(t) \tag{5.43}$$

The general solution to this equation is the sum of the homogeneous solution and the particular solution, which is

$$u(t) = Ae^{-2Dt/m} + u_p \tag{5.44}$$

in which A is a constant of integration that depends on both the initial conditions and the initial values of the particular solution u_p. The particular solution depends in turn on the form of the wavelet. Having found the total field $u(t)$ as well as the scattered field $v(t) = u(t) - u_0(t)$ at the location of the mass, we can proceed to compute the total field elsewhere as the sum of the scattered and incident fields, that is,

$$\begin{aligned} u(x,t) &= v(t - x/C_r) + U(t + x/C_r) & x > 0 \\ &= u(t - x/C_r) - U(t - x/C_r) + U(t + x/C_r) \end{aligned} \tag{5.45}$$

$$\begin{aligned} u(x,t) &= v(t + x/C_r) + U(t + x/C_r) & x < 0 \\ &= u(t + x/C_r) \end{aligned} \tag{5.46}$$

Notice that if $m = 0$, then $u = u_0$ and the particular solution is identical to the incident field, while the exponential term in the homogeneous solution decays instantly to zero. Hence, no interaction takes place, as expected.

Example 2: Rod Impinging onto a Rigid Surface
Consider an initially unstressed rod of finite length L impinging head on onto a flat, perfectly rigid surface, and doing so with axial velocity V. Disregarding gravitational effects, a shock wave ensues that propagates through the rod and after a brief instant, the rod bounces off. We now wish to determine the stresses in the rod and the duration of contact with the flat surface.

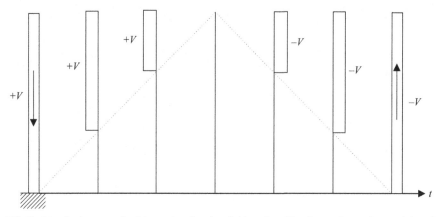

Figure 5.8. Rod impinging onto (and bouncing from) a rigid surface. The figure shows the particle velocity at various times during impact.

At the instant of the collision, the bottom end of the rod abruptly changes velocity from V to 0, which elicits an axial force $N = \rho C_r A V$ that equals the ground reaction. A compression wavefront then begins to move up through the rod with velocity C_r. The rod section below the wavefront remains quiescent (i.e., does not move) and is subjected to a compressed state with axial force N, while above the wavefront, the rod is unstressed and still falling with velocity V. Figuratively speaking, the upper part does not yet *know* that the bottom part has collided. Eventually, the wavefront reaches the upper end, and because the axial stress there must be zero, a tensile wavefront is reflected downward with velocity C_r that cancels the incident compressive wavefront. The portion above the reflected wavefront goes thus into an unstressed state, while the particles there move *upwards* with velocity V (i.e., are already bouncing off). Eventually, the reflected wavefront reaches the bottom, at which point in time the entire rod is again stress free, and all particles are moving up with velocity V. At that instant, the rod loses contact with the surface. Hence, the total contact time is the time that it takes for the incident and reflected wavefronts to move up and down the rod, that is $T = 2L/C_r$, which equals the fundamental period of the rod. Figure 5.8 illustrates the particle velocities in the rod at seven instants in time. The rod segments that lie below the dashed line do not move and are in compression.

Notice that in the collision phenomenon described in the preceding text, the rod fully recovers its kinetic energy after the impact, which means that energy is conserved. However, if we were to carry out an actual experiment, we would surely hear a noise associated with the impact, and a distinctive ringing after the rod bounced off, even if the rod were to hit the surface at a perfectly normal angle. This is direct evidence that the model proposed above is only an approximation. The vibrations in the rod, which are transmitted through the air as audible noise (i.e., energy), are caused by a variety of sources. These include the finite rigidity of the surface at the point of impact as well as the presence of waves in the rod other than the simple compressive waves elaborated on in the preceding.

5.2 Waves in Layered Media via Spectral Elements

Having discussed the propagation of waves in simple systems, we can now elaborate on the much more complicated case of waves in unbounded, horizontally layered media. For this purpose, we make use herein of wave-based elements or *spectral elements*, which provide *exact* solutions expressed in terms of *dynamic stiffness matrices* for each individual layer, which are valid no matter how thick the layers may be (for further details see also chapter 10 in Kausel, 2006[1]).

Let us examine first the case of a full, homogeneous, isotropic, elastic space. A full space has no boundaries, so it cannot sustain *guided waves*, nor does it exhibit any resonances or preferred directions. Hence, in the absence of sources, only *body waves* in the form of longitudinal (dilatational, primary, pressure or P) waves and transverse (secondary, shear or S) waves can propagate along arbitrary directions in such a full space. Depending on the coordinate system being used, these P and S waves could take the form of plane waves, cylindrical waves, or spherical waves. A full space has no preferred propagation modes, which means that S–P waves of any frequency and any wavenumber are admissible.

By contrast, a homogeneous half-space has a free boundary that admits a nondispersive guided wave, namely the Rayleigh wave. In addition, it may support body waves, among which we distinguish between the *SV–P waves* (vertically polarized S and P waves, which exhibit particle motions in the vertical plane defined by the normal to the surface and the direction of propagation), and *SH waves* (horizontally polarized shear waves involving particle motions in horizontal planes parallel to the free surface and perpendicular to the plane of propagation). In general, when a body wave hits the free surface, it both reflects and also converts partially to other waves, a phenomenon referred to as *mode conversion*. In particular, an SV wave reflects as an SV wave and also converts partially to a reflected P wave, provided the angle of incidence is not too shallow, that is, it is greater than some critical angle. The addition of one or more layers to the half-space greatly complicates the picture. Now a host of guided waves will exist, which are dispersive; these consist of SV–P modes (generalized Rayleigh waves and Stoneley waves) and of SH modes (generalized Love waves and torsional waves). In addition, complex branches for evanescent modes and leaky modes will exist as well.

A very important – indeed fundamental – observation that should be made about waves in horizontally layered media is that the wave spectra for SV–P and SH waves do *not* depend on whether plane strain waves or cylindrical waves are being considered, even if the displacements patterns elicited by such waves are not identical. For example, guided torsional (cylindrical SH) waves in a layer over an elastic half-space exhibit *exactly* the same dispersion characteristics as plane Love waves in that *same medium*; that is, they propagate at exactly the same speed for a given frequency. Thus, it suffices for us to consider in detail only the two plane strain cases for SH and SV–P waves, which we then generalize without much ado to cylindrical waves.

5.2.1 SH Waves and Generalized Love Waves

Consider a horizontally stratified soil consisting of layers that are laterally homogenous. Using a right-handed Cartesian coordinates with x from left to right and z up, the equation of motion for SH waves within a layer is of the form

$$\rho \ddot{u}_y = b_y + \mu \left(\frac{\partial^2 u_y}{\partial x^2} + \frac{\partial^2 u_y}{\partial z^2} \right) \tag{5.47}$$

[1] E. Kausel, *Fundamental Solutions in Elastodynamics* (Cambridge: Cambridge University Press, 2006).

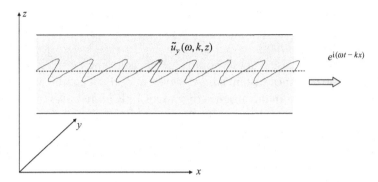

Figure 5.9. Single-layer displacements in the frequency-wavenumber domain.

where u_y is the *antiplane* displacement caused by SH waves (i.e., along the y-direction into the paper), b_y is the body force, and ρ, μ are the mass density and shear modulus of the layer. Applying a Fourier transform in x and t of the form

$$\tilde{u}_y(k,\omega) = \int_{-\infty}^{+\infty} \int_{-\infty}^{+\infty} u_y(x,t)e^{-i(\omega t - kx)}dx\,dt \tag{5.48}$$

and a similar transform on the body load, we obtain

$$-\omega^2 \rho \tilde{u}_y = \tilde{b}_y + \mu\left(-k^2 \tilde{u}_y + \frac{\partial^2 \tilde{u}_y}{\partial z^2}\right) \tag{5.49}$$

This is the wave equation expressed in the frequency–wavenumber domain. A visualization of the transformed quantities is depicted in Figure 5.9. In the absence of body sources, this equation can be written as a differential equation in z:

$$\frac{\partial^2 \tilde{u}_y}{\partial z^2} + \left(k_S^2 - k^2\right)\tilde{u}_y = 0 \tag{5.50}$$

with

$$\boxed{k_S = \frac{\omega}{C_S}}, \qquad C_S = \sqrt{\frac{\mu}{\rho}} \tag{5.51}$$

whose solution is

$$\tilde{u}_y = A_1\,e^{-ksz} + A_2\,e^{ksz} = A_1\,e^{-i\beta z} + A_2\,e^{i\beta z} \tag{5.52}$$

with A_1, A_2 being arbitrary constants, and

$$s = \sqrt{1 - \left(\frac{k}{k_S}\right)^2}, \qquad ks = i\,\beta \tag{5.53}$$

$$\boxed{\beta = \sqrt{k_S^2 - k^2}} \qquad \begin{cases} \mathrm{Re}\,\beta \geq 0 \\ \mathrm{Im}\,\beta \leq 0 \end{cases} = \text{vertical wavenumber} \tag{5.54}$$

From the preceding we obtain the shearing stresses in horizontal planes as

$$\tilde{\tau}_{yz} = \mu \frac{\partial \tilde{u}_y}{\partial z} = i\mu\beta \left(A_1 e^{-i\beta z} - A_2 e^{i\beta z} \right) \tag{5.55}$$

Consider now a single layer of thickness h as a *free body*, and assign (temporarily) the subindices 1, 2 to the upper and lower surfaces, respectively. To preserve dynamic equilibrium, we apply external tractions p_1, p_2 at these surfaces so as to match the internal stresses at these locations. Setting the origin of coordinates at the lower horizon (at which $z = 0$), writing the displacement and stresses in matrix format, and evaluating these at the upper and lower external surfaces, we obtain

$$\begin{Bmatrix} \tilde{u}_1 \\ \tilde{u}_2 \end{Bmatrix} = \begin{Bmatrix} e^{-i\beta h} & e^{i\beta h} \\ 1 & 1 \end{Bmatrix} \begin{Bmatrix} A_1 \\ A_2 \end{Bmatrix} \tag{5.56}$$

and

$$\begin{Bmatrix} \tilde{p}_1 \\ \tilde{p}_2 \end{Bmatrix} = \begin{Bmatrix} \tilde{\tau}_1 \\ -\tilde{\tau}_2 \end{Bmatrix} = i\mu\beta \begin{Bmatrix} e^{-i\beta h} & -e^{i\beta h} \\ -1 & 1 \end{Bmatrix} \begin{Bmatrix} A_1 \\ A_2 \end{Bmatrix} \tag{5.57}$$

Here we reversed the sign of the internal stress component at the lower interface so as to express this stress as an external traction. Eliminating the integration constants A_1, A_2 between these two expressions, we obtain after brief algebra

$$\begin{Bmatrix} \tilde{p}_1 \\ \tilde{p}_2 \end{Bmatrix} = \frac{i\mu\beta}{e^{i\beta h} - e^{-i\beta h}} \begin{Bmatrix} e^{i\beta h} + e^{-i\beta h} & -2 \\ -2 & e^{i\beta h} + e^{-i\beta h} \end{Bmatrix} \begin{Bmatrix} \tilde{u}_1 \\ \tilde{u}_2 \end{Bmatrix} \tag{5.58}$$

which can be written as

$$\begin{Bmatrix} \tilde{p}_1 \\ \tilde{p}_2 \end{Bmatrix} = \frac{\mu\beta}{\sin\beta h} \begin{Bmatrix} \cos\beta h & -1 \\ -1 & \cos\beta h \end{Bmatrix} \begin{Bmatrix} \tilde{u}_1 \\ \tilde{u}_2 \end{Bmatrix} \tag{5.59}$$

This equation can be written in compact form as

$$\tilde{\mathbf{p}}_l = \mathbf{K}_l \tilde{\mathbf{u}}_l \tag{5.60}$$

with

$$\boxed{\mathbf{K}_l = \mu\beta \begin{Bmatrix} \cot\beta h & -\sin^{-1}\beta h \\ -\sin^{-1}\beta h & \cot\beta h \end{Bmatrix}} \tag{5.61}$$

being the symmetric *stiffness matrix* of the lth layer. This relationship is *exact*, and is valid for any frequency ω, horizontal wavenumber k, and layer thickness h.

Before moving on to consider an ensemble of layers in a stratified soil, we first examine the special case of a homogeneous elastic half-space, that is, of an infinitely deep layer. In this case $A_2 = 0$ (i.e., no waves moving up), in which case

$$\tilde{u}_y = A_1 e^{i\beta z} \tag{5.62}$$

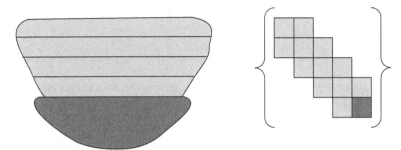

Figure 5.10. Construction of stiffness matrix by overlapping layer matrices.

and

$$p_y = \tilde{\tau}_{yz} = i\mu\beta e^{i\beta z} A_1 \tag{5.63}$$

which leads immediately to

$$p_y = i\mu\beta \tilde{u}_y \tag{5.64}$$

Hence

$$\boxed{K_H = i\mu\beta} \tag{5.65}$$

is the stiffness (impedance) of the half-space, as seen from its upper surface.

We can now generalize the preceding results to a stack of N parallel layers that may or not be underlain by an elastic half-space. To obtain the stiffness matrix for the global system of layers, it suffices for us to overlap the stiffness matrices of the individual layers at the appropriate interface locations (analogous to the assembly of the global stiffness matrix of a structural system obtained by overlapping member stiffness matrices); see Figure 5.10.

Renumbering the layer interfaces from the top down, the final result is a matrix equation of the form

$$\tilde{p} = K\tilde{u} \tag{5.66}$$

$$\tilde{u} = \begin{bmatrix} \tilde{u}_1 & \tilde{u}_2 & \cdots & \tilde{u}_{N+1} \end{bmatrix}^T = \tilde{u}(\omega,k) \tag{5.67}$$

$$\tilde{p} = \tilde{p}(\omega,k) \tag{5.68}$$

in which K is the symmetric, block-tridiagonal, global stiffness matrix; $\tilde{p}$ is the vector of external interface tractions (i.e., sources); and $\tilde{u}$ is the vector of interface displacements. We shall discuss the use of these matrices in the ensuing, but before doing so, we first provide (without proof) the expressions needed to obtain the displacements and stresses in the interior of any layer and/or half-space by means of the so-called *analytic continuation*.

With reference to Figure 5.11, these are as follows.

(a) Layer

$$\boxed{\tilde{u}(\zeta) = \Delta u \frac{\sin\frac{1}{2}\zeta\beta h}{\sin\frac{1}{2}\beta h} + u_m \frac{\cos\frac{1}{2}\zeta\beta h}{\cos\frac{1}{2}\beta h}} \qquad \boxed{\tilde{\tau}_{yz}(\zeta) = \mu\beta \left\{ \Delta u \frac{\cos\frac{1}{2}\zeta\beta h}{\sin\frac{1}{2}\beta h} - u_m \frac{\sin\frac{1}{2}\zeta\beta h}{\cos\frac{1}{2}\beta h} \right\}} \tag{5.69}$$

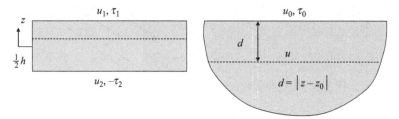

Figure 5.11. Analytic continuation in layer and half-space.

in which

$$\Delta u = \tfrac{1}{2}\big(\tilde{u}_1 - \tilde{u}_2\big), \qquad u_m = \tfrac{1}{2}\big(\tilde{u}_1 + \tilde{u}_2\big), \qquad \zeta = 2z/h \tag{5.70}$$

($z = 0$ at mid-height, $\zeta = \pm 1$ at top and bottom)

(b) Half-space

$$\tilde{u} = \tilde{u}_0\, e^{-i\beta d} \tag{5.71}$$

$$\tilde{\tau} = \tilde{\tau}_0\, e^{-i\beta d} \qquad d \geq 0 = \text{depth below half-space interface} \tag{5.72}$$

provided that there are no sources within the half-space.

Having obtained the global equilibrium equation, we proceed to describe its uses. The following three problems can be addressed by means of the stiffness matrix formulation:

- Normal modes: Determine the generalized Love modes in a system of layers.
- Source problem: Apply a dynamic source in some layer or in the half-space.
- Wave amplification: Given a plane wave that impinges the layers from the half-space underneath, determine the motions anywhere in the layers.

(A) Normal Modes

These are obtained by considering free waves, that is, waves in the absence of sources. This is accomplished by setting the external tractions (source) vector $\tilde{p}$ to zero; that is,

$$\mathbf{K}\tilde{u} = 0 \tag{5.73}$$

A nontrivial solution exists only if the determinant of $\mathbf{K}$ vanishes, that is $|\mathbf{K}| = 0$. To illustrate matters, consider the case of a single elastic layer underlain by an elastic half-space. The global stiffness matrix is

$$\mathbf{K}_{\text{system}} = \begin{Bmatrix} \mu_1\beta_1 \cot\beta_1 h_1 & -\mu_1\beta_1 \sin^{-1}\beta_1 h_1 \\ -\mu_1\beta_1 \sin^{-1}\beta_1 h_1 & \mu_1\beta_1 \cot\beta_1 h_1 + i\mu_2\beta_2 \end{Bmatrix} \tag{5.74}$$

$$\beta_l = h\sqrt{k_l^2 - k^2} \qquad k_l = \frac{\omega}{C_l} \qquad l = 1,2 \tag{5.75}$$

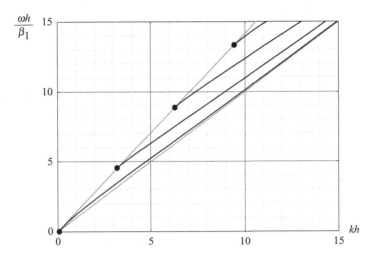

Figure 5.12. Dispersion spectrum for Love waves.

in which $l = 1$ refers to the layer, and $l = 2$ refers to the half-space. Setting the determinant of this matrix to zero, we obtain

$$\cot \beta_1 h_1 \left(\cot \beta_1 h_1 + \frac{i \mu_2 \beta_2}{\mu_1 \beta_1} \right) - \sin^{-2} \beta_1 h_1 = 0 \tag{5.76}$$

which reduces to

$$\tan h_1 \sqrt{k_1^2 - k^2} = \frac{\mu_2}{\mu_1} \frac{\sqrt{k^2 - k_2^2}}{\sqrt{k_1^2 - k^2}} \tag{5.77}$$

This is the classical characteristic equation for Love waves, whose solution (Figure 5.12) is well known. Having obtained the characteristic eigenvalues k, the *modal shapes* are obtained by introducing the eigenvalues found into the characteristic equation, setting arbitrarily the first component to 1, and solving for the remaining components (in this case, only one). Hence,

$$\phi = \left\{ \begin{array}{c} 1 \\ \cos \beta_1 h_1 \end{array} \right\} = \left\{ \begin{array}{c} 1 \\ \cos h_1 \sqrt{k_1^2 - k^2} \end{array} \right\} \tag{5.78}$$

In general, solving the transcendental eigenvalue problem $|\mathbf{K}| = 0$ for an arbitrarily layered system is very difficult, especially so when both real and complex roots are sought. Although tedious and error-prone search techniques can be used for this purpose, a much more convenient and very robust technique is the Thin-Layer Method (TLM), which resorts to a discretization of the displacement field in the direction of layering. Further details on the TLM can be found in Barbosa and Kausel[2] and especially in the references therein.

[2] J. Barbosa and E. Kausel, "The thin-layer method in a cross-anisotropic 3D space," *Int. J. Numer. Methods Eng.*, 89, 2012, 537–560.

(B) Source Problem

This is carried out by first casting the source in terms of interface tractions, solving for the displacements, and then carrying out an inverse Fourier transform into space–time. We illustrate the procedure summarized in Box 5.1 by means of a simple example, namely a single layer underlain by an elastic half-space that contains an antiplane line source at elevation z_0. Refer also to the first two boxed equations of the previous pages giving the impedance of the layer and half-space as well as Figures 5.10 and 5.11.

Box 5.1   Summary of steps in integral transform method for SH waves

i) Source at depth $d = |z_0|$ in the half-space: $p_y(x, z_0, t) = \delta(x)\,\delta(t)$

ii) Fourier transform of source: $\tilde{p}_y(\omega, k, z_0) = 1$ ⟶

iii) Equilibrium equation: $\mathbf{K}\tilde{\mathbf{u}} = \tilde{\mathbf{p}}$

or in full

$$\begin{bmatrix} \beta_1\mu_1 \cot\beta_1 h & -\beta_1\mu_1/\sin\beta_1 h & 0 \\ -\beta_1\mu_1/\sin\beta_1 h & \beta_1\mu_1\cot\beta_1 h + \beta_2\mu_2\cot\beta_2 d & -\beta_2\mu_2/\sin\beta_2 d \\ 0 & -\beta_2\mu_2/\sin\beta_2 d & \beta_2\mu_2(\cot\beta_2 d + i) \end{bmatrix} \begin{Bmatrix} \tilde{u}_{y1} \\ \tilde{u}_{y2} \\ \tilde{u}_y \end{Bmatrix} = \begin{Bmatrix} 0 \\ 0 \\ 1 \end{Bmatrix} ⟵$$

iv) Solve for displacements for a dense set of k, ω: $\tilde{\mathbf{u}} = \mathbf{K}^{-1}\tilde{\mathbf{p}}$

v) Carry out an inverse Fourier transform

$$\mathbf{u}(x, t) = \left(\tfrac{1}{2\pi}\right)^2 \int_{-\infty}^{+\infty}\int_{-\infty}^{+\infty} \tilde{\mathbf{u}}(k, \omega)\, e^{i(\omega t - kx)}\, dk\, d\omega$$

It should be observed that the preceding outline glosses over many issues, such as the numerical method used to evaluate the inverse transform, the wavenumber step, or the highest wavenumber used in the computation of the improper integrals. We shall elaborate briefly on these issues in Section 5.2.4, although a complete treatment is beyond the scope of this book.

(C) Wave Amplification Problem

Consider a layered system underlain by an elastic half-space with shear wave velocity C_r ("rock") that is subjected to plane waves of amplitude A that propagate upwards in the half-space at some angle ϑ with respect to the vertical, as shown on the left in Figure 5.13. In the absence of reflections and feedback from the layers (i.e., if the half-space were instead a full space, as shown in Figure 5.13 on the right), this incident wave would elicit displacements

$$u_y^* = A e^{-i(k_x x + k_z z)} \equiv A e^{-ik_z z}\left(e^{-ikx}\right) \tag{5.79}$$

$$k_x \equiv k = \frac{\omega}{C_r}\sin\vartheta \tag{5.80}$$

$$k_z = \frac{\omega}{C_r}\cos\vartheta \tag{5.81}$$

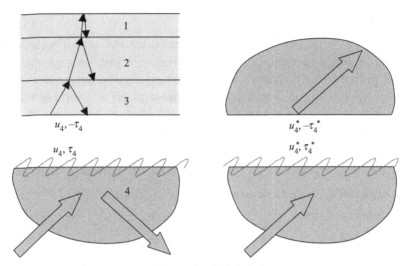

Figure 5.13. Layers over half-space versus full space (exploded view).

We refer to these displacements as belonging to the *free field*, a fact that we emphasize by means of a superscript star. Evaluating the free field displacement at the surface of the half-space (for which we conveniently take $z = 0$), and discarding the exponential term in x (because it is common to all N layers, so it drops out), we obtain $u_{N+1}^* = A$. We shall now show that the solution for the layers can be obtained by applying a fictitious source at the half-space interface of the form

$$\tilde{p}_{N+1} = K_{\text{full}} u_{N+1}^* = K_{\text{full}} A \tag{5.82}$$

in which $K_{\text{full}} = K_{\text{lower}} + K_{\text{upper}} = 2i\beta_r \mu_r$ is the impedance of an imagined homogeneous *full* space, as seen from the elevation of the half-space interface with the layers.

With reference to Figure 5.13, consider a system of three layers underlain by an elastic half-space. On the left in this figure, we show an "exploded" view in which the layers are separated as a free body from the half-space. On the right is an imagined reference system consisting of a full, homogenous space, and again, we separate the upper part from the lower. Clearly, in the latter case, the incident wave produces no reflections or refractions, so the wave continues unperturbed to the upper part. Comparing next the lower parts on the left with that on the right, we observe that they are geometrically and materially *identical*, and that they are also subjected to *exactly* the same source, yet the displacements and stresses on the surface are still different. Clearly, it must be the *difference* in stresses on the surface – a secondary source – that is causing the *difference* in displacements, and giving rise to the downgoing wave on the left. Hence, these differences must be related by the stiffness of the half-space as

$$\tau_4 - \tau_4^* = K_{\text{lower}} \left(u_4 - u_4^* \right) = i\beta_{\text{full}} \mu_{\text{full}} \left(u_4 - u_4^* \right) \tag{5.83}$$

On the other hand, the upper half-space on the right is free from sources, so it too must obey a proportionality condition between stresses and displacements:

$$-\tau_4^* = K_{\text{upper}} u_4^* = i\beta_4 \mu_4 u_4^* \tag{5.84}$$

Combining these two expressions, we obtain

$$-\tau_4 = -K_{\text{lower}}u_4 + \left(K_{\text{lower}} + K_{\text{upper}}\right)u_4^* = -K_{\text{lower}}u_4 + K_{\text{full}}u_4^* \qquad (5.85)$$

Introducing this result into the equilibrium equation for the upper layers and moving the term in $K_{\text{lower}}u_4$ to the left hand side, we finally obtain the global system equation

$$\begin{Bmatrix} K_{11}^1 & K_{12}^1 & 0 & 0 \\ K_{21}^1 & K_{22}^1 + K_{11}^2 & K_{12}^2 & 0 \\ 0 & K_{21}^2 & K_{22}^2 + K_{11}^3 & K_{12}^3 \\ 0 & 0 & K_{21}^3 & K_{22}^3 + K_{\text{lower}} \end{Bmatrix} \begin{Bmatrix} u_1 \\ u_2 \\ u_3 \\ u_4 \end{Bmatrix} = \begin{Bmatrix} 0 \\ 0 \\ 0 \\ K_{\text{full}}u_4^* \end{Bmatrix} \qquad (5.86)$$

in which $K_{full} = 2i\beta_4\mu_4$ is the impedance of the full space, and $u_4^* = A$ is again the amplitude of the incident wave, and

$$\beta_\ell = \sqrt{k_\ell^2 - k^2} = \sqrt{\left(\frac{\omega}{C_\ell}\right)^2 - \left(\frac{\omega}{C_4}\sin\vartheta\right)^2} = \frac{\omega}{C_\ell}\sqrt{1 - \left(\frac{C_\ell}{C_4}\right)^2 \sin^2\vartheta}, \qquad \ell = 1,\dots 4 \qquad (5.87)$$

Example: One-Dimensional Amplification of SH Waves

Consider a homogeneous layer of arbitrary thickness h, modulus G_{soil}, and mass density ρ_{soil} that is underlain by a homogeneous half-space of shear modulus G_{rock} and mass density ρ_{rock}. This system is subjected to vertically propagating SH waves (i.e., $\vartheta = 0, k = 0$ so $\beta_\ell = k_\ell$). These waves get partially transmitted to the layer, get partially reflected, and there is also some wave amplification. The motion amplitude is specified at rock outcrop, that is, at the location of the soil–rock interface before the soil is "added" (or taken into account), which results in a free rock surface. The incident amplitude is $A_{\text{SH}} = \frac{1}{2} \equiv u_{\text{rock}}^*$, which at rock outcrop would produce a motion of unit amplitude, that is, $u_{\text{outcrop}}^* = 2u_{\text{rock}}^* = 1$ (the reflected wave at rock outcrop has the same amplitude as the incident wave, and both add up at the outcrop surface). From the preceding, we have then

$$\begin{Bmatrix} K_{11}^1 & K_{12}^1 \\ K_{21}^1 & K_{22}^1 + K_{\text{lower}} \end{Bmatrix} \begin{Bmatrix} u_{\text{surf}} \\ u_{\text{rock}} \end{Bmatrix} = \begin{Bmatrix} 0 \\ K_{\text{full}}A_{\text{SH}} \end{Bmatrix} = \begin{Bmatrix} 0 \\ K_{\text{rock}} \end{Bmatrix} \qquad (5.88)$$

because $K_{\text{full}} = 2K_{\text{rock}}$. Substituting the expressions for the impedance matrix of the layer and the scalar impedance of the half-space for the data given, we obtain after brief algebra

$$\eta\frac{G_{\text{soil}}}{h} \begin{Bmatrix} \cot\eta & -\dfrac{1}{\sin\eta} \\ -\dfrac{1}{\sin\eta} & \cot\eta + i\dfrac{\rho_{\text{rock}}C_{\text{rock}}}{\rho_{\text{soil}}C_{\text{soil}}} \end{Bmatrix} \begin{Bmatrix} u_{\text{surf}} \\ u_{\text{rock}} \end{Bmatrix} = \begin{Bmatrix} 0 \\ i\omega\rho_{\text{rock}}C_{\text{rock}} \end{Bmatrix}, \qquad \eta = \frac{\omega h}{C_{\text{soil}}} \qquad (5.89)$$

We now define the *impedance contrast* as the ratio

$$\boxed{\chi = \frac{\rho_{\text{soil}}\,C_{\text{soil}}}{\rho_{\text{rock}}C_{\text{rock}}}} \qquad (5.90)$$

in which case the preceding simplifies into

$$
\begin{Bmatrix} \cos \eta & -1 \\ -1 & \cos \eta + i \dfrac{1}{\chi} \sin \eta \end{Bmatrix} \begin{Bmatrix} u_{\text{surf}} \\ u_{\text{rock}} \end{Bmatrix} = \begin{Bmatrix} 0 \\ i \dfrac{\sin \eta}{\chi} \end{Bmatrix}
\tag{5.91}
$$

the solution of which is

$$
\begin{Bmatrix} u_{\text{surf}} \\ u_{\text{rock}} \end{Bmatrix} = \frac{1}{\cos \eta + i \chi \sin \eta} \begin{Bmatrix} 1 \\ \cos \eta \end{Bmatrix}
\tag{5.92}
$$

The transfer function from rock outcrop to the free surface is then

$$
H_{\text{surf\,loutcrop}} = \frac{1}{\cos \eta + i \chi \sin \eta}, \qquad \left| H_{\text{surf\,loutcrop}} \right| = \frac{1}{\sqrt{\cos^2 \eta + \chi^2 \sin^2 \eta}}
\tag{5.93}
$$

At the resonant frequencies of the soil layer (if it were on rigid base), $\cos \eta = 0$, $\sin \eta = \pm 1$, at which point the (maximum) amplification is $\left| H \right|_{\max} = \chi^{-1}$. We see that this maximum amplification is similar to that of a shear beam on fixed base with uniform modal damping $\xi = \frac{1}{2} \chi$. Thus, the impedance ratio is a direct measure of *radiation damping*, that is, of earthquake energy that is fed back into the rock. A plot of the amplification function in terms of the frequency ratio ω / ω_1 for $\chi = 0.2$ (i.e., a fraction of damping of $\xi = 0.10$) is given in Figure 5.14. **Note:** See also the section in Chapter 7.4 entitled *Location of the Control Motion*.

5.2.2 SV-P Waves and Generalized Rayleigh Waves

We consider next the more complicated case of vertically polarized S and P waves. In general, SV–P problems are similar to those of SH waves, but much more involved, more difficult to solve, involve double the number of degrees of freedom and double the bandwidth, and the wave spectra are far more elaborate. Still, since the developments for SV–P

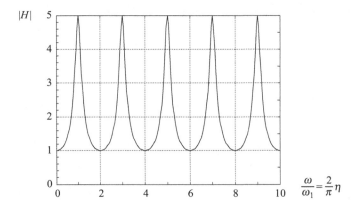

Figure 5.14. Amplification of SH waves with respect to rock outcrop, single layer over elastic half-space.

waves closely parallel the methods used for SH waves, we present only a sketch with the most essential parts. We begin with the in-plane dynamic equilibrium equation, which is

$$
\left\{ \begin{matrix} b_x \\ b_z \end{matrix} \right\} + \left[\begin{Bmatrix} \lambda+2\mu & 0 \\ 0 & \mu \end{Bmatrix} \frac{\partial^2}{\partial x^2} + \begin{Bmatrix} 0 & \lambda+\mu \\ \lambda+\mu & 0 \end{Bmatrix} \frac{\partial^2}{\partial x \partial z} \right.
$$
$$
\left. + \begin{Bmatrix} \mu & 0 \\ 0 & \lambda+2\mu \end{Bmatrix} \frac{\partial^2}{\partial z^2} - \begin{Bmatrix} \rho & 0 \\ 0 & \rho \end{Bmatrix} \frac{\partial^2}{\partial t^2} \right] \left\{ \begin{matrix} u_x \\ u_z \end{matrix} \right\} = \left\{ \begin{matrix} 0 \\ 0 \end{matrix} \right\} \tag{5.94}
$$

where we are assuming the vertical z coordinate and the displacement u_z to be positive when upward. Applying a spatiotemporal Fourier transform, and – to attain *symmetric* stiffness matrices – adding an imaginary factor $-\mathrm{i} = -\sqrt{-1}$ to the vertical components, we obtain

$$
\left\{ \begin{matrix} \tilde{b}_x \\ -\mathrm{i}\tilde{b}_z \end{matrix} \right\} + \left[\begin{Bmatrix} \mu & 0 \\ 0 & \lambda+2\mu \end{Bmatrix} \frac{\partial^2}{\partial z^2} + k \begin{Bmatrix} 0 & \lambda+\mu \\ \lambda+\mu & 0 \end{Bmatrix} \frac{\partial}{\partial z} + \rho\omega^2 \begin{Bmatrix} 1 & 0 \\ 0 & 1 \end{Bmatrix} \right.
$$
$$
\left. -k^2 \begin{Bmatrix} \lambda+2\mu & 0 \\ 0 & \mu \end{Bmatrix} \right] \left\{ \begin{matrix} \tilde{u}_x \\ -\mathrm{i}\tilde{u}_z \end{matrix} \right\} = \left\{ \begin{matrix} 0 \\ 0 \end{matrix} \right\} \tag{5.95}
$$

Solving the homogeneous equation (i.e., no body loads), we obtain

$$
\left\{ \begin{matrix} \tilde{u}_x \\ -\mathrm{i}\tilde{u}_z \end{matrix} \right\} = \left\{ \begin{matrix} 1 & -s & 1 & s \\ -p & 1 & p & 1 \end{matrix} \right\} \begin{bmatrix} e^{-\mathrm{i}\alpha z} & & & \\ & e^{-\mathrm{i}\beta z} & & \\ & & e^{\mathrm{i}\alpha z} & \\ & & & e^{\mathrm{i}\beta z} \end{bmatrix} \left\{ \begin{matrix} A_1 \\ A_2 \\ A_3 \\ A_4 \end{matrix} \right\} \tag{5.96}
$$

in which

$$
\alpha = -\mathrm{i}kp = \sqrt{k_P^2 - k^2} \qquad \begin{cases} \mathrm{Re}\,\alpha \geq 0 \\ \mathrm{Im}\,\alpha \leq 0 \end{cases} \tag{5.97}
$$

$$
\beta = -\mathrm{i}ks = \sqrt{k_S^2 - k^2} \qquad \begin{cases} \mathrm{Re}\,\beta \geq 0 \\ \mathrm{Im}\,\beta \leq 0 \end{cases} \tag{5.98}
$$

$$
p = \sqrt{1 - (k_P / k)^2} \qquad \begin{cases} \mathrm{Re}\,kp \geq 0 \\ \mathrm{Im}\,kp \geq 0 \end{cases} \tag{5.99}
$$

$$
s = \sqrt{1 - (k_S / k)^2} \qquad \begin{cases} \mathrm{Re}\,ks \geq 0 \\ \mathrm{Im}\,ks \geq 0 \end{cases} \tag{5.100}
$$

$$
k_P = \omega / C_P \tag{5.101}
$$

$$
k_S = \omega / C_S \tag{5.102}
$$

By differentiation, the stresses in horizontal planes are (observe the convenience factor $-\mathrm{i}$ added to the vertical stress component)

$$
\left\{ \begin{matrix} \tilde{\tau}_{xz} \\ -i\,\tilde{\tau}_{zz} \end{matrix} \right\} = 2k\mu \left\{ \begin{matrix} -p & \frac{1}{2}(1+s^2) & p & \frac{1}{2}(1+s^2) \\ \frac{1}{2}(1+s^2) & -s & \frac{1}{2}(1+s^2) & s \end{matrix} \right\} \left[\begin{matrix} e^{-i\alpha z} & & & \\ & e^{-i\beta z} & & \\ & & e^{i\alpha z} & \\ & & & e^{i\beta z} \end{matrix} \right] \left\{ \begin{matrix} A_1 \\ A_2 \\ A_3 \\ A_4 \end{matrix} \right\}
$$

$$(5.103)$$

Applying these two results to a single free layer, evaluating both at the top and bottom surfaces of the layer, and eliminating the amplitudes, we obtain

$$
\left\{ \begin{matrix} \tilde{p}_{x1} \\ -i\,\tilde{p}_{z1} \\ \hdashline \tilde{p}_{x2} \\ -i\,\tilde{p}_{z2} \end{matrix} \right\} = \left\{ \begin{matrix} \tilde{\tau}_{xz1} \\ -i\,\tilde{\tau}_{zz1} \\ \hdashline -\tilde{\tau}_{xz2} \\ -i\left(-\tilde{\tau}_{zz2}\right) \end{matrix} \right\} = \left\{ \begin{matrix} K_{11} & K_{12} & \vdots & K_{13} & K_{14} \\ K_{21} & K_{22} & \vdots & K_{23} & K_{24} \\ \hdashline K_{31} & K_{32} & \vdots & K_{33} & K_{34} \\ K_{41} & K_{42} & \vdots & K_{43} & K_{44} \end{matrix} \right\} \left\{ \begin{matrix} \tilde{u}_{x1} \\ -i\,\tilde{u}_{z1} \\ \hdashline \tilde{u}_{x2} \\ -i\,\tilde{u}_{z2} \end{matrix} \right\}
$$

$$(5.104)$$

or

$$
\left\{ \begin{matrix} \tilde{\mathbf{p}}_1 \\ \tilde{\mathbf{p}}_2 \end{matrix} \right\} = \left[\begin{matrix} \mathbf{K}_{11} & \mathbf{K}_{12} \\ \mathbf{K}_{21} & \mathbf{K}_{22} \end{matrix} \right] \left\{ \begin{matrix} \tilde{\mathbf{u}}_1 \\ \tilde{\mathbf{u}}_2 \end{matrix} \right\}
$$

$$(5.105)$$

which relates the external tractions to the interface displacements through a symmetric stiffness matrix. The elements of this stiffness matrix are as given in Box 5.2.[3]

Box 5.2

$$C_p = \cosh kph \qquad\qquad\qquad\qquad S_p = \sinh kph \qquad (5.106)$$

$$C_s = \cosh ksh \qquad\qquad\qquad\qquad S_s = \sinh ksh \qquad (5.107)$$

$$D = 2(1 - C_p C_s) + \left(\frac{1}{ps} + ps \right) S_p S_s \qquad (5.108)$$

$$\mathbf{K}_{11} = 2k\mu \left[\frac{1-s^2}{2D} \left\{ \begin{matrix} \frac{1}{s}(C_p S_s - ps\, C_s S_p) & 1 - C_p C_s + ps\, S_p S_s \\ 1 - C_p C_s + ps\, S_p S_s & \frac{1}{p}(C_s S_p - ps\, C_p S_s) \end{matrix} \right\} + \frac{1+s^2}{2} \left\{ \begin{matrix} 0 & 1 \\ 1 & 0 \end{matrix} \right\} \right] \qquad (5.109)$$

$$\mathbf{K}_{22} = \text{same as } \mathbf{K}_{11}, \text{ with off-diagonal signs reversed}$$

$$\mathbf{K}_{12} = 2k\mu \left[\frac{1-s^2}{2D} \left\{ \begin{matrix} \frac{1}{s}(ps\, S_p - S_s) & C_p - C_s \\ -(C_p - C_s) & \frac{1}{p}(ps\, S_s - S_p) \end{matrix} \right\} \right], \qquad \mathbf{K}_{21} = \mathbf{K}_{12}^T \qquad (5.110)$$

Also useful is the inverse of the coupling submatrix $\mathbf{K}_{12}$,

$$
\mathbf{K}_{12}^{-1} = \frac{1}{2k\mu} \left[\frac{2}{1-s^2} \left\{ \begin{matrix} \frac{1}{p}(ps\, S_s - S_p) & -(C_p - C_s) \\ C_p - C_s & \frac{1}{s}(ps\, S_p - S_s) \end{matrix} \right\} \right]
$$

$$(5.111)$$

---

[3] Kausel (2006).

The half-space stiffness matrix is obtained by discarding the terms for upgoing waves. The final result is

$$\mathbf{K}_{\text{lower}} = 2k\mu \left[\frac{1-s^2}{2(1-ps)} \begin{Bmatrix} p & -1 \\ -1 & s \end{Bmatrix} + \begin{Bmatrix} 0 & 1 \\ 1 & 0 \end{Bmatrix} \right] \qquad \text{Lower half-space} \qquad (5.112)$$

which satisfies

$$\Delta = ps - \left(\frac{1+s^2}{2}\right)^2 = \det(\mathbf{K}) = \text{Rayleigh function} \qquad (5.113)$$

In addition

$$\mathbf{K}_{\text{lower}}^{-1} = \frac{1}{2k\mu\Delta} \begin{Bmatrix} s\frac{1}{2}(1-s^2) & ps-\frac{1}{2}(1-s^2) \\ ps-\frac{1}{2}(1-s^2) & p\frac{1}{2}(1-s^2) \end{Bmatrix} \qquad (5.114)$$

The stiffness matrix for an upper half-space $\mathbf{K}_{\text{upper}}$ is the same as that of a lower, but with the off-diagonal signs reversed; that is,

$$\mathbf{K}_{\text{upper}} = 2k\mu \left[\frac{1-s^2}{2(1-ps)} \begin{Bmatrix} p & 1 \\ 1 & s \end{Bmatrix} - \begin{Bmatrix} 0 & 1 \\ 1 & 0 \end{Bmatrix} \right] \qquad (5.115)$$

Finally, the stiffness matrix (i.e., impedance) of a full space, as seen from any given horizon, is

$$\mathbf{K}_{\text{full}} = \mathbf{K}_{\text{lower}} + \mathbf{K}_{\text{upper}} = 2k\mu \frac{1-s^2}{1-ps} \begin{Bmatrix} p & 0 \\ 0 & s \end{Bmatrix} \qquad (5.116)$$

It should be observed that the elements of the stiffness matrix are assembled with hyperbolic functions of complex argument in both the numerator and the denominator, and although their ratio is finite, the individual terms may grow exponentially. To avoid potential numerical problems associated with underflow, overflow, and severe cancellation errors, it is necessary to make the following substitutions. Let

$$\begin{aligned} C_p &= \cosh kph = \cosh(a+ib) = \cosh a\cos b + i\sinh a\sin b \\ &= e^a \tfrac{1}{2}\left[(1+e^{-2a})\cos b + i(1-e^{-2a})\sin b\right] \\ &\underset{\text{def}}{=} e^a C_1 \end{aligned} \qquad (5.117)$$

Similarly

$$S_p = e^a S_1 \qquad (5.118)$$

$$C_s = \cosh ksh = \cosh(c+id) = e^c C_2 \qquad (5.119)$$

$$S_s = e^c S_2 \qquad (5.120)$$

Hence, an individual term such as the first one in $\mathbf{K}_{11}$ would be computed as

$$\begin{aligned} \frac{C_p S_s}{D} &= \frac{e^{a+c} C_1 S_2}{2(1-e^{a+c}C_1 C_2) + \left(\frac{1}{ps}+ps\right)e^{a+c}S_1 S_2} \\ &= \frac{C_1 S_2}{2(e^{-(a+c)} - C_1 C_2) + \left(\frac{1}{ps}+ps\right)S_1 S_2} \end{aligned} \qquad (5.121)$$

Observe that the exponential growth factor has canceled out, and the ratio is numerically stable.

As in the antiplane case, the global system matrices are obtained by overlapping the layer stiffness matrices, including the lower and/or upper half-spaces, if any. This results in a symmetric, block-tridiagonal system stiffness matrix. Additional useful expressions for the element of the stiffness matrices in the limit of zero wavenumber, zero frequency, or both can be found in Kausel (2006), op cit. After solving for the displacements, it is necessary to remove the additional imaginary factor applied to the vertical components, which was introduced to attain symmetry.

Normal Modes

As in the antiplane case, this is achieved by setting the determinant of the global matrix to zero. We illustrate this by means of the classical problem of Rayleigh waves. For a half-space without any added layers, this implies

$$\det(\mathbf{K}_{\text{lower}}) = \Delta = ps - \left(\frac{1+s^2}{2}\right)^2 = 0 \tag{5.122}$$

With the substitution $C_R = \omega / k$ (= speed of Rayleigh waves) we obtain the characteristic equation for Rayleigh waves

$$\sqrt{1-\left(\frac{C_R}{C_P}\right)^2}\sqrt{1-\left(\frac{C_R}{C_S}\right)^2} - \left(1-\frac{1}{2}\left(\frac{C_R}{C_S}\right)^2\right)^2 = 0 \tag{5.123}$$

which after rationalization leads to a cubic equation in C_R/C_S. This equation has three roots, one of which gives the speed of Rayleigh waves, and the other two are either real or complex, and do not represent physically meaningful waves. A very close solution for the celerity of Rayleigh waves that is valid for any Poisson's ratio is given by the polynomial approximation

$$C_R / C_s = 0.874 + 0.197v - 0.056v^2 - 0.027v^3 \tag{5.124}$$

5.2.3 Stiffness Matrix Method in Cylindrical Coordinates

It can be shown that when a horizontally layered system is described in cylindrical coordinates r, θ, z, and the Fourier transform of Cartesian coordinates is replaced by a Hankel (i.e., Fourier–Bessel) transform, the resulting equilibrium equations in the frequency–wavenumber domain are *identical* to a combination of the SH and SV–P system equations, except that the vertical components do *not* carry an imaginary unit factor. Thus, the traction–displacement equation in the frequency–radial wavenumber domain for a single layer is of the form

$$
\begin{Bmatrix} \tilde{p}_{r1} \\ \tilde{p}_{z1} \\ \tilde{p}_{r2} \\ \tilde{p}_{z2} \\ \tilde{p}_{\theta 1} \\ \tilde{p}_{\theta 2} \end{Bmatrix} = \begin{Bmatrix} \begin{array}{cccc|cc} K_{11}^{SVP} & K_{12}^{SVP} & K_{13}^{SVP} & K_{14}^{SVP} & 0 & 0 \\ K_{21}^{SVP} & K_{22}^{SVP} & K_{23}^{SVP} & K_{24}^{SVP} & 0 & 0 \\ K_{31}^{SVP} & K_{32}^{SVP} & K_{33}^{SVP} & K_{34}^{SVP} & 0 & 0 \\ K_{41}^{SVP} & K_{42}^{SVP} & K_{43}^{SVP} & K_{44}^{SVP} & 0 & 0 \\ \hline 0 & 0 & 0 & 0 & K_{11}^{SH} & K_{12}^{SH} \\ 0 & 0 & 0 & 0 & K_{21}^{SH} & K_{22}^{SH} \end{array} \end{Bmatrix} \begin{Bmatrix} \tilde{u}_{r1} \\ \tilde{u}_{z1} \\ \tilde{u}_{r2} \\ \tilde{u}_{z2} \\ \tilde{u}_{\theta 1} \\ \tilde{u}_{\theta 2} \end{Bmatrix} \qquad (5.125)
$$

Observe that the "in-plane" components are uncoupled from the "antiplane" components, so the global stiffness matrix necessarily decouples into two *separate* equations. It follows that when the normal modes in cylindrical coordinates are sought via $\det(\mathbf{K}) = 0$, one obtains exactly the same roots as $\det(\mathbf{K}_{SV-P}) = 0$ *and* $\det(\mathbf{K}_{SH}) = 0$. Hence, the wavenumber spectra for cylindrical waves, say, torsional guided waves, are *identical* to those in plane strain. Furthermore, these eigenvalue problems do not depend on the circumferential index n used to accomplish the Fourier expansion in the azimuth θ. In summary, finding the normal modes in cylindrical coordinates involves exactly the same eigenvalue problems as in the two plane strain cases.

For completeness, we provide also the necessary formulas and describe the computational steps for the case of *3-D* sources (say a point source) at some location in the medium. These are as follows.

(a) Express loads in ω–k via Hankel transform.

$$
\tilde{\mathbf{b}}_n(k,z,\omega) = a_n \int_0^\infty r \, \mathbf{C}_n \int_0^{2\pi} \mathbf{T}_n \int_{-\infty}^{+\infty} \mathbf{b}(r,\theta,z,t) \, e^{-i\omega t} \, dt \, d\theta \, dr \qquad (5.126)
$$

$$
a_n = \begin{cases} \dfrac{1}{2\pi} & n = 0 \\ \dfrac{1}{\pi} & n \neq 0 \end{cases} \qquad (5.127)
$$

$$
\mathbf{T}_n = \mathrm{diag}\left\{ \begin{pmatrix} \cos n\theta \\ \sin n\theta \end{pmatrix}, \begin{pmatrix} -\sin n\theta \\ \cos n\theta \end{pmatrix}, \begin{pmatrix} \cos n\theta \\ \sin n\theta \end{pmatrix} \right\} \quad \text{(either the upper or the lower entries)} \qquad (5.128)
$$

$$
\mathbf{C}_n = \begin{Bmatrix} J'_n & \dfrac{n}{kr} J_n & 0 \\ \dfrac{n}{kr} J_n & J'_n & 0 \\ 0 & 0 & J_n \end{Bmatrix} \qquad (5.129)
$$

$$
J_n = J_n(kr) = \text{Bessel function} \qquad (5.130)
$$

$$
J'_n = \frac{dJ_n}{d(kr)} \qquad (5.131)
$$

(b) Solve for the displacements.

$$\tilde{\mathbf{u}}_n = \mathbf{K}^{-1}\tilde{\mathbf{p}}_n \qquad \text{(note that } \mathbf{K} \text{ is } \textit{not} \text{ a function of } n\text{)} \tag{5.132}$$

(c) Obtain displacements in the spatial domain r, θ by inverse Hankel transform.

$$\mathbf{u}(r,\theta,z,t) = \frac{1}{2\pi}\int_{-\infty}^{+\infty} e^{i\omega t}\sum_{n=0}^{\infty}\mathbf{T}_n\int_0^{\infty} k\,\mathbf{C}_n\tilde{\mathbf{u}}_n(k,z,\omega)\,dk\,d\omega \tag{5.133}$$

Example
To illustrate matters, consider the classical problem of a point load applied tangentially at the surface of an elastic half-space (Chao's problem[4]). In cylindrical coordinates, the load can be expressed as

$$\mathbf{p}_1 = \begin{Bmatrix} p_r \\ p_\theta \\ p_z \end{Bmatrix}_{(1)} = \frac{1}{2\pi}\begin{Bmatrix} \cos\theta \\ -\sin\theta \\ 0 \end{Bmatrix}\frac{\delta(r)}{r} = \begin{bmatrix} \cos\theta & 0 & 0 \\ 0 & -\sin\theta & 0 \\ 0 & 0 & \cos\theta \end{bmatrix}\begin{bmatrix}1\\1\\0\end{bmatrix}\frac{\delta(r)}{2\pi r} = \mathbf{T}_1\begin{Bmatrix}1\\1\\0\end{Bmatrix}\frac{\delta(r)}{2\pi r} \tag{5.134}$$

We begin by transforming this into the wavenumber domain via a Hankel transform:

$$\tilde{\mathbf{p}}_1 = \begin{Bmatrix} \tilde{p}_r \\ \tilde{p}_\theta \\ \tilde{p}_z \end{Bmatrix}_{(1)} = \int_0^\infty \begin{bmatrix} J_1' & \frac{1}{kr}J_1 & 0 \\ \frac{1}{kr}J_1 & J_1' & 0 \\ 0 & 0 & J_1 \end{bmatrix}\begin{Bmatrix}1\\1\\0\end{Bmatrix}\frac{\delta(r)}{2\pi}\,dr = \int_0^\infty \begin{Bmatrix}J_0\\J_0\\0\end{Bmatrix}\frac{\delta(r)}{2\pi}\,dr = \frac{1}{2\pi}\begin{Bmatrix}1\\1\\0\end{Bmatrix} \tag{5.135}$$

Observe that because the load involves only the $n = 1$ azimuthal component, it follows that the displacements will exist solely for that component.

Next, as in the plane strain case, we solve separately for the SV–P and SH components of the displacements. This involves multiplication of the inverse of the half-space impedances (stiffnesses) by the loads in the frequency–wavenumber domain just obtained:

$$\begin{Bmatrix}\tilde{u}_r\\\tilde{u}_z\end{Bmatrix} = \frac{1}{2k\mu\Delta}\begin{Bmatrix} s\frac{1}{2}(1-s^2) & ps-\frac{1}{2}(1-s^2) \\ ps-\frac{1}{2}(1-s^2) & p\frac{1}{2}(1-s^2)\end{Bmatrix}\frac{1}{2\pi}\begin{Bmatrix}1\\0\end{Bmatrix} \tag{5.136}$$

$$\tilde{u}_\theta = \frac{1}{2\mu ks}\left(\frac{1}{2\pi}\right) \tag{5.137}$$

Thereafter, we apply an inverse Hankel transform

$$\mathbf{u} = \begin{Bmatrix} u_r \\ u_\theta \\ u_z \end{Bmatrix} = \mathbf{T}_1\int_0^\infty \begin{bmatrix} J_1' & \frac{1}{kr}J_1 & 0 \\ \frac{1}{kr}J_1 & J_1' & 0 \\ 0 & 0 & J_1 \end{bmatrix}\begin{Bmatrix} \tilde{u}_r \\ \tilde{u}_\theta \\ \tilde{u}_z \end{Bmatrix}_{(1)} k\,dk \tag{5.138}$$

Substituting the preceding results for the various components into the integral, we obtain ultimately

[4] C. C. Chao, "Dynamical response of an elastic half-space to tangential surface loading," *J. Appl. Mechan.*, 27, 1960, 559–567.

$$u_r = (\cos\theta)\left[\frac{1}{4\pi\mu}\int_0^\infty \frac{s^{\frac{1}{2}}(1-s^2)}{\Delta}\left(J_0(kr) - \frac{J_1(kr)}{kr}\right)dk + \frac{1}{2\pi\mu}\int_0^\infty \frac{1}{s}\frac{J_1(kr)}{kr}dk\right] \tag{5.139}$$

$$u_\theta = (-\sin\theta)\left[\frac{1}{4\pi\mu}\int_0^\infty \frac{s^{\frac{1}{2}}(1-s^2)}{\Delta}\frac{J_1(kr)}{kr}dk + \frac{1}{2\pi\mu}\int_0^\infty \frac{1}{s}\left(J_0(kr) - \frac{J_1(kr)}{kr}\right)dk\right] \tag{5.140}$$

$$u_z = (\cos\theta)\left[\frac{1}{4\pi\mu}\int_0^\infty \frac{ps - \frac{1}{2}(1+s^2)}{\Delta}J_1(kr)dk\right] \tag{5.141}$$

These expressions can be integrated analytically via contour integration, as Chao did for the particular case of $v = 0.25$, or for any arbitrary Poisson's ratio and any load direction as given by Kausel (2012[5]). A numerical solution involving a moderate computational effort is also possible, even if it that requires some care to give proper treatment to the singularity at the Rayleigh pole (i.e., $\Delta = 0$ when $k = k_R = \omega/C_R$), decide on the wavenumber sampling rate and how high the cutoff frequency should be, worry also about the tails of integral after that limit, and so forth).

5.2.4 Accurate Integration of Wavenumber Integrals

In the preceding three sections on the Stiffness Matrix Method (SMM) we learned how to calculate the response functions for layered media when harmonic sources are applied somewhere. Ultimately, we saw that this task required evaluation of improper integrals over either horizontal or radial wavenumbers. In most cases, such integrals are not amenable to closed form or analytical integration, so one must necessarily rely on numerical methods.

We also mentioned in passing that such improper integrals are fraught with difficulties that must adequately be dealt with, lest the results be highly inaccurate or even totally wrong. Although we cannot cover exhaustively all aspects of this problem herein, we shall provide nonetheless a summary of the important issues and suggest some remedies. These issues are as follows:

- The integrals are improper, that is, their upper limit is infinitely large. In addition, the kernels of the integrals decay relatively slowly with the wavenumber k, especially when the source and the receiver are at the same elevation. Thus, it behooves to decide appropriately on the maximum wavenumber, say, k_{max}, and at the same time provide a strategy to circumvent mere truncation so as to capture the contribution of the tail to that integral above k_{max}.
- In addition, the layers strongly couple the various interfaces, but at sufficiently high wavenumbers this coupling becomes negligible. One must then decide on how large the needed truncation wavenumber k_{max} should be.
- The integrands are highly wavy, and exhibit sharp peaks in the vicinity of the wavenumbers associated with the normal modes at the given frequency. One then needs to be able to estimate a priori how many such peaks can be expected.

---

[5] E. Kausel, "Lamb's problem at its simplest," *Proc. R. Soc. London A*, 2012, RSPA-20120462

- The integrals must be discretized by an appropriate choice of the wavenumber step. This step has to be small enough to properly model the sharp peaks in the integrands.

We discuss briefly each of these issues in turn.

Maximum Wavenumber for Truncation and Layer Coupling

We begin by showing that at sufficiently high wavenumber, the flexibility functions (the integrands) behave quasistatically (i.e., as if $\omega = 0$), and furthermore, that the layer interfaces decouple as if the interface belonged to two infinitely deep half-spaces (an upper and a lower half-space) with material properties equal to those of the two layers meeting at that interface. Thus, the layer stiffness (or impedance) matrices attain a block-diagonal structure, that is, $\mathbf{K}_{12} \to \mathbf{O}$ so that $\mathbf{K} \to [\mathbf{K}_{11}, \mathbf{O}; \mathbf{O}, \mathbf{K}_{22}]$. The simplest way to demonstrate this is by considering the *dispersion equations* for S and P waves for any of the layers, while omitting for simplicity the layer subindex. These equations are then

$$k_x^2 + k_{P_z}^2 = k_P^2 \equiv \left(\frac{\omega}{C_P}\right)^2 \qquad k_x^2 + k_{S_z}^2 = k_S^2 \equiv \left(\frac{\omega}{C_s}\right)^2 \tag{5.142}$$

where we have added an additional subscript P or S to the vertical wavenumbers so as to distinguish one from the other. Indeed, only the horizontal wavenumber is common to both, that is, $k_{P_x} = k_{S_x} = k_x \equiv k$. From here, we obtain

$$k_{P_z} = \pm\sqrt{k_P^2 - k^2} \qquad\qquad k_{S_z} = \pm\sqrt{k_S^2 - k^2} \tag{5.143}$$

We define

$$kp = -ik_{P_z} = \sqrt{k^2 - k_P^2} \qquad\qquad s = -ik_{S_z} = \sqrt{k^2 - k_S^2} \tag{5.144}$$

Hence, the propagation of P or S waves within a specific layer is given by expressions of the form

$$A_P \exp\left[i(\omega t - kx)\right]\exp\left(\mp kp|z|\right), \qquad A_S \exp\left[i(\omega t - kx)\right]\exp\left(\mp ks|z|\right) \tag{5.145}$$

where A_S, A_P are arbitrary amplitudes (they do not matter here) and the sign of the exponential depends on whether the wave moves up or down. If we focus attention on the last factor and evaluate it at $|z| = H$ (where z is measured from the interface being considered), and ask ourselves what value the horizontal wavenumber $k > k_S > k_P$ must have so that a wave that emanates from one interface will have decayed to nearly zero upon reaching the neighboring interface at a distance H, that is, for what value of k does the exponential term becomes negligibly small. Clearly this factor is given by

$$\exp\left(-ks|z|\right) \to \exp\left(-ksH\right) = \exp\left(-H\sqrt{k^2 - k_S^2}\right) < \varepsilon = 10^{-m} \tag{5.146}$$

(e.g., $m = 2$, or $m = 3$ will do). The solution is then

$$H\sqrt{k^2 - k_S^2} > m\ln 10 \qquad \text{or} \qquad k > \sqrt{\left(\frac{m\ln 10}{H}\right)^2 + k_S^2} \tag{5.147}$$

For $m = 2, 2\ln 10 = 4.61 \approx 4.71 = \frac{3}{2}\pi$ this would imply for the current layer

$$k > \sqrt{\left(\frac{\frac{3}{2}\pi}{H}\right)^2 + \left(\frac{\omega}{C_s}\right)^2} \tag{5.148}$$

Although a similar limit exists for the P-wave term, it need not be considered herein because $k_S > k_P$. The actual maximum wavenumber used in a computation depends on the shear wave velocities of the various layers as well as their thicknesses, but for practical purposes, it is easiest to choose the maximum for all of the layers on the basis of the minimum shear wave velocity and the minimum thickness, that is, $\min(C_{s\ell}), \min(H_\ell)$, say

$$\boxed{k_{\max} = \max\left\{\begin{array}{c} \sqrt{\left(\dfrac{\frac{3}{2}\pi}{\min H_\ell}\right)^2 + \left(\dfrac{\omega}{\min C_{s\ell}}\right)^2} \\[2ex] \dfrac{2\omega}{\min C_{s\ell}} \end{array}\right.} \tag{5.149}$$

Observe that the maximum wavenumber is a function of the current frequency.

Static Asymptotic Behavior: Tail of Integrals

At high wavenumbers not only do the impedance matrices decouple into block-diagonal form, but they also converge to the static solution, that is, $ks \to k, kp \to k$. Moreover, at a specific interface, the static SVP and SH impedances of two dissimilar upper (U) and lower (L) half-spaces with shear moduli G_U, G_L and Poisson's ratios ν_U, ν_L are

$$\mathbf{K}_{\text{SVP}} = 2k\left[\frac{G_L}{1 + a_L^2}\begin{Bmatrix} 1 & a_L^2 \\ a_L^2 & 1 \end{Bmatrix} + \frac{G_U}{1 + a_U^2}\begin{Bmatrix} 1 & -a_U^2 \\ -a_U^2 & 1 \end{Bmatrix}\right], \qquad K_{\text{SH}} = k(G_L + G_U) \tag{5.150}$$

$$a_L^2 = \frac{1 - 2\nu_L}{2 - 2\nu_L}, \qquad a_U^2 = \frac{1 - 2\nu_U}{2 - 2\nu_U} \tag{5.151}$$

(**Note**: the matrices above are checkerboard symmetric–antisymmetric with respect to positive and negative values of k (the off-diagonal terms change sign, the diagonal terms do not).

The inverses of these stiffnesses are

$$\mathbf{F}_{\text{SVP}} = \mathbf{K}_{\text{SVP}}^{-1} = \frac{1}{k}\begin{Bmatrix} f_{11} & f_{13} \\ f_{31} & f_{33} \end{Bmatrix}, \qquad F_{\text{SH}} = K_{\text{SH}}^{-1} = \frac{1}{k}f_{22} \tag{5.152}$$

with

$$f_{11} = f_{33} = \frac{1}{\Delta}\left[\frac{G_L}{1 + a_L^2} + \frac{G_U}{1 + a_U^2}\right] \tag{5.153}$$

$$f_{22} = 1/(G_L + G_U) \tag{5.154}$$

$$f_{31} = f_{13} = -\frac{1}{\Delta}\left[\frac{a_L^2 G_L}{1 + a_L^2} - \frac{a_U^2 G_U}{1 + a_U^2}\right] \tag{5.155}$$

$$\Delta = 2\left[\left(\frac{G_L}{1 + a_L^2} + \frac{G_U}{1 + a_U^2}\right)^2 - \left(\frac{a_L^2 G_L}{1 + a_L^2} - \frac{a_U^2 G_U}{1 + a_U^2}\right)^2\right] \tag{5.156}$$

As luck would have it, one can also compute the *exact* static solution due to line loads (2-D) and point loads (3-D) applied at the interface of two half-spaces, as will be shown. We begin with the-3-D case.

3-D: The *static* response due to *point loads* at the interface of two half-spaces is as follows:

Horizontal point load	Vertical point load
$u_{rx} = (\cos\theta)\frac{1}{2\pi} f_{22}$	$u_{rz} = -\frac{1}{2\pi} f_{13}$
$u_{\theta x} = (-\sin\theta)\frac{1}{2\pi} f_{11}$	$u_{\theta z} = 0$
$u_{zx} = (\cos\theta)\frac{1}{2\pi} f_{31}$	$u_z = \frac{1}{2\pi} f_{33}$

Setting $G_U = 0$, it can readily be shown that these expressions agree with the classical Cerruti and Boussinesq solutions for point loads applied onto the surface of an elastic half-space. The procedure to follow is then as follows:

- Subtract the static flexibilities $\mathbf{F}_{SVP}$ and F_{SH} given previously from the diagonal blocks of the dynamic flexibility matrices, using the appropriate material properties for each pair of layers (which here play the roles of "upper" and "lower" half-spaces). Do this for each and every wavenumber up to k_{max}. The net effect is that the tails will have vanished.
- Do numerically the wavenumber integrals (here a Hankel transform) up to k_{max} with the system impedance matrix from which the static terms have been subtracted.
- Add back the *exact* static solution just given.

2-D: Elastic half-spaces, whether layered or not, do not have a proper static response due to line loads, because a static load causes the displacements to diverge everywhere, that is, they are infinitely large. Thus, the procedure just described for point loads is not applicable to line loads. Fortunately, we can instead evaluate the tails themselves:

$$\frac{1}{2\pi}\int_{k_{max}}^{\infty} \mathbf{F}_{SVP}e^{-ikx}dk = \frac{1}{2\pi}\begin{Bmatrix} f_{11} & f_{13} \\ f_{31} & f_{33} \end{Bmatrix}\int_{k_{max}}^{\infty}\frac{e^{-ikx}}{k}dk$$
$$= (-)\frac{1}{2\pi}\begin{Bmatrix} f_{11} & f_{13} \\ f_{31} & f_{33} \end{Bmatrix}\left[\text{Ci}(k_{max}x) + i\left(\tfrac{\pi}{2} - \text{Si}(k_{max}x)\right)\right] \tag{5.157}$$

$$\frac{1}{2\pi}\int_{k_{max}}^{\infty} F_{SH}e^{-ikx}dk = (-)\frac{1}{2\pi}f_{22}\left[\text{Ci}(k_{max}x) + i\left(\tfrac{\pi}{2} - \text{Si}(k_{max}x)\right)\right] \tag{5.158}$$

where Ci, Si are the cosine-integral and sine-integral functions, for which effective routines exist. Alternatively, one can also express this in terms of the exponential integral Ei.

Thus, all that is required is to carry out the numerical integration up to k_{max} and then add the tails given above. Although these change from layer to layer, the factor in the integrals does not change, so the tem in square brackets need be evaluated only once.

Wavenumber Step

The wavenumber step Δk is a rather delicate matter. It is influenced by all of the following considerations:

- A discrete wavenumber summation automatically implies a periodic source with a spatial period of $L = 2\pi / \Delta k$. Hence, the step must be small enough to achieve a good separation to those "neighboring sources." If one is *not* interested in the Green's functions in the frequency domain themselves, but only care to use these as tools to obtain the time response, then the separation need only be as large as the distance traveled by the fastest waves from the neighboring source to the farthest receiver (maximum range) up to the maximum time of interest. But if the actual objective is the Green's functions in the frequency domain, one generally needs a larger distance to prevent contamination, in which case one relies on damping to attenuate the contribution of those neighboring sources. An alternative is also to use complex frequencies (see Chapter 6, Section 6.6.14).
- Depending on the method used to compute the wavenumber integrals, it is necessary to use a sufficient number of points to resolve the oscillations of the trigonometric or Bessel functions that multiply the kernels, which have arguments $kr \rightarrow k_{max}r_{max}$. Hence, the higher the frequency or the farther the range r_{max} (the largest distance to the receivers), the more points are needed.
- The step must also be small enough so that all poles (or nearly all of the poles) are properly resolved. By and large, as the frequency increases, so does the number of poles. In practice, we do not know ahead of time how many poles exist at any given frequency, but we can estimate these by determining the frequencies at which the stratified soil has resonances when the waves move vertically, that is, $k = 0$. (**Note:** in principle, a layered half-space has no resonant frequencies, but it exhibits strong near-resonances similar to those of plates and strata). This estimation follows from a fairly simple and inexpensive calculation involving the layered stratum obtained by making the half-space rigid, and the free layered plate obtained by removing that half-space. Suppose that these "resonant" frequencies are $\omega_1, \omega_2 \cdots$. Then, the number of poles for any wavenumber is the number of poles for $k = 0$ that exist below the frequency being considered, plus one. This is because at each cutoff frequency, we gain one more pole.

For example, for a layer on an elastic half-space, and for SH waves, the layer will have resonances at frequencies that lie between those of the stratum and the free plate (Love modes). Choosing the latter, these cutoff frequencies are $\omega_j = j\pi C_s / H/2$, so the number of poles estimated is $N = 2H\omega/(\pi C_s) + 1$, so at any given frequency and on average, the distance between peaks is on the order of k_{max}/N, and one would need to use just a small fraction of this quantity, say $\Delta k = 0.005 k_{max}/N$

One final observation: unless the response is needed at many equidistant receiver stations, it is generally not convenient to use either the FFT or some kind of FHT (Fast-Hankel transform) to carry out the integrations over wavenumbers. This is because a direct numerical integration for the response at a few receiving stations is usually more economical than using any of the fast transforms, and because that allows also using uneven spacing of the receivers. In addition, it allows for uneven wavenumber steps, say smaller in the vicinity of resonances, and fewer elsewhere (i.e. adaptive integration). For example, in an FFT approach for integration in 2-D space, the spatial spacing will be $\Delta x = \pi / k_{max} = \frac{1}{2} L / N$, where N is the number of steps in the wavenumber integration, and L is the spatial period of the loads. Most likely then there will exist many points at which the response will needlessly be computed when using the FFT.

Other enhancements to the numerical evaluation of the wavenumber integrals, such as using the exponential window method (Chapter 6, Section 6.6.14), and/or resorting to special integration methods (such as Filon's) to take into account the oscillations of the trigonometric or Bessel functions, will be left to readers to explore.

6 Numerical Methods

6.1 Normal Modes by Inverse Iteration

Although there exist many well-established methods and routines for solving the eigenvalue problems occurring in structural dynamics, the inverse iteration or *Stodola–Vianello* method has proven to be particularly simple and easy to program. This method is especially convenient when only the fundamental mode, or the first few lower modes, are desired. However, it may fail if the system has repeated eigenvalues, such as in a building with identical torsional and lateral frequencies. However, the main reason for presenting this numerical tool here is for its great didactic value, helping as it does in the learning of eigenvalue problems in general. At the same time, we can then avoid the use of canned routines and black boxes, at least for a while. For serious work on eigenvalue problems, however, readers should consider the specialized literature, particularly the documentation of the widely available *LINPACK* procedures, and the relevant chapters in the excellent book *Numerical Recipes: The Art of Scientific Computing* by W. H. Press, S. A. Teukolsky, W. T. Vetterling, and B. P. Flannery.

6.1.1 Fundamental Mode

The Stodola–Vianello method is based on a very simple computational scheme. If $\mathbf{v}_0$ is an arbitrary starting vector, then the iteration

$$\boxed{\mathbf{K}\mathbf{v}_{k+1} = \mathbf{M}\mathbf{v}_k} \qquad k = 0, 1, 2, \ldots \qquad (6.1)$$

converges to the fundamental mode of the system represented by the stiffness and matrices $\mathbf{K}$ and $\mathbf{M}$. From a physical point of view, the procedure has a simple interpretation, if the right-hand side in Eq. 6.1 is seen as an external, static load, to which the structure is responding and deforming. For example, in the case of a simply supported beam, and if all elements of the starting vector are 1, then the initial load is proportional to the weight of the beam. After the first iteration, the forces will begin approaching a parabolic distribution, with larger distributed forces near the center. Eventually, the shape of the load will stabilize, and so will the deformation. At that point, the iteration will have converged.

To prove mathematically why inverse iteration converges to the first mode, consider the modal expansion of the starting vector:

$$\mathbf{v}_0 = \mathbf{\Phi}\mathbf{c} = \sum_{j=1}^{n} c_j \, \boldsymbol{\phi}_j \tag{6.2}$$

Of course, at first we know neither the modal coordinates of the starting vector, nor the modes themselves, but that does not prevent us from writing the expansion equation as if we did know these quantities. After the first iteration, the expansion then changes to

$$\mathbf{v}_1 = \sum_{j=1}^{n} c_j \mathbf{K}^{-1}\mathbf{M}\,\boldsymbol{\phi}_j = \sum_{j=1}^{n} \frac{c_j}{\omega_j^2} \mathbf{K}^{-1}\mathbf{K}\,\boldsymbol{\phi}_j = \sum_{j=1}^{n} \frac{c_j}{\omega_j^2}\boldsymbol{\phi}_j \tag{6.3}$$

and in general

$$\mathbf{v}_k = \sum_{j=1}^{n} \frac{c_j}{\omega_j^{2k}}\boldsymbol{\phi}_j = \frac{c_1}{\omega_1^{2k}}\left[\boldsymbol{\phi}_1 + \frac{c_2}{c_1}\left(\frac{\omega_1}{\omega_2}\right)^{2k}\boldsymbol{\phi}_2 + \cdots + \frac{c_n}{c_1}\left(\frac{\omega_1}{\omega_n}\right)^{2k}\boldsymbol{\phi}_n \right] \tag{6.4}$$

Since the frequency ratios in this equation are all less than 1, when raised to the $2k$th power they become smaller and smaller as the iteration progresses. Hence, the iteration converges to the first term, namely

$$\mathbf{v}_k \xrightarrow[k\to\infty]{} \frac{c_1}{\omega_1^{2k}}\boldsymbol{\phi}_1 \tag{6.5}$$

Thus, the iteration converges to a vector that is proportional to the first mode. Since eigenvectors are defined only up to an arbitrary scaling factor, the iteration indeed converges to the first mode, at least in principle.

As described, however, this procedure still has one problem: as the iteration progresses, the magnitude of $\mathbf{v}_k$ becomes either larger and larger, or smaller and smaller, depending on whether the fundamental eigenvalue ω_1 is smaller or larger than 1 in numerical value. To avoid this problem, we normalize each iteration by some appropriate scaling factor. Common choices for this factor are the largest element, the vector norm, or the quadratic form with the mass matrix, that is, $\mathbf{v}_k^T\mathbf{M}\mathbf{v}_k$. The last option is usually the best, as it may be available as a by-product of the mixed Rayleigh–Schwarz quotients used to check for convergence.

Example

Consider the close-coupled, three-mass system shown in Figure 6.1.

The exact solution for the fundamental mode can be shown to be given by

$$\omega_1 = 2\sqrt{\frac{k}{m}}\sin\frac{\pi}{12} = 0.517\sqrt{\frac{k}{m}} \tag{6.6}$$

$$\phi_{i1} = \cos\frac{\pi(i-1)}{6} \tag{6.7}$$

where $i = 1$ corresponds to the top mass. Hence, the exact first mode is

$$\boldsymbol{\phi}_1 = \begin{Bmatrix} 1.000 \\ 0.866 \\ 0.500 \end{Bmatrix} \tag{6.8}$$

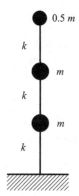

Figure 6.1. 3-DOF system.

The matrices required for the computation are as follows:

$$\mathbf{M} = m \begin{bmatrix} \frac{1}{2} & 0 & 0 \\ 0 & 1 & 0 \\ 0 & 0 & 1 \end{bmatrix} \qquad \mathbf{K} = k \begin{bmatrix} 1 & -1 & 0 \\ -1 & 2 & -1 \\ 0 & -1 & 2 \end{bmatrix} \qquad \mathbf{K}^{-1} = \frac{1}{k} \begin{bmatrix} 3 & 2 & 1 \\ 2 & 2 & 1 \\ 1 & 1 & 1 \end{bmatrix} \tag{6.9}$$

(**Note**: here we use $\mathbf{K}^{-1}$ for convenience. In a program, we would solve a system of equations.) We choose an initial vector in the form of a straight line, and use it to compute the Rayleigh quotient:

$$\mathbf{v}_0 = \begin{Bmatrix} 3 \\ 2 \\ 1 \end{Bmatrix} \equiv 3 \begin{Bmatrix} 1 \\ 0.667 \\ 0.333 \end{Bmatrix} \qquad R_{00} = \frac{\mathbf{v}_0^T \mathbf{K} \mathbf{v}_0}{\mathbf{v}_0^T \mathbf{M} \mathbf{v}_0} = 0.315 \frac{k}{m} \tag{6.10}$$

which yields an estimated fundamental frequency $\omega_1 \approx \sqrt{R_{00}} = 0.562\sqrt{\frac{k}{m}}$. This value is only 8.7% larger than the true frequency. The first iteration is then

$$\mathbf{v}_1' = \frac{m}{k} \begin{bmatrix} 3 & 2 & 1 \\ 2 & 2 & 1 \\ 1 & 1 & 1 \end{bmatrix} \begin{bmatrix} \frac{1}{2} & 0 & 0 \\ 0 & 1 & 0 \\ 0 & 0 & 1 \end{bmatrix} \begin{Bmatrix} 3 \\ 2 \\ 1 \end{Bmatrix} = \frac{m}{2k} \begin{Bmatrix} 19 \\ 16 \\ 9 \end{Bmatrix} = \frac{19m}{2k} \begin{Bmatrix} 1 \\ 0.842 \\ 0.473 \end{Bmatrix} \tag{6.11}$$

Using this iterated value to compute the mixed Rayleigh quotient defined earlier, we obtain an improved estimate for the frequency of $\omega_1 \approx \sqrt{R_{01}} = 0.523\sqrt{\frac{k}{m}}$, which is only 1.1% larger than the true value. As can be seen, the iteration converges very fast. However, the fundamental eigenvalue will generally converge faster than the corresponding eigenvector. Comparison of the normalized components of the first iteration (the last column) with those of the true eigenvector shows general agreement, but the values are still off by some 5%. Notice also that the ratio of any two corresponding eigenvector components between consecutive iterations is also a coarse approximation to the eigenvalue. For example, the ratio of the first component is $\frac{(3)(2)}{19}\frac{k}{m}$, whose square root yields an estimated frequency $\omega_1 \approx 0.562\sqrt{\frac{k}{m}}$.

6.1.2 Higher Modes: Gram–Schmidt Sweeping Technique

Assume we have already computed the fundamental mode by inverse iteration, and that we wish to obtain the second mode. For this purpose, we begin as before with an arbitrary starting vector, which we express in terms of the modes:

$$\mathbf{v}_0 = c_1 \phi_1 + c_2 \phi_2 + \cdots c_n \phi_n \tag{6.12}$$

Since we know the first eigenvector, we can subtract (or sweep out, or *rinse*) its contribution to the starting vector, that is,

$$\mathbf{v}_0' = \mathbf{v}_0 - c_1 \phi_1 = c_2 \phi_2 + \cdots c_n \phi_n \tag{6.13}$$

Now, from the modal expansion theorem, we can compute the first modal coordinate as

$$c_1 = \frac{\phi_1^T \mathbf{M} \mathbf{v}_0}{\phi_1^T \mathbf{M} \phi_1} \tag{6.14}$$

so that

$$\mathbf{v}_0' = \mathbf{v}_0 - \frac{\phi_1^T \mathbf{M} \mathbf{v}_0}{\phi_1^T \mathbf{M} \phi_1} \phi_1 \tag{6.15}$$

is a starting vector that does *not* contain any contribution from the first mode. Hence, the inverse iteration with the rinsed starting vector should now converge toward the second mode. However, because of numerical round-off errors, small contributions in the first mode are reintroduced as iteration progresses. If these errors were not corrected, they would eventually grow and cause the computation to, once more, converge to the fundamental mode. To prevent this situation, it is necessary to rinse after each iteration. Of course, it will also be necessary to scale the iterated vectors to prevent their growth or diminution.

6.1.3 Inverse Iteration with Shift by Rayleigh Quotient

Consider the eigenvalue problem

$$\mathbf{K} \phi = \omega^2 \mathbf{M} \phi \tag{6.16}$$

If we *define* $\omega^2 = \lambda + R$, where R is any arbitrary number (generally, an a priori estimate of ω_j^2 by means of the Rayleigh quotient), then we can rewrite the eigenvalue problem in modified form as

$$(\mathbf{K} - R\mathbf{M}) \phi = \lambda \mathbf{M} \phi \tag{6.17}$$

or

$$\bar{\mathbf{K}} \phi = \lambda \mathbf{M} \phi \tag{6.18}$$

which is an eigenvalue problem in λ with modified stiffness matrix $\bar{\mathbf{K}} = \mathbf{K} - R\mathbf{M}$ (notice that this matrix could be nearly singular). Clearly, this eigenvalue problem has the same

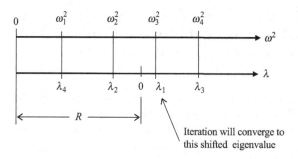

Figure 6.2. Eigenvalue shift.

eigenvectors as the original one, and its eigenvalues are shifted. Hence, the inverse iteration will converge to the eigenvector associated with the smallest modified eigenvalue,
that is, to the smallest λ in absolute value. For example, if the shift R is an approximation
to the third eigenvalue (as shown in Figure 6.2), then the iteration will converge to that
mode. The convergence rate is controlled by the ratio λ_1 / λ_2; the smaller this ratio (in
absolute value) the faster the convergence. This will occur if the original shift is already
close to an eigenvalue.

The inverse iteration with shift by Rayleigh quotient can be accomplishes according to
the following scheme:

Inverse iteration $\qquad\qquad (\mathbf{K} - R_k\mathbf{M})\mathbf{v}'_{k+1} = \mathbf{M}\mathbf{v}_k$ $\qquad\qquad$ (6.19)

Mixed Rayleigh quotient $\qquad \lambda_{k+1} = \dfrac{\mathbf{v}_k^T\mathbf{M}\mathbf{v}_k}{\mathbf{v}_k^T\mathbf{M}\mathbf{v}'_{k+1}}$ $\qquad\qquad$ (6.20)

Rescaling $\qquad\qquad\qquad \mathbf{v}_{k+1} = \lambda_{k+1}\mathbf{v}'_{k+1}$ $\qquad\qquad$ (6.21)

Eigenvalue $\qquad\qquad\qquad R_{k+1} = R_k + \lambda_{k+1}$ $\qquad\qquad$ (6.22)

Defining the vector increment $\Delta\mathbf{v}_{k+1} = \mathbf{v}_{k+1} - \mathbf{v}_k$, it is a simple matter to prove that the
three steps above can be expressed together in a single matrix equation as

$$\begin{Bmatrix} \mathbf{K} - R_k\mathbf{M} & -\mathbf{M}\mathbf{v}_k \\ -\mathbf{v}_k^T\mathbf{M} & \mathbf{O} \end{Bmatrix} \begin{Bmatrix} \Delta\mathbf{v}_{k+1} \\ \lambda_{k+1} \end{Bmatrix} = \begin{Bmatrix} -(\mathbf{K} - R_k\mathbf{M})\mathbf{v}_k \\ 0 \end{Bmatrix} \qquad (6.23)$$

Notice that the second equations implies $\mathbf{v}_k^T\mathbf{M}\,\Delta\mathbf{v}_{k+1} = 0$, that is, the change to the current
vector is orthogonal to that vector with respect to the mass matrix.

Advantages

- Much faster convergence
- No need for repeated rinsing
- Much more robust, as iteration is less affected by errors in eigenvalues already found

Disadvantages

- Shifted stiffness matrix is different for each mode. This implies that the triangulariza-
  tion (i.e., the Cholesky decomposition) of this matrix must be repeated for each mode.
- Shifted matrix may be ill conditioned (but this can be handled explicitly).

- For repeated or closely spaced eigenvalues, special handling is required, because the iterated eigenvectors may rotate in the plane defined by the eigenvectors corresponding to the multiple (or close) eigenvalues, which causes orthogonality to fail. This is true also for conventional inverse iteration, and can be handled explicitly.

The inverse iteration scheme is similar to the one presented before, except that rinsing is only required at the beginning of each iteration, and that for each mode, a new shifted stiffness matrix must be computed. The shift is determined from the Rayleigh quotient obtained with the rinsed starting vector. The eigenvalues may, or may not appear in natural order.

6.1.4 Improving Eigenvectors after Inverse Iteration

It can be shown that eigenvalues in the inverse iteration method converge much faster to their exact values than the components of the eigenvectors. Indeed, eigenvalues are often surprisingly accurate, while eigenvectors can still exhibit considerable error. Fortunately, once the modes have been found, it is often possible to achieve a substantial improvement in the eigenvectors by a simple reorthogonalization process. The cleaning procedure is as follows.

Let $\tilde{\Phi}$ be the matrix containing the computed modes. Because of unavoidable errors, these modes will not be the exact modes Φ, but will contain some errors $\mathbf{E}$. In the normal case of nonrepeated eigenvalues, the modes will span the full n-dimensional space, in which case the error matrix can be expressed in terms of the modes. Hence, the approximate modes will be of the form

$$\tilde{\Phi} = \Phi + \mathbf{E} = \Phi + \Phi\mathbf{C} \tag{6.24}$$

In general, the coefficient matrix $\mathbf{C} = \{c_{ij}\}$ will be nonsymmetric and fully populated. However, because all modes are known only up to a multiplication constant, it follows that the true modes do not contribute any error in their own direction. Hence all of the diagonal elements vanish, that is, $c_{ii} = 0$. Consider now the orthogonality condition with respect to the mass matrix:

$$\begin{aligned}\tilde{\Phi}^T \mathbf{M} \tilde{\Phi} &= (\mathbf{C}^T \Phi^T + \Phi^T)\mathbf{M}(\Phi + \Phi\mathbf{C}) \\ &= \mathbf{C}^T \Phi^T \mathbf{M}\Phi\mathbf{C} + \mathbf{C}^T \Phi^T \mathbf{M}\Phi + \Phi^T \mathbf{M}\Phi\mathbf{C} + \Phi^T \mathbf{M}\Phi\end{aligned} \tag{6.25}$$

Assuming that the error terms are small, we proceed to neglect the first term in Eq. 6.25, because it is quadratic in the coefficients c_{ij}. Hence, the components of the above matrix expression are of the form

$$\tilde{\mu}_{ij} = \mu_i c_{ji} + \mu_j c_{ij} + \mu_i \delta_{ij} \tag{6.26}$$

in which δ_{ij} is the Kronecker delta, and μ_j is the modal mass for the jth mode. Specializing Eq. 6.26 for $i = j$ and taking into account that $c_{jj} = 0$, we conclude that $\mu_j \approx \tilde{\mu}_{jj} = \tilde{\phi}_j^T \mathbf{M} \tilde{\phi}_j$. Of course, this is only an approximation, because we neglected the quadratic terms. It follows that for $i \neq j$

$$\tilde{\mu}_{ij} = \tilde{\mu}_{ii} c_{ji} + \tilde{\mu}_{jj} c_{ij} \tag{6.27}$$

By a similar development with the second orthogonality condition involving the stiffness matrix, we obtain

$$\tilde{\kappa}_{ij} = \tilde{\kappa}_{ii} c_{ji} + \tilde{\kappa}_{jj} c_{ij} = \omega_i^2 \tilde{\mu}_{ii} c_{ji} + \omega_j^2 \tilde{\mu}_{jj} c_{ij} \qquad (6.28)$$

in which $\tilde{\kappa}_{ij} = \tilde{\phi}_i^T \mathbf{K} \tilde{\phi}_j$. Combining Eqs. 6.27 and 6.28, we obtain a system of two equations in two unknowns c_{ij}, c_{ji}. The final result is

$$c_{ij} = \frac{\omega_i^2 \tilde{\mu}_{ij} - \tilde{\kappa}_{ij}}{\tilde{\mu}_{jj} \left(\omega_i^2 - \omega_j^2 \right)} \qquad c_{ji} = \frac{\omega_j^2 \tilde{\mu}_{ij} - \tilde{\kappa}_{ij}}{\tilde{\mu}_{ii} \left(\omega_j^2 - \omega_i^2 \right)} \qquad \mathbf{\Phi} = \tilde{\mathbf{\Phi}} - \tilde{\mathbf{\Phi}} \mathbf{C} = \tilde{\mathbf{\Phi}} (\mathbf{I} - \mathbf{C}) \qquad (6.29)$$

We observe that $\tilde{\mu}_{ij} = \tilde{\mu}_{ji}$ and $\tilde{\kappa}_{ij} = \tilde{\kappa}_{ji}$. In addition, if two modes are close, the error terms can be large indeed; this is a consequence of the fact that any vector in the plane defined by the eigenvectors for repeated eigenvalues is also an eigenvector.

Notice that the error coefficients depend only on pairs of eigenvectors. Hence, the correction can be carried out *on the fly* as an integral part of the inverse iteration process. In principle, the correction may involve a substantial computational effort, but if carried out on the fly, it may help reduce the number of inverse iterations.

6.1.5 Inverse Iteration for Continuous Systems

As will be seen in the ensuing sections, the application of the Rayleigh–Ritz approach and of weighted residual methods to obtain the vibration frequencies and modes of continuous systems always results in approximate solutions whose accuracy depends on how closely the trial functions can describe the actual modal functions. To improve this accuracy, it is possible in some cases to modify these trial functions iteratively by means of an inverse iteration scheme that closely parallels the method used for discrete systems.

To illustrate this concept, let us consider the example of a homogeneous, simply supported bending beam. The dynamic equilibrium equations for free vibration and the boundary conditions are

$$EI \frac{\partial^4 \phi}{\partial x^4} = \omega^2 \rho A \phi \qquad \phi(0) = \phi''(0) = \phi(L) = \phi''(L) = 0 \qquad (6.30)$$

in which $\phi = \phi(x)$ is the modal shape for the fundamental mode, which we shall pretend to be unknown. To obtain this mode by successive approximations, we change this *eigenvalue problem* into an iterative *boundary value problem* (in essence, a static problem) of the form

$$EI \frac{\partial^4 \psi_{k+1}}{\partial x^4} = \rho A \, \psi_k \qquad k = 0, 1, 2 \ldots \qquad (6.31)$$

which we initiate with a starting, single trial function approximation $\psi_0(x)$ chosen by us. For example, we can use as initial guess for the first mode the parabola

$$\phi \approx \psi_0(x) = 4 \left(1 - \frac{x}{L} \right) \frac{x}{L} \qquad (6.32)$$

which satisfies only the essential boundary conditions. Hence

$$\frac{\partial^4 \psi_1}{\partial x^4} = \frac{4\rho A}{EI}\left(1 - \frac{x}{L}\right)\frac{x}{L} \tag{6.33}$$

The initial guess for the frequency follows from the Rayleigh quotient

$$R = \frac{\int_0^L EI\left(\psi_0''\right)^2 dx}{\int_0^L \rho A\, \psi_0^2 dx} = 120\frac{EI}{\rho A L^4} \qquad \text{compared to the exact} \quad \omega_1^2 = \frac{\pi^4 EI}{\rho A L^4} \tag{6.34}$$

The error in the eigenvalue is $120/\pi^4 - 1 = 0.232$ or 23% (i.e., about 12% of the frequency). We now proceed to integrate according to the iterative scheme, and obtain

$$\psi_1 = \frac{4\rho A}{EI}\left(\tfrac{1}{120}\xi^5 - \tfrac{1}{360}\xi^6\right) + \tfrac{1}{6}C_1\xi^3 + \tfrac{1}{2}C_2\xi^2 + C_3\xi + C_4. \tag{6.35}$$

Imposing boundary conditions, we obtain $C_2 = C_4 = 0, C_1 = -\tfrac{1}{3}\rho A/EI, C_3 = \tfrac{1}{30}\rho A/EI$, so

$$\psi_1 = \frac{\rho A}{90EI}\left(3 - 5\xi^2 + 3\xi^4 - \xi^5\right)\xi \tag{6.36}$$

which can be rescaled arbitrarily to

$$\psi_1 = \left(3 - 5\xi^2 + 3\xi^4 - \xi^5\right)\xi \tag{6.37}$$

The new Rayleigh quotient is

$$R = \frac{\int_0^L EI\left(\psi_1''\right)^2 dx}{\int_0^L \rho A\, \psi_1^2 dx} = \frac{531960}{5461}\frac{EI}{\rho A L^4} = 97.411\frac{EI}{\rho A L^4} \tag{6.38}$$

This result is virtually exact, since $97.411/\pi^4 = 1.00002$.

6.2 Method of Weighted Residuals

The objective in these methods is to reduce the system of partial differential equations characterizing the continuous systems into a system of ordinary differential equations. In effect, we attempt to reduce the continuous systems into discrete ones – ideally, with the fewest number of degrees of freedom (DOF), to allow hand calculations. With this goal in mind, we begin by making the following observation: All of the continuous systems considered previously involve partial differential equations of the general form

$$\mathfrak{M}_{2\mu}\ddot{\mathbf{u}} + \mathfrak{K}_{2\kappa}\mathbf{u} = \mathbf{b}(\mathbf{x},t) \tag{6.39}$$

in which $\mathfrak{M}, \mathfrak{K}$ are mass and stiffness *differential operator matrices* of orders $2\mu, 2\kappa$ respectively (with $\mu < \kappa$). Furthermore, for *each* boundary point, there are κ boundary conditions of the form

$$\mathfrak{B}_i\mathbf{u}(\mathbf{x}_b,t) = \mathbf{0} \qquad i = 1,2,\ldots\kappa \tag{6.40}$$

that is, the *number* of boundary conditions at each boundary point x_b equals half the order of the differential equation. These boundary conditions involve differential operations of orders not exceeding $2\kappa - 1$.

The boundary conditions can be classified into the following two fundamental groups:

- *Essential*, geometric or Dirichlet boundaries: These are boundaries at which geometric conditions, such as displacements or rotations, are prescribed. They involve differential operations whose orders are in the range $[0, \kappa - 1]$.
- *Natural*, additional or Neumann boundaries: These are boundaries at which stresses, forces, or moments are prescribed. They involve differential operations whose orders are in the range $[\kappa, 2\kappa - 1]$.

In conjunction with the boundary conditions, the differential operators $\mathfrak{M}, \mathfrak{K}$ satisfy two important mathematical relations, namely they are *self-adjoint* and *positive semidefinite*. The essential aspects of these properties are as follows. Let $\mathbf{v}(\mathbf{x}), \mathbf{w}(\mathbf{x})$ be two distinct, *arbitrary* test functions satisfying the boundary conditions, but *not* necessarily the differential equation. The operators satisfy then the following conditions (the proof can be obtained with integration by parts): $\mathfrak{K}$

Self-adjoint property:

$$\int_V \mathbf{v}^T \mathfrak{M} \mathbf{w} \, dV = \int_V \mathbf{w}^T \mathfrak{M} \mathbf{v} \, dV \quad \text{and} \quad \int_V \mathbf{v}^T \mathfrak{K} \mathbf{w} \, dV = \int_V \mathbf{w}^T \mathfrak{K} \mathbf{v} \, dV \tag{6.41}$$

which is analogous to the symmetry property of matrices: a matrix $\mathbf{A}$ is symmetric if and only if the equality $\mathbf{x}^T \mathbf{A} \mathbf{y} = \mathbf{y}^T \mathbf{A} \mathbf{x}$ holds for arbitrary nonzero vectors $\mathbf{x}, \mathbf{y}$.

Positive semidefinite property

$$\int_V \mathbf{v}^T \mathfrak{M} \mathbf{v} \, dV \geq 0 \quad \text{and} \quad \int_V \mathbf{v}^T \mathfrak{K} \mathbf{v} \, dV \geq 0 \tag{6.42}$$

If the equal sign is satisfied only when $\mathbf{v} = \mathbf{0}$, then the operator is said to be *positive definite* (instead of semidefinite). This property relates to the nonnegativity of the kinetic and strain energies. Now, in the method of *weighted residuals*, we assume that an approximate solution can be written as

$$\mathbf{u} = \boldsymbol{\psi}_1 q_1 + \cdots \boldsymbol{\psi}_n q_n = \sum_{j=1}^n \boldsymbol{\psi}_j(\mathbf{x}) q_j(t) = \boldsymbol{\psi} \mathbf{q} \tag{6.43}$$

$$\boldsymbol{\psi} = \begin{bmatrix} \boldsymbol{\psi}_1 & \cdots & \boldsymbol{\psi}_n \end{bmatrix} \quad \text{and} \quad \mathbf{q} = \begin{Bmatrix} q_1 \\ \vdots \\ q_n \end{Bmatrix} \tag{6.44}$$

in which the $\boldsymbol{\psi}_j = \boldsymbol{\psi}_j(\mathbf{x})$ are <u>known</u> *trial* functions, which are chosen arbitrarily by the analyst on the basis of what he or she knows about the problem, and the $q_j = q_j(t)$ are unknown functions of time (to be determined). The number n of these functions is also chosen arbitrarily – the fewest possible. The trial functions must satisfy *all* boundary conditions,

$$\mathfrak{B}_i \boldsymbol{\psi}_j = \mathbf{0} \quad i = 1, 2, \ldots \kappa, \quad j = 1, 2, \ldots n \tag{6.45}$$

and they must have continuous derivatives up to order 2κ (in jargon: they must have continuity $C_{2\kappa}$). We next substitute the trial solution into the differential equation, and obtain

$$\mathfrak{M}_{2\mu}\mathbf{\Psi}\ddot{\mathbf{q}} + \mathfrak{K}_{2\kappa}\mathbf{\Psi}\mathbf{q} = \mathbf{b}(\mathbf{x},t) + \mathbf{r}(\mathbf{x},t) \tag{6.46}$$

in which $\mathbf{r}$ is the *residual*. This residual arises because the trial solution almost certainly does *not* satisfy the differential equation – if it did, then the trial solution would be the exact solution, which is unlikely. This residual can be interpreted physically as external body forces, which are necessary to force the system to vibrate in the pattern implied by the trial solution. To dispose of this residual, we choose appropriate weighting function $\mathbf{W}(\mathbf{x})$ (as described later), multiply the previous equation by the transpose of this function, and require the integral of the weighted residual $\mathbf{W}^T\mathbf{r}$ over the entire body to be zero:

$$\int_V \mathbf{W}^T[\mathfrak{M}_{2\mu}\mathbf{\Psi}\ddot{\mathbf{q}} + \mathfrak{K}_{2\kappa}\mathbf{\Psi}\mathbf{q} - \mathbf{b}]\,dV = \int_V \mathbf{W}^T\mathbf{r}\,dV = 0 \tag{6.47}$$

in which V is the volume, area, or length of the body in question. Since $\mathbf{q}$ is *not* a function of *space*, it follows that this equation can be written as

$$\mathfrak{M}\ddot{\mathbf{q}} + \mathfrak{K}\mathbf{q} = \mathbf{p} \tag{6.48}$$

in which

$$\mathbf{M} = \int_V \mathbf{W}^T\mathfrak{M}_{2\mu}\mathbf{\Psi}\,dV \qquad = \text{mass matrix} \tag{6.49}$$

$$\mathbf{K} = \int_V \mathbf{W}^T\mathfrak{K}_{2\kappa}\mathbf{\Psi}\,dV \qquad = \text{stiffness matrix} \tag{6.50}$$

$$\mathbf{p}(t) = \int_V \mathbf{W}^T\mathbf{b}\,dV \qquad = \text{load vector} \tag{6.51}$$

There exist several weighted residual methods, the implementation of which depends on how the weighting functions are chosen. We describe briefly several of these methods, but consider only one in detail, namely the Galerkin method. For further details see, for example, Crandall.[1]

Why Zero Work by Residual Forces?

You may perhaps wonder why the virtual work done by the time-varying residual forces should be zero. The answer is that in general, they must not. Indeed, before we require this condition to be true, the undetermined parameters $\mathbf{q}$ are still only that, namely undetermined. However, the very moment that *we* impose the requirement that the work done by the residual should be zero, we are in fact imposing restrictions on the admissible solutions, so only a well-defined set of $\mathbf{q}$ values will satisfy the virtual work equation. In other words, it is *we* who make the determination that the virtual work done by the residual must be zero.

---

[1] S. H. Crandall, *Engineering Analysis: A Survey of Numerical Procedures* (New York: McGraw–Hill, 1956).

6.2.1 Point Collocation

Here, the residual is forced to be zero at n arbitrary, discrete points in the body (or equivalently, the weights are chosen in the form of Dirac delta functions). This does *not* make the solution exact at these points, however.

6.2.2 Sub-domain

In this method, the body (the "domain") is divided arbitrarily into n parts (or cells, or sub-domains); the integral of the residual over each part is then forced to be zero. This is equivalent to choosing weighting functions that are unity in a given sub-domain, and zero elsewhere.

6.2.3 Least Squares

Here, the weighting functions are chosen equal to the residual, and the integral of the squared error ($\mathbf{r}^T\mathbf{r}$) over the entire body is minimized (instead of making it zero).

6.2.4 Galerkin

In this alternative, the weighting functions are chosen equal to the trial functions, $\mathbf{W} \equiv \mathbf{\Psi}$. Again, the trial functions must satisfy all boundary conditions, and they must have $C_{2\kappa}$ continuity. The required matrices are in this case:

$$\mathbf{M} = \int_V \mathbf{\Psi}^T \,\mathfrak{M}_{2\mu}\, \mathbf{\Psi}\, dV \qquad = \text{mass matrix}$$
(6.52)

$$\mathbf{K} = \int_V \mathbf{\Psi}^T \,\mathfrak{K}_{2\kappa}\, \mathbf{\Psi}\, dV \qquad = \text{stiffness matrix}$$
(6.53)

$$\mathbf{p}(t) = \int_V \mathbf{\Psi}^T \,\mathbf{b}\, dV \qquad = \text{load vector}$$
(6.54)

In the following, we demonstrate the Galerkin method by means of various examples.

Example 1

Consider a uniform, simply supported bending beam of length L as shown in Figure 6.3. There is a time-varying concentrated load acting at the center, at which point we also take the origin of coordinates. The load can be represented mathematically as a distributed load of intensity $b(x,t) = \delta(x)p(t)$. Although this problem could be solved *exactly* by taking trial functions from the family of functions $\cos(j\pi x / L)$ (which for odd j happen to be the exact solutions for this case), we shall pretend ignorance of this fact. Instead, we shall seek an approximate solution with a single trial function ($n = 1$), namely the elastica of a beam under a uniformly distributed (static) load (which we can take out of a handbook), and which we assume to be a good approximation for our problem. This function is of the form

$$\psi_1 = \tfrac{1}{5}\left[5 - 6\left(\tfrac{2x}{L}\right)^2 + \left(\tfrac{2x}{L}\right)^4\right] \equiv \tfrac{1}{5}[5 - 6\zeta^2 + \zeta^4], \qquad -1 \le \zeta \le 1$$
(6.55)

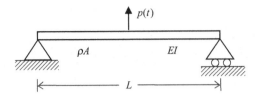

Figure 6.3. Simply supported beam.

For convenience, we have scaled this function so that it has a value of 1 at the center (i.e., at $x = 0$). This function satisfies automatically the boundary conditions as well as the required C_4 continuity condition. Notice also that the elastica of a beam under a concentrated load at the center could not be used here, since its fourth derivative is discontinuous at the midpoint and is zero elsewhere. Considering that we take only one trial function and that the differential equation involves only 1 DOF per point in the beam, we obtain here a single-DOF (SDOF) system. Defining $\zeta = 2x/L$, we have:

$$m = \int_{-L/2}^{L/2} \rho A \psi_1^2 \, dx = \frac{\rho A L}{25} \int_0^1 \left[5 - 6\zeta^2 + \zeta^4 \right]^2 d\zeta \tag{6.56}$$
$$= 0.504 \rho A L$$

$$k = \int_{-L/2}^{L/2} EI \psi_1 \frac{\partial^4 \psi_1}{\partial x^2} \, dx = \frac{384 EI}{25 L^3} \int_0^1 (5 - 6\zeta^2 + \zeta^4) \, d\zeta \tag{6.57}$$
$$= 49.152 \frac{EI}{L^3}$$

$$p(t) = \int_{-L/2}^{L/2} p(t) \, \delta(x) \, dx \tag{6.58}$$

These values should be compared with those provided by the heuristic approach, which would have given coefficients 0.500 for the mass and 48.000 for the stiffness. (A word of caution: if we had *not* normalized the function to 1 at the center, the above coefficients would have been different by a constant factor, but this would not have changed the final solution for the characteristic frequencies or for the displacements).

The values for the beam's fundamental frequency are then

Exact result: $$\omega_1 = \pi^2 \sqrt{\frac{EI}{\rho A L^4}} = 9.867 \sqrt{\frac{EI}{\rho A L^4}} \tag{6.59}$$

Galerkin: $$\omega_1 = \sqrt{\frac{49.152}{0.504}} \sqrt{\frac{EI}{\rho A L^4}} = 9.875 \sqrt{\frac{EI}{\rho A L^4}} \tag{6.60}$$

Heuristic method: $$\omega_1 = \sqrt{\frac{48}{0.5}} \sqrt{\frac{EI}{\rho A L^4}} = 9.798 \sqrt{\frac{EI}{\rho A L^4}} \tag{6.61}$$

As can be seen, the two numerical estimates are very close to the exact value.

Example 2
We consider once more a uniform beam of length L subjected to a concentrated load at the center, but we *add* a lumped mass at $x = \frac{1}{3} L$ (Figure 6.4). This problem is exceedingly difficult to solve rigorously, so we use Galerkin instead. This time, however, we use *two*

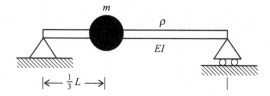

Figure 6.4. Beam with lumped mass.

trial functions, choosing for this purpose the exact expressions for the first two modes of the beam *without* the mass:

$$\psi_1 = \sin\frac{\pi x}{L} \qquad \psi_2 = \sin\frac{2\pi x}{L} \tag{6.62}$$

Hence,

$$\Psi = \left[\sin\frac{\pi x}{L} \quad \sin\frac{2\pi x}{L}\right] \qquad \frac{\partial^4 \Psi}{\partial x^4} = \left(\frac{\pi}{L}\right)^4 \left[\sin\frac{\pi x}{L} \quad 16\sin\frac{2\pi x}{L}\right] \tag{6.63}$$

On the other hand, the mass per unit length as well as the load for this beam can be represented as

$$\rho A(x) = \rho A + m\,\delta(x - \tfrac{1}{3}L) \tag{6.64}$$

$$b(x,t) = p(t)\delta\!\left(x - \tfrac{1}{2}L\right) \tag{6.65}$$

The mass matrix is then:

$$\mathbf{M} = \int_0^L \begin{bmatrix} \sin\frac{\pi x}{L} \\ \sin\frac{2\pi x}{L} \end{bmatrix} (\rho A + m\,\delta(x - \tfrac{1}{3}L)) \left[\sin\frac{\pi x}{L} \quad \sin\frac{2\pi x}{L}\right] dx$$

$$= \rho A \int_0^L \left\{ \begin{matrix} \sin^2\frac{\pi x}{L} & \sin\frac{\pi x}{L}\sin\frac{2\pi x}{L} \\ \sin\frac{\pi x}{L}\sin\frac{2\pi x}{L} & \sin^2\frac{2\pi x}{L} \end{matrix} \right\} dx + m\left\{ \begin{matrix} \sin^2\frac{\pi}{3} & \sin\frac{\pi}{3}\sin\frac{2\pi}{3} \\ \sin\frac{\pi}{3}\sin\frac{2\pi}{3} & \sin^2\frac{2\pi}{3} \end{matrix} \right\}$$

$$= \frac{\rho A L}{2}\left\{ \begin{matrix} 1 & \\ & 1 \end{matrix} \right\} + \frac{3m}{4}\left\{ \begin{matrix} 1 & 1 \\ 1 & 1 \end{matrix} \right\} \tag{6.66}$$

Similarly, the stiffness matrix is

$$\mathbf{K} = \int_0^L \begin{bmatrix} \sin\frac{\pi x}{L} \\ \sin\frac{2\pi x}{L} \end{bmatrix} EI\frac{\pi^4}{L^4}\left[\sin\frac{\pi x}{L} \quad 16\sin\frac{2\pi x}{L}\right] dx$$

$$= \frac{EI\pi^4}{L^4} \int_0^L \left\{ \begin{matrix} \sin^2\frac{\pi x}{L} & 16\sin\frac{\pi x}{L}\sin\frac{2\pi x}{L} \\ \sin\frac{\pi x}{L}\sin\frac{2\pi x}{L} & 16\sin^2\frac{2\pi x}{L} \end{matrix} \right\} dx$$

$$= \frac{EI}{L^3}\frac{\pi^4}{2}\left\{ \begin{matrix} 1 & \\ & 16 \end{matrix} \right\} \tag{6.67}$$

Finally, the load vector is

$$
\begin{aligned}
\mathbf{p} &= \int_0^L \begin{bmatrix} \sin\frac{\pi x}{L} \\ \sin\frac{2\pi x}{L} \end{bmatrix} p(t)\,\delta(x - \tfrac{1}{2}L)\,dx \\
&= p(t)\begin{bmatrix} 1 \\ 0 \end{bmatrix}
\end{aligned}
\tag{6.68}
$$

We have now reduced this problem to one involving only 2 DOF.

6.3 Rayleigh–Ritz Method

This method is intimately related to Galerkin's approach, and is based on the principle of virtual displacements (**Note**: in the ensuing, δ denotes a "variation" or virtual perturbation and *not* a Dirac delta function):

$$
\delta W_i = \int_V \delta\mathbf{u}^T (\mathbf{b} - \rho\ddot{\mathbf{u}})\,dV + \oint_{S_p} \delta\mathbf{u}_n^T \mathbf{t}_n\,dS
\tag{6.69}
$$

The left-hand side represents the *internal* virtual work, while the right-hand side contains the virtual work done by the *external* forces acting on the body (i.e., the integral in V) and the external tractions acting on that part of the boundary where stresses are prescribed (i.e., on S_p). For consistency with the previous material, we shall assume "homogeneous" (i.e., zero force) boundary conditions here, which eliminates the last term:

$$
\delta W_i = \int_V \delta\mathbf{u}^T (\mathbf{b} - \rho\ddot{\mathbf{u}})\,dV
\tag{6.70}
$$

The internal virtual work is as follows:

Taut string:

$$
\delta W_i = \int_0^L \delta u'\, T\, u'\,dx
\tag{6.71}
$$

Shear beam:

$$
\delta W_i = \int_0^L \delta\gamma\, GA_s\,\gamma\,dx = \int_0^L \delta u'\, GA_s\, u'\,dx
\tag{6.72}
$$

Bending beam:

$$
\delta W_i = \int_0^L \delta\theta'\, EI\,\theta'\,dx = \int_0^L \delta u''\, EI\, u''\,dx
\tag{6.73}
$$

Plate:

$$
\delta W_i = \iint_A D\left[(\nabla^2 \delta u)(\nabla^2 u) - (1-v)\left(\frac{\partial^2 \delta u}{\partial x^2}\frac{\partial^2 u}{\partial y^2} + \frac{\partial^2 \delta u}{\partial y^2}\frac{\partial^2 u}{\partial x^2} - 2\frac{\partial^2 \delta u}{\partial x\,\partial y}\frac{\partial^2 u}{\partial x\,\partial y} \right) \right] dx\,dy
\tag{6.74}
$$

Solid:

$$
\delta W_i = \int_V \delta\boldsymbol{\sigma}^T \boldsymbol{\varepsilon}\,dV
\tag{6.75}
$$

in which $\boldsymbol{\sigma}, \boldsymbol{\varepsilon}$ are the stress and strain vectors at a point.

As in the weighted residual methods, we assume a solution of the form

$$\mathbf{u} = \boldsymbol{\psi}_1 q_1 + \cdots \boldsymbol{\psi}_n q_n = \sum_{j=1}^{n} \boldsymbol{\psi}_j(\mathbf{x}) q_j(t) = \boldsymbol{\Psi}\mathbf{q} \tag{6.76}$$

which implies

$$\delta \mathbf{u}^T = \delta \mathbf{q}^T \boldsymbol{\Psi}^T \tag{6.77}$$

As an example of how the virtual work equation is accomplished, we consider the particular case of a bending beam:

$$\int_0^L \delta \mathbf{q}^T \boldsymbol{\Psi}''^T EI \, \boldsymbol{\Psi}'' \mathbf{q} \, dx = \int_0^L \delta \mathbf{q}^T \boldsymbol{\Psi}^T (\mathbf{b} - \rho A \, \boldsymbol{\Psi} \ddot{\mathbf{q}}) \, dx \tag{6.78}$$

That is,

$$\delta \mathbf{q}^T \left[\left\{ \int_0^L \boldsymbol{\Psi}^T \rho A \, \boldsymbol{\Psi} \, dx \right\} \ddot{\mathbf{q}} + \left\{ \int_0^L \boldsymbol{\Psi}''^T EI \, \boldsymbol{\Psi}'' \, dx \right\} \mathbf{q} - \int_0^L \boldsymbol{\Psi}^T \mathbf{b} \, dx \right] = 0 \tag{6.79}$$

The unknown functions $\mathbf{q}$ are then obtained by requiring this equation to be valid for arbitrary variations $\delta \mathbf{q}$:

$$\left\{ \int_0^L \boldsymbol{\Psi}^T \boldsymbol{\Psi} \rho A \, dx \right\} \ddot{\mathbf{q}} + \left\{ \int_0^L \boldsymbol{\Psi}''^T \boldsymbol{\Psi}'' EI \, dx \right\} \mathbf{q} = \int_0^L \boldsymbol{\Psi}^T \mathbf{b} \, dx \tag{6.80}$$

which can be expressed more simply as

$$\mathbf{M}\ddot{\mathbf{q}} + \mathbf{K}\mathbf{q} = \mathbf{p}(t) \tag{6.81}$$

6.3.1 Boundary Conditions and Continuity Requirements in Rayleigh–Ritz

Unlike in weighted residual methods, the trial functions in Rayleigh–Ritz only have to satisfy the *essential* (or geometric) boundary conditions, and in addition, they only have to have C_κ continuity. Hence, it is substantially easier to choose appropriate trial functions in this method.

However, if the trial functions satisfy *all* boundary conditions and they have the *full* $C_{2\kappa}$ continuity, then the Rayleigh–Ritz method is the same as Galerkin. The proof is based on integration by parts. Because of the stronger requirements on the trial functions, Galerkin's method is said to represent the *strong form* of the variational formulation, while the Rayleigh–Ritz approach is said to represent the *weak form*. (From a practical point of view, however, the Rayleigh–Ritz approach is "stronger" in the sense that one can get away with coarser functions; however, these designations are now firmly established.)

Example

We repeat the first example in Section 6.2.4, that is, a uniform beam, but we now use a much simpler trial function, namely a second-order parabola (also, we set the origin at the left support):

$$u = \psi_1 q_1 \tag{6.82}$$

$$\psi_1 = 4\left(1 - \frac{x}{L}\right)\frac{x}{L} \tag{6.83}$$

This function has the required C_2 continuity, and it satisfies only the geometric boundary conditions. Its second derivative is

$$\psi_1'' = -\frac{8}{L^2} \tag{6.84}$$

After we apply the Rayleigh–Ritz equation for beams already presented, we obtain

$$16\rho A\int_0^L \left(1 - \frac{x}{L}\right)^2 \left(\frac{x}{L}\right)^2 dx\, \ddot{q}_1 + \frac{64EI}{L^4}\int_0^L dx\, q_1 = 4\int_0^L \left(1 - \frac{x}{L}\right)\frac{x}{L}\delta(x - \tfrac{1}{2}L)\, dx\, p(t) \tag{6.85}$$

That is,

$$\left(\frac{8}{15}\rho AL\right)\ddot{q}_1 + \left(\frac{64EI}{L^3}\right)q_1 = p(t) \tag{6.86}$$

The estimated resonant frequency of the beam is then

$$\omega_1 = \sqrt{\frac{(64)(15)}{8}}\sqrt{\frac{EI}{\rho AL^4}} = 10.954\sqrt{\frac{EI}{\rho AL^4}} \tag{6.87}$$

Clearly, this result is not as accurate as those obtained previously with the Galerkin and heuristic methods, but the trial function is certainly much simpler. (If we had used instead the elastica for the distributed load, the result would have been exactly as before.)

6.3.2 Rayleigh–Ritz versus Galerkin

If in the equation stating the principle of virtual displacements we integrate by parts the left-hand side containing the internal virtual work, and we do so without discarding the boundary terms, we will arrive at an alternative expression for this principle in the form

$$\int_V \delta\mathbf{u}^T[\mathfrak{M}_{2\mu}\ddot{\mathbf{u}} + \mathfrak{K}_{2\kappa}\mathbf{u} - \mathbf{b}]dV = \oint_{S_p} \delta\mathbf{u}_n^T(\mathbf{t}_n - \boldsymbol{\sigma}_n)dS \tag{6.88}$$

in which $\boldsymbol{\sigma}_n$ are the normal and tangential components of the *internal* stresses developing on that part of the boundary where tractions $(\mathbf{t}_n)$ are prescribed (i.e., on S_p). Requiring equation 6.88 to be valid for arbitrary variation $\delta\mathbf{u}$ leads immediately to the differential equation for the system together with the equilibrium conditions at the boundary S_p. If, as before, we assume homogeneous boundary conditions $\mathbf{t}_n = \mathbf{0}$ (which is not really a restriction, as external tractions there could still be included via singular terms in $\mathbf{b}$), then this equation can be written as

$$\int_V \delta\mathbf{u}^T[\mathfrak{M}_{2\mu}\ddot{\mathbf{u}} + \mathfrak{K}_{2\kappa}\mathbf{u}]dV + \oint_{S_p} \delta\mathbf{u}_n^T\boldsymbol{\sigma}_n\, dS = \int_V \delta\mathbf{u}^T\mathbf{b}\, dV \tag{6.89}$$

which is fully equivalent to the Rayleigh–Ritz equation given previously, and could be solved in like manner by assuming a combination of trial functions. If these trial functions are chosen so that they satisfy *all* homogeneous boundary conditions, then $\sigma_n = \mathbf{0}$ at all points on the boundary, the surface integral vanishes, and Eq. 6.89 becomes identical to Galerkin's. On the other hand, if the trial functions satisfy only the essential boundary conditions, then the surface integral does not vanish, and must be accounted for explicitly. This task can be accomplished by expressing the boundary stresses in terms of the trial functions. In some cases, this alternative form of Rayleigh–Ritz might be preferable to the previous one, particularly if we use simple trial functions with at most C_κ continuity, because then many terms in $\mathfrak{K}_{2\kappa}\mathbf{u}$ will be zero, making the integration easier. We illustrate this concept with the same beam example presented previously, and using the same crude trial function:

$$\psi_1 = 4\left(1-\frac{x}{L}\right)\frac{x}{L} \qquad \psi_1' = \frac{4}{L}\left(1-\frac{2x}{L}\right) \qquad \psi_1'' = -\frac{8}{L^2} \tag{6.90}$$

$$\oint_{S_p} \delta\mathbf{u}_n^T \sigma_n \, dS \equiv \delta u' E I u''\big|_0^L = \delta q_1 \, \psi_1' \, E I \, \psi_1'' q_1 \big|_0^L$$

$$= \delta q_1 \frac{4}{L}\left[\left(1-\frac{2L}{L}\right)\left(\frac{-8EI}{L^2}\right)-\left(1-\frac{0}{L}\right)\left(\frac{-8EI}{L^2}\right)\right]q_1$$

$$= \delta q_1 \frac{64EI}{L^3} q_1 \tag{6.91}$$

$$\mathfrak{K}_{2\kappa}\Psi \equiv EI\frac{\partial^4 \psi_1}{\partial x^4} = 0 \tag{6.92}$$

The mass term is as before. As can be seen, the boundary term now gives the same value that the internal virtual work gave previously, while the stiffness differential term is identically zero. Hence, we obtain exactly the same result as before.

This example also shows that if the trial functions in Galerkin do not satisfy all boundary conditions, nonsensical results will be obtained, since the important contribution of the boundary term will be missing.

6.3.3 Rayleigh–Ritz versus Finite Elements

In Rayleigh–Ritz, we use global trial functions that are defined over the entire body. By contrast, in finite elements we use interpolation functions between adjacent nodal points. The correspondence is

$$\Psi \quad \rightarrow \quad \text{interpolation functions}$$
$$\mathbf{q} \quad \rightarrow \quad \text{nodal displacements}$$

Other than that, the two methods are basically the same (at least when using a displacement formulation). The advantage of Rayleigh–Ritz over finite elements is that it permits computations with fewer trial functions, which is accomplished at the expense of having to divine appropriate functions.

6.3.4 **Rayleigh–Ritz Method for Discrete Systems**

Consider a discrete system with a large number N of DOF. Its dynamic equilibrium equation can be written in the homogeneous form:

$$\mathbf{M\ddot{u}} + \mathbf{C\dot{u}} + \mathbf{Ku} - \mathbf{p} = \mathbf{0} \tag{6.93}$$

It is often possible for us to reduce substantially the size of this system, particularly if we are interested only in the behavior and participation of the lower modes. For this purpose, we arbitrarily choose a set of $n < N$ trial vectors, or *assumed modes*, or *Ritz vectors*, $\mathbf{\Psi} = \{\psi_1 \quad \psi_2 \quad \cdots \quad \psi_n\}$ that we think may be reasonable approximations to some of the actual modes. At this stage we need not worry about the orthogonality of these trial vectors, but only about their appropriateness to describe in sufficient detail the deformation of the system, and that they be linearly independent. Let $q_j(t)$ be as yet *undetermined parameters* with which we form a trial solution of the form

$$\mathbf{u}(t) = \mathbf{\Psi q}(t) = \sum_{j=1}^{n} \psi_j \, q_j(t) \tag{6.94}$$

When we substitute this trial solution into the dynamic equilibrium equation 6.93, in most cases it will not satisfy it exactly at any time, because the assumed modes are not necessarily solutions to this system. In that case, the equation will produce a *residual* $\mathbf{r}(t)$ that we can interpret as unbalanced nodal loads, or as additional external loads that we must apply to deform the system in the shape of the assumed modes:

$$\mathbf{M\Psi\ddot{q}} + \mathbf{C\Psi\dot{q}} + \mathbf{K\Psi q} - \mathbf{p} = \mathbf{r} \tag{6.95}$$

To dispose of this residual, we resort to the principle of virtual displacements. We accomplish this by applying an arbitrary set of virtual displacements $\delta\mathbf{u} = \mathbf{\Psi}\delta\mathbf{q}$ to the system and requiring the virtual work done by the residual to be zero, that is, $\delta\mathbf{u}^T\mathbf{r} = 0$. This gives us

$$\delta\mathbf{q}^T \mathbf{\Psi}^T \left[\mathbf{M\Psi\ddot{q}} + \mathbf{C\Psi\dot{q}} + \mathbf{K\Psi q} - \mathbf{p} \right] = 0 \tag{6.96}$$

For this equation to be true no matter what the $\delta\mathbf{q}$ are requires that the other factor be zero at all times, that is,

$$\mathbf{\Psi}^T \left[\mathbf{M\Psi\ddot{q}} + \mathbf{C\Psi\dot{q}} + \mathbf{K\Psi q} - \mathbf{p} \right] = \mathbf{0} \tag{6.97}$$

This leads us immediately to the reduced system equation

$$\mathbf{\widehat{M}\ddot{q}} + \mathbf{\widehat{C}\dot{q}} + \mathbf{\widehat{K}q} = \mathbf{\hat{p}} \tag{6.98}$$

in which

$$\mathbf{\widehat{M}} = \mathbf{\Psi}^T \mathbf{M\Psi} \qquad \mathbf{\widehat{C}} = \mathbf{\Psi}^T \mathbf{C\Psi} \qquad \mathbf{\widehat{K}} = \mathbf{\Psi}^T \mathbf{K\Psi} \qquad \mathbf{\hat{p}} = \mathbf{\Psi}^T \mathbf{p} \tag{6.99}$$

are reduced system matrices of size $n \times n$, which are symmetric and generally fully populated. Thus, the normal modes for this reduced system follow from

$$\mathbf{\widehat{K}x}_j = \omega_j^2 \, \mathbf{\widehat{M}x}_j \tag{6.100}$$

Approximations to the n first actual modes of the system are then obtained by multiplication of the reduced modes by the trial vectors, that is,

$$\boldsymbol{\Phi} \approx \boldsymbol{\Psi} \mathbf{X}, \qquad \mathbf{X} = \{\mathbf{x}_1 \quad \cdots \quad \mathbf{x}_n\} \tag{6.101}$$

If the number n of trial vectors is not too large, the reduced system may offer significant computational savings when solved by standard methods, such as modal superposition.

Example

Consider a close-coupled structure with five masses whose stiffness and mass matrices are

$$\mathbf{K} = k \begin{Bmatrix} 1 & -1 & & & \\ -1 & 2 & -1 & & \\ & -1 & 2 & -1 & \\ & & -1 & 2 & -1 \\ & & & -1 & 2 \end{Bmatrix} \qquad \mathbf{M} = \frac{m}{2} \begin{Bmatrix} 1 & & & & \\ & 2 & & & \\ & & 2 & & \\ & & & 2 & \\ & & & & 2 \end{Bmatrix} \tag{6.102}$$

We wish to reduce this structure to only 2 DOF, which can be accomplished by means of two Ritz vectors. To this effect, we choose two vectors that we know are crude approximations to the first and second modes:

$$\boldsymbol{\Psi} = \begin{Bmatrix} 5 & -2 \\ 4 & -1 \\ 3 & 0 \\ 2 & 1 \\ 1 & 1 \end{Bmatrix} \tag{6.103}$$

The reduced stiffness and mass matrices are then

$$\widehat{\mathbf{K}} = \boldsymbol{\Psi}^T \mathbf{K} \boldsymbol{\Psi} = k \begin{Bmatrix} 5 & -2 \\ -2 & 4 \end{Bmatrix} \qquad \widehat{\mathbf{M}} = \boldsymbol{\Psi}^T \mathbf{M} \boldsymbol{\Psi} = \frac{m}{2} \begin{Bmatrix} 85 & -12 \\ -12 & 10 \end{Bmatrix} \tag{6.104}$$

Solving the reduced eigenvalue problem, we obtain the eigenvectors

$$\mathbf{x}_1 = \begin{Bmatrix} 1 \\ 0.3943 \end{Bmatrix} \qquad \mathbf{x}_2 = \begin{Bmatrix} 0.1004 \\ 1 \end{Bmatrix} \tag{6.105}$$

which imply

$$\phi_1 \approx \begin{Bmatrix} 5 \\ 4 \\ 3 \\ 2 \\ 1 \end{Bmatrix} + 0.3943 \begin{Bmatrix} -2 \\ -1 \\ 0 \\ 1 \\ 1 \end{Bmatrix} = \begin{Bmatrix} 4.2114 \\ 3.6057 \\ 3.0000 \\ 2.3943 \\ 1.3943 \end{Bmatrix} \qquad \phi_2 \approx 0.1004 \begin{Bmatrix} 5 \\ 4 \\ 3 \\ 2 \\ 1 \end{Bmatrix} + \begin{Bmatrix} -2 \\ -1 \\ 0 \\ 1 \\ 1 \end{Bmatrix} = \begin{Bmatrix} -1.4981 \\ -0.5985 \\ 0.3011 \\ 1.2007 \\ 1.1004 \end{Bmatrix} \tag{6.106}$$

and renormalizing with respect to the top elements

$$\phi_1 \approx \begin{Bmatrix} 1.0000 \\ 0.8562 \\ 0.7124 \\ 0.5685 \\ 0.3311 \end{Bmatrix}, \quad \phi_2 \approx \begin{Bmatrix} 1.0000 \\ 0.3995 \\ -0.2010 \\ -0.8015 \\ -0.7345 \end{Bmatrix} \tag{6.107}$$

Also, the eigenvalues for the reduced system are

$$\omega_1 = 0.3239\sqrt{\frac{k}{m}}, \quad \omega_2 = 0.9295\sqrt{\frac{k}{m}} \tag{6.108}$$

By contrast, the *exact* first two frequencies and modal shapes of the five-mass system are

$$\omega_1 = 0.3128\sqrt{\frac{k}{m}}, \quad \omega_2 = 0.9080\sqrt{\frac{k}{m}} \tag{6.109}$$

$$\phi_1 \approx \begin{Bmatrix} 1.0000 \\ 0.9511 \\ 0.8090 \\ 0.5878 \\ 0.3090 \end{Bmatrix}, \quad \phi_2 \approx \begin{Bmatrix} 1.0000 \\ 0.5878 \\ -0.3090 \\ -0.9511 \\ -0.8090 \end{Bmatrix} \tag{6.110}$$

As can be seen, the frequencies of the reduced system are reasonably close, while the modal shapes are less so, particularly the second mode. These could be improved by means of a generalization of the inverse iteration method referred to as the *subspace iteration method*[2] which basically consists in carrying out iterations with several vectors simultaneously.

6.3.5 Trial Functions versus True Modes

In the vast number of cases where the Rayleigh–Ritz method, or the Assumed Modes Method of Section 6.4, is used, the trial functions almost certainly will *not* coincide with the modes. If fortuitously they did, this would mean that the discrete stiffness and mass matrices obtained would be diagonal. But the reverse is not true: diagonal mass and stiffness matrices do *not* guarantee that the modes are exact. All it says it that the trial functions are orthogonal with respect to the mass and stiffness operators. A simple example is that of a simply supported, homogeneous beam for which the true modes are known to be sine functions. Say we used instead the deflected shapes due to static uniform load and to static linearly varying load as trial functions, namely

$$\psi_1 = (1-\xi^2)(5-\xi^2) \qquad \text{shape due to uniform load} \quad -1 \le \xi \le 1, \, \xi = 2x/L \tag{6.111}$$

$$\psi_2 = \xi(1-\xi^2)(7-3\xi^2) \qquad \text{shape due to linear load } q(x) = \xi \tag{6.112}$$

---

[2] K. J. Bathe, *The Finite Element Method*.

where $x = 0$ at the center, then both the integrals of $\rho AL \int_{-1}^{+1} \psi_1 \psi_2 \, d\xi = 0$ and $EI / L^3 \int_{-1}^{+1} \psi_1'' \psi_2'' \, d\xi = 0$ vanish, that is, they are orthogonal to each other. It follows that orthogonality of the trial functions per se is not enough to discriminate true modes from merely good trial functions that satisfy all boundary conditions. The residue in the differential equation would be needed to reveal that.

6.4 Discrete Systems via Lagrange's Equations

Lagrange's equations provide a very convenient means to reduce complicated continuous systems into discrete models with a finite number of DOF, which can then be solved using conventional methods. As in the method of weighted residuals previously described, the strategy relies on making educated assumptions about how the system is able to deform, that is, using *trial functions* or *assumed modes* to describe the spatial variation of the motion. Unlike weighted residuals, however, the use of Lagrange's equations only requires that the trial functions satisfy the *essential* boundary conditions. The trial function must also be sufficiently continuous, that is, they must have C_κ continuity (see Section 6.2). Thus, the method is comparable in flexibility and power to the Rayleigh–Ritz approach.

6.4.1 Assumed Modes Method

In a nutshell, the assumed modes method consists in the following steps:

- Make assumptions about how the structure will deform using "generalized coordinates"
- Choose a convenient number of *trial functions* or assumed modes, $\psi_j(x)$, which are functions of space only. These *must* satisfy the geometric boundary conditions (displacement and/or rotations). If they also happen to satisfy the *force (or natural) boundary conditions*, still better results are obtained. *You* choose both the shape and number of these functions, ideally as close as possible to what you believe the true modes to be. The more trial functions, the better, but then more work is involved. Surprisingly good results are often obtained with just one function.
- Express the displacements of the structure in terms of these assumed modes times unknown parameters q_j ("generalized coordinates"), which are functions of time only.
- Apply Lagrange's equations. This leads to a discrete system that has as many DOF as the number of independent parameters chosen.

The method is best explained by means of examples, but before considering these, we provide a brief explanation of our notation for partial derivatives.

6.4.2 Partial Derivatives

Consider a quadratic form (a scalar!) with a *symmetric* square matrix $\mathbf{A}$ (where $\mathbf{A} = \mathbf{\Psi}^T \mathbf{\Psi}$ or $\mathbf{A} = \mathbf{\Psi}''^T \mathbf{\Psi}''$):

$$f = \tfrac{1}{2} \mathbf{q}^T \mathbf{A} \mathbf{q} \tag{6.113}$$

The partial derivative of f with respect to q_j (the result of which is a scalar!) is

$$
\begin{aligned}
\frac{\partial f}{\partial q_j} = \frac{\partial}{\partial q_j}\left(\tfrac{1}{2}\mathbf{q}^T\mathbf{A}\mathbf{q}\right) &= \tfrac{1}{2}\frac{\partial \mathbf{q}^T}{\partial q_j}\mathbf{A}\mathbf{q} + \tfrac{1}{2}\mathbf{q}^T\mathbf{A}\frac{\partial \mathbf{q}}{\partial q_j} \\
&= \tfrac{1}{2}\frac{\partial \mathbf{q}^T}{\partial q_j}\left(\mathbf{A}+\mathbf{A}^T\right)\mathbf{q} \\
&= \frac{\partial \mathbf{q}^T}{\partial q_j}\mathbf{A}\mathbf{q} = \mathbf{e}_j^T\mathbf{A}\mathbf{q}
\end{aligned}
\tag{6.114}
$$

in which $\mathbf{e}_j$ is a vector whose jth element is 1, and all others are zero. The second line in Eq. 6.114 follows because the second term on the first line is a scalar, and the transpose of a scalar equals the scalar itself. In addition, $\mathbf{A}+\mathbf{A}^T = 2\mathbf{A}$ because the matrix is symmetric. Writing all of the partial derivatives in matrix form, we obtain

$$
\frac{\partial f}{\partial \mathbf{q}} = \tfrac{1}{2}\frac{\partial \mathbf{q}^T\mathbf{A}\mathbf{q}}{\partial \mathbf{q}} = \left\{\frac{\partial f}{\partial q_j}\right\} = \left\{\mathbf{e}_j^T\right\}\mathbf{A}\mathbf{q} = \mathbf{I}\mathbf{A}\mathbf{q} \equiv \mathbf{A}\mathbf{q}
\tag{6.115}
$$

Thus, we have managed to express a system of partial derivatives in compact form.

6.4.3 Examples of Application

Example 1: Beam with Concentrated Mass

Consider a uniform, simply supported beam with bending stiffness EI, mass per unit length ρA, and total length L. At a distance $a = \tfrac{1}{4}L$ from the left support is a mass m attached, which oscillates together with the beam (Figure 6.5). Formulate the equations of motion using Lagrange's equations.

This continuous system has infinitely many DOF characterized by transcendental equations with no closed-form solution. However, the displacements of the beam can be expressed with as much accuracy as desired by means of the finite expansion

$$
u(x,t) = \sum_{j=1}^{N} \psi_j(x)q_j(t) = \mathbf{\Psi}\mathbf{q} \equiv \mathbf{q}^T\mathbf{\Psi}^T
\tag{6.116}
$$

$$
\mathbf{\Psi} = \left\{\psi_1 \quad \cdots \quad \psi_N\right\}, \qquad \mathbf{q} = \left\{\begin{array}{c} q_1 \\ \vdots \\ q_N \end{array}\right\}
\tag{6.117}
$$

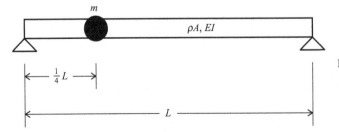

Figure 6.5. Beam with lumped mass.

in which the ψ_j are user chosen, known *trial* functions of space that *must* satisfy the beam's geometric (or essential) boundary conditions. In addition, they may, but need not satisfy the natural (force) boundary conditions (i.e., those in shear and/or bending moment). The number N of functions is also chosen by the analyst. The family of functions must form a "complete set" in the sense that when $N \to \infty$, one must be able to describe any arbitrary deformation $u(x,t)$. From the expansions in Eqs. 6.116 and 6.117, we can write in compact form

$$\dot{u}^2 = \dot{\mathbf{q}}^T \mathbf{\Psi}^T \mathbf{\Psi} \dot{\mathbf{q}}, \qquad (u'')^2 = \mathbf{q}^T \mathbf{\Psi}''^T \mathbf{\Psi}'' \mathbf{q} \tag{6.118}$$

The kinetic and potential energies are then

$$K = \tfrac{1}{2}\left(m\dot{u}^2_{x=\frac{1}{4}L} + \int_0^L \rho A \dot{u}^2 \, dx\right) = \tfrac{1}{2}\dot{\mathbf{q}}^T \left[m\left(\mathbf{\Psi}^T \mathbf{\Psi}\right)_{x=\frac{1}{4}L} + \rho A \int_0^L \mathbf{\Psi}^T \mathbf{\Psi} \, dx\right] \dot{\mathbf{q}} \tag{6.119}$$

$$V = \tfrac{1}{2}\int_0^L EI\,(u'')^2 \, dx = \tfrac{1}{2} EI\, \mathbf{q}^T \left[\int_0^L \mathbf{\Psi}''^T \mathbf{\Psi}'' \, dx\right] \mathbf{q} \tag{6.120}$$

the partial derivatives of which are

$$\frac{\partial K}{\partial \mathbf{q}} = \mathbf{0} \tag{6.121}$$

$$\frac{d}{dt}\left(\frac{\partial K}{\partial \dot{\mathbf{q}}}\right) = \left[m\left(\mathbf{\Psi}^T \mathbf{\Psi}\right)_{x=\frac{1}{4}L} + \rho A \int_0^L \mathbf{\Psi}^T \mathbf{\Psi} \, dx\right] \ddot{\mathbf{q}} = \mathbf{M}\ddot{\mathbf{q}} \tag{6.122}$$

$$\frac{\partial V}{\partial \mathbf{q}} = EI\left[\int_0^L \mathbf{\Psi}''^T \mathbf{\Psi}'' \, dx\right] \mathbf{q} \tag{6.123}$$

The Lagrange equations are then

$$\frac{d}{dt}\left(\frac{\partial K}{\partial \dot{\mathbf{q}}}\right) - \frac{\partial K}{\partial \mathbf{q}} + \frac{\partial V}{\partial \mathbf{q}} = \left[m\left(\mathbf{\Psi}^T \mathbf{\Psi}\right)_{x=\frac{1}{4}L} + \rho A \int_0^L \mathbf{\Psi}^T \mathbf{\Psi} \, dx\right] \ddot{\mathbf{q}} + EI\left[\int_0^L \mathbf{\Psi}''^T \mathbf{\Psi}'' \, dx\right]\mathbf{q} = \mathbf{0} \tag{6.124}$$

Hence, the mass and stiffness matrices for the beam with a lumped mass are

$$\mathbf{M} = m\left(\mathbf{\Psi}^T \mathbf{\Psi}\right)_{x=\frac{1}{4}L} + \rho A \int_0^L \mathbf{\Psi}^T \mathbf{\Psi} \, dx \tag{6.125}$$

$$\mathbf{K} = EI\left[\int_0^L \mathbf{\Psi}''^T \mathbf{\Psi}'' \, dx\right] \tag{6.126}$$

In particular, the product of the trial functions gives

$$\mathbf{\Psi}^T \mathbf{\Psi} = \left\{\begin{matrix} \psi_1 \\ \vdots \\ \psi_N \end{matrix}\right\} \left\{\psi_1 \quad \cdots \quad \psi_N\right\} = \left\{\begin{matrix} \psi_1^2 & \cdots & \psi_1 \psi_N \\ \vdots & \ddots & \vdots \\ \psi_N \psi_1 & \cdots & \psi_N^2 \end{matrix}\right\} \tag{6.127}$$

Here we choose $\psi_j = \sin(j\pi x/L)$, which satisfy *all* boundary conditions, that is, both the essential (geometric) boundary conditions and the additional (natural, or force) boundary conditions. These trial functions happen to be the exact modes of the beam when $m = 0$, that is, for a uniform beam without the lumped mass.

Substituting the trial function 6.127 into the integrals 6.125 and 6.126 and carrying out the required operations, we obtain

$$
\int_0^L \begin{Bmatrix} \psi_1^2 & \cdots & \psi_1\psi_N \\ \vdots & \ddots & \vdots \\ \psi_N\psi_1 & \cdots & \psi_N^2 \end{Bmatrix} dx = \int_0^L \begin{bmatrix} \sin^2\pi\frac{x}{L} & \sin\pi\frac{x}{L}\sin\pi\frac{2x}{L} & \cdots & \sin\pi\frac{x}{L}\sin\pi\frac{Nx}{L} \\ \sin\pi\frac{2x}{L}\sin\pi\frac{x}{L} & \sin^2\pi\frac{2x}{L} & \cdots & \sin\pi\frac{2x}{L}\sin\pi\frac{Nx}{L} \\ \vdots & \vdots & \ddots & \vdots \\ \sin\pi\frac{Nx}{L}\sin\pi\frac{x}{L} & \sin\pi\frac{Nx}{L}\sin\pi\frac{2x}{L} & \cdots & \sin^2\pi\frac{Nx}{L} \end{bmatrix} dx
$$

$$
= \frac{L}{2}\begin{Bmatrix} 1 & 0 & \cdots & 0 \\ 0 & 1 & \cdots & \vdots \\ \vdots & \vdots & \ddots & 0 \\ 0 & \cdots & 0 & 1 \end{Bmatrix} \tag{6.128}
$$

$$
\int_0^L \begin{Bmatrix} \psi_1''^2 & \cdots & \psi_1''\psi_N'' \\ \vdots & \ddots & \vdots \\ \psi_N''\psi_1'' & \cdots & \psi_N''^2 \end{Bmatrix} dx = \left(\frac{\pi}{L}\right)^4 \int_0^L \begin{Bmatrix} \sin^2\pi\frac{x}{L} & \cdots & N\sin\pi\frac{x}{L}\sin\pi\frac{Nx}{L} \\ \vdots & \ddots & \vdots \\ \sin\pi\frac{Nx}{L}\sin\pi\frac{x}{L} & \cdots & N^2\sin^2\pi\frac{Nx}{L} \end{Bmatrix} dx
$$

$$
= \frac{\pi^4}{2L^3}\begin{bmatrix} 1 & 0 & \cdots & 0 \\ 0 & 4 & \ddots & \vdots \\ \vdots & \ddots & \ddots & 0 \\ 0 & \cdots & 0 & N^2 \end{bmatrix} \tag{6.129}
$$

Also, $\sin j\pi\frac{x}{L}\big|_{x=\frac14 L} = \sin j\frac{\pi}{4} = \frac12\sqrt{2}, 1, \frac12\sqrt{2}, 0, -\frac12\sqrt{2}, -1, \cdots$. For $N = 4$, this gives

$$
(\mathbf{\Psi}^T\mathbf{\Psi})_{x=\frac14 L} = \begin{Bmatrix} \left(\frac12\sqrt{2}\right)^2 & \frac12\sqrt{2}\times 1 & \frac12\sqrt{2}\times\frac12\sqrt{2} & \frac12\sqrt{2}\times 0 \\ & 1^2 & 1\times\frac12\sqrt{2} & 1\times 0 \\ & & \left(\frac12\sqrt{2}\right)^2 & \frac12\sqrt{2}\times 0 \\ \text{Symm} & & & 0^2 \end{Bmatrix} \tag{6.130}
$$

$$
= \frac12\begin{Bmatrix} 1 & \sqrt{2} & 1 & 0 \\ \sqrt{2} & 1 & \sqrt{2} & 0 \\ 1 & \sqrt{2} & 1 & 0 \\ 0 & 0 & 0 & 0 \end{Bmatrix}
$$

Hence

$$\mathbf{M} = \frac{m}{2}\begin{Bmatrix} 1 & \sqrt{2} & 1 & 0 \\ \sqrt{2} & 1 & \sqrt{2} & 0 \\ 1 & \sqrt{2} & 1 & 0 \\ 0 & 0 & 0 & 0 \end{Bmatrix} + \frac{\rho A L}{2}\begin{Bmatrix} 1 & 0 & \cdots & 0 \\ 0 & 1 & \cdots & \vdots \\ \vdots & \vdots & \ddots & 0 \\ 0 & \cdots & 0 & 1 \end{Bmatrix} \tag{6.131}$$

$$\mathbf{K} = \frac{\pi^4 EI}{2L^3}\begin{Bmatrix} 1 & 0 & \cdots & 0 \\ 0 & 4 & \ddots & \vdots \\ \vdots & \ddots & \ddots & 0 \\ 0 & \cdots & 0 & N^2 \end{Bmatrix} \tag{6.132}$$

Finally, the equation of motion is

$$\mathbf{M\ddot{q}} + \mathbf{Kq} = \mathbf{0} \tag{6.133}$$

After solving this equation for $\mathbf{q}(t)$ we obtain the actual motion of the beam by computing $\mathbf{u} = \mathbf{\Psi q}$.

Example 2: Oscillator Mounted on a Bending Beam
A 1-DOF oscillator is mounted on a simply supported, homogeneous bending beam of stiffness EI, mass density ρ, and cross section A, as shown in Figure 6.6. In addition, there is dynamic force acting at the ¾ point. We identify three points at which these elements are located or connected with indices α, β, γ, as marked (α = mass, β = attachment point of spring, and γ = point of application of load. We use Greek letters, to avoid confusion with the indices for the assumed modes). The system is constrained to move only in the vertical direction.

In principle, this problem involves a continuous body with distributed mass (the beam), so it has infinitely many DOF. However, we can still obtain excellent approximations by making assumptions about how the body will deform. In essence, we postulate that the deformation of the system can be described in terms of a linear combination of freely chosen, independent *trial functions*, which are scaled by as yet unknown generalized

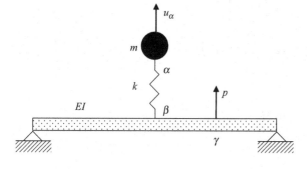

Figure 6.6. Beam with mounted oscillator.

coordinates q_i. Clearly, the more parameters we choose, the better the answer will be, but also the more effort will be required to obtain and solve the equations of motion.

In this problem, *we* choose four generalized coordinates as follows:

Motion of beam : $\quad u(x,t) = q_1(t)\sin\frac{\pi x}{L} + q_2(t)\sin\frac{2\pi x}{L} + q_3(t)\sin\frac{3\pi x}{L}$ \hfill (6.134)

Motion of oscillator : $\quad u_\alpha = q_4(t)$ \hfill (6.135)

These functions satisfy, as required, the essential (geometric) boundary conditions, and also the natural (i.e., stress) boundary conditions. These functions happen to be the *exact* first three vibration modes of a simply supported bending beam (our beam *without* the oscillator). Other choices would have been possible. For example, we could have chosen only two functions equal to the deflected shapes associated with *static* forces applied at $x = \frac{1}{2}L$ (where the oscillator is attached) and $x = \frac{3}{4}L$ (where the force acts). These would have been cubic parabolas with discontinuities at the points of application of the loads (which entails much more work...).

In terms of the generalized coordinates, the motion at the three physical points α, β, γ is

$$u_\alpha \equiv q_4 \tag{6.136}$$

$$\begin{aligned} u_\beta &= q_1 \sin\frac{\pi}{L}\frac{L}{2} + q_2 \sin\frac{2\pi}{L}\frac{L}{2} + q_2 \sin\frac{3\pi}{L}\frac{L}{2} \\ &= q_1 - q_3 \end{aligned} \tag{6.137}$$

$$\begin{aligned} u_\gamma &= q_1 \sin\frac{\pi}{L}\frac{3L}{4} + q_2 \sin\frac{2\pi}{L}\frac{3L}{4} + q_3 \sin\frac{3\pi}{L}\frac{3L}{4} \\ &= \tfrac{\sqrt{2}}{2}q_1 - q_2 + \tfrac{\sqrt{2}}{2}q_3 \end{aligned} \tag{6.138}$$

Thus, the deformation of the spring is

$$\begin{aligned} e &= u_\alpha - u_\beta = q_4 - (q_1 - q_3) \\ &= -q_1 + q_3 + q_4 \end{aligned} \tag{6.139}$$

We are now ready to derive the Lagrange equations for this problem.

Kinetic energy : $\quad T = \tfrac{1}{2}m\dot{q}_4^2 + \tfrac{1}{2}\int_0^L \dot{u}^2 \rho A\,dx$ \hfill (6.140)

Elastic potential : $\quad V = \tfrac{1}{2}k(q_3 - q_1)^2 + \tfrac{1}{2}\int_0^L EI\left(\frac{\partial^2 u}{\partial x^2}\right)^2 dx$ \hfill (6.141)

External work : $\quad W_e = p\,u_\gamma = p\left(\tfrac{\sqrt{2}}{2}q_1 - q_2 + \tfrac{\sqrt{2}}{2}q_3\right)$ \hfill (6.142)

But

$$\begin{aligned} \int_0^L \dot{u}^2 \rho A\,dx &= \rho A\int_0^L \left(\dot{q}_1(t)\sin\frac{\pi x}{L} + \dot{q}_2(t)\sin\frac{2\pi x}{L} + \dot{q}_3(t)\sin\frac{3\pi x}{L}\right)^2 dx \\ &= \tfrac{1}{2}M\left(\dot{q}_1^2 + \dot{q}_2^2 + \dot{q}_3^2\right) \end{aligned} \tag{6.143}$$

in which $M = \rho A L$ is the mass of the beam. Also,

$$\int_0^L EI \, (u'')^2 \, dx = EI \int_0^L \left[\left(\tfrac{\pi}{L} \right)^2 q_1(t) \sin \tfrac{\pi x}{L} + \left(\tfrac{2\pi}{L} \right)^2 q_2(t) \sin \tfrac{2\pi x}{L} + \left(\tfrac{3\pi}{L} \right)^2 q_2(t) \sin \tfrac{3\pi x}{L} \right]^2 dx$$

$$= \frac{\pi^4 EI}{2 L^3} (q_1^2 + 16 q_2^2 + 81 q_3^2) \tag{6.144}$$

The fact that the cross-products $q_1 q_2$, and so forth do *not* contribute here to the integrals is the convenient result of how we chose the trial functions: the modes of the beam alone satisfy the orthogonality condition. Hence

$$T = \tfrac{1}{2} M (\dot{q}_1^2 + \dot{q}_2^2 + \dot{q}_3^2) + \tfrac{1}{2} m \dot{q}_4^2 \qquad V = \frac{\pi^4 EI}{2 L^3} (q_1^2 + 16 q_2^2 + 81 q_3^2) + \tfrac{1}{2} k (q_1 - q_3 - q_4)^2$$

$$\tag{6.145}$$

$$\frac{\partial T}{\partial \dot{q}_1} = M \dot{q}_1 \qquad \frac{\partial T}{\partial \dot{q}_2} = M \dot{q}_2 \qquad \frac{\partial T}{\partial \dot{q}_3} = M \dot{q}_3 \qquad \frac{\partial K}{\partial \dot{q}_4} = m \dot{q}_4 \tag{6.146}$$

and

$$\frac{d}{dt} \frac{\partial T}{\partial \dot{q}_1} = M \ddot{q}_1 \qquad \frac{d}{dt} \frac{\partial T}{\partial \dot{q}_2} = M \ddot{q}_2 \qquad \frac{d}{dt} \frac{\partial T}{\partial \dot{q}_3} = M \ddot{q}_3 \qquad \frac{d}{dt} \frac{\partial T}{\partial \dot{q}_4} = m \ddot{q}_4 \tag{6.147}$$

$$\frac{\partial V}{\partial q_1} = \frac{\pi^4 EI}{L^3} q_1 + k (q_1 - q_3 - q_4) \qquad \frac{\partial V}{\partial q_2} = \frac{16 \pi^4 EI}{L^3} q_2 \tag{6.148}$$

$$\frac{\partial V}{\partial q_3} = \frac{81 \pi^4 EI}{L^3} q_3 + k (-q_1 + q_3 + q_4) \qquad \frac{\partial V}{\partial q_4} = k (-q_1 + q_3 + q_4) \tag{6.149}$$

$$\frac{\partial W_e}{\partial q_1} = \frac{\sqrt{2}}{2} p \qquad \frac{\partial W_e}{\partial q_2} = -p \qquad \frac{\partial W_e}{\partial q_3} = \frac{\sqrt{2}}{2} p \qquad \frac{\partial W_e}{\partial q_4} = 0 \tag{6.150}$$

The Lagrange's equations are then

$$\frac{d}{dt} \frac{\partial T}{\partial \dot{q}_i} + \frac{\partial V}{\partial q_i} = \frac{\partial W_e}{\partial q_i} \qquad i = 1, 2, 3, 4 \tag{6.151}$$

In matrix form, and with the shorthand $k_b = \pi^4 EI / L^3$, the equations of motion are then

$$
\begin{bmatrix} M & & & \\ & M & & \\ & & M & \\ & & & m \end{bmatrix}
\begin{Bmatrix} \ddot{q}_1 \\ \ddot{q}_2 \\ \ddot{q}_3 \\ \ddot{q}_4 \end{Bmatrix}
+
\begin{bmatrix} k + k_b & & -k & -k \\ & 16 k_b & & \\ -k & & k + 81 k_b & k \\ -k & & k & k \end{bmatrix}
\begin{Bmatrix} q_1 \\ q_2 \\ q_3 \\ q_4 \end{Bmatrix}
=
\begin{Bmatrix} \frac{\sqrt{2}}{2} p \\ -p \\ \frac{\sqrt{2}}{2} p \\ 0 \end{Bmatrix}
\tag{6.152}
$$

which is a 4-DOF system of the form

$$\mathbf{M} \ddot{\mathbf{q}} + \mathbf{K} \mathbf{q} = \mathbf{p}(t) \tag{6.153}$$

Notice that the second DOF is uncoupled from the others. This is because the second assumed mode does not interact with the spring-mass system, so we recover the true second mode of the beam. If we set m and k to zero (i.e., *no* oscillator) and also get rid of the last row and column (since $q_4 = 0$), we recover the *exact* solution for the simply supported beam. After solving this equation for $\mathbf{q}(t)$, we can use the assumed modes expansion to find the displacements of the beam.

To test this problem, let's use some numbers. Assume $m = M$, $k = k_b$, and $p = 0$. Hence, this is a free vibration problem of the form

$$k \begin{bmatrix} 2 & & -1 & -1 \\ & 16 & & \\ -1 & & 82 & 1 \\ -1 & & 1 & 1 \end{bmatrix} \begin{Bmatrix} q_1 \\ q_2 \\ q_3 \\ q_4 \end{Bmatrix}_j = m\,\omega_j^2 \begin{Bmatrix} q_1 \\ q_2 \\ q_3 \\ q_4 \end{Bmatrix}_j \tag{6.154}$$

whose eigenvalues and modal shapes are (using MATLAB®)

$$\omega_1^2 = 0.3807\,k/m \qquad \omega_2^2 = 2.5942\,k/m \qquad \omega_3^2 = 16\,k/m \qquad \omega_4^2 = 82.0252\,k/m \tag{6.155}$$

$$\mathbf{Q} = \{q_{ij}\} = \begin{Bmatrix} 0.5236 & 0.8519 & 0 & -0.0127 \\ 0 & 0 & 1 & \\ -0.0040 & 0.0173 & 0 & 0.9998 \\ 0.8519 & -0.5235 & 0 & 0.0125 \end{Bmatrix} \tag{6.156}$$

Let's compare the above eigenvalues against the first three eigenvalues of the beam without the oscillator, which are

$$\omega_1^2 = k/m, \qquad \omega_2^2 = 16k/m, \qquad \omega_3^2 = 81\,k/m \tag{6.157}$$

We see that the third root matches exactly the second eigenvalue of the beam without the oscillator, while the fourth root is close the third eigenvalue. Thus, the oscillator basically changed the first root of the continuous beam into two roots, one that is smaller and another larger than the root of the beam alone.

To obtain the (approximated) modes of vibration of the beam and/or oscillator, it suffices to apply once more the assumed modal expansion. For example, for the beam we would have

$$\phi_j(x) = \sum_{i=1}^{3} q_{ij}\psi_i = \sum_{i=1}^{3} q_{ij}\sin\frac{i\pi}{L} \qquad j = 1,2,3,4 \tag{6.158}$$

or in full

$$\phi_{b1}(x) = 0.5236\sin\tfrac{\pi x}{L} - 0.0040\sin\tfrac{3\pi x}{L} \tag{6.159}$$

$$\phi_{b2}(x) = 0.8519\sin\tfrac{\pi x}{L} + 0.0173\sin\tfrac{3\pi x}{L} \tag{6.160}$$

$$\phi_{b3}(x) = 1.0\sin\tfrac{2\pi x}{L} \tag{6.161}$$

$$\phi_{b4}(x) = -0.0127\sin\tfrac{\pi x}{L} + 0.9998\sin\tfrac{3\pi x}{L} \tag{6.162}$$

while the oscillator modes are

$$\phi_{o1} = 0.8519, \qquad \phi_{o2} = -0.5235, \qquad \phi_{o3} = 0., \qquad \phi_{o4} = 0.0125 \tag{6.163}$$

For example, a vibration that excited only the first mode would be of the form

$$u_b(x,t) = \left(0.5236 \sin \tfrac{\pi x}{L} - 0.0040 \sin \tfrac{3\pi x}{L}\right) \sin \omega_1 t \qquad \text{for the beam} \tag{6.164}$$

$$u_o(t) = 0.8519 \sin \omega_1 t \qquad\qquad\qquad\qquad \text{for the oscillator} \tag{6.165}$$

6.4.4 What If Some of the Discrete Equations Remain Uncoupled?

It may well occur that after introducing the trial function one fortuitously arrives at a discrete model where one or more DOF in both the mass and stiffness matrices remain uncoupled from the other DOF. This could perhaps mean that one has been rather fortunate and somehow guessed a trial function for the uncoupled DOF which happens to be an *exact* mode of the system. But more likely, it may mean that just by chance one has chosen pairs of trial function that are linear combinations of modes without modal overlap. To illustrate this concept, it suffices to consider an originally discrete system to which we apply the Rayleigh–Ritz approach with the objective of reducing the number of DOF (see Section 6.3.4). Say the original system has $N = 5$ DOF, which we wish to reduce to 2, and that the original eigenvalue problem is given by

$$\mathbf{K}\phi_j = \omega_j^2 \mathbf{M}\phi_j, \qquad j = 1,2,\ldots 5 \tag{6.166}$$

Suppose next that we choose to reduce this into a 2-DOF system, and use for this purpose the two trial functions ψ_1, ψ_2, which in principle are arbitrary. This would mean that the transformed 2×2 stiffness matrix is

$$\bar{\mathbf{K}} = \mathbf{\Psi}^T \mathbf{K} \mathbf{\Psi} = \begin{Bmatrix} \psi_1^T \mathbf{K} \psi_1 & \psi_1^T \mathbf{K} \psi_2 \\ \psi_2^T \mathbf{K} \psi_1 & \psi_2^T \mathbf{K} \psi_2 \end{Bmatrix} \tag{6.167}$$

and a similar expression for the transformed mass matrix. But as luck would have it, say that we hit the jackpot and fortuitously chose for the trial function the exact combinations

$$\psi_1 = \alpha_1 \phi_1 + \alpha_2 \phi_2, \qquad \psi_2 = \alpha_3 \phi_3 + \alpha_4 \phi_4 + \alpha_5 \phi_5 \tag{6.168}$$

where the α_j are arbitrary coefficients; also, observe that there is *no* modal overlap between the two trial functions. In that case, and considering orthogonality, the transformed stiffness matrix is

$$\bar{\mathbf{K}} = \begin{Bmatrix} \alpha_1 \kappa_1 + \alpha_2 \kappa_2 & 0 \\ 0 & \alpha_3 \kappa_3 + \alpha_4 \kappa_4 + \alpha_5 \kappa_5 \end{Bmatrix} \tag{6.169}$$

where the $\kappa_j = \phi_j^T \mathbf{K} \phi_j$. A similar expression applies also to the mass matrix. Clearly, in this case the reduced stiffness and mass matrices are uncoupled, in which case the estimated uncoupled frequencies will be

$$\bar{\omega}_1^2 = \frac{\alpha_1 \kappa_1 + \alpha_2 \kappa_2}{\alpha_1 \mu_1 + \alpha_2 \mu_2}, \qquad \bar{\omega}_2^2 = \frac{\alpha_3 \kappa_3 + \alpha_4 \kappa_4 + \alpha_5 \kappa_5}{\alpha_3 \mu_3 + \alpha_4 \mu_4 + \alpha_5 \mu_5} \tag{6.170}$$

These are weighted averages of the true frequencies of the system, and such that

$$\omega_1^2 \leq \overline{\omega}_1^2 \leq \omega_2^2 \qquad \text{and} \qquad \omega_3^2 \leq \overline{\omega}_2^2 \leq \omega_5^2 \tag{6.171}$$

That is, the estimates may or not be close to any two true frequencies. Hence, uncoupled equations only imply that the trial functions happen to be mutually orthogonal, but give no indications as to the accuracy of their choice. They can, but need not resemble any true mode.

6.5 Numerical Integration in the Time Domain

By and large, the available numerical solution methods can be grouped into two broad classes, namely the *time domain* methods, and the *frequency domain* methods. The former operate directly on the equations of motion, provide the response at discrete *time steps*, and may generally be used for both linear and nonlinear problems. The latter, on the other hand, involve the use of complex algebra and the application of Fourier series and Fourier transforms. While frequency domain solutions are (in principle) restricted to linear systems, they allow the application of powerful numerical techniques, including the now well-known *Fast Fourier Transform* (FFT) algorithm. Also, they provide a simple mechanism for modeling hysteretic damping, so they are widely used in practice. We shall consider first the solution strategy in the time domain and defer the application of Fourier methods to Section 6.6.

As we have seen in earlier chapters, the analysis of structural vibration problems with discrete models entails the solution of systems of ordinary, second-order differential equations. For example, a linearly *elastic* structural system subjected to dynamic loads producing only small displacements and rotations is characterized by the *linear* system

$$\mathbf{M\ddot{u}} + \mathbf{C\dot{u}} + \mathbf{Ku} = \mathbf{p}(t) \tag{6.172}$$

By contrast, when structural materials exhibit *inelastic* behavior, the internal forces in the structure do depend on the deformation *path* (i.e., on the deformation time history). In that case, the governing equations must be expressed in the general nonlinear form

$$\mathbf{M\ddot{u}} + \mathbf{C\dot{u}} + \mathbf{f}(\mathbf{u},t) = \mathbf{p}(t) \tag{6.173}$$

with $\mathbf{f}(\mathbf{u},t)$ denoting the time-varying, path-dependent *internal* forces. In most cases, these equations must be solved with the aid of numerical integration methods. This is so because the sources have an arbitrary variation in time, because the equations are non-linear, or simply because the structural complexity or the nature of damping is such that only numerical solutions are practical.

All solution methods in the time domain involve some kind of approximations to the equations of motion, be they *physical* or *mathematical* approximations. The physical approximations are usually based on making simplifying assumptions on how the forcing function and/or the response vary with time, while the mathematical approximations rely on replacing the differential equation with difference equations and/or using numerical integration methods. The former are targeted specifically at the second-order differential equations that characterize problems in structural dynamics, while the latter are generic in the sense that they can be applied to any type of differential equations. To be useful, all methods must satisfy some *stability* and *accuracy* conditions, a subject that we shall discuss in a separate section.

In all numerical methods in the time domain, the time variable t is discretized or sampled at small intervals Δt, which are usually (but not always) held constant for the duration of the analysis. When we say *small*, we mean small in comparison to the dynamic characteristics of the system being solved and the dominant frequencies (or more precisely, periods) of the excitation. A discrete instant in time is then given by $t_i = i \Delta t$, with i being an integer counter; hence, $\Delta t = t_{i+1} - t_i$ The discretized equations of motion are

$$\mathbf{M}\ddot{\mathbf{u}}_i + \mathbf{C}\dot{\mathbf{u}}_i + \mathbf{K}\mathbf{u}_i = \mathbf{p}_i \qquad \text{linear problem} \tag{6.174}$$

$$\mathbf{M}\ddot{\mathbf{u}}_i + \mathbf{C}\dot{\mathbf{u}} + \mathbf{f}_i = \mathbf{p}_i \qquad \text{nonlinear problem} \tag{6.175}$$

In the nonlinear case, the restoring force $\mathbf{f}_i = \mathbf{f}_i(\mathbf{u}_i, \mathbf{u}_{i-1}, \ldots \mathbf{u}_0)$ is a function of the system's past deformation history. At each instant in time, the response of the system is completely defined by the displacement and velocity vectors $\mathbf{u}_i$ and $\dot{\mathbf{u}}_i$, because the forcing function is known at all times, and the acceleration can be deduced directly from the differential equation. In some methods, the displacement and velocity vectors are combined into a single entity referred to as the *state vector*.

6.5.1 Physical Approximations to the Forcing Function

In this first type of numerical methods, the forcing function is assumed to be constant, or to vary linearly within each small time step, over the course of which the equations of motion are solved *exactly*. Hence, these methods can be applied only to *linear systems*. Obviously, in the case of MDOF system, exact solutions can be found only to each of the modal equations of motion, so these methods are restricted to modal superposition. For this reason, we consider here only SDOF systems, for which this class of methods is intended. An example is the computation of earthquake response spectra. Consider a specific time interval $[t_i, t_{i+1}]$, and assume that during this interval the forcing function can be approximated as

$$p(t) = (1-\alpha)\breve{p}_i + \alpha \left[p_i + \frac{\Delta p_i}{\Delta t}(t - t_i) \right] \tag{6.176}$$

in which $\breve{p}_i$ is a constant, representative value of the forcing function in the interval being considered, $\Delta p_i = p_{i+1} - p_i$ is the load increment during this interval, and α is an arbitrary weighting factor. Choosing $\alpha = 0$ together with $\breve{p}_i = \frac{1}{2}(p_i + p_{i+1})$, we obtain the *rectangular force method*, while if we take $\alpha = 1$, we have the *trapezoidal force method*, for which the forcing function varies linearly between time points. In either case, the forcing function is of the form

$$p(t) = a + b(t - t_i) \tag{6.177}$$

in which the values of a, b follow by simple inspection of the expression for $p(t)$:

$$a = (1-\alpha)\breve{p}_i + \alpha p_i \qquad b = \alpha \frac{\Delta p_i}{\Delta t} \tag{6.178}$$

The equation of motion for the SDOF system (or the modal equation) is then

$$m\ddot{u} + c\dot{u} + ku = a + b(t - t_i) \tag{6.179}$$

The exact solution to this differential equation can be found, as usual, by considering a homogeneous solution with initial conditions $u_i, \dot{u}_i$, and a particular solution $u_p = A + B(t - t_i)$. Substituting u_p into the differential equation, we find that the constants A, B are related to a, b as

$$A = \frac{1}{k}\left[a - \frac{2\xi}{\omega_n}b\right] \quad \text{and} \quad B = \frac{b}{k} \tag{6.180}$$

Hence, the complete solution is

$$u(t) = e^{-\xi\omega_n(t-t_i)}\left\{(u_i - u_{pi})\cos\omega_d(t - t_i) + \frac{(\dot{u}_i - \dot{u}_{pi}) + \xi\omega_n(u_i - u_{pi})}{\omega_d}\sin\omega_d(t - t_i)\right\} + u_p(t) \tag{6.181}$$

$$\dot{u}(t) = e^{-\xi\omega_n(t-t_i)}\left\{(\dot{u}_i - \dot{u}_{pi})\cos\omega_d(t - t_i) - \frac{\omega_n}{\omega_d}\left[\xi(\dot{u}_i - \dot{u}_{pi}) + \omega_n(u_i - u_{pi})\right]\sin\omega_d(t - t_i)\right\} + \dot{u}_p(t) \tag{6.182}$$

wherein, of course, $t_i \le t \le t_{i+1}$. With the auxiliary definitions

$$S = \frac{\omega_n}{\omega_d}e^{-\xi\omega_n\Delta t}\sin\omega_d\Delta t \tag{6.183}$$

$$C = e^{-\xi\omega_n\Delta t}\cos\omega_d\Delta t \tag{6.184}$$

the solution at the end of the current time step, which is the beginning of the next step, can be written compactly as

$$\left\{\begin{array}{c} u_{i+1} \\ \frac{1}{\omega_n}\dot{u}_{i+1} \end{array}\right\} = \left\{\begin{array}{cc} C + \xi S & S \\ -S & C - \xi S \end{array}\right\}\left\{\begin{array}{c} u_i - u_{pi} \\ \frac{1}{\omega_n}(\dot{u}_i - \dot{u}_{pi}) \end{array}\right\} + \left\{\begin{array}{c} u_{p,i+1} \\ \frac{1}{\omega_n}\dot{u}_{p,i+1} \end{array}\right\} \tag{6.185}$$

with $u_{pi} = A, \dot{u}_{pi} = B, u_{p,i+1} = A + B\Delta t$, and $\dot{u}_{p,i+1} = B$. We define next the dimensionless matrices

$$\mathbf{A} = \left\{\begin{array}{cc} a_{11} & a_{12} \\ a_{21} & a_{22} \end{array}\right\} \equiv \left\{\begin{array}{cc} C + \xi S & S \\ -S & C - \xi S \end{array}\right\} \tag{6.186}$$

$$\mathbf{B} = \left\{\begin{array}{cc} b_{11} & b_{12} \\ b_{21} & b_{22} \end{array}\right\} \equiv \left\{\begin{array}{cc} -a_{11} + \frac{1}{\omega_n\Delta t}[a_{12} + 2\xi(1 - a_{11})] & 1 - \frac{1}{\omega_n\Delta t}[a_{12} + 2\xi(1 - a_{11})] \\ -a_{21} - \frac{1}{\omega_n\Delta t}[2\xi a_{21} + 1 - a_{22}] & \frac{1}{\omega_n\Delta t}[2\xi a_{21} + 1 - a_{22}] \end{array}\right\} \tag{6.187}$$

Also, we write the state and load vectors as

$$\mathbf{z}_i = \left\{\begin{array}{c} u_i \\ \frac{1}{\omega_n}\dot{u}_i \end{array}\right\} \quad \mathbf{w}_i = \frac{\alpha}{k}\left\{\begin{array}{c} p_i \\ p_{i+1} \end{array}\right\} \quad \breve{\mathbf{w}}_i = \frac{1-\alpha}{k}\left\{\begin{array}{c} 1 - a_{11} \\ -a_{21} \end{array}\right\}\breve{p}_i \tag{6.188}$$

The iterative solution to the equations of motion is then

$$\mathbf{z}_{i+1} = \mathbf{A}\mathbf{z}_i + \mathbf{B}\mathbf{w}_i + \breve{\mathbf{w}}_i \tag{6.189}$$

Notice that if the time step is taken to be identical at all times, the coefficient matrices $\mathbf{A}$, $\mathbf{B}$ will not change in the course of the integration, so they would need be computed only once. This will be true in most practical cases, as there is rarely much reason to consider varying time steps in a linear problem, except perhaps in the context of an adaptive integration scheme.

At any instant in time, the acceleration could be obtained, if desired, from the differential equation and the state vector. In matrix form, this relationship is

$$\ddot{u}_i = \omega_n^2 \left(\tfrac{1}{k} p_i - \begin{bmatrix} 1 & 2\xi \end{bmatrix} \mathbf{z}_i \right) \tag{6.190}$$

In the seismic case, we use the iterative integration formula with $p(t) = -m\ddot{u}_g(t)$ and the state vector $\mathbf{z}_i$ representing the *relative* motions. The absolute acceleration follows then from

$$\ddot{u}_i = -\omega_n^2 \begin{bmatrix} 1 & 2\xi \end{bmatrix} \mathbf{z}_i \tag{6.191}$$

In most cases, the integration will be carried out with only the pure versions of either the rectangular or trapezoidal methods, that is, with $\alpha = 0$ or $\alpha = 1$. While higher order integration schemes could be developed, it should be remembered that these would be based on an exact linear solution to the equation of motion for a SDOF system, so they would be applicable only to linear systems. The additional complications probably do not warrant the additional effort.

6.5.2 Physical Approximations to the Response

This class of methods is applicable to both linear and nonlinear systems. In general, the time step in these methods can be changed as integration progresses, but to avoid notational congestion, we will use a subindex for the time step only when required.

Constant Acceleration Method

As the name implies, in this method we assume the acceleration to have some constant, representative value in each time interval:

$$\ddot{\mathbf{u}}(t) = (1 - \alpha)\ddot{\mathbf{u}}_i + \alpha\ddot{\mathbf{u}}_{i+1} \tag{6.192}$$

with α being an arbitrary constant (usually ½). Since the acceleration is constant, upon integration we obtain

$$\dot{\mathbf{u}}_{i+1} = \dot{\mathbf{u}}_i + \ddot{\mathbf{u}}\,\Delta t \tag{6.193}$$

$$\mathbf{u}_{i+1} = \mathbf{u}_i + \dot{\mathbf{u}}_i\,\Delta t + \tfrac{1}{2}\ddot{\mathbf{u}}\,\Delta t^2 \tag{6.194}$$

The last term in this expression is obtained by substituting the differential equation of motion into the expression for $\ddot{\mathbf{u}}(t)$. If $\alpha \neq 0$, the method is *implicit*, unless the system is linear, in which case it could be converted into an *explicit* method. However, the latter option is not practical except when analyzing SDOF systems. In the implicit case, we have to compute $\mathbf{f}_{i+1}$ iteratively by assuming values for $\mathbf{u}_{i+1}$. As will be seen, the constant acceleration method is a special case of Newmark's method.

Linear Acceleration Method

Here, the acceleration is assumed to vary linearly between time steps:

$$\ddot{\mathbf{u}} = \ddot{\mathbf{u}}_i + (\ddot{\mathbf{u}}_{i+1} - \ddot{\mathbf{u}}_i)(t - t_i) / \Delta t \tag{6.195}$$

Then

$$\dot{\mathbf{u}}_{i+1} = \dot{\mathbf{u}}_i + \tfrac{1}{2}\ddot{\mathbf{u}}_i\,\Delta t + \tfrac{1}{2}\ddot{\mathbf{u}}_{i+1}\,\Delta t \tag{6.196}$$

$$\mathbf{u}_{i+1} = \mathbf{u}_i + \dot{\mathbf{u}}_i\,\Delta t + \tfrac{1}{3}\ddot{\mathbf{u}}_i\,\Delta t^2 + \tfrac{1}{6}\ddot{\mathbf{u}}_{i+1}\,\Delta t^2 \tag{6.197}$$

We have again an implicit method. If the system is linear, the equations could once more be written in explicit form, but for MDOF systems this option would require matrix inversions (or equivalently, a Gaussian elimination), unless the implicit damping forces could be neglected. The linear acceleration method is also a special case of Newmark's method.

Newmark's β Method

Newmark's β method is perhaps the most widely used numerical integration method in structural dynamics. This method is based on the following ansatz:

$$\dot{\mathbf{u}}_{i+1} = \dot{\mathbf{u}}_i + (1 - \alpha)\ddot{\mathbf{u}}_i\,\Delta t + \alpha\ddot{\mathbf{u}}_{i+1}\,\Delta t \tag{6.198}$$

$$\mathbf{u}_{i+1} = \mathbf{u}_i + \dot{\mathbf{u}}_i\,\Delta t + (\tfrac{1}{2} - \beta)\ddot{\mathbf{u}}_i\,\Delta t^2 + \beta\ddot{\mathbf{u}}_{i+1}\,\Delta t^2 \tag{6.199}$$

That is,

$$\mathbf{u}_{i+1} = \mathbf{u}_i + \dot{\mathbf{u}}_i\,\Delta t + (\tfrac{1}{2} - \beta)\Delta t^2\,\mathbf{M}^{-1}\left[\mathbf{p}_i - \mathbf{C}\dot{\mathbf{u}}_i - \mathbf{f}_i\right] + \beta\Delta t^2\,\mathbf{M}^{-1}\left[\mathbf{p}_{i+1} - \mathbf{C}\dot{\mathbf{u}}_{i+1} - \mathbf{f}_{i+1}\right] \tag{6.200}$$

$$\dot{\mathbf{u}}_{i+1} = \dot{\mathbf{u}}_i + (1 - \alpha)\Delta t\,\mathbf{M}^{-1}\left[\mathbf{p}_i - \mathbf{C}\dot{\mathbf{u}}_i - \mathbf{f}_i\right] + \alpha\Delta t\,\mathbf{M}^{-1}\left[\mathbf{p}_{i+1} - \mathbf{C}\dot{\mathbf{u}}_{i+1} - \mathbf{f}_{i+1}\right] \tag{6.201}$$

For $\alpha = \tfrac{1}{2}$ and $\beta = \tfrac{1}{6}$, we obtain the *linear acceleration method*, while if we choose $\beta = \tfrac{1}{2}(1 - \alpha)$, we recover the *constant acceleration method*. For $\alpha = 0$, the method is unstable. In most cases, this is an *implicit* method. In the case of linear systems, $\mathbf{f}_i = \mathbf{K}\mathbf{u}_i$. The equations can then be changed into an explicit form. To see how this is done, we write the linear state equations in matrix form as

$$\begin{Bmatrix} \mathbf{M} + \beta\Delta t^2\,\mathbf{K} & \beta\Delta t^2\,\mathbf{C} \\ \alpha\Delta t\,\mathbf{K} & \mathbf{M} + \alpha\Delta t\,\mathbf{C} \end{Bmatrix} \begin{Bmatrix} \mathbf{u}_{i+1} \\ \dot{\mathbf{u}}_{i+1} \end{Bmatrix} = \begin{Bmatrix} \mathbf{M} - (\tfrac{1}{2} - \beta)\Delta t^2\,\mathbf{K} & \Delta t\left[\mathbf{M} - (\tfrac{1}{2} - \beta)\Delta t\,\mathbf{C}\right] \\ -(1 - \alpha)\Delta t\,\mathbf{K} & \mathbf{M} - (1 - \alpha)\Delta t\,\mathbf{C} \end{Bmatrix} \begin{Bmatrix} \mathbf{u}_i \\ \dot{\mathbf{u}}_i \end{Bmatrix} + $$
$$+ \begin{Bmatrix} (\tfrac{1}{2} - \beta)\Delta t^2\,\mathbf{I} & \beta\Delta t^2\,\mathbf{I} \\ (1 - \alpha)\Delta t\,\mathbf{I} & \alpha\Delta t\,\mathbf{I} \end{Bmatrix} \begin{Bmatrix} \mathbf{p}_i \\ \mathbf{p}_{i+1} \end{Bmatrix} \tag{6.202}$$

A naïve approach to solve for the state vector at step $i + 1$ would be to multiply this matrix equation by the inverse of the matrix on the left. However, this would destroy the narrow-bandedness of the equations (if any). A better alternative is to find at the start of the computation the triangular form of this matrix and use it to efficiently solve at each instant in time the system of equations.

In the particular case of *linearly elastic SDOF systems*, Eq. 6.202 reduces to

$$
\begin{Bmatrix} \alpha & -\beta\omega_n\Delta t \\ \alpha\omega_n\Delta t & 1+\alpha\Delta t\,2\xi\omega_n \end{Bmatrix} \begin{Bmatrix} u_{i+1} \\ \frac{1}{\omega_n}\dot{u}_{i+1} \end{Bmatrix} = \begin{Bmatrix} \alpha-(\frac{1}{2}\alpha-\beta)\omega_n^2\Delta t^2 & -[\beta-\alpha+(\alpha-2\beta)\xi\omega_n\Delta t]\omega_n\Delta t \\ -(1-\alpha)\omega_n\Delta t & 1-2(1-\alpha)\xi\omega_n\Delta t \end{Bmatrix}
$$

$$
\begin{Bmatrix} u_i \\ \frac{1}{\omega_n}\dot{u}_i \end{Bmatrix} + \frac{1}{k}\begin{Bmatrix} (\frac{1}{2}\alpha-\beta)\omega_n^2\Delta t^2 & 0 \\ (1-\alpha)\omega_n\Delta t & \alpha\omega_n\Delta t \end{Bmatrix}\begin{Bmatrix} p_i \\ p_{i+1} \end{Bmatrix}
$$

(6.203)

which can be solved for the state vector

$$
\mathbf{z}_{i+1} = \mathbf{A}\mathbf{z}_i + \mathbf{B}\mathbf{w}_i
$$

(6.204)

$$
\mathbf{z}_i = \begin{Bmatrix} u_i \\ \frac{1}{\omega_n}\dot{u}_i \end{Bmatrix} \qquad \mathbf{w}_i = \frac{1}{k}\begin{Bmatrix} p_i \\ p_{i+1} \end{Bmatrix}
$$

(6.205)

with matrices

$$
\mathbf{A} = \frac{1}{d}\begin{Bmatrix} d-\omega_n^2\Delta t^2\left[\frac{1}{2}+(\alpha-2\beta)\xi\omega_n\Delta t\right] & \omega_n\Delta t\left[1+(2\alpha-1)\xi\omega_n\Delta t-2(\alpha-2\beta)\xi^2\omega_n^2\Delta t^2\right] \\ -\omega_n\Delta t\left[1-\frac{1}{2}(\alpha-2\beta)\omega_n^2\Delta t^2\right] & d-2\xi\omega_n\Delta t-\omega_n^2\Delta t^2\left[\alpha-(\alpha-2\beta)\xi\omega_n\Delta t\right] \end{Bmatrix}
$$

(6.206)

$$
\mathbf{B} = \frac{\omega_n\Delta t}{d}\begin{Bmatrix} \omega_n\Delta t\left[\frac{1}{2}-\beta+(\alpha-2\beta)\xi\omega_n\Delta t\right] & \beta\omega_n\Delta t \\ 1-\alpha-\frac{1}{2}(\alpha-2\beta)\omega_n^2\Delta t^2 & \alpha \end{Bmatrix}
$$

(6.207)

in which $d = 1+2\alpha\xi\omega_n\Delta t+\beta\omega_n^2\Delta t^2$

Impulse Acceleration Method

In this *explicit* method, it is assumed that the velocity has a constant average value at each time step:

$$
\dot{\mathbf{u}}_{avg} = \frac{\mathbf{u}_i-\mathbf{u}_{i-1}}{\Delta t_{i-1}}+\ddot{\mathbf{u}}_i\,\frac{\Delta t_{i-1}+\Delta t_i}{2} = \dot{\mathbf{u}}_i+\frac{1}{2}\ddot{\mathbf{u}}_i\,\Delta t_i
$$

(6.208)

$$
\dot{\mathbf{u}}_{i+1} = \dot{\mathbf{u}}_{avg}+\frac{1}{2}\ddot{\mathbf{u}}_{i+1}\Delta t_i = \dot{\mathbf{u}}_i+\frac{1}{2}(\ddot{\mathbf{u}}_i+\ddot{\mathbf{u}}_{i+1})\Delta t_i
$$

(6.209)

$$
\mathbf{u}_{i+1} = \mathbf{u}_i+\dot{\mathbf{u}}_{avg}\Delta t_i
$$

(6.210)

From Eq. 6.208,

$$\dot{\mathbf{u}}_i = \frac{\mathbf{u}_i - \mathbf{u}_{i-1}}{\Delta t_{i-1}} + \tfrac{1}{2} \ddot{\mathbf{u}}_i \, \Delta t_{i-1} \tag{6.211}$$

Substituting into this expression the differential equation

$$\ddot{\mathbf{u}}_i = \mathbf{M}^{-1}(\mathbf{p}_i - \mathbf{C}\dot{\mathbf{u}} - \mathbf{f}_i) \tag{6.212}$$

we obtain the *velocity predictor*

$$(\mathbf{I} + \tfrac{1}{2}\Delta t_{i-1}\mathbf{M}^{-1}\mathbf{C})\dot{\mathbf{u}}_i = \frac{\mathbf{u}_i - \mathbf{u}_{i-1}}{\Delta t_{i-1}} + \tfrac{1}{2}\Delta t_{i-1}\mathbf{M}^{-1}(\mathbf{p}_i - \mathbf{f}_i) \tag{6.213}$$

which for known values of the right-hand side can be solved for $\dot{\mathbf{u}}_i$. Also, by combining Eqs. 6.208 and 6.210, we obtain finally the *displacement predictor*

$$\mathbf{u}_{i+1} = \mathbf{u}_i\left(1 + \frac{\Delta t_i}{\Delta t_{i-1}}\right) - \mathbf{u}_{i-1}\frac{\Delta t_i}{\Delta t_{i-1}} + \tfrac{1}{2}\ddot{\mathbf{u}}_i\,\Delta t_i(\Delta t_{i-1} + \Delta t_i) \tag{6.214}$$

Application of the impulse acceleration method to a nonlinear system is not difficult, as the method is *explicit*. The solution would proceed along the following iterative lines:

- Knowing $\mathbf{u}_i, \mathbf{u}_{i-1},\dots$ compute $\mathbf{f}_i$
- Use the velocity predictor equation to solve for $\dot{\mathbf{u}}_i$
- Use the differential equation to find $\ddot{\mathbf{u}}_i$
- Use the displacement predictor equation to find $\mathbf{u}_{i+1}$

In the somewhat odd case in which $\mathbf{u}_0 = \dot{\mathbf{u}}_0 = \mathbf{p}_0 = \mathbf{0}$, the method yields $\mathbf{u}_1 = \mathbf{0}$. Hence, to start the method, it may be necessary to use another formula for the first step. If viscous damping forces can be neglected, or if they are mass-proportional, then solving for $\dot{\mathbf{u}}_i$ is particularly simple.

In the special case of SDOF systems, the velocity predictor can be written directly as

$$\dot{u}_i = \frac{1}{(1 + \tfrac{1}{2}\Delta t_{i-1}\,c/m)}\left[\frac{u_i - u_{i-1}}{\Delta t_{i-1}} + \frac{\Delta t_{i-1}(p_i - f_i)}{2m}\right] \tag{6.215}$$

6.5.3 Methods Based on Mathematical Approximations

Numerical methods for the solution of *initial value problems* (propagation problems) have been mainly developed for (systems of) first-order differential equations of the form

$$\dot{\mathbf{u}} = \mathbf{f}(\mathbf{u},t) \tag{6.216}$$

However, higher order differential equations can always be expressed in the form of systems of first order differential equations, for which similar procedures may be used. This is accomplished by simply adding the trivial equation $\dot{\mathbf{u}} - \mathbf{v} = \mathbf{0}$ and writing the differential equation as $\mathbf{M}\dot{\mathbf{v}} + \mathbf{C}\mathbf{v} + \mathbf{f}(\mathbf{u},t) = \mathbf{0}$.

On the other hand, it is also possible in some cases to derive methods directly for higher order equations. We will consider initially the solution of the first-order differential equation, then extending these solutions to our particular second-order differential equation in structural dynamics. Later on, we will also derive special formulas for the second-order case, and will show how the methods based on physical approximations can be related to the more general mathematical schemes.

Multistep Methods for First-Order Differential Equations

This class of methods is closely related to what in statistical analysis is referred to as ARMA models, which are tools to understand and predict future values in a series. In general, ARMA models consist of two parts, namely the finite differences (or Auto-Regressive) part, and the numerical integration (or Moving Average) part.

Finite Differences Formulas

Consider the first-order differential equation at some discrete instant

$$\dot{\mathbf{u}}_i = \mathbf{f}(\mathbf{u}_i, t_i) \equiv \mathbf{f}_i \tag{6.217}$$

The derivative $\dot{\mathbf{u}}_i$ can be expressed in terms of the function evaluated at different values of t using finite differences formulas. This is equivalent to assuming that $\mathbf{u}$ can be approximated over a given time interval by a polynomial expression. For example, consider the following three cases.

Forward Difference

$$\dot{\mathbf{u}}_i = \frac{\mathbf{u}_{i+1} - \mathbf{u}_i}{t_i - t_{i-1}} = \frac{\mathbf{u}_{i+1} - \mathbf{u}_i}{\Delta t_i} \tag{6.218}$$

$$\boxed{\mathbf{u}_{i+1} = \mathbf{u}_i + \Delta t_i\, \mathbf{f}(\mathbf{u}_i, t_i)} \qquad \text{Euler's Method} \tag{6.219}$$

Central Difference

$$\dot{\mathbf{u}}_i = \frac{\mathbf{u}_{i+1} - \mathbf{u}_{i-1}}{t_{i+1} - t_{i-1}} = \frac{\mathbf{u}_{i+1} - \mathbf{u}_{i-1}}{\Delta t_i + \Delta t_{i-1}} \tag{6.220}$$

$$\boxed{\mathbf{u}_{i+1} = \mathbf{u}_{i-1} + (\Delta t_i + \Delta t_{i-1})\,\mathbf{f}(\mathbf{u}_i, t_i)} \tag{6.221}$$

Backward Difference

$$\dot{\mathbf{u}}_i = \frac{\mathbf{u}_i - \mathbf{u}_{i-1}}{t_i - t_{i-1}} = \frac{\mathbf{u}_i - \mathbf{u}_{i-1}}{\Delta t_{i-1}} \tag{6.222}$$

$$\boxed{\mathbf{u}_i = \mathbf{u}_{i-1} + \Delta t_{i-1}\, \mathbf{f}(\mathbf{u}_i, t_i)} \tag{6.223}$$

The first two formulas are of an *explicit* type, while the third one is an *implicit* formula. If $\mathbf{f}(\mathbf{u}, t)$ is linear in $\mathbf{u}$, an explicit expression can be derived for $\mathbf{u}_i$. Otherwise it will be necessary to assume a value for $\mathbf{u}_i$, compute $\mathbf{f}(\mathbf{u}_i, t_i)$, find the corresponding $\mathbf{u}_i$, and iterate.

Forward – Two-Step Backward Difference

Many different formulas can be obtained using various finite differences approximations to $\dot{\mathbf{u}}_i$. Say, for instance, that we want to express $\dot{\mathbf{u}}_i$ in terms of $\mathbf{u}_{i+1}, \mathbf{u}_i, \mathbf{u}_{i-1}$, and $\mathbf{u}_{i-2}$, and that the time step is constant. Using Taylor series expansions, we can write

$$\mathbf{u}_{i+1} = \mathbf{u}_i + \Delta t\, \dot{\mathbf{u}}_i + \tfrac{1}{2!}\Delta t^2\, \ddot{\mathbf{u}}_i + \tfrac{1}{3!}\Delta t^3\, \dddot{\mathbf{u}}_i + \tfrac{1}{4!}\Delta t^4\, \ddddot{\mathbf{u}}_i + \cdots \tag{6.224}$$

$$\mathbf{u}_i = \mathbf{u}_i \tag{6.225}$$

$$\mathbf{u}_{i-1} = \mathbf{u}_i - \Delta t\, \dot{\mathbf{u}}_i + \tfrac{1}{2!}\Delta t^2\, \ddot{\mathbf{u}}_i - \tfrac{1}{3!}\Delta t^3\, \dddot{\mathbf{u}}_i + \tfrac{1}{4!}\Delta t^4\, \ddddot{\mathbf{u}}_i - \cdots \tag{6.226}$$

$$\mathbf{u}_{i-2} = \mathbf{u}_i - 2\Delta t\, \dot{\mathbf{u}}_i + \tfrac{4}{2!}\Delta t^2\, \ddot{\mathbf{u}}_i - \tfrac{8}{3!}\Delta t^3\, \dddot{\mathbf{u}}_i + \tfrac{16}{4!}\Delta t^4\, \ddddot{\mathbf{u}}_i - \cdots \tag{6.227}$$

We next wish to find a set of coefficients $a_1, \ldots a_4$ such that the weighted sum $a_1\mathbf{u}_{i+1} + a_2\mathbf{u}_i + a_3\mathbf{u}_{i-1} + a_4\mathbf{u}_{i-2}$ yields the best possible approximation to $\dot{\mathbf{u}}_i$. This is achieved combining Eqs. 6.224 to 6.227 in such a way that the resulting expression for $\dot{\mathbf{u}}_i$ contains only derivatives of order higher than $\ddot{\mathbf{u}}$ (in this example). We make thus

$$\begin{bmatrix} 1 & 1 & 1 & 1 \\ 1 & 0 & -1 & -2 \\ 1 & 0 & 1 & 1 \\ 1 & 0 & -1 & -8 \end{bmatrix} \begin{Bmatrix} a_1 \\ a_2 \\ a_3 \\ a_4 \end{Bmatrix} = \begin{Bmatrix} 0 \\ \Delta t^{-1} \\ 0 \\ 0 \end{Bmatrix} \quad \begin{array}{l} \text{(coefficient of } \mathbf{u}_i) \\ \text{(coefficient of } \dot{\mathbf{u}}_i) \\ \text{(coefficient of } \ddot{\mathbf{u}}_i) \\ \text{(coefficient of } \dddot{\mathbf{u}}_i) \end{array} \tag{6.228}$$

The solution of this system of four equations is

$$a_1 = \frac{1}{3\Delta t} \qquad a_2 = \frac{1}{2\Delta t} \qquad a_3 = \frac{-1}{\Delta t} \qquad a_4 = \frac{1}{6\Delta t} \tag{6.229}$$

Hence

$$\dot{\mathbf{u}}_i = \tfrac{1}{6\Delta t}\left[2\mathbf{u}_{i+1} + 3\mathbf{u}_i - 6\mathbf{u}_{i-1} + \mathbf{u}_{i-2}\right] + O(\Delta t^3) \tag{6.230}$$

The last term in Eq. 6.230 implies that the finite difference approximation for this example is of the order of Δt^3. We can thus derive the explicit formula

$$\mathbf{u}_{i+1} = -\tfrac{3}{2}\mathbf{u}_i + 3\mathbf{u}_{i-1} - \tfrac{1}{2}\mathbf{u}_{i-2} + 3\Delta t\, \mathbf{f}(\mathbf{u}_i, t_i) \tag{6.231}$$

It can be shown, however, that this formula is numerically unstable, and thus useless.

Park's Method (Three-Step Backward Difference)

A closely related formula is the three-step backward difference

$$\dot{\mathbf{u}}_i = \tfrac{1}{6\Delta t}\left[10\mathbf{u}_i - 15\mathbf{u}_{i-1} + 6\mathbf{u}_{i-2} - \mathbf{u}_{i-3}\right] + O(\Delta t^3) \tag{6.232}$$

which leads to Park's unconditionally stable, implicit method:

$$\mathbf{u}_i = \tfrac{3}{2}\mathbf{u}_{i-1} - \tfrac{3}{5}\mathbf{u}_{i-2} + \tfrac{1}{10}\mathbf{u}_{i-3} - \tfrac{3}{5}\Delta t\, \mathbf{f}(\mathbf{u}_i, t_i) \tag{6.233}$$

Numerical Integration Formulas

If we integrate the equation $\dot{\mathbf{u}} = \mathbf{f}(\mathbf{u},t)$ over an interval $[t_i, t_{i+k}]$, we obtain

$$\mathbf{u}_{i+k} = \mathbf{u}_i + \int_{t_i}^{t_{i+k}} \mathbf{f}(\mathbf{u},t)\,dt \tag{6.234}$$

Using now any numerical integration method to evaluate the integral, or assuming a polynomial expression for $\mathbf{f}$ in the range $t_i \le t \le t_{i+k}$, the integration can readily be performed. For instance, taking $k = 1$ and assuming the function to be constant $\breve{\mathbf{f}}_i$, we obtain

$$\mathbf{u}_{i+1} = \mathbf{u}_i + \Delta t_i \breve{\mathbf{f}} \tag{6.235}$$

The constant value $\breve{\mathbf{f}}_i$ can be taken as

$$\breve{\mathbf{f}}_i = (1-\alpha)\mathbf{f}_i + \alpha\mathbf{f}_{i+1} \qquad 0 \le \alpha \le 1 \tag{6.236}$$

Hence

$$\mathbf{u}_{i+1} = \mathbf{u}_i + (1-\alpha)\Delta t_i\,\mathbf{f}_i + \alpha\Delta t_i\,\mathbf{f}_{i+1} \tag{6.237}$$

For $\alpha = 0$, this numerical integration scheme is explicit; otherwise it is implicit. More general formulas can be derived using other polynomial expansions to evaluate the integral, such as Simpson's method.

Difference and Integration Formulas

It can be seen that both the finite differences and the integration formulas can be generalized by an equation of the form (for constant Δt)

$$\boxed{\alpha_k\mathbf{u}_{i+k} + \cdots + \alpha_1\mathbf{u}_{i+1} + \alpha_0\mathbf{u}_i = \Delta t\left[\beta_k\,\mathbf{f}_{i+k} + \cdots + \beta_1\,\mathbf{f}_{i+1} + \beta_0\,\mathbf{f}_i\right]} \tag{6.238}$$

where we assume that $\alpha_k \ne 0$ and $|\alpha_0| + |\beta_0| \ne 0$. If $\beta_k = 0$, we obtain an explicit formula, otherwise the expression is implicit. This formula represents a kth-order method.

Multistep Methods for Second-Order Differential Equations

(a) General Case $\ddot{\mathbf{u}} = \mathbf{f}(\mathbf{u},\dot{\mathbf{u}},t)$
The equation of motion for a problem in structural dynamics can be expressed as

$$\ddot{\mathbf{u}} = \mathbf{M}^{-1}\left[\mathbf{p}(t) - \mathbf{C}\dot{\mathbf{u}} - \mathbf{f}(\mathbf{u},t)\right] \tag{6.239}$$

which is of the form $\ddot{\mathbf{u}} = \mathbf{f}(\mathbf{u},\dot{\mathbf{u}},t)$. This equation can be reduced to a system of first-order differential equations by it writing in matrix form as

$$\begin{Bmatrix}\dot{\mathbf{u}}\\\dot{\mathbf{v}}\end{Bmatrix} = \begin{bmatrix}\mathbf{O} & \mathbf{I}\\\mathbf{O} & -\mathbf{M}^{-1}\mathbf{C}\end{bmatrix}\begin{Bmatrix}\mathbf{u}\\\mathbf{v}\end{Bmatrix} + \begin{Bmatrix}\mathbf{0}\\\mathbf{M}^{-1}\left[\mathbf{p}(t)-\mathbf{f}(\mathbf{u},t)\right]\end{Bmatrix} \tag{6.240}$$

which is of the form $\dot{\mathbf{z}} = \mathbf{f}(\mathbf{z},t)$, with $\mathbf{z}$ being the state vector. Hence, the methods described previously for first-order differential equations could be used to solve the problem at hand.

(b) Special Case $\ddot{\mathbf{u}} = \mathbf{f}(\mathbf{u},t)$

If the system has no damping, the differential equation is said to be *special*. It is then possible to derive directly a multistep method to compute $\mathbf{u}$ without evaluating $\dot{\mathbf{u}}$. The general form of these methods is

$$\boxed{\alpha_k \mathbf{u}_{i+k} + \cdots + \alpha_1 \mathbf{u}_{i+1} + \alpha_0 \mathbf{u}_i = \Delta t^2 \left[\beta_k \mathbf{f}_{i+k} + \cdots + \beta_1 \mathbf{f}_{i+1} + \beta_0 \mathbf{f}_i \right]} \tag{6.241}$$

6.5.4 Runge–Kutta Type Methods

Given the first-order equation

$$\dot{\mathbf{u}} = \mathbf{f}(\mathbf{u},t) \tag{6.242}$$

the Runge–Kutta type methods are based on the ansatz

$$\mathbf{u}_{i+1} = \mathbf{u}_i + \Delta t\, \varphi(\mathbf{u}_i, t_i, \Delta t) \tag{6.243}$$

in which φ is formed with a family of functions $\mathbf{g}_1, \mathbf{g}_2, \ldots$, as will be seen. This family of functions satisfies the following rules:

1. $\mathbf{f}(\mathbf{u},t)$ belongs to φ
2. If $\mathbf{g}_1(\mathbf{u},t,\Delta t)$ and $\mathbf{g}_2(\mathbf{u},t,\Delta t)$ belong to φ, and if $a_1 = a_1(\Delta t), a_2 = a_2(\Delta t), \ldots b_1 = b_1(\Delta t)$, $b_2 = b_2(\Delta t), \ldots$ are arbitrary functions of Δt, then the functions

$$\mathbf{g}_3(\mathbf{u},t,\Delta t) = a_1\, \mathbf{g}_1(\mathbf{u},t,\Delta t) + a_2\, \mathbf{g}_2(\mathbf{u},t,\Delta t) \tag{6.244}$$

$$\mathbf{g}_4(\mathbf{u},t,\Delta t) = \mathbf{g}_1 \left[\mathbf{u} + b_1 \Delta t\, \mathbf{g}_2(\mathbf{u},t,\Delta t),\ t + b_2 \Delta t, \Delta t \right] \tag{6.245}$$

belong to φ.

A simple and usual way to generate the function using these rules is by taking constants $a_1, a_2, \ldots, b_1, b_2, \ldots$ and forming the set of functions

$$\mathbf{g}_1 = \mathbf{f}(t,\mathbf{u}) \tag{6.246}$$

$$\mathbf{g}_2 = \mathbf{f}(\mathbf{u} + b_1 \Delta t\, \mathbf{g}_1,\ t + b_2 \Delta t) \tag{6.247}$$

$$\mathbf{g}_3 = \mathbf{f}(\mathbf{u} + b_3 \Delta t\, \mathbf{g}_2,\ t + b_4 \Delta t) \tag{6.248}$$

and so forth
and

$$\varphi = a_1\, \mathbf{g}_1 + a_2\, \mathbf{g}_2 + \cdots a_r\, \mathbf{g}_r \tag{6.249}$$

The constants are then determined by comparing $\varphi(u_i, t_i, \Delta t)$ with the Taylor series expansion of the forward difference, namely

$$\frac{1}{\Delta t}(\mathbf{u}_{i+1} - \mathbf{u}_i) = \dot{\mathbf{u}}_i + \frac{\Delta t}{2!}\ddot{\mathbf{u}}_i + \frac{\Delta t^2}{3!}\dddot{\mathbf{u}}_i + \cdots \tag{6.250}$$

and trying to make equal as many terms as possible. The right-hand side of this expansion contains the terms

$$\dot{\mathbf{u}}_i = \mathbf{f}_i = \mathbf{f}(\mathbf{u}_i, t_i) \tag{6.251}$$

$$\ddot{\mathbf{u}}_i = \frac{\partial \mathbf{f}}{\partial \mathbf{u}} \frac{\partial \mathbf{u}}{\partial t} + \frac{\partial \mathbf{f}}{\partial t} = \left\{ \frac{\partial f_m}{\partial u_n} \right\}\Bigg|_i \dot{\mathbf{u}}_i + \dot{\mathbf{f}}_i = \mathbf{J}_i \mathbf{f}_i + \dot{\mathbf{f}}_i \tag{6.252}$$

and so on for the higher derivatives (the indices m, n refer to the rows and columns of $\mathbf{J}_i$, respectively). Hence

$$\frac{1}{\Delta t}(\mathbf{u}_{i+1} - \mathbf{u}_i) = \mathbf{f}_i + \frac{\Delta t}{2!}(\mathbf{J}_i \mathbf{f}_i + \dot{\mathbf{f}}_i) + \cdots = \boldsymbol{\varphi} = a_1 \mathbf{g}_1 + a_2 \mathbf{g}_2 + \cdots \tag{6.253}$$

We illustrate these concepts with some specific examples.

Euler's Method

The simplest possible alternative would be to use only *one* function:

$$\boldsymbol{\varphi} = \frac{1}{\Delta t}(\mathbf{u}_{i+1} - \mathbf{u}_i) = \mathbf{f}(\mathbf{u}_i, t_i) + O(\Delta t) \tag{6.254}$$

so that

$$\boxed{\mathbf{u}_{i+1} = \mathbf{u}_i + \Delta t \, \mathbf{f}(\mathbf{u}_i, t_i)} \tag{6.255}$$

Thus, Euler's method is a first-order method, which can be shown to be unstable.

Improved and Modified Euler Methods

Here, we use *two* functions for $\boldsymbol{\varphi}$, that is, $\boldsymbol{\varphi} = a_1 \mathbf{g}_1 + a_2 \mathbf{g}_2$:

$$\boldsymbol{\varphi} = \frac{1}{\Delta t}(\mathbf{u}_{i+1} - \mathbf{u}_i) = a_1 \mathbf{f}(t_i, \mathbf{u}_i) + a_2 \mathbf{f}[\mathbf{u}_i + b_1 \Delta t \, \mathbf{f}(t_i, \mathbf{u}_i), \, t_i + b_2 \Delta t] \tag{6.256}$$

Next, we expand the last term in Taylor series

$$\mathbf{f}[\mathbf{u}_i + b_1 \Delta t \, \mathbf{f}(t_i, \mathbf{u}_i), \, t_i + b_2 \Delta t] = \mathbf{f}_i + \Delta t \left(b_1 \mathbf{J}_i \mathbf{f}_i + b_2 \dot{\mathbf{f}}_i \right) + \cdots \tag{6.257}$$

and compare the result for $\boldsymbol{\varphi}$ with the Taylor series expansion of the forward difference

$$\frac{1}{\Delta t}(\mathbf{u}_{i+1} - \mathbf{u}_i) = \mathbf{f}_i + \frac{\Delta t}{2!}(\mathbf{J}_i \mathbf{f}_i + \dot{\mathbf{f}}_i) + \cdots = a_1 \mathbf{f}_i + a_2 \left[\mathbf{f}_i + \Delta t \left(b_1 \mathbf{J}_i \mathbf{f}_i + b_2 \dot{\mathbf{f}}_i \right) + \cdots \right] \tag{6.258}$$

It follows that

$$a_1 + a_2 = 1 \qquad a_2 b_1 = \tfrac{1}{2} \qquad a_2 b_2 = \tfrac{1}{2} \tag{6.259}$$

We have thus a second-order method by taking $a_1 = 1 - a_2, b_1 = b_2 = 1/(2a_2)$.

In particular, if $a_1 = a_2 = \tfrac{1}{2}$ and $p_1 = p_2 = 1$

$$\boxed{\mathbf{u}_{i+1} = \mathbf{u}_i + \tfrac{1}{2}\Delta t \left[\mathbf{f}_i + \mathbf{f}(t_{i+1}, \mathbf{u}_i + \Delta t \, \mathbf{f}_i) \right]} \tag{6.260}$$

which is the *improved Euler method* (Figure 6.7, top).

For $a_1 = 0, a_2 = 1, p_1 = p_2 = \tfrac{1}{2}$

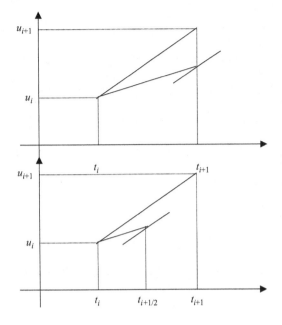

Figure 6.7. Improved and modified Euler methods.

$$\boxed{\mathbf{u}_{i+1} = \mathbf{u}_i + \Delta t \left[\mathbf{f}(t_i + \tfrac{1}{2}\Delta t, \mathbf{u}_i + \tfrac{1}{2}\Delta t \, \mathbf{f}_i) \right]}$$ (6.261)

which is the *modified Euler method* (Figure 6.7, bottom).

The Normal Runge–Kutta Method

The method normally called *Runge–Kutta* is a fourth-order method obtained by selecting the first four functions of the set

$$\boldsymbol{\varphi} = \tfrac{1}{6}(\mathbf{g}_1 + 2\mathbf{g}_2 + 2\mathbf{g}_3 + \mathbf{g}_4)$$ (6.262)

where

$$\mathbf{g}_1 = \mathbf{f}(t_i, \mathbf{u}_i)$$ (6.263)

$$\mathbf{g}_2 = \mathbf{f}(t_i + \tfrac{1}{2}\Delta t, \mathbf{u}_i + \tfrac{1}{2}\Delta t \, \mathbf{g}_1)$$ (6.264)

$$\mathbf{g}_3 = \mathbf{f}(t_i + \tfrac{1}{2}\Delta t, \mathbf{u}_i + \tfrac{1}{2}\Delta t \, \mathbf{g}_2)$$ (6.265)

$$\mathbf{g}_4 = \mathbf{f}(t_i + \Delta t, \mathbf{u}_i + \Delta t \, \mathbf{g}_3)$$ (6.266)

The solution is then

$$\mathbf{u}_{i+1} = \mathbf{u}_i + \tfrac{1}{6}\Delta t (\mathbf{g}_1 + 2\mathbf{g}_2 + 2\mathbf{g}_3 + \mathbf{g}_4)$$ (6.267)

The application of Runge–Kutta type methods to higher order equations is again done by replacing them by a system of first-order equations. In the case of *second-order differential equations*,

$$\ddot{\mathbf{u}} = \mathbf{f}(\mathbf{u}, \dot{\mathbf{u}}, t)$$ (6.268)

we would write

$$\begin{Bmatrix} \dot{\mathbf{u}} \\ \dot{\mathbf{v}} \end{Bmatrix} = \begin{Bmatrix} \mathbf{v}(t) \\ \mathbf{f}(\mathbf{u},\mathbf{v},t) \end{Bmatrix} \tag{6.269}$$

The family of functions is now

$$\mathbf{g}_1 = \mathbf{v}_i \qquad\qquad \mathbf{h}_1 = \mathbf{f}(\mathbf{u}_i, \mathbf{v}_i, t_i) \tag{6.270}$$

$$\mathbf{g}_2 = \mathbf{v}_i + \tfrac{1}{2}\Delta t\,\mathbf{h}_1 \qquad \mathbf{h}_2 = \mathbf{f}(\mathbf{u}_i + \tfrac{1}{2}\Delta t\,\mathbf{g}_1, \mathbf{g}_2, t_i + \tfrac{1}{2}\Delta t) \tag{6.271}$$

$$\mathbf{g}_3 = \mathbf{v}_i + \tfrac{1}{2}\Delta t\,\mathbf{h}_2 \qquad \mathbf{h}_3 = \mathbf{f}(\mathbf{u}_i + \tfrac{1}{2}\Delta t\,\mathbf{g}_2, \mathbf{g}_3, t_i + \tfrac{1}{2}\Delta t) \tag{6.272}$$

$$\mathbf{g}_4 = \mathbf{v}_i + \Delta t\,\mathbf{h}_3 \qquad \mathbf{h}_4 = \mathbf{f}(\mathbf{u}_i + \Delta t\,\mathbf{g}_3, \mathbf{g}_4, t_i + \Delta t) \tag{6.273}$$

and the solution is

$$\mathbf{u}_{i+1} = \mathbf{u}_i + \tfrac{1}{6}\Delta t\left(\mathbf{g}_1 + 2\mathbf{g}_2 + 2\mathbf{g}_3 + \mathbf{g}_4\right) \tag{6.274}$$

$$\mathbf{v}_{i+1} = \mathbf{v}_i + \tfrac{1}{6}\Delta t\left(\mathbf{h}_1 + 2\mathbf{h}_2 + 2\mathbf{h}_3 + \mathbf{h}_4\right) \tag{6.275}$$

6.5.5 Stability and Convergence Conditions for Multistep Methods

When a general difference-integration formula such as that presented earlier

$$\alpha_k \mathbf{u}_{i+k} + \cdots + \alpha_1 \mathbf{u}_{i+1} + \alpha_0 \mathbf{u}_i = \Delta t\left[\beta_k\,\mathbf{f}_{i+k} + \cdots + \beta_1\,\mathbf{f}_{i+1} + \beta_0\,\mathbf{f}_i\right] \tag{6.276}$$

is adopted, requirements of *stability* and *convergence* must be met by the coefficients α, β. The motivation for the stability condition stems from the fact that we are replacing a differential equation by a *difference equation*, and we are thus introducing into the general solution of the *homogeneous* equation a set of artificial roots, which may cause the solution to grow without bounds. On the other hand, to be useful, a numerical method must not only be stable, but it must also provide accurate results, and converge in the limit of an infinitesimal step size to the exact solution. Thus, it must also satisfy a convergence or consistency condition. Details on these conditions can be found in textbooks on numerical methods, such as those by Henrici[3] and Hildebrand.[4] It should be noted, however, that although these conditions are shown to be necessary for stability and convergence, they do not guarantee from a practical point of view the adequacy of a method. For this reason, they are not discussed in detail herein.

Conditional and Unconditional Stability of Linear Systems

As stated previously, the conditions of stability and consistency referred to earlier, although necessary, are not sufficient to guarantee the adequacy of a numerical method. For linear systems, on the other hand, it is possible to determine a priori the conditions

---

[3] P. Henrici, *Discrete Variable Methods in Ordinary Differential Equations* (New York: Wiley, 1962).

[4] F. B. Hildebrand, *Introduction to Numerical Analysis* (New York: McGraw-Hill, 1972).

that a numerical integration formula must obey so that the solution will not grow without bounds. Consider for this purpose the expression

$$\alpha_k \mathbf{z}_{i+k} + \cdots + \alpha_1 \mathbf{z}_{i+1} + \alpha_0 \mathbf{z}_i = \Delta t \left[\beta_k \mathbf{f}_{i+k} + \cdots + \beta_1 \mathbf{f}_{i+1} + \beta_0 \mathbf{f}_i \right] \tag{6.277}$$

in which

$$\mathbf{z}_j = y_j \qquad \text{for a first-order differential equation (a scalar)} \tag{6.278}$$

$$\mathbf{z}_j = \left\{ y_j \quad \dot{y}_j \right\}^T \qquad \text{for a second-order differential equation (a vector)} \tag{6.279}$$

and so on for higher order differential equations. In the case of an nth-order *linear* differential equation with constant coefficients, the function $\mathbf{f}_i$ is of the form

$$\dot{\mathbf{z}} = \mathbf{f}(t,\mathbf{z}) = \mathbf{A}\mathbf{z} + \mathbf{b}(t) \tag{6.280}$$

with $\mathbf{A}$ being an n by n square matrix of coefficients, and $\mathbf{b}$ is the loading function. Thus, the integration formula can be written as

$$(\alpha_k \mathbf{I} - \beta_k \Delta t \, \mathbf{A})\mathbf{z}_{i+k} + \cdots + (\alpha_0 \mathbf{I} - \beta_0 \Delta t \, \mathbf{A})\mathbf{z}_i = \Delta t \left[\beta_k \mathbf{b}(t_k) + \cdots + \beta_0 \mathbf{b}(t_0) \right] \tag{6.281}$$

with $\mathbf{I}$ being the identity matrix. With the definition $\mathbf{C}_j = \alpha_j \mathbf{I} - \beta_j \Delta t \, \mathbf{A}$, this equation can be written as

$$\mathbf{C}_k \mathbf{z}_{i+k} + \cdots + \mathbf{C}_0 \mathbf{z}_i = \Delta t \left[\beta_k \mathbf{b}(t_k) + \cdots + \beta_0 \mathbf{b}(t_0) \right] \tag{6.282}$$

We consider in particular the homogeneous equation

$$\mathbf{C}_k \mathbf{z}_{i+k} + \cdots + \mathbf{C}_0 \mathbf{z}_i = \mathbf{0} \tag{6.283}$$

which admits solutions of the form

$$\mathbf{z}_j = \lambda^j \mathbf{z}_0 \tag{6.284}$$

Substituting this expression into the homogeneous equation, and dividing by $\lambda^i \neq 0$, we obtain

$$\left\{ \sum_{j=0}^{k} \lambda^j \mathbf{C}_j \right\} \mathbf{z}_0 = \mathbf{0} \tag{6.285}$$

which has a nontrivial solution only if the determinant of the term in parentheses vanishes; that is,

$$\left| \sum_{j=0}^{k} \lambda^j \mathbf{C}_j \right| = 0 \tag{6.286}$$

This equation constitutes an nth-order eigenvalue problem of kth degree, with λ being the eigenvalues and $\mathbf{z}_0$ being an eigenvector. By introducing appropriate trivial equations, the equation for λ can also be expressed as the linear eigenvalue problem

$$\begin{Bmatrix} \begin{bmatrix} \mathbf{C}_{k-1} & \cdots & \mathbf{C}_1 & \mathbf{C}_0 \\ \mathbf{I} & \mathbf{0} & \mathbf{0} & \cdots \\ \vdots & \ddots & \ddots & \mathbf{0} \\ \mathbf{0} & \cdots & \mathbf{I} & \mathbf{0} \end{bmatrix} \begin{bmatrix} \mathbf{z}_{k-1} \\ \vdots \\ \mathbf{z}_1 \\ \mathbf{z}_0 \end{bmatrix} \end{Bmatrix} = \lambda \begin{Bmatrix} \begin{bmatrix} -\mathbf{C}_k & \mathbf{0} & \cdots & \mathbf{0} \\ \mathbf{0} & \mathbf{I} & \ddots & \vdots \\ \vdots & \ddots & \ddots & \mathbf{0} \\ \mathbf{0} & \cdots & \mathbf{0} & \mathbf{I} \end{bmatrix} \begin{bmatrix} \mathbf{z}_{k-1} \\ \vdots \\ \mathbf{z}_1 \\ \mathbf{z}_0 \end{bmatrix} \end{Bmatrix} \tag{6.287}$$

In general, the eigenvalues from this equation will be complex numbers. Thus, they are most conveniently written in polar form

$$\lambda = r\, e^{i\theta} \qquad r = |\lambda| \qquad \theta = \arg(\lambda) \tag{6.288}$$

In particular, the eigenvalue having the largest absolute value is referred to as the *spectral radius*:

$$\rho = r_{\max} \tag{6.289}$$

It is easy to see now that in order to have a stable solution to the homogeneous equation, the spectral radius must be less than or equal to 1 ($\rho \le 1$); otherwise the solution will grow unbounded as j (i.e., time) increases. On the other hand, the spectral radius is a function of the time step, that is $\rho = \rho(\Delta t)$. If $\rho(\Delta t) \le 1$ for arbitrary values of the time step, the method is said to be *unconditionally stable*. Conversely, if $\rho(\Delta t) > 1$ when Δt exceeds a certain threshold, the method is only *conditionally stable*. If the spectral radius is larger than 1 no matter how small the time step, the method is *unstable*, and cannot be used to integrate numerically the equations at hand.

To illustrate these concepts, consider the case of a linear SDOF system subjected to support (i.e., seismic) motion:

$$\dot{\mathbf{z}} = \begin{Bmatrix} \dot{v} \\ \dot{y} \end{Bmatrix} = \begin{Bmatrix} 0 & 1 \\ -\omega_n^2 & -2\xi\omega_n \end{Bmatrix} \begin{Bmatrix} v \\ y \end{Bmatrix} + \begin{Bmatrix} 0 \\ -\ddot{u}_g \end{Bmatrix} = \mathbf{A}\mathbf{z} + \mathbf{b} \tag{6.290}$$

Using Euler's method, the integration formula is

$$\mathbf{z}_{i+1} = \mathbf{z}_i + \Delta t\, \mathbf{A}\mathbf{z}_i + \Delta t\, \mathbf{b}_i \tag{6.291}$$

Thus, the eigenvalue problem is $\det[\lambda\mathbf{I} - (\mathbf{I} + \Delta t\, A)] = 0$

$$\left| \begin{Bmatrix} \lambda - 1 & -\Delta t \\ \omega_n^2 \Delta t & \lambda - 1 + 2\xi\omega_n\Delta t \end{Bmatrix} \right| = 0 \tag{6.292}$$

which yields

$$\lambda_{1,2} = 1 - \xi\omega_n\Delta t \pm i\omega_n\Delta t \sqrt{1 - \xi^2} \tag{6.293}$$

Both roots have absolute value

$$\rho = |\lambda| = \sqrt{1 - 2\xi\omega_n\Delta t + \omega_n^2\Delta t^2} \tag{6.294}$$

which is less than 1 only if $\omega_n\,\Delta t \le 2\xi$; that is, if $\Delta t \le 2\xi / \omega_n$. In the case of an undamped system, $\xi = 0$, implying $\Delta t = 0$. Thus, Euler's method would not be stable, and could not be used to integrate the equations of motion.

Using similar analyses in connection with an undamped system, it can be shown that the constant acceleration method is unconditionally stable, whereas Newmark's β method is only conditionally stable for given values of the parameters α, β.

While the stability condition is demonstrated only for a linear system, the methods are assumed to be adequate also for nonlinear problems, although difficulties could arise if the system exhibits strain hardening.

In the case of linear systems with more than 1 DOF, it can be argued that the direct solution of the equations of motion is equivalent to the simultaneous solution of the modal equations. Since each modal equation is the same as that of a SDOF system, the stability condition is controlled by a statement of the form $\omega_j \Delta t \leq$ [some limit] for $j = 1, 2, \ldots n$, with n being the number or DOF. Thus, it is the highest frequency (shortest period) present in the system that controls the stability of the particular integration scheme used. Since this frequency can be quite high in the case of structures with many DOF, the necessary time step could turn out to be very small, even if the higher modes do not contribute to the response of the system. This can be avoided using unconditionally stable algorithms, such as the Wilson θ method, although this may sometimes involve sacrifices in the accuracy of the computation.

6.5.6 Stability Considerations for Implicit Integration Schemes

When an unconditionally stable integration method is used in *implicit* form (e.g., in connection with nonlinear problems), it may in some cases cease to be stable. To see why this is so, let us consider once more the free vibration problem of an undamped SDOF system, which we solve with the constant acceleration method (i.e., Newmark's β method with $\alpha = \frac{1}{2}, \beta = \frac{1}{4}$). When used in its *explicit* form, this method is known to be unconditionally stable. The iteration is again

$$m\ddot{u}_{i+1} + ku_{i+1} = 0 \tag{6.295}$$

or

$$m\ddot{u}_{i+1} + k\left[u_i + \dot{u}_i\Delta t + \tfrac{1}{4}\ddot{u}_i\,\Delta t^2 + \tfrac{1}{4}\ddot{u}_{i+1}\,\Delta t^2\right] = 0 \tag{6.296}$$

With the definition $f_i = k\left[u_i + \dot{u}_i\Delta t + \tfrac{1}{4}\ddot{u}_i\,\Delta t^2\right]$, Eq. 6.296 can be expressed as

$$m\ddot{u}_{i+1} + \tfrac{1}{4}k\Delta t^2\,\ddot{u}_{i+1} = -f_i \tag{6.297}$$

We define the auxiliary variable $a = \ddot{u}_{i+1}$ (the converged value of the acceleration), and denote with a_j the tentative values for this variable in the course of the implicit iteration scheme (in other words, $a \equiv a_\infty$). The implicit iteration is then given by

$$ma_{j+1} = -\tfrac{1}{4}k\Delta t^2 a_j - f_i \tag{6.298}$$

or

$$a_{j+1} + \tfrac{1}{4}\omega_n^2\Delta t^2 a_j = -\tfrac{1}{m}f_i \tag{6.299}$$

whose error is $\varepsilon_j = a_j - a$. The stability requirement for this iteration follows from an examination of the roots of the equation

$$z^{j+1} + \tfrac{1}{4}\omega_n^2 \Delta t^2 z^j = 0 \tag{6.300}$$

which has the single root $z = -(\tfrac{1}{2}\omega_n \Delta t)^2$. Hence, the iteration is stable only if the absolute value of this root does not exceed 1, that is, if

$$\tfrac{1}{2}\omega_n \Delta t < 1 \qquad \text{or} \qquad \frac{\pi}{T_n}\Delta t < 1 \tag{6.301}$$

with $T_n = 2\pi / \omega_n$ being the natural period. This implies the stability condition

$$\Delta t < \frac{T_n}{\pi} \tag{6.302}$$

Hence, the implicit iteration is only *conditionally stable*, in contrast to the explicit scheme, which is unconditionally stable.

6.6 Fundamentals of Fourier Methods

Following the development of efficient digital computers, and particularly after the discovery of the *Fast Fourier Transform* (FFT) algorithm by Cooley and Tuckey in 1965, Fourier transform methods have found extensive use by engineers and scientists as tools to obtain numerical solutions to systems of linear differential equations. While the mathematical theory was fully developed and available well before computers were invented, the procedures and basic jargon in current use are due primarily to electrical engineers, who used Fourier methods in the context of problems of electrical oscillations in circuits. Thus, it is customary in structural dynamics to use such terms as *transients* to refer to a temporary changes in dynamic response caused by abruptly applied loads, or mechanical *impedance* to refer to the relationship between forces and displacements (or between pressures and velocities in the case of acoustical problems).

In the sections that follow, we present the basic concepts needed to understand the manipulation of functions with the aid of Fourier methods. We elaborate on the following tools in the Fourier family:

- Fourier transform
- Fourier series
- Discrete Fourier transform
- Discrete Fourier series

In addition to these, there exist also the real-valued cosine and sine transforms and/or series, which are sometimes used for symmetric (even) and antisymmetric (odd) functions, respectively. However, as these are special cases of the methods listed previously, we need not study them separately.

6.6.1 Fourier Transform

Let $f(t)$ be a real, single-valued, piecewise continuous function of t, defined in the interval $[t_1, t_2]$ (Figure 6.8). Such function is said to be *time limited*. Within this interval, it has only

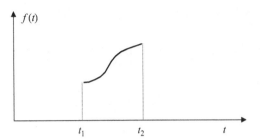

Figure 6.8. Function with compact support.

a finite number of discontinuities, a finite number of maxima and minima, and it satisfies the so-called *Dirichlet* condition

$$\int_{t_1}^{t_2} |f(t)| dt = \text{finite} \tag{6.303}$$

Most functions likely to be used in applied physics satisfy these conditions. Consider next a continuous frequency parameter ω, and define the integral transform

$$\tilde{f}(\omega) = \int_{t_1}^{t_2} f(t) e^{-i\omega t} dt \tag{6.304}$$

This equation establishes a one-to-one correspondence between the real-valued function $f(t)$ in the *time domain*, and a complex valued function $\tilde{f}(\omega)$ in the *frequency domain*. Clearly, the latter has complex-conjugate properties with respect to positive and negative values of frequency, that is $\tilde{f}(-\omega) = \tilde{f}^c(\omega)$. Hence, the real/imaginary parts of $\tilde{f}$ are symmetric/antisymmetric, respectively, with respect to frequency.

If outside the interval $[t_1, t_2]$ the function vanishes, or it is *defined* as zero (for example, by means of a rectangular "window" or box function), then we could obviously change the lower and upper limits to negative and positive infinity without changing the result of the integral. We would then write

$$\boxed{\tilde{f}(\omega) = \int_{-\infty}^{+\infty} f(t) e^{-i\omega t} dt} \tag{6.305}$$

This expression gives the *Fourier transform* (FT) of the function $f(t)$. It can be shown that the required inversion formula is

$$\boxed{f(t) = \frac{1}{2\pi} \int_{-\infty}^{+\infty} \tilde{f}(\omega) e^{i\omega t} d\omega} \tag{6.306}$$

which is the *inverse Fourier transform*. It has the same form as the direct transform, except for the sign of the exponential term, and the $\frac{1}{2\pi}$ in front. Within the interval $[t_1, t_2]$ used to define the Fourier transform, this inversion formula recovers the original function $f(t)$, while outside of the interval it yields zero.

Example: The Box Function

Let $f(t) = a$ if $t_1 \le t \le t_2$

        $f(t) = 0$ otherwise $\tag{6.307}$

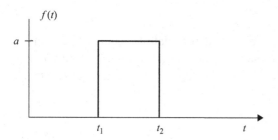

Figure 6.9. Box function.

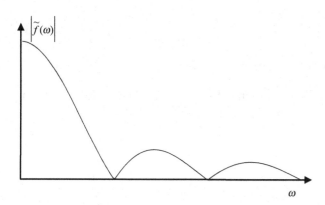

Figure 6.10. Box function, amplitude of Fourier transform.

as shown in Figure 6.9. The Fourier transform is then

$$
\tilde{f}(\omega) = a \int_{t_1}^{t_2} e^{-i\omega t} dt = a \frac{e^{-i\omega t_2} - e^{-i\omega t_1}}{-i\omega}
$$
$$
= a t_d e^{-i\omega t_c} \frac{\sin\theta}{\theta}
$$

(6.308)

with

$$t_c = \tfrac{1}{2}(t_1 + t_2) \qquad = \text{center of the box function} \tag{6.309}$$

$$t_d = t_2 - t_1 \qquad = \text{duration (or width) of the box function} \tag{6.310}$$

$$\theta = \tfrac{1}{2}\omega t_d \qquad = \text{dimensionless frequency} \tag{6.311}$$

The Fourier transform can be expressed also in terms of its absolute value and phase angle (Figure 6.10); that is,

$$
\tilde{f}(\omega) = a t_d \left| \frac{\sin\theta}{\theta} \right| e^{-i\varphi}
$$

(6.312)

in which

$$
\varphi = \begin{cases}
\theta \dfrac{2t_c}{t_d} \pm 2k\pi & \text{if } \sin\theta > 0 \\[2ex]
\theta \dfrac{2t_c}{t_d} \pm (2k-1)\pi & \text{if } \sin\theta < 0
\end{cases}
$$

(6.313)

with k being an arbitrary integer.

6.6.2 **Fourier Series**

Consider once more the function $f(t)$ of the previous section, defined again in the interval $[t_1, t_2]$ outside of which the function vanishes, and assume that this interval is contained in another, larger interval $[t_0, t_3]$ such that $[t_0 \leq t_1 < t_2 \leq t_3]$ and $f(t) = 0$ for $t_0 \leq t < t_1$, $t_2 < t \leq t_3$, see Figure 6.11.

Denote the parameters

$$T = t_3 - t_0 \qquad = \text{period of the series} \tag{6.314}$$

$$\Delta\omega = \frac{2\pi}{T} \qquad = \text{frequency step} \tag{6.315}$$

$$\omega_j = j\,\Delta\omega \qquad = \text{discrete frequency (with } j \text{ being an integer)} \tag{6.316}$$

Using the *same* transformed values $\tilde{f}(\omega)$ considered previously for the Fourier transform, that is,

$$\boxed{\tilde{f}(\omega) = \int_{t_1}^{t_2} f(t)\, e^{-i\,\omega t}\, dt} \tag{6.317}$$

then the infinite summation

$$\boxed{f(t) = \frac{1}{2\pi} \sum_{j=-\infty}^{\infty} \tilde{f}(\omega_j)\, e^{i\,\omega_j t}\, \Delta\omega} \tag{6.318}$$

constitutes now the *discrete Fourier series* representation of the function $f(t)$. Within the interval $[t_0, t_3]$, it converges once more to the original function, except that in the neighborhood of discontinuities, it converges to the average of the two function values at the discontinuity, and then overshoots the discontinuity by some finite amount. These local deviations are known as the *Gibbs* phenomenon. Outside of that interval, this formula periodically replicates the original function in $[t_0, t_3]$ at intervals T.

Comparison of the expression for the Fourier series with that of the Fourier transform shows that the former can be interpreted as the numerical integration of the latter by means of the rectangular rule, with integration step $\Delta\omega$. It is a rather remarkable fact that within the original interval, both converge to the same function, and that the main consequence of a discretization of the integral is simply rendering the function periodic.

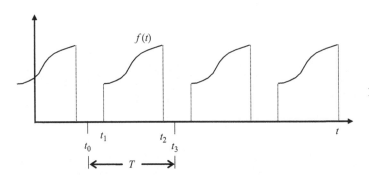

Figure 6.11. Periodic function.

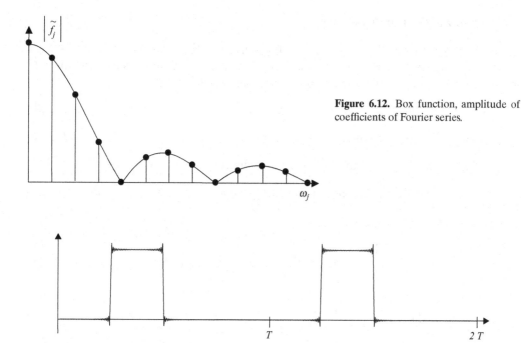

Figure 6.12. Box function, amplitude of coefficients of Fourier series.

Figure 6.13. Periodic box function by Fourier series with 512 terms.

A corollary of this observation is that Fourier integrals in the computer always represent periodic functions, never time-limited functions, as the integrals must by necessity be approximated by summations and are thus always discrete.

Example: The Periodic Box Function

To illustrate these matters, consider once more the box function of the previous section, which we now assume to be periodic, and because of that the Fourier amplitude spectrum is now discrete, as shown in Figure 6.12. Assume also the following parameters: $a\,t_d = 1, t_0 = 0, t_1 = 2, t_2 = 4$, and $t_3 = 8$. These imply in turn $t_d = 2, T = t_3 - t_0 = 8$ and $t_c = \frac{1}{2}(t_1 + t_2) = 3$. Also, we define the dimensionless time, frequency, and frequency step $\tau = 2t\,/\,t_d, \theta_j = \frac{1}{2}\omega_j\,t_d = j\,\Delta\theta$, and $\Delta\theta = \frac{1}{2}\Delta\omega t_d = \frac{1}{2}\frac{2\pi}{T}t_d = \frac{1}{4}\pi$, respectively. The dimensionless transformation formulae are then

$$\tilde{f}(\theta_j) = e^{-\mathrm{i}\theta_j\tau_c}\,\frac{\sin\theta_j}{\theta_j} \qquad = \text{coefficients of FS (identical to FT)} \qquad (6.319)$$

$$f(\tau) = \frac{\Delta\theta}{2\pi}\sum_{j=-\infty}^{\infty}\frac{\sin\theta_j}{\theta_j}e^{\mathrm{i}\theta_j(\tau-\tau_c)} \qquad = \text{Fourier series} \qquad (6.320)$$

The result of applying the Fourier series with a finite number of terms is shown in Figure 6.13. As can be seen, the Fourier series is not time limited, but is instead *periodic*. Also, it is a *continuous* function. The Gibbs phenomenon is clearly visible; in addition, there is some ringing in the neighborhood of the discontinuity due to the truncation of the Fourier series at some finite frequency ($j_{\max} = 2^9 = 512$; a higher cutoff frequency, say $j_{\max} = 2^{12} = 4096$, reduces this ringing substantially).

6.6.3 Discrete Fourier Transform

The *discrete Fourier transform* can be visualized best as a *Fourier series* in which the time and frequency spaces have been interchanged. Thus, the time domain is now the discrete space, while the frequency domain is continuous and periodic. Let

$$\Delta t \qquad\qquad = \text{time step} \tag{6.321}$$

$$t_k = k\,\Delta t \qquad = \text{discrete time (}k\text{ is an integer)} \tag{6.322}$$

$$f_k \equiv f(t_k) \qquad = \text{values of function or } signal \text{ at discrete times ("samples")} \tag{6.323}$$

$$v = \pi/\Delta t \qquad = Nyquist\,(\text{folding, or cutoff) frequency), in rad/s} \tag{6.324}$$

The appropriate formula are now

$$\boxed{\tilde{F}(\omega) = \sum_{k=-\infty}^{\infty} f_k\, e^{-i\omega t_k}\,\Delta t} \qquad \text{Discrete Fourier transform (DFT)} \tag{6.325}$$

$$\boxed{f_k = \frac{1}{2\pi}\int_{-v}^{+v} \tilde{F}(\omega)\, e^{i\omega t_k}\, d\omega} \qquad \text{Inverse discrete Fourier transform (IDFT)} \tag{6.326}$$

We have used a capitalized symbol $\tilde{F}$ for the DFT, because its values are not identical to those of the Fourier transform, even if the samples correspond to those of an actual continuous function. This is because of the so-called *aliasing* phenomenon, by way of which high-frequency components (i.e., those with periods shorter than twice the time step) get "impersonated" by lower frequencies, as will be seen later on.

Example: The Discrete Box Function (Figure 6.14)

$$\text{Let} \quad \begin{aligned} f_k &= a & \text{if} & & k_1 < k < k_2 \\ f_k &= \tfrac{1}{2}a & \text{if} & & k = k_1 \quad \text{or} \quad k = k_2 \\ f_k &= 0 & \text{otherwise} \end{aligned} \tag{6.327}$$

Hence

$$\tilde{F}(\omega) = a\Delta t\left\{\tfrac{1}{2}e^{-i\omega\Delta t k_1} + e^{-i\omega\Delta t(k_1+1)} + \cdots + e^{-i\omega\Delta t(k_2-1)} + \tfrac{1}{2}e^{-i\omega\Delta t k_2}\right\} \tag{6.328}$$

or with the auxiliary variable $z = e^{-i\omega\Delta t}$,

$$\tilde{F}(\omega) = a\Delta t\left\{\tfrac{1}{2}z^{k_1} + z^{k_1+1} + \cdots + z^{k_2-1} + \tfrac{1}{2}z^{k_2}\right\} \tag{6.329}$$

If $\omega\Delta t = 2\pi n$ (where n is an integer), or what is the same, when $\omega = 2nv$, then $z = 1$, in which case the summation in z is simply $k_2 - k_1$. In particular, for $n = 0$, we have

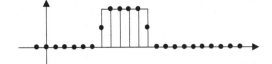

Figure 6.14. Discrete box function.

$$\tilde{F}(0) = a\,\Delta t(k_2 - k_1) = a\,t_d \tag{6.330}$$

On the other hand, when $z \ne 1$, the summation can be shown to be given by

$$\tilde{F}(\omega) = a\,\Delta t\,\frac{z+1}{z-1}(z^{k_2} - z^{k_1}) \tag{6.331}$$

Defining the auxiliary variables

$$t_c = \tfrac{1}{2}(k_1 + k_2)\Delta t \qquad = \text{center of discrete box function} \tag{6.332}$$

$$t_d = (k_2 - k_1)\Delta t \qquad = \text{width of discrete box function} \tag{6.333}$$

$$\theta = \tfrac{1}{2}\omega t_d \qquad = \text{dimensionless frequency} \tag{6.334}$$

we obtain the DFT after some algebra as

$$\tilde{F}(\omega) = a\,\Delta t\,\cot\tfrac{1}{2}\omega\Delta t\,\sin\theta\,e^{-i\omega t_c} \tag{6.335}$$

If we now apply the IDFT to this expression, we recover the original discrete box function. The DFT is a periodic function with period 2ν (the period of z). The base period or *band* of this periodic function extends over the frequency range $-\nu \le \omega \le \nu$, with ν being the Nyquist frequency. In general, it is only this band that has physical meaning. Thus, we say that the discrete function or signal is *band limited*.

Comparison of the above DFT with the Fourier transform of a continuous box function of equal center and width reveals that they are similar, but not identical. In the Nyquist band, the ratio of these two functions is

$$\frac{\tilde{F}(\omega)}{\tilde{f}(\omega)} = \frac{\Delta t\,\theta\cot\tfrac{1}{2}\omega\Delta t}{t_d} = \frac{\tfrac{1}{2}\omega\Delta t}{\tan\tfrac{1}{2}\omega\Delta t} = \frac{\tfrac{\pi}{2}\tfrac{\omega}{\nu}}{\tan\tfrac{\pi}{2}\tfrac{\omega}{\nu}} \tag{6.336}$$

which decays monotonically from 1 to 0 in the range $0 \le \omega \le \nu$.

6.6.4 Discrete Fourier Series

The discrete Fourier series is the periodic counterpart of the DFT, that is, the signal in the time domain is both discrete and periodic. In general, when a function is discrete in one domain, it must be periodic in the other, and vice versa. Thus, the DFS must be discrete and periodic.

Before continuing, we first define the *split-summation*

$$\overline{\sum_{j=M}^{N}} a_j = \tfrac{1}{2}a_M + a_{M+1} + \cdots + a_{N-1} + \tfrac{1}{2}a_N \tag{6.337}$$

The bar across the summation stipulates that the first and last elements must be halved.

Assume next that we have a discrete and periodic function or signal, and that the first period is composed of $2N$ equal intervals Δt containing $2N + 1$ points. In the interest of greater generality, we will allow for periodic discontinuities, so that the function values of the first and last point need not be equal. The length of one period is then $T = 2N\,\Delta t$, which we shall assume is aligned with the origin (begins at zero time).

Let

$$\Delta t = \frac{T}{2N} \qquad = \text{time step} \tag{6.338}$$

$$\Delta\omega = \frac{v}{N} = \frac{2\pi}{T} \qquad = \text{frequency step} \tag{6.339}$$

$$t_k = k\,\Delta t \qquad = \text{discrete time } (0 \le k \le 2N) \tag{6.340}$$

$$\omega_j = j\,\Delta\omega \qquad = \text{discrete frequency} (-N \le j \le N) \tag{6.341}$$

$$\omega_j t_k = \frac{\pi}{N} jk \tag{6.342}$$

The transformation formulae are then

$$\boxed{\tilde{F}_j = \sum_{k=0}^{2N} f_k\, e^{-i\omega_j t_k}\,\Delta t} \qquad \text{Discrete Fourier series (DFS)} \tag{6.343}$$

$$\boxed{f_k = \frac{1}{2\pi}\sum_{j=-N}^{N} \tilde{F}_j\, e^{i\omega_j t_k}\,\Delta\omega} \qquad \text{Inverse discrete Fourier series (IDFS)} \tag{6.344}$$

If the function $f(t)$ from which the discrete samples f_k were taken is such that that the first and last values are equal, that is, $f_0 \equiv f_{2N}$ (e.g., when these values are both zero, such as in the case of the box function that we used in the previous section), then the split summations in Eqs. 6.343 and 6.344 can be changed into conventional summations:

$$\tilde{F}_j = \Delta t \sum_{k=0}^{2N-1} f_k\, e^{-i\pi jk/N} \tag{6.345}$$

$$f_k = \frac{\Delta\omega}{2\pi}\sum_{j=-N}^{N-1} \tilde{F}_j\, e^{i\pi jk/N} = \frac{\Delta\omega}{2\pi}\sum_{j=-N+1}^{N} \tilde{F}_j\, e^{i\pi jk/N} \tag{6.346}$$

which are the expressions usually found in textbooks on Fourier methods. The advantage of the split summations over the conventional summations is that they allow for discontinuities between periods, and the end values $\tilde{F}_N, \tilde{F}_{-N}$ need not be real, but can be complex conjugates. Also, because they treat end points equally, they are in some sense more "symmetric."

Example 1: Discrete, Periodic Box Function
This is exactly the same case we examined in the previous section. Since $0 < k_1 < k_2 < 2N$, it follows that the coefficients of the DFS are the same as those of the DFT, namely $\tilde{F}_j = \tilde{F}(\omega_j)$. The only difference is that now these values occur at discrete intervals (i.e., ω is no longer continuous):

$$\tilde{F}_j = a\,\Delta t\,\cot\frac{j\pi}{N}\sin\frac{j\pi}{N}\frac{k_2-k_1}{2}\exp(-i\frac{j\pi}{N}\frac{k_2+k_1}{2}) \tag{6.347}$$

The IDFS recovers the original box function, and in addition, it is periodic.

Example 2: The Discrete Sawtooth Function (Figure 6.15)
Let

$$f_k = a\frac{t_k}{T} = a\frac{k}{2N} \qquad \text{for} \qquad 0 \le k \le 2N \tag{6.348}$$

Notice that this function will be "discontinuous" at the end of the period. The DFS is then

$$\tilde{F}_j = \frac{a\,\Delta t}{2N}\sum_{k=0}^{2N} k\, e^{-i\frac{\pi}{N}jk} \tag{6.349}$$

Defining $z = e^{-i\frac{\pi}{N}j}$, the split sum is

$$S = \tfrac{1}{2}0z^0 + 1z + 2z^2 + \cdots + kz^k + \cdots + \tfrac{1}{2}(2N)z^{2N} \tag{6.350}$$

When $j = 0$, then $z = 1$, in which case the summation is simply

$$S = 1 + 2 + \cdots + (2N-1) + \tfrac{1}{2}2N = 2N^2 \tag{6.351}$$

To evaluate this summation when $z \ne 0$, we consider instead the auxiliary summation (using the formula from the previous section, with $k_1 \to 0$ and $k_2 \to 2N$):

$$Q = \tfrac{1}{2}z^0 + z + z^2 + \cdots + z^k + \cdots + \tfrac{1}{2}z^{2N} = \frac{z+1}{z-1}(z^{2N}-1) \tag{6.352}$$

Comparing term for term the summations for S and Q, it becomes obvious that

$$S = \frac{dQ}{dz} = \frac{d}{dz}\left[\frac{z+1}{z-1}(z^{2N}-1)\right] = \frac{2}{(z-1)^2}\left[N(z^2-1)z^{2N-1} - z^{2N} + 1\right] \tag{6.353}$$

But $z^{2N} = e^{-i2\pi j} = 1$, so that

$$S = 2N\,z^{-1}\frac{z+1}{z-1} = 2i\,N\,e^{i\frac{\pi}{N}j}\cot\frac{\pi}{2N}j \tag{6.354}$$

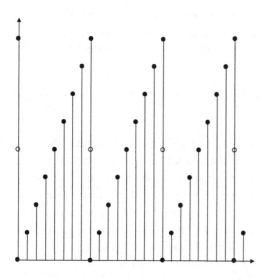

Figure 6.15. Discrete sawtooth function.

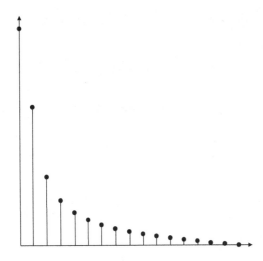

Figure 6.16. Fourier series coefficients of discrete sawtooth function.

Hence, we obtain finally (with $T = 2N \Delta t$)

$$
\tilde{F}_j =
\begin{cases}
\frac{1}{2} a t_p & \text{if} \quad j = 0 \\
i \frac{1}{2N} a t_p \, e^{i\frac{\pi}{N}j} \cot \frac{\pi}{2N}j & \text{if} \quad 0 < j \le N
\end{cases}
\tag{6.355}
$$

In the positive Nyquist range, the absolute value of the DFS can be written as

$$
|\tilde{F}_j| = \frac{aT}{\pi j} \frac{\frac{\pi}{2N}j}{\tan \frac{\pi}{2N}j} \qquad 1 \le j \le N
\tag{6.356}
$$

which are shown in Figure 6.16. Eqs. 6.355 and 6.356 indicate that at low frequencies and starting from zero, the amplitudes decay approximately as ½, $1/\pi$, $1/2\pi$,.... This is illustrated in Figure 6.16.

6.6.5 The Fast Fourier Transform

The FFT is not another member of the Fourier family, but is simply a superbly efficient algorithm to compute the discrete Fourier series of Section 6.6.4. A conventional term by term evaluation of the DFT equation for a discrete function f_k consisting of $2N$ samples, namely

$$
\tilde{F}_j = \Delta t \sum_{k=0}^{2N-1} f_k \, e^{-i\pi jk/N}
\tag{6.357}
$$

requires on the order of $4N^2$ operations; in contrast, the FFT algorithm requires only on the order of $2N \log_2 N$ operations, which is substantially less. Many computer versions of this procedure also take advantage of the complex conjugate symmetry of the transform of real functions, returning only the transformed values in the positive Nyquist range consisting of $N+1$ complex points ($0 \le j \le N$). Since $N+1$ complex words occupy the same storage space as $2N+2$ real words, it is possible to replace the original function consisting of $2N$ points by the transform while displacing essentially the same memory space. Thus, the FFT often destroys the original data and replaces it with the transformed data.

The FFT transformation is typically made with the assumption that $\Delta t = 1, \Delta \omega = 2\pi$. As a result, a double application of the algorithm returns the original array multiplied by $2N$. For a description of the algorithm itself, readers are referred to the specialized literature. Computer routines are also readily available, including in MATLAB®.

6.6.6 Orthogonality Properties of Fourier Expansions

The complex exponentials used for the Fourier expansions in the previous sections satisfy some important orthogonality conditions. These conditions can be used to prove the inversion formulae.

(a) Fourier Transform

In this case, we *define* the orthogonality conditions as

$$\frac{1}{2\pi} \int_{-\infty}^{\infty} e^{\pm i(\omega - \omega_0)t} \, dt = \delta(\omega - \omega_0) \tag{6.358}$$

$$\frac{1}{2\pi} \int_{-\infty}^{\infty} e^{\pm i \omega(t - t_0)} \, d\omega = \delta(t - t_0) \tag{6.359}$$

in which $\delta(\ )$ is the Dirac delta function. We emphasize the word "define," because these integrals do not really converge. Nonetheless, when they are used as argument in some other integrals, they *behave* like Dirac delta functions, a property that mathematicians call a *distribution*. Hence, we both abbreviate and interpret these integrals as Dirac delta functions.

(b) Fourier Series

Let $\omega_j = j \frac{2\pi}{T}$ and $\omega_n = n \frac{2\pi}{T}$ then

$$\frac{1}{T} \int_0^T e^{i(\omega_j - \omega_n)t} \, dt = \delta_{jn} = \begin{cases} 1 & \text{if } j = n \\ 0 & \text{else} \end{cases} \tag{6.360}$$

in which δ_{jn} is the Kronecker delta (the discrete counterpart to the Dirac delta function).

(c) Discrete Fourier Series

Let $\omega_j t_k = \frac{\pi}{N} j k$, then

$$\frac{1}{2N} \sum_{j=0}^{2N} e^{i \omega_j (t_k - t_l)} = \delta_{kl} = \begin{cases} 1 & \text{if } k = l \\ 0 & \text{else} \end{cases} \tag{6.361}$$

$$\frac{1}{2N} \sum_{j=-N}^{N} e^{i(\omega_j - \omega_n)t_k} = \delta_{jn} = \begin{cases} 1 & \text{if } j = n \\ 0 & \text{else} \end{cases} \tag{6.362}$$

Also

$$\frac{1}{2N} \sum_{j=0}^{2N-1} e^{i \omega_j (t_k - t_l)} = \delta_{kl} = \begin{cases} 1 & \text{if } k = l \\ 0 & \text{else} \end{cases} \tag{6.363}$$

$$\frac{1}{2N} \sum_{j=-N+1}^{N} e^{i(\omega_j - \omega_n)t_k} = \delta_{jn} = \begin{cases} 1 & \text{if } j = n \\ 0 & \text{else} \end{cases} \qquad (6.364)$$

6.6.7 Fourier Series Representation of a Train of Periodic Impulses

In the *time domain*, we consider an infinite train of Dirac delta impulses at regular intervals Δt

$$\hat{\delta}(t) = \sum_{k=-\infty}^{\infty} \delta(t - t_k) \qquad (6.365)$$

Since this is a periodic function with period Δt, we can represent it – at least formally – by means of a Fourier series. We obtain the coefficients of that series by integration over the fundamental period:

$$D_j = D(\Omega_j) = \int_0^{\Delta t} \delta(t) e^{-i\Omega_j t} dt = 1 \qquad (6.366)$$

where, to avoid confusion, we have used an alternate symbol Ω to denote frequencies. Hence,

$$\hat{\delta}(t) = \frac{1}{2\pi} \Delta\Omega \sum_{j=-\infty}^{\infty} (1) e^{i\Omega_j t} \qquad (6.367)$$

with discrete frequencies $\Omega_j = j\,\Delta\Omega$, $\Delta\Omega = \frac{2\pi}{\Delta t} = 2\nu$. It follows that

$$\boxed{\hat{\delta}(t) \equiv \sum_{k=-\infty}^{\infty} \delta(t - t_k) = \frac{1}{\Delta t} \sum_{j=-\infty}^{\infty} e^{i2\nu jt}} \qquad (6.368)$$

By an entirely analogous development, we can also obtain the formal representation of a train of impulses in the *frequency domain*, spaced at regular intervals $\Delta\omega = \frac{2\pi}{T}$

$$\boxed{\hat{\delta}(\omega) \equiv \sum_{j=-\infty}^{\infty} \delta(\omega - \omega_j) = \frac{1}{\Delta\omega} \sum_{n=-\infty}^{\infty} e^{-i\omega nT}} \qquad (6.369)$$

Both of the previous representations must be understood in the sense of *distributions*.

6.6.8 Wraparound, Folding, and Aliasing

Let $f(t)$ be a real-valued continuous function of t which, for the sake of generality, we now define as extending over the entire real axis. Assuming that it satisfies the Dirichlet conditions, then its Fourier transform exists. Hence, we have the Fourier transform pair

$$\boxed{\tilde{f}(\omega) = \int_{-\infty}^{+\infty} f(t) e^{-i\omega t} dt} \quad \text{and} \quad \boxed{f(t) = \frac{1}{2\pi} \int_{-\infty}^{+\infty} \tilde{f}(\omega) e^{i\omega t} d\omega} \qquad (6.370)$$

Consider next the *periodic* function $g(t)$ which is obtained by evaluating the Fourier series associated with the discrete values $\tilde{f}_j = \tilde{f}(\omega_j)$ of the transform. Expressing these values with the help of a train of impulses as defined in the previous section, we can write

$$\tilde{g}(\omega) = \tilde{f}(\omega)\Delta\omega \sum_{j=-\infty}^{\infty} \delta(\omega-\omega_j) = \tilde{f}(\omega) \sum_{n=-\infty}^{\infty} e^{-i\omega nT} \tag{6.371}$$

The inverse Fourier transform of the left part of this equation is then

$$g(t) = \frac{1}{2\pi}\int_{-\infty}^{\infty}\left[\tilde{f}(\omega)\Delta\omega \sum_{j=-\infty}^{\infty} \delta(\omega-\omega_j)\right]d\omega = \frac{\Delta\omega}{2\pi} \sum_{j=-\infty}^{\infty} \tilde{f}_j\, e^{i\omega_j t} \tag{6.372}$$

which yields the conventional Fourier series representation for $g(t)$. To determine the relationship that $g(t)$ has with the original function $f(t)$, we Fourier-invert the last term on the right in the expression for $\tilde{g}(\omega)$, which results in

$$g(t) = \frac{1}{2\pi}\int_{-\infty}^{+\infty}\tilde{g}(\omega)e^{i\omega t}\,d\omega = \sum_{n=-\infty}^{\infty}\frac{1}{2\pi}\int_{-\infty}^{\infty}\tilde{f}(\omega)e^{i\omega(t-nT)}\,d\omega \tag{6.373}$$

That is,

$$\boxed{g(t) = \sum_{n=-\infty}^{\infty} f(t-nT)} \tag{6.374}$$

Thus, each periodic segment of the Fourier series $g(t)$ can be interpreted as resulting from an overlap or *wraparound* of all segments of length $T = 2\pi/\Delta\omega$ of the original, non-periodic function. Graphically speaking, it is as if the function had been drawn on paper, rolled into a cylinder of circumference T, and the curves on the overlapping sheets added together. If, as we assumed in the previous sections, the original function is nonzero only in the interval $0 \le t \le T$, then $g(t) = f(t)$, except that now it is also periodic.

Consider next what happens when we discretize the function $f(t)$ in the time domain (i.e., when we sample the function at discrete intervals Δt). If we use for this purpose a train of impulses in the time domain, we can write

$$F(t) = f(t)\Delta t \sum_{k=-\infty}^{\infty} \delta(t-t_k) = f(t) \sum_{j=-\infty}^{\infty} e^{i2vjt} \tag{6.375}$$

Now, the Fourier transform of the left part of this equation is simply

$$\tilde{F}(\omega) = \int_{-\infty}^{\infty} F(t)e^{-i\omega t}\,dt = \Delta t \sum_{k=-\infty}^{\infty}\int_{-\infty}^{\infty}\left[f(t)\,\delta(t-t_k)e^{-i\omega t}\right]dt = \Delta t \sum_{k=-\infty}^{\infty} f_k\, e^{-i\omega t_k} \tag{6.376}$$

which agrees with the discrete Fourier transform of the samples. Alternatively, the Fourier transform for the equivalent expression on the right leads to

$$\tilde{F}(\omega) = \sum_{j=-\infty}^{\infty}\int_{-\infty}^{\infty} f(t)e^{-i(\omega-2vj)t}\,dt \tag{6.377}$$

That is,

$$\boxed{\tilde{F}(\omega) = \sum_{j=-\infty}^{\infty} \tilde{f}(\omega-2vj)} \tag{6.378}$$

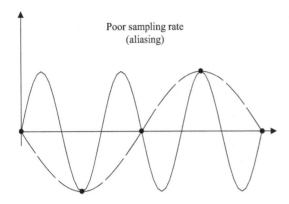

Poor sampling rate
(aliasing)

Figure 6.17. Aliasing due to poor sampling.

Thus, the Fourier spectrum of the discrete samples does not agree with that of the continuous function. Instead, it is a periodic spectrum, which recurs at frequency intervals $2v$, with the frequency range $-v \leq \omega \leq v$ constituting the *physical* band of that spectrum. Within this band, the spectral amplitudes for the discrete signal are related to those for the continuous signal by the *aliasing* equation 6.378, which can be visualized as an overlap of all spectral runs of length $2v$. Because of the complex conjugate symmetry of the spectrum, the aliasing equation can also be interpreted as a sequence of folds of the true spectrum about integer multiples of the Nyquist frequency, reversing for each *odd* fold the sign of the imaginary parts, and adding the results.

The aliasing phenomenon also has a physical interpretation: the components in $f(t)$ with frequencies higher than the Nyquist frequency v are *folded* or aliased into components with frequencies within the Nyquist band. In other words, their energy gets "converted" into low-frequency energy, as illustrated schematically in Figure 6.17. Examples of this phenomenon can be found in the stationary patterns seen under stroboscopic light; the slow moving (or even reversing) wheels of a carriage in a Western film; or in digital audio recordings, the transformation into audible noise of the normally inaudible ultrasonic components produced by instruments. However, if the original signal is band-limited and this band lies within the Nyquist range, then no aliasing takes place when sampling the signal at least as fast as $\Delta t = \pi / v$.

The equations in this section demonstrate also that a signal *cannot* be both time limited *and* band limited. If a signal is time limited, it must by necessity be band unlimited, and if it is band limited, then it must be time unlimited (but see also the Ricker wavelet in Chapter 9, Section 9.3.5). Of course, a signal can be both time-unlimited and band unlimited.

6.6.9 Trigonometric Interpolation and the Fundamental Sampling Theorem

The DFS can be utilized to interpolate discrete data sets while avoiding addition of high-frequency components. Formally, this goal can be achieved by allowing the time variable in the inverse DFS to vary continuously, or to attain values corresponding to noninteger multiples of the time step. In practice, interpolation is achieved by adding after the Nyquist frequency an arbitrarily long string of *trailing* zeros to the DFS, and redefining correspondingly the number of data points to reflect this addition. When the augmented

DFS is then inverted, the resulting array will contain the interpolated data points. In the audio industry, this technique is referred to as *oversampling*.

For example, suppose that we have an earthquake record consisting of $2N = 1000$ points, sampled at intervals of $\Delta t = 0.02$ s, and that we wish to interpolate it to have samples at $\Delta t = 0.01$ s. The initial Nyquist (folding or cutoff) frequency, in cycles per second, is $0.5/\Delta t = 25$ Hz. We begin by computing the initial DFS, which contains 501 complex points between 0 and 25 Hz. We then augment this transform with 500 complex zeros, which in effect redefine the Nyquist frequency to 50 Hz. The inverse transform of this augmented spectrum then yields an interpolated array with 2000 data points. Hence

- Addition of trailing zeros in the frequency domain = interpolation in the time domain
- Addition of trailing zeros in the time domain = interpolation in the frequency domain

The interpolation operation described can also be carried out directly in the time domain by means of the *fundamental sampling theorem*, whose derivation is mercifully brief. Indeed, from the equations for the DFT, we have

$$F(t) = \frac{1}{2\pi}\int_{-v}^{+v} \tilde{F}(\omega)\,e^{i\,\omega t_k}\,d\omega = \frac{1}{2\pi}\Delta t \sum_{k=-\infty}^{\infty}\int_{-v}^{+v} f_k\,e^{i\,\omega(t-t_k)}d\omega \tag{6.379}$$

Integrating the last expression on the right and expressing the result in terms of trigonometric functions, we obtain immediately

$$\boxed{F(t) = \sum_{k=-\infty}^{\infty} f_k\,\frac{\sin v(t-t_k)}{v(t-t_k)}} \tag{6.380}$$

which permits finding interpolated values $F(t)$ directly from the known samples f_k of $f(t)$. Notice that since $v(t_{k+1} - t_k) = v\Delta t = \pi$, the interpolation function is zero when $t = t_n$ coincides with one of the samples (say the nth), and is 1 for the sample itself; thus $F(t_k) = f(t_k)$. However, in general $F(t) \neq f(t)$. If the original signal is band-limited within the Nyquist band defined by the sampling rate, then it is indeed true that $F(t) \equiv f(t)$ for any t. In that case, the discrete samples are enough to recover the complete original signal. The fundamental sampling theorem then states that "twice the time step – the Nyquist period – must be equal or shorter than the shortest period contained in the signal," if no information in the signal is to be lost.

In the case of a DFS obtained by discretization of the Fourier spectrum of the DFT, or what is equivalent, obtained by wrapping-around runs of length $2N$ of the discrete signal, the interpolation equation is obtained as follows:

$$F(t) = \frac{\Delta t\,\Delta\omega}{2\pi}\sum_{j=-N}^{N}\sum_{k=-\infty}^{\infty} f_k\,e^{-i\,\omega_j t_k}\,e^{i\,\omega_j t} = \frac{1}{2N}\sum_{k=-\infty}^{\infty}\sum_{j=-N}^{N} f_k\,e^{i\,\omega_j(t-t_k)} \tag{6.381}$$

The inner split sum on the right can be evaluated by the same method used earlier in the example on the sawtooth function. The final result is

$$F(t) = \sum_{k=-\infty}^{\infty} f_k\,\frac{\frac{1}{2N}\sin v(t-t_k)}{\tan\frac{1}{2N}v(t-t_k)} \tag{6.382}$$

It can be shown that this equation implies

$$F_k = F(t_k) = \sum_{n=-\infty}^{\infty} f_{k+2nN} \tag{6.383}$$

which is again the overlapped (wrapped-around) discrete function.

6.6.10 Smoothing, Filtering, Truncation, and Data Decimation

A smoothing operation consists in modifying, reducing, or setting to zero the high frequency components in the DFT of a signal. The resulting inverse transform yields thus a function that is similar to the original function, but has less sharp variations and corners. More generally, we can also apply appropriate weighting functions or *filters* to arbitrary frequency bands within the Fourier spectrum, to achieve certain desirable effects. Common among these are the low-pass, high-pass, and band-pass filters.

If the transform is instead truncated, the effect is again a smoothing operation, but the new time increment is larger (in effect, this constitutes the reverse of interpolation). Using developments similar to those we have used in the previous sections, it can be shown that the time histories associated with inversion of Fourier spectra that have been truncated at a frequency $\omega = v$ are given by the following expressions:

$$h_1(t) = \frac{v}{\pi} \int_{-\infty}^{\infty} f(\tau) \left[\frac{\sin v(t-\tau)}{v(t-\tau)} \right] d\tau \qquad \text{Truncated Fourier transform} \tag{6.384}$$

$$h_2(t) = \frac{1}{T} \int_0^T g(\tau) \left[\frac{\sin v(t-\tau)}{\tan \frac{\pi}{T}(t-\tau)} \right] d\tau \qquad \text{Truncated Fourier series} \tag{6.385}$$

Another possible manipulation could consist in deleting data points in the DFT. For example, we could *decimate* every other complex point. In effect, this action would reduce the number of data points, changing the period of the signal in the time domain, but not the time step. In most cases, decimation constitutes an action of last resort, a drastic operation that nearly always degrades the signal.

6.6.11 Mean Value

The zero-frequency (or DC) component of a Fourier series is a real number that is proportional to the mean or average value $\bar{f}$ of the signal in the interval 0 to T. This is so because for zero frequency,

$$\tilde{f}_0 = \int_0^T f(t)\, dt = T \bar{f} \tag{6.386}$$

Hence, if we set this value to zero, we shift the whole function vertically without altering its shape.

We mention also in passing that in a DFS or DFT, the spectral amplitude at the Nyquist frequency is also real and proportional to the sum of the signal with alternating signs, that is, $f_0 - f_1 + f_2 - f_3 + \cdots$.

6.6.12 Parseval's Theorem

Parseval's theorem is a useful and important property of Fourier methods that we shall state without proof.[5] If $x(t)$ and $y(t)$ are two real and continuous functions of t defined in the interval $[0, T]$, whose Fourier transforms are $\tilde{x}(\omega)$ and $\tilde{y}(\omega)$, then

$$\int_0^T x(t)\, y(t)\, dt = \frac{1}{2\pi}\int_{-\infty}^{\infty} \tilde{x}(\omega)\, \tilde{y}^c(\omega)\, d\omega = \frac{1}{2\pi}\int_{-\infty}^{\infty} \tilde{x}^c(\omega)\, \tilde{y}(\omega)\, d\omega \qquad (6.387)$$

in which a superscript c denotes the complex conjugate. In particular, if $x(t) \equiv y(t)$, then

$$\int_0^T x^2(t)\, dt = \frac{1}{2\pi}\int_{-\infty}^{\infty} |\tilde{x}(\omega)|^2 \, d\omega \qquad (6.388)$$

In the case of a *Fourier series*, the corresponding expressions are

$$\int_0^T x(t)\, y(t)\, dt = \frac{1}{2\pi} \sum_{j=-\infty}^{\infty} \tilde{x}_j\, \tilde{y}_j^c\, \Delta\omega = \frac{1}{2\pi} \sum_{j=-\infty}^{\infty} \tilde{x}_j^c\, \tilde{y}_j\, \Delta\omega \qquad (6.389)$$

$$\int_0^T x^2(t)\, dt = \frac{1}{2\pi} \sum_{j=-\infty}^{\infty} |\tilde{x}_j|^2 \, \Delta\omega \qquad (6.390)$$

in which $\frac{\Delta\omega}{2\pi} = \frac{1}{T}$. It is a remarkable fact that the change of the Fourier integral into a summation did not affect the result of the operation.

Finally, Parseval's equations for the *discrete Fourier series* are

$$\sum_{k=0}^{2N} x_k\, y_k\, \Delta t = \frac{1}{2\pi} \sum_{j=-N}^{N} \tilde{X}_j\, \tilde{Y}_j^c\, \Delta\omega \qquad (6.391)$$

$$\sum_{0}^{2N} x_k^2\, \Delta t = \frac{1}{2\pi} \sum_{-N}^{N} |X_j|^2\, \Delta\omega \qquad (6.392)$$

with $\tilde{X}_j = \tilde{X}(\omega_j)$ and $\tilde{Y}_j = \tilde{Y}(\omega_j)$ being the aliased spectral amplitudes. The split summation in Eqs. 6.391 and 6.392 can often be expressed in terms of conventional summations from 0 to $2N-1$, and from $-N+1$ to $+N$. Care must be exercised, however, when the spectral value at the Nyquist frequency is a complex number, a situation that could have arisen as a result of manipulations like multiplication by a transfer function. The imaginary part of that element must then be set to zero.

---

[5] Due to Parseval (1799) but sometimes referred to also as Plancherel's theorem (1910).

6.6.13 Summary of Important Points

	Time domain	Frequency domain
FT	$f(t) = \frac{1}{2\pi} \int_{-\infty}^{+\infty} \tilde{f}(\omega) e^{i\omega t} d\omega$	$\tilde{f}(\omega) = \int_{-\infty}^{+\infty} f(t) e^{-i\omega t} dt$
FS	$g(t) = \frac{1}{2\pi} \sum_{j=-\infty}^{\infty} \tilde{f}(\omega_j) e^{i\omega_j t} \Delta\omega$	$\tilde{f}(\omega) = \int_0^T f(t) e^{-i\omega t} dt$
DFT	$f_k = \frac{1}{2\pi} \int_{-v}^{+v} \tilde{F}(\omega) e^{i\omega t_k} d\omega$	$\tilde{F}(\omega) = \sum_{k=-\infty}^{\infty} f_k e^{-i\omega t_k} \Delta t$
DFS	$f_k = \frac{1}{2\pi} \sum_{j=-N}^{N} \tilde{F}_j e^{i\omega_j t_k} \Delta\omega$	$\tilde{F}_j = \sum_{k=0}^{2N} f_k e^{-i\omega_j t_k} \Delta t$

Time limited	$\Rightarrow$	Unlimited band
Unlimited time	$\Leftarrow$	Band limited
Discrete	$\Leftrightarrow$	Periodic, aliased
Periodic	$\Leftrightarrow$	Discrete

	FT	FS	DFT	DFS
Highest frequency (Hz)	∞	∞	$1/2\Delta t$	$1/2\Delta t$
Time domain periodic?	No	Yes, T	No	Yes, $T = 2N\Delta t$
Frequency domain periodic?	No	No	Yes, $1/\Delta t$	Yes, $1/\Delta t$
Spectral value	Continuous	Discrete	Continuous	Discrete
Nature	Exact	Exact	Aliased	Aliased

6.6.14 Frequency Domain Analysis of Lightly Damped or Undamped Systems

It is well known that conventional numerical methods based on the FFT algorithm cannot be applied to the analysis of undamped systems, because of the singularities at the resonant frequencies of the system. Although such singularities do not exist in lightly damped systems, it is still necessary to include a sufficient number of points so as to resolve accurately the transfer functions in the neighborhood of the natural frequencies. Also, it is necessary to add at the end of the force time history a quiet zone of *trailing zeros* of sufficient duration so as to damp out the free vibration terms and avoid *wraparound*. This duration is thus a function of the fundamental period of the system and the amount of damping, and can be very large for lightly damped systems. For undamped systems, the free vibration terms will never decay and, therefore, the standard application of the FFT algorithm is no longer possible.

A powerful general approach to obtain solutions with the FFT method for undamped or lightly damped systems is provided by the *Exponential Window Method* (EWM) described in the ensuing, which ultimately can be regarded as a numerical implementation of the *Laplace Transform*. Although this method has been used in signal processing and in seismology, it is virtually unknown in structural dynamics. In the sections that follow, we first describe succinctly the theoretical basis of this method for arbitrary

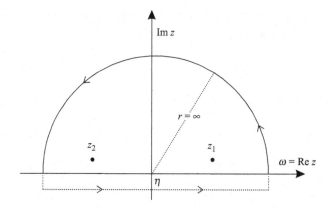

Figure 6.18. Fourier inversion by contour integration along a shifted path.

dynamical systems, and then illustrate its application to a continuous system. A more complete account of this method can be found in Kausel and Roësset (1992)[6] and in Hall and Beck (1993).[7]

Exponential Window Method: The Preferred Tool

As we have already seen, the response of a lightly damped or undamped MDOF (or continuous) system follows from the inverse Fourier transform

$$u(t) = \frac{1}{2\pi} \int_{-\infty}^{\infty} H(\omega)\,\tilde{p}(\omega)\,e^{i\omega t} d\omega \qquad (6.393)$$

in which $H(\omega)$ is the transfer function at an arbitrary elevation, and $\tilde{p}(\omega)$ is the Fourier transform of the excitation $p(t)$. A formal analytical evaluation of this integral can be accomplished by contour integration in the complex frequency plane, using a complex frequency z, as depicted schematically in Figure 6.18:

$$u(t) = \frac{1}{2\pi} \oint H(z)\,\tilde{p}(z)\,e^{izt} dz \qquad (6.394)$$

The choice of the integration path depends on the sign of t: for positive times, the exponential term is bounded in the upper complex half-plane, while for negative times, it is bounded in the lower. Since both the excitation and the vibrating system are causal, it follows that the lower half-plane cannot contain any poles (see also Section 9.2, "Functions of complex variables: a brief introduction"). On the other hand, the value of the contour integral depends only on the poles enclosed by the integration path. Hence, for positive times, the contour can be taken along a path that runs parallel to the real axis at some arbitrary distance η underneath it, and closed in the upper half-plane with a circle of infinite radius, as shown in Figure 6.18. Invoking standard arguments of contour integration,

[6] E. Kausel and J. M. Roësset (1992), "Frequency domain analysis of undamped systems," *Journal of Engineering Mechanics*, ASCE, 118 (4), 721–734

[7] J. F. Hal, and J. L. Beck (1993), "Linear system response by DFT: analysis of a recent modified method," *Earthquake Engineering and Structural Dynamics*, 22, 599–615.

it can be shown that the integral along the infinite circle vanishes, and only the integral along the path $z = \omega - i\eta$ remains. Hence, the Fourier integral is equivalent to

$$u(t) = \frac{1}{2\pi} \int_{-\infty}^{\infty} H(\omega - i\eta)\, \tilde{p}(\omega - i\eta)\, e^{i(\omega - i\eta)t} d\omega \qquad (6.395)$$

Since η does not depend on ω, it follows that the response is given by

$$u(t) = e^{\eta t} \left[\frac{1}{2\pi} \int_{-\infty}^{\infty} H(\omega - i\eta)\, \tilde{p}(\omega - i\eta)\, e^{i\omega t} d\omega \right] \qquad (6.396)$$

with

$$\tilde{p}(\omega - i\eta) = \int_{0}^{t_d} \left[e^{-\eta t} p(t) \right] e^{-i\omega t} dt \qquad (6.397)$$

The transfer function H for complex frequency $z = \omega - i\eta$ is just one of the components of the vector $\tilde{\mathbf{u}}$ obtained from the solution of the well-known equilibrium equation in the frequency domain

$$(\mathbf{K} + i z \mathbf{C} - z^2 \mathbf{M})\tilde{\mathbf{u}} = \tilde{\mathbf{p}} \qquad (6.398)$$

This equation differs from the classical equation in structural dynamics only in that ω is replaced by z. This system will not exhibit singularities along the axis of integration, even if $\mathbf{C}$ vanishes (i.e., for undamped systems). Hence, to compute the response in the frequency domain, it suffices to

- Compute the FFT of the excitation, modified by a decaying exponential window
- Evaluate the transfer functions for complex frequency
- Compute the inverse FFT of the product of the previous two quantities
- Modify the result by a rising exponential window

While in theory any arbitrary factor $\eta > 0$ could be used, in practice the choice of this number is limited by the finite precision with which the computations are made. Indeed, the value of the rising exponential term at the end of the window is $\exp(\eta T) = \exp(2\pi h/\Delta\omega)$. Once this value exceeds some three to four orders of magnitude, numerical errors develop, particularly at large times. Numerical experiments indicate that good results are obtained with the simple rule of thumb $\eta = \Delta\omega$ (i.e., the imaginary component of the complex frequency z equals the frequency step), in which case $\exp(\eta T) = \exp(2\pi) = 535 = 10^{2.73}$. Consequently, this choice depresses wraparound by almost three orders of magnitude (i.e., by $20 \times 2.73 = 55$ dB). Just about as good is also $\eta = \frac{1}{2}\Delta\omega$.

Observe that the contour integral 6.394 can alternatively be written in terms of the Laplace parameter $s = i(\omega - i\eta)$, which causes a 90 degree counterclockwise rotation of the axes such that the new integration path along a parallel to the vertical axis coincides with the Bromwich contour lying to the right of all poles and then enclosing these with a large circle "around infinity" for $s \ll 0$. Thus, the EWM is in effect a numerical implementation of the Laplace transform.

The method described applies equally well to continuous systems with an infinite number of resonant frequencies. Consider, for example, an undamped homogeneous cantilever shear beam of length L, uniform cross section A, shear modulus G, and mass density ρ.

This beam is subjected to a concentrated load $p(t)$ applied at the tip. From a solution of the differential equation for this problem by the method of characteristics, it is known that the response velocity $\dot{u}(x,t)$ in the beam consists of a pulse with the same shape as $p(t)$ that travels along the beam with velocity $c = \sqrt{G/\rho}$. This pulse repeatedly reflects at the two extremes of the shear beam, changing polarity every time it impinges on the fixed end. In the frequency domain, on the other hand, the solution for the displacement and velocity involves the transfer functions

$$H_u(\omega,x) = \frac{L}{GA}\frac{\sin \alpha x}{\alpha L \cos \alpha L} \quad \text{and} \quad H_{\dot{u}}(\omega,x) = \frac{i}{\rho c A}\frac{\sin \alpha x}{\cos \alpha L} \tag{6.399}$$

in which $\alpha = \omega/c$. The solution of these equations with the exponential window method requires replacing ω by $z = \omega - i\eta$.

We present next two examples. The first is a simple 1-DOF system, which although uncomplicated illustrates rather dramatically the power of the EWM, and the second one is a continuous system that illustrates the superb accuracy that can be achieved.

Example 1
Consider an undamped 1-DOF system subjected to a delayed step load of infinite duration with the following properties: Fundamental frequency $= f_n = 1.25$ Hz (i.e., period $T_n = 0.8$ s); mass $m = 5$ kg; and stiffness $k = 4\pi^2 m f_n^2 = 308.425$ N/m. The step load begins with a delay of $t_0 = 1$ s, after which it remains constant at $p_0 = 1,000$. The static deflection caused by the load is $u_s = p_0 / k = 3.242$. We choose a total duration for the response of $t_p = 5.12$ s, which is slightly more than six periods; this is also the implied period of the response in the context of the standard frequency inversion method. The implied frequency step is $\Delta f = 1/t_p = 0.195$ s. We also choose $N = 64$ points for the FFT, which means that the time step is $\Delta t = 5.12/64 = 0.08$ s, and the Nyquist frequency is $f_{max} = 0.5/\Delta t = 6.25$ Hz, that is, five times the fundamental frequency. Also, there are $\frac{1}{2}N = 32$ frequency steps between zero frequency and the Nyquist frequency. Since the system has no damping, the transient never decays. The exact solution is

$$u = \begin{cases} 0 & t < t_0 \\ u_s\left(1 - \cos\left[\omega_n\left(t - t_0\right)\right]\right) & t \geq t_0 \end{cases}, \quad t_0 = 1 \tag{6.400}$$

We apply the EWM with an imaginary part $\eta = \frac{1}{2}\Delta\omega = \pi\Delta f$, that is, half of the frequency step. Figure 6.19 shows the absolute value of the transfer function for this case, which exhibits a moderate peak, but not an infinite resonance, and this despite the absence of damping. Furthermore, although this figure displays both markers at the discrete frequencies as well as a dashed spline to increase the visibility, in reality only the discrete points are used in the calculation, and this without doing any interpolation whatsoever. In fact, one should avoid the temptation of increasing the resolution near the peak, for that would significantly deteriorate the results.

Figure 6.20 shows the response at the discrete time steps as dots. The continuous curve is the *exact* solution. As can be seen, the agreement is perfect, and this despite the rather coarse time step used and what would appear to be an even coarser transfer function. Of course, repeating this example with either 128 or 256 points would have given results that are just as accurate as those shown herein, but with double or four times the resolution.

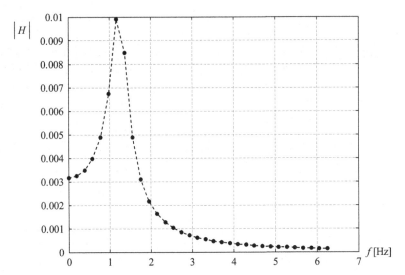

Figure 6.19. EWM, transfer function for example 1. Only the discrete values are used.

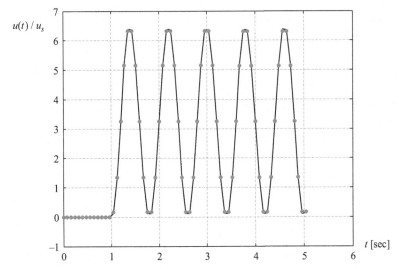

Figure 6.20. EWM, response function for example 1. The continuous line is the exact solution, while the markers show the discrete solution.

If we used only 64 points, it was to show how good the results are despite the coarseness of the sampling.

Example 2

We consider next an undamped, continuous, cantilever shear beam with the following properties.

Shear wave velocity $C_S = 5$ m/ms; mass density 2 Mg/m³; length $L = 0.5$ m, and cross-section $A = 10^{-4}$m. It is subjected at its free end to a triangular pulse of amplitude $p_0 = 1$, duration $t_d = 0.02$ ms, and the response is observed at the center $x = 0.25$ m. This system has

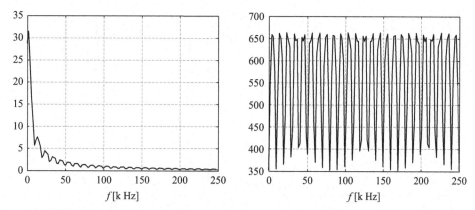

Figure 6.21. EWM, displacement, and velocity transfer functions for example 2.

a fundamental frequency $f_n = \frac{c_s}{4L} = 2.5\,\text{kHz}$ and higher resonant frequencies $f_n = (2n-1)f_1$, that is, periods $T_1 = 0.4$ ms, $T_n = \frac{1}{2n-1}T_1$.

For the EWM, we choose $N = 256$ points for the FFT and a time increment $\Delta t = 0.002$ ms. With these data, the width of the Fourier window is $t_p = N\Delta t = 0.512$ ms, the load is sampled at 11 points, the Nyquist frequency is $f_{\max} = 0.5/\Delta t = 250$ kHz, and the frequency step is $\Delta f = 1/t_p = 0.1953$ kHz. We also choose the imaginary component of frequency to be equal to the frequency step, that is, $\eta = \Delta\omega = 2\pi\Delta f$. Up to the Nyquist frequency, the shear beam contains 50 resonant peaks spaced at 5-kHz intervals. Hence, there are some 25 frequency steps between the resonant peaks. Figure 6.21 depicts both the displacement and velocity transfer functions, which to the untrained eye might appear to be unacceptably coarse, and not looking like anything that would readily be recognized as a transfer function, and especially so the velocity's.

The triangular pulse elicits a wavelet of spatial width $t_d / C_s = 0.004$ m which propagates down, reflects at the base with opposite polarity, then rebounds upwards, and on reaching the top reflects there with the same negative polarity. Upon a further reflection at the bottom and then again at the top, the whole pattern starts repeating itself. That is, the duration of the response pattern equals the fundamental period, which in turn equals twice the travel time of shear waves down and up again. The exact solution for this problem can easily be obtained with the method of characteristics, which would demonstrate that the agreement with the numerical solution is virtually perfect.

Figure 6.22 shows the displacement and velocity response at the center of the beam. The computed results are supremely good, to the point that they could not be distinguished from the exact analytical solution, except for some mild ringing at the very end of the time window in the displacement. Rather remarkable is the faithful reproduction of the sharp temporal discontinuities, and in the case of the displacement time histories, the large differences between initial and final values. This is in contrast to a conventional implementation of the FFT method for which the response is periodic, so initial and final values must agree. Thus, this shows that trailing zeros are not necessary in this method. We emphasize again that NO interpolation is used for the transfer functions, and indeed, that none should be used, lest the results deteriorate. At the same time, the user should not at all be concerned about the resolution of the peaks. The method takes care of the details.

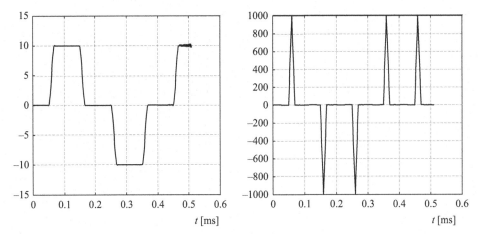

Figure 6.22. EWM, displacement (left) and velocity (right) observed at the center of a shear beam due to triangular pulse at the free end. The time axis is in milliseconds.

In summary, the steps to be taken by the analyst are the following:

- Decide on the duration T needed for the response functions (that is, the time window if interest), This automatically defines the frequency step $\Delta f = 1/T$ (or $\Delta \omega = 2\pi/T$). Choose also the imaginary component of frequency, say $\eta = \frac{1}{2}\Delta\omega = \pi/T$. You should be largely unconcerned with the resolution of the transfer functions, that is, about how small (or large) $\Delta\omega$ actually is. In particular, do not worry about the resolution of the resonant peaks.
- Decide on the time step Δt needed to define adequately both the excitation and the response. For the load, it must be a fraction of the load duration and of sufficient resolution to capture the variation in time of the load. For the response, it should be a fraction of the shortest natural period of vibration that is thought to influence the response, that is, $\Delta t < T_{\min}$, which can easily be estimated.
- Decide on the number of points in the Fourier transform, $N \geq \mathrm{round}(T/\Delta t)$. where "round" is the rounding function that changes the fraction into an integer. Many modern implementations of the FFT work just as well with an arbitrary number of points, so that powers of two are no longer necessary, for example in Matlab. Thus, no trailing zeros are necessary.
- Apply the method and obtain the response function, but under no circumstances should you try to interpolate the transfer functions.

6.7 Fundamentals of Finite Elements

We begin with a brief review of the numerical integration methods widely used in finite-element analyses, and then go on with some fundamentals concepts in the theory together with explicit expressions for the stiffness and mass matrices for some commonly used finite elements.

6.7.1 Gaussian Quadrature

Assume we wish to estimate numerically the integral

$$I = \int_a^b f(x)\,dx \tag{6.401}$$

We begin making an ansatz for $f(x)$ of the form

$$f(x) = \sum_{j=1}^n f_j L_j(x) + p(x)\sum_{k=0}^\infty \beta_k x^k \tag{6.402}$$

with arbitrary n and $f_j = f(x_j)$, with x_j, β_j being the as yet unknown, but appropriate integration points and weights. Also,

$$L_j(x) = \frac{\prod\limits_{\substack{k\neq j}}^n (x - x_k)}{\prod\limits_{\substack{k\neq j}}^n (x_j - x_k)} \tag{6.403}$$

is the Lagrangian polynomial of degree $n-1$. In addition,

$$p(x) = \prod_{k=1}^n (x - x_k) = (x - x_1)(x - x_2)\cdots(x - x_n) \tag{6.404}$$

is a complete polynomial of degree n. Observe that

$$L_i(x_j) = \delta_{ij} = \begin{cases} 1 & i = j \\ 0 & i \neq j \end{cases} \quad \text{and} \quad p(x_i) = 0 \tag{6.405}$$

which means that the Lagrangian polynomial vanishes at each of the $x = x_k \neq x_j$, and equals 1 when $x = x_j$. Hence

$$I = \int_a^b f(x)\,dx = \sum_{j=1}^n f_j \int_a^b L_j(x)\,dx + \sum_{k=0}^\infty \beta_k \int_a^b x^k p(x)\,dx \tag{6.406}$$

We now proceed to choose integration stations $x_1, x_2 \ldots x_n$ such that they satisfy the condition

$$\boxed{\int_a^b x^k p(x)\,dx = 0} \quad k = 0, 1, \ldots n-1 \tag{6.407}$$

This implies that the sought integral has the form

$$I = \int_a^b f(x)\,dx = \sum_{j=1}^n \alpha_j f_j + R_n \approx \sum_{j=1}^n \alpha_j f_j \tag{6.408}$$

where

$$\alpha_j = \int_a^b L_j(x)\,dx \quad = \text{integration weights} \tag{6.409}$$

$$R_n = \sum_{k=0}^{\infty} \beta_k \int_a^b x^k p(x)\, dx \qquad = \text{Residual} \tag{6.410}$$

The residue vanishes, that is, $R_n = 0$, whenever $f(x)$ is a polynomial of degree no higher that $2n-1$, because in that case, $\beta_k = 0$ for $k \geq 2n$. Hence, the above formula is *exact* for any polynomial of degree up to $2n-1$, even though the computation requires only n function values $f_j, j = 1,\dots n$.

Normalization

Setting

$$x = x_0 + \xi c = \frac{a+b}{2} + \xi \frac{b-a}{2} \qquad \begin{aligned} x_0 &= \tfrac{1}{2}(a+b) &= \text{midpoint} \\ c &= \tfrac{1}{2}(b-a) &= \text{half-length of interval} \end{aligned} \tag{6.411}$$

Hence

$$I = \int_a^b f(x)\, dx = c \int_{-1}^{+1} f(\xi)\, d\xi \tag{6.412}$$

so that

$$I = \tfrac{1}{2}(b-a) \sum_{j=1}^n w_j f(x_j), \qquad x_j = \tfrac{1}{2}(a+b) + \tfrac{1}{2}(b-a)\xi_j, \qquad w_j = \int_{-1}^{+1} L_j(\xi)\, d\xi \tag{6.413}$$

To focus ideas, we now proceed to determine the particular values of the integration points and weights when $n = 1$ and $n = 2$.
$n = 1$:

$$p(\xi) = \xi - \xi_1 \quad ` \quad L_1(\xi) = 1 \tag{6.414}$$

$$\int_{-1}^{+1} p(\xi)\, d\xi = \left(\tfrac{1}{2}\xi^2 - \xi_1 \xi\right)\big|_{-1}^{+1} = -2\xi_1 = 0 \qquad \xi_1 = 0 \qquad x_1 = \tfrac{1}{2}(a+b) \tag{6.415}$$

$$w_1 = \int_{-1}^{+1} L_1(\xi)\, d\xi = \xi\big|_{-1}^{+1} = 2 \tag{6.416}$$

Hence

$$I \approx \tfrac{1}{2}(b-a)\left[w_1 f(x_1)\right] = (b-a) f_1\left(\tfrac{1}{2}(b+a)\right) \tag{6.417}$$

$n = 2$:

$$p(x) = (x - x_1)(x - x_2) \tag{6.418}$$

$$\int_{-1}^{+1}(\xi - \xi_1)(\xi - \xi_2)\, d\xi = \int_a^b \left(\xi^2 - (\xi_1 + \xi_2)\xi + \xi_1\xi_2\right) d\xi = \tfrac{2}{3} + 2\xi_1\xi_2 = 0 \tag{6.419}$$

$$\int_{-1}^{+1}\xi(\xi-\xi_1)(\xi-\xi_2)d\xi=\int_a^b\left(\xi^2-(\xi_1+\xi_2)\xi^2+\xi_1\xi_2\xi\right)d\xi=-\tfrac{2}{3}(\xi_1+\xi_2)=0 \tag{6.420}$$

$$\text{Hence}\quad \xi_1\xi_2=-\tfrac{1}{3},\quad \xi_1+\xi_2=0\qquad \xi_1=-\frac{\sqrt{3}}{3},\quad \xi_2=\frac{\sqrt{3}}{3} \tag{6.421}$$

$$L_1(\xi)=\frac{\xi-\xi_2}{\xi_1-\xi_2}\qquad w_1=\int_{-1}^{+1}\frac{\xi-\xi_2}{\xi_1-\xi_2}d\xi=1 \tag{6.422}$$

$$L_2(\xi)=\frac{\xi-\xi_1}{\xi_2-\xi_1}\qquad w_2=\int_{-1}^{+1}\frac{\xi-\xi_1}{\xi_2-\xi_1}d\xi=1 \tag{6.423}$$

$$x_1=\tfrac{1}{2}(a+b)-\tfrac{1}{2}(a+b)\tfrac{\sqrt{3}}{3}\qquad x_2=\tfrac{1}{2}(a+b)+\tfrac{1}{2}(a+b)\tfrac{\sqrt{3}}{3} \tag{6.424}$$

$$I\approx\tfrac{1}{2}(b-a)\left[w_1 f(x_1)+w_2 f(x_2)\right]=\tfrac{1}{2}(b-a)(f_1+f_2) \tag{6.425}$$

More generally, for $n\le 6$, we obtain the results given in Table 6.1.

Gaussian Quadrature

$$\boxed{I=\tfrac{1}{2}(b-a)\sum_{j=1}^{n}w_j f(x_j)}\qquad \boxed{x_j=\tfrac{1}{2}(a+b)+\tfrac{1}{2}(b-a)\xi_j}\qquad \boxed{w_j=\int_{-1}^{+1}L_j(\xi)d\xi} \tag{6.426}$$

Table 6.1. Gaussian stations and weights

N	ξ_j	w_j
1	0.	2.
2	$\pm 0.57735\,02691\,89626=\pm\tfrac{1}{3}\sqrt{3}$	1.
3	$\pm 0.77459\,66692\,41483=\pm\sqrt{\tfrac{3}{5}}$	$0.55555\,55555\,55556=\tfrac{5}{9}$
	0.	$0.88888\,88888\,88889=\tfrac{8}{9}$
4	$\pm 0.86113\,63115\,94053$	$0.34785\,48451\,37454$
	$=\pm\sqrt{\tfrac{3}{7}+\tfrac{2}{7}\sqrt{\tfrac{6}{5}}}$	$=\tfrac{1}{2}-\tfrac{1}{36}\sqrt{30}$
	$\pm 0.33998\,10435\,84856$	$0.65214\,51548\,62546$
	$=\pm\sqrt{\tfrac{3}{7}-\tfrac{2}{7}\sqrt{\tfrac{6}{5}}}$	$=\tfrac{1}{2}+\tfrac{1}{36}\sqrt{30}$
5	$\pm 0.90617\,98459\,38664$	$0.23692\,68850\,56186$
	$\pm 0.53846\,93101\,05683$	$0.47862\,86704\,99366$
	0.	$0.56888\,88888\,88889$
6	$\pm 0.93246\,95142\,03152$	$0.17132\,44923\,79170$
	$\pm 0.66120\,93864\,66265$	$0.36076\,15730\,48139$
	$\pm 0.23861\,91860\,83197$	$0.46791\,39345\,72691$

Example 1

$$I = \int_1^3 \ln x \, dx = x (\ln x - 1)\big|_1^3 = 3 \ln 3 - 2 = 1.2958 \qquad (6.427)$$

For $n = 2$, that is, a linear expansion, which is exact for polynomials up to order $2n - 1 = 3$, the result is

$$x_1 = 2 - \tfrac{\sqrt{3}}{3} = 1.42265, \qquad x_2 = 2 + \tfrac{\sqrt{3}}{3} = 2.57735 \qquad (6.428)$$

$$I \approx 1 \times \left[\ln x_1 + \ln x_2\right] = 1.2993 \qquad (3\% \text{ error}) \qquad (6.429)$$

For $n = 3$, that is, a quadratic expansion, which is exact for polynomials up to order $2n - 1 = 5$, we obtain

$$x_1 = 2 - \sqrt{\tfrac{3}{5}} = 1.22540, \qquad x_2 = 2, \qquad x_3 = 2 + \sqrt{\tfrac{3}{5}} = 2.77459 \qquad (6.430)$$

$$I \approx 1 \times \left[\tfrac{5}{9} \ln x_1 + \tfrac{8}{9} \ln x_2 + \tfrac{5}{9} \ln x_3\right] = 1.2960 \qquad (0.2\ \% \text{ error}) \qquad (6.431)$$

Example 2
Similar to Example 2, but with different interval:

$$I = \int_0^2 \ln x \, dx = x (\ln x - 1)\big|_0^2 = 2 (\ln 2 - 1) = -0.6137 \qquad (6.432)$$

Observe that the integrand is *singular* at the left end of the integration interval. The numerical results are as follows:

n	I	*Error* (%)
2	−0.40546	34
3	−0.50905	17
4	−0.55076	10

Figure 6.23 shows the error as function of $1/n$, the red dashed line being from the tabulation above, and the black line for a fitted cubic parabola. In principle, when $n \to \infty$, one should get "exact" results, but of course this neglects errors due to the finite precision. Still, the plot seems to show that even with $n = 10$ Gaussian points, the error would still be on the order of 2%. There are ways to improve the integration by changing the criterion for the weights, as described in the well-known book *Numerical Recipes*.

6.7.2 Integration in the Plane

By and large, formulas for numerical integration over two- and three-dimensional domains are not as well developed as those for one-dimensional domains. Still, one can use repeatedly the formulas for 1-D domains, as shown next, or use specially developed quadrature formulas for specific shapes, such as triangles or tetrahedra.

Consider a function $z = f(x, y)$ that we wish to integrate over some area in the 2-D plane. Toward this purpose, we consider it given in terms of curvilinear coordinates ξ, η, as shown in Figure 6.24.

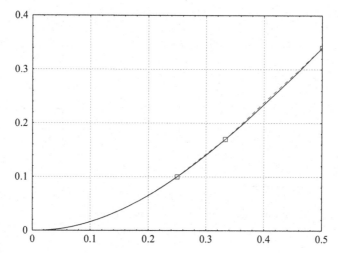

Figure 6.23. Error when using Gaussian quadrature to integrate a logarithmic function.

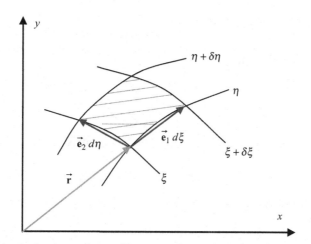

Figure 6.24. Integration in the plane using curvilinear coordinates.

The position vector to some arbitrary point is then

$$\vec{\mathbf{r}} = x\,\hat{\mathbf{i}} + y\,\hat{\mathbf{j}}, \qquad d\vec{\mathbf{r}} = dx\,\hat{\mathbf{i}} + dy\,\hat{\mathbf{j}} \tag{6.433}$$

In curvilinear coordinates with base vectors $\vec{\mathbf{e}}_1, \vec{\mathbf{e}}_2$

$$\vec{\mathbf{e}}_1 d\xi = \left(\frac{\partial x}{\partial \xi}\hat{\mathbf{i}} + \frac{\partial y}{\partial \xi}\hat{\mathbf{j}}\right)d\xi, \qquad \vec{\mathbf{e}}_2 d\eta = \left(\frac{\partial x}{\partial \eta}\hat{\mathbf{i}} + \frac{\partial y}{\partial \eta}\hat{\mathbf{j}}\right)d\eta \tag{6.434}$$

The elementary area contained by neighboring curves is then

$$dA = \left(\vec{\mathbf{e}}_1 \times \vec{\mathbf{e}}_1 \cdot \hat{\mathbf{k}}\right)d\xi d\eta \tag{6.435}$$

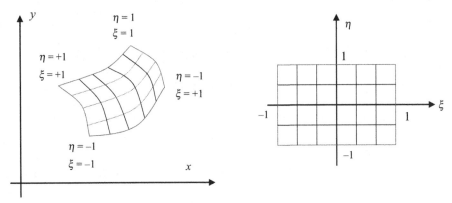

Figure 6.25. Mapping a curvilinear quadrilateral into a normalized rectangle.

with

$$\vec{e}_1 \times \vec{e}_1 \cdot \hat{\mathbf{k}} = \begin{vmatrix} \dfrac{\partial x}{\partial \xi} & \dfrac{\partial y}{\partial \xi} & 0 \\[6pt] \dfrac{\partial x}{\partial \eta} & \dfrac{\partial y}{\partial \eta} & 0 \\[6pt] 0 & 0 & 1 \end{vmatrix} = \begin{vmatrix} \dfrac{\partial x}{\partial \xi} & \dfrac{\partial y}{\partial \xi} \\[6pt] \dfrac{\partial x}{\partial \eta} & \dfrac{\partial y}{\partial \eta} \end{vmatrix} = |\mathbf{J}| \qquad = \text{the Jacobian} \tag{6.436}$$

$$\mathbf{J} = \left\{ \begin{matrix} \dfrac{\partial x}{\partial \xi} & \dfrac{\partial y}{\partial \xi} \\[6pt] \dfrac{\partial x}{\partial \eta} & \dfrac{\partial y}{\partial \eta} \end{matrix} \right\} = \mathbf{J}(\xi, \eta) \qquad = \text{Jacobi matrix} \tag{6.437}$$

Hence, applying Gaussian quadrature, the integral over some domain can be approximated by the double sum

$$\iint f(x,y)\,dx\,dy = \iint f(\xi,\eta)|\mathbf{J}|\,d\xi\,d\eta \approx \sum_{i=1}^{n}\sum_{j=1}^{n} w_i w_j J_{ij} f_{ij} \tag{6.438}$$

with

$$J_{ij} = \left| \mathbf{J}(\xi_i, \eta_j) \right| \tag{6.439}$$

The mapping of the curved space to a normalized space is as shown in Figure 6.25.

(a) Integral over a Rectangular Area (Figure 6.26)

$$x = \tfrac{1}{2}(a_1 + b_1) + \tfrac{1}{2}(b_1 - a_1)\xi \qquad\qquad y = \tfrac{1}{2}(a_2 + b_2) + \tfrac{1}{2}(b_2 - a_2)\eta \tag{6.440}$$

$$\frac{\partial x}{\partial \xi} = \tfrac{1}{2}(b_1 - a_1) \qquad\qquad \frac{\partial y}{\partial \xi} = 0 \tag{6.441}$$

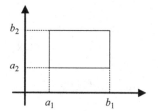

Figure 6.26. Rectangular area.

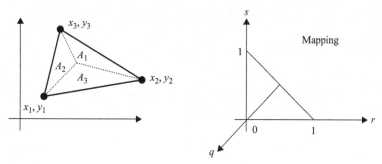

Figure 6.27 Mapping of triangular area.

$$\frac{\partial x}{\partial \eta} = 0 \qquad\qquad \frac{\partial y}{\partial \eta} = \frac{1}{2}(b_2 - a_2) \tag{6.442}$$

$$|\mathbf{J}| = \frac{1}{4}(b_1 - a_1)(b_2 - a_2) = \frac{1}{4}A \qquad A = \text{area} \tag{6.443}$$

$$I = \iint f(x,y)\,dx\,dy = \frac{1}{4}A \iint f(\xi,\eta)\,d\xi\,d\eta = \frac{1}{4}A \sum_{i=1}^{m}\sum_{j=1}^{n} w_i w_j f_{ij} \tag{6.444}$$

(b) Integral over a Triangular Area (Figure 6.27)

Triangular Coordinates

Consider a point x, y within the triangle and connect it with each of the vertices. This separates the triangle into subtriangles with areas A_1, A_2, A_3 such that $A = A_1 + A_2 + A_3$. Define

$$q = \frac{A_1}{A} \qquad r = \frac{A_2}{A} \qquad s = \frac{A_3}{A} \qquad q + r + s = 1 \tag{6.445}$$

The arbitrary point x, y in the triangle can then be expressed as

$$x = qx_1 + rx_2 + sx_3 = x_1 + (x_2 - x_1)r + (x_3 - x_1)s \tag{6.446}$$

$$y = qy_1 + ry_2 + sy_3 = y_1 + (y_2 - y_1)r + (y_3 - y_1)s \tag{6.447}$$

So

$$J = |\mathbf{J}| = (x_2 - x_1)(y_3 - y_1) - (x_3 - x_1)(y_2 - y_1)$$
$$= 2A \tag{6.448}$$

With the affine transformation of the triangle mapped into the "standard" triangle, the integration can be expressed as

$$I = \iint f(x,y)\,dx\,dy = I = \int_0^1 \int_0^{1-r} f(r,s) J\,dr\,ds = 2A \int_0^1 \int_0^{1-r} f(r,s)\,dr\,ds \tag{6.449}$$

which in turn can be reduced to

$$I = \iint f(x,y)\,dx\,dy = (2A)\tfrac{1}{2}\sum_{j=1}^n w_j f(r_j,s_j) = A\sum_{j=1}^n w_j f(r_j,s_j) \tag{6.450}$$

where the r_j, s_j and w_j are the coordinates of the Gaussian points and weights obtained from Table 6.2, and $f(r_j,s_j) = f(x_{gj},y_{gj})$, with x_{gj},y_{gj} inferred from the above formulas.

$$\boxed{I = A\sum_{j=1}^n w_j f(r_j,s_j)} \tag{6.451}$$

$$\boxed{x_j = x_1 + (x_2 - x_1)r_j + (x_3 - x_1)s_j} \qquad \boxed{y_j = y_1 + (y_2 - y_1)r_j + (y_3 - y_1)s_j}$$

(c) Curvilinear Triangle (Figure 6.28)

This case is similar to the previous one, even if somewhat more complicated. It too can be solved by means of triangular coordinates, expressing the coordinates at some arbitrary point in terms of the nodal coordinates at the vertices and edges as

$$x = q(2q-1)x_1 + r(2r-1)x_2 + s(2s-1)x_3 + 4(qr\,x_4 + rs\,x_5 + sq\,x_6) \tag{6.452}$$

$$y = q(2q-1)y_1 + r(2r-1)y_2 + s(2s-1)y_3 + 4(qr\,y_4 + rs\,y_5 + sq\,y_6) \tag{6.453}$$

Hence, with $q + r + s = 1$, we have then

$$\frac{\partial x}{\partial r} = (1-4q)x_1 + (4r-1)x_2 + 4(q-r)x_4 + 4s\,x_5 - 4s\,x_6 \tag{6.454}$$

$$\frac{\partial y}{\partial r} = (1-4q)y_1 + (4r-1)y_2 + 4(q-r)y_4 + 4s\,y_5 - 4s\,y_6 \tag{6.455}$$

$$\frac{\partial x}{\partial s} = (1-4q)x_1 + (4s-1)x_3 - 4r\,x_4 + 4r\,x_5 + 4(q-s)x_6 \tag{6.456}$$

$$\frac{\partial y}{\partial s} = (1-4q)y_1 + (4s-1)y_3 - 4r\,y_4 + 4r\,y_5 + 4(q-s)y_6 \tag{6.457}$$

Table 6.2. Gaussian quadrature over a triangular area

j	r_j	s_j	w_j
1	$0.16666\ 66666\ 66667 = \frac{1}{6}$	r_1	$0.33333\ 33333\ 33333$
2	$0.66666\ 66666\ 66667 = \frac{2}{3}$	r_1	w_1
$n = 3$	r_1	r_2	w_1
1	$0.10128\ 65073\ 23456$	r_1	$0.12593\ 91805\ 44827$
2	$0.79742\ 69853\ 53087$	r_1	w_1
3	r_1	r_2	w_1
4	$0.47014\ 20641\ 05115$	r_6	$0.13239\ 41527\ 88506$
5	r_4	r_4	w_4
6	$0.05971\ 58717\ 89770$	r_4	w_4
$n = 7$	$0.33333\ 33333\ 33333$	r_7	$0.22500\ 00000\ 00000$
1	$0.06513\ 01029\ 02216$	r_1	$0.05334\ 72356\ 08839$
2	$0.86973\ 97941\ 95568$	r_1	w_1
3	r_1	r_2	w_1
4	$0.31286\ 54960\ 04875$	r_6	$0.07711\ 37608\ 90257$
5	$0.63844\ 41885\ 69809$	r_4	w_4
6	$0.04869\ 03154\ 25316$	r_5	w_4
7	r_5	r_6	w_4
8	r_4	r_5	w_4
9	r_6	r_4	w_4
10	$0.26034\ 59660\ 79038$	r_{10}	$0.17561\ 52574\ 33204$
11	$0.47930\ 80678\ 41923$	r_{10}	w_{10}
12	r_{10}	r_{11}	w_{10}
$n = 13$	$0.33333\ 33333\ 33333$	r_{13}	$-0.14957\ 00444\ 67670$

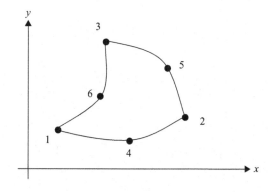

Figure 6.28. Curvilinear triangular element.

$$J = |\mathbf{J}| = \frac{\partial x}{\partial r}\frac{\partial y}{\partial s} - \frac{\partial x}{\partial s}\frac{\partial y}{\partial r} \tag{6.458}$$

$$I = \iint f(x,y)\,dx\,dy = \tfrac{1}{2}\sum_{j=1}^{n} w_j\, J_j\, f_j \tag{6.459}$$

$$J_j = J(r_j, s_j), \qquad f_j = f(r_j, s_j) \tag{6.460}$$

where again the weights are taken from Table 6.2.

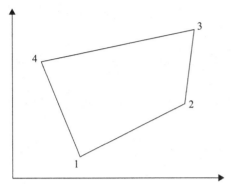

Figure 6.29. Quadrilateral element.

(d) Quadrilateral (Figure 6.29)

$$x = \tfrac{1}{4}\big[(1-\xi)(1-\eta)x_1 + (1+\xi)(1-\eta)x_2 + (1+\xi)(1+\eta)x_3 + (1-\xi)(1+\eta)x_4\big] = \mathbf{\Phi x} \quad (6.461)$$

$$y = \tfrac{1}{4}\big[(1-\xi)(1-\eta)y_1 + (1+\xi)(1-\eta)y_2 + (1+\xi)(1+\eta)y_3 + (1-\xi)(1+\eta)y_4\big] = \mathbf{\Phi y} \quad (6.462)$$

where

$$\mathbf{\Phi} = \tfrac{1}{4}\big[(1-\xi)(1-\eta) \quad (1+\xi)(1-\eta) \quad (1+\xi)(1+\eta) \quad (1-\xi)(1+\eta)\big] \quad (6.463)$$

$$\mathbf{x} = \big[x_1 \quad x_2 \quad x_3 \quad x_4\big]^T, \qquad \mathbf{y} = \big[y_1 \quad y_2 \quad y_3 \quad y_4\big]^T \quad (6.464)$$

Hence

$$\frac{\partial x}{\partial \xi} = \tfrac{1}{4}\big[-(1-\eta)x_1 + (1-\eta)x_2 + (1+\eta)x_3 - (1+\eta)x_4\big] = \mathbf{\Phi}_\xi \mathbf{x} \quad (6.465)$$

$$\frac{\partial y}{\partial \xi} = \tfrac{1}{4}\big[-(1-\eta)y_1 + (1-\eta)y_2 + (1+\eta)y_3 - (1+\eta)y_4\big] = \mathbf{\Phi}_\xi \mathbf{y} \quad (6.466)$$

$$\frac{\partial x}{\partial \eta} = \tfrac{1}{4}\big[-(1-\xi)x_1 - (1+\xi)x_2 + (1+\xi)x_3 + (1-\xi)x_4\big] = \mathbf{\Phi}_\eta \mathbf{x} \quad (6.467)$$

$$\frac{\partial y}{\partial \eta} = \tfrac{1}{4}\big[-(1-\xi)y_1 - (1+\xi)y_2 + (1+\xi)y_3 + (1-\xi)y_4\big] = \mathbf{\Phi}_\eta \mathbf{y} \quad (6.468)$$

$$J = |\mathbf{J}| = \frac{\partial x}{\partial \xi}\frac{\partial y}{\partial \eta} - \frac{\partial x}{\partial \eta}\frac{\partial y}{\partial \xi} \quad (6.469)$$

$$I = \iint f(x,y)\,dx\,dy = \sum_{i=1}^{m}\sum_{j=1}^{n} w_i w_j J_{ij} f_{ij} \quad (6.470)$$

Table 6.3. Quadratic interpolation functions

	8-noded element	9-noded element
φ_1	$-\frac{1}{4}(\xi-1)(\eta-1)(1+\xi+\eta)$	$\frac{1}{4}(\xi-1)(\eta-1)\xi\eta$
φ_2	$\frac{1}{4}(\xi+1)(\eta-1)(1-\xi+\eta)$	$\frac{1}{4}(\xi+1)(\eta-1)\xi\eta$
φ_3	$\frac{1}{4}(\xi+1)(\eta+1)(\xi+\eta-1)$	$\frac{1}{4}(\xi+1)(\eta+1)\xi\eta$
φ_4	$\frac{1}{4}(\xi-1)(\eta+1)(1+\xi-\eta)$	$\frac{1}{4}(\xi-1)(\eta+1)\xi\eta$
φ_5	$\frac{1}{2}(1-\xi^2)(1-\eta)$	$\frac{1}{2}(1-\xi^2)(\eta-1)\eta$
φ_6	$\frac{1}{2}(1+\xi)(1-\eta^2)$	$\frac{1}{2}(\xi+1)(1-\eta^2)\xi$
φ_7	$\frac{1}{2}(1-\xi^2)(1+\eta)$	$\frac{1}{2}(1-\xi^2)(\eta+1)\eta$
φ_8	$\frac{1}{2}(1-\xi)(1-\eta^2)$	$\frac{1}{2}(\xi-1)(1-\eta^2)\xi$
φ_9	–	$(1-\xi^2)(1-\eta^2)$

where the Gaussian points and their weights are taken from Table 6.1.

(e) Curvilinear Quadrilateral

$$x = \sum_{k=1}^{N} \varphi_k x_k = \mathbf{\Phi}\mathbf{x}, \qquad y = \sum_{k=1}^{N} \varphi_k y_k = \mathbf{\Phi}\mathbf{y} \qquad \varphi_k(\xi,\eta) = \text{interpolation functions} \quad (6.471)$$

Observe that

$$\sum_{k=1}^{N} \varphi_k = 1 \qquad \text{and} \qquad \varphi_k(\xi_i,\eta_j) = \delta_{ij} = \begin{cases} 1 & i=j \\ 0 & i \neq j \end{cases} \tag{6.472}$$

where the indices i, j refer to the nodal numbers and their coordinates. The partial derivatives are shown in Tables 6.3, 6.4 and 6.5.

The rest of the formulation is as before.

Inadmissible Shapes

The areas of the elements to be integrated must be convex and their edges cannot intersect. Figure 6.30 shows examples of shapes that are not acceptable:

6.7.3 Finite Elements via Principle of Virtual Displacements

Consider an elastic body subjected to dynamic forces. With reference to the principle of virtual displacements, it is known that the following relationship applies:

$$\iiint_V \delta\mathbf{u}^T \ddot{\mathbf{u}} \rho \, dV + \iiint_V \delta\boldsymbol{\varepsilon}^T \mathbf{D}\boldsymbol{\varepsilon} \, dV = \iiint_V \delta\mathbf{u}^T \mathbf{b} \, dV + \iint_A \delta\mathbf{u}^T \mathbf{q} \, dA \tag{6.473}$$

where $\delta\mathbf{u}, \delta\boldsymbol{\varepsilon}$ are the virtual displacements and strain at a point; $\mathbf{u}, \boldsymbol{\varepsilon}$ are the actual displacement and strain vectors at some point and time; $\mathbf{D}$ is the constitutive matrix; and the right-hand side represents the virtual work done by the external forces in the body and on the surface.

Table 6.4. Derivatives of quadratic interpolation functions with respect to ξ

	Φ_ξ	
	8-noded element	9-noded element
$\varphi_{1,\xi}$	$\frac{1}{4}(1-\eta)(2\xi+\eta)$	$\frac{1}{4}(1-2\xi)(1-\eta)\eta$
$\varphi_{2,\xi}$	$\frac{1}{4}(1-\eta)(2\xi-\eta)$	$-\frac{1}{4}(1+2\xi)(1-\eta)\eta$
$\varphi_{3,\xi}$	$\frac{1}{4}(1+\eta)(2\xi+\eta)$	$\frac{1}{4}(1+2\xi)(1+\eta)\eta$
$\varphi_{4,\xi}$	$\frac{1}{4}(1+\eta)(2\xi-\eta)$	$-\frac{1}{4}(1-2\xi)(1+\eta)\eta$
$\varphi_{5,\xi}$	$-\xi(1-\eta)$	$\xi\eta(1-\eta)$
$\varphi_{6,\xi}$	$\frac{1}{2}(1-\eta^2)$	$\frac{1}{2}(1+2\xi)(1-\eta^2)$
$\varphi_{7,\xi}$	$-\xi(1+\eta)$	$-\xi\eta(1+\eta)$
$\varphi_{8,\xi}$	$-\frac{1}{2}(1-\eta^2)$	$-\frac{1}{2}(1-2\xi)(1-\eta^2)$
$\varphi_{9,\xi}$	$-$	$-2\xi(1-\eta^2)$

Table 6.5. Derivatives of quadratic interpolation functions with respect to η

	Φ_η	
	8-noded element	9-noded element
$\varphi_{1,\eta}$	$\frac{1}{4}(1-\xi)(\xi+2\eta)$	$\frac{1}{4}\xi(1-\xi)(1-2\eta)$
$\varphi_{2,\eta}$	$-\frac{1}{4}(1+\xi)(\xi-2\eta)$	$-\frac{1}{4}\xi(1+\xi)(1-2\eta)$
$\varphi_{3,\eta}$	$\frac{1}{4}(1+\xi)(\xi+2\eta)$	$\frac{1}{4}\xi(1+\xi)(1+2\eta)$
$\varphi_{4,\eta}$	$-\frac{1}{4}(1-\xi)(\xi-2\eta)$	$-\frac{1}{4}\xi(1-\xi)(1+2\eta)$
$\varphi_{5,\eta}$	$-\frac{1}{2}(1-\xi^2)$	$-\frac{1}{2}(1-\xi^2)(1-2\eta)$
$\varphi_{6,\eta}$	$-(1+\xi)\eta$	$-\xi(1+\xi)\eta$
$\varphi_{7,\eta}$	$\frac{1}{2}(1-\xi^2)$	$\frac{1}{2}(1-\xi^2)(1+2\eta)$
$\varphi_{8,\eta}$	$-(1-\xi)\eta$	$\xi(1-\xi)\eta$
$\varphi_{9,\eta}$	$-$	$-2(1-\xi^2)\eta$

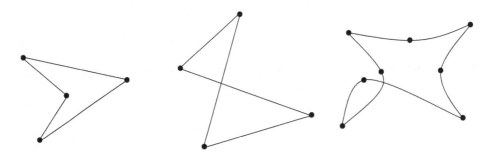

Figure 6.30. Inadmissible shapes.

Applying this principle to a small, finite domain defined by some arbitrarily shaped finite element, we express the displacements $\mathbf{u}$ within the finite element in terms of the displacements at the nodes $\mathbf{U}$ as

$$\mathbf{u} = \mathbf{\Phi U}, \quad \mathbf{\Phi} = \mathbf{\Phi}(x, y), \quad \mathbf{U} = \begin{Bmatrix} \mathbf{u}_1 \\ \vdots \\ \mathbf{u}_N \end{Bmatrix} = \text{nodal displacements} \tag{6.474}$$

where $\mathbf{\Phi}$ is the matrix of interpolation functions. Also, the strain vector can be expressed in terms of the displacements as

$$\varepsilon = \mathbf{Lu} = \mathbf{L\Phi U} = \mathbf{CU}, \quad \mathbf{C} = \mathbf{L\Phi} \tag{6.475}$$

where $\mathbf{L}$ is a differential operator matrix which can readily be constructed. For example, in Cartesian coordinates and for plane-strain conditions, this matrix is of the form

$$\mathbf{L} = \begin{Bmatrix} \frac{\partial}{\partial x} & 0 \\ 0 & \frac{\partial}{\partial y} \\ \frac{\partial}{\partial y} & \frac{\partial}{\partial x} \end{Bmatrix}, \quad \mathbf{D} = \begin{Bmatrix} \lambda + 2\mu & \lambda & \lambda & 0 & 0 & 0 \\ \lambda & \lambda + 2\mu & \lambda & 0 & 0 & 0 \\ \lambda & \lambda & \lambda + 2\mu & 0 & 0 & 0 \\ 0 & 0 & 0 & \mu & 0 & 0 \\ 0 & 0 & 0 & 0 & \mu & 0 \\ 0 & 0 & 0 & 0 & 0 & \mu \end{Bmatrix} \tag{6.476}$$

where λ, μ are the Lamé constants. Hence

$$\delta \varepsilon^T = \delta \mathbf{u}^T \mathbf{L}^T = \delta \mathbf{U}^T \mathbf{C}^T \tag{6.477}$$

So

$$\delta \mathbf{U}^T \left[\iiint_{Elem} \mathbf{\Phi}^T \mathbf{\Phi} \rho \, dV \right] \ddot{\mathbf{U}} + \delta \mathbf{U}^T \left[\iiint_V \mathbf{C}^T \mathbf{D} \mathbf{C} \, dV \right] \mathbf{U} = \delta \mathbf{U}^T \left[\iiint_V \mathbf{\Phi}^T \mathbf{b} \, dV + \iint_A \mathbf{\Phi}^T \mathbf{q} \, dA \right] \tag{6.478}$$

Requiring this expression to be valid for any arbitrary virtual displacement $\delta \mathbf{U}$, we are led to the equation

$$\mathbf{M}_{elem} \ddot{\mathbf{U}} + \mathbf{K}_{elem} \mathbf{U} = \mathbf{P} \tag{6.479}$$

where

$$\mathbf{M}_{elem} = \iiint_{Elem} \mathbf{\Phi}^T \mathbf{\Phi} \rho \, dV \qquad = \text{the element's consistent mass matrix} \tag{6.480}$$

$$\mathbf{K}_{elem} = \iiint_V \mathbf{C}^T \mathbf{D} \mathbf{C} \, dV \qquad = \text{the element's stiffness matrix} \tag{6.481}$$

$$\mathbf{P} = \iiint_V \mathbf{\Phi}^T \mathbf{b} \, dV + \iint_A \mathbf{\Phi}^T \mathbf{q} \, dA \qquad = \text{the consistent nodal load vector} \tag{6.482}$$

Thus, we have obtained the means to derive the elastic and inertia properties of the finite element in terms of its geometry and number of nodes.

The interpolation polynomials $\boldsymbol{\Phi} = \{\varphi_k\}$ must satisfy the following requirements.

(a) Consistency

The interpolation functions satisfy $\varphi_k(x_i, y_j) = \delta_{ij}$, which implies that

$$\mathbf{x}_k \equiv \boldsymbol{\Phi}(\mathbf{x}_k)\mathbf{X}, \qquad \mathbf{X} = \begin{Bmatrix} \mathbf{x}_1 \\ \vdots \\ \mathbf{x}_N \end{Bmatrix} \tag{6.483}$$

and

$$\mathbf{u}_k \equiv \boldsymbol{\Phi}(\mathbf{x}_k)\mathbf{U} \tag{6.484}$$

This means that the interpolated displacements must coincide with the nodal displacements whenever $\mathbf{x} = \mathbf{x}_k$, that is, when evaluated at a node.

(b) Conformity

The displacements must be continuous along any of the element's edges, and they can depend only on the displacements of the nodes that lie on that edge. The continuity condition applies also to the first spatial derivatives if these appear as *geometric* boundary conditions (beam, plates, shells). Elements that fail these conditions are said to be *nonconforming* or *incompatible* (still, they are often used for plate bending).

(c) Rigid Body Test

The element must be able to execute rigid body translations and rotations without eliciting elastic forces, that is, $\mathbf{K}_{\text{elem}}\mathbf{U}_{\text{rigid body}} = \mathbf{0}$. In general, this is satisfied if

$$\mathbf{L}\mathbf{u}_{\text{rigid body}} = \mathbf{L}\boldsymbol{\Phi}\mathbf{U}_{\text{rigid body}} = \mathbf{0} \tag{6.485}$$

at any arbitrary point within the element.

(d) Convergence (Patch Test)

As the elements in the finite element grid are made to be smaller and smaller, the strains within each element begin to approach a constant value. Hence, the interpolation functions must be able to represent such a constant strain state. This ability is verified by means of the "patch test," during which a body discretized into finite elements is subjected to a deformation state that causes constant strain everywhere. An example is shown in Figure 6.31.

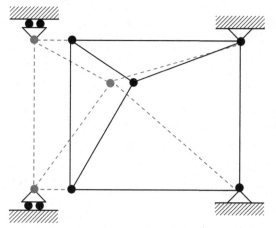

Figure 6.31. Patch test.

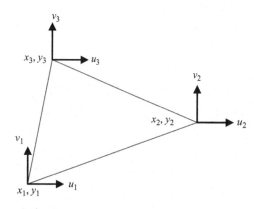

Figure 6.32. Triangular Plane Strain Element.

6.7.4 Plate Stretching Elements (Plane Strain)

(a) Triangular Elements (Figure 6.32)

We begin with a linear ansatz for the displacements of the form

$$u = a_1 + a_2 x + a_3 y = \boldsymbol{\psi}\, \mathbf{a}, \qquad \boldsymbol{\psi} = \begin{bmatrix} 1 & x & y \end{bmatrix} \tag{6.486}$$

$$v = b_1 + b_2 x + b_3 y = \boldsymbol{\psi}\, \mathbf{b} \tag{6.487}$$

with as yet undetermined coefficients a_j, b_j. These equations must be valid at each of the three vertices, which implies

$$\begin{Bmatrix} u_1 \\ u_2 \\ u_3 \end{Bmatrix} = \begin{bmatrix} 1 & x_1 & y_1 \\ 1 & x_2 & y_2 \\ 1 & x_3 & y_3 \end{bmatrix} \begin{Bmatrix} a_1 \\ a_2 \\ a_3 \end{Bmatrix}, \qquad \begin{Bmatrix} v_1 \\ v_2 \\ v_3 \end{Bmatrix} = \begin{bmatrix} 1 & x_1 & y_1 \\ 1 & x_2 & y_2 \\ 1 & x_3 & y_3 \end{bmatrix} \begin{Bmatrix} b_1 \\ b_2 \\ b_3 \end{Bmatrix} \tag{6.488}$$

or in matrix form

$$\mathbf{U} = \boldsymbol{\Psi}\mathbf{a}, \qquad \mathbf{V} = \boldsymbol{\Psi}\mathbf{b}, \qquad \boldsymbol{\Psi} = \begin{Bmatrix} 1 & x_1 & y_1 \\ 1 & x_2 & y_2 \\ 1 & x_3 & y_3 \end{Bmatrix} = \begin{Bmatrix} \boldsymbol{\psi}_1 \\ \boldsymbol{\psi}_2 \\ \boldsymbol{\psi}_3 \end{Bmatrix} \tag{6.489}$$

Hence

$$\mathbf{a} = \boldsymbol{\Psi}^{-1}\mathbf{U} \qquad \mathbf{b} = \boldsymbol{\Psi}^{-1}\mathbf{V} \tag{6.490}$$

Also, the analytical inverse of $\boldsymbol{\Psi}$ is

$$\boldsymbol{\Psi}^{-1} = \frac{1}{2A} \begin{Bmatrix} x_2 y_3 - x_3 y_2 & x_3 y_1 - x_1 y_3 & x_1 y_2 - x_2 y_1 \\ y_2 - y_3 & y_3 - y_1 & y_1 - y_2 \\ x_3 - x_2 & x_1 - x_3 & x_2 - x_1 \end{Bmatrix} \tag{6.491}$$

where

$$2A = \sum_{j=1}^{3} \left(x_j y_{j+1} - x_{j+1} y_j \right) = x_1\left(y_2 - y_3 \right) + x_2\left(y_3 - y_1 \right) + x_3\left(y_1 - y_2 \right) \tag{6.492}$$

is twice the area of the triangle (with $x_4 \equiv x_1, y_4 \equiv y_1$). Hence

$$\mathbf{u} = \begin{Bmatrix} \boldsymbol{\psi} & \mathbf{0} \\ \mathbf{0} & \boldsymbol{\psi} \end{Bmatrix} \begin{Bmatrix} \mathbf{a} \\ \mathbf{b} \end{Bmatrix} = \begin{Bmatrix} \boldsymbol{\psi} & \mathbf{0} \\ \mathbf{0} & \boldsymbol{\psi} \end{Bmatrix} \begin{Bmatrix} \boldsymbol{\Psi}^{-1}\mathbf{U} \\ \boldsymbol{\Psi}^{-1}\mathbf{V} \end{Bmatrix} = \begin{Bmatrix} \boldsymbol{\Gamma} & \mathbf{0} \\ \mathbf{0} & \boldsymbol{\Gamma} \end{Bmatrix} \begin{Bmatrix} \mathbf{U} \\ \mathbf{V} \end{Bmatrix} \tag{6.493}$$

where

$$\boldsymbol{\Gamma} = \boldsymbol{\psi}\,\boldsymbol{\Psi}^{-1} = \tfrac{1}{2A}\begin{bmatrix} \gamma_1 & \gamma_2 & \gamma_3 \end{bmatrix} \tag{6.494}$$

$$\gamma_1 = \left(x - x_3 \right)\left(y_2 - y_3 \right) + \left(y - y_3 \right)\left(x_3 - x_2 \right) \tag{6.495}$$

$$\gamma_2 = \left(x - x_1 \right)\left(y_3 - y_1 \right) + \left(y - y_1 \right)\left(x_1 - x_3 \right) \tag{6.496}$$

$$\gamma_3 = \left(x - x_2 \right)\left(y_1 - y_2 \right) + \left(y - y_2 \right)\left(x_2 - x_1 \right) \tag{6.497}$$

Since each term in $\gamma_1, \gamma_2, \gamma_3$ is formed with differences of coordinates, it follows that we could add any arbitrary shift to the origin without affecting these differences. This means that the product $\boldsymbol{\Gamma} = \boldsymbol{\psi}\,\boldsymbol{\Psi}^{-1}$ does not depend on the placement of the origin of coordinates.

The strain vector is now

$$\boldsymbol{\varepsilon} = \mathbf{L}\mathbf{u} = \begin{Bmatrix} \frac{\partial}{\partial x} & 0 \\ 0 & \frac{\partial}{\partial y} \\ \frac{\partial}{\partial y} & \frac{\partial}{\partial x} \end{Bmatrix} \begin{bmatrix} \boldsymbol{\Gamma} & \mathbf{0} \\ \mathbf{0} & \boldsymbol{\Gamma} \end{bmatrix} \begin{Bmatrix} \mathbf{U} \\ \mathbf{V} \end{Bmatrix} = \boldsymbol{\Phi} \begin{Bmatrix} \mathbf{U} \\ \mathbf{V} \end{Bmatrix} \tag{6.498}$$

where

$$\boldsymbol{\Phi} = \frac{1}{2A} \begin{Bmatrix} y_2 - y_3 & y_3 - y_1 & y_1 - y_2 & 0 & 0 & 0 \\ 0 & 0 & 0 & x_3 - x_2 & x_1 - x_3 & x_2 - x_1 \\ x_3 - x_2 & x_1 - x_3 & x_2 - x_1 & y_2 - y_3 & y_3 - y_1 & y_1 - y_2 \end{Bmatrix} = \frac{1}{2A} \begin{Bmatrix} \mathbf{Y} & \mathbf{0} \\ \mathbf{0} & \mathbf{X} \\ \mathbf{X} & \mathbf{Y} \end{Bmatrix} \tag{6.499}$$

where

$$\mathbf{X} = \{x_3 - x_2 \quad x_1 - x_3 \quad x_2 - x_1\}, \quad \mathbf{Y} = \{y_2 - y_3 \quad y_3 - y_1 \quad y_1 - y_2\} \tag{6.500}$$

With

$$\mathbf{D} = \begin{cases} \lambda + 2\mu & \lambda & 0 \\ \lambda & \lambda + 2\mu & 0 \\ 0 & 0 & \mu \end{cases} \tag{6.501}$$

the element stiffness matrix is then

$$\mathbf{K} = \iint_A \mathbf{\Phi}^T \mathbf{D} \mathbf{\Phi} \, dA = A \, \mathbf{\Phi}^T \mathbf{D} \mathbf{\Phi} = (\lambda + 2\mu) \begin{cases} \mathbf{Y}^T \mathbf{Y} & \mathbf{O} \\ \mathbf{O} & \mathbf{X}^T \mathbf{X} \end{cases} + \mu \begin{cases} \mathbf{X}^T \mathbf{X} & \mathbf{X}^T \mathbf{Y} \\ \mathbf{Y}^T \mathbf{X} & \mathbf{Y}^T \mathbf{Y} \end{cases} \tag{6.502}$$

It can be shown that this stiffness matrix satisfies both the rigid body condition and the ability to model uniform strain (i.e., the patch test); also, it is compatible.

On the other hand, the consistent mass matrix is

$$\mathbf{M} = \rho \iint_A \begin{cases} \mathbf{\Gamma}^T \mathbf{\Gamma} & \mathbf{0} \\ \mathbf{0} & \mathbf{\Gamma}^T \mathbf{\Gamma} \end{cases} dA = \begin{cases} \mathbf{m} & \mathbf{0} \\ \mathbf{0} & \mathbf{m} \end{cases}, \quad \mathbf{m} = \frac{\rho A}{12} \begin{cases} 2 & 1 & 1 \\ 1 & 2 & 1 \\ 1 & 1 & 2 \end{cases} \tag{6.503}$$

(b) Rectangular Element

Consider a rectangular element whose sides are parallel to the *local* axes x, y. Like in the triangle, we start by assuming a displacement field, but somewhat more general:

$$u = a_1 + a_2 x + a_3 y + a_4 xy = \boldsymbol{\psi} \mathbf{a}, \quad \boldsymbol{\psi} = \begin{bmatrix} 1 & x & y & xy \end{bmatrix} \tag{6.504}$$

$$v = b_1 + b_2 x + b_3 y + b_4 xy = \boldsymbol{\psi} \mathbf{b} \tag{6.505}$$

Once again, this expansion must be satisfied at the vertices (nodes) of the rectangle, which requires the two conditions

$$\begin{Bmatrix} u_1 \\ u_2 \\ u_3 \\ u_4 \end{Bmatrix} = \begin{bmatrix} 1 & x_1 & y_1 & x_1 y_1 \\ 1 & x_2 & y_2 & x_2 y_2 \\ 1 & x_3 & y_3 & x_3 y_3 \\ 1 & x_4 & y_4 & x_4 y_4 \end{bmatrix} \begin{Bmatrix} a_1 \\ a_2 \\ a_3 \\ a_4 \end{Bmatrix}, \quad \begin{Bmatrix} v_1 \\ v_2 \\ v_3 \\ v_4 \end{Bmatrix} = \begin{bmatrix} 1 & x_1 & y_1 & x_1 y_1 \\ 1 & x_2 & y_2 & x_2 y_2 \\ 1 & x_3 & y_3 & x_3 y_3 \\ 1 & x_4 & y_4 & x_4 y_4 \end{bmatrix} \begin{Bmatrix} b_1 \\ b_2 \\ b_3 \\ b_4 \end{Bmatrix} \tag{6.506}$$

So

$$\mathbf{U} = \mathbf{\Psi} \mathbf{a}, \quad \mathbf{V} = \mathbf{\Psi} \mathbf{b}, \quad \mathbf{u}(x, t) = \boldsymbol{\psi} \mathbf{\Psi}^{-1} \mathbf{U}, \quad \mathbf{v}(x, t) = \boldsymbol{\psi} \mathbf{\Psi}^{-1} \mathbf{V} \tag{6.507}$$

Using the symbolic tool in MATLAB®, it can be shown that if we replace $x \to x - x_0$, $x_j \to x_j - x_0$ and $y \to y - y_0$, $y_j \to y_j - y_0$, the products above do not change; that is, they are independent of x_0, y_0. This implies that the formulation is independent of the choice for the origin of the coordinate system. In particular, we can choose the origin at the left lower corner of the rectangular element of size $a \times b$, so $0 \le x \le a, 0 \le y \le b$, in which case

$$
\Psi = \begin{Bmatrix} 1 & 0 & 0 & 0 \\ 1 & a & 0 & 0 \\ 1 & a & b & ab \\ 1 & 0 & b & 0 \end{Bmatrix}, \qquad \Psi^{-1} = \frac{1}{ab} \begin{Bmatrix} ab & 0 & 0 & 0 \\ -b & b & 0 & 0 \\ -a & 0 & 0 & a \\ 1 & -1 & 1 & -1 \end{Bmatrix}, \tag{6.508}
$$

and

$$
\Gamma = \psi \Psi^{-1} = \frac{1}{ab} \left[(a-x)(b-y) \quad x(b-y) \quad xy \quad (a-x)y \right] \tag{6.509}
$$

So

$$
\mathbf{m} = \rho \iint_A \Gamma^T \Gamma \, dA = \rho \frac{ab}{36} \begin{bmatrix} 4 & 2 & 1 & 2 \\ 2 & 4 & 2 & 1 \\ 1 & 2 & 4 & 2 \\ 2 & 1 & 2 & 4 \end{bmatrix}, \qquad \mathbf{M} = \begin{Bmatrix} \mathbf{m} & \mathbf{o} \\ \mathbf{o} & \mathbf{m} \end{Bmatrix} \tag{6.510}
$$

Also

$$
\boldsymbol{\varepsilon} = \mathbf{L}\mathbf{u} = \begin{Bmatrix} \frac{\partial}{\partial x} & 0 \\ 0 & \frac{\partial}{\partial y} \\ \frac{\partial}{\partial y} & \frac{\partial}{\partial x} \end{Bmatrix} \begin{Bmatrix} \Gamma & \mathbf{0} \\ \mathbf{0} & \Gamma \end{Bmatrix} \begin{Bmatrix} \mathbf{U} \\ \mathbf{V} \end{Bmatrix} = \Phi \begin{Bmatrix} \mathbf{U} \\ \mathbf{V} \end{Bmatrix} \tag{6.511}
$$

where

$$
\Phi = \frac{1}{ab} \begin{Bmatrix} -(b-y) & b-y & y & -y & 0 & 0 & 0 & 0 \\ 0 & 0 & 0 & 0 & -(a-x) & -x & x & a-x \\ -(a-x) & -x & x & a-x & -(b-y) & b-y & y & -y \end{Bmatrix} \tag{6.512}
$$

Hence,

$$
\mathbf{K} = \iint_A \Phi^T \mathbf{D} \Phi \, dA = \mathbf{K}_1 + \mathbf{K}_2 + \mathbf{K}_3 \tag{6.513}
$$

where

$$
\mathbf{K}_1 = \mu \begin{Bmatrix} \mathbf{A} & \mathbf{O} \\ \mathbf{O} & \mathbf{B} \end{Bmatrix}, \qquad \mathbf{K}_2 = (\lambda + 2\mu) \begin{Bmatrix} \mathbf{B} & \mathbf{O} \\ \mathbf{O} & \mathbf{A} \end{Bmatrix}, \qquad \mathbf{K}_3 = \begin{Bmatrix} \mathbf{O} & \mathbf{C} \\ \mathbf{C}^T & \mathbf{O} \end{Bmatrix} \tag{6.514}
$$

with

$$
\mathbf{A} = \frac{1}{6} \frac{a}{b} \begin{Bmatrix} 2 & 1 & -1 & -2 \\ 1 & 2 & -2 & -1 \\ -1 & -2 & 2 & 1 \\ -2 & -1 & 1 & 2 \end{Bmatrix} \qquad \mathbf{B} = \frac{1}{6} \frac{b}{a} \begin{Bmatrix} 2 & -2 & -1 & 1 \\ -2 & 2 & 1 & -1 \\ -1 & 1 & 2 & -2 \\ 1 & -1 & -2 & 2 \end{Bmatrix} \qquad \mathbf{O} = \text{null} \tag{6.515}
$$

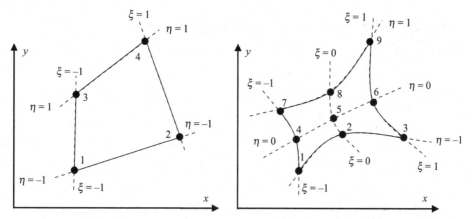

Figure 6.33. Isoparametric linear and quadratic quadrilaterals, with dimensionless coordinates.

$$\mathbf{C} = \tfrac{1}{4}(\lambda+\mu) \begin{Bmatrix} 1 & 0 & -1 & 0 \\ 0 & -1 & 0 & 1 \\ -1 & 0 & 1 & 0 \\ 0 & 1 & 0 & -1 \end{Bmatrix} + \tfrac{1}{4}(\lambda-\mu) \begin{Bmatrix} 0 & 1 & 0 & -1 \\ -1 & 0 & 1 & 0 \\ 0 & -1 & 0 & 1 \\ 1 & 0 & -1 & 0 \end{Bmatrix} \tag{6.516}$$

Observe that all horizontal displacements are listed first, followed by all vertical displacements.

6.7.5 Isoparametric Elements

Isoparametric finite elements are obtained by using the same expansions for the displacements as for the coordinates at an interior point (i.e., using the same interpolants).

Plane Strain Curvilinear Quadrilaterals

$$x = \mathbf{NX}, \qquad y = \mathbf{NY}, \qquad \mathbf{X}^T = \begin{bmatrix} x_1 & \cdots & x_N \end{bmatrix}, \qquad \mathbf{Y}^T = \begin{bmatrix} y_1 & \cdots & y_N \end{bmatrix} \tag{6.517}$$

$$u = \mathbf{NU}, \qquad v = \mathbf{NV}, \qquad \mathbf{U}^T = \begin{bmatrix} u_1 & \cdots & u_N \end{bmatrix}, \qquad \mathbf{V}^T = \begin{bmatrix} v_1 & \cdots & v_N \end{bmatrix} \tag{6.518}$$

where $\mathbf{N}$ is a row vector of interpolation functions (detailed expressions for both linear and quadratic elements are given later on). With reference to Figure 6.33, Eqs. 6.517 and 6.518 can be written compactly as

$$\mathbf{u} = \begin{Bmatrix} u \\ v \end{Bmatrix} = \begin{bmatrix} \mathbf{N} & \mathbf{O} \\ \mathbf{O} & \mathbf{N} \end{bmatrix} \begin{Bmatrix} \mathbf{U} \\ \mathbf{V} \end{Bmatrix} = \begin{bmatrix} \mathbf{N} & \mathbf{O} \\ \mathbf{O} & \mathbf{N} \end{bmatrix} \mathcal{U}, \qquad \mathcal{U} = \begin{Bmatrix} \mathbf{U} \\ \mathbf{V} \end{Bmatrix} \tag{6.519}$$

$$\begin{Bmatrix} \dfrac{\partial}{\partial x} \\ \dfrac{\partial}{\partial y} \end{Bmatrix} = \mathbf{J}^{-1} \begin{Bmatrix} \dfrac{\partial}{\partial \xi} \\ \dfrac{\partial}{\partial \eta} \end{Bmatrix}, \qquad \mathbf{J} = \begin{Bmatrix} \dfrac{\partial x}{\partial \xi} & \dfrac{\partial y}{\partial \xi} \\ \dfrac{\partial x}{\partial \eta} & \dfrac{\partial y}{\partial \eta} \end{Bmatrix} = \begin{Bmatrix} \mathbf{N}_\xi \mathbf{X} & \mathbf{N}_\xi \mathbf{Y} \\ \mathbf{N}_\eta \mathbf{X} & \mathbf{N}_\eta \mathbf{Y} \end{Bmatrix} \tag{6.520}$$

where $\mathbf{N}_\xi = \frac{\partial}{\partial \xi}\mathbf{N}$ and $\mathbf{N}_\eta = \frac{\partial}{\partial \eta}\mathbf{N}$. On the other hand, the strains are

$$\varepsilon = \mathbf{L}\mathbf{u} = \left\{ \begin{array}{c} \dfrac{\partial u}{\partial x} \\[2mm] \dfrac{\partial v}{\partial y} \\[2mm] \dfrac{\partial v}{\partial y} + \dfrac{\partial u}{\partial x} \end{array} \right\} = \mathbf{J}^{-1} \left\{ \begin{array}{c} \mathbf{N}_\xi \mathbf{U} \\[1mm] \mathbf{N}_\eta \mathbf{V} \\[1mm] \mathbf{N}_\eta \mathbf{U} + \mathbf{N}_\xi \mathbf{V} \end{array} \right\} = \mathbf{J}^{-1} \left[\begin{array}{cc} \mathbf{N}_\xi & \mathbf{0} \\ \mathbf{0} & \mathbf{N}_\eta \\ \mathbf{N}_\eta & \mathbf{N}_\xi \end{array} \right] \left\{ \begin{array}{c} \mathbf{U} \\ \mathbf{V} \end{array} \right\} = \mathcal{B}\mathcal{U}$$

$$(6.521)$$

where

$$\mathcal{B} = \mathbf{J}^{-1} \left\{ \begin{array}{cc} \mathbf{N}_\xi & \mathbf{0} \\ \mathbf{0} & \mathbf{N}_\eta \\ \mathbf{N}_\eta & \mathbf{N}_\xi \end{array} \right\} \tag{6.522}$$

On the other hand, the differential area is

$$dA = |\mathbf{J}| \, d\xi \, d\eta \tag{6.523}$$

Hence, the element stiffness and mass matrices are (here still grouped by DOF, not by nodes)

$$\mathbf{K} = \iint_A \mathcal{B}^T \mathbf{D} \mathcal{B} \, |\mathbf{J}| \, d\xi \, d\eta \tag{6.524}$$

$$\mathbf{M} = \iint_A \rho \left\{ \begin{array}{cc} \mathbf{N} & \mathbf{0} \\ \mathbf{0} & \mathbf{N} \end{array} \right\}^T \left\{ \begin{array}{cc} \mathbf{N} & \mathbf{0} \\ \mathbf{0} & \mathbf{N} \end{array} \right\} |\mathbf{J}| \, dA = \rho \iint_A \left\{ \begin{array}{cc} \mathbf{N}^T \mathbf{N} & \mathbf{O} \\ \mathbf{O} & \mathbf{N}^T \mathbf{N} \end{array} \right\} |\mathbf{J}| \, d\xi \, d\eta \tag{6.525}$$

Interpolation Functions for Linear and Quadratic Elements

The interpolation functions for linear and quadratic elements are as given next. Observe that $\mathbf{N}_1$ is an auxiliary vector used in the definition of $\mathbf{N}$, and that the order of the elements in these interpolation functions define the nodal sequence shown schematically in Figure 6.34.

Linear Expansion

$$\begin{aligned} \mathbf{N} &= \tfrac{1}{2}\left[(1-\eta)\mathbf{N}_1 \quad (1+\eta)\mathbf{N}_1 \right] \\ \mathbf{N}_1 &= \tfrac{1}{2}\left[(1-\xi) \quad (1+\xi) \right] \end{aligned} \qquad -1 \le \xi \le 1, \, -1 \le \eta \le 1, \tag{6.526}$$

Quadratic Expansion

$$\begin{aligned} \mathbf{N} &= \tfrac{1}{2}\left[\eta(\eta-1)\mathbf{N}_1 \quad 2(1-\eta)^2 \mathbf{N}_1 \quad \eta(\eta+1)\mathbf{N}_1 \right] \\ \mathbf{N}_1 &= \tfrac{1}{2}\left[\xi(\xi-1) \quad 2(1-\xi)^2 \quad \xi(\xi+1) \right] \end{aligned} \qquad -1 \le \xi \le 1, \, -1 \le \eta \le 1, \tag{6.527}$$

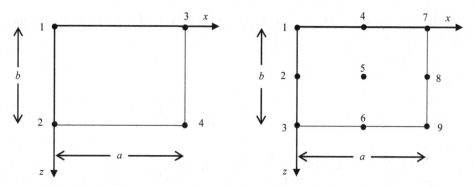

Figure 6.34. Linear (4-noded) and quadratic (9-noded) rectangular SH and SVP elements. Direction of vertical axis is irrelevant for SH element. Matrix elements ordered according to nodal sequences shown.

Rectangular 4-Noded SH Element (Linear Expansion)

Displacements and nodal forces are perpendicular to the plane. Refer also to Figure 6.34.

$$\mathbf{K}_1 = \frac{a}{b}\begin{bmatrix} 1 & -1 \\ -1 & 1 \end{bmatrix}, \qquad \mathbf{K}_2 = \frac{b}{a}\begin{bmatrix} 2 & 1 \\ 1 & 2 \end{bmatrix} \tag{6.528}$$

$$\mathbf{K} = \frac{G}{6}\left(\begin{bmatrix} 2\mathbf{K}_1 & \mathbf{K}_1 \\ \mathbf{K}_1 & 2\mathbf{K}_1 \end{bmatrix} + \begin{bmatrix} \mathbf{K}_2 & -\mathbf{K}_2 \\ -\mathbf{K}_2 & \mathbf{K}_2 \end{bmatrix} \right) \tag{6.529}$$

$$\mathbf{M} = \begin{bmatrix} 2\mathbf{M}_1 & \mathbf{M}_1 \\ \mathbf{M}_1 & 2\mathbf{M}_1 \end{bmatrix}, \qquad \mathbf{M}_1 = \frac{\rho}{36}ab\begin{bmatrix} 2 & 1 \\ 1 & 2 \end{bmatrix} \tag{6.530}$$

Rectangular 9-Noded SH Element (Quadratic Expansion)

Displacements and nodal forces are perpendicular to the plane. Refer also to Figure 6.34.

$$\mathbf{K}_1 = \frac{a}{b}\begin{bmatrix} 7 & -8 & 1 \\ -8 & 16 & -8 \\ 1 & -8 & 7 \end{bmatrix}, \qquad \mathbf{K}_2 = \frac{b}{a}\begin{bmatrix} 4 & 2 & -1 \\ 2 & 16 & 2 \\ -1 & 2 & 4 \end{bmatrix} \tag{6.531}$$

$$\mathbf{K} = \frac{G}{90}\left(\begin{bmatrix} 4\mathbf{K}_1 & 2\mathbf{K}_1 & -\mathbf{K}_1 \\ 2\mathbf{K}_1 & 16\mathbf{K}_1 & 2\mathbf{K}_1 \\ -\mathbf{K}_1 & 2\mathbf{K}_1 & 4\mathbf{K}_1 \end{bmatrix} + \begin{bmatrix} 7\mathbf{K}_2 & -8\mathbf{K}_2 & \mathbf{K}_2 \\ -8\mathbf{K}_2 & 16\mathbf{K}_2 & -8\mathbf{K}_2 \\ \mathbf{K}_2 & -8\mathbf{K}_2 & 7\mathbf{K}_2 \end{bmatrix} \right). \tag{6.532}$$

$$\mathbf{M} = \begin{bmatrix} 4\mathbf{M}_2 & 2\mathbf{M}_2 & -\mathbf{M}_2 \\ 2\mathbf{M}_2 & 16\mathbf{M}_2 & 2\mathbf{M}_2 \\ -\mathbf{M}_2 & 2\mathbf{M}_2 & 4\mathbf{M}_2 \end{bmatrix}, \qquad \mathbf{M}_2 = \frac{\rho ab}{900}\begin{bmatrix} 4 & 2 & -1 \\ 2 & 16 & 2 \\ -1 & 2 & 4 \end{bmatrix} \tag{6.533}$$

Horizontal–vertical displacements and nodal forces are defined in the plane of the element. Vertical forces and displacements are positive down. The nodal displacement vectors are grouped by nodes, that is, $u_1, v_1, u_2, v_2 \cdots u_4, v_4$. Also, $\mathbf{I}$ is the 2×2 identity matrix. Refer also to Figure 6.34.

$$\alpha_1 = G \qquad \alpha_2 = \lambda + 2G \qquad \alpha_3 = \lambda - G, \qquad \alpha_4 = \lambda + G \tag{6.534}$$

$$\mathbf{A} = \frac{a}{b} \begin{bmatrix} \alpha_1 & 0 \\ 0 & \alpha_2 \end{bmatrix}, \qquad \mathbf{B} = \frac{b}{a} \begin{bmatrix} \alpha_2 & 0 \\ 0 & \alpha_1 \end{bmatrix} \qquad \mathbf{C} = \begin{bmatrix} 0 & -\alpha_3 \\ \alpha_3 & 0 \end{bmatrix} \qquad \mathbf{D} = \begin{bmatrix} 0 & \alpha_4 \\ \alpha_4 & 0 \end{bmatrix} \tag{6.535}$$

$$\mathbf{M} = \begin{bmatrix} 2\mathbf{M}_1 & \mathbf{M}_1 \\ \mathbf{M}_1 & 2\mathbf{M}_1 \end{bmatrix}, \qquad \mathbf{M}_1 = \frac{\rho ab}{36} \begin{bmatrix} 2\mathbf{I} & \mathbf{I} \\ \mathbf{I} & 2\mathbf{I} \end{bmatrix}, \qquad \mathbf{K}_1 = \begin{bmatrix} \mathbf{A} & -\mathbf{A} \\ -\mathbf{A} & \mathbf{A} \end{bmatrix} \tag{6.536}$$

$$\mathbf{K}_2 = \begin{bmatrix} 2\mathbf{B} & \mathbf{B} \\ \mathbf{B} & 2\mathbf{B} \end{bmatrix}, \qquad \mathbf{K}_3 = \frac{3}{2} \begin{bmatrix} \mathbf{D} & \mathbf{C} \\ \mathbf{C}^T & -\mathbf{D} \end{bmatrix}, \qquad \mathbf{K}_4 = \frac{3}{2} \begin{bmatrix} \mathbf{C} & \mathbf{D} \\ -\mathbf{D} & \mathbf{C}^T \end{bmatrix} \tag{6.537}$$

$$\mathbf{K} = \frac{1}{6} \left(\begin{bmatrix} 2\mathbf{K}_1 & \mathbf{K}_1 \\ \mathbf{K}_1 & 2\mathbf{K}_1 \end{bmatrix} + \begin{bmatrix} \mathbf{K}_2 & -\mathbf{K}_2 \\ -\mathbf{K}_2 & \mathbf{K}_2 \end{bmatrix} + \begin{bmatrix} \mathbf{K}_3 & \mathbf{K}_4^T \\ \mathbf{K}_4 & -\mathbf{K}_3 \end{bmatrix} \right) \tag{6.538}$$

To reverse the direction of z, simply flip up the figure, which moves nodes 2, 4 to the top. Alternatively, maintain the nodal order and multiply even rows and columns of $\mathbf{K}$ by -1.

Rectangular 9-Noded SVP Element (Quadratic Expansion)

Horizontal–vertical displacements and nodal forces are defined in the plane of the element. Vertical forces and displacements are positive down, as indicated by the arrow. The nodal displacement vectors are grouped by nodes, that is, $u_1, v_1, u_2, v_2 \cdots u_9, v_9$. Refer also to Figure 6.34.

$\mathbf{A}, \mathbf{B}, \mathbf{C}, \mathbf{D}, \mathbf{I}$ as shown in the previous section

$$\mathbf{M} = \frac{\rho ab}{900} \begin{bmatrix} 4\mathbf{M}_2 & 2\mathbf{M}_2 & -\mathbf{M}_2 \\ 2\mathbf{M}_2 & 16\mathbf{M}_2 & 2\mathbf{M}_2 \\ -\mathbf{M}_2 & 2\mathbf{M}_2 & 4\mathbf{M}_2 \end{bmatrix}, \qquad \mathbf{M}_2 = \begin{bmatrix} 4\mathbf{I} & 2\mathbf{I} & -\mathbf{I} \\ 2\mathbf{I} & 16\mathbf{I} & 2\mathbf{I} \\ -\mathbf{I} & 2\mathbf{I} & 4\mathbf{I} \end{bmatrix} \tag{6.539}$$

$$\mathbf{K}_1 = \begin{bmatrix} 7\mathbf{A} & -8\mathbf{A} & \mathbf{A} \\ -8\mathbf{A} & 16\mathbf{A} & -8\mathbf{A} \\ \mathbf{A} & -8\mathbf{A} & 7\mathbf{A} \end{bmatrix}, \qquad \mathbf{K}_2 = \begin{bmatrix} 4\mathbf{B} & 2\mathbf{B} & -\mathbf{B} \\ 2\mathbf{B} & 16\mathbf{B} & 2\mathbf{B} \\ -\mathbf{B} & 2\mathbf{B} & 4\mathbf{B} \end{bmatrix} \tag{6.540}$$

$$\mathbf{K}_3 = \frac{5}{2} \begin{bmatrix} 3\mathbf{D} & 4\mathbf{C} & -\mathbf{C} \\ 4\mathbf{C}^T & \mathbf{O} & 4\mathbf{C} \\ -\mathbf{C}^T & 4\mathbf{C}^T & -3\mathbf{D} \end{bmatrix}, \qquad \mathbf{K}_4 = \frac{5}{2} \begin{bmatrix} 3\mathbf{C} & 4\mathbf{D} & -\mathbf{D} \\ -4\mathbf{D} & \mathbf{O} & 4\mathbf{D} \\ \mathbf{D} & -4\mathbf{D} & -3\mathbf{C} \end{bmatrix} \tag{6.541}$$

$$\mathbf{K} = \frac{1}{90} \left(\begin{bmatrix} 4\mathbf{K}_1 & 2\mathbf{K}_1 & -\mathbf{K}_1 \\ 2\mathbf{K}_1 & 16\mathbf{K}_1 & 2\mathbf{K}_1 \\ -\mathbf{K}_1 & 2\mathbf{K}_1 & 4\mathbf{K}_1 \end{bmatrix} \begin{bmatrix} 7\mathbf{K}_2 & -8\mathbf{K}_2 & \mathbf{K}_2 \\ -8\mathbf{K}_2 & 16\mathbf{K}_2 & -8\mathbf{K}_2 \\ \mathbf{K}_2 & -8\mathbf{K}_2 & 7\mathbf{K}_2 \end{bmatrix} \begin{bmatrix} 3\mathbf{K}_3 & 4\mathbf{K}_4^T & -\mathbf{K}_4^T \\ 4\mathbf{K}_4 & \mathbf{O} & 4\mathbf{K}_4^T \\ -\mathbf{K}_4 & 4\mathbf{K}_4 & -3\mathbf{K}_3 \end{bmatrix} \right) \tag{6.542}$$

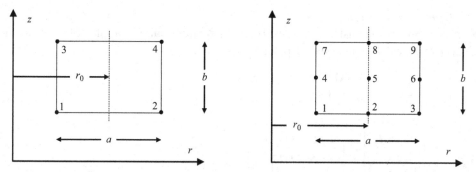

Figure 6.35. Linear (4-noded) and quadratic (9-noded) rectangular elements in cylindrical coordinate. Matrix elements ordered according to nodal sequences shown.

To reverse the direction of z, simply flip up Figure 6.34, which moves nodes 3, 6, 9 to the top, and 1, 4, 7 to the bottom. Alternatively, maintain the nodal order and multiply even rows and columns of $\mathbf{K}$ by -1 (i.e., reverse signs in checkerboard fashion).

Cylindrical Coordinates

Displacement and Strains in Cylindrical Coordinates

$$\mathbf{u} = \sum_{n=0}^{\infty} \begin{Bmatrix} \cos n\theta & 0 & 0 \\ 0 & -\sin n\theta & 0 \\ 0 & 0 & \cos n\theta \end{Bmatrix} \begin{Bmatrix} \bar{u}_r \\ \bar{u}_\theta \\ \bar{u}_z \end{Bmatrix}, \qquad \bar{\mathbf{u}}_n = \begin{Bmatrix} \bar{u}_r \\ \bar{u}_\theta \\ \bar{u}_z \end{Bmatrix}_n \tag{6.543}$$

where n is the index in the Fourier expansion. Typically, only $n = 0$ (axisymmetric loads, such as vertical or torsional loads) and $n = 1$ (lateral x-loads) are needed. The overbar is a reminder that it is a Fourier component. The strains can similarly be expanded in Fourier series, and they are

$$\bar{\boldsymbol{\varepsilon}}_n = \begin{bmatrix} \bar{\varepsilon}_r & \bar{\varepsilon}_\theta & \bar{\varepsilon}_z & \bar{\gamma}_{\theta z} & \bar{\gamma}_{rz} & \bar{\gamma}_{r\theta} \end{bmatrix}_n^T \tag{6.544}$$

The actual strains $\boldsymbol{\varepsilon}$ involve summations that are analogous to those for the displacements, except that $\begin{bmatrix} \bar{\varepsilon}_r & \bar{\varepsilon}_\theta & \bar{\varepsilon}_z & \bar{\gamma}_{rz} \end{bmatrix}^T$ change as $(\cos n\theta)$, and $\begin{bmatrix} \bar{\gamma}_{\theta z} & \bar{\gamma}_{r\theta} \end{bmatrix}^T$ change as $(-\sin\theta)$.

$$\bar{\boldsymbol{\varepsilon}}_n = \left(\mathbf{L}_r \frac{\partial}{\partial r} + \frac{n}{r} \bar{\mathbf{L}}_\theta + \mathbf{L}_z \frac{\partial}{\partial z} + \mathbf{L}_1 \frac{1}{r} \right) \bar{\mathbf{u}}_n, \qquad \bar{\boldsymbol{\sigma}}_n = \mathbf{D} \bar{\boldsymbol{\varepsilon}}_n \tag{6.545}$$

$$\mathbf{L}_r = \begin{Bmatrix} 1 & 0 & 0 \\ 0 & 0 & 0 \\ 0 & 0 & 0 \\ 0 & 0 & 0 \\ 0 & 0 & 1 \\ 0 & 1 & 0 \end{Bmatrix} \qquad \bar{\mathbf{L}}_\theta = \begin{Bmatrix} 0 & 0 & 0 \\ 0 & -1 & 0 \\ 0 & 0 & 0 \\ 0 & 0 & 1 \\ 0 & 0 & 0 \\ 1 & 0 & 0 \end{Bmatrix}, \qquad \mathbf{L}_z = \begin{Bmatrix} 0 & 0 & 0 \\ 0 & 0 & 0 \\ 0 & 0 & 1 \\ 0 & 1 & 0 \\ 1 & 0 & 0 \\ 0 & 0 & 0 \end{Bmatrix} \qquad \mathbf{L}_1 = \begin{Bmatrix} 0 & 0 & 0 \\ 1 & 0 & 0 \\ 0 & 0 & 0 \\ 0 & 0 & 0 \\ 0 & 0 & 0 \\ 0 & -1 & 0 \end{Bmatrix} \tag{6.546}$$

Finite-Element Expansion

The finite-element expansion for an isoparametric quadrilateral, applicable to both the nodal coordinates and the nodal displacements is

$$\mathbf{x} = \begin{Bmatrix} r \\ z \end{Bmatrix} = \begin{Bmatrix} \mathbf{NR} \\ \mathbf{NZ} \end{Bmatrix} = \begin{bmatrix} \mathbf{N} & \mathbf{O} \\ \mathbf{O} & \mathbf{N} \end{bmatrix} \begin{Bmatrix} \mathbf{R} \\ \mathbf{Z} \end{Bmatrix} \tag{6.547}$$

$$\bar{\mathbf{u}}_n = \begin{Bmatrix} \bar{u}_r \\ \bar{u}_\theta \\ \bar{u}_z \end{Bmatrix}_n = \begin{Bmatrix} \mathbf{N}\bar{\mathbf{U}}_r \\ \mathbf{N}\bar{\mathbf{U}}_\theta \\ \mathbf{N}\bar{\mathbf{U}}_z \end{Bmatrix} = \begin{bmatrix} \mathbf{N} & \mathbf{O} & \mathbf{O} \\ \mathbf{O} & \mathbf{N} & \mathbf{O} \\ \mathbf{O} & \mathbf{O} & \mathbf{N} \end{bmatrix} \begin{Bmatrix} \bar{\mathbf{U}}_r \\ \bar{\mathbf{U}}_\theta \\ \bar{\mathbf{U}}_z \end{Bmatrix} = \mathbf{Q}\bar{\mathbf{U}}_n \tag{6.548}$$

$$\bar{\mathbf{U}}_n = \begin{Bmatrix} \bar{\mathbf{U}}_r \\ \bar{\mathbf{U}}_\theta \\ \bar{\mathbf{U}}_z \end{Bmatrix}_n = \text{nodal disp.}, \qquad \mathbf{Q} = \begin{Bmatrix} \mathbf{N} & \mathbf{O} & \mathbf{O} \\ \mathbf{O} & \mathbf{N} & \mathbf{O} \\ \mathbf{O} & \mathbf{O} & \mathbf{N} \end{Bmatrix} = \text{interpolation matrix} \tag{6.549}$$

where for the quadrilaterals with the nodal sequence shown

$$\mathbf{R} = \begin{bmatrix} r_1 & r_2 & \cdots & r_N \end{bmatrix}^T \qquad \mathbf{Z} = \begin{bmatrix} z_1 & z_2 & \cdots & z_N \end{bmatrix}^T \qquad N = 4 \quad \text{or} \quad N = 9 \tag{6.550}$$

$$\bar{\mathbf{U}}_j = \begin{bmatrix} u_{j1} & u_{j2} & \cdots & u_{jN} \end{bmatrix}^T, \qquad j = r, \theta, z \tag{6.551}$$

Linear Expansion (N = 4)

$$\begin{aligned} \mathbf{N}_1 &= \tfrac{1}{2}\begin{bmatrix} (1-\xi) & (1+\xi) \end{bmatrix} \\ \mathbf{N} &= \tfrac{1}{2}\begin{bmatrix} (1-\eta)\mathbf{N}_1 & (1+\eta)\mathbf{N}_1 \end{bmatrix} \end{aligned} \qquad -1 \le \xi \le 1, -1 \le \eta \le 1, \tag{6.552}$$

Quadratic Expansion (N = 9)

$$\begin{aligned} \mathbf{N}_1 &= \tfrac{1}{2}\begin{bmatrix} \xi(\xi-1) & 2(1-\xi)^2 & \xi(\xi+1) \end{bmatrix} \\ \mathbf{N} &= \tfrac{1}{2}\begin{bmatrix} \eta(\eta-1)\mathbf{N}_1 & 2(1-\eta)^2\mathbf{N}_1 & \eta(\eta+1)\mathbf{N}_1 \end{bmatrix} \end{aligned} \qquad -1 \le \xi \le 1, -1 \le \eta \le 1, \tag{6.553}$$

Also

$$\mathbf{N}_\xi = \frac{\partial \mathbf{N}}{\partial \xi}, \qquad \mathbf{N}_\eta = \frac{\partial \mathbf{N}}{\partial \eta} \tag{6.554}$$

$$\mathbf{J}(\xi, \eta) = \begin{Bmatrix} \dfrac{\partial r}{\partial \xi} & \dfrac{\partial z}{\partial \xi} \\ \dfrac{\partial r}{\partial \eta} & \dfrac{\partial z}{\partial \eta} \end{Bmatrix} = \begin{Bmatrix} \mathbf{N}_\xi \mathbf{R} & \mathbf{N}_\xi \mathbf{Z} \\ \mathbf{N}_\eta \mathbf{R} & \mathbf{N}_\eta \mathbf{Z} \end{Bmatrix} = \text{Jacobi matrix} \tag{6.555}$$

where $\mathbf{J}$ is the Jacobi matrix and $J = |\mathbf{J}|$ is the Jacobian. For a linear expansion, the Jacobian is

$$J = |\mathbf{J}| = \tfrac{1}{8}\Big\{\big[(r_2 - r_1)(z_4 - z_3) - (r_4 - r_3)(z_2 - z_1)\big]\xi$$
$$+\big[(r_4 - r_2)(z_3 - z_1) - (r_3 - r_1)(z_4 - z_2)\big]\eta \qquad (6.556)$$
$$+(r_4 - r_1)(z_3 - z_2) + (r_2 - r_3)(z_4 - z_1)\Big\}$$

Rectangular Elements

For a rectangular element in cylindrical coordinates and a linear expansion, the Jacobi matrix is

$$\mathbf{J}(\xi,\eta) = \begin{Bmatrix} \tfrac{1}{2}a & 0 \\ 0 & \tfrac{1}{2}b \end{Bmatrix}, \qquad J = |\mathbf{J}| = \tfrac{1}{4}ab \qquad = \text{Jacobian} \qquad (6.557)$$

Hence, if $\bar{u}_j = \bar{u}_j(r,z)$ is any displacement component (i.e., radial, tangential, or vertical), then

$$\begin{bmatrix} \dfrac{\partial \bar{u}_j}{\partial r} \\ \dfrac{\partial \bar{u}_j}{\partial z} \end{bmatrix} = \mathbf{J}^{-1}\begin{bmatrix} \dfrac{\partial \bar{u}_j}{\partial \xi} \\ \dfrac{\partial \bar{u}_j}{\partial \eta} \end{bmatrix} = \begin{bmatrix} \dfrac{2}{a}\dfrac{\partial \bar{u}_j}{\partial \xi} \\ \dfrac{2}{b}\dfrac{\partial \bar{u}_j}{\partial \eta} \end{bmatrix} = \begin{bmatrix} \dfrac{2}{a}\mathbf{N}_\xi \bar{\mathbf{U}}_j \\ \dfrac{2}{b}\mathbf{N}_\eta \bar{\mathbf{U}}_j \end{bmatrix}, \qquad j = r,\theta,z \qquad (6.558)$$

That is,

$$\boxed{\bar{u}_{j,r} = \tfrac{2}{a}\mathbf{N}_\xi \bar{\mathbf{U}}_j}, \qquad\qquad \boxed{\bar{u}_{j,z} = \tfrac{2}{b}\mathbf{N}_\eta \bar{\mathbf{U}}_j} \qquad (6.559)$$

where

$$\bar{\mathbf{U}}_j = \begin{bmatrix} u_{j1} & u_{j2} & \cdots & u_{jN} \end{bmatrix}^T, \qquad j = r,\theta,z \qquad (6.560)$$

Stiffness Matrix

The stress tensor (represented as vector here) is

$$\bar{\sigma}_n = \mathbf{D}\bar{\varepsilon}_n = \mathbf{DLQ}\bar{\mathbf{U}}_n = \mathbf{D}\Big(\mathbf{L}_r\mathbf{Q}_r + \tfrac{n}{r}\bar{\mathbf{L}}_\theta\mathbf{Q} + \mathbf{L}_z\mathbf{Q}_z + \tfrac{1}{r}\mathbf{L}_1\mathbf{Q}\Big)\bar{\mathbf{U}}_n \qquad (6.561)$$

where

$$\mathbf{Q}_r = \tfrac{\partial}{\partial r}\mathbf{Q} = \tfrac{2}{a}\mathbf{Q}_\xi \qquad \mathbf{Q}_z = \tfrac{\partial}{\partial z}\mathbf{Q} = \tfrac{2}{b}\mathbf{Q}_\eta. \qquad (6.562)$$

Define also

$$\mathbf{D}_{\alpha\beta} = \mathbf{L}_\alpha^T \mathbf{D}\mathbf{L}_\beta \qquad (6.563)$$

where

$$
\mathbf{D}_{rr} = \left\{ \begin{matrix} \lambda+2\mu & 0 & 0 \\ 0 & \mu & 0 \\ 0 & 0 & \mu \end{matrix} \right\} \quad
\bar{\mathbf{D}}_{\theta\theta} = \left[\begin{matrix} \mu & 0 & 0 \\ 0 & \lambda+2\mu & 0 \\ 0 & 0 & \mu \end{matrix} \right] \quad
\mathbf{D}_{zz} = \left\{ \begin{matrix} \mu & 0 & 0 \\ 0 & \mu & 0 \\ 0 & 0 & \lambda+2\mu \end{matrix} \right\} \quad (6.564)
$$

$$
\bar{\mathbf{D}}_{r\theta} = \bar{\mathbf{D}}_{\theta r}^{T} = \left\{ \begin{matrix} 0 & -\lambda & 0 \\ \mu & 0 & 0 \\ 0 & 0 & 0 \end{matrix} \right\} \quad
\mathbf{D}_{rz} = \mathbf{D}_{zr}^{T} = \left\{ \begin{matrix} 0 & 0 & \lambda \\ 0 & 0 & 0 \\ \mu & 0 & 0 \end{matrix} \right\} \quad
\bar{\mathbf{D}}_{\theta z} = \bar{\mathbf{D}}_{z\theta}^{T} = \left\{ \begin{matrix} 0 & 0 & 0 \\ 0 & 0 & -\lambda \\ 0 & \mu & 0 \end{matrix} \right\}
$$
$$
(6.565)
$$

$$
\mathbf{D}_{r1} = \mathbf{D}_{1r} = \left\{ \begin{matrix} \lambda & 0 & 0 \\ 0 & -\mu & 0 \\ 0 & 0 & 0 \end{matrix} \right\} \quad
\bar{\mathbf{D}}_{\theta 1} = \bar{\mathbf{D}}_{1\theta}^{T} = -\left\{ \begin{matrix} 0 & \mu & 0 \\ \lambda+2\mu & 0 & 0 \\ 0 & 0 & 0 \end{matrix} \right\} \quad
\mathbf{D}_{z1} = \mathbf{D}_{1z}^{T} = \left\{ \begin{matrix} 0 & 0 & 0 \\ 0 & 0 & 0 \\ \lambda & 0 & 0 \end{matrix} \right\}
$$
$$
(6.566)
$$

$$
\mathbf{D}_{11} = \left\{ \begin{matrix} \lambda+2\mu & 0 & 0 \\ 0 & \mu & 0 \\ 0 & 0 & 0 \end{matrix} \right\} \qquad \text{(Matrices with overbars have been formed with } \bar{\mathbf{L}}_{\theta}) \quad (6.567)
$$

Then

$$
\delta\bar{\sigma}^{T}\varepsilon = \delta\bar{\mathbf{U}}^{T}\left(\mathbf{Q}_{r}^{T}\mathbf{L}_{r}^{T} + \frac{n}{r}\mathbf{Q}^{T}\bar{\mathbf{L}}_{\theta}^{T} + \mathbf{Q}_{z}^{T}\mathbf{L}_{z}^{T} + \frac{1}{r}\mathbf{Q}^{T}\mathbf{L}_{1}^{T} \right)^{T} \mathbf{D}\left(\mathbf{L}_{r}\mathbf{Q}_{r} + \frac{n}{r}\bar{\mathbf{L}}_{\theta}\mathbf{Q} + \mathbf{L}_{z}\mathbf{Q}_{z} + \frac{1}{r}\mathbf{L}_{1}\mathbf{Q} \right)\bar{\mathbf{U}}
$$

$$
= \delta\bar{\mathbf{U}}^{T}\left[\mathbf{H}_{rr} + \frac{n}{r}\left(\mathbf{H}_{r\theta} + \mathbf{H}_{\theta r} \right) + \left(\mathbf{H}_{rz} + \mathbf{H}_{zr} \right) + \mathbf{H}_{zz} + \frac{1}{r}\left(\mathbf{H}_{r1} + \mathbf{H}_{1r} \right) + \frac{1}{r}\left(\mathbf{H}_{z1} + \mathbf{H}_{1z} \right) \right.
$$

$$
\left. + \left(\frac{n}{r} \right)^{2}\mathbf{H}_{\theta\theta} + \frac{n}{r}\left(\mathbf{H}_{\theta z} + \mathbf{H}_{z\theta} \right) + \frac{n}{r^{2}}\left(\mathbf{H}_{\theta 1} + \mathbf{H}_{1\theta} \right) + \mathbf{H}_{11}\left(\frac{1}{r} \right)^{2} \right]\bar{\mathbf{U}} \quad (6.568)
$$

where

$$
\mathbf{H}_{rr} = \mathbf{Q}_{r}^{T}\mathbf{D}_{rr}\mathbf{Q}_{r}, \qquad \mathbf{H}_{\theta\theta} = \mathbf{Q}^{T}\mathbf{D}_{\theta\theta}\mathbf{Q}, \qquad \mathbf{H}_{zz} = \mathbf{Q}_{z}^{T}\mathbf{D}_{zz}\mathbf{Q}_{z} \qquad (6.569)
$$

$$
\mathbf{H}_{rz} = \mathbf{Q}_{r}^{T}\mathbf{D}_{rz}\mathbf{Q}_{z}, \qquad \mathbf{H}_{r1} = \mathbf{Q}_{r}^{T}\mathbf{D}_{r1}\mathbf{Q}, \qquad \mathbf{H}_{z1} = \mathbf{Q}_{z}^{T}\mathbf{D}_{z1}\mathbf{Q} \qquad (6.570)
$$

$$
\mathbf{H}_{r\theta} = \mathbf{Q}_{r}^{T}\mathbf{D}_{r\theta}\mathbf{Q}, \qquad \mathbf{H}_{\theta z} = \mathbf{Q}^{T}\mathbf{D}_{\theta z}\mathbf{Q}_{z}, \qquad \mathbf{H}_{\theta 1} = \mathbf{Q}^{T}\mathbf{D}_{\theta 1}\mathbf{Q} \qquad (6.571)
$$

$$
\mathbf{H}_{11} = \mathbf{Q}^{T}\mathbf{D}_{11}\mathbf{Q} \qquad (6.572)
$$

$$
\mathbf{H}_{1r} = \mathbf{H}_{r1}^{T} \qquad \mathbf{H}_{z\theta} = \mathbf{H}_{\theta z}^{T} \qquad \mathbf{H}_{\theta r} = \mathbf{H}_{r\theta}^{T} \qquad (6.573)
$$

$$
\mathbf{H}_{zr} = \mathbf{H}_{rz}^{T}, \qquad \mathbf{H}_{1z} = \mathbf{H}_{z1}^{T} \qquad \mathbf{H}_{1\theta} = \mathbf{H}_{\theta 1}^{T} \qquad (6.574)
$$

Although each of the above matrices is of size 12×12, they can be written compactly in terms of $3\times3 = 9$ matrices of size 4×4 each, identical in structure to $\mathbf{D}_{\alpha\beta}$, but with elements multiplied by some appropriate matrix $\Xi_{\alpha\beta} = \mathbf{N}_\alpha^T \mathbf{N}_\beta$, and $\mathbf{O}$ being the null matrix of order 4×4. For example,

$$\mathbf{H}_{rr} = \mathbf{Q}_r^T \mathbf{D}_{rr} \mathbf{Q}_r = \begin{Bmatrix} (\lambda + 2\mu)\Xi_{rr} & \mathbf{O} & \mathbf{O} \\ \mathbf{O} & \mu\Xi_{rr} & \mathbf{O} \\ \mathbf{O} & \mathbf{O} & \mu\Xi_{rr} \end{Bmatrix}, \quad \Xi_{rr} = \mathbf{N}_r^T \mathbf{N}_r \tag{6.575}$$

which is constructed by multiplication of the elements of $\mathbf{D}_{rr}$ by Ξ_{rr}. Similarly,

$$\mathbf{H}_{rz} = \mathbf{Q}_r^T \mathbf{D}_{rz} \mathbf{Q}_z = \begin{Bmatrix} \mathbf{O} & \mathbf{O} & \lambda\Xi_{rz} \\ \mathbf{O} & \mathbf{O} & \mathbf{O} \\ \mu\Xi_{rz} & \mathbf{O} & \mathbf{O} \end{Bmatrix}, \quad \Xi_{rz} = \mathbf{N}_r^T \mathbf{N}_z \tag{6.576}$$

and so on. The stiffness matrix is then (this omits the integration in θ)

$$\begin{aligned} \mathbf{K} &= \iint_A (\mathbf{LQ})^T \mathbf{D}(\mathbf{LQ}) r \, dr \, dz = \int_{-1}^{+1}\int_{-1}^{+1} (\mathbf{LQ})^T \mathbf{D}(\mathbf{LQ}) r J \, d\xi \, d\eta \\ &= \int_{-1}^{+1}\int_{-1}^{+1} (\mathbf{LQ})^T \mathbf{D}(\mathbf{LQ}) r J \, d\xi \, d\eta \\ &= \int_{-1}^{+1}\int_{-1}^{+1} \Big\{ r\big[\mathbf{H}_{rr} + (\mathbf{H}_{rz} + \mathbf{H}_{zr}) + \mathbf{H}_{zz} \big] + \big[(\mathbf{H}_{r1} + \mathbf{H}_{1r}) + (\mathbf{H}_{z1} + \mathbf{H}_{1z}) + \tfrac{1}{r}\mathbf{H}_{11} \big] \\ &\quad + n\big[(\mathbf{H}_{r\theta} + \mathbf{H}_{\theta r}) + (\mathbf{H}_{\theta z} + \mathbf{H}_{z\theta}) + \tfrac{1}{r}(\mathbf{H}_{\theta 1} + \mathbf{H}_{1\theta}) \big] + \tfrac{1}{r}n^2 \mathbf{H}_{\theta\theta} \Big\} J \, d\xi \, d\eta \end{aligned} \tag{6.577}$$

Since the material parameters are constant within each element, we see that it suffices to examine the integrals in $\Xi_{\alpha\beta}$, and not those of $\mathbf{H}_{\alpha\beta}$. These integrals are as follows.

Define

$$\boxed{\alpha = \frac{a}{2r_0}}, \quad \boxed{\beta = \frac{b}{a}} \quad \boxed{L = \ln\frac{1+\alpha}{1-\alpha}} \tag{6.578}$$

(a) Linear Expansion

$$\mathbf{A}_{rr} = \int_{-1}^{+1}\int_{-1}^{+1} r\Xi_{rr} J \, d\xi \, d\eta = r_0 \frac{\beta}{6} \left\{ \begin{array}{cc:cc} 2 & -2 & 1 & -1 \\ -2 & 2 & -1 & 1 \\ \hdashline 1 & -1 & 2 & -2 \\ -1 & 1 & -2 & 2 \end{array} \right\} \tag{6.579}$$

$$\mathbf{A}_{rz} = \int_{-1}^{+1}\int_{-1}^{+1} r\Xi_{rz} J \, d\xi \, d\eta = r_0 \frac{1}{12} \left[3\left\{ \begin{array}{cc:cc} 1 & 1 & -1 & -1 \\ -1 & -1 & 1 & 1 \\ \hdashline 1 & 1 & -1 & -1 \\ -1 & -1 & 1 & 1 \end{array} \right\} + \alpha\left\{ \begin{array}{cc:cc} -1 & 1 & 1 & -1 \\ 1 & -1 & -1 & 1 \\ \hdashline -1 & 1 & 1 & -1 \\ 1 & -1 & -1 & 1 \end{array} \right\} \right] \tag{6.580}$$

$$\mathbf{A}_{zz} = \int_{-1}^{+1}\int_{-1}^{+1} r\Xi_{zz} J \, d\xi \, d\eta = r_0 \frac{1}{6\beta} \left[\left\{ \begin{array}{cc:cc} 2 & 1 & -2 & -1 \\ 1 & 2 & -1 & -2 \\ \hdashline -2 & -1 & 2 & 1 \\ -1 & -2 & 1 & 2 \end{array} \right\} + \alpha\left\{ \begin{array}{cc:cc} -1 & 0 & 1 & 0 \\ 0 & 1 & 0 & -1 \\ \hdashline 1 & 0 & -1 & 0 \\ 0 & -1 & 0 & 1 \end{array} \right\} \right] \tag{6.581}$$

$$\mathbf{A}_{rt} = \int_{-1}^{+1}\int_{-1}^{+1} \Xi_{rt} J \, d\xi d\eta = r_0 \frac{1}{6}\alpha\beta \left\{ \begin{array}{cc:cc} -2 & -2 & -1 & -1 \\ 2 & 2 & 1 & 1 \\ \hdashline -1 & -1 & -2 & -2 \\ 1 & 1 & 2 & 2 \end{array} \right\} \tag{6.582}$$

$$\mathbf{A}_{z1} = \int_{-1}^{+1}\int_{-1}^{+1} \Xi_{z1} J \, d\xi d\eta = r_0 \frac{\alpha}{6} \left\{ \begin{array}{cc:cc} -2 & -1 & -2 & -1 \\ -1 & -2 & -1 & -2 \\ \hdashline 2 & 1 & 2 & 1 \\ 1 & 2 & 1 & 2 \end{array} \right\} \tag{6.583}$$

$$\mathbf{A}_{tz} = \int_{-1}^{+1}\int_{-1}^{+1} \Xi_{z1} J \, d\xi d\eta = r_0 \frac{\alpha}{6} \left\{ \begin{array}{cc:cc} -2 & -1 & 2 & 1 \\ -1 & -2 & 1 & 2 \\ \hdashline -2 & -1 & 2 & 1 \\ -1 & -2 & 1 & 2 \end{array} \right\} \tag{6.584}$$

$$\mathbf{A}_{tt} = \int_{-1}^{+1}\int_{-1}^{+1} \Xi_{tt} J \, d\xi d\eta = r_0 \frac{\beta}{12\alpha} \left\{ \begin{array}{cc:cc} 2a_1 & 2a_2 & a_1 & a_2 \\ 2a_2 & 2a_3 & a_2 & a_3 \\ \hdashline a_1 & a_2 & 2a_1 & 2a_2 \\ a_2 & a_3 & 2a_2 & 2a_3 \end{array} \right\}, \qquad \begin{array}{l} a_1 = (\alpha+1)^2 L - 2\alpha(1+2\alpha) \\ a_2 = (\alpha^2-1)L + 2\alpha \\ a_3 = (\alpha-1)^2 L - 2\alpha(1-2\alpha) \end{array} \tag{6.585}$$

Observe that

$$\lim_{\alpha\to 1} a_1 = \infty, \qquad \lim_{\alpha\to 1} a_2 = 2, \qquad \lim_{\alpha\to 1} a_3 = 2 \tag{6.586}$$

These matrices are of the form

$$\mathbf{A}_{rr} = \begin{Bmatrix} 2\mathbf{B}_{rr} & \mathbf{B}_{rr} \\ \mathbf{B}_{rr} & 2\mathbf{B}_{rr} \end{Bmatrix}, \qquad\qquad \mathbf{B}_{rr} = r_0 \frac{\beta}{6}\begin{Bmatrix} 1 & -1 \\ -1 & 1 \end{Bmatrix} \tag{6.587}$$

$$\mathbf{A}_{rz} = \begin{Bmatrix} \mathbf{B}_{rz} & -\mathbf{B}_{rz} \\ \mathbf{B}_{rz} & -\mathbf{B}_{rz} \end{Bmatrix}, \qquad\qquad \mathbf{B}_{rz} = r_0 \frac{1}{12}\left[3\begin{Bmatrix} 1 & 1 \\ -1 & -1 \end{Bmatrix} + \alpha\begin{Bmatrix} -1 & 1 \\ 1 & -1 \end{Bmatrix} \right] \tag{6.588}$$

$$\mathbf{A}_{zz} = \begin{Bmatrix} \mathbf{B}_{zz} & -\mathbf{B}_{zz} \\ -\mathbf{B}_{zz} & \mathbf{B}_{zz} \end{Bmatrix}, \qquad\qquad \mathbf{B}_{zz} = \frac{r_0}{6\beta}\left[\begin{Bmatrix} 2 & 1 \\ 1 & 2 \end{Bmatrix} + \alpha\begin{Bmatrix} -1 & 0 \\ 0 & 1 \end{Bmatrix} \right] \tag{6.589}$$

$$\mathbf{A}_{rt} = \begin{Bmatrix} 2\mathbf{B}_{rt} & \mathbf{B}_{rt} \\ \mathbf{B}_{rt} & 2\mathbf{B}_{rt} \end{Bmatrix}, \qquad\qquad \mathbf{B}_{rt} = \frac{r_0\alpha\beta}{6}\begin{Bmatrix} -1 & -1 \\ 1 & 1 \end{Bmatrix} \tag{6.590}$$

$$\mathbf{A}_{z1} = \begin{Bmatrix} -\mathbf{B}_{z1} & -\mathbf{B}_{z1} \\ \mathbf{B}_{z1} & \mathbf{B}_{z1} \end{Bmatrix}, \qquad\qquad \mathbf{B}_{z1} = \frac{r_0\alpha}{6}\begin{Bmatrix} 2 & 1 \\ 1 & 2 \end{Bmatrix} \tag{6.591}$$

$$\mathbf{A}_{tz} = \begin{Bmatrix} -\mathbf{B}_{tz} & \mathbf{B}_{tz} \\ -\mathbf{B}_{tz} & \mathbf{B}_{tz} \end{Bmatrix}, \qquad\qquad \mathbf{B}_{tz} = \frac{r_0\alpha}{6}\begin{Bmatrix} 2 & 1 \\ 1 & 2 \end{Bmatrix} \tag{6.592}$$

$$\mathbf{A}_{tt} = \begin{Bmatrix} 2\mathbf{B}_{tt} & \mathbf{B}_{tt} \\ \mathbf{B}_{tt} & 2\mathbf{B}_{tt} \end{Bmatrix}, \qquad\qquad \mathbf{B}_{tt} = \frac{r_0\beta}{12\alpha} \begin{bmatrix} a_1 & a_2 \\ a_2 & a_3 \end{bmatrix} \tag{6.593}$$

where

$$\begin{aligned} a_1 &= (\alpha+1)^2 L - 2\alpha(1+2\alpha) \\ a_2 &= (\alpha^2-1)L + 2\alpha \\ a_3 &= (\alpha-1)^2 L - 2\alpha(1-2\alpha) \end{aligned} \tag{6.594}$$

Also,

$$\mathbf{A}_{r1} = \mathbf{A}_{rt}, \mathbf{A}_{t1} = \mathbf{A}_{11} = \mathbf{A}_{tt} \tag{6.595}$$

To construct the element stiffness matrix, insert the above matrices into the corresponding material matrices, expanding the zeros to match the size of the $\mathbf{A}_{\alpha\beta}$ matrices. For example,

$$\mathbf{H}_{rz} + \mathbf{H}_{zr} \rightarrow \begin{Bmatrix} \mathbf{0} & \mathbf{0} & \lambda\mathbf{A}_{rz} \\ \mathbf{0} & \mathbf{0} & \mathbf{0} \\ \mu\mathbf{A}_{rz} & \mathbf{0} & \mathbf{0} \end{Bmatrix} + \begin{Bmatrix} \mathbf{0} & \mathbf{0} & \lambda\mathbf{A}_{rz} \\ \mathbf{0} & \mathbf{0} & \mathbf{0} \\ \mu\mathbf{A}_{rz} & \mathbf{0} & \mathbf{0} \end{Bmatrix}^T \tag{6.596}$$

and so on. Observe that some terms must still be multiplied by either n (those in $\mathbf{A}_{r\theta}, \mathbf{A}_{\theta z}$) or n^2 (those in $\mathbf{A}_{\theta\theta}, \mathbf{A}_{\theta1}$), see the integral defining $\mathbf{K}$ given earlier in this section.

Mass Matrix
For each DOF, the mass matrix is

$$\mathbf{M} = \begin{Bmatrix} 2\mathbf{M}_0 & \mathbf{M}_0 \\ \mathbf{M}_0 & 2\mathbf{M}_0 \end{Bmatrix}, \qquad\qquad \mathbf{M}_0 = r_0 \frac{\rho ab}{36} \begin{Bmatrix} 2-\alpha & 1 \\ 1 & 2+\alpha \end{Bmatrix} \tag{6.597}$$

(b) Quadratic Expansion
The component matrices can be written compactly as

$$\mathbf{A}_{rr} = r_0 \frac{\beta}{90} \begin{bmatrix} 4\mathbf{B}_{rr} & 2\mathbf{B}_{rr} & -\mathbf{B}_{rr} \\ 2\mathbf{B}_{rr} & 16\mathbf{B}_{rr} & 2\mathbf{B}_{rr} \\ -\mathbf{B}_{rr} & 2\mathbf{B}_{rr} & 4\mathbf{B}_{rr} \end{bmatrix}, \quad \mathbf{B}_{rr} = \begin{Bmatrix} 7 & -8 & 1 \\ -8 & 16 & -8 \\ 1 & -8 & 7 \end{Bmatrix} + \alpha \begin{bmatrix} -4 & 4 & 0 \\ 4 & 0 & -4 \\ 0 & -4 & 4 \end{bmatrix} \tag{6.598}$$

$$\mathbf{A}_{rz} = r_0 \frac{1}{180} \begin{bmatrix} 3\mathbf{B}_{rz} & -4\mathbf{B}_{rz} & \mathbf{B}_{rz} \\ 4\mathbf{B}_{rz} & \mathbf{0} & -4\mathbf{B}_{rz} \\ -\mathbf{B}_{rz} & 4\mathbf{B}_{rz} & -3\mathbf{B}_{rz} \end{bmatrix}, \quad \mathbf{B}_{rz} = \begin{Bmatrix} 15 & 20 & -5 \\ -20 & 0 & 20 \\ 5 & -20 & -15 \end{Bmatrix} + \alpha \begin{bmatrix} -11 & -8 & -1 \\ 12 & 16 & 12 \\ -1 & -8 & -11 \end{bmatrix} \tag{6.599}$$

$$\mathbf{A}_{zz} = r_0 \frac{1}{90\beta} \begin{bmatrix} 7\mathbf{B}_{zz} & -8\mathbf{B}_{zz} & \mathbf{B}_{zz} \\ -8\mathbf{B}_{zz} & 16\mathbf{B}_{zz} & -8\mathbf{B}_{zz} \\ \mathbf{B}_{zz} & -8\mathbf{B}_{zz} & 7\mathbf{B}_{zz} \end{bmatrix}, \quad \mathbf{B}_{zz} = \begin{Bmatrix} 4 & 2 & -1 \\ 2 & 16 & 2 \\ -1 & 2 & 4 \end{Bmatrix} + \alpha \begin{bmatrix} -3 & -2 & 0 \\ -2 & 0 & 2 \\ 0 & 2 & 3 \end{bmatrix} \tag{6.600}$$

$$\mathbf{A}_{r1} = r_0 \frac{\alpha\beta}{90} \begin{Bmatrix} 4\mathbf{B}_{r1} & 2\mathbf{B}_{r1} & -\mathbf{B}_{r1} \\ 2\mathbf{B}_{r1} & 16\mathbf{B}_{r1} & 2\mathbf{B}_{r1} \\ -\mathbf{B}_{r1} & 2\mathbf{B}_{r1} & 4\mathbf{B}_{r1} \end{Bmatrix}, \qquad \mathbf{B}_{r1} = \begin{Bmatrix} -3 & -4 & 1 \\ 4 & 0 & -4 \\ -1 & 4 & 3 \end{Bmatrix} \tag{6.601}$$

$$\mathbf{A}_{z1} = r_0 \frac{\alpha}{90} \begin{Bmatrix} -3\mathbf{B}_{z1} & -4\mathbf{B}_{z1} & \mathbf{B}_{z1} \\ 4\mathbf{B}_{z1} & \mathbf{O} & -4\mathbf{B}_{z1} \\ -\mathbf{B}_{z1} & 4\mathbf{B}_{z1} & 3\mathbf{B}_{z1} \end{Bmatrix}, \qquad \mathbf{B}_{z1} = \begin{Bmatrix} 4 & 2 & -1 \\ 2 & 16 & 2 \\ -1 & 2 & 4 \end{Bmatrix} \tag{6.602}$$

Define

$$b_1 = 3(1+\alpha)^2 L - 2\alpha(2\alpha^3 + 4\alpha^2 + 6\alpha + 3) \tag{6.603}$$

$$b_2 = 3(1-\alpha^2)L - 2\alpha(3 - 2\alpha^2) \tag{6.604}$$

$$b_3 = 4\left[3(1-\alpha^2)^2 L - 2\alpha(3 - 5\alpha^2)\right] \tag{6.605}$$

$$b_4 = 3(1-\alpha)^2 L + 2\alpha(2\alpha^3 - 4\alpha^2 + 6\alpha - 3) \tag{6.606}$$

Then

$$\mathbf{A}_{11} = \begin{Bmatrix} 4\mathbf{B} & 2\mathbf{B} & -\mathbf{B} \\ 2\mathbf{B} & 16\mathbf{B} & 2\mathbf{B} \\ -\mathbf{B} & 2\mathbf{B} & 4\mathbf{B} \end{Bmatrix}, \qquad \mathbf{B} = \frac{r_0 \beta}{180\alpha^3} \begin{Bmatrix} b_1 & -2(1+\alpha)b_2 & b_2 \\ -2(1+\alpha)b_2 & b_3 & -2(1-\alpha)b_2 \\ b_2 & -2(1-\alpha)b_2 & b_4 \end{Bmatrix} \tag{6.607}$$

$$\mathbf{M} = \begin{Bmatrix} 4\mathbf{M}_0 & 2\mathbf{M}_0 & -\mathbf{M}_0 \\ 2\mathbf{M}_0 & 16\mathbf{M}_0 & 2\mathbf{M}_0 \\ -\mathbf{M}_0 & 2\mathbf{M}_0 & 4\mathbf{M}_0 \end{Bmatrix}, \qquad \mathbf{M}_0 = r_0 \frac{\rho ab}{900} \left[\begin{Bmatrix} 4 & 2 & -1 \\ 2 & 16 & 2 \\ -1 & 2 & 4 \end{Bmatrix} + \alpha \begin{Bmatrix} -3 & -2 & 0 \\ -2 & 0 & 2 \\ 0 & 2 & 3 \end{Bmatrix} \right] \tag{6.608}$$

In expanded form, the matrices for the quadratic case are

$$\mathbf{A}_{r1} = r_0 \frac{\alpha\beta}{90} \begin{Bmatrix} -12 & -16 & 4 & \vdots & -6 & -8 & 2 & \vdots & 3 & 4 & -1 \\ 16 & 0 & -16 & \vdots & 8 & 0 & -8 & \vdots & -4 & 0 & 4 \\ -4 & 16 & 12 & \vdots & -2 & 8 & 6 & \vdots & 1 & -4 & -3 \\ \cdots & & & & & & & & & & \\ -6 & -8 & 2 & \vdots & -48 & -64 & 16 & \vdots & -6 & -8 & 2 \\ 8 & 0 & -8 & \vdots & 64 & 0 & -64 & \vdots & 8 & 0 & -8 \\ -2 & 8 & 6 & \vdots & -16 & 64 & 48 & \vdots & -2 & 8 & 6 \\ \cdots & & & & & & & & & & \\ 3 & 4 & -1 & \vdots & -6 & -8 & 2 & \vdots & -12 & -16 & 4 \\ -4 & 0 & 4 & \vdots & 8 & 0 & -8 & \vdots & 16 & 0 & -16 \\ 1 & -4 & -3 & \vdots & -2 & 8 & 6 & \vdots & -4 & 16 & 12 \end{Bmatrix} \tag{6.609}$$

$$
\mathbf{A}_{rr} = r_0 \frac{\beta}{90}
\left\{
\begin{array}{ccc:ccc:ccc}
28 & -32 & 4 & 14 & -16 & 2 & -7 & 8 & -1 \\
-32 & 64 & -32 & -16 & 32 & -16 & 8 & -16 & 8 \\
4 & -32 & 28 & 2 & -16 & 14 & -1 & 8 & -7 \\
\hdashline
14 & -16 & 2 & 112 & -128 & 16 & 14 & -16 & 2 \\
-16 & 32 & -16 & -128 & 256 & -128 & -16 & 32 & -16 \\
2 & -16 & 14 & 16 & -128 & 112 & 2 & -16 & 14 \\
\hdashline
-7 & 8 & -1 & 14 & -16 & 2 & 28 & -32 & 4 \\
8 & -16 & 8 & -16 & 32 & -16 & -32 & 64 & -32 \\
-1 & 8 & -7 & 2 & -16 & 14 & 4 & -32 & 28
\end{array}
\right\}
$$

$$
+ r_0 \frac{\alpha\beta}{90}
\left[
\begin{array}{ccc:ccc:ccc}
-16 & 16 & 0 & -8 & 8 & 0 & 4 & -4 & 0 \\
16 & 0 & -16 & 8 & 0 & -8 & -4 & 0 & 4 \\
0 & -16 & 16 & 0 & -8 & 8 & 0 & 4 & -4 \\
\hdashline
-8 & 8 & 0 & -64 & 64 & 0 & -8 & 8 & 0 \\
8 & 0 & -8 & 64 & 0 & -64 & 8 & 0 & -8 \\
0 & -8 & 8 & 0 & -64 & 64 & 0 & -8 & 8 \\
\hdashline
4 & -4 & 0 & -8 & 8 & 0 & -16 & 16 & 0 \\
-4 & 0 & 4 & 8 & 0 & -8 & 16 & 0 & -16 \\
0 & 4 & -4 & 0 & -8 & 8 & 0 & -16 & 16
\end{array}
\right]
$$

(6.610)

$$
\mathbf{A}_{rz} = r_0 \frac{1}{180}
\left\{
\begin{array}{ccc:ccc:ccc}
45 & 60 & -15 & -60 & -80 & 20 & 15 & 20 & -5 \\
-60 & 0 & 60 & 80 & 0 & -80 & -20 & 0 & 20 \\
15 & -60 & -45 & -20 & 80 & 60 & 5 & -20 & -15 \\
\hdashline
60 & 80 & -20 & 0 & 0 & 0 & -60 & -80 & 20 \\
-80 & 0 & 80 & 0 & 0 & 0 & 80 & 0 & -80 \\
20 & -80 & -60 & 0 & 0 & 0 & -20 & 80 & 60 \\
\hdashline
-15 & -20 & 5 & 60 & 80 & -20 & -45 & -60 & 15 \\
20 & 0 & -20 & -80 & 0 & 80 & 60 & 0 & -60 \\
-5 & 20 & 15 & 20 & -80 & -60 & -15 & 60 & 45
\end{array}
\right\}
$$

$$
+ r_0 \frac{\alpha}{180}
\left[
\begin{array}{ccc:ccc:ccc}
-33 & -24 & -3 & 44 & 32 & 4 & -11 & -8 & -1 \\
36 & 48 & 36 & -48 & -64 & -48 & 12 & 16 & 12 \\
-3 & -24 & -33 & 4 & 32 & 44 & -1 & -8 & -11 \\
\hdashline
-44 & -32 & -4 & 0 & 0 & 0 & 44 & 32 & 4 \\
48 & 64 & 48 & 0 & 0 & 0 & -48 & -64 & -48 \\
-4 & -32 & -44 & 0 & 0 & 0 & 4 & 32 & 44 \\
\hdashline
11 & 8 & 1 & -44 & -32 & -4 & 33 & 24 & 3 \\
-12 & -16 & -12 & 48 & 64 & 48 & -36 & -48 & -36 \\
1 & 8 & 11 & -4 & -32 & -44 & 3 & 24 & 33
\end{array}
\right]
$$

(6.611)

$$
\mathbf{A}_{zz} = \frac{r_0}{90\beta}
\begin{Bmatrix}
28 & 14 & -7 & -32 & -16 & 8 & 4 & 2 & -1 \\
14 & 112 & 14 & -16 & -128 & -16 & 2 & 16 & 2 \\
-7 & 14 & 28 & 8 & -16 & -32 & -1 & 2 & 4 \\
-32 & -16 & 8 & 64 & 32 & -16 & -32 & -16 & 8 \\
-16 & -128 & -16 & 32 & 256 & 32 & -16 & -128 & -16 \\
8 & -16 & -32 & -16 & 32 & 64 & 8 & -16 & -32 \\
4 & 2 & -1 & -32 & -16 & 8 & 28 & 14 & -7 \\
2 & 16 & 2 & -16 & -128 & -16 & 14 & 112 & 14 \\
-1 & 2 & 4 & 8 & -16 & -32 & -7 & 14 & 28
\end{Bmatrix}
$$

$$
+ \frac{r_0\,\alpha}{90\beta}
\begin{Bmatrix}
-21 & -14 & 0 & 24 & 16 & 0 & -3 & -2 & 0 \\
-14 & 0 & 14 & 16 & 0 & -16 & -2 & 0 & 2 \\
0 & 14 & 21 & 0 & -16 & -24 & 0 & 2 & 3 \\
24 & 16 & 0 & -48 & -32 & 0 & 24 & 16 & 0 \\
16 & 0 & -16 & -32 & 0 & 32 & 16 & 0 & -16 \\
0 & -16 & -24 & 0 & 32 & 48 & 0 & -16 & -24 \\
-3 & -2 & 0 & 24 & 16 & 0 & -21 & -14 & 0 \\
-2 & 0 & 2 & 16 & 0 & -16 & -14 & 0 & 14 \\
0 & 2 & 3 & 0 & -16 & -24 & 0 & 14 & 21
\end{Bmatrix}
$$

(6.612)

$$
\mathbf{A}_{z1} = r_0\,\frac{\alpha}{90}
\begin{Bmatrix}
-12 & -6 & 3 & -16 & -8 & 4 & 4 & 2 & -1 \\
-6 & -48 & -6 & -8 & -64 & -8 & 2 & 16 & 2 \\
3 & -6 & -12 & 4 & -8 & -16 & -1 & 2 & 4 \\
16 & 8 & -4 & 0 & 0 & 0 & -16 & -8 & 4 \\
8 & 64 & 8 & 0 & 0 & 0 & -8 & -64 & -8 \\
-4 & 8 & 16 & 0 & 0 & 0 & 4 & -8 & -16 \\
-4 & -2 & 1 & 16 & 8 & -4 & 12 & 6 & -3 \\
-2 & -16 & -2 & 8 & 64 & 8 & 6 & 48 & 6 \\
1 & -2 & -4 & -4 & 8 & 16 & -3 & 6 & 12
\end{Bmatrix}
$$

(6.613)

Also, the elements of the $\mathbf{A}_{11}$ matrix are

$$
a_{11} = \frac{r_0\beta}{45\alpha^3}\left[3(1+\alpha)^2 L - 2\alpha(2\alpha^3 + 4\alpha^2 + 6\alpha + 3)\right]
$$

(6.614)

$$
a_{21} = -\frac{2 r_0\beta(1+\alpha)}{45\alpha^3}\left[3(1-\alpha^2)L - 2\alpha(3-2\alpha^2)\right]
$$

(6.615)

$$
a_{31} = \frac{r_0\beta}{45\alpha^3}\left[3(1-\alpha^2)L - 2\alpha(3-2\alpha^2)\right]
$$

(6.616)

$$a_{41} = \frac{r_0 \beta}{90\alpha^3} \left[3(1+\alpha)^2 L - 2\alpha(2\alpha^3 + 4\alpha^2 + 6\alpha + 3) \right] \tag{6.617}$$

$$a_{51} = -\frac{r_0 \beta(1+\alpha)}{45\alpha^3} \left[3(1-\alpha^2)L - 2\alpha(3 - 2\alpha^2) \right] \tag{6.618}$$

$$a_{61} = \frac{r_0 \beta}{90\alpha^3} \left[3(1-\alpha^2)L - 2\alpha(3 - 2\alpha^2) \right] \tag{6.619}$$

$$a_{71} = -\frac{r_0 \beta}{180\,\alpha^3} \left[3(1+\alpha)^2 L - 2\alpha(2\alpha^3 + 4\alpha^2 + 6\alpha + 3) \right] \tag{6.620}$$

$$a_{81} = \frac{r_0 \beta(1+\alpha)}{90\alpha^3} \left[3(1-\alpha^2)L - 2\alpha(3 - 2\alpha^2) \right] \tag{6.621}$$

$$a_{91} = -\frac{r_0 \beta}{180\alpha^3} \left[3(1-\alpha^2)L - 2\alpha(3 - 2\alpha^2) \right] \tag{6.622}$$

$$a_{22} = \frac{4r_0 \beta}{45\alpha^3} \left[3(1-\alpha^2)^2 L - 2\alpha(3 - 5\alpha^2) \right] \tag{6.623}$$

$$a_{32} = -\frac{2r_0 \beta(1-\alpha)}{45\alpha^3} \left[3(1-\alpha^2)L - 2\alpha(3 - 2\alpha^2) \right] \tag{6.624}$$

$$a_{42} = -\frac{r_0 \beta(1+\alpha)}{45\alpha^3} \left[3(1-\alpha^2)L - 2\alpha(3 - 2\alpha^2) \right] \tag{6.625}$$

$$a_{52} = \frac{2r_0 \beta}{45\alpha^3} \left[3(1-\alpha^2)^2 L - 2\alpha(3 - 5\alpha^2) \right] \tag{6.626}$$

$$a_{62} = -\frac{r_0 \beta(1-\alpha)}{45\alpha^3} \left[3(1-\alpha^2)L - 2\alpha(3 - 2\alpha^2) \right] \tag{6.627}$$

$$a_{72} = \frac{r_0 \beta(1+\alpha)}{90\alpha^3} \left[3(1-\alpha^2)L - 2\alpha(3 - 2\alpha^2) \right] \tag{6.628}$$

$$a_{82} = -\frac{r_0 \beta}{45\alpha^3} \left[3(1-\alpha^2)^2 L - 2\alpha(3 - 5\alpha^2) \right] \tag{6.629}$$

$$a_{92} = \frac{r_0 \beta(1-\alpha)}{90\alpha^3} \left[3(1-\alpha^2)L - 2\alpha(3 - 2\alpha^2) \right] \tag{6.630}$$

$$a_{33} = \frac{r_0 \beta}{45\alpha^3} \left[3(1-\alpha)^2 L + 2\alpha(2\alpha^3 - 4\alpha^2 + 6\alpha - 3) \right] \qquad (6.631)$$

$$a_{43} = \frac{r_0 \beta}{90\alpha^3} \left[3(1-\alpha^2) L - 2\alpha(3 - 2\alpha^2) \right] \qquad (6.632)$$

$$a_{53} = -\frac{r_0 \beta (1-\alpha)}{45\alpha^3} \left[3(1-\alpha^2) L - 2\alpha(3 - 2\alpha^2) \right] \qquad (6.633)$$

$$a_{63} = \frac{r_0 \beta}{90\alpha^3} \left[3(1-\alpha)^2 L + 2\alpha(2\alpha^3 - 4\alpha^2 + 6\alpha - 3) \right] \qquad (6.634)$$

$$a_{73} = -\frac{r_0 \beta}{180\alpha^3} \left[3(1-\alpha^2) L - 2\alpha(3 - 2\alpha^2) \right] \qquad (6.635)$$

$$a_{83} = \frac{r_0 \beta (1-\alpha)}{90\alpha^3} \left[3(1-\alpha^2) L - 2\alpha(3 - 2\alpha^2) \right] \qquad (6.636)$$

$$a_{93} = -\frac{r_0 \beta}{180\alpha^3} \left[3(1-\alpha)^2 L + 2\alpha(2\alpha^3 - 4\alpha^2 + 6\alpha - 3) \right] \qquad (6.637)$$

$$a_{44} = \frac{4r_0 \beta}{45\alpha^3} \left[3(1+\alpha)^2 L - 2\alpha(2\alpha^3 + 4\alpha^2 + 6\alpha + 3) \right] \qquad (6.638)$$

$$a_{54} = -\frac{8r_0 \beta (1+\alpha)}{45\alpha^3} \left[3(1-\alpha^2) L - 2\alpha(3 - 2\alpha^2) \right] \qquad (6.639)$$

$$a_{64} = \frac{4r_0 \beta}{45\alpha^3} \left[3(1-\alpha^2) L - 2\alpha(3 - 2\alpha^2) \right] \qquad (6.640)$$

$$a_{74} = \frac{r_0 \beta}{90\alpha^3} \left[3(1+\alpha)^2 L - 2\alpha(2\alpha^3 + 4\alpha^2 + 6\alpha + 3) \right] \qquad (6.641)$$

$$a_{84} = -\frac{r_0 \beta (1+\alpha)}{45\alpha^3} \left[3(1-\alpha^2) L - 2\alpha(3 - 2\alpha^2) \right] \qquad (6.642)$$

$$a_{94} = \frac{r_0 \beta}{90\alpha^3} \left[3(1-\alpha^2) L - 2\alpha(3 - 2\alpha^2) \right] \qquad (6.643)$$

$$a_{55} = \frac{16 r_0 \beta}{45\alpha^3} \left[3(1-\alpha^2)^2 L - 2\alpha(3 - 5\alpha^2) \right] \qquad (6.644)$$

$$a_{65} = -\frac{8r_0 \beta (1-\alpha)}{45\alpha^3} \left[3(1-\alpha^2) L - 2\alpha(3 - 2\alpha^2) \right] \qquad (6.645)$$

$$a_{75} = -\frac{r_0\beta(1+\alpha)}{45\alpha^3}\left[3(1-\alpha^2)L - 2\alpha(3-2\alpha^2)\right] \tag{6.646}$$

$$a_{85} = \frac{2r_0\beta}{45\alpha^3}\left[3(1-\alpha^2)^2 L - 2(3-5\alpha^2)\right] \tag{6.647}$$

$$a_{95} = -\frac{r_0\beta(1-\alpha)}{45\alpha^3}\left[3(1-\alpha^2)L - 2\alpha(3-2\alpha^2)\right] \tag{6.648}$$

$$a_{66} = \frac{4r_0\beta}{45\alpha^3}\left[3(1-\alpha)^2 L + 2\alpha(2\alpha^3 - 4\alpha^2 + 6\alpha - 3)\right] \tag{6.649}$$

$$a_{76} = \frac{r_0\beta}{90\alpha^3}\left[3(1-\alpha^2) - 2\alpha(3-2\alpha^2)\right] \tag{6.650}$$

$$a_{86} = -\frac{r_0\beta(1-\alpha)}{45\alpha^3}\left[3(1-\alpha^2)L - 2\alpha(3-2\alpha^2)\right] \tag{6.651}$$

$$a_{96} = \frac{r_0\beta}{90\alpha^3}\left[3(1-\alpha)^2 + 2\alpha(2\alpha^3 - 4\alpha^2 + 6\alpha - 3)\right] \tag{6.652}$$

Consistent Nodal Loads

For rectangular elements, the body loads contribute consistent nodal forces

$$\mathbf{P} = \iint \left\{ \begin{bmatrix} \mathbf{N}^T & \mathbf{0} & \mathbf{0} \\ \mathbf{0} & \mathbf{N}^T & \mathbf{0} \\ \mathbf{0} & \mathbf{0} & \mathbf{N}^T \end{bmatrix} \begin{bmatrix} q_r \\ q_\theta \\ q_z \end{bmatrix} \right\} r\, dr\, dz = \left\{ \begin{matrix} \mathbf{p}_r \\ \mathbf{p}_\theta \\ \mathbf{p}_z \end{matrix} \right\} \tag{6.653}$$

so for each coordinate direction $j = r, \theta, z$, we must determine an integral of the form

$$\mathbf{p}_j = \iint \mathbf{N}^T q_j \; r\, dr\, dz \tag{6.654}$$

Expressing the load in terms of the interpolation functions $\mathbf{p}_j = \mathbf{N}\mathbf{q}_j$, where $\mathbf{q}_j$ is the vector that contains the values which the body load attains at the nodes, then

$$\mathbf{p}_j = \left\{ \iint \mathbf{N}^T \mathbf{N} \; r\, dr\, dz \right\} \mathbf{q}_j = \frac{1}{\rho}\left\{ \iint \rho \mathbf{N}^T \mathbf{N} \; r\, dr\, dz \right\} \mathbf{q}_j = \frac{1}{\rho}\mathbf{M}\mathbf{q}_j \tag{6.655}$$

That is, the consistent nodal forces can be obtained from the mass matrix $\mathbf{M}$ associated with each DOF by appropriate summation with the nodal values of the body load.

Loads that are *uniformly* distributed on either the upper or lower surface of the element can be idealized as body loads of intensity q/b acting on an element that is infinitesimally thin in the vertical direction that is, $b \to 0$, in which case the consistent nodal loads of the upper and lower nodes merge into one, and the vertical thickness factor b cancels out. This is accomplished by adding up all submatrices of the mass matrix as will be shown, and defining the uniform load as $\mathbf{q} = q\,\mathbf{1}$, where $\mathbf{1}$ is a vector composed of ones.

For a linear expansion, this results in the following:

$$\mathbf{p} = \frac{q}{\rho b}[\mathbf{I} \quad \mathbf{I}]\begin{Bmatrix} 2\mathbf{M}_0 & \mathbf{M}_0 \\ \mathbf{M}_0 & 2\mathbf{M}_0 \end{Bmatrix}\begin{bmatrix} \mathbf{I} \\ \mathbf{I} \end{bmatrix}\mathbf{1} = \frac{6q}{\rho b}\mathbf{M}_0\mathbf{1}$$

$$= q\frac{ar_0}{6}\begin{bmatrix} 2-\alpha & 1 \\ 1 & 2+\alpha \end{bmatrix}\begin{bmatrix} 1 \\ 1 \end{bmatrix}, \qquad \boxed{\alpha = \frac{a}{2r_0}}$$

(6.656)

That is,

$$\boxed{\mathbf{p} = q\frac{ar_0}{6}\begin{bmatrix} 3-\alpha \\ 3+\alpha \end{bmatrix}}$$

(6.657)

which gives the nodal loads at the two nodes that define the side of length a. In particular, if $r_1 = 0$, $r_2 = a$, $r_0 = \frac{1}{2}a$, $\alpha = 1$, then

$$\mathbf{p} = q\frac{a^2}{6}\begin{bmatrix} 1 \\ 2 \end{bmatrix} \qquad 2\pi(p_1 + p_2) = 2\pi(q\frac{1}{2}a^2) = \pi a^2 q$$

(6.658)

which is the correct result for the total load acting on a disk of radius a.

Similarly, for a quadratic expansion

$$\frac{q}{\rho b}\{\mathbf{I} \quad \mathbf{I} \quad \mathbf{I}\}\begin{Bmatrix} 4\mathbf{M}_0 & 2\mathbf{M}_0 & -\mathbf{M}_0 \\ 2\mathbf{M}_0 & 16\mathbf{M}_0 & 2\mathbf{M}_0 \\ -\mathbf{M}_0 & 2\mathbf{M}_0 & 4\mathbf{M}_0 \end{Bmatrix}\begin{bmatrix} \mathbf{I} \\ \mathbf{I} \\ \mathbf{I} \end{bmatrix} = \frac{30q}{\rho b}\mathbf{M}_0$$

(6.659)

$$\mathbf{p} = \frac{30q}{\rho b}\mathbf{M}_0\mathbf{1} = q\frac{ar_0}{30}\begin{Bmatrix} 4-3\alpha & 2(1-\alpha) & -1 \\ 2(1-\alpha) & 16 & 2(1+\alpha) \\ -1 & 2(1+\alpha) & 4+3\alpha \end{Bmatrix}\begin{bmatrix} 1 \\ 1 \\ 1 \end{bmatrix}$$

(6.660)

That is,

$$\boxed{\mathbf{p} = q\frac{ar_0}{6}\begin{Bmatrix} 1-\alpha \\ 4 \\ 1+\alpha \end{Bmatrix}}$$

(6.661)

which gives the nodal forces at the three nodes lying on the horizontal edge at which the load is applied. Again, if we consider a uniform load acting on a solid disk, we recover the total load $\pi a^2 q$.

A more general method is given next.

Consistent Nodal Forces for Inclined and Curved Edges

We now proceed to develop the consistent loads for the more general case where the load is distributed along a nonhorizontal edge as shown in Figure 6.36. The consistent nodal load is now of the form

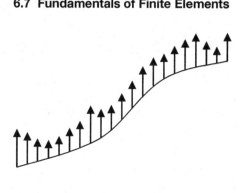

Figure 6.36. Surface tractions distributed on a curvilinear edge.

$$\mathbf{p} = \int_{s_1}^{s_2} \mathbf{N}^T q(s) \, r \, ds \tag{6.662}$$

where the integral extends along the edge with curvilinear coordinate s, and $q(s)$ is a surface traction. We have then

$$r = \mathbf{N}\mathbf{R} = \mathbf{R}^T\mathbf{N}^T \qquad z = \mathbf{N}\mathbf{Z} = \mathbf{Z}^T\mathbf{N}^T, \qquad q = \mathbf{N}\mathbf{q} = \mathbf{q}^T\mathbf{N}^T \tag{6.663}$$

$$\mathbf{p} = \left\{ \int_{s_1}^{s_2} \mathbf{N}^T\mathbf{N} \, r \, ds \right\} \mathbf{q} \tag{6.664}$$

$$ds = \sqrt{dr^2 + dz^2} = d\xi \sqrt{\left(\tfrac{dr}{d\xi}\right)^2 + \left(\tfrac{dz}{d\xi}\right)^2} = d\xi \sqrt{\left(\mathbf{N}_\xi \mathbf{R}\right)^2 + \left(\mathbf{N}_\xi \mathbf{Z}\right)^2} \tag{6.665}$$

(a) Linear Expansion

$$\mathbf{N} = \tfrac{1}{2}\left[(1-\xi) \quad (1+\xi) \right], \qquad \mathbf{N}^T\mathbf{N} = \tfrac{1}{4} \begin{Bmatrix} (1-\xi)^2 & 1-\xi^2 \\ 1-\xi^2 & (1+\xi)^2 \end{Bmatrix} \tag{6.666}$$

$$\begin{aligned} r &= \tfrac{1}{2}\left[(1-\xi)r_1 + (1+\xi)r_2 \right] = \tfrac{1}{2}(r_1 + r_2) + \tfrac{1}{2}(r_2 - r_1)\xi \\ &= r_0 + \tfrac{1}{2}a\xi \\ &= r_0(1 + \alpha\xi) \end{aligned} \tag{6.667}$$

$$\mathbf{N}_\xi = \tfrac{1}{2}[-1 \quad 1] \quad dr = \mathbf{N}_\xi \mathbf{R} \; d\xi = \tfrac{1}{2}(r_2 - r_1) \, d\xi, \quad dz = \mathbf{N}_\xi \mathbf{Z} \; d\xi = \tfrac{1}{2}(z_2 - z_1) \, d\xi \tag{6.668}$$

$$ds = \tfrac{1}{2} d\xi \sqrt{(r_2 - r_1)^2 + (z_2 - z_1)^2} = \tfrac{1}{2} c \, d\xi \tag{6.669}$$

where c is the length of the edge where the load is applied. Then

$$\begin{aligned} \mathbf{p} &= \left\{ \tfrac{1}{8} c r_0 \int_{-1}^{+1} \begin{Bmatrix} (1-\xi)^2 & 1-\xi^2 \\ 1-\xi^2 & (1+\xi)^2 \end{Bmatrix} (1+\alpha\xi) \, d\xi \right\} \mathbf{q} \\ &= \frac{c r_0}{6} \begin{Bmatrix} 2-\alpha & 1 \\ 1 & 2+\alpha \end{Bmatrix} \mathbf{q} = \frac{c r_0}{6} \begin{Bmatrix} (2-\alpha)q_1 + q_2 \\ q_1 + (2+\alpha)q_2 \end{Bmatrix} \end{aligned} \tag{6.670}$$

If we set $q_1 = q_2 = q$, we recover the results obtained earlier by the ad hoc method, that is,

$$\mathbf{p} = q \frac{cr_0}{6} \begin{Bmatrix} 3 - \alpha \\ 3 + \alpha \end{Bmatrix} \tag{6.671}$$

Also, for loads that vary linearly from the axis, say $q(r) = q_R r / R$, the consistent load vector is

$$\mathbf{p} = \frac{q_R}{R} \frac{cr_0}{6} \begin{Bmatrix} (2 - \alpha)r_1 + r_2 \\ r_1 + (2 + \alpha)r_2 \end{Bmatrix} = \frac{q_R}{R} \frac{cr_0^2}{6} \begin{Bmatrix} 3 - 2\alpha + \alpha^2 \\ 3 + 2\alpha + \alpha^2 \end{Bmatrix} \tag{6.672}$$

which can be useful to represent overturning moments (rocking, torsion, etc.).

(b) Quadratic Expansion

$$\mathbf{N} = \tfrac{1}{2} \left[\xi(\xi - 1) \quad 2(1 - \xi^2) \quad \xi(\xi + 1) \right] \tag{6.673}$$

$$\mathbf{N}^T \mathbf{N} = \tfrac{1}{4} \begin{Bmatrix} \xi^2(\xi - 1)^2 & 2\xi(\xi - 1)(1 - \xi^2) & \xi^2(\xi^2 - 1) \\ 2\xi(\xi - 1)(1 - \xi^2) & 4(1 - \xi^2)^2 & 2\xi(\xi + 1)(1 - \xi^2) \\ \xi^2(\xi^2 - 1) & 2\xi(\xi + 1)(1 - \xi^2) & \xi^2(\xi + 1)^2 \end{Bmatrix} \tag{6.674}$$

$$\begin{aligned} r &= \tfrac{1}{2} \left[\xi(\xi - 1)r_1 + 2(1 - \xi^2)r_2 + \xi(\xi + 1)r_3 \right] \\ &= \left[r_2 + \tfrac{1}{2}(r_3 - r_1)\xi + \tfrac{1}{2}(r_1 - 2r_2 + r_3)\xi^2 \right] \end{aligned} \tag{6.675}$$

$$\mathbf{N}_\xi = \tfrac{1}{2} \left[2\xi - 1 \quad -4\xi \quad 2\xi + 1 \right] \tag{6.676}$$

$$\begin{aligned} \mathbf{N}_\xi \mathbf{R} &= \tfrac{1}{2} \left[(2\xi - 1)r_1 - 4\xi r_2 + (2\xi + 1)r_3 \right] \\ &= \tfrac{1}{2}(r_3 - r_1) + (r_1 - 2r_2 + r_3)\xi \end{aligned} \tag{6.677}$$

$$\begin{aligned} \mathbf{N}_\xi \mathbf{Z} &= \tfrac{1}{2} \left[(2\xi - 1)z_1 - 4\xi z_2 + (2\xi + 1)z_3 \right] \\ &= \tfrac{1}{2}(z_3 - z_1) + (z_1 - 2z_2 + z_3)\xi \end{aligned} \tag{6.678}$$

$$\begin{aligned} ds &= d\xi \sqrt{\left[\tfrac{1}{2}(r_3 - r_1) + (r_1 - 2r_2 + r_3)\xi \right]^2 + \left[\tfrac{1}{2}(z_3 - z_1) + (z_1 - 2z_2 + z_3)\xi \right]^2} \\ &= d\xi \sqrt{\left[\tfrac{1}{2}a + (r_1 - 2r_2 + r_3)\xi \right]^2 + \left[\tfrac{1}{2}b + (z_1 - 2z_2 + z_3)\xi \right]^2} \end{aligned} \tag{6.679}$$

Because of the square root term, the integration will be virtually impossible to carry out analytically, but we can still do so numerically via Gaussian quadrature.

On the other hand, the two differences $e_r = r_1 - 2r_2 + r_3 = (r_3 - r_2) - (r_2 - r_1)$ and $e_z = z_1 - 2z_2 + z_3 = (z_3 - z_2) - (z_2 - z_1)$ measure the eccentricity of the intermediate node from the center. Clearly, if r_2 is close to r_1, then $e_r \to a$ and if r_2 is close to r_3, then $e_r \to -a$. Hence, the ratio $\varepsilon_r = e_r / a$ is less than one in absolute value. Likewise, the ratio $\varepsilon_z = e_z / b$ is also less than 1 in absolute value. To simplify matters, we shall assume in the ensuing that $\varepsilon_r = \varepsilon_z = \varepsilon$, which implies that the three nodes lie on a straight line (i.e., the edge is straight). Also, the common ratio satisfies $-1 \leq \varepsilon \leq 1$

$$ds = d\xi \tfrac{1}{2}\sqrt{(a^2+b^2)(1+2\varepsilon\xi)^2} = \tfrac{1}{2}c|1+2\varepsilon\xi|d\xi, \qquad c = \sqrt{a^2+b^2} \tag{6.680}$$

Simpler still, we shall now make the additional assumption that $\varepsilon = 0$, that is, that the middle node is at the center of a straight edge. Then $ds = \tfrac{1}{2}cd\xi$, and

$$r = r_0 + \tfrac{1}{2}a\xi = r_0(1+\alpha\xi) \tag{6.681}$$

So

$$\int_{s_1}^{s_3} \mathbf{N}^T\mathbf{N}\, r\, ds = \frac{cr_0}{8}\int_{-1}^{+1}\begin{Bmatrix} \xi^2(\xi-1)^2 & 2\xi(\xi-1)(1-\xi^2) & \xi^2(\xi^2-1) \\ 2\xi(\xi-1)(1-\xi^2) & 4(1-\xi^2)^2 & 2\xi(\xi+1)(1-\xi^2) \\ \xi^2(\xi^2-1) & 2\xi(\xi+1)(1-\xi^2) & \xi^2(\xi+1)^2 \end{Bmatrix}(1+\alpha\xi)\, d\xi$$

$$= \frac{cr_0}{30}\begin{Bmatrix} 4-3\alpha & 2(1-\alpha) & -1 \\ 2(1-\alpha) & 16 & 2(1+\alpha) \\ -1 & 2(1+\alpha) & 4+3\alpha \end{Bmatrix} \tag{6.682}$$

which agrees perfectly with the result obtained earlier using the ad hoc method. Hence, for an arbitrary load distribution, we obtain

$$\mathbf{p} = \frac{cr_0}{30}\begin{Bmatrix} 4-3\alpha & 2(1-\alpha) & -1 \\ 2(1-\alpha) & 16 & 2(1+\alpha) \\ -1 & 2(1+\alpha) & 4+3\alpha \end{Bmatrix}\begin{Bmatrix} q_1 \\ q_2 \\ q_3 \end{Bmatrix} = \frac{cr_0}{30}\begin{Bmatrix} (4-3\alpha)q_1+2(1-\alpha)q_2-q_3 \\ 2(1-\alpha)q_1+16q_2+2(1+\alpha)q_3 \\ (4+3\alpha)q_3+2(1+\alpha)q_2-q_1 \end{Bmatrix} \tag{6.683}$$

For a uniform load $q_1 = q_2 = q_3 = q$

$$\mathbf{p} = q\frac{cr_0}{6}\begin{Bmatrix} 1-\alpha \\ 4 \\ 1+\alpha \end{Bmatrix} \tag{6.684}$$

and for a linearly distributed load of the form $q = q_R\, r/R$

$$\mathbf{p} = \frac{q_R}{R}\frac{cr_0}{30}\begin{Bmatrix} (4-3\alpha)r_1+2(1-\alpha)r_2-r_3 \\ 2(1-\alpha)r_1+16r_2+2(1+\alpha)r_3 \\ (4+3\alpha)r_3+2(1+\alpha)r_2-r_1 \end{Bmatrix} = q_R\frac{r_0}{R}\frac{cr_0}{30}\begin{Bmatrix} 5(1-2\alpha)+3\alpha^2 \\ 4(5+\alpha^2) \\ 5(1+2\alpha)+3\alpha^2 \end{Bmatrix} \tag{6.685}$$

which for a solid cylinder with $r_0 = \frac{1}{2}a$, $R = a$, $\alpha = 1$, $c = a$ and a vertical force with $n = 1$ yields

$$\mathbf{p} = q_R \frac{a^2}{60} \begin{Bmatrix} -1 \\ 12 \\ 9 \end{Bmatrix}, \qquad \text{moment} = M = \pi q_R \frac{a^3}{60} \left(-1 \times 0 + 12 \times \tfrac{1}{2} + 9 \times 1 \right) = \frac{\pi q_R a^3}{4} \qquad (6.686)$$

which is the correct result. Observe that in this case, the force placed on node 1 (i.e., the axis) has no effect, and could be set to zero.

7 Earthquake Engineering and Soil Dynamics

7.1 Stochastic Processes in Soil Dynamics

A stochastic process is a random function of some continuous variable, usually time. For example, the acceleration time history of an earthquake at some location is a stochastic process because its intensity, duration, frequency content, and temporal evolution depend on nondeterministic parameters, such as the distance to the fault, the characteristics and length of the fault rupture, the travel path and type of the seismic waves, and so on. Thus, no two earthquakes will ever be perfectly alike, not even when both are associated with one and the same fault and are recorded at exactly the same station and with the same instrument. More generally, a *stochastic field* is a multidimensional random function of multiple variables such as the spatial location and time. An example is the spatial distribution of the three components of earthquake motion over some well-defined region, say a valley or a city. Any random process may be of either finite or infinite duration.

7.1.1 Expectations of a Random Process

As explained by Crandall,[1] a random process is defined by a family – or ensemble – of time histories that are the outcome $x(t)$ of some experiment or observation, say different earthquakes at a site, each of which constitutes a *sample*. The value attained by x at any time t is a random variable with some underlying – even if unknown – first-order probability density function $p_x = p_{x(t)}(x)$, which defines the likelihood of x lying in some interval at that instant in time. This process is associated with statistical properties referred to as moments. Formally, these are

$$m_x(t) = E[x] = \int_{-\infty}^{+\infty} x\, p_{x(t)}\, dx \qquad \text{Mean (first moment)} \tag{7.1}$$

$$E\left[x^2(t)\right] = \int_{-\infty}^{+\infty} x^2\, p_{x(t)}\, dx \qquad \text{Mean square (second moment)} \tag{7.2}$$

---

[1] S. H. Crandall and W. D. Mark, *Random Vibration in Mechanical Systems* (New York and London: Academic Press, 1963).

$$\text{var}(t) = E\left[(x - m_x)^2\right] = \int_{-\infty}^{+\infty} (x - m_x)^2 \, p_{x(t)} \, dx \qquad \text{Variance} \qquad (7.3)$$
$$= E\left[x^2\right] - m_x^2$$

$$\sigma_x(t) = \sqrt{\text{var}(t)} \qquad\qquad\qquad\qquad \text{Standard deviation} \qquad (7.4)$$

In addition, if $x_1 = x(t_1), x_2 = x(t_2)$ and $p(x_1, x_2)$ is the joint probability density function, then

$$R(t_1, t_2) = E\left[x_1 \, x_2\right] = \int_{-\infty}^{+\infty}\int_{-\infty}^{+\infty} x_1 \, x_2 \, p(x_1, x_2, t) dx_1 \, dx_2 \qquad \text{Autocorrelation function} \qquad (7.5)$$

$$C(t_1, t_2) = E\left[(x_1 - \bar{x}_1)(x_2 - \bar{x}_2)\right] \qquad\qquad \text{Covariance function} \qquad (7.6)$$

$$\rho(t_1, t_2) = \frac{C(t_1, t_2)}{\sigma_1 \sigma_2} = \frac{E\left[(x_1 - \bar{x}_1)(x_2 - \bar{x}_2)\right]}{\sqrt{E\left[(x_1 - \bar{x}_1)^2\right] E\left[(x_2 - \bar{x}_2)^2\right]}} \qquad \text{Correlation function} \qquad (7.7)$$

It can be shown that $-1 \le \rho(t_1, t_2) \le 1$. If $\rho = 1$, the samples at t_1, t_2 are perfectly correlated (i.e., the samples are linearly related in the form $x_1 = a + bx_2$ with $b > 0$); if $\rho = -1$, the samples are negatively correlated; and if $\rho = 0$, the values at t_1, t_2 are uncorrelated.

7.1.2 Functions of Random Variable

In the case of a function of random variable $u = f(x)$ whose probability density function is p_u, it is shown in books on probability theory that $p_u \, du = p_x \, dx$. Let $u = f\left[x(t)\right]$ and $w = g\left[x(t)\right]$ be two deterministic functions of the random variable $x(t)$ – for example, the strains and stresses elicited by an earthquake at some location. Thus, $u(t), w(t)$ will also be random processes whose joint probability density function can be written symbolically as $p_{uw} = p(u, w)$. If $u_1 = u(t_1), w_2 = w(t_2)$, then the cross-correlation function of these processes is

$$C_{uw}(t_1, t_2) = E\left[u_1 \, w_2\right] = \int_{-\infty}^{+\infty}\int_{-\infty}^{+\infty} u_1 \, w_2 \, p(u_1, w_2) \, du_1 \, dw_2 \qquad (7.8)$$

The cross-covariance function is defined as

$$\sigma_{uw}(t_1, t_2) = E\left[(u - \bar{u})(w - \bar{w})\right] = \int_{-\infty}^{+\infty} (u_1 - \bar{u})(w_2 - \bar{w}) \, p_{uw}(u_1, w_2) \, du_1 \, dw_2 \qquad (7.9)$$

and the cross-correlation function is

$$\rho(t_1, t_2) = \frac{E\left[(u_1 - \bar{u})(w_2 - \bar{w})\right]}{\sqrt{E\left[(u_1 - \bar{u})^2\right] E\left[(w_2 - \bar{w})^2\right]}} = \frac{\sigma_{uw}}{\sigma_u \sigma_w} \qquad (7.10)$$

7.1.3 Stationary Processes

A *stationary process* in one in which the underlying probability density functions do not depend on time. In this case, the mean and the variance are constants, that is,

$$\bar{u} = E\big[u(t)\big] = \int_{-\infty}^{+\infty} u\, p(u)\, du \tag{7.11}$$

$$\sigma_u^2 = E\Big[\big(u(t)-\bar{u}\big)^2\Big] = E\big[u^2(t)\big] - \bar{u}^2 = \int_{-\infty}^{+\infty} (u-\bar{u})^2\, p(u)\, du \tag{7.12}$$

On the other hand, the cross-correlation and autocorrelation functions still depend on the time difference $\tau = t_2 - t_1$:

$$C_{uw}(\tau) = C_{wu}(-\tau) = E\big[u(t)w(t+\tau)\big] \tag{7.13}$$

$$R_u(\tau) = R_u(-\tau) = E\big[u(t)u(t+\tau)\big] \tag{7.14}$$

Observe the reversal of subscripts in the cross-correlation function for negative time lags.

7.1.4 Ergodic Processes

In many practical situations, stochastic processes are not only *stationary*, but they are also assumed to be *ergodic*, which means that any one sample of infinite duration is statistically representative of the ensemble as a whole. Hence, the expectations for the ensembles obtained in terms of probability density functions can be replaced by temporal averages over any one sample as follows:

$$\bar{u} = \lim_{T\to\infty} \frac{1}{T} \int_{-T/2}^{+T/2} u(t)\, dt \qquad\qquad \text{Mean} \tag{7.15}$$

$$\sigma_u^2 = \lim_{T\to\infty} \frac{1}{T} \int_{-T/2}^{+T/2} \big[u(t)-\bar{u}\big]^2\, dt \qquad\qquad \text{Variance} \tag{7.16}$$

$$R(\tau) = R(-\tau) = \lim_{T\to\infty} \frac{1}{T} \int_{-T/2}^{+T/2} u(t)u(t+\tau)\, dt \qquad \text{Autocorrelation} \tag{7.17}$$

$$\sigma_{uu}(\tau) = \lim_{T\to\infty} \frac{1}{T} \int_{-T/2}^{+T/2} \big[u(t)-\bar{u}\big]\big[u(t+\tau)-\bar{u}\big]\, dt \qquad \text{Autocovariance} \tag{7.18}$$

$$\sigma_{uu}(0) = \sigma_u^2 \qquad\qquad \text{Square of standard deviation} \tag{7.19}$$

$$\sigma_{uw}(\tau) = \lim_{T\to\infty} \frac{1}{T} \int_{-T/2}^{+T/2} \big[u(t)-\bar{u}\big]\big[w(t+\tau)-\bar{w}\big]\, dt \qquad \text{Cross-covariance} \tag{7.20}$$

$$\rho_{uw}(\tau) = \rho_{wu}(-\tau) = \frac{\sigma_{uw}(\tau)}{\sigma_u \sigma_w} \qquad\qquad \text{Cross-correlation function} \tag{7.21}$$

7.1.5 Spectral Density Functions

The Fourier transform of the autocovariance function is the spectral density function

$$S_{uu}(\omega) = \frac{1}{2\pi} \int_{-\infty}^{+\infty} \sigma_{uu}(\tau) e^{-i\omega t}\, d\tau \tag{7.22}$$

which is a purely real function of frequency because the autocovariance is an even function of the time delay. The spectral density function admits the inverse transform

$$\sigma_{uu}(\tau) = \int_{-\infty}^{+\infty} S_{uu}(\omega) e^{i\omega\tau}\, d\omega \tag{7.23}$$

This pair of equations is known as the *Wiener–Khinchin* (or Khintchine) relationship. Notice the reversal in the position of the $0.5/\pi$ factor in the Fourier transforms, which runs counter to the usual convention in Fourier transform pairs. This it is done purely for reasons of convenience: the integral of the spectral density function then equals the variance

$$\sigma_u^2 = \sigma_{uu}(0) = \int_{-\infty}^{+\infty} S_{uu}(\omega) d\omega = \text{var}(u) \tag{7.24}$$

Still, one could use instead the conventional definition of Fourier transformation without detriment in the analyses, provided that one is consistent in all operations with the so defined spectral density function. In practice, the most-often used definition is that of the one-sided spectral density function $\bar{S}_u(\omega) = 2 S_u(\omega)$, which allows the one-sided integrals

$$\sigma_{uu}(\tau) = \int_0^\infty \bar{S}_u(\omega) \cos(\omega\tau) d\omega \tag{7.25}$$

$$\sigma_u^2 = \int_0^\infty \bar{S}_u(\omega) d\omega \tag{7.26}$$

On the other hand, the Fourier transform of the cross-covariance function yields the cross-spectral density function, and its inverse recovers the cross-covariance function

$$S_{uw}(\omega) = \frac{1}{2\pi} \int_{-\infty}^{+\infty} \sigma_{uw}(\tau) e^{-i\omega\tau} d\tau \tag{7.27}$$

$$\sigma_{uw}(\tau) = \int_{-\infty}^{+\infty} S_{uw}(\omega) e^{i\omega\tau} d\omega \tag{7.28}$$

Unlike the spectral density function that is real, the cross-spectral density function is generally complex.

7.1.6 Coherence Function

The coherence function is defined as

$$\gamma_{uw}^2(\omega) = \frac{|S_{uw}(\omega)|^2}{S_{uu}(\omega) S_{ww}(\omega)}, \qquad 0 \le \gamma_{uw} \le 1 \tag{7.29}$$

The coherence function gives a measure of the linear dependence between two signals as a function of frequency. If $\gamma_{uw} = 1$ for all frequencies, the two signals are perfectly correlated. If $\gamma_{uw} < 0.5$ or thereabout in either all frequencies or in some frequency band, the signals are poorly correlated in that band, and if $\gamma_{uw} = 0$, the signals are uncorrelated, that is, they are *incoherent*.

7.1.7 Estimation of Spectral Properties

In engineering situations, the tools of spectral estimation are often used to make unbiased assessment of the behavior of physical systems. An unavoidable fact of life is, however, that all signals in any experiment or observation will always be contaminated with some degree of noise, which tarnishes the true nature to some unknown extent. This is the reason for using instead sets of observations, and not just a single observation. To see why, consider a set of n observations for some experiment, and begin by writing each measured signal in the form

$$u_j(t) = c_j \, \tilde{u}(t)_j + \varepsilon_j(t), \qquad j = 1, 2, \ldots n \tag{7.30}$$

where u_j is the actual measurement and $\tilde{u}_j$ is the true signal to be measured. Also, c_j is a constant that may vary from observation to observation – it is difficult to repeat experiments with exactly the same intensity – and ε_j is the unavoidable noise. The average of these observations is then

$$\frac{1}{n}\sum_1^n u_j(t) = C\,\tilde{u}(t) + \frac{1}{n}\sum_1^n \varepsilon_j(t), \qquad C = \frac{1}{n}\sum_1^n c_j \tag{7.31}$$

Because the signals of a *linear* system are perfectly coherent and repeatable while the noise is (thought to be) incoherent between experiments, the average of the latter will be quite small; indeed it will drop linearly in proportion with the number of samples. Hence, at any given time, the averaged signal will have an improved signal to noise ratio. A similar proof can be used to demonstrate that the contribution of the noise to the variance (i.e., to the standard deviation) drops rapidly with the number of samples. The proof hinges on the fact that the noise is *uncorrelated* to the signal – this is the very *definition* of the noise – which causes the cross-expectation of the signal and the noise to vanish.

Consider now a set (i.e., family) of two stochastic processes, say $u_j(t), w_j(t), j = 1, 2, \ldots n$, each of which has finite duration. The number n of samples (i.e., observations) in each set is generally as small as possible (say 5 or 10), but enough to diminish any errors and noise that may exist in the measurement of the signals, and thus strengthen these in turn. Their Fourier transforms are

$$U_j(\omega) = \int_{-\infty}^{+\infty} u_j(t)e^{-i\omega t}\,dt, \qquad W_j(\omega) = \int_{-\infty}^{+\infty} w_j(t)e^{-i\omega t}\,dt \tag{7.32}$$

The infinite limits in the integrals in Eq. 7.32 are simply a matter of notational convenience: we thus avoid having to specify the individual intervals in which the signals do not vanish. However, since these time histories have finite duration, they cannot possibly be ergodic, but it is customary to assume that they can still be regarded as such in order for their spectral properties to be estimated directly from the samples.

Using the set of observations, we *define* average spectral density functions as follows. (**Note:** A superscript star denotes a complex conjugate.)

$$G_{uu}(\omega) = \frac{1}{2n\pi}\sum_{j=1}^n U_j(\omega)\,U_j^*(\omega)$$
$$= \frac{1}{2n\pi}\sum_{j=1}^n |U_j(\omega)|^2 \tag{7.33}$$

$$G_{ww}(\omega) = \frac{1}{2n\pi}\sum_{j=1}^n W_j(\omega)\,W_j^*(\omega)$$
$$= \frac{1}{2n\pi}\sum_{j=1}^n |W_j(\omega)|^2 \tag{7.34}$$

$$G_{uw}(\omega) = \frac{1}{2n\pi} \sum_{j=1}^{n} U_j(\omega) W_j^*(\omega) \qquad (7.35)$$

$$\gamma_{uw}^2(\omega) = \frac{|G_{uw}(\omega)|^2}{|G_{uu}(\omega)||G_{ww}(\omega)|} = \frac{\left|\sum_{j=1}^{n} U_j(\omega) W_j^*(\omega)\right|^2}{\sum_{j=1}^{n} |U_j(\omega)|^2 \sum_{j=1}^{n} |W_j(\omega)|^2} \qquad (7.36)$$

Observe that in the numerator, the absolute value is taken after the sum has been evaluated. Notice thus that if $n = 1$ is used, the implied coherence is seemingly perfect, that is,

$$\gamma_{uw}(\omega) = \frac{|U_1(\omega) W_1^*(\omega)|}{|U_1(\omega)||W_1(\omega)|} = \frac{|U_1(\omega)||W_1(\omega)|}{|U_1(\omega)||W_1(\omega)|} \equiv 1 \qquad (7.37)$$

However, when using multiple samples, the coherence function will not equal 1, because as already mentioned, the signals have noise. In other words, the ratios $H_j(\omega) = W_j / U_j$ may not all be the same for each sample.

More generally, an estimation of the complex *transfer function* of a linear system relating an "input" $u(t)$ to some "output" $w(t)$ can be obtained as follows:

$$H_{wu}(\omega) = \frac{G_{wu}}{G_{uu}} = \frac{G_{uw}^*}{G_{uu}} \qquad (7.38)$$

Finally, the *signal to noise ratio* is defined as

$$\frac{S}{N}(\omega) = \frac{\gamma^2}{1-\gamma^2} \qquad (7.39)$$

which provides a measure of the quality of the signal at some frequency.

A word in closing: This brief summary necessarily glosses over much detail in spectral data estimation, especially data conditioning (e.g., sliding Hanning windows) and filtering (e.g., Wiener filtering). Readers are referred to the specialized literature on the subject.

Example: Spectral Analysis of Surface Waves (SASW)

Consider an elastic half-space subjected to a disturbance at some location on its surface. This excitation elicits waves that at even fairly close range are already dominated by Rayleigh waves. This is because such waves have the slowest decay due to geometric spreading. If the half-space is homogeneous and the source is small in comparison to the range where the motions are measured, such motions will attain the form

$$u(r,t) \approx \frac{A}{\sqrt{r}} f(t - r/C_R) \qquad (7.40)$$

where A is some appropriate amplitude, C_R is the velocity of Rayleigh waves – which is roughly 90% to 95% of the shear wave velocity and is independent of the frequency, or equivalently, independent of the wavelength – and $f(t)$ depends in some way on the source function (e.g., Lamb's problem). Hence, the temporal Fourier transform is

$$U(r,\omega) = \frac{A}{\sqrt{r}} e^{-\mathrm{i}\,\omega r/C_R} F(\omega) \tag{7.41}$$

where $F(\omega)$ is the Fourier transform of $f(t)$, and r/C_R is the travel time of Rayleigh waves over the distance r. Consider two receivers placed at r_1, $r_2 = r_1 + L$ that record the motions of the waves traveling underneath. To connect to the previous material, these motions could be denoted as $u_j \equiv u_j(r_1,t)$ and $w_j \equiv u_j(r_2,t)$, with $j = 1,2\dots n$ denoting the experiment number, except that the motion at the first station r_1 is not truly an *input* to the *output* at r_2, but just another observation. In the absence of noise, the ratio of their Fourier spectra (i.e., the transfer function) is

$$H_{21}(\omega) = \frac{U(r_2,\omega)}{U(r_1,\omega)} = e^{\mathrm{i}\omega(r_2 - r_1)/C_R} \sqrt{\frac{r_2}{r_1}} \tag{7.42}$$

whose theoretical amplitude and phase are

$$|H_{21}| = \sqrt{\frac{r_2}{r_1}}, \qquad \phi = \arg(H_{21}) = \omega L/C_R, \qquad L = r_2 - r_1 \tag{7.43}$$

Thus, the phase grows linearly with frequency. This is an indication that Rayleigh waves are nondispersive, that is, that their speed is constant. Hence, in principle we could measure the speed of Rayleigh waves as the slope of the phase spectrum

$$C_R = \omega L / \phi(\omega) \tag{7.44}$$

We mention in passing that the estimation of the phase spectrum $\phi(\omega)$ usually relies on a technique known as *phase unwrapping*, which extends continuously the angle inferred from a complex quantity initially constrained to lie in the range of $-\pi$ to π (i.e., -180 to 180 degrees). Abrupt jumps in the phase angle are thus eliminated.

Even in the ideal case of a homogeneous half-space, however, the measured signals will be contaminated by noise, which means that the phase spectrum will not be quite linear. A set of measurements with various signals will then be necessary to minimize the noise, yet it may well be found that the coherence $\gamma(\omega)$ for the set of signals exhibits high values only in some limited frequency band. For example, the coherence could be low for low frequencies if the energy of the source is poor at such frequencies (i.e., if $F(\omega)$ is low). Alternatively, the coherence could be low for high frequencies if the waves were subjected to scattering and dispersion due to local, small-scale irregularities in the soil that solely affect waves of short wavelength. Either way, the measured speed of Rayleigh waves will be reliable only in the frequency band at which the coherence function exhibits reasonably high values, say in excess of 0.9. If the system is strongly nonlinear such that the experiments are not quite repeatable (which not the case in this example), then the coherence could be low throughout.

Consider next a half-space in which the shear modulus increases gradually with depth. It can be shown that such a medium is able to propagate various kinds of guided (surface) waves whose speeds change with frequency. Among these, the fundamental Rayleigh mode may well dominate the motion again for sources near the surface, but this time it will be a dispersive wave. To a first approximation, the speed of such waves is a function of the *penetration* (i.e., the depth of significant motion), which in turn depends on the frequency. Waves of long wavelength (i.e., low frequency) penetrate the most and are the fastest, while waves of short wavelength (high frequency) penetrate the least, and thus are the slowest. A simple rule of thumb posits that the penetration of Rayleigh waves is on the order of one third of the wavelength observed on the surface, that is, $d = \frac{1}{3}\lambda$, with $\lambda = C_R(f)/f = 2\pi C_R(\omega)/\omega$. Hence, a spectral decomposition of the motion observed at two stations will result in a generally nonlinear phase spectrum from which the phase velocity of Rayleigh waves as a function of frequency could be inferred. This would allow one in turn to estimate the shear wave velocities as a function of depth, which is the principle underlying Stokoe's widely used Spectral Analysis of Surface Waves method, usually referred to by its acronym of SASW. In that method, a set of signals is used to minimize the noise using sources adequate to elicit waves with sufficient energy in the frequency bands of interest, and placing the receivers at a distance L which is comparable to the wavelengths of interest.

7.1.8 Spatial Coherence of Seismic Motions

It has long been known that when an earthquake is sensed at some distant receiver, the so-called P or pressure waves are the first to arrive, followed by the S or shear waves and then shortly thereafter by the surface (Rayleigh) waves, with the whole quake then rolling off into some coda that results from scattered and reverberant waves as well as noise. Thus, the recorded motions are seen to exhibit strong temporal variability. More recently, after detailed evaluations of numerous seismic records obtained by means of arrays of strong motion instruments and these motions were compared within some local neighborhood for any given event, it was found that they exhibit random spatial characteristics that remained evident even at small distances. That is, the motions observed in some neighborhood during an earthquake defined a random field in space-time, with probabilistic properties that depended on various physical parameters, among which the material properties and seismic wave velocities near the surface play an important role. Stiff media – implying long wavelengths – exhibit relatively mild spatial incoherence while soft media (i.e., short wavelengths) show them more clearly. Such random characteristics are found not only at the ground surface, but exist also at depth, and they play an important role in the seismic design of spatially extended structures, which is why we consider the topic herein, even if only very lightly.

Coherency Function Based on Statistical Analyses of Actual Earthquake Motions

Statistical studies carried out in past years on the spatial variation and coherence of seismic motions have been obtained from extensive data sets acquired with arrays of strong motion instruments. In a nutshell, the coherence functions have emanated from analyses similar to those outlined in Section 7.1.7, after combining the results of many seismic events at any given site as well as integrating the data from different sites. An example is the set of coherency functions presented in one of several EPRI reports,[2] which are too

---

[2] Electric Power Research Institute (EPRI), Report 1014101, December 2006, Program on Technology Innovation: "Spatial coherency models for soil–structure interaction."

complicated and lengthy to be included herein. More importantly, it would seem to us that those empirical coherency functions are likely to significantly overestimate the degree of spatial incoherence because they aggregate different seismic events with rather disparate source characteristics, even at one and the same site.

To clarify this point, consider the following mental experiment: Suppose that there is an ideal site, say a deep, homogeneous alluvial soil deposit, on the surface of which, at some fixed distance d, there are two recording stations A, B. During a first seismic event, the epicenter lies behind station A along the straight line AB and elicits motions on the surface dominated by a single wave mode, which means that the motions on the surface are perfect, delayed replicas, that is, $u_B(t) = u_A(t - t_{AB})$, where $t_{AB} = d/C_1$ is the temporal delay, d is the distance and C_1 is the apparent phase velocity (or celerity). By mental construction, this motion has an absolutely perfect spatial coherence. Months (or years) later, another earthquake takes place at this same site, but this time the epicenter is symmetrically located beyond station B, so motions are first observed at station B and a little later on at A. By itself, this event also exhibits *perfect coherence*. But if we now mix these two events with the statistical tools used in the EPRI report referred to earlier, the coherence will no longer be perfect, as can easily be shown. Let the two events produce motions

$$U_{jA}(\omega) = E_j |H_{jA}| \exp i\phi_{jA}, \qquad U_{jB}(\omega) = E_j |H_{jB}| \exp i\phi_{jB}, \qquad j = 1, 2 \tag{7.45}$$

where $E_j = E_j(\omega)$ is the Fourier spectrum of the source for the two seismic events (which need not be equal), H_{jA}, H_{jB} from the source to the receivers, and ϕ_{jA}, ϕ_{jB} are the phase angles of the transfer functions. By mental construction (i.e., symmetry), these two satisfy $|H_{jA}| = |H_{jb}| \equiv |H|$, $\phi_{1B} = \phi_{1A} - \Delta\phi_{AB}$, $\phi_{2B} = \phi_{2A} + \Delta\phi_{AB}$, where $\Delta\phi_{AB} = \omega t_{AB} = \omega d / C$. Hence

$$\begin{aligned} S_{AB}(\omega) &= U_{1A}(\omega)U_{1B}^*(\omega) + U_{2A}(\omega)U_{2B}^*(\omega) \\ &= |E_1|^2 |H|^2 \exp(-i\Delta\phi_{AB}) + |E_2|^2 |H|^2 \exp(+i\Delta\phi_{AB}) \\ &= |H|^2 \left[\left(|E_1|^2 + |E_2|^2 \right) \cos\phi_{AB} + i\left(|E_1|^2 - |E_2|^2 \right) \sin\phi_{AB} \right] \end{aligned} \tag{7.46}$$

$$\begin{aligned} S_{AA}(\omega) &= U_{1A}(\omega)U_{1A}^*(\omega) + U_{2A}(\omega)U_{2A}^*(\omega) \\ &= |H|^2 \left(|E_1|^2 + |E_2|^2 \right) \end{aligned} \tag{7.47}$$

$$\begin{aligned} S_{BB}(\omega) &= U_{1B}(\omega)U_{1B}^*(\omega) + U_{2B}(\omega)U_{2B}^*(\omega) \\ &= |H|^2 \left(|E_1|^2 + |E_2|^2 \right) \end{aligned} \tag{7.48}$$

so

$$\begin{aligned} \gamma &= \frac{|H|^2 \left[\left(|E_1|^2 + |E_2|^2 \right) \cos\phi_{AB} + i\left(|E_1|^2 - |E_2|^2 \right) \sin\phi_{AB} \right]}{|H|^2 \left(|E_1|^2 + |E_2|^2 \right)} \\ &= \cos\phi_{AB} + i\frac{|E_1|^2 - |E_2|^2}{|E_1|^2 + |E_2|^2} \sin\phi_{AB} \end{aligned} \tag{7.49}$$

If the earthquakes have similar power, then $|E_1| \approx |E_2|$, in which case

$$\gamma = \cos\phi_{AB} = \cos\frac{\omega d}{C} \tag{7.50}$$

which is identical to the factor on page 2-2 of the EPRI report, middle of the last paragraph, and this despite the fact that each of the imagined quakes herein has *perfect* coherence.

More generally, spatial incoherence arises from various uncertainties in the wave content (i.e., phase velocities) at any given site because of multiple travel paths, wave scattering, nonparallel or undulating layers, nonlinear effects and the like. That is, the incoherence arises because the wave content itself is complex. It has nothing to do with our ability – or lack thereof – to predict the degree of similarity of seismic motions at a site observed at two near points during some specific earthquake.

Wave Model for Random Field

We present next a simple mechanistic (numerical) model that allows making analytical predictions on the spatial variability of seismic motions at a given site. The model reckons that the motion observed on the surface results from multiple trains of plane waves emanating simultaneously from underneath along different travel paths and multiple directions.[3] Thus, it aims to mimic the effects of multiple wave paths, refraction and scattering, and has the merit of being based on physical considerations and not simply on earthquake statistics inferred from dense arrays. It can be justified further by the fact that, given *any* arbitrary stochastic wave field prescribed at the free surface, it is *always* possible (at least in principle) to carry out a wavenumber decomposition of that motion field (i.e., a Fourier transform in space–time) and arrive at the precise plane wave decomposition that gave rise to that very stochastic wave field, and thus is indistinguishable from it, at least in a statistical sense. This mapping into a wave field is useful because – at least in the vicinity of the surface – it allows a rational means to model and project coherence functions from the surface to points immediately underneath, provided such points do not lie too deep within the soil mass. We present a succinct summary in the ensuing.

Simple Cross-Spectrum for SH Waves

Consider the particular case of wave motion at the surface of a layered soil that results from horizontally polarized shear waves (SH waves) that are incident on the surface from the medium underneath. This means that only one displacement component $u = u(x,t)$ will be observed on the surface, which thus defines a scalar *random field*. Assume further that the motion in the near surface layers of the stratified medium consists of N trains of stationary plane SH waves of the form

$$u(x,t) = \sum_{j=1}^{N} f_j(t - s_j x) \tag{7.51}$$

where $s_j = \sin\theta_j / C_S$ is the apparent slowness of the jth wave train propagating in the uppermost layer with angle θ_j with respect to the vertical, and C_S is the shear wave velocity in that upper layer. The cross-expectation between two observation points (A, B) with coordinates x_A, x_B is then

---

[3] E. Kausel and A. Pais, "Stochastic deconvolution of earthquake motions," *J. Eng. Mech. ASCE*, 113(2), 1987, 266–277.

$$C_{AB}(\tau) = E\left[\sum_{i=1}^{N}\sum_{j=1}^{N} f_i(t - s_i x_A) f_j(t + \tau - s_j x_B)\right]$$

$$= \sum_{i=1}^{N}\sum_{j=1}^{N} E\left[f_i(t - s_i x_A) f_j(t + \tau - s_j x_B)\right]$$

$$= \sum_{i=1}^{N}\sum_{j=1}^{N}\left[R_{ij}(\tau - s_j x_B + s_i x_A)\right] \qquad (7.52)$$

where the R_{ij} are the auto- and cross-correlation functions for each pair $f_i f_j$. If we assume further that the N wave trains are *uncorrelated*, then the coupling terms R_{ij} vanish, in which case

$$C_{AB}(\tau) = \sum_{j=1}^{N} R_{jj}(\tau - s_j(x_B - x_A))$$

$$= \sum_{j=1}^{N} R_{jj}(\tau - s_j d) \qquad (7.53)$$

where $d = x_B - x_A$ is the separation between the observation points. The cross-spectral density function is then

$$S_{AB}(\omega) = \sum_{j=1}^{N} e^{-i s_j \omega d} S_{jj}(\omega) \qquad (7.54)$$

and in particular, when $A = B$ and $d = 0$,

$$S_{AA}(\omega) = S_{BB}(\omega) = \sum_{j=1}^{N} S_{jj}(\omega) \qquad (7.55)$$

The coherence function for points on the surface is then

$$\gamma = \frac{S_{AB}}{\sqrt{S_{AA} S_{BB}}} = \frac{\sum_{j=1}^{N} e^{-i \sin\theta_j \frac{\omega d}{C_s}} S_{jj}}{\sum_{j=1}^{N} S_{jj}} = \frac{\sum_{j=1}^{N} S_{jj} \cos\left(\sin\theta_j \frac{\omega d}{C_s}\right) - i\sum_{j=1}^{N} S_{jj} \sin\left(\theta_j \frac{\omega d}{C_s}\right)}{\sum_{j=1}^{N} S_{jj}} \qquad (7.56)$$

We can now generalize the results to a set of wave trains that vary continuously with the angle of incidence, assuming each train to be statistically independent as previously assumed for two wave trains. This is consistent with the usual model where earthquakes are assumed to be white noise, and thus lacking in autocorrelation. The summation then goes over into an integral for which the expression is

$$\boxed{\gamma = \frac{\int_{-\frac{1}{2}\pi}^{\frac{1}{2}\pi} S_\theta \, e^{-i \sin\theta \frac{\omega d}{C_s}} \, d\theta}{\int_{-\frac{1}{2}\pi}^{\frac{1}{2}\pi} S_\theta \, d\theta}} \qquad (7.57)$$

Example: Quarter Space Noise

For example, consider a quarter space noise where uncorrelated waves arrive with equal spectral amplitudes S_θ within the dihedron $0 \le \theta \le \frac{1}{2}\pi$. In that case, the integral can be evaluated in closed form and results in

$$\gamma = J_0\left(\Omega_{AB}\right) - i\,\mathbf{H}_0\left(\Omega_{AB}\right), \qquad \Omega_{AB} = \frac{\omega d}{C_S} = \frac{\omega\left(x_B - x_A\right)}{C_S} \tag{7.58}$$

where $J_0, \mathbf{H}_0$ are, respectively, the Bessel and Struve-H functions of order zero. Its absolute value is

$$\left|\gamma\right| = \sqrt{J_0^2\left(\Omega_{AB}\right) + \mathbf{H}_0^2\left(\Omega_{AB}\right)} \tag{7.59}$$

which decays steadily with Ω_{AB}. A plot of this function is shown in Figure 7.1, which can also be approximated closely by

$$y = \exp\left[-\frac{1}{2\pi^2}\Omega_{AB}^2\right] = \exp\left[-2\left(d\,/\,\lambda\right)^2\right], \qquad \lambda = C_s\,/\,f = 2\pi C_s\,/\,\omega \tag{7.60}$$

as shown by the dashed line in Figure 7.1. We see that in this case, the coherence is basically nil when the separation equals the characteristic wavelength, that is, $d = \left|x_B - x_A\right| = \lambda$. This result is perfectly consistent with empirical findings. For example, for a soil with $C_S = 1{,}000$ m/s, a separation $d = 25$ m, and a frequency $f = 20$ Hz, then $\lambda = 50$ m, $d\,/\,\lambda = \frac{1}{2}$, then $\left|\gamma\right| \sim 0.61$.

On the other hand, the coherence function given in a 2006 EPRI report (eq. 2-1 and Fig. 2-1 therein) is provided without any reference to soil parameters. A decent match with the preceding results can still be obtained by choosing $C_S \sim 628 = 200\pi$ m/s, for which $\omega\xi\,/\,C_S = 2\pi \times 20 \times 25\,/\,(2\pi \times 100) = 5$ and $\left|\gamma\right| \sim 0.25$.

Although in the previous example the various wave trains arriving at different angles had all the same power spectral density function $S\left(\omega\right)$, the formulation allows for an

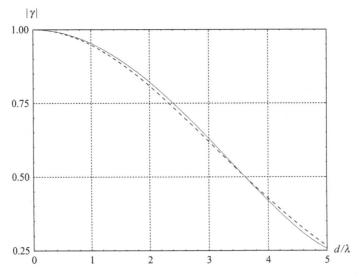

Figure 7.1. Coherency function of quarter space noise.

arbitrary variation with the angle of incidence of that function. But despite the homogeneity of S_θ with θ, we still arrived at a stochastic model where the seismic motions at the surface were imperfectly correlated. Why should this be so? Because the various rays arrive at the surface at different times – indeed, they propagate with different phase velocities – so when they are added together, their summations at the various receivers do not coincide in time. Moreover, if the rays (wave trains) were white noise (or nearly so), then each of them would be uncorrelated with itself, and therefore it would also be uncorrelated with neighboring rays, and this too leads to incoherence.

We have thus shown that even an elementary wave propagation model can lead to stochastic fields with coherence functions rather similar to those based on statistics alone. This opens the door to methods that can be used to extrapolate the random field at the surface inferred by statistical (i.e., empirical) means to points within the soil mass, even if that soil were layered. It suffices to deconvolve the surface spectra by means of the transfer functions from the surface to the depth below, as roughly described in the next paragraph.

Stochastic Deconvolution

We now proceed to project the spectral properties from the surface to some depth z, a process which we refer to as *stochastic deconvolution*, but do so only for the simpler case when the motions consist purely of SH waves, and then again we provide only a sketch. The general case of motions that result from any kind of wave types can be found in Zendagui et al.[4]

As is well known, when a random process is passed through a linear filter, such as a vibrating structure responding to a random source, then its input and output spectral densities are related through the squared absolute value of the transfer functions. More generally, this also applies to a random field involving temporal and spatial variables. Indeed, when the motions at all elevations are expressed in the so-called frequency–wavenumber (ω, k) domain, one can deconvolve the motions to any elevation, as shown in Chapter 5, Section 5.2.1 in this book. All that is required is to find the transfer functions in the frequency wavenumber domain.

In the particular case of SH wave motions in a homogeneous half-space (or in the topmost layer of a stratified system), the transfer function from the surface to a depth z below the surface is given by the simple expression $H(\omega, k) = \cos\left(z\sqrt{k_s^2 - k^2}\right)$, where $k_s = \omega / C_s$ is the wavenumber for shear waves, and k is the horizontal wavenumber. In the frequency–wavenumber domain, the coherence function at depth is then obtained as follows:

a. Decide on the spectral characteristics on the free ground, which will be assumed to be known. For example, let's say that the coherence functions at the surface can be approximated as

$$\gamma = \exp\left(-\alpha^2 k_s^2 x^2\right) \tag{7.61}$$

[4] D. Zendagui, M. K. Berrah, and E. Kausel, "Stochastic deamplification of spatially varying seismic motions," *Soil Dyn. Earthquake Eng.*, 18, 1999, 409–421.

b. Compute the spatial Fourier transform at the surface

$$\tilde{\gamma}(\omega,k,0) = \int_{-\infty}^{+\infty} \gamma(\omega,x)e^{ikx}dx \rightarrow \int_{-\infty}^{+\infty} \exp(-\alpha^2\,k_S^2 x^2)e^{ikx}dx = \frac{1}{\alpha k_S \sqrt{\pi}}\exp\left(-\left(\frac{k}{2\alpha k_S}\right)^2\right) \quad (7.62)$$

c. Compute the coherence function in the frequency–wavenumber domain at depth z as

$$\tilde{\gamma}(\omega,k,z) = \left|H(\omega,k)\right|^2 \tilde{\gamma}(\omega,k,0) = \frac{1}{\alpha k_S \sqrt{\pi}}\exp\left(-\left(\frac{\kappa}{2\alpha}\right)^2\right)\cos^2\left(k_S z\sqrt{1-\kappa^2}\right) \quad (7.63)$$

$$\kappa = k/k_S$$

d. Invert the result into the spatial domain

$$\begin{aligned}
\gamma(\omega,x,z) &= \frac{1}{\alpha k_S \sqrt{\pi}}\frac{1}{2\pi}\int_{-k_S}^{+k_S} \exp\left(-\left(\frac{\kappa}{2\alpha}\right)^2\right)\cos^2\left(k_S z\sqrt{1-\kappa^2}\right)e^{-ikx}dk \\
&= \frac{1}{\alpha\sqrt{\pi}}\frac{1}{2\pi}\int_{-1}^{+1}\cos^2\left[k_S z\sqrt{1-\kappa^2}\right]e^{-\left(\frac{\kappa}{2\alpha}\right)^2}e^{-i\kappa(k_S x)}d\kappa
\end{aligned} \quad (7.64)$$

where we now omit wavenumbers $k > k_S$ (or $\kappa > 1$) in the inverse Fourier transform because these correspond to inhomogeneous (evanescent) waves. Thus, this dispenses with the situation where the square root term of the transfer function turns imaginary. The last integral must be integrated numerically, but that poses no problems. In particular, at $x = 0$ we obtain

$$\gamma(\omega,0,z) = \frac{1}{\alpha\sqrt{\pi}}\frac{1}{\pi}\int_0^1\left[\cos\left(k_S z\sqrt{1-\kappa^2}\right)e^{-\left(\frac{\kappa}{2\alpha}\right)}\right]^2 d\kappa < \frac{1}{\alpha\sqrt{\pi}}\frac{1}{\pi} \quad (7.65)$$

which is less than 1 in value provided $\alpha > \pi^{-\frac{3}{2}} = 0.179$. This is the coherence function between a point on the surface at $x = z = 0$ and a point at depth at $x = 0$, $z > 0$.

7.2 Earthquakes, and Measures of Quake Strength

As the earth surface moves gradually because of plate tectonics, ground deformations build up in the earth crust, and strain energy begins to accumulate locally near a fault. However, this process cannot continue forever, for there are limits on how much strain energy can be stored in the ground before the rock will fail. When that limit is reached, the fault breaks in some region and slips, as a result of which most of the accumulated strain energy is converted into heat, some is expended in fracturing the material, and the remainder is released in the form of seismic waves. The longer the fault (i.e., the longer the fault break), the more energy is released, although this energy emanates neither from a single point in space nor is it released during a brief instant. Hence, any measure of earthquake size will relate to the length of the fracture zone and to the intensity of ground motions recorded at some distance to the fault.

In a nutshell, there exist two common ways of measuring the strength of earthquakes and their effects on structures, namely the *magnitude* and the *intensity*. The first is an intrinsic property of any given earthquake and is a measure of the total energy released, which does not depend on the place on earth from which it is inferred, while the second is an empirical description of observed effects on people and on manmade structures, so it changes with distance to the source, soil conditions, and the like. The ensuing paragraphs provide a very basic description of each.

7.2.1 Magnitude

As stated earlier, the earthquake magnitude is a scale providing a rough measure of the strain energy released by the fracture of the earth crust along a fault during an earthquake. The earliest such scale is the so-called *Richter magnitude* scale, which while now outdated and no longer in use by seismologists remains firmly anchored in the news media, which invariable refer to any estimated earthquake magnitude simply as an "intensity on the Richter scale." The Richter magnitude was originally defined as

> *the logarithm of the maximum trace of a Wood–Anderson instrument of [some specific physical characteristics] placed at 100 km from the epicenter...*

Since no instrument was ever placed at exactly that distance, methods were also provided to correct for various epicentral distances. However, this scale saturated quickly for strong seismic events and was generally deemed to be unsatisfactory. For this reason, seismologists developed various alternative magnitude scales, of which the most important and frequently used is the *moment magnitude*. To explain roughly what this scale means, we use next a very simple fault model (perhaps too simple for a seismology course, but certainly enough to illustrate concepts). Figure 7.2 shows a schematic view of the elastic rebound region containing the fracture zone before and after the earthquake. The liberated energy is expended as frictional dissipation in the fault plane, fracturing of the rock nearby, and in the form of seismic waves.

Seismic Moment

Consider a fault with a vertical plane that intersects the surface, see Figure 7.2. This highly idealized fault is rectangular in shape and the fracture zone has a width a, length b, and depth c, as shown in the figure. Thus, the fracture zone has a volume $V = abc$, and the fault area is $A = bc$. Over time, strain accumulates in this volume, producing internal shearing stresses τ on either side of the fault which, in the aggregate, are equivalent to a total shearing force $F = \tau A$. This pair of forces has a moment arm a and produces a net moment (or torque)

$$M = Fa = \tau Aa = \tau\, abc \qquad (7.66)$$

On the other hand, the shear strain deforms the volume so that the right side is displaced with respect to the left side by some amount Δu, which produces a shear strain $\gamma = \Delta u / a$, so the shear stress is $\tau = \mu \gamma = \mu\, \Delta u / a$, where μ is the shear modulus. Hence,

$$M = \mu \frac{\Delta u}{a}\, abc = \mu\, bc\, \Delta u = \mu\, A\, \Delta u \qquad (7.67)$$

Changing M into the standard notation M_0, we obtain

$$\boxed{M_0 = \mu\, A\, \Delta u} \qquad (7.68)$$

This is the formula normally used by seismologists to define the *seismic moment*, where μ is a representative measure of the shear rigidity of the rock, A is the estimated area of the fault, and Δu is the observed fault slip, usually estimated after the earthquake.

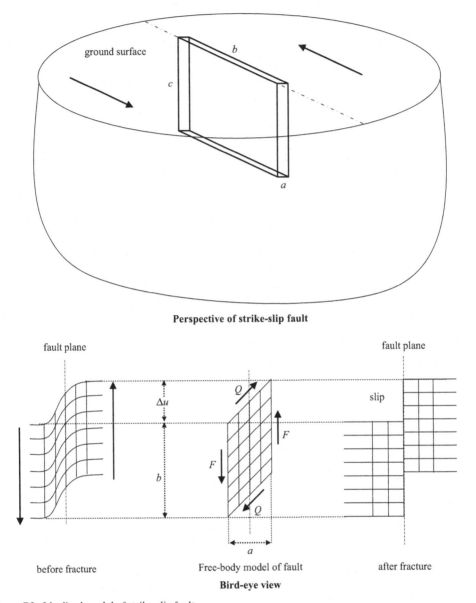

Figure 7.2. Idealized model of strike-slip fault.

On the other hand, there exists also a pair of transverse shearing forces acting at the upper and lower ends in the fault's free body diagram, each of which equal $Q = \tau ac$, and these exert a moment in the opposite direction $M = \tau abc$ which equals the moment already found, and both of these are released when the fault breaks. That this must be so follows from the fundamental physical fact that immediately prior to the ground breaking, the fractured area most certainly *was* in perfect static equilibrium, neither moving nor rotating. Thus, across the fault and prior to its breaking, there *must* have existed both normal and shearing stresses such that not only the sum of forces acting on the fractured area added up to zero, but also the total moment of these forces. Hence, the true equivalent system must necessarily include two equal magnitude torques acting in opposite directions, which is what seismologists refer to as a *double couple*, and this affects in turn how the wave energy is radiated and scattered into the surrounding medium. Ultimately, the net effect of the earthquake, at least as seen by an observer at some distance from the fault, is similar to that caused by a point double couple with reversed sign, the change of which is necessary to model the *release* of internal stresses due to fracture.

Of course, this very simplistic model glosses over many important aspects, such as the fact that both the shear modulus and the confining pressure holding the fault together by friction change with depth, that the fault may not intersect the surface and does not break everywhere simultaneously but the fracture propagates at some finite speed, that the direction of the slip may not be horizontal or the fault plane may not be vertical, and so forth.

Moment Magnitude

The moment magnitude is defined as

$$\boxed{M_W = \tfrac{2}{3}\log_{10} M_0 - 10.7} \tag{7.69}$$

where M_0 is the seismic moment in dynes-cm, the subscript W refers to mechanical *work*, and the constants are chosen so as to achieve some degree of consistency with the old Richter scale. The inverse transformation is then

$$M_0 = 10^{\frac{3}{2}(M_W + 10.7)} \tag{7.70}$$

which means that an increase of two steps in moment magnitude corresponds to an increase in the seismic moment by a factor $10^3 = 1,000$, so one step corresponds to $10^{3/2} = \sqrt{1000} = 31.6$. Since the energy released can be shown to be proportional to the seismic moment, this means that each step of increase in the magnitude scale corresponds to an increase in the mechanical power of the earthquake by a factor 31.6.

In principle, the magnitude scale has no upper limit, but it is believed that magnitudes greater than 10 are exceedingly unlikely to be ever observed, for the simple reason that there are physical limits on how much strain can be accumulated in the earth's crust before it breaks. The largest earthquake observed anywhere in the world during recent historical times is the 1960 Valdivia (Chile) earthquake, which topped the magnitude scale at about 9.5 and had a fault break of some 1,000 km. At the other extreme, small-magnitude earthquakes equal to or less than zero are also possible, but they represent ground vibrations that are indistinguishable from ambient noise caused by miscellaneous sources, such as wind, and induced micro-earthquakes such as rock bursts in mines, building demolitions, or explosions in quarry blasts. Thus, the practical range of moment magnitude is somewhere between 1 and 10.

7.2.2 Seismic Intensity

Another common measure of seismic severity is expressed in terms of the earthquake intensity, which is a qualitative, subjective description of the effects of earthquakes on structures and people. As such, for any given moment magnitude, it is highly dependent on the distance to the causative fault, the geological and morphological characteristics of

Table 7.1. Masonry quality

Masonry type	Description
A	Good workmanship, mortar, and design; reinforced, especially laterally, and bound together by using steel, concrete, etc.; designed to resist lateral forces.
B	Good workmanship and mortar; has reinforcement, but not designed in detail to resist lateral forces.
C	Ordinary workmanship and mortar; no extreme weaknesses like failing to tie in at corners, but neither reinforced nor designed against horizontal forces
D	Weak materials, such as adobe; poor mortar; low standards of workmanship; weak horizontally.

Table 7.2. Mercalli Intensity Scale modified by Richter

1	Not felt. Marginal and long-period of large earthquakes.
2	Felt by persons at rest, especially on upper floors, or favorably placed.
3	Felt indoors. Hanging objects swing. Vibration like passing of light trucks. Duration estimated. May not be recognized as an earthquake.
4	Hanging objects swing. Vibration like passing of heavy trucks; or sensation of a jolt like a heavy ball striking the walls. Standing motor cars rock. Windows, dishes, doors rattle. Glasses clink. Crockery clashes. In the upper range of 4, wooden walls and frames crack.
5	Felt outdoors; direction estimated. Sleepers wakened. Liquids disturbed, some spilled. Small unstable objects displaced or upset. Doors swing, close, open. Shutters, pictures move. Pendulum clocks stop, start, change rate.
6	Felt by all. Many frightened and run outdoors. Persons walk unsteadily. Windows, dishes, glassware broken. Knickknacks, books, and so on, off shelves. Pictures off walls. Furniture moved or overturned. Weak plaster and masonry *D* cracked. Small bells ring (church, school). Trees, bushes shaken visibly, or heard to rustle.
7	Difficult to stand. Noticed by drivers of motor cars. Hanging objects quiver. Furniture broken. Damage to masonry *D* including cracks. Weak chimneys broken at roof line. Fall of plaster, loose bricks, stones, tiles, cornices, unbraced parapets, and architectural ornaments. Some cracks in masonry *C*. Waves on ponds; water turbid with mud. Small slides and caving in along sand or gravel banks. Large bells ring. Concrete irrigation ditches damaged.
8	Steering of motor cars affected. Damage to masonry *C*; partial collapse. Some damage to masonry *B*, none to masonry *A*. Fall of stucco and some masonry walls. Twisting, fall of chimneys, factory stacks, monuments, towers, elevated tanks. Frame houses moved on foundations if not bolted down; loose panel walls thrown out. Decayed piling broken off. Branches broken from trees. Changes in flow or temperature of springs and walls. Cracks in wet ground and on steep slopes.
9	General panic. Masonry *D* destroyed; masonry *C* heavily damaged, sometimes with complete collapse; masonry *B* seriously damaged. General damage to foundations. Framed structures, if not bolted, shifted off foundations. Frames racked. Conspicuous cracks in ground. In alluviated areas sand and mud ejected, earthquake fountains, sand craters.
10	Most masonry and frame structures destroyed with their foundations. Some well-built wooden structures and bridges destroyed. Serious damage to dams, dikes, embankments. Large landslides. Water thrown on banks of canals, rivers, lakes, etc. Sand and mud shifted horizontally on beaches and flat land. Rails bent slightly.
11	Rails bent greatly. Underground pipelines completely out of service
12	Damage nearly total. Large rock masses displaced. Lines of sight and level distorted. Objects thrown into the air.

Table 7.3. Concise modern version of the Modified Mercalli Intensity Scale

Mercalli Intensity	Witness Observations
I	Felt by very few people; barely noticeable.
II	Felt by a few people, especially on upper floors.
III	Noticeable indoors, especially on upper floors, but may not be recognized as an earthquake.
IV	Felt by many indoors, few outdoors. May feel like heavy truck passing by.
V	Felt by almost everyone, some people awakened. Small objects moved. Trees and poles may shake.
VI	Felt by everyone. Difficult to stand. Some heavy furniture moved, some plaster falls. Chimneys may be slightly damaged.
VII	Slight to moderate damage in well built, ordinary structures. Considerable damage to poorly built structures. Some walls may fall.
VIII	Little damage in specially built structures. Considerable damage to ordinary buildings, severe damage to poorly built structures. Some walls collapse.
IX	Considerable damage to specially built structures, buildings shifted off foundations. Ground cracked noticeably. Wholesale destruction. Landslides.
X	Most masonry and frame structures and their foundations destroyed. Ground badly cracked. Landslides. Wholesale destruction.
XI	Total damage. Few, if any, structures standing. Bridges destroyed. Wide cracks in ground. Waves seen on ground.
XII	Total damage. Waves seen on ground. Objects thrown up into air.

the travel path of the seismic waves to the observation site, to the local soil conditions, the type and quality of constructions, the materials used, the size and height of buildings, and so on. Still, it is a useful measure of the destructiveness of any given earthquake. Maps showing contours of equal seismic intensity near the source of an earthquake are referred to as *isoseismal maps*.

The most widespread intensity scale is the 1902 *Modified Mercalli Intensity* scale, or simply the *MM scale*. In a now classical piece of work on earthquake intensity and its relation with magnitude, Gutenberg and Richter (1942 and 1956) began by defining structural quality so as to avoid verbose and repetitive descriptions within their intensity table. Thus, they denoted the quality of masonry, brick or other materials with the letters *A, B, C, D* listed in Table 7.1. It should be noted that these types have no connection with the conventional classes *A, B, C* currently used in construction.

7.2.3 Seismic Risk: Gutenberg–Richter Law

In general, in any earthquake-prone region large earthquakes tend to occur much less frequently than small earthquakes. This is, of course, because larger earthquakes demand more time to build up the necessary energy, and that accumulation is likely to be interrupted – and thus halted – by smaller quakes. Based on historical evidence, Gutenberg and Richter have suggested that the frequency with which earthquakes take place obey an empirical law of the form

$$\log_{10} N = a - bM \qquad \text{or} \qquad \log N = A - BM \tag{7.71}$$

Alternatively,

$$\boxed{N = 10^{a-bM} = c \times 10^{-bM}} \qquad \text{or} \qquad \boxed{N = e^{A-BM} = C \times e^{-BM}} \tag{7.72}$$

where N is the number of earthquakes per year whose magnitude equals or exceeds the magnitude M, and a, b are constants, with $b \approx 1$ (or equivalently, $B \approx 2.3$). The parameter c (or C) equals the product of the size and the average seismicity of the ground area under consideration. For reasonably large areas, these parameters can reliably be estimated.

While reasonably robust, this frequency law overestimates the frequency of extremely large events (of which $M = 10$ seems to be an upper limit), and also predicts an infinite number of vanishingly small events $M < 0$ which in the aggregate would release an infinite amount of energy, an impossibility. Still, it produces useful results in the range of engineering interest.

A similar law can also be written for the *epicentral intensity* I_0.

Having the magnitude, however, is not enough to conduct seismic risk studies. One also needs a relationship between magnitude and local seismic intensity. This is accomplished by means of the so-called *attenuation laws*. A number of such laws have been proposed, again from empirical evidence, and many have the form

$$I = \begin{cases} I_0 & D < D_0 \\ I_0 + a - b \log D & D > D_0 \end{cases} \tag{7.73}$$

where a, b are again constants, D is the epicentral distance, and D_0 is a minimum such distance. For example, for northeastern sites in the United States, $a = 4.9$, $b = 2.1$, $D_0 = 10$ miles. Alternatively, instead of attenuation laws, isoseismal and seismic risk maps can be used instead, such as ASCE-SEI 7-05, or the U.S. Department of Defense document *Unified Facilities Criteria, Structural Load Data*, UFC 3-310-01, May 25, 2005 (including changes as of December 2007).

7.2.4 Direction of Intense Shaking

With reference to Figure 7.3, we consider herein the problem of finding the horizontal direction for which the seismic motion is most intense. We shall accomplish this by employing elementary concepts of random vibration. Let $x(t), y(t)$ be the displacements produced at the site by an earthquake, which we can idealize as zero-mean random processes, that is, $E[x] = 0$, $E[y] = 0$, with the operator E being the expectation. This is the same as saying that the average acceleration in either direction is zero, so the ground comes to rest after the earthquake is over. The instantaneous motion in some arbitrary, *fixed* direction α is then

$$r(t) = x(t) \cos \alpha + y(t) \sin \alpha \tag{7.74}$$

for which the first and second-order expectations are

$$E[r] = E[x]\cos \alpha + E[y] \sin \alpha = 0 \tag{7.75}$$

$$\begin{aligned} E[r^2] &= E\left[(x \cos \alpha + y \sin \alpha)^2\right] \\ &= E[x^2]\cos^2 \alpha + 2E[x\,y]\sin \alpha \cos \alpha + E[y^2]\sin^2 \alpha \end{aligned} \tag{7.76}$$

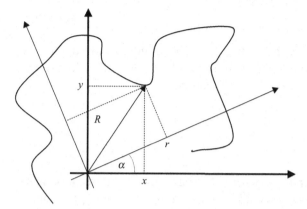

Figure 7.3. Principal axes of ground motion.

Thus, the motion is zero mean in any direction. We define next the shorthand $E[x^2] = S_x^2$, $E[y^2] = S_y^2$, $E[xy] = \rho_{xy} S_x S_y$ to denote the variances and covariance, respectively, with ρ_{xy} being the *coefficient of cross-correlation*. In terms of these quantities, the mean square response in direction α is

$$S_\alpha^2 = S_x^2 \cos^2 \alpha + 2\rho_{xy} S_x S_y \sin \alpha \cos \alpha + S_y^2 \sin^2 \alpha \tag{7.77}$$

The directions of maximum and minimum response follow from the condition

$$\frac{\partial S_\alpha^2}{\partial \alpha} = -2S_x^2 \cos \alpha \sin \alpha + 2\rho_{xy} S_x S_y (\cos^2 \alpha - \sin^2 \alpha) + 2S_y^2 \sin \alpha \cos \alpha = 0 \tag{7.78}$$

which yields

$$\tan 2\alpha = \frac{2\rho_{xy} S_x S_y}{S_x^2 + S_y^2} \tag{7.79}$$

In particular, if the two components of motion are uncorrelated, then $\alpha = 0$ and $\alpha = \frac{1}{2}\pi$, which constitute, respectively, the directions of maximum and minimum intensity of ground shaking – or the other way around.

You should not confuse $r(t)$ for the ground motion along a *fixed* direction with the orbital motion $R(t,\theta)$ of the ground along a constantly changing direction. The magnitude and instantaneous position angle of this orbital motion is

$$R(t) = \sqrt{x^2 + y^2} \qquad \theta = \arctan \frac{y}{x} \tag{7.80}$$

Inasmuch as R is never negative, it follows that its mean value must be positive, that is, $E[R] > 0$, unlike the motions along fixed directions, which are zero mean. On the other hand, the mean squared value is

$$E[R^2] = E[x^2] + E[y^2] = S_x^2 + S_y^2 > 0 \tag{7.81}$$

In particular, if the two earthquake components are equally intense and their cross-correlation is zero, then the earthquake has the same intensity in *any* direction. Hence, it follows that

$$E\left[R^2\right] = 2\,S_x^2 \tag{7.82}$$

This does not contradict the finding that the motion has uniform intensity. In the former case, the direction is fixed, while in the latter, it continuously changes direction, so the value obtained is not representative of any specific direction.

7.3 Ground Response Spectra

7.3.1 Preliminary Concepts

The seismic response spectrum is a plot depicting the maximum response of a single degree of freedom (SDOF) system with some fraction of critical damping to a given earthquake excitation. It is plotted either as a function of the natural frequency or the natural period of that system. The principal application of response spectra lies in engineering design, because when a structure is modeled as a SDOF system, all internal forces are synchronous with, and proportional to, the response. Hence, the maximum values of those internal forces, which are needed to design the structure, are attained when the response is maximum.

Consider first an undamped system:

$$m\ddot{u} + kv = 0 \tag{7.83}$$

If we divide by the mass and separate the two terms, we obtain

$$\ddot{u} = -\omega_n^2 v \tag{7.84}$$

Hence, we reach the important conclusion that *the absolute acceleration is at all times proportional to the relative displacement*. In other words, except for a factor, the time histories of these two response quantities are identical. In particular, the maximum values are also proportional and occur simultaneously, that is,

$$\left|\ddot{u}_{\max}\right| = \omega_n^2 \left|v_{\max}\right| \tag{7.85}$$

Consider next a damped system:

$$m\ddot{u} + c\dot{v} + kv = 0 \tag{7.86}$$

or

$$\ddot{u} = -(2\xi\omega_n \dot{v} + \omega_n^2 v) \tag{7.87}$$

Clearly, when the relative displacement is maximum, the relative velocity must be zero, so the acceleration at that point in time is

$$\left|\ddot{u}\right| = \omega_n^2 \left|v_{\max}\right| \tag{7.88}$$

Although this acceleration is *not* the maximum absolute acceleration, it is a very close approximation for it, if the damping is moderate or small. In that case the difference can be neglected. In general, the acceleration predicted by the previous equation is referred to as the *pseudo-acceleration*. As shown before, it is exactly the maximum acceleration when the system has no damping.

Figure 7.4 illustrates schematically a shaking table onto which a set of oscillators has been attached, each with a distinct natural frequency $\omega_1, \omega_2, \cdots \omega_n$ (in growing order of frequencies), but all possessing the same fraction of critical damping ξ. Plotted at the top are the time histories for relative displacements versus time for each oscillator, which at some

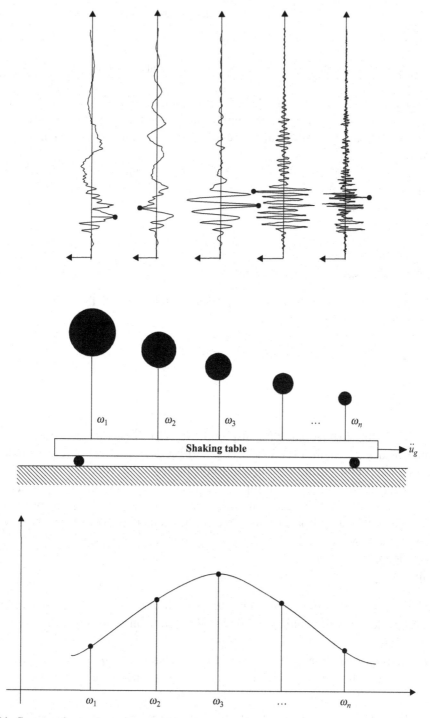

Figure 7.4. Concept of ground response spectrum.

moment in time attain some maximum elongation, even if not all simultaneously. Plotted at the bottom are then those maximum values versus the frequency of the oscillators. This constitutes the response spectrum for relative displacements.

In the light of the previous developments, we introduce the following definitions:

$$S_d \equiv S_d(\omega_n,\xi) = |v_{max}| \qquad = \textit{Spectral displacement}\,(\text{max. relative displacement}) \qquad (7.89)$$

$$S_v = \omega_n S_d \qquad\qquad\quad = \textit{Pseudo-velocity} \qquad\qquad\qquad\qquad\qquad (7.90)$$

$$S_a = \omega_n^2 S_d = \omega_n S_v \qquad = \textit{Pseudo-acceleration}\,(\text{spectral acceleration}) \qquad (7.91)$$

Unlike the pseudo-acceleration, the pseudo-velocity has some relationship to, but is *not* a close approximation for, the maximum relative velocity, particularly for low frequencies where S_v tends to zero but the relative velocity tends to the maximum ground velocity.

A plot of either the spectral displacement or spectral acceleration versus the frequency of the SDOF system defines then the desired response spectrum for the seismic acceleration record considered. Clearly, different fractions of critical damping will lead also to different spectra.

7.3.2 Tripartite Response Spectrum

With reference to Figure 7.5 and from the spectral definitions given previously, we take the logarithms and obtain immediately

$$\log S_v = \log S_d + \log \omega_n \qquad\qquad\qquad\qquad\qquad\qquad\qquad\qquad (7.92)$$

$$\log S_v = \log S_v \qquad\qquad\qquad\qquad\qquad\qquad\qquad\qquad\qquad\qquad (7.93)$$

$$\log S_v = \log S_a - \log \omega_n \qquad\qquad\qquad\qquad\qquad\qquad\qquad\qquad (7.94)$$

Hence, in a doubly logarithmic plot of pseudo-velocity versus frequency, lines of *constant* relative displacement, pseudo-velocity, and pseudo-acceleration are straight lines with inclinations of +45, 0 and –45 degrees to the horizontal. It follows that in such a plot, all three spectral quantities can be represented simultaneously, and this is achieved by superimposing a logarithmic mesh that is inclined at 45 degrees to the principal grid. Such a plot is referred to as a tripartite plot. Accelerations increase then from the lower left to the upper right, while displacements increase from the lower right to the upper left, as shown in Figure 7.6.

A tripartite spectrum has two *asymptotes*:

- At low frequencies, the spectrum tends to the *maximum ground displacement*, because very flexible structures cannot respond to the ground motion, and remain essentially motionless.
- At high frequencies, the spectrum tends to the *maximum ground acceleration*, because very stiff structures hardly deform and follow the ground motion at all times.

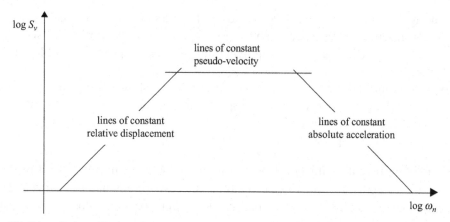

Figure 7.5. Tripartite logarithmic scale.

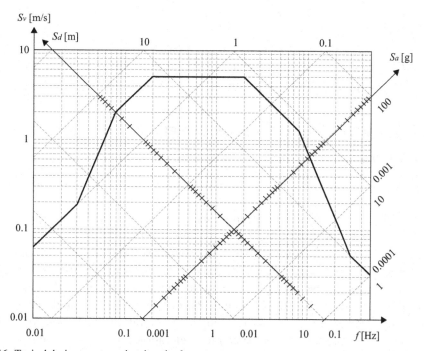

Figure 7.6. Typical design spectrum in tripartite format.

7.3.3 Design Spectra

Response spectra of actual earthquakes are very jagged, particularly for low values of damping. This means that the response can change substantially with very small changes in natural frequency of the SDOF, shifting from a peak to a valley of the spectrum or vice versa. Since two earthquakes with the same general characteristics will not have the peaks and valleys of the spectra coinciding, it is not realistic to design for the spectrum of a single earthquake motion. It is thus customary in engineering practice to use idealized, smooth spectra for design, which are normalized to a unit peak ground acceleration. In actual designs, these spectra must be scaled by the site's design peak acceleration. The rules of construction for the spectra have been developed by computing the spectra for many actual earthquake records, and applying statistical envelopes to ensembles of these actual spectra.

A typical idealized design spectrum for zero damping is shown in the Figure 7.6. While it does not represent any particular design spectrum, it is based on rules of thumb similar to those used in actual seismic codes.

The rules used in this particular design spectrum are

- An ideal earthquake with a peak acceleration of $1g$ has a peak velocity of 1 m/s and a peak displacement of 1 m, which constitutes the easy-to-remember 1–1–1 rule. These define three straight lines of constant displacement, constant velocity and constant acceleration that form the starting point for the design spectrum.
- The maximum displacement magnification is a factor 3.
- The maximum velocity magnification factor depends on the fraction of damping ξ considered according to the following empirical formula $A_v = 4 / (1 + 20\xi)$. This means that for $\xi = 0$, $A_v = 4$.
- The maximum acceleration magnification is a factor 5.
- Transition lines from the ground's lines to the magnified lines and back are drawn at the frequency pairs $0.03 \rightarrow 0.08$ Hz and $8 \rightarrow 30$ Hz

7.3.4 Design Spectrum in the Style of ASCE/SEI-7-05

Design Earthquake

The ASCE/SEI-7-05 seismic design code designates the design earthquake as the Maximum Considered Earthquake and refers to it by means of the acronym MCE. It defines it as the earthquake whose strength has a 2% probability of being exceeded in 50 years, which translates into an earthquake with a return period of 2475 years (i.e., approximately $2,500 = 50/0.02$ years). The earthquake itself is given in terms of a ground response spectrum similar to the one depicted in Figure 7.7, which in turn is fully defined by the parameters prescribed by the code, and as summarized briefly in the ensuing.

Specifically, the code defines the MCE in terms of two separate ground response spectrum parameters that it refers to as the Mapped Acceleration Parameters, or MAP for short. These are S_S = the short period spectral acceleration and S_1 = the spectral acceleration at $T = 1$ seconds. The code also presents a table of coefficients F_a, F_v that depend on ground conditions and the strength of the earthquake, but in the case of stiff ground or rock, these are simply $F_a = F_v = 1$, in which case the two regional parameters are not independent, but satisfy the functional relationship $S_S = 2.5 S_1$. We shall assume in the ensuing that this simplifying relationship holds.

Transition Periods

The ASCE/SEI-7-05 code defines the following transition points:

$$T_0 = \tfrac{1}{5} S_{D1} / S_{DS} \rightarrow 0.08 \tag{7.95}$$

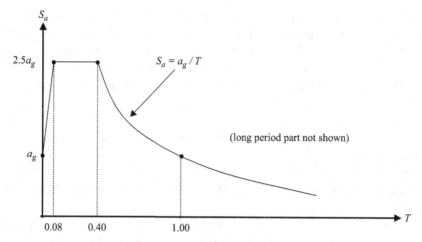

Figure 7.7. Design response spectrum in the style of ASCE/SEI-7-05.

$$T_S = S_{D1} / S_{DS} \rightarrow 0.4 \tag{7.96}$$

$$T_L \sim 6 \rightarrow 12 \qquad \text{Long periods at which the spectral } \textit{displacement} \text{ is constant} \tag{7.97}$$

Implied Ground Motion Parameters

When the simplifying assumption $S_S = 2.5\,S_1$ holds, the ASCE/SEI-7-05 design response spectrum implies the following peak ground motion parameters on rock (where $g = 9.8 =$ acceleration of gravity, in m/s^2)

Ground acceleration $\qquad a_g = S_a\left(T = 0\right) = 0.4\,S_{DS} = S_{D1}$ $\tag{7.98}$

Ground displacement
$$d_g = \lim_{T \to \infty}\left[\left(\frac{T}{2\pi}\right)^2 S_a\,g\right]$$
$$= \frac{\left(S_{D1}g\right)T_L}{4\pi^2} \approx \frac{S_{D1}\,T_L}{4} = \frac{a_g\,T_L}{4} \sim 1.5 a_g \rightarrow 3 a_g \tag{7.99}$$

That is, the spectrum contains the implicit rule that "a 1g earthquake produces a ground displacement anywhere from 1.5 m to 3 m. This is consistent with Newmark's old rule of thumb which stated that "a 1g earthquake produces a peak velocity of 36 in/s and a displacement of 4 ft."

7.3.5 MDOF Systems: Estimating Maximum Values from Response Spectra

In a design office where structural components are dimensioned so as to resist maximum credible forces, it is often not necessary to know exactly how a given physical parameter, such as the acceleration of some floor, the shear in a column, or the bending stresses in a beam, evolve in time in response to a seismic event. Instead, it may suffice to estimate such maximum values by means of either actual or design response spectra.

If the structure can be assumed to have normal modes, then a modal decomposition will decouple the structure into a set of SDOF equations, namely one for each of the modes in the system. It is then possible to ascertain exactly what the maximum response in each mode will be, since the response spectrum, by its very definition, provides the maximum response for an SDOF system subjected to a given ground motion. However, the global maximum for the physical parameter in question cannot be obtained simply as the addition of the modal maxima, because these maxima will not all occur simultaneously. Hence, we must resort to statistical arguments to combine the modal maxima.

To keep the explanations simple, we present the concept here by means of an example involving a structure that has only lateral DOF, and is subjected to an earthquake with a single horizontal component.

Consider a closely coupled, four-story lumped mass structure that has only one translational DOF at each elevation, as shown in Figure 7.8. We assume that the modal frequencies and modal shapes for this structure are known to us. This structure deforms laterally in response to an earthquake for which the response spectrum is also known. Suppose now that we wish to estimate the maximum shearing force F_{max} in the third spring from the top (i.e., in spring k_3). There are two equivalent alternatives available to us to obtain this force: 1) from the deformation of that spring, and 2) from the global equilibrium of the free body that we obtain by cutting through the spring in question. We consider each of these options in turn:

$$F(t) = k_3 \left(u_3 - u_4 \right) = k_3 \left(v_3 - v_4 \right) \tag{7.100}$$

and

$$F(t) = -\left[m_1 \ddot{u}_1 + m_2 \ddot{u}_2 + m_3 \ddot{u}_3 \right] = -\sum_{i=1}^{3} m_i \ddot{u}_i \tag{7.101}$$

in which the u_i, v_i are the absolute and relative displacements, respectively. If we express the first of these in terms of a modal superposition, we obtain

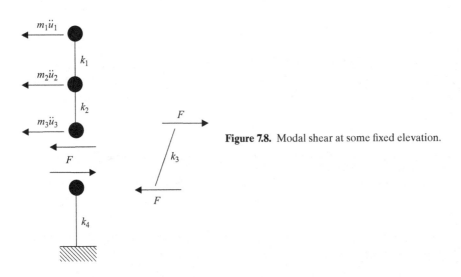

Figure 7.8. Modal shear at some fixed elevation.

$$F(t) = k_3 \left(\sum_{j=1}^{N} \left(\varphi_{3j} - \varphi_{4j} \right) q_j(t) \right) = \sum_{j=1}^{N} F_j(t) \qquad (7.102)$$

in which the modal response function is the solution to

$$\mu_j \ddot{q}_j + \eta_j \dot{q}_j + \kappa_j q_j = -\mu_j \gamma_j \ddot{u}_g \qquad j = 1, \ldots n \qquad (7.103)$$

Clearly, the maximum value of $q(t)$ must be $q_{max} = |\gamma_j| S_d(\omega_j, \xi_j) \equiv |\gamma_j| S_{dj}$, so the maximum modal shear is

$$\overline{F}_j = \max F_j(t) = k_3 |\varphi_{3j} - \varphi_{4j}| |\gamma_j| S_{dj} \qquad (7.104)$$

Since the maximum modal shears do not all occur simultaneously, it follows that the maximum total shear has the (conservative) upper bound known as the *sum of the absolute values* (SAV), that is,

$$F_{max} \leq \sum_{j=1}^{N} \overline{F}_j = k_3 \sum_{j=1}^{N} |\varphi_{3j} - \varphi_{4j}| |\gamma_j| S_{dj} \qquad \text{SAV} \qquad (7.105)$$

Using concepts from random vibration, which start from the assumption that the earthquake is a zero-mean, broad band, white noise random process, and the use of mathematical models for the probabilities of exceedance of threshold values, it is argued that the modal responses $F_j(t)$ are *statistically independent*, and the combined maximum can be estimated from the variance of the stochastic process. This leads in turn to the square root of the sum of the squares rule, or SRSS rule,

$$F_{max} \approx \sqrt{\sum_{j=1}^{N} \overline{F}_j^2} = k_3 \sqrt{\sum_{j=1}^{N} |\varphi_{3j} - \varphi_{4j}|^2 |\gamma_j|^2 S_{dj}^2} \qquad \text{SRSS} \qquad (7.106)$$

This widely used rule gives reasonably close estimations, provided that the modes are *well separated*, that is, that the frequencies of any two modes are not close to each other. If, however, this is not the case, then the SRSS may lead to unconservative estimations of response. The reason is that if two distinct modes, for example a translational and a torsional mode, have the same frequencies (and, of course, equal damping ratios), then the modal responses are proportional to each other at all instants in time; thus, their maxima will occur simultaneously. This can be seen by observing that, except for the modal participation factor, the modal equations are identical. This has led to the development of alternative rules to the SRSS, of which one of the most widely used is the Wilson–Der Kiureghian Complete Quadratic Combination rule, or CQC method (sometimes referred to as the CSM rule, in reference to Closely Spaced Modes):

$$F_{max} \approx \sqrt{\sum_{i=1}^{N} \sum_{j=1}^{N} \rho_{ij} |\overline{F}_i \overline{F}_j|} \qquad \text{CQC} \qquad (7.107)$$

in which the modal cross-correlation coefficients ρ_{ij} are given by

$$\rho_{ij} = \frac{8\sqrt{\xi_i \xi_j} (\xi_i + r\xi_j) r^{3/2}}{(1 - r^2)^2 + 4\xi_i \xi_j (1 + r^2) r + 4(\zeta_i^2 + \xi_j^2) r^2} \qquad (7.108)$$

with $r = \omega_j / \omega_i$ being the tuning ratio. You should observe that this expression yields exactly the same result when r is replaced by $1/r$. The implication is that $\rho_{ij} = \rho_{ji}$. Also, the diagonal terms correspond to $\rho_{jj} = 1$, so if the cross-modal terms are disregarded, we recover the SRSS rule.

We return now to the alternative expression for the shear based on global equilibrium of inertia forces. If the structure is undamped, we can write the dynamic equilibrium equation as

$$\mathbf{M}\ddot{\mathbf{u}} + \mathbf{K}\mathbf{v} = \mathbf{0} \tag{7.109}$$

which in terms of the modal response for the relative displacements is

$$\mathbf{M}\ddot{\mathbf{u}} = -\mathbf{K}\mathbf{v} = -\mathbf{K}\boldsymbol{\Phi}\mathbf{q} = -\mathbf{M}\boldsymbol{\Phi}\boldsymbol{\Omega}^2\mathbf{q} \tag{7.110}$$

Hence

$$\ddot{\mathbf{u}} = -\boldsymbol{\Phi}\boldsymbol{\Omega}^2\mathbf{q} \tag{7.111}$$

It follows that the acceleration of the ith floor is

$$\ddot{u}_i(t) = -\sum_{j=1}^{N} \varphi_{ij}\, \omega_j^2\, q_j(t) \tag{7.112}$$

whose modal maximum is

$$\bar{\ddot{u}}_{ij} = \left| \varphi_{ij}\, \omega_j^2\, \gamma_j\, S_{dj} \right| = \left| \varphi_{ij}\, \gamma_j\, S_{aj} \right| \tag{7.113}$$

with $S_{aj} = S_a\left(\omega_j, \xi_j\right)$ being the spectral acceleration (pseudo-acceleration) at the modal frequency and damping. In terms of the modes, the shearing force is then

$$F(t) = \sum_{i=1}^{3} m_i \sum_{j=1}^{N} \varphi_{ij}\, \omega_j^2\, q_j = \sum_{j=1}^{N}\sum_{i=1}^{3} m_i\, \varphi_{ij}\, \omega_j^2\, q_j = \sum_{j=1}^{N} F_j(t) \tag{7.114}$$

so that the maximum shear by the SRSS rule is

$$F_{\max} \approx \sqrt{\sum_{j=1}^{N} \bar{F}_j^2} = \sqrt{\sum_{j=1}^{N} \left| \sum_{i=1}^{3} m_i\, \varphi_{ij} \right|^2 \left| \gamma_j \right|^2 S_{aj}^2} \tag{7.115}$$

Common Error in Modal Combination

You should carefully observe that in all of these rules, the statistical combination is *always the last step*. Hence, it is *not* appropriate for you to first compute the SRSS for the displacements and then use these to estimate some other effect, such as the shear (as in this example). This means that the last of the following three formulas is most definitely wrong:

$$\bar{v}_3 \approx \sqrt{\sum_{j=1}^{N} \left| \varphi_{3j} \right|^2 \left| \gamma_j \right|^2 S_{dj}^2} \qquad \bar{v}_4 \approx \sqrt{\sum_{j=1}^{N} \left| \varphi_{4j} \right|^2 \left| \gamma_j \right|^2 S_{dj}^2} \tag{7.116}$$

$$F_{\text{max}} = k_3(\bar{v}_3 - \bar{v}_4) \qquad wrong!!! \tag{7.117}$$

In summary, you must first compute any effect at the level of the modes, and only then apply the SRSS or CQC rules. Otherwise, the results may be unpredictable, either being too conservative, or worse, very unconservative. The latter can happen when cancellations occur as a result of the loss of the sign in the displacements after application of the SRSS rule.

As an additional example, suppose we wish to compute the maximum axial stress in a column. If $M(t)$ and $N(t)$ are the bending moment and axial stress in the column, and W, A are the section modulus and the cross section, respectively, then the instantaneous maximum stress is

$$\sigma(t) = \frac{M(t)}{W} + \frac{N(t)}{A} \tag{7.118}$$

The correct and incorrect modal combinations are then

$$\bar{\sigma} = \sqrt{\sum_{j=1}^{N} \bar{\sigma}_j^2} = \sqrt{\sum_{j=1}^{N} \left(\frac{M_j}{W} + \frac{N_j}{A}\right)^2} \qquad \text{Correct} \tag{7.119}$$

$$\bar{\sigma} = \frac{\bar{M}}{W} + \frac{\bar{N}}{A} = \frac{1}{W}\sqrt{\sum_{j=1}^{N} \bar{M}_j^2} + \frac{1}{A}\sqrt{\sum_{j=1}^{N} \bar{N}_j^2} \qquad \text{Incorrect} \tag{7.120}$$

A word in closing: One often finds that in the design of high-rise buildings, engineers determine without much ado the maximum floor accelerations by means of the SRSS rule, which they then multiply by the floor masses to estimate the maximum D'Alembert forces. From this point on, these inertia forces are treated as *static* forces to analyze the structure as a whole, say to determine bending moments or shears throughout the structure. Although this practice may be convenient, it is intrinsically *wrong*. With some luck, however, they might get away with it when the response is dominated by the fundamental mode, which may not be the case for tall buildings.

General Case: Response Spectrum Estimation for Complete Seismic Environment

We generalize now the rules we presented previously and consider the case of a general 3-D structure that is subjected to an earthquake with multiple components (say, east–west and north–south). If we focus attention onto one specific point in the structure and consider its relative response in each of the two horizontal directions, say x and y, we can then write the response as

$$u_x(t) = u_{xx}(t) + u_{yx}(t) \tag{7.121}$$

$$u_y(t) = u_{yx}(t) + u_{yy}(t) \tag{7.122}$$

in which $u_{xx}(t)$ is the response in direction x due to the earthquake in direction x, and so forth. Thus, the first subindex identifies the direction of response, while the second identifies the direction of the earthquake. Using modal decomposition, we can write these two components of displacements as

$$u_x(t) = \sum_{j=1}^{N} \left[u_{xx}^j(t) + u_{xy}^j(t) \right] \equiv \sum_{j=1}^{N} u_x^j(t) \tag{7.123}$$

$$u_y(t) = \sum_{j=1}^{N} \left[u_{yx}^j(t) + u_{yy}^j(t) \right] \equiv \sum_{j=1}^{N} u_y^j(t) \tag{7.124}$$

Hence, the maximum response in either direction estimated with the CQC method is

$$\bar{u}_x \approx \sqrt{\sum_{i=1}^{N}\sum_{j=1}^{N} \rho_{ij}\, \bar{u}_x^i\, \bar{u}_x^j} = \sqrt{\sum_{i=1}^{N}\sum_{j=1}^{N} \rho_{ij} \left(\overline{u_{xx}^i + u_{xy}^i} \right)\left(\overline{u_{xx}^j + u_{xy}^j} \right)} \tag{7.125}$$

$$\bar{u}_y \approx \sqrt{\sum_{i=1}^{N}\sum_{j=1}^{N} \rho_{ij}\, \bar{u}_y^i\, \bar{u}_y^j} = \sqrt{\sum_{i=1}^{N}\sum_{j=1}^{N} \rho_{ij} \left(\overline{u_{yx}^i + u_{yy}^i} \right)\left(\overline{u_{yx}^j + u_{yy}^j} \right)} \tag{7.126}$$

If the east–west and north–south components of the earthquake are *statistically independent*, then so are the modal responses, which means that the two modal maxima can be combined by the SRSS rule, that is,

$$\bar{u}_x^j \approx \sqrt{\left(\bar{u}_{xx}^j \right)^2 + \left(\bar{u}_{xy}^j \right)^2} \tag{7.127}$$

Roughly speaking, two random processes are said to be statistically independent when knowledge of the first gives no clue as to what the second should be. This assumption is actually quite good, because for any real earthquake, it is always possible to find two orthogonal directions for which the motion components are indeed statistically independent. The method to accomplish this is very similar to that used to find principal stresses by means of Mohr's circle, with the variance and covariance playing the role of tensile and shearing stresses, respectively. It follows that

$$\bar{u}_x \approx \sqrt{\sum_{i=1}^{N}\sum_{j=1}^{N} \rho_{ij} \left[\left(\bar{u}_{xx}^i \right)^2 + \left(\bar{u}_{xy}^i \right)^2 \right]\left[\left(\bar{u}_{xx}^j \right)^2 + \left(\bar{u}_{xy}^j \right)^2 \right]} \tag{7.128}$$

Assume next that the motion u_x corresponds to the lth DOF in the structure. If so, then the maximum relative displacement and absolute acceleration in the jth mode at that point are

$$\bar{v}_{xx}^j = \left| \varphi_{lj}\, \gamma_{xj} \right| S_{dj}^x \qquad \bar{v}_{xy}^j = \left| \varphi_{lj}\, \gamma_{yj} \right| S_{dj}^y \tag{7.129}$$

$$\ddot{\bar{u}}_{xx}^j = \left| \varphi_{lj}\, \gamma_{xj} \right| S_{aj}^x \qquad \ddot{\bar{u}}_{xy}^j = \left| \varphi_{lj}\, \gamma_{yj} \right| S_{aj}^y \tag{7.130}$$

in which γ_{xj}, γ_{yj} are the participation factors for the jth mode due to the seismic motion in direction x and y, and $S_{dj}^x = S_d^x(\omega_j, \xi_j)$, $S_{dj}^y = S_d^y(\omega_j, \xi_j)$ are the response spectra for these two earthquake components. Finally, in the absence of further information to the contrary, it is often reasonable to assume that the two earthquake components have similar (or even identical) response spectra, an assumption that does *not* contradict the fact that they may be statistically independent. If so, then the maximum relative displacement in direction x is

$$\bar{v}_x \approx \sqrt{\sum_{i=1}^{N}\sum_{j=1}^{N} \rho_{ij} |\varphi_{li}||\varphi_{lj}| \sqrt{\left(\gamma_{xi}^2 + \gamma_{yi}^2\right)} \sqrt{\left(\gamma_{xi}^2 + \gamma_{yi}^2\right)} S_{di}\, S_{dj}} \qquad (7.131)$$

The relative displacement and absolute acceleration for well-separated modes are then

$$\bar{v}_x \approx \sqrt{\sum_{i=1}^{N} \varphi_{lj}^2 \left(\gamma_{xi}^2 + \gamma_{yi}^2\right) S_{dj}^2} \qquad \bar{\bar{u}}_x \approx \sqrt{\sum_{i=1}^{N} \varphi_{lj}^2 \left(\gamma_{xi}^2 + \gamma_{yi}^2\right) S_{aj}^2} \qquad (7.132)$$

7.4 Dynamic Soil–Structure Interaction

7.4.1 General Considerations

Soil–structure interaction – or SSI for short – is a broad discipline in Applied Mechanics that is concerned with the development and investigation of theoretical methods and practical tools for the analysis of dynamically loaded structures, taking into consideration the flexibility and dynamic properties of the supporting soil. A concise review on the history of SSI can be found in Kausel (2010).[5]

SSI also touches on other interdisciplinary problems in geomechanics and seismology, such as earthquake source modeling, soil amplification, and wave scattering elicited by local geologic conditions (modification of the seismic signal by canyons, valleys, soft soil deposits, etc.), development of generalized constitutive equations for the inelastic modeling of porous multiphase media (i.e., nonlinear material models), soil liquefaction, fluid–structure interaction (e.g., water reservoirs, dams, tanks), seismic isolation and vibration absorption, numerical methods (energy-absorbing boundaries to model infinite media by means of finite elements, etc.), and many more.

Broadly speaking, the theory of SSI deals with two related, but distinct issues. On the one hand are problems involving *external* loads, that is, with problems where the dynamic excitation is applied directly onto the structure. Examples are (unbalanced) reciprocating machines on elastic foundation, railroad tracks loaded by fast moving trains, tall buildings subjected to wind loads, or radar tracking stations responding to operational loads. On the other hand, the design of massive structures in seismic areas and of underground facilities resistant to blast loads required extension of the theory to *internal* loads, that is, to dynamic excitations and sources applied within the soil mass. In the first situation, interaction effects arise solely as a result of external and inertial forces being transmitted to the ground, a phenomenon that is now referred to as *inertial interaction*. The mechanical energy thereby transmitted to the ground and scattered away from the structure in the form of stress waves is called *radiation damping*. In the second case, additional interaction effects also arise for earthquake loads because the stiffer structural foundation cannot conform to the distortions of the soil elicited by the incident seismic waves. The structure – or inclusion – acts like an opaque or reflective object in the path if the incident seismic rays that then produces a scattered wave field that modifies locally the motion

---

[5] E. Kausel, "Early history of soil–structure interaction," *Soil Dyn. Earthquake Eng.,* 30, 2010, 822–832.

in the vicinity of the foundation. This effect is now referred to as *kinematic interaction*. Finally, the motion that the ground would have experienced if neither the soil had been excavated nor the structure erected is normally called the *free-field* problem.

It is generally accepted that SSI effects are negligible when the soil is very stiff. When this is indeed the case, then a conventional seismic analysis of the structure at hand can be carried out without consideration of SSI effects. On the other hand, for soft to interme-diately firm soils or rock – having a shear wave velocity of less than, say 700 m/s – these interaction effects can no longer be ignored. A proper model should then account for the subgrade flexibility and the coupled interaction between the soil and the structures.

Few, if any, structures are perfectly symmetric with respect to vertical planes. As a result, the response of structures is inherently 3-D. Nonetheless, many solution techniques assume that the structure has some form of symmetry, which can have important effects on the computed response. Simplifications are also frequently introduced in the modeling of the geometric boundary conditions of the soil–structure system, which can again play a decisive role on the system's dynamic behavior. This is particularly true when material nonlinear effects and/or 3-D effects are accounted for in the soil.

The greatest difficulty, however, lies in the determination of the mechanical properties of the soil underneath and in the general vicinity of the structure. These properties depend not only on the spatial coordinates (sediments of sand or clay; soft layers; boulders, etc.), but also on time (changing phreatic surface; excavation and construction sequence; settle-ment of structure and consolidation of soil; effect of prior earthquakes; loading/unload-ing; etc.). In addition, the accuracy with which these properties can be determined in the laboratory by means of *undisturbed samples* (an oxymoron!) or in situ by seismic testing (cross-hole etc.) is very limited. Thus, it is difficult indeed to determine experimentally the parameters needed for any "true" nonlinear model.

In addition, soils are not homogeneous, and their properties generally vary in both the horizontal direction and with depth. However, due to geological sedimentation processes, the soil properties vary more rapidly in the vertical than in the horizontal direction, and the idealization of the supporting soil as a set of horizontal layers is often adequate.

Seismic Excitation (Free-Field Problem)

The free field problem addresses the evaluation or estimation of the ground motion at the site before excavation or erection of any structure. The methods and solution techniques for this problem are closely related to those used by seismologists, although in most cases rather simple 1-D wave propagation models are used in the context of waves propagating vertically in horizontally layered soils of finite or infinite depth.

Kinematic Interaction

This refers to the so-called wave passage or traveling wave problem, the scattering of seismic rays by a rigid inclusion, or the *tau effect*. In the case of an ideally rigid foun-dation, kinematic interaction depends only on the geometry of the foundation, the soil configuration near the foundation, and the travel path of the seismic excitation across the soil–structure interface. The interaction effects observed in many structures are mostly the result of this phenomenon. Examples are buried pipes, tunnels, deeply embedded buildings or buried structures, long bridges resting on distant piers, torsion of buildings caused by surface waves, and so forth. This topic will be taken up in more detail later on.

Inertial Interaction

This is the classical SSI problem involved in machine vibrations, shallowly embedded buildings subjected to earthquakes, and so forth. The solution techniques range from "exact" analytical solutions for very simple systems, to the semi-analytical and fully numerical solutions using finite element procedures for complex systems. Again, more details will be seen later on.

7.4.2 Modeling Considerations

Continuum Solutions versus Finite Elements

Basically, there are two alternatives to solve for SSI effects:

- In the first approach, the mathematical model of the soil and the structure is based on discrete methods such as finite elements or finite differences. This method is usually referred to as the complete, one-pass or *direct approach.*
- Alternatively, it is possible to model the subgrade by stiffness or impedance functions that can be interpreted as sets of springs, dashpots, and masses. The structure is again modeled with finite elements or regular linear members. This approach is normally referred to as the spring method, substructure method, or the *three-step solution.*

As will be shown later on, these two methods are mathematically equivalent, provided consistent assumptions are made. Each of these methods has its own advantages and limitations, but many of the modeling considerations in the solution of the SSI problem are common to both. In particular, one of the most important ingredients in either model is the definition of the design earthquake, and the identification of the wave mechanism giving rise to the *free field* ground motion, that is, the motion that would take place if no structure were present.

Finite Element Discretization

Finite element (and finite difference) techniques have some well-established restrictions for acceptable accuracy. To reproduce adequately the propagation of waves through the continuum, the size of the elements should not be larger than about $\frac{1}{6}\lambda_{\min}$ to $\frac{1}{8}\lambda_{\min}$, that is, a fraction of the smallest wave length of interest. This wave length is given by $\lambda_{\min} = C_s T_{\min}$, where $T_{\min}$ represents the smallest period of interest and C_s is the typical shear wave velocity of the soil. In addition, the refinement of the finite elements underneath the footing could be dictated by additional considerations such as strain gradients. Indeed, the number of elements must be sufficient to achieve acceptable resolution in those regions where the strains are large and expected to change rapidly with distance.

For solution procedures in the time domain, it is usually advantageous to work with diagonal (lumped) mass matrices. Frequency domain solutions, on the other hand, can be performed either with lumped or consistent matrices. A combination of both, typically $\frac{2}{3}$ consistent and $\frac{1}{3}$ lumped has been found to decrease numerical dispersion effects and improve significantly the accuracy of the dynamic responses.

Boundary Conditions

Another important consideration is the selection of the appropriate boundary conditions in the finite model to simulate the semi-infinite extent of the soil and the radiation of waves into the outer region.[6] For a half-space, this radiation takes place in all directions, but for a layered stratum resting on stiff rock, radiation can only take place laterally and for frequencies higher than the first natural frequency of the stratum.

The typical finite element *island* of soil will be delimited by at least two boundaries, one at the bottom and one (or more) at the sides (i.e., lateral boundaries). For shallow strata over stiff rock, the bottom boundary is often idealized as a rigid interface at bedrock, at which the displacements are specified. For deep soil strata over rock, or when there is a deep layer of soil without a clear, sharp change in elastic properties, it becomes necessary to define the bottom boundary at some arbitrary depth that is at least two foundation diameters down into the soil, to ensure that the waves generated by the vibration of the foundation are significantly attenuated before being reflected at this boundary.

Both the bottom and lateral boundaries can be simulated numerically by means of numerical devices of varying degrees of sophistication.

Viscous Boundaries

This is the simplest and least accurate of the absorbing boundaries, indeed just an emergency measure to be used in the absence of better alternatives. In this simple alternative, attempts are made to absorb the waves radiating away from the structure by means of viscous dampers that are occasionally supplemented by springs and masses. These procedures are based on 1-D wave propagation theory in rods – for which they are exact – and assume both a pattern of waves (S, P, or R waves) as well as their angle of incidence on the boundary. Therefore, they are only approximate in two or three dimensions and they also break down at low frequencies, for dampers have no static impedance.

Paraxial Boundaries

Paraxial boundaries are mathematical artifacts that improve considerably on the wave absorbing capacity of viscous boundaries, even if they are hardly more complicated than the latter.[7] These boundaries are designed to absorb waves impinging on the boundary not only at normal incidence, but also waves that arrive at oblique angles, provided the incidence angles do not deviate substantially form normality; thus the name *paraxial*, which means "close to the (normal) axis."

---

[6] E. Kausel, "Local transmitting boundaries," *J. Eng. Mech.*, *117*(6), 1988, 1011–1027.

[7] E. Kausel, "Physical interpretation and stability of paraxial boundaries," *Bull. Seismol. Soc. Am.*, 82 (2), 1992, 898–913.

Consistent Boundaries

Transmitting boundaries exist that reproduce the soil region beyond the boundaries as if the finite element grid had been extended without any limit. These artifacts are based on an exact solution to the wave propagation problem in a layered medium, and are thus exact in the finite element sense. For this reason, they can be placed immediately in contact with the irregular part of the structure modeled with finite elements, so there is no need for any transition region. Although only defined properly in the frequency domain, these boundaries are extremely effective and work stupendously even for static problems.

Boundary Elements

The Boundary Element Method (BEM) is often used to model exterior regions, but although technically accurate, it is hampered by the great computational expense that it entails for 3-D models. For this reason, it is used for the most part only in 2-D models, and then again it is restricted to homogenous exterior media, although layered media could be considered as well if one were to use appropriate *fundamental solutions* (or Green's functions) to construct the boundary.

Perfectly Matched Layers

Perfectly Matched Layers (PMLs) is a relatively new mathematical technique that seems to work very well indeed[8]. PMLs are based on a complex-valued mapping and stretching of space, and have thus no simple physical interpretation. In essence, the finite element model is surrounded by a few layers of finite elements whose dimensions are progressively more complex-valued with distance to the transition horizon. It can be shown that in the limit of a fine grid, PMLs are virtually perfect absorbers of waves.

7.4.3 Solution Methods

Direct Approach

We review in the ensuing the methods available for the solution of soil–structure interaction problems, and discuss their similarities and differences.[9]

In the direct approach, the structure (or structures) and the surrounding soil are analyzed together. The excitation is given in the form of a base motion, or in the form of equivalent lateral forces applied onto the structures. Finite elements and regular linear members are normally used to model the different components of the system. Finite difference schemes may also be used to model the soil, although this procedure is less frequently used in practice.

The direct approach has two main advantages:

a. It allows to solve a true nonlinear dynamic problem, where superposition is no longer valid, accounting both for the nonlinear effects in the soil amplification problem (variations of properties with depth) and in the interaction problem (variations of

---

[8] E. Kausel and J. Barbosa, "PMLs: A direct approach," *Int. J. Numer. Methods Eng.*, 90, 2012, 343–352.

[9] E. Kausel and J. M. Roësset (1974), "Soil–structure interaction problems for nuclear containment structures," in *Electric Power and the Civil Engineer*, Proceedings of the ASCE Power Division Specialty Conference held in Boulder, Colorado, on August 12–14, 1974, pp. 469–498.

properties both in depth and with horizontal distance). This type of solution is, how-ever, rarely used.

b. It allows including the effect of the flexibility of the mat, and its exact connection to the structure.

The main disadvantage of the direct solution is its relative computational cost, since a large number of DOF are treated simultaneously, namely those representing the soil and those of the structures. We shall elaborate in much greater detail on the direct solution in a later section.

Superposition Theorem

With reference to Figure 7.9, the general equations of motion for the soil–structure system considered in the direct approach can be written in matrix form as:

$$\mathbf{M\ddot{u} + C\dot{y} + Ky = 0} \tag{7.133}$$

where $\mathbf{y}$ is a vector of relative displacements, $\mathbf{u}$ the vector of absolute accelerations, and $\mathbf{y} = \mathbf{u} - \mathbf{u}_g$, where $\mathbf{u}_g$ is a generalized ground acceleration vector. It is possible, alterna-tively, to write this equation in the form of 2 equations

$$\mathbf{M\ddot{u}_1 + C\dot{y}_1 + Ky_1 = 0} \tag{7.134}$$

$$\mathbf{M\ddot{y}_2 + C\dot{y}_2 + Ky_2 = -M_2\ddot{u}_1} \tag{7.135}$$

where $\mathbf{u}_1 = \mathbf{y}_1 + \mathbf{u}_g, \mathbf{u} = \mathbf{u}_1 + \mathbf{y}_2, \mathbf{y} = \mathbf{y}_1 + \mathbf{y}_2$, and $\mathbf{M} = \mathbf{M}_1 + \mathbf{M}_2$. $\mathbf{M}_1$ represents the mass of the system excluding the mass of the structure, while $\mathbf{M}_2$ represents exclusively the mass of the structure. ($\mathbf{M}_1, \mathbf{M}_2$ are conveniently filled with zeros to match the dimensions of $\mathbf{M}$). The equivalence of these two equations with the first one is demonstrated by simple addition.

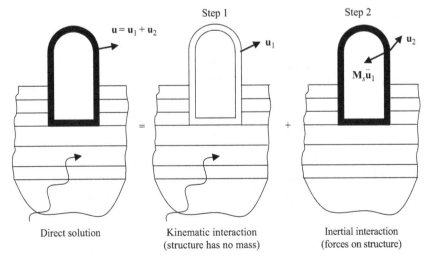

Figure 7.9. Superposition theorem.

In Eq. 7.134, the response of the *massless structure* is found first, and is referred to as the *kinematic interaction* or *wave passage*, and less frequently, as the *tau effect*. The results are then used in Eq. 7.135, which defines the *inertial interaction*, and that is solved by application of fictitious inertia forces applied to the structure alone (Figure 7.9). At this time the soil could be replaced, if so desired, by appropriate impedance functions that account for layering and embedment effects.

Three-Step Approach

Whenever the foundation–structure system can be considered to be very rigid, it is possible to remove the structure altogether from the above Eq. 7.134 for kinematic interaction and replace it by an infinitely rigid, massless foundation. This is legitimate because the structure in this step acts as a rigid body without mass. That equation then describes the solution for a massless rigid foundation subjected to the specified seismic environment, which elicits in that foundation up to six components of motion, namely three translations and three rotations.

On the other hand, the vector $\mathbf{y}_2$ can be regarded as the displacements relative to a fictitious support, while $\mathbf{u}_1$ is the equivalent support motion. For a rigid foundation (slab and lateral walls if the structure is embedded) it is therefore valid to break the solution into the following three steps (see Figure 7.10):

a. *Kinematic interaction*: Determination of the motion of the massless rigid foundation when subjected to the same input motion as the total solution. For surface foundations, the motion is usually equal to that of the free surface, but it could also contain rotations (rocking and torsion) when the waves do not propagate vertically. For an embedded foundation it will yield in general both translations and rotations, the solution of which generally requires the use of the same numerical techniques described for the direct solution. This fact would make the procedure not attractive since one could equally well include the structure at little extra cost. However, it is possible to obtain excellent approximations to the kinematic interaction problem by means of the *Iguchi method* that will be expounded in detail later on.

b. *Foundation impedances*: Determination of the frequency-dependent subgrade stiffness for the relevant DOF, which account for layering and embedment effects. This

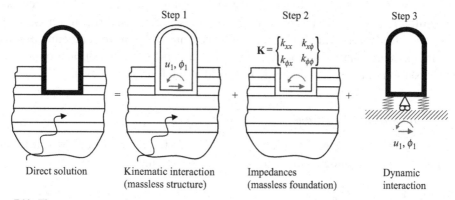

Figure 7.10. Three-step approach.

step corresponds formally to a dynamic condensation of the DOF of the soil. It yields the *soil impedance matrix* whose components are the so called soil "springs." In practice, one never carries out a condensation, but directly obtains the impedances by other means. Each stiffness coefficient is of the form $k_0(1+2i\xi)(k+ia_0c)$, where k_0 is the static stiffness, ξ is a measure of the internal damping in the soil (of a hysteric nature) and a_0 is the dimensionless frequency $\omega R/C_s$, where ω is the circular frequency of the motion and excitation, R is a characteristic dimension (e.g., the radius) of the foundation slab, and C_s is an arbitrary reference shear wave velocity. Also, the functions $k = k(a_0)$ and $c = c(a_0)$ are frequency dependent coefficients, normalized with respect to the static stiffness. The coefficient c is related to the energy loss by radiation. Commercial computer programs as well as numerical solutions for the impedances of circular and rectangular footings lying on (or embedded in) either uniform or layered half-spaces are available in the literature.

   c. *Inertial interaction*: Computation of the response of the real structure supported on frequency dependent soil "springs" (i.e., impedances) and subjected at the base of these "springs" to the motion computed in the first step.

The only approximation involved in this approach concerns the deformability of the structural foundation. If it were rigid, the solution of this procedure would be identical to that of the direct approach – assuming of course consistent definitions of the motion, impedances and similar numerical procedures.

The substructure method has the advantage of being substantially less time consuming when approximations are used for the kinematic interaction problem. It allows, therefore, conducting more parametric studies, and the accuracy of each step is subject to better control. Of particular importance is the possibility in this method to make use of symmetry or cylindrical conditions if the foundation meets these requirements, even if the structure does not – which is a frequent situation. The coupling between the corresponding terms will come in naturally in the third step.

From a practical standpoint, the procedure has an additional advantage when the kinematic interaction phase is used to *define* the seismic motion components (translation, rotations and torsion), whatever the physical arguments against this type of specification. In this manner, undesirable deamplification of certain frequency components following the use of 1-D amplification theory can be avoided while achieving a fully 3-D representation of the structure. It is important to mention again that a solution consistent with the direct approach will involve both rotations and translations in an embedded massless rigid foundation, and that the impedances must account for embedment effects. Furthermore, the translation is not, in general, equal to the control motion, nor is it equal to the translation of the subgrade in the free field at the foundation level.

Approximate Stiffness Functions

Good approximations to the impedances (or stiffness functions) of cylindrical foundations embedded in a homogeneous half-space can be obtained as follows.[10]

---

[10] A. Pais and E. Kausel, "Approximate formulas for dynamic stiffnesses of rigid foundations," *Soil Dyn. Earthquake Eng.*, 7(4), 1988, 213–227.

Static Values

In the ensuing formulas, G is the shear modulus of the soil underneath the mat; R is the radius of the foundation; E is the depth of embedment; and v is Poisson's ratio.

$$K_v^0 = \frac{4GR}{1-v}\left(1+0.54\frac{E}{R}\right) \qquad \text{Vertical (heaving)} \tag{7.136}$$

$$K_h^0 = \frac{8GR}{2-v}\left(1+\frac{E}{R}\right) \qquad \text{Swaying (horizontal)} \tag{7.137}$$

$$K_{hr}^0 = K_h^0 R\left(\frac{2}{5}\frac{E}{R}-0.03\right) \qquad \text{Coupling (horizontal-rocking)} \tag{7.138}$$

$$K_r^0 = \frac{8GR^3}{3(1-v)}\left(1+\frac{2}{3}\frac{E}{R}+0.58\left(\frac{E}{R}\right)^3\right) \qquad \text{Rocking (rotation about horizontal axis)} \tag{7.139}$$

$$K_t^0 = \frac{16GR^3}{3}\left(1+\frac{8}{3}\frac{E}{R}\right) \qquad \text{Torsion (rotation about vertical axis)} \tag{7.140}$$

Dynamic Impedances

In all modes of motion the impedances are of the form $Z = K^0\left(k+\mathrm{i}\,a_0 c\right)(1+2\mathrm{i}\,\xi)$, where K^0 is one of the static values given above, ξ is the fraction of material damping, and k,c (with an added subindex to identify the mode) are stiffness functions as follows:

$$k_v = 1, \qquad c_v = \frac{\pi}{K_v^0/GR}\left[\alpha+2\frac{E}{R}\right] \tag{7.141}$$

$$k_h = 1, \qquad c_h = \frac{\pi}{K_h^0/GR}\left[1+(1+\alpha)\frac{E}{R}\right], \qquad \alpha = \frac{C_L}{C_T} = \sqrt{\frac{2(1-v)}{1-2v}} \le 2.5 \tag{7.142}$$

$$k_{hr} = k_h \qquad c_{hr} = c_h \tag{7.143}$$

$$k_r = 1-\frac{0.35 a_0^2}{1+a_0^2}, \qquad b_r = \frac{2}{1+E/R} \tag{7.144}$$

$$c_r = \frac{\pi}{K_r^0/GR^3}\left[\frac{1}{4}\alpha+\frac{E}{R}+\frac{1}{3}(1+\alpha)\left(\frac{E}{R}\right)^3\right]\frac{a_0^2}{b_r+a_0^2}+0.84(1+\alpha)\left(\frac{E}{R}\right)^{2.5}\frac{b_r}{b_r+a_0^2} \tag{7.145}$$

$$k_t = 1-\frac{0.35 a_0^2}{1+a_0^2}, \qquad b_t = \frac{1}{0.37+0.87(E/R)^{2/3}} \tag{7.146}$$

$$c_t = \frac{\pi}{2K_t^0/GR^3}\left(1+4\frac{E}{R}\right)\frac{a_0^2}{b_t+a_0^2}, \tag{7.147}$$

Additional approximations for rectangular (prismatic) foundations can be found in the reference given in the footnote to this section.

7.4.4 **Direct Formulation of SSI Problems**

The Substructure Theorem

Consider a structure made up of an arbitrary combination of linear members, which may include plates or shells. We assume this structure to occupy the full 3-D space, with each node having up to 6 DOF, namely three translations and three rotations, so this is a discrete system with a large but finite number of degrees of freedom. We further carry out a mental experiment wherein we imagine the subgrade to be modeled by means of an obscenely large – indeed nearly infinitely large – finite element grid where we imagine the boundaries to be so far away that their presence has absolutely no effect on the structural neighborhood, that is, reflections at those far boundaries can be neglected (alternatively, we may assume that the far boundaries are perfect absorbers of waves). In addition, we assume that the seismic source is prescribed somewhere within this finite element mesh although its actual details and location need not be known to us. In lieu of a precise description of the source, we shall pretend to know how to solve the free-field problem when the structure is absent. Although in principle this might follow from that very same finite element grid within which we applied the source, we assume instead that we are able to obtain that solution by some other means. We rush to add that if no such alternative solution method for the free field problem existed, then we would not be able to solve the SSI problem to begin with.

Consider now the free field problem and assume that in the light of our previous mental experiment, the nodal vector of displacements $\mathbf{u}_b^*$ and internal stresses $\mathbf{p}_b^*$ in the neighborhood of the location where the structure will be standing are known to us. The subscript b identifies the *base*, that is, the DOF along the soil–structure interface, and the star reminds us that these forces and displacements correspond to the free field problem. Proceed now to extract the prismatic soil portion where the structure will be standing (i.e., the excavated soil), and imagine the soil prism as a free body in space. To maintain equilibrium of this new two-body system, we apply tractions $\mathbf{p}_b^*$ onto the free body and tractions $-\mathbf{p}_b^*$ onto the now open interface of the soil. Also, because of geometric compatibility, both the open interface and the outer surface of the prism will experience the same free-field displacements $\mathbf{u}_b^*$. In principle, the two separate bodies describe exactly the same problem as the original, undivided free-field problem, so nothing has really changed.

Consider next a similar situation, but this time with the soil prism replaced by the structure as a free body in space. We will then observe that the interaction forces and the displacements are different, namely $\mathbf{p}_b$ and $\mathbf{u}_b$, and this is so because of SSI effects. If we next compare the outer regions in cases 1 (free field) and 2 (with the structure), we see that the two outer regions are *identical*, and contain exactly the same source. Thus, if we carry out a mental "subtraction" of these two problems, we see that the remote source cancels out (i.e., it "disappears") and the displacement and stress field acting on the open interface between the soil and the structure change into $\Delta\mathbf{u}_b = \mathbf{u}_b - \mathbf{u}_b^*$ and $\Delta\mathbf{p}_b = -(\mathbf{p}_b - \mathbf{p}_b^*) = \mathbf{p}_b^* - \mathbf{p}_b$, respectively. At this point we reason that the difference in displacements at the interface (and beyond) must be the result of the difference in tractions, that is, we regard $\Delta\mathbf{p}_b$ as a bona fide source that acts in the absence of sources elsewhere. Hence, we can legitimately relate these two fields through the impedance matrix $\mathbf{Z} = \mathbf{Z}(\omega)$ of the outer region, which in continuation of our mental experiment we imagine was obtained by a static condensation of all *exterior* degrees of freedom (but of course, in reality $\mathbf{Z}$ is obtained by other means). If so, then

$$\mathbf{p}_b^* - \mathbf{p}_b = \mathbf{Z}\left(\mathbf{u}_b - \mathbf{u}_b^*\right), \qquad \text{so} \qquad \mathbf{p}_b = -\mathbf{Z}\,\mathbf{u}_b + \mathbf{Z}\mathbf{u}_b^* + \mathbf{p}_b^* \tag{7.148}$$

On the other hand, if a subscript s identifies the degrees of freedom in the structure other than those along the soil–structure interface, then dynamic equilibrium in the frequency domain of the complete structure as a free body in space demands that

$$\begin{Bmatrix} \mathbf{Z}_{ss} & \mathbf{Z}_{sb} \\ \mathbf{Z}_{bs} & \mathbf{Z}_{bb} \end{Bmatrix} \begin{Bmatrix} \mathbf{u}_s \\ \mathbf{u}_b \end{Bmatrix} = \begin{Bmatrix} \mathbf{0} \\ \mathbf{p}_b \end{Bmatrix} = \begin{Bmatrix} \mathbf{0} \\ -\mathbf{Z}\mathbf{u}_b + \mathbf{Z}\mathbf{u}_b^* + \mathbf{p}_b^* \end{Bmatrix} \tag{7.149}$$

where the elements $\mathbf{Z}_{ij}$ are the conventional impedances of the structure, that is, they are dynamic stiffness matrices of the form $\mathbf{Z}_{ij} = \mathbf{K}_{ij} + i\omega\mathbf{C}_{ij} - \omega^2\mathbf{M}_{ij}$. Taking the first element of the load vector to the left-hand side, it follows that

$$\boxed{\begin{bmatrix} \mathbf{Z}_{ss} & \mathbf{Z}_{sb} \\ \mathbf{Z}_{bs} & \mathbf{Z}_{bb} + \mathbf{Z} \end{bmatrix} \begin{Bmatrix} \mathbf{u}_s \\ \mathbf{u}_b \end{Bmatrix} = \begin{Bmatrix} \mathbf{0} \\ \mathbf{Z}\mathbf{u}_b^* + \mathbf{p}_b^* \end{Bmatrix}} \qquad \text{Substructure theorem} \tag{7.150}$$

which constitutes an extremely important result that we refer to as the *substructure theorem*.[11] It shows that

- The excitation in the SSI problem is fully defined by the seismic environment in the free field in the neighborhood of the structure, namely before excavation and erection of that structure.
- The SSI problem for the complete dynamic system is defined in terms of equivalent, fictitious forces $\mathbf{Z}\mathbf{u}_b^* + \mathbf{p}_b^*$ applied solely along the soil–structure interface. It should be carefully noticed that the $\mathbf{p}_b^*$ are the tractions acting on the extracted *soil island* and not on the cavity. The latter are equal and opposite to the former. Alternatively, we can interpret these loads in terms of an effective ground motion $\tilde{\mathbf{u}}_g^* = \mathbf{u}_b^* + \mathbf{Z}^{-1}\mathbf{p}_b^*$ "traveling" underneath the "soil springs" $\mathbf{Z}$, in which case the right-hand side changes into $\mathbf{Z}\tilde{\mathbf{u}}_g^*$.
- The elastic, dynamic, and radiation effects of the soil beyond the boundaries (i.e., the entire outer world) can be captured completely in terms of an impedance matrix $\mathbf{Z} = \mathbf{Z}(\omega)$ defined at the soil–structure boundary, and this matrix is simply added to the structural impedance matrix.

Other than the linearity of the soil, this result is *exact* and makes no approximations. Thus, this theorem is extremely useful because it demonstrates how to rigorously prescribe the seismic motion to a discrete model of the soil–structure system. In the ensuing sections, we make use of this result and elaborate on the formulation needed for infinitely rigid foundations, that is, when the motion of the whole foundation can be described by no more than 6 DOF.

---

[11] E. Kausel, R. V. Whitman, J. P. Morray, and F. Elsabee, "The spring method for embedded foundations," *Nucl. Eng. Design*, 48, 1978, 377–392.

SSI Equations for Structures with Rigid Foundation

We assume again that the structure is supported by, and partially embedded in, a compliant soil of arbitrary properties whose dynamic stiffness characteristics we assume to be captured through a frequency-dependent impedance matrix $\mathbf{Z}$ defined at the soil–structure interface. This impedance matrix has as many DOF as there are DOF along that interface, that is, along the base, and can be visualized as having been obtained via a hypothetical, refined, discrete model of the soil in the frequency domain in which all DOF away from the soil–structure interface have been removed by static condensation. For the sake of generality, we shall assume here that external forces, seismic or not, are applied to any DOF throughout the structure, including the soil–structure interface.

Denoting with a subscript s all the DOF in the *structure* not in contact with the soil, and with a subscript b the degrees of freedom at the *base* of the structure in contact with that soil, then for an undamped structure the dynamic equilibrium equation in the frequency domain can be written as

$$\begin{bmatrix} \mathbf{K}_{ss} - \omega^2 \mathbf{M}_{ss} & \mathbf{K}_{sb} - \omega^2 \mathbf{M}_{sb} \\ \mathbf{K}_{bs} - \omega^2 \mathbf{M}_{bs} & \mathbf{K}_{bb} - \omega^2 \mathbf{M}_{bb} + \mathbf{Z} \end{bmatrix} \begin{Bmatrix} \mathbf{u}_s \\ \mathbf{u}_b \end{Bmatrix} = \begin{Bmatrix} \mathbf{p}_s \\ \mathbf{p}_b \end{Bmatrix} \tag{7.151}$$

in which $\mathbf{M}_{ss}, \dots \mathbf{M}_{bb}$ are the submatrices of the consistent or lumped mass matrix for the super-structure and $\mathbf{K}_{ss}, \dots \mathbf{K}_{bb}$ define the complete structural stiffness matrix. For now, and to keep matters simple, we shall assume the system to be undamped, but later on we shall add either viscous or hysteretic damping, or both.

In principle, if we consider input forcing functions of unit magnitude (either forces or appropriately defined seismic motions), we can solve the dynamic equilibrium equation for a dense set of frequencies, a process that yields the transfer functions for the system. Thereafter, as is usual, we convolve these transfer functions with the actual excitation time histories via the fast Fourier transform algorithm and obtain the desired response functions. However, we shall consider in the ensuing the special case where the base – but not the soil – can be assumed to be infinitely rigid.

We begin by defining the rigid-body matrix (here written in transposed form)

$$\mathbf{E} = \begin{Bmatrix} \mathbf{I} & \mathbf{O} & \mathbf{I} & \mathbf{O} & \cdots & \mathbf{I} & \mathbf{O} \\ \mathbf{T}_1^T & \mathbf{I} & \mathbf{T}_2^T & \mathbf{I} & \cdots & \mathbf{T}_n^T & \mathbf{I} \end{Bmatrix}^T = \begin{Bmatrix} \mathbf{E}_s \\ \mathbf{E}_b \end{Bmatrix} \tag{7.152}$$

in which $\mathbf{I}$ is a 3×3 identity matrix, $\mathbf{O}$ is a 3×3 null matrix, and the $\mathbf{T}_i, i = 1, \dots n$ matrices are

$$\mathbf{T}_i = \begin{Bmatrix} 0 & z_i - z_0 & -(y_i - y_0) \\ -(z_i - z_0) & 0 & x_i - x_0 \\ y_i - y_0 & -(x_i - x_0) & 0 \end{Bmatrix} \tag{7.153}$$

with n being the number of nodes, x_i, y_i, z_i are the coordinates for the ith mass point under consideration, and x_0, y_0, z_0 are the coordinates of an *arbitrary* point on the foundation with respect to which the soil stiffnesses and seismic motions are defined. This point is usually taken at the location where the vertical axis intersects the soil at the geometric center of the foundation, if such center exists. Thus, $\mathbf{E}$ has six columns and $6n$ rows, with n being the total number of nodes in the structure, including the foundation. This matrix is easily partitioned into structural nodes and base nodes, as shown earlier.

Assuming now the structural foundation to be perfectly rigid, the motion for the DOF on the soil–structure interface can be expressed as

$$\mathbf{u}_b = \mathbf{E}_b \mathbf{u}_f \tag{7.154}$$

in which $\mathbf{u}_f$ is the 6×1 displacement vector of the foundation, which consists of the three translations and three rotations of the base, measured relative to the reference point x_0, y_0, z_0.

Let's digress briefly and visualize the structure as a complete free body in space. We see then that the system stiffness matrix must satisfy the rigid-body condition

$$\begin{Bmatrix} \mathbf{K}_{ss} & \mathbf{K}_{sb} \\ \mathbf{K}_{bs} & \mathbf{K}_{bb} \end{Bmatrix} \begin{Bmatrix} \mathbf{E}_s \\ \mathbf{E}_b \end{Bmatrix} = \begin{Bmatrix} \mathbf{O} \\ \mathbf{O} \end{Bmatrix} \tag{7.155}$$

This is because rigid body translations and/or rotations of the structure as a whole produce no internal deformations, and thus, such motions require no external forces. It follows that

$$\begin{aligned} \mathbf{K}_{sb}\mathbf{E}_b = -\mathbf{K}_{ss}\mathbf{E}_s \quad &\rightarrow \quad \mathbf{E}_s^T \mathbf{K}_{sb}\mathbf{E}_b = -\mathbf{E}_s^T \mathbf{K}_{ss}\mathbf{E}_s \\ \mathbf{K}_{bb}\mathbf{E}_b = -\mathbf{K}_{bs}\mathbf{E}_s \quad &\rightarrow \quad \mathbf{E}_b^T \mathbf{K}_{bb}\mathbf{E}_b = -\mathbf{E}_b^T \mathbf{K}_{bs}\mathbf{E}_s \end{aligned} \tag{7.156}$$

Taking into account the symmetry of the stiffness matrix as a whole, we infer

$$\begin{aligned} \mathbf{E}_b^T \mathbf{K}_{bb}\mathbf{E}_b &= -\mathbf{E}_b^T \mathbf{K}_{bs}\mathbf{E}_s = -\left(\mathbf{E}_b^T \mathbf{K}_{sb}\mathbf{E}_b \right)^T \\ &= \left(\mathbf{E}_s \mathbf{K}_{ss}\mathbf{E}_s \right)^T = \mathbf{E}_s \mathbf{K}_{ss}\mathbf{E}_s \end{aligned} \tag{7.157}$$

Similar relationships would hold also for a damping matrix, were it to exist. If we now substitute the preceding equations into our original equilibrium equation, after brief algebra we obtain

$$\left[\begin{Bmatrix} \mathbf{K}_{ss} & -\mathbf{K}_{ss}\mathbf{E}_s \\ -\mathbf{E}_s^T \mathbf{K}_{ss} & \mathbf{E}_s^T \mathbf{K}_{ss}\mathbf{E}_s + \mathbf{T}_b^T \mathbf{Z}\mathbf{T}_b \end{Bmatrix} - \omega^2 \begin{Bmatrix} \mathbf{M}_{ss} & \mathbf{M}_{sb}\mathbf{E}_b \\ \mathbf{E}_b^T \mathbf{M}_{bs} & \mathbf{E}_b^T \mathbf{M}_{bb}\mathbf{E}_b \end{Bmatrix} \right] \begin{Bmatrix} \mathbf{u}_s \\ \mathbf{u}_f \end{Bmatrix} = \begin{Bmatrix} \mathbf{p}_s \\ \mathbf{T}_b^T \mathbf{p}_b \end{Bmatrix} \tag{7.158}$$

We define next two 6×6 matrices $\mathbf{M}_f, \mathbf{Z}_f$ together with an effective foundation load vector $\mathbf{p}_f$ as

$$\boxed{\mathbf{M}_f = \mathbf{E}_b^T \mathbf{M}_{bb}\mathbf{E}_b} \qquad = \text{foundation inertia} \tag{7.159}$$

$$\boxed{\mathbf{Z}_f = \mathbf{E}_b^T \mathbf{Z}\mathbf{E}_b} \qquad = \text{frequency-dependent foundation impedances} \tag{7.160}$$

$$\boxed{\mathbf{p}_f = \mathbf{E}_b^T \mathbf{p}_b} \qquad = \text{total external forces and moments on foundation} \tag{7.161}$$

With these definitions, our transformed equilibrium equation simplifies into

$$\begin{Bmatrix} \mathbf{K}_{ss} - \omega^2 \mathbf{M}_{ss} & -\left(\mathbf{K}_{ss}\mathbf{E}_s + \omega^2 \mathbf{M}_{sb}\mathbf{E}_b \right) \\ -\left(\mathbf{E}_s^T \mathbf{K}_{ss} + \omega^2 \mathbf{E}_b^T \mathbf{M}_{bs} \right) & \mathbf{Z}_f - \omega^2 \mathbf{M}_f + \mathbf{E}_s^T \mathbf{K}_{ss}\mathbf{E}_s \end{Bmatrix} \begin{Bmatrix} \mathbf{u}_s \\ \mathbf{u}_f \end{Bmatrix} = \begin{Bmatrix} \mathbf{p}_s \\ \mathbf{p}_f \end{Bmatrix} \tag{7.162}$$

which requires only the stiffness and mass matrices for the superstructure as if the structure were supported by an infinitely rigid soil.

Defining the impedance matrix of the structure as

$$\mathbf{Z}_s = \mathbf{K}_{ss} - \omega^2 \mathbf{M}_{ss} \tag{7.163}$$

and solving from the foregoing for the two unknowns, we obtain

$$\mathbf{u}_f = \left[\mathbf{Z}_f - \omega^2 \mathbf{M}_f + \mathbf{E}_s^T \mathbf{K}_{ss} \mathbf{E}_s - \left(\mathbf{E}_s^T \mathbf{K}_{ss} + \omega^2 \mathbf{E}_b^T \mathbf{M}_{bs} \right) \mathbf{Z}_s^{-1} \left(\mathbf{K}_{ss} \mathbf{E}_s + \omega^2 \mathbf{M}_{sb} \mathbf{E}_b \right) \right]^{-1} \times$$
$$\left[\mathbf{p}_f + \left(\mathbf{E}_s^T \mathbf{K}_{ss} + \omega^2 \mathbf{E}_b^T \mathbf{M}_{bs} \right) \mathbf{Z}_s^{-1} \mathbf{p}_s \right] \tag{7.164}$$
$$\mathbf{u}_s = \mathbf{Z}_s^{-1} \left[\mathbf{p}_s + \left(\mathbf{K}_{ss} \mathbf{E}_s + \omega^2 \mathbf{M}_{sb} \mathbf{E}_b \right) \right] \mathbf{u}_f$$

In most practical cases, the inertial coupling terms $\mathbf{M}_{sb}$ of the structure with the base can be disregarded, and indeed, for lumped mass systems they are identically zero, that is, $\mathbf{M}_{sb} = \mathbf{O}$. Neglecting these terms and considering the identity $\mathbf{E}_s^T \mathbf{K}_{ss} \left(\mathbf{I} - \mathbf{Z}_s^{-1} \mathbf{K}_{ss} \right) \mathbf{E}_s = -\omega^2 \mathbf{E}_s^T \mathbf{K}_{ss} \mathbf{Z}_s^{-1} \mathbf{M}_{ss} \mathbf{E}_s$, Eq. 7.164 ultimately simplifies to

$$\boxed{\begin{aligned} \mathbf{u}_f &= \left[\mathbf{Z}_f - \omega^2 \mathbf{M}_f - \omega^2 \mathbf{E}_s^T \mathbf{K}_{ss} \mathbf{Z}_s^{-1} \mathbf{M}_{ss} \mathbf{E}_s \right]^{-1} \left[\mathbf{p}_f + \mathbf{E}_s^T \mathbf{K}_{ss} \mathbf{Z}_s^{-1} \mathbf{p}_s \right] \\ \mathbf{u}_s &= \mathbf{Z}_s^{-1} \left(\mathbf{p}_s + \mathbf{K}_{ss} \mathbf{E}_s \mathbf{u}_f \right) \end{aligned}} \tag{7.165}$$

Although Eq. 7.165 could be solved directly and without much ado, a better way is described in Section 7.4.5.

7.4.5 SSI via Modal Synthesis in the Frequency Domain

We proceed next to provide a very convenient solution to the simplified dynamic equilibrium equation just presented by recourse to modal synthesis in the frequency domain, that is, by expressing the superstructure in terms of its normal modes on fixed base. This is advantageous because it allows modeling some very general types of structures via external programs, such as finite elements, which by themselves need not include SSI capabilities, but whose extensive library of structural elements can't (and shouldn't) be replicated in an SSI program.

Consider the eigenvalue problem for the normal modes of the structure on *fixed* base expressed in matrix form, that is,

$$\mathbf{K}_{ss} \mathbf{\Phi} = \mathbf{M}_{ss} \mathbf{\Phi} \mathbf{\Omega}^2 \tag{7.166}$$

This system satisfies the orthogonality conditions

$$\mathbf{\Phi}^T \mathbf{M}_{ss} \mathbf{\Phi} = \mathfrak{M} = \mathrm{diag}\left(\mu_j \right) \qquad = \text{modal mass} \tag{7.167}$$

$$\begin{aligned} \mathbf{\Phi}^T \mathbf{K}_{ss} \mathbf{\Phi} &= \mathfrak{K} = \mathrm{diag}\left(\kappa_j \right) \\ &= \mathfrak{M} \mathbf{\Omega}^2 = \mathrm{diag}\left(\mu_j \, \omega_j^2 \right) \end{aligned} \qquad = \text{modal stiffness} \tag{7.168}$$

If none of the modal frequencies of the structure on rigid base vanish (i.e., it contains no rigid-body modes), then the modal matrix spans the full multidimensional space, in which case $\mathbf{\Phi}^{-1}$ exists. If so, then

$$\mathbf{M}_{ss} = \mathbf{\Phi}^{-T} \mathfrak{M} \mathbf{\Phi}^{-1}, \qquad \text{and} \qquad \mathbf{K}_{ss} = \mathbf{\Phi}^{-T} \mathfrak{M} \mathbf{\Omega}^2 \, \mathbf{\Phi}^{-1} \tag{7.169}$$

Also, $\mathbf{Z}_s^{-1} = \mathbf{\Phi}\,\mathfrak{M}^{-1}\left(\mathbf{\Omega}^2 - \omega^2\mathbf{I}\right)^{-1}\mathbf{\Phi}^T$, which we write compactly as

$$\boxed{\mathbf{Z}_s^{-1} = \mathbf{\Phi}\,\mathfrak{M}^{-1}\mathfrak{D}\,\mathbf{\Phi}^T} \tag{7.170}$$

where

$$\mathfrak{D} = \left(\mathbf{\Omega}^2 - \omega^2\mathbf{I}\right)^{-1} = \mathrm{diag}\left\{\frac{1}{\omega_j^2 - \omega^2}\right\} \tag{7.171}$$

More generally, if the structural system has damping, then $\mathfrak{D}$ must be replaced by

$$\boxed{\mathfrak{D} = \mathrm{diag}\left\{\frac{1}{\omega_j^2 - \omega^2 + 2\mathrm{i}\,\xi_j\,\omega_j\,\omega}\right\}} \quad \text{for viscous damping} \tag{7.172}$$

$$\boxed{\mathfrak{D} = \mathrm{diag}\left\{\frac{1}{\omega_j^2 - \omega^2 + 2\mathrm{i}\,\xi_j\,\omega_j^2\,\mathrm{sgn}(\omega)}\right\}} \quad \text{for hysteretic damping} \tag{7.173}$$

$$\mathrm{sgn}(\omega) = \begin{cases} 1 & \omega > 0 \\ 0 & \omega = 0 \\ -1 & \omega < 0 \end{cases} \tag{7.174}$$

These equations imply that the damping is prescribed at the level of the modes, or that the damping matrix is of the proportional type.

We define also the matrix of participation factors $\mathbf{\Gamma}$ as the modal coordinates of the rigid body matrix $\mathbf{E}_s$, that is,

$$\boxed{\mathbf{E}_s = \mathbf{\Phi}\mathbf{\Gamma}}, \qquad \boxed{\mathbf{\Gamma} = \mathbf{\Phi}^{-1}\mathbf{E}_s = \mathfrak{M}^{-1}\mathbf{\Phi}^T\mathbf{M}\mathbf{E}_s} \tag{7.175}$$

which satisfies

$$\mathbf{\Gamma}^T\mathbf{M}\mathbf{\Gamma} = \left(\mathbf{E}_s^T\mathbf{\Phi}^{-T}\right)\mathfrak{M}\left(\mathbf{\Phi}^{-1}\mathbf{E}_s\right) = \mathbf{E}_s^T\left(\mathbf{\Phi}^{-T}\mathfrak{M}\mathbf{\Phi}^{-1}\right)\mathbf{E}_s = \mathbf{E}_s^T\mathbf{M}_{ss}\mathbf{E}_s \tag{7.176}$$

which we write compactly as

$$\boxed{\mathbf{M}_s \underset{def}{=} \mathbf{\Gamma}^T\mathbf{M}\mathbf{\Gamma} = \mathbf{E}_s^T\mathbf{M}_{ss}\mathbf{E}_s} \tag{7.177}$$

Hence, the quadratic form of the modal participation factors together with the modal mass equals the total mass of the structure $\mathbf{M}_s = \mathbf{E}_s^T\mathbf{M}_{ss}\mathbf{E}_s$, a 6×6 matrix of masses and products of inertia.

Case 1: Modal Synthesis with Structural Loads

From the equations in Section 7.4.4 as well as the modal expressions, we have

$$\begin{aligned}
\mathbf{E}_s^T\mathbf{K}_{ss}\mathbf{K}_d^{-1} &= \left(\mathbf{\Gamma}^T\mathbf{\Phi}^T\right)\left(\mathbf{\Phi}^{-T}\mathfrak{M}\mathbf{\Omega}^2\mathbf{\Phi}^{-1}\right)\left(\mathbf{\Phi}\mathfrak{M}^{-1}\mathfrak{D}\mathbf{\Phi}^T\right) = \mathbf{\Gamma}^T\mathbf{\Omega}^2\mathfrak{D}\mathbf{\Phi}^T \\
&= \mathbf{\Gamma}^T\left(\mathbf{\Omega}^2 - \omega^2\mathbf{I} + \omega^2\mathbf{I}\right)\mathfrak{D}\mathbf{\Phi}^T = \mathbf{\Gamma}^T\mathbf{\Phi}^T + \omega^2\mathbf{\Gamma}^T\mathfrak{D}\mathbf{\Phi}^T \\
&= \mathbf{E}_s^T + \omega^2\mathbf{\Gamma}^T\mathfrak{D}\mathbf{\Phi}^T
\end{aligned} \tag{7.178}$$

$$
\begin{aligned}
\mathbf{E}_s^T \mathbf{K}_{ss} \mathbf{Z}_s^{-1} \mathbf{M}_{ss} \mathbf{E}_s &= \left(\mathbf{\Gamma}^T \mathbf{\Omega}^2 \mathfrak{D} \mathbf{\Phi}^T \right) \mathbf{M}_{ss} \left(\mathbf{\Phi} \mathbf{\Gamma} \right) = \mathbf{\Gamma}^T \mathbf{\Omega}^2 \mathfrak{D} \mathfrak{M} \, \mathbf{\Gamma} \\
&= \mathbf{\Gamma}^T \left(\mathbf{\Omega}^2 - \omega^2 \mathbf{I} + \omega^2 \mathbf{I} \right) \mathfrak{D} \mathfrak{M} \, \mathbf{\Gamma} = \mathbf{\Gamma}^T \left(\mathbf{\Omega}^2 - \omega^2 \mathbf{I} + \omega^2 \mathbf{I} \right) \mathfrak{D} \mathfrak{M} \, \mathbf{\Gamma} \\
&= \mathbf{\Gamma}^T \mathfrak{M} \mathbf{\Gamma} + \omega^2 \mathbf{\Gamma}^T \mathfrak{D} \mathfrak{M} \mathbf{\Gamma} \\
&= \mathbf{M}_s + \omega^2 \mathbf{\Gamma}^T \mathfrak{D} \mathfrak{M} \mathbf{\Gamma}
\end{aligned}
\tag{7.179}
$$

Also, we define the dimensionless transfer function matrix and the structural modal vector

$$
\boxed{\mathfrak{H} = \omega^2 \mathfrak{D} + \mathbf{I} = \operatorname{diag} \left\{ \frac{\omega_j^2 + 2 \mathrm{i} \, \xi_j \omega_j \omega}{\omega_j^2 + 2 \mathrm{i} \, \xi_j \omega_j \omega - \omega^2} \right\}}
\qquad \text{Viscous damping} \tag{7.180}
$$

$$
\boxed{\mathfrak{H} = \omega^2 \mathfrak{D} + \mathbf{I} = \operatorname{diag} \left\{ \frac{\omega_j^2 + 2 \mathrm{i} \, \xi_j \omega_j^2 \operatorname{sgn}(\omega)}{\omega_j^2 + 2 \mathrm{i} \, \xi_j \omega_j^2 \operatorname{sgn}(\omega) - \omega^2} \right\}}
\qquad \text{Hysteretic damping} \tag{7.181}
$$

$$
\mathfrak{P}_s = \mathbf{\Phi}^T \mathbf{p}_s \tag{7.182}
$$

Hence, making use of the various relationships given above, we obtain an equivalent form of the simplified equilibrium equation given at the end of Section 7.4.4, that is,

$$
\begin{aligned}
\mathbf{u}_f &= \left[\mathbf{Z}_f - \omega^2 \left(\mathbf{M}_f + \mathbf{M}_s \right) - \omega^4 \mathbf{\Gamma}^T \mathfrak{M} \mathfrak{D} \, \mathbf{\Gamma} \right]^{-1} \left[\mathbf{p}_f + \mathbf{E}_s^T \mathbf{p}_s + \omega^2 \mathbf{\Gamma}^T \mathfrak{D} \mathfrak{P}_s \right] \\
\mathbf{u}_s &= \mathbf{\Phi} \mathfrak{D} \mathfrak{M}^{-1} \mathfrak{P}_s + \left(\mathbf{E}_s + \omega^2 \mathbf{\Phi} \mathfrak{D} \mathbf{\Gamma} \right) \mathbf{u}_f
\end{aligned}
\tag{7.183}
$$

or alternatively, using also the definitions for $\mathbf{E}_s, \mathbf{M}_s, \mathfrak{H}, \mathfrak{P}_s$, we finally obtain

$$
\boxed{
\begin{aligned}
\mathbf{u}_f &= \left[\mathbf{Z}_f - \omega^2 \left(\mathbf{M}_f + \mathbf{\Gamma}^T \mathfrak{M} \mathfrak{H} \mathbf{\Gamma} \right) \right]^{-1} \left[\mathbf{p}_f + \mathbf{\Gamma}^T \mathfrak{H} \mathfrak{P}_s \right] \\
\mathbf{u}_s &= \mathbf{\Phi} \mathfrak{D} \mathfrak{M}^{-1} \mathfrak{P}_s + \mathbf{\Phi} \mathfrak{H} \mathbf{\Gamma} \mathbf{u}_f
\end{aligned}
}
\tag{7.184}
$$

If the modes have been normalized, then $\mathfrak{M} = \mathbf{I}$ is the identity matrix. Observe also that $\mathbf{M}_f + \mathbf{M}_s = \mathbf{M}_0$ is the 6×6 total mass matrix of the structure as if it were a rigid body (relative to the reference point).

Case 2: Support Motion

We consider next the case of seismic motions prescribed at the base of the foundation in the absence of structural forces. It suffices to express the 6×1 vector of support motions $\mathbf{u}_g$ (three translations and three rotations) in terms of a fictitious load applied onto the foundation, that is, $\mathbf{p}_f = \mathbf{Z}_f \mathbf{u}_g$, with $\mathbf{p}_s = \mathbf{0}$, in which case our equation for $\mathbf{u}_f, \mathbf{u}_s$ just given simplifies into

$$
\boxed{
\begin{aligned}
\mathbf{u}_f &= \left[\mathbf{Z}_f - \omega^2 \left(\mathbf{M}_f + \mathbf{\Gamma}^T \mathfrak{M} \mathfrak{H} \mathbf{\Gamma} \right) \right]^{-1} \mathbf{Z}_f \mathbf{u}_g \\
\mathbf{u}_s &= \mathbf{\Phi} \mathfrak{H} \mathbf{\Gamma} \mathbf{u}_f
\end{aligned}
}
\tag{7.185}
$$

with the symbols as previously defined. It should be observed also that the transfer functions $\mathfrak{H}$ from ground displacement to absolute structural displacement are identical to the transfer functions from ground acceleration to absolute structural acceleration. Hence, when the ground motion is given, as usual, in terms of accelerations, exactly the same solution equations apply, except for the physical interpretation of the output, which will be absolute accelerations and not displacements.

Partial Modal Summation

When not all modes are present or have been computed, the modal summation will be incomplete. However, the contribution of the missing modes can be accounted for in at least an approximate fashion by assuming those modes to have very large natural frequencies, which in turn implies that their behavior is quasistatic. In that case, the contribution of the missing modes can be accomplished by computing the residual modal participation, using for this purpose the available modes as well as the static deformations, as will be shown.

Denote with subindex 1 the matrices and vectors associated with the known (available) modes and with subindex 2 the corresponding entities for those that are not available. Partitioning the modal matrix, the participation factor matrix and the *modal* masses into known and unknown modes, we obtain

$$\Phi = \{\Phi_1 \quad \Phi_2\}, \qquad \Gamma = \begin{Bmatrix} \Gamma_1 \\ \Gamma_2 \end{Bmatrix}, \qquad \mathfrak{M} = \begin{Bmatrix} \mathfrak{M}_1 & \mathbf{O} \\ \mathbf{O} & \mathfrak{M}_2 \end{Bmatrix} \tag{7.186}$$

Also, since the frequencies of the unknown modes are assumed to be very large, the dynamic amplification factors for those modes tend asymptotically to zero in proportion to the inverse squared frequencies, that is,

$$\mathfrak{D}_2 \to \Omega_2^{-2}, \qquad \mathfrak{D} \to \begin{Bmatrix} \mathfrak{D}_1 & \mathbf{O} \\ \mathbf{O} & \Omega_2^{-2} \end{Bmatrix} \tag{7.187}$$

It follows that

$$\Phi\mathfrak{D}\mathfrak{M}^{-1}\mathfrak{P}_s \to \Phi_1\mathfrak{D}_1\mathfrak{M}_1^{-1}\mathfrak{P}_1 + \Phi_2\Omega_2^{-2}\mathfrak{M}_2^{-1}\mathfrak{P}_2 \tag{7.188}$$

Now, from the eigenvalue problem together with the orthogonality conditions,

$$\mathbf{K}_{ss}^{-1} = \Phi\Omega^{-2}\mathfrak{M}^{-1}\Phi^T = \Phi_1\Omega_1^{-2}\mathfrak{M}_1^{-1}\Phi_1^T + \Phi_2\Omega_2^{-2}\mathfrak{M}_2^{-1}\Phi_2^T \tag{7.189}$$

so

$$\Phi_2\Omega_2^{-2}\mathfrak{M}_2^{-1}\Phi_2^T = \mathbf{K}_{ss}^{-1} - \Phi_1\Omega_1^{-2}\mathfrak{M}_1^{-1}\Phi_1^T \tag{7.190}$$

Post-multiplying this by the load vector, we obtain

$$\Phi_2\Omega_2^{-2}\mathfrak{M}_2^{-1}\mathfrak{P}_2 = \mathbf{K}_{ss}^{-1}\mathbf{p}_s - \Phi_1\Omega_1^{-2}\mathfrak{M}_1^{-1}\mathfrak{P}_1 \tag{7.191}$$

Combining this with the preceding three equations, we infer

$$\Phi\mathfrak{D}\mathfrak{M}^{-1}\mathfrak{P}_s \to \mathbf{K}_{ss}^{-1}\mathbf{p}_s + \Phi_1\mathfrak{B}_1\mathfrak{M}_1^{-1}\mathfrak{P}_1 \tag{7.192}$$

where

$$\mathfrak{B}_1 = \mathfrak{D}_1 - \Omega_1^{-2} = \mathrm{diag}\left\{\frac{1}{\omega_j^2 - \omega^2 + 2i\xi_j\omega\omega_j} - \frac{1}{\omega_j^2}\right\} \tag{7.193}$$

That is,

$$\boxed{\mathfrak{B}_1 = \mathrm{diag}\left\{\frac{r_j^2 - 2\mathrm{i}\xi_j r_j}{\left(1 - r_j^2 + 2\mathrm{i}\xi_j r_j\right)\omega_j^2}\right\},} \qquad r_j = \frac{\omega}{\omega_j} \tag{7.194}$$

(for the known or available modes). On the other hand,

$$\mathbf{M}_s = \mathbf{E}_s^T \mathbf{M} \mathbf{E}_s = \mathbf{\Gamma}^T \mathfrak{M} \mathbf{\Gamma} \mathbf{\Gamma} = \mathbf{\Gamma}_1^T \mathfrak{M}_1 \mathbf{\Gamma}_1 + \mathbf{\Gamma}_2^T \mathfrak{M}_2 \mathbf{\Gamma}_2 \tag{7.195}$$

Hence

$$\mathbf{\Gamma}_2^T \mathfrak{M}_2 \mathbf{\Gamma}_2 = \mathbf{M}_s - \mathbf{\Gamma}_1^T \mathfrak{M}_1 \mathbf{\Gamma}_1 \tag{7.196}$$

Also,

$$\mathbf{E}_s = \mathbf{\Phi}\mathbf{\Gamma}, \qquad \mathbf{E}_s^T = \mathbf{\Gamma}^T \mathbf{\Phi}^T = \mathbf{\Gamma}_1^T \mathbf{\Phi}_1^T + \mathbf{\Gamma}_2^T \mathbf{\Phi}_2^T \tag{7.197}$$

So

$$\mathbf{\Gamma}_2^T \mathbf{\Phi}_2^T = \mathbf{E}_s^T - \mathbf{\Gamma}_1^T \mathbf{\Phi}_1^T \tag{7.198}$$

and

$$\mathbf{\Gamma}_2^T \mathfrak{P}_2 = \left(\mathbf{E}_s^T - \mathbf{\Gamma}_1^T \mathbf{\Phi}_1^T\right)\mathbf{p}_s = \mathbf{E}_s^T \mathbf{p}_s - \mathbf{\Gamma}_1^T \mathfrak{P}_1 \tag{7.199}$$

Now, if $\Omega_2 \to \infty$, then $\mathfrak{H}_2 \to \mathbf{I}$ and $\mathfrak{H} = \begin{bmatrix} \mathfrak{H}_1 & \mathfrak{H}_2 \end{bmatrix} \to \begin{bmatrix} \mathfrak{H}_1 & \mathbf{I} \end{bmatrix}$. Hence

$$\begin{aligned}\mathbf{\Gamma}^T \mathfrak{M} \mathfrak{H} \mathbf{\Gamma} &\to \mathbf{\Gamma}_1^T \mathfrak{M}_1 \mathfrak{H}_1 \mathbf{\Gamma}_1 + \mathbf{\Gamma}_2^T \mathfrak{M}_2 \mathbf{\Gamma}_2 \\ &= \mathbf{\Gamma}_1^T \mathfrak{M}_1 \mathfrak{H}_1 \mathbf{\Gamma}_1 - \mathbf{\Gamma}_1^T \mathfrak{M}_1 \mathbf{\Gamma}_1 + \mathbf{M}_s\end{aligned} \tag{7.200}$$

or

$$\mathbf{\Gamma}^T \mathfrak{M} \mathfrak{H} \mathbf{\Gamma} \to \mathbf{M}_s + \mathbf{\Gamma}_1^T \mathfrak{M}_1 \left(\mathfrak{H}_1 - \mathbf{I}\right)\mathbf{\Gamma}_1 = \mathbf{M}_s + \omega^2 \mathbf{\Gamma}_1^T \mathfrak{M}_1 \mathfrak{D}_1 \mathbf{\Gamma}_1 \tag{7.201}$$

Also,

$$\begin{aligned}\mathbf{\Gamma}^T \mathfrak{H} \mathfrak{P}_s &\to \mathbf{\Gamma}_1^T \mathfrak{H}_1 \mathfrak{P}_1 + \mathbf{\Gamma}_2^T \mathfrak{P}_2 \\ &= \mathbf{\Gamma}_1^T \mathfrak{H}_1 \mathfrak{P}_1 - \mathbf{\Gamma}_1^T \mathfrak{P}_1 + \mathbf{E}_s^T \mathbf{p}_s\end{aligned} \tag{7.202}$$

or

$$\mathbf{\Gamma}^T \mathfrak{H} \mathfrak{P}_s \to \mathbf{E}_s^T \mathbf{p}_s + \mathbf{\Gamma}_1^T \left(\mathfrak{H}_1 - \mathbf{I}\right)\mathfrak{P}_1 = \mathbf{E}_s^T \mathbf{p}_s + \omega^2 \mathbf{\Gamma}_1^T \mathfrak{D}_1 \mathfrak{P}_1 \tag{7.203}$$

Finally,

$$\begin{aligned}\mathbf{\Phi} \mathfrak{H} \mathbf{\Gamma} &\to \mathbf{\Phi}_1 \mathfrak{H}_1 \mathbf{\Gamma}_1 + \mathbf{\Phi}_2 \mathbf{\Gamma}_2 \\ &= \mathbf{\Phi}_1 \mathfrak{H}_1 \mathbf{\Gamma}_1 - \mathbf{\Phi}_1 \mathbf{\Gamma}_1 + \mathbf{E}_s\end{aligned} \tag{7.204}$$

or

$$\mathbf{\Phi} \mathfrak{H} \mathbf{\Gamma} \to \mathbf{E}_s + \mathbf{\Phi}_1 \left(\mathfrak{H}_1 - \mathbf{I}\right)\mathbf{\Gamma}_1 = \mathbf{E}_s + \omega^2 \mathbf{\Phi}_1 \mathfrak{D}_1 \mathbf{\Gamma}_1 \tag{7.205}$$

Hence,

$$
\boxed{
\begin{aligned}
\mathbf{u}_f &= \left[\mathbf{K}_f - \omega^2 \left(\mathbf{M}_f + \mathbf{M}_s + \omega^2\, \boldsymbol{\Gamma}_1^T \mathfrak{M}_1 \mathfrak{D}_1 \boldsymbol{\Gamma}_1 \right) \right]^{-1} \times \\
&\qquad \left[\mathbf{K}_f \mathbf{u}_g + \mathbf{p}_f + \mathbf{E}_s^T \mathbf{p}_s + \omega^2 \boldsymbol{\Gamma}_1^T \mathfrak{D}_1 \mathfrak{P}_1 \right] \\
\mathbf{u}_s &= \mathbf{K}_{ss}^{-1} \mathbf{p}_s + \boldsymbol{\Phi}_1 \mathfrak{B}_1 \mathfrak{M}_1^{-1} \mathfrak{P}_1 + \left(\mathbf{E}_s + \omega^2 \boldsymbol{\Phi}_1 \mathfrak{D}_1 \boldsymbol{\Gamma}_1 \right) \mathbf{u}_f
\end{aligned}
}
\tag{7.206}
$$

which gives the response of the soil–structure system in terms of the known properties. Observe that

$$
\mathbf{K}_{ss}^{-1} \mathbf{p}_s + \boldsymbol{\Phi}_1 \mathfrak{B}_1 \mathfrak{M}_1^{-1} \mathfrak{P}_1 = \left(\mathbf{K}_{ss}^{-1} - \boldsymbol{\Phi}_1 \boldsymbol{\Omega}_1^2 \mathfrak{M}_1^{-1} \boldsymbol{\Phi}_1^T \right) \mathbf{p}_s + \boldsymbol{\Phi}_1 \mathfrak{D}_1 \mathfrak{M}_1^{-1} \boldsymbol{\Phi}_1^T \mathbf{p}_s
\tag{7.207}
$$

The term in parentheses $\mathfrak{F} = \mathbf{K}_{ss}^{-1} - \boldsymbol{\Phi}_1 \boldsymbol{\Omega}_1^2 \mathfrak{M}_1^{-1} \boldsymbol{\Phi}_1^T$ adds the *quasistatic* contribution of the missing modes to the structural flexibility. If all modes are present, then $\mathfrak{F} = \mathbf{0}$, all DOF are active, and these equations reduce the equations of Section 7.4.4. At the other extreme, if *all* modes are missing, then

$$
\begin{aligned}
\mathbf{u}_f &= \left[\mathbf{K}_f - \omega^2 \left(\mathbf{M}_f + \mathbf{M}_s \right) \right]^{-1} \left[\mathbf{K}_f \mathbf{u}_g + \mathbf{p}_f + \mathbf{E}_s^T \mathbf{p}_s \right] \\
\mathbf{u}_s &= \mathbf{K}_{ss}^{-1} \mathbf{p}_s + \mathbf{E}_s \, \mathbf{u}_f
\end{aligned}
\tag{7.208}
$$

which is the dynamic response of a quasirigid structure on a compliant soil. Thus, the formulation is consistent.

What If the Modes Occupy Only a Subspace?

In some cases the structure on fixed base may have fewer DOF in each node than the foundation upon which it rests. If so, then the modal count will certainly be incomplete, even when *all* of the modes are included in the analysis. For example, a building idealized as a shear beam has only lateral DOF. If that building were supported by a foundation on rocking and vertical springs, then the interaction with the ground would elicit both rotations and vertical motions in every structural node, and these would activate inertial effects in the structure that would feedback onto the foundation and affect its motion. Thus, it behooves to include the inertial properties of the structure for all active DOF of the combined soil–structure system, and not just the mass used in the eigenvalue analysis. In the example of the beam, this would correspond to the rotational and vertical inertial properties relative to the base. It is easy to see that this situation can be reduced to that of Section 7.4.4 by assuming the missing modes – say, the vertical and rotational oscillations of the beam – to be modes with infinite frequencies. In this case, the quantities with subscript 1 refer to the modes available while $\mathbf{M}_s$ contains only the mass that is missing.

For example, consider a shear beam of height H, mass density ρ, effective lateral and vertical inertial areas A_x, A_z, and rotational moment of inertia I_y. This beam is upright at $x = 0$, $z > 0$, has a finite lateral stiffness but is infinitely stiff vertically as well as in rotation (i.e., bending). The beam rests on a foundation with 3 DOF, namely lateral, vertical and rotational motion in the plane, that is, u_x, u_z, θ_y. Thus, the rigid body matrix $\mathbf{E}_s$ (here continuous) and the distributed structural mass matrix are of the form

$$\mathbf{E}_s = \begin{bmatrix} \mathbf{e}_x & \mathbf{e}_z & \mathbf{e}_\theta \end{bmatrix} = \begin{bmatrix} 1 & 0 & z \\ 0 & 1 & -x \\ 0 & 0 & 1 \end{bmatrix} \rightarrow \begin{bmatrix} 1 & 0 & z \\ 0 & 1 & 0 \\ 0 & 0 & 1 \end{bmatrix}, \quad \mathbf{M}(z) = \begin{bmatrix} \rho A_x & 0 & 0 \\ 0 & \rho A_z & 0 \\ 0 & 0 & \rho I_y \end{bmatrix} \quad (7.209)$$

The contribution of the lateral modes of the structure to the foundation mass is then

$$\mathbf{M}_s = \int_0^H \mathbf{E}_s^T \mathbf{M} \mathbf{E}_s \, dz = \int_0^H \begin{bmatrix} 1 & 0 & 0 \\ 0 & 1 & 0 \\ z & -x & 1 \end{bmatrix} \begin{bmatrix} \rho A_x & 0 & 0 \\ 0 & \rho A_z & 0 \\ 0 & 0 & \rho I_y \end{bmatrix} \begin{bmatrix} 1 & 0 & z \\ 0 & 1 & -x \\ 0 & 0 & 1 \end{bmatrix} dz \quad (7.210)$$

That is,

$$\mathbf{M}_s = \rho \int_0^H \begin{bmatrix} A_x & 0 & z \\ 0 & A_z & 0 \\ z & 0 & I_y + A_x z^2 \end{bmatrix} dz \quad (7.211)$$

which yields

$$\mathbf{M}_s = \begin{Bmatrix} m_x & 0 & \frac{1}{2} m_x H \\ 0 & m_z & 0 \\ \frac{1}{2} m_x H & 0 & J_y + \frac{1}{3} m_x H^2 \end{Bmatrix}, \quad m_x = \rho A_x H, \quad m_z = \rho A_z H, \quad J_y = \rho I_y H \quad (7.212)$$

The available lateral modes automatically contribute the inertial terms in m_x, while the missing inertial terms m_z, J_y remain to be added by hand to the foundation matrix $\mathbf{M}_f$. It suffices then to augment the foundation mass matrix with the inertial contribution of the superstructure for the DOF which are inactive in the structure, but made active by SSI. Of course, in most cases the lateral and vertical inertias are equal, $m_x = m_z$, but here we had to make a distinction between these two to see more clearly what was missing.

Then again, a structure may actually have nodes that have as many (or even more) DOF than the soil–structure system, yet it may possess rigid members that establish kinematic constraints between some of those DOF. If so, then these will have been eliminated through master-slave conditions, in which case the system will be deficient in its inertial properties – at least as far as SSI is concerned. An example is a simple frame consisting of two inclined (nonparallel) columns connected by a horizontal girder, two masses with translational and rotation al inertia lumped at the intersection of the columns and the girder, and the system rests on a foundation that allows both horizontal and vertical translation as well as rotation. At first, this system has the same number of DOF in each of the two structural nodes as the foundation, that is, a total of 6 DOF. However, if the columns and the girder are axially rigid, this couples the motions of the nodes, even if it does not eliminate their motion in any of the three directions. Thus, the nodes will continue to translate and rotate in all directions, but such motions are no longer independent of each other. In fact, the system now has only a single master

DOF, and the remaining 5 are slaves. Hence, this system has now only one mode and not six. Furthermore, that master DOF could be the lateral displacement of the center of the girder, or its rotation. By the time that the modes are computed and supplied to the SSI program implementing the equations in these chapters, that program has no way of knowing what choices were made earlier concerning the definition of the master DOF, and indeed, what kinematic constraints were used and where. Still, this situation can be remedied by supplying to the program the full mass matrix *before* kinematic condensation, as if each node still had the full count of DOF, and then assume that the modes associated with rigid conditions have infinite frequencies. We are then led back to exactly the same method used in Section 7.4.4.

Member Forces

In a design office it may be of interest to determine not only the motions of the nodes in the structure, including SSI effects, but also the forces in the structural members. When the structure is given in terms of linear members, the information about connectivity and member properties is known, in which case the forces in the members can be computed directly from the nodal displacements. However, when the structure is given in terms of its normal modes and the motion is modified for SSI effects, this is not the case, since the member information is not directly available to the SSI program. Thus, it behooves to either augment the information available to the SSI program so that such forces can be computed, or to return to the structural program the displacements computed by the SSI program, and proceed with its built-in capabilities to assess member forces.

In the first alternative, one arrives at the total, absolute motions of the system $\mathbf{u}_f, \mathbf{u}_s$ with respect to an inertial reference system. The first step in the calculation of the member forces is to compute the motion of the structure relative to the base, that is,

$$\mathbf{y}_s(t) = \mathbf{u}_s - \mathbf{E}_s \mathbf{u}_f \tag{7.213}$$

This removes the rigid-body motion components, which do not elicit structural deformations, but that can affect the computation of forces in those members which are directly connected to the foundation. The member forces are then of the form

$$\mathbf{F} = \mathfrak{R}\,\mathbf{y} \tag{7.214}$$

where $\mathfrak{R}$ is the rigidity matrix relating member forces to nodal displacements. Conceptually it is similar to the stiffness matrix $\mathbf{K}_{ss}$, but differs from it in that the member stiffnesses have not been overlapped, and also makes an automatic transformation of the global displacements of the structural nodes into local member deformations. This matrix is a system property that does not depend on either the normal modes or on the mass matrix. A simple example will suffice to illustrate matters.

Suppose the structure consists of a single beam-column that is inclined with respect to the vertical. The stiffness matrix of the beam, in *local* member coordinates, is of the form

$$\mathbf{K}_L = \frac{EI}{L^3}\begin{bmatrix} 0 & 0 & 0 & 0 & 0 & 0 \\ 0 & 12 & 6L & 0 & -12 & 6L \\ 0 & 6L & 4L^2 & 0 & -6L & 2L^2 \\ 0 & 0 & 0 & 0 & 0 & 0 \\ 0 & -12 & -6L & 0 & 12 & -6L \\ 0 & 6L & 2L^2 & 0 & -6L & 4L^2 \end{bmatrix} + \frac{EA}{L}\begin{bmatrix} 1 & 0 & 0 & -1 & 0 & 0 \\ 0 & 0 & 0 & 0 & 0 & 0 \\ 0 & 0 & 0 & 0 & 0 & 0 \\ -1 & 0 & 0 & 1 & 0 & 0 \\ 0 & 0 & 0 & 0 & 0 & 0 \\ 0 & 0 & 0 & 0 & 0 & 0 \end{bmatrix} = \begin{Bmatrix} \mathbf{K}_{AA} & \mathbf{K}_{AB} \\ \mathbf{K}_{BA} & \mathbf{K}_{BB} \end{Bmatrix}$$

(7.215)

This member stiffness matrix relates local nodal forces to local nodal displacements.

The member AB has nodal coordinates $\mathbf{x}_A, \mathbf{x}_B$ with which we can decide how the local coordinates relate to the global coordinates. Say the start node is A and the end node is B, which define the orientation of the member in the plane. Its length and direction cosines with respect to the x, z axes are

$$L = \sqrt{(x_B - x_A)^2 + (z_B - z_A)^2}, \qquad c = \frac{x_B - x_A}{L}, \qquad s = \frac{z_B - z_A}{L} \tag{7.216}$$

Hence, if u_L, w_L are the axial and transverse displacements in local member coordinates and u_G, w_G are the horizontal and vertical components in Cartesian coordinates, then

$$u_L = c\, u_G + s\, w_G, \qquad w_L = -s\, u_G + c\, w_G \tag{7.217}$$

Defining the displacement gradients

$$\begin{aligned} \Delta u_L &= u_L^B - u_L^A = \left(u_G^B - u_G^A\right)c + \left(w_G^B - w_G^A\right)s \\ \Delta \phi_L^{AB} &= \frac{w_L^B - w_L^A}{L} = \frac{-\left(u_G^B - u_G^A\right)s + \left(w_G^B - w_G^A\right)c}{L} \end{aligned} \tag{7.218}$$

then the internal member forces are

$$\begin{aligned} N^B &= \frac{EA}{L}\Delta u_L \\ S^B &= \frac{2EI}{L^2}\left(6\Delta\phi_L^{AB} - 2\theta^A - 2\theta^B\right) \\ M^A &= \frac{2EI}{L}\left(2\theta^A + \theta^B - 3\Delta\phi_L^{AB}\right) \\ M^B &= \frac{2EI}{L}\left(\theta^A + 2\theta^B - 3\Delta\phi_L^{AB}\right) \end{aligned} \tag{7.219}$$

where N^B, S^B are the axial and shearing forces as seen from the ending node, and M^A, M^B are the end moments. The above can be used to infer the member rigidity matrix $\mathfrak{R}_{AB}$ with which the local member forces can be evaluated and passed on, as may be the case, to the SSI program, should that program have this capability.

7.4.6 The Free-Field Problem: Elements of 1-D Soil Amplification

As briefly alluded to earlier, the free-field problem deals with the assessment of the seismic environment imagined to exist at a site before the soil has been excavated and any structure erected. Figure 7.11 shows a schematic view of this problem: a seismic fault

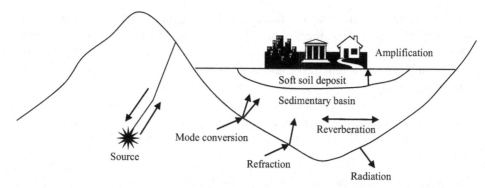

Figure 7.11. Free-field problem.

ruptures and elicit waves which travel along multiple paths to the site, encountering variegated soil and topographic conditions and ultimately arriving at the location of interest, which may or not contain buildings extraneous to the structure of interest but not yet erected, that is, they just happen to be already there.

Clearly, the characteristics of the free-field motions depend on a multitude of factors such as earthquake magnitude, epicentral distance, attenuation laws, large-scale as well as local geology, and particularly, the local soil properties, including inelastic behavior of the soil overburden. Most engineering studies of local amplification assume, however, that the soil is horizontally layered and exhibits depth-dependent viscoelastic properties, that it has an infinite lateral extent, and that the motion in the free field consist of plane shear waves propagating at some angle, usually vertically. An obvious shortcoming of this 1-D amplification model is that it assumes that only one type of waves gives rise to the local seismic motion. Still, more general 2-D wave propagation analyses with SV and P waves at arbitrary angles show amplification functions not very different from those predicted by the 1-D theory, except for unusual combinations of waves. In such models, material nonlinear effects are most often incorporated only qualitatively by using soil properties (i.e., moduli and damping) that are consistent with the levels of strain expected to develop during the seismic event. The mathematical tools needed to carry out soil amplification analyses in horizontally layered media can be found in Chapter 5, Section 5.2, so they need not be repeated herein.

In some cases, the assumption of an infinite horizontal extent of the soil layers may not be warranted, because trapping and focusing of seismic energy can occur as a result of lateral boundaries or dipping layers. For example, reverberations may be observed in narrow alluvial valleys bounded by firm rock. Also, the assumption of a single wave type (i.e., plane body waves) may lead to artifacts associated with standing waves, whereas actual motions are caused by a mix of many wave types.

It has become accepted practice in the seismic analysis for SSI effects to define design earthquakes on the basis of smooth response spectra, which will supposedly envelop over the frequency range of interest the response of SDOF oscillators to any credible ground motion, and are thus believed to constitute a safe basis for the design of such systems. Sets of these response spectra, and rules to construct them, have been derived from the normalized records of actual ground motions using statistical analyses. It is also customary to generate artificial or synthetic earthquake time histories (accelerograms) whose spectra envelope the design spectra everywhere. Although a pragmatic approach, these synthetic quakes have little physical significance and they are especially problematic when nonlinear effects are taken into consideration, for they greatly exaggerate the deformations in the soil and thus introduce unrealistic levels of strain-softening as well as excessive values of hysteretic damping.

In the light of these considerations, and being an intrinsically difficult problem, in most cases significant simplifications are made and a number of assumptions are taken for granted in the definition of the free-field problem, of which the following are the most common:

- The soil properties change only in the depth direction, yet not horizontally. In other words, the material characteristics either change gradually with depth, or the soil exhibits well-defined horizontal layers.
- Inelastic soil effects are assumed to be sufficiently moderate that either linear or quasilinear models will suffice to take into account the inelasticity.
- The seismic environment in the soil is the result of *plane waves* that emanate from the depths underneath, and is associated with a well-defined horizontal phase velocity.
- In most cases (indeed, almost universally) the waves are assumed to propagate along a vertical path. This is what is normally referred to as 1-D soil amplification involving horizontally polarized shear waves, or SH waves. The implication is that the apparent horizontal phase velocity is infinitely large, and motions in horizontal planes are uniform. This greatly simplifies the mathematics and gets rid of mode conversions.
- The characteristic design motion is defined independently of the free field problem, and is chosen on the basis of the soil type (firm soil, soft alluvium, sand, etc.), structural location (on flat or sloping terrain, etc.), location of known active faults and potential sources, recent and ancient historical seismicity at the site (previous earthquakes, sand boils, dislocations in the stratigraphy), and not in the least also on the prevailing seismic codes and regulations. All of these are used to define the characteristics of the design earthquake, which include maximum acceleration, duration, frequency content, and so forth.
- The design motion, which is often referred to as the *control motion*, is assumed to take place at a well-defined location, usually at the surface of the soil or at some nearby outcropping of rock. This point is typically referred to as the *control point*.
- If the soil is very rigid, say there is crystalline rock, then the soil amplification effects are usually neglected and the control motion is defined directly at the surface of the soil.
- If the soil is not rigid, then there are various choices as to where the control motion will take place, as described in the ensuing.

A most important point in the solution of the free field problem is the definition of *where* the control motion is assumed to take place, that is, where the *control point* is. There are at least four different possibilities, each of which leads to drastically different motions within and on the soil and hence in the structure:

a. The control motion is specified at the free surface of the soil deposit, without any structure. It implies basically that amplification effects by the soil are already included in the selected spectra. This is probably the safest and most rational choice for the control point when the soils are not soft.

b. The control motion is specified at the level of the foundation but far away from the structure, so that any effects of structure or excavation may be neglected. Its physical implications are difficult to interpret since the motion at any level within the soil mass must depend on the material characteristics throughout the soil profile. Its meaning is further confused if several adjacent structures are founded at different levels. This choice is the least optimal and usually results in spurious resonances at frequencies associated with the short soil column above the foundation level.

c. The control motion is specified at some hypothetical outcropping of rock, and without any structure. This choice raises some questions as to the exact definition of rock, and about the characteristics of the design earthquake, which ought to be consistent with firm rock. It will usually result in substantial amplification of motions.

d. The motion is specified at "bedrock." Although simple, this option is not as logical as any of the previous ones, since the motion "deep down there" is generally unknown, and radiation damping is neglected. We strongly recommend against this choice.

For an elastic material and any specified class of waves, we can readily *deconvolve* the control motions implied by these four choices and obtain the motions anywhere else in the medium, and in the process determine a unique relationship between the various control points. This will demonstrate that the motions observed at any fixed location depend greatly on the choice of the control point. Thus, that point must then be chosen wisely and with great care.

Effect of Location of Control Motion in 1-D Soil Amplification

Consider a layered, elastic system consisting of $n-1$ layers underlain by an elastic half-space ("rock") that is subjected to vertically propagating shear waves. The amplitudes of the incident and reflected waves in the elastic half-space, which give rise to the motion in the layers, are designated with A, B, respectively. We define as rock outcrop the soil configuration where all soil layers are removed (or do not exist), and only the rock remains.

Numbering the layer interfaces from the top down, then the mathematical soil amplification problem at some arbitrary frequency ω will predict motions in the frequency domain $u_j(\omega)$, $j = 1, 2, \ldots n$. In addition, the *control motion* will be denoted as $u_{cont}(\omega)$. The situation is as illustrated schematically in Figure 7.12. In all cases, the transfer functions will be assumed to be determinable with the tools given in Chapter 5, Section 5.2.

The transfer function from *rock outcrop* to any arbitrary layer is formally defined as the ratio of the observed motion u_j to the input motion at rock outcrop u_R, that is,

$$\boxed{T_{jR} = \frac{u_j}{u_R}}, \qquad j = 1, 2, \ldots n, \qquad T_{RR} = 1, \quad A = \frac{1}{2}, \qquad u_j = T_{jR} u_R = T_{jR} u_{cont} \qquad (7.220)$$

Observe that u_n is the motion at the soil–rock interface, that is, at the *rock within*, which differs from the motion at *rock outcrop*. We now consider other alternative locations for the control motion.

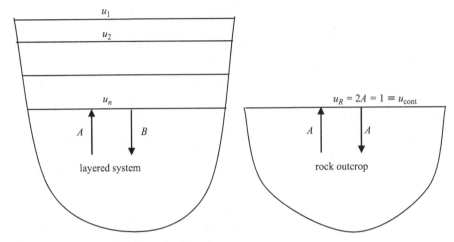

Figure 7.12. Layered system, incident and reflected waves.

Control Motion Specified at Some Layer Within

Next consider the situation where the control motion is specified somewhere else, namely at the kth layer *within*, that is, $u_{\text{cont}} = u'_k = 1$, which elicits motions throughout which are now denoted with single primes (to distinguish these from the case where the control motion is defined at rock outcrop). Then

$$\boxed{T_{jk} = \frac{T_{jR}}{T_{kR}}}, \qquad u'_j = T_{jk}u'_k = \frac{T_{jR}}{T_{kR}}u'_k, \qquad u'_j = \frac{T_{jR}}{T_{kR}}u_{\text{cont}} \tag{7.221}$$

and at the elevation of the rock interface

$$u'_n = T_{nk}u'_k = \frac{T_{nR}}{T_{kR}}u'_k \qquad\qquad = \text{rock within} \tag{7.222}$$

$$u'_R = \frac{T_{RR}}{T_{kR}}u'_k = \frac{1}{T_{kR}}u'_k = \frac{1}{T_{kR}}u_{\text{cont}} \qquad = \text{rock outcrop} \tag{7.223}$$

The incident amplitude is now not $A = \tfrac{1}{2}$ but instead

$$A' = \frac{\tfrac{1}{2}u'_k}{T_{kR}} = \frac{1}{2T_{kR}} = \tfrac{1}{2}T_{Rk} \tag{7.224}$$

In particular, when the control motion is specified at the *free surface*, then $k = 1$ and

$$T_{j1} = \frac{T_{jR}}{T_{1R}}, \qquad T_{n1} = \frac{T_{nR}}{T_{1R}}, \qquad T_{R1} = \frac{1}{T_{1R}} \tag{7.225}$$

It should be noted that this alternative suffers from one severe inconsistency: the motions at locations above the control horizon do not at all depend on the motions, or even the material properties of the soil, below the control horizon. Clearly, this violates physics, because the motion observed anywhere is most certainly the result of waves that previously propagated through the deepest layers and were surely affected by them. But we pretend to know better about the effect of those deeper layers, and thus assume that transmission, reflection and amplification effects below the control horizon are already built into the control motion.

Control Motion Specified at Some Outcropping Layer

Imagine next that the control motion is specified at the kth layer *outcropping*, which means that we must first physically remove the upper $k-1$ layers, that is, those above the control elevation so as to define the transfer functions without those upper layers. To distinguish this case and that in the previous section, we shall use tildes. Then for the system without the upper $k-1$ layers, we have

$$\tilde{T}_{jk} = \frac{\tilde{T}_{jR}}{\tilde{T}_{kR}}, \qquad j \geq k \tag{7.226}$$

and in particular, at rock outcrop

$$\tilde{T}_{Rk} = \frac{\tilde{T}_{RR}}{\tilde{T}_{kR}} = \frac{1}{\tilde{T}_{kR}}, \qquad \tilde{A} = \frac{\tilde{u}_k}{2\tilde{T}_{kR}} = \frac{1}{2\tilde{T}_{kR}} = \tfrac{1}{2}\tilde{T}_{Rk} = \text{amplitude of incident wave} \tag{7.227}$$

Hence, the motions in the *original* layered system with all n layers will be

$$\boxed{\tilde{u}_j = T_{jR}\tilde{T}_{Rk}\tilde{u}_k = \frac{T_{jR}}{\tilde{T}_{kR}} u_{\text{cont}}}, \qquad j = 1,2,\dots n \tag{7.228}$$

which is clearly different from the motion elicited by a control motion specified *within*.
In summary:

Location control motion	Transfer function from control surface to output interface	Ratio of incident wave to control motion
Rock outcrop	T_{jR}	$\tfrac{1}{2}$
Rock within	$T_{jn} = T_{jR}/T_{nR} = T_{jR}T_{Rn}$	$\tfrac{1}{2}T_{Rn} = \tfrac{1}{2}/T_{nR}$
Layer within	$T_{jk} = T_{jR}/T_{kR} = T_{jR}T_{Rk}$	$\tfrac{1}{2}T_{Rk} = \tfrac{1}{2}/T_{kR}$
Free surface	$T_{j1} = T_{jR}/T_{1R} = T_{jR}T_{R1}$	$\tfrac{1}{2}T_{R1} = \tfrac{1}{2}/T_{1R}$

Observe that with the sole exception of the case where the motion is prescribed directly at rock outcropping (namely our initial reference case), *in no other case will the wave amplitudes at a hypothetical rock outcropping be simply half of the control motion.*

We observe that the location of the control motion will have an enormous effect on the resulting motions within the layered profile. When the motion is prescribed at rock outcrop, one talks properly of soil amplification, whereas when the control motion is prescribed at the free surface, then the process of figuring out what motion must exist within to obtain that motion at the surface is referred to a *deconvolution*. Although there exist excellent engineering reasons for why the surface should be the preferred control point ("less danger of things going wrong" between the control point and the structure), it must be added that deconvolution constitutes an ill-posed mathematical problem, because it involves a body where two intrinsically incompatible boundary conditions are imposed simultaneously. On the one hand we prescribe the displacements (or accelerations) at the free surface and at the same time we prescribe the stresses, namely zero. Another way of looking at this is to realize that we pretend to know what the motion is at the surface (whatever the soil properties), even though that motion ought to depend on the mechanical properties of the soil underneath.

7.4.7 Kinematic Interaction of Rigid Foundations

The term *kinematic interaction*, or alternatively *wave passage*, is used to refer to the interaction and response of a massless, rigid foundation of arbitrary shape to some arbitrary seismic environment, as illustrated in Figure 7.13. It can be shown that this response can be used as an effective support motion to the superstructure resting on this foundation, but now mounted on (generally frequency-dependent) soil impedances in lieu of the soil. This motion accounts for both soil amplification and the wave scattering caused by the inability of the rigid foundation to accommodate the ground deformations elicited by the seismic waves

As we have seen, when the structure is modeled via some discrete model of arbitrary refinement, the seismic response in the frequency domain obeys an equation of the form

$$\begin{bmatrix} \mathbf{Z}_{ss} & \mathbf{Z}_{sb} \\ \mathbf{Z}_{bs} & \mathbf{Z}_{bb} + \mathbf{Z} \end{bmatrix} \begin{Bmatrix} \mathbf{u}_s \\ \mathbf{u}_b \end{Bmatrix} = \begin{Bmatrix} \mathbf{0} \\ \mathbf{Z}\mathbf{u}_b^* + \mathbf{p}_b^* \end{Bmatrix} \tag{7.229}$$

where the subindex s refers to the DOF in the *structure* away from the soil interface, the subindex b refers to the DOF at the *base* in contact with the soil, and $\mathbf{Z} = \mathbf{Z}(\omega)$ is the frequency dependent, fully populated, impedance matrix of the soil, as seen from the soil–foundation interface after the DOF in the soil have been condensed out. Also, $\mathbf{u}_b^*$ is the *free-field motion* observed at the soil–structure interface, that is, the motion observed there before the soil is excavated and any structure is erected, and $\mathbf{p}_b^*$ represents the net forces in the free-field acting on the soil *to be excavated*, that is, they are the discrete counterpart of the internal stresses acting in the free-field along the soil–structure interface. Clearly, if we could not solve the free-field problem, then the SSI problem could not be solved either. Thus, we assume $\mathbf{u}_b^*, \mathbf{p}_b^*$ to be known. Observe that for a foundation without any embedment, the free-field forces vanish, that is, $\mathbf{p}_b^* = \mathbf{0}$.

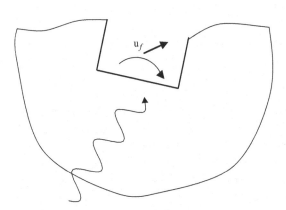

Figure 7.13. Kinematic interaction.

When the foundation is very rigid, or indeed, when it is perfectly rigid, then the motion of the foundation obeys a kinematic constraint of the form

$$\mathbf{u}_b = \mathbf{E}_b \, \mathbf{u}_f \qquad (7.230)$$

in which $\mathbf{E}_b$ is the rigid-body matrix for the DOF along the soil–structure interface, as defined as in Section 7.4.4, that is,

$$\mathbf{E}_b = \left\{ \begin{array}{c} \mathbf{T}_1 \\ \vdots \\ \mathbf{T}_j \\ \vdots \end{array} \right\} \qquad (7.231)$$

is a rigid-body transformation matrix assembled with the nodal transformation matrices

$$\mathbf{T}_j = \begin{bmatrix} 1 & 0 & 0 & 0 & z_j - z_0 & -(y_j - y_0) \\ 0 & 1 & 0 & -(z_j - z_0) & 0 & x_j - x_0 \\ 0 & 0 & 1 & y_j - y_0 & -(x_j - x_0) & 0 \end{bmatrix} \qquad (7.232)$$

in which x_j, y_j, z_j are the coordinates of the jth node along the soil–structure interface, x_0, y_0, z_0 are the coordinates of the (arbitrary) reference point at which the motion of the rigid foundation is being observed, and

$$\mathbf{u}_f = \left\{ u_{fx} \quad u_{fy} \quad u_{fz} \quad \vartheta_{fx} \quad \vartheta_{fy} \quad \vartheta_{fz} \right\}^T \qquad (7.233)$$

is the vector of foundation motion, which contains the three translations and three rotations of the foundation. Hence, making use of these various equations, we ultimately obtain

$$\begin{Bmatrix} \mathbf{Z}_{ss} & \mathbf{Z}_{sb}\mathbf{E}_b \\ \mathbf{E}_b^T\mathbf{Z}_{bs} & \mathbf{E}_b^T\mathbf{Z}_{bb}\,\mathbf{E}_b + \mathbf{E}_b^T\,\mathbf{Z}\,\mathbf{E}_b \end{Bmatrix} \begin{Bmatrix} \mathbf{u}_s \\ \mathbf{u}_f \end{Bmatrix} = \begin{Bmatrix} \mathbf{0} \\ \mathbf{E}_b^T\left(\mathbf{Z}\mathbf{u}_b^* + \mathbf{p}_b^*\right) \end{Bmatrix} \qquad (7.234)$$

Next, we define

$$\mathbf{Z}_f = \mathbf{E}_b^T\,\mathbf{Z}\,\mathbf{E}_b \qquad (7.235)$$

as the 6×6 matrix of frequency-dependent foundation impedances (the "soil springs and dashpots"), relative to the reference point. In most cases, this matrix will either be *directly* available from results in the literature, or it can be determined with an appropriate external program, or it can even be *estimated*, in which case one can bypass the determination of $\mathbf{Z}$. Hence, with an appropriate re-definition of the structural terms, we can write

$$\begin{bmatrix} \mathbf{Z}_{ss} & \mathbf{Z}_{sf} \\ \mathbf{Z}_{fs} & \mathbf{Z}_{ff} + \mathbf{Z}_f \end{bmatrix} \begin{Bmatrix} \mathbf{u}_s \\ \mathbf{u}_f \end{Bmatrix} = \begin{Bmatrix} \mathbf{0} \\ \mathbf{E}_b^T\mathbf{Z}\mathbf{u}_b^* + \mathbf{E}_b^T\mathbf{p}_b^* \end{Bmatrix} \qquad (7.236)$$

Iguchi's Approximation, General Case

On the right-hand side of this last equation, we recognize the second loading term as the overall resultant of the free-field forces and moments acting on the excavated soil, which can readily be computed. However, evaluation of the first term would actually require knowledge of $\mathbf{Z}$, which in many practical cases is not available. To circumvent this difficulty, we resort to an excellent approximation proposed by Iguchi[12] which consists in setting $\mathbf{u}_b^* = \mathbf{E}_b\,\mathbf{u}_f^*$, that is, expressing the free-field displacements along the soil–structure interface in terms of an as yet unknown rigid body motion vector $\mathbf{u}_f^*$ such that $\mathbf{E}_b^T\mathbf{Z}\mathbf{u}_b^* = \mathbf{E}_b^T\mathbf{Z}\mathbf{E}_b\,\mathbf{u}_f^* = \mathbf{Z}_f\mathbf{u}_f^*$, in which case

$$\begin{bmatrix} \mathbf{Z}_{ss} & \mathbf{Z}_{sf} \\ \mathbf{Z}_{fs} & \mathbf{Z}_{ff}+\mathbf{Z}_f \end{bmatrix}\begin{Bmatrix} \mathbf{u}_s \\ \mathbf{u}_f \end{Bmatrix} = \begin{Bmatrix} \mathbf{0} \\ \mathbf{Z}_f\,\mathbf{u}_f^* + \mathbf{p}_f^* \end{Bmatrix} \tag{7.237}$$

with $\mathbf{p}_f^* = \mathbf{E}_b^T\mathbf{p}_b^*$ being a 6×1 vector containing the resulting forces and moments acting on the excavated soil, and $\mathbf{u}_f^*$ being a 6×1 vector whose components remain to be determined. From dynamic equilibrium considerations in the excavated soil (i.e., Newton's law), we know also that the aggregated force resultants $\mathbf{p}_f^*$ of the tractions acting on the external surface of the excavation must necessarily equal the net resultant of the inertia forces acting on the body of excavated soil, that is,

$$\mathbf{p}_f^* = \iiint_V \mathbf{T}^T\,\ddot{\mathbf{u}}^*\,\rho\,dV \tag{7.238}$$

with $\ddot{\mathbf{u}}^*$ being the absolute free-field acceleration at some point within the excavated soil, and $\mathbf{T} = \mathbf{T}(\mathbf{x})$ is a matrix similar to the $\mathbf{T}_j$ defied earlier, but replacing the nodal position $\mathbf{x}_j$ with the generic position $\mathbf{x}$ within the soil mass. In the light of the identity $\ddot{\mathbf{u}}^* = -\omega^2\mathbf{u}^*$, this reduces to

$$\mathbf{p}_f^* = \iint_A \mathbf{T}^T\mathbf{p}_b^*\,dA = -\omega^2\iiint_V \mathbf{T}^T\,\mathbf{u}^*\,\rho\,dV \tag{7.239}$$

Considering that $\mathbf{u}^*$ is a function of position that is not of the rigid-body type, we see that it cannot readily be extracted out of the volume integral. However, this volume integral might still be simpler to evaluate than the surface integral, because the latter requires computing the up-to-nine internal stresses and projecting these onto the external surface of the excavation, that is, accounting for the direction of the normal, whereas the displacement vector has only three components that are aligned with the global coordinates.

Concerning the still unknown 6×1 rigid body motion vector $\mathbf{u}_f^*$, we compute it as the spatial average of the free-field motion along the contact surface whose deviation $\Delta(\mathbf{x}) = \mathbf{u}_b^* - \mathbf{T}\mathbf{u}_f^*$ is optimal in the least square sense, that is, $\min\iint_A \Delta^T\Delta\,dA$, with A being the contact area of the structure and the soil, and $\mathbf{u}_b^* = \mathbf{u}_b^*(\mathbf{x}), \mathbf{T} = \mathbf{T}(\mathbf{x})$, with $\mathbf{x}$ being a generic point on the soil interface. As is well known, the least square optimum is obtained by taking partial derivatives with respect to the unknown components of $\mathbf{u}_f^*$, that is,

$$\frac{\partial}{\partial\mathbf{u}_f^*}\iint_A \Delta^T\Delta\,dA = \mathbf{0} \tag{7.240}$$

[12] M. Iguchi, "An approximate analysis of input motions for rigid embedded foundations," *Trans. Arch. Inst. Jpn.*, 315 (May), 1982, 61–75.

That is,

$$\iint_A \mathbf{T}^T \left(\mathbf{u}_b^* - \mathbf{T}\mathbf{u}_f^* \right) dA = 0 \tag{7.241}$$

or

$$\left\{ \iint_A \mathbf{T}^T \mathbf{T}\, dA \right\} \mathbf{u}_f^* = \iint_A \mathbf{T}^T \mathbf{u}_b^*\, dA \tag{7.242}$$

We define

$$\mathbf{H} = \iint_A \mathbf{T}^T \mathbf{T}\, dA = \begin{Bmatrix} A & 0 & 0 & 0 & S_z & -S_y \\ 0 & A & 0 & -S_z & 0 & S_x \\ 0 & 0 & A & S_y & -S_x & 0 \\ 0 & -S_z & 0 & I_{xx} & -I_{xy} & -I_{xz} \\ S_z & 0 & -S_x & -I_{yx} & I_{yy} & -I_{yz} \\ -S_y & S_x & 0 & -I_{zx} & -I_{zy} & I_{zz} \end{Bmatrix} \tag{7.243}$$

as the matrix with the area properties of the soil–structure interface, in which A is the total contact area, S_j are the static moments of area, and the I_{ij} are the moments and products of inertia of that area, all of which are with respect to the reference point. Hence

$$\mathbf{u}_f^* = \mathbf{H}^{-1} \iint_A \mathbf{T}^T \mathbf{u}_b^*\, dA \tag{7.244}$$

Choosing this reference point as the centroid of the contact area (not of the excavated volume!) and assuming the principal axes of this area to coincide in orientation with the global coordinate axes (e.g., a doubly symmetric section), then the static moments and products of inertia vanish and the matrix is purely diagonal, that is,

$$\mathbf{H} = \text{diag}\{ A \quad A \quad A \quad I_x \quad I_y \quad I_z \} \tag{7.245}$$

Hence $\mathbf{H}^{-1} = \text{diag}\{ A^{-1} \quad A^{-1} \quad A^{-1} \quad I_x^{-1} \quad I_y^{-1} \quad I_z^{-1} \}$. Our problem reduces then to solving

$$\begin{bmatrix} \mathbf{Z}_{ss} & \mathbf{Z}_{sf} \\ \mathbf{Z}_{fs} & \mathbf{Z}_{ff} + \mathbf{Z}_f \end{bmatrix} \begin{Bmatrix} \mathbf{u}_s \\ \mathbf{u}_f \end{Bmatrix} = \begin{Bmatrix} \mathbf{0} \\ \mathbf{Z}_f \mathbf{u}_f^* + \mathbf{p}_f^* \end{Bmatrix} \tag{7.246}$$

with

$$\mathbf{p}_f^* = \iint_A \mathbf{T}^T \mathbf{p}_b^*\, dA = -\omega^2 \iiint_V \mathbf{T}^T \mathbf{u}^* \rho\, dV \tag{7.247}$$

and $\mathbf{u}_f^*$ as defined in Eq. 7.244. Observe that $\mathbf{p}_f^*$ are the net forces and moments with which the free-field motion acts at the reference point onto the excavated soil (not on the excavation, which would be $-\mathbf{p}_f^*$). Finally, all of the preceding can also be written as

$$\begin{bmatrix} \mathbf{Z}_{ss} & \mathbf{Z}_{sf} \\ \mathbf{Z}_{fs} & \mathbf{Z}_{ff} + \mathbf{Z}_f \end{bmatrix} \begin{Bmatrix} \mathbf{u}_s \\ \mathbf{u}_f \end{Bmatrix} = \begin{Bmatrix} \mathbf{0} \\ \mathbf{Z}_f \mathbf{u}_0^* \end{Bmatrix} \tag{7.248}$$

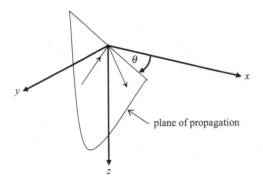

Figure 7.14. SH wave ray in azimuthal plane.

where

$$
\boxed{
\begin{aligned}
\mathbf{u}_0^* &= \mathbf{H}^{-1} \iint_A \mathbf{T}^T \mathbf{u}_b^* \, dA + \mathbf{K}_f^{-1} \iint_A \mathbf{T}^T \mathbf{p}_b^* \, dA \\
&= \mathbf{H}^{-1} \iint_A \mathbf{T}^T \mathbf{u}_b^* \, dA - \omega^2 \, \mathbf{K}_f^{-1} \iiint_V \mathbf{T}^T \mathbf{u}^* \, \rho \, dV
\end{aligned}
}
\tag{7.249}
$$

is the effective ground motion (or "support motion") that must be prescribed at the reference support point $\mathbf{x}_0$ "underneath" the foundation $\mathbf{Z}_f$ that is attached at that point.

Iguchi Approximation for Cylindrical Foundations Subjected to SH Waves

The next several pages present an application of Iguchi's method to cylindrical foundations embedded in a homogeneous half-space when excited by SH waves. The material in these pages has been inferred from an appendix in a report by Pais and Kausel (1985),[13] which contains also the solution for SV–P waves as well as the extension to prismatic (rectangular) foundations, the detail of which are too lengthy and complicated to be included herein. We begin by reviewing the basic equations for SH waves propagating in a homogeneous half-space.

Plane SH Waves in a Homogeneous Half-Space

Consider a homogeneous elastic half-space subjected to plane SH waves, as shown in Figures 7.14 and 7.15. We choose a right-handed Cartesian coordinate system whose origin is at the surface, the axis x is from left to right, y is toward the viewer, and z is down. Assume then an SH wave propagating in the plane x, z which is polarized in direction y. In the vertical plane, this wave propagates at an angle ϕ with respect to the vertical, which is positive when the wave has positive horizontal phase velocity. Hence

$$
k = \frac{\omega}{C_s} \sin \phi = k_s \sin \phi \quad = \text{horizontal wavenumber} \tag{7.250}
$$

$$
n = \frac{\omega}{C_s} \cos \phi = k_s \cos \phi \quad = \text{vertical wavenumber} \tag{7.251}
$$

---

[13] A. Pais and E. Kausel, "Stochastic response of foundations," *MIT Research Report R85-6*, Department of Civil Engineering, Cambridge, MA, 1985. A PDF copy is available online.

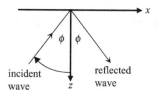

Figure 7.15. Reflection of SH wave.

$$\boxed{u_y = e^{i(\omega t - kx)}\cos nz}$$ (7.252)

which is normalized so that at the surface, $u_y = 1$. The shearing stresses are

$$\boxed{\begin{aligned}\tau_{xy} &= G\frac{\partial u_y}{\partial x}\\ &= -ikGe^{i(\omega t - kx)}\cos nz\end{aligned}}$$ (7.253)

and

$$\boxed{\begin{aligned}\tau_{yz} &= G\frac{\partial u_y}{\partial z}\\ &= -nGe^{i(\omega t - kx)}\sin nz\end{aligned}}$$ (7.254)

More generally, if the wave propagates at a *clockwise*-positive azimuth θ with respect to the x-axis, then

$$k_x = k\cos\theta, \qquad k_y = k\sin\theta$$ (7.255)

in which case

$$\boxed{u_x = e^{i(\omega t - k_x x - k_y y)}\left(-\sin\theta\right)\cos nz}$$ (7.256)

$$\boxed{u_y = e^{i(\omega t - k_x x - k_y y)}\left(\cos\theta\right)\cos nz}$$ (7.257)

In the ensuing, we shall need the following integrals:

$$\int_0^{2\pi}\cos m\theta\, e^{\pm ia\cos\theta}\,d\theta = 2\pi(\pm i)^m J_m(a), \qquad \int_0^{2\pi}\sin m\theta\, e^{\pm ia\cos\theta}\,d\theta = 0$$ (7.258)

$$\int_0^{2\pi} e^{\pm ikR\cos\theta}\,d\theta = 2\pi J_0(kR), \qquad \int_0^{2\pi}\cos\theta\, e^{\pm ikR\cos\theta}\,d\theta = \pm 2\pi i J_1(kR)$$ (7.259)

$$\int_0^R J_0(kr)\,r\,dr = R^2\frac{J_1(kR)}{kR}, \qquad \int_0^R J_1(kr)\,r^2\,dr = R^3\frac{J_2(kR)}{kR}$$ (7.260)

Geometric Properties

$$e = \frac{E}{R} \qquad\qquad \text{Embedment ratio}$$ (7.261)

$$A = \pi R^2\left(1 + 2e\right) \qquad \text{Contact area}$$ (7.262)

$$H = R\frac{e^2}{1+2e} \qquad\qquad \text{Height of centroid of contact area} \qquad (7.263)$$

$$h = \frac{H}{E} = \frac{e}{1+2e} \qquad\qquad \text{Dimensionless height of centroid} \qquad (7.264)$$

$$I_x = I_y = \pi R^4 \left(\tfrac{1}{4} + e - e^2 h + \tfrac{2}{2}e^3\right) \quad \begin{array}{l}\text{Moment of inertia, any horizontal} \\ \text{axis through centroid}\end{array} \qquad (7.265)$$

$$I_z = \pi R^4 \left(\tfrac{1}{2} + 2e\right) \qquad\qquad \text{Moment of inertia about vertical axis (torsion)} \quad (7.266)$$

$$\mathbf{H} = \operatorname{diag}\left\{A \quad A \quad A \quad I_x \quad I_y \quad I_z\right\} \quad \text{Matrix of area properties} \qquad (7.267)$$

Free-Field Motion Components at Arbitrary Point, Zero Azimuth

$$\mathbf{u}^* = \mathbf{T}^T \mathbf{u}^* = \begin{Bmatrix} 1 & 0 & 0 \\ 0 & 1 & 0 \\ 0 & 0 & 1 \\ 0 & -\left(z_j - z_0\right) & y_j - y_0 \\ z_j - z_0 & 0 & -\left(x_j - x_0\right) \\ -\left(y_j - y_0\right) & x_j - x_0 & 0 \end{Bmatrix} \begin{bmatrix} 0 \\ u_y^* \\ 0 \end{bmatrix} = \begin{Bmatrix} 0 \\ 1 \\ 0 \\ -\left(z_j - z_0\right) \\ 0 \\ x_j - x_0 \end{Bmatrix} u_y^* \qquad (7.268)$$

where

$$x_0 = 0, \qquad z_0 = E - H, \qquad (7.269)$$

$$\boxed{u_y^* = e^{-ikx}\cos nz} \qquad (7.270)$$

in which

$$k = k_s \sin\phi, \qquad n = k_s \cos\phi, \qquad k_s = \omega / C_s \qquad (7.271)$$

Surface Integrals

$$\iint_A \mathbf{T}^T \mathbf{u}^* \, dA = \iint_A \begin{Bmatrix} 0 \\ 1 \\ 0 \\ -\left(z - z_0\right) \\ 0 \\ x - x_0 \end{Bmatrix} u_y^* \, dA = \begin{Bmatrix} 0 \\ K_1 \\ 0 \\ K_2 \\ 0 \\ K_3 \end{Bmatrix} \qquad (7.272)$$

This involves the following three integrals, which extend over both the base and the sides of the cylinder (see also Figure 7.16).

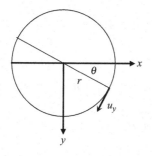

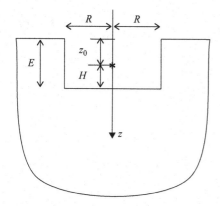

Figure 7.16. Cylindrical foundation.

First Integral

$$\iint_A e^{-ikx} \cos nz \, dA = \cos nE \int_0^R r \int_0^{2\pi} e^{-ikr\cos\theta} \, d\theta \, dr + R \int_0^{2\pi} e^{-ikR\cos\theta} \int_0^E \cos nz \, dz \, d\theta$$

$$= 2\pi \left[\cos nE \int_0^R J_0(kr) \, r \, dr + R \frac{\sin nE}{n} J_0(kR) \right] \tag{7.273}$$

$$\boxed{K_1 = 2\pi R^2 \left[\cos nE \, \frac{J_1(kR)}{kR} + e \frac{\sin nE}{nE} J_0(kR) \right]} \tag{7.274}$$

Second Integral

$$-\iint_A (z - z_0) e^{-ikx} \cos nz \, dA$$

$$= -\left[(E - z_0) \cos nE \int_0^R r \int_0^{2\pi} e^{-ikr\cos\theta} \, d\theta \, dr + R \int_0^{2\pi} e^{-ikR\cos\theta} \int_0^E (z - z_0) \cos nz \, dz \, d\theta \right]$$

$$= -2\pi \left[H \cos nE \int_0^R J_0(kr) \, r \, dr + R J_0(kR) \left(H \frac{\sin nE}{n} + \frac{1}{n^2}(\cos nE - 1) \right) \right] \tag{7.275}$$

$$\boxed{K_2 = -2\pi R^3 e \left[h \cos nE \, \frac{J_1(kR)}{kR} + e \left(h \frac{\sin nE}{nE} - \frac{1}{2} \left(\frac{\sin \frac{1}{2}nE}{\frac{1}{2}nE} \right)^2 \right) J_0(kR) \right]} \tag{7.276}$$

Third Integral

$$\iint_A x e^{-ikx} \cos nz \, dA$$

$$= \cos nE \int_0^R r^2 \int_0^{2\pi} \cos\theta \, e^{-ikr\cos\theta} \, d\theta \, dr + R^2 \int_0^{2\pi} \cos\theta \, e^{-ikR\cos\theta} \int_0^E \cos nz \, dz \, d\theta$$

$$= -2\pi i \left[\cos nE \int_0^R J_1(kr) \, r^2 \, dr + R^2 E J_1(kR) \frac{\sin nE}{nE} \right] \tag{7.277}$$

$$\boxed{K_3 = -2\pi i R^3 \left[\cos nE \, \frac{J_2(kR)}{kR} + e \, J_1(kR) \frac{\sin nE}{nE} \right]} \tag{7.278}$$

Volume Integrals

$$\iiint_V \mathbf{T}^T \mathbf{u}^* dV = \iiint_V \begin{Bmatrix} 0 \\ 1 \\ 0 \\ -(z-z_0) \\ 0 \\ x-x_0 \end{Bmatrix} u_y^* \, dV = \begin{Bmatrix} 0 \\ K_4 \\ 0 \\ K_5 \\ 0 \\ K_6 \end{Bmatrix}, \quad x_0 = 0, \quad z_0 = E-H, \quad u_y^* = e^{i(\omega t - kx)} \cos nz \tag{7.279}$$

This involves the following three volume integrals.

First Integral

$$\iiint_V e^{-ikx} \cos nz \, dV = \int_0^R r \int_0^{2\pi} e^{-ikr\cos\theta} d\theta \int_0^E \cos nz \, dz \, dr$$

$$= 2\pi \frac{\sin nE}{n} \int_0^R r J_0(kr) \, dr \tag{7.280}$$

$$\boxed{K_4 = 2\pi R^3 \frac{\sin nE}{nE} \frac{J_1(kR)}{kR} e} \tag{7.281}$$

Second Integral

$$-\iiint_V (z-z_0) e^{-ikx} \cos nz \, dV = -\int_0^R r \int_0^{2\pi} e^{-ikr\cos\theta} d\theta \int_0^E (z-z_0)\cos nz \, dz \, dr$$

$$= -2\pi \left(H \frac{\sin nE}{n} - \frac{1}{2} E^2 \frac{\sin^2 \frac{1}{2} nE}{\left(\frac{1}{2} nE\right)^2} \right) \int_0^R r J_0(kr) \, dr \tag{7.282}$$

$$\boxed{K_5 = -2\pi R^4 \left[h \frac{\sin nE}{nE} - \frac{1}{2} \left(\frac{\sin \frac{1}{2} nE}{\frac{1}{2} nE} \right)^2 \right] \frac{J_1(kR)}{kR} e^2} \tag{7.283}$$

Third Integral

$$\iiint_V x e^{-ikx} \cos nz \; dV = \int_0^R r^2 \int_0^{2\pi} \cos\theta \, e^{-ikr\cos\theta} d\theta \int_0^E \cos nz \, dz \, dr$$

$$= -2\pi i \frac{\sin nE}{n} \int_0^R r^2 J_1(kr) \, dr \qquad (7.284)$$

$$\boxed{K_6 = -2\pi i R^4 \frac{\sin nE}{nE} \frac{J_2(kR)}{kR} e} \qquad (7.285)$$

Effective Motions

$$\mathbf{u}_0^* = \mathbf{H}^{-1} \iint_A \mathbf{T}^T \mathbf{u}^* \, dA - a_0^2 \frac{G}{R^2} \mathbf{K}_{\bar{f}}^{-1} \iiint_V \mathbf{T}^T \mathbf{u}^* \, dV \quad , \qquad a_0 = \frac{\omega R}{C_S} \qquad (7.286)$$

$$= \mathbf{u}_1^* + \mathbf{u}_2^*$$

with $\mathbf{K}_{\bar{f}}^{-1} = \left(\mathbf{K}_f^0\right)^{-1}$ being the impedance matrix relative to the centroid.

Components of $\mathbf{u}_1^*$

$$u_{x1}^* = u_{z1}^* = \vartheta_{y1}^* = 0 \qquad (7.287)$$

$$\boxed{u_{y1}^* = \frac{2}{1+2e}\left[\cos nE \frac{J_1(kR)}{kR} + e\frac{\sin nE}{nE}J_0(kR)\right]} \qquad (7.288)$$

$$\boxed{R\,\vartheta_{x1}^* = \frac{-2}{\frac{1}{4}+e-e^2h+\frac{2}{3}e^3}\left[h\,e\,\cos nE \frac{J_1(kR)}{kR} + \left(h\frac{\sin nE}{nE} - \frac{1}{2}\left(\frac{\sin\frac{1}{2}nE}{\frac{1}{2}nE}\right)^2\right)e^2 J_0(kR)\right]} \quad (7.289)$$

$$\boxed{R\,\vartheta_{z1}^* = \frac{-2i}{\frac{1}{2}+2e}\left[\cos nE \frac{J_2(kR)}{kR} + e\,J_1(kR)\frac{\sin nE}{nE}\right]} \qquad (7.290)$$

Components of $\mathbf{u}_2^*$

For a cylindrical foundation embedded in homogeneous elastic half-space, the impedance matrix relative to the centroid of the contact area is of the form

$$\mathbf{K}_{\bar{f}}^0 = GR \left\{ \begin{array}{cccccc} k_{hh}^0 & 0 & 0 & 0 & -R\,k_{hr}^0 & 0 \\ 0 & k_{hh}^0 & 0 & R\,k_{hr}^0 & 0 & 0 \\ 0 & 0 & k_{vv}^0 & 0 & 0 & 0 \\ 0 & R\,k_{hr}^0 & 0 & R^2\,k_{rr}^0 & 0 & 0 \\ -R\,k_{hr}^0 & 0 & 0 & 0 & R^2\,k_{rr}^0 & 0 \\ 0 & 0 & 0 & 0 & 0 & R^2\,k_{tt}^0 \end{array} \right\} \qquad (7.291)$$

in which $k_{hh}^0, k_{hr}^0, k_{rr}^0, k_{tt}^0$ are dimensionless, complex stiffness functions that depend on the dimensionless frequency $a_0 = \omega R / C_S$ as well as on the aspect ratio e and damping ξ. The signs of the coupling terms in the above matrix have been chosen consistent with the directions of positive rotations used in this document. The stiffness coefficients relative to the centroid at a distance H above the base are then related to those at the base as follows:

$$k_{hh}^0 = k_{hh}, \quad k_{hr}^0 = k_{hr} - h\,k_{hh}, \quad k_{rr}^0 = k_{rr} - 2h\,k_{hr} + h^2 k_{hh}, \quad k_{tt}^0 = k_{tt} \tag{7.292}$$

Omitting an implied zero superscript, the inverse of the impedance matrix is the flexibility matrix

$$\mathbf{F} = \left(\mathbf{K}_f^0\right)^{-1} = \frac{1}{GR}
\begin{Bmatrix}
f_{hh} & 0 & 0 & 0 & \frac{1}{R}f_{hr} & 0 \\
0 & f_{hh} & 0 & -\frac{1}{R}f_{hr} & 0 & 0 \\
0 & 0 & f_{vv} & 0 & 0 & 0 \\
0 & -\frac{1}{R}f_{hr} & 0 & \frac{1}{R^2}f_{rr} & 0 & 0 \\
\frac{1}{R}f_{hr} & 0 & 0 & 0 & \frac{1}{R^2}f_{rr} & 0 \\
0 & 0 & 0 & 0 & 0 & \frac{1}{R^2}f_{tt}
\end{Bmatrix}
\tag{7.293}$$

where

$$f_{hh} = \frac{k_{rr}}{\Delta}, \quad f_{rr} = \frac{k_{hh}}{\Delta}, \quad f_{hr} = \frac{k_{hr}}{\Delta}, \quad \Delta = k_{hh}k_{rr} - k_{hr}^2, \quad f_{vv} = k_{vv}^{-1}, \quad f_{tt} = k_{tt}^{-1} \tag{7.294}$$

$$\mathbf{u}_2^* = -a_0^2
\begin{Bmatrix}
f_{hh} & 0 & 0 & 0 & \frac{1}{R}f_{hr} & 0 \\
0 & f_{hh} & 0 & -\frac{1}{R}f_{hr} & 0 & 0 \\
0 & 0 & f_{vv} & 0 & 0 & 0 \\
0 & -\frac{1}{R}f_{hr} & 0 & \frac{1}{R^2}f_{rr} & 0 & 0 \\
\frac{1}{R}f_{hr} & 0 & 0 & 0 & \frac{1}{R^2}f_{rr} & 0 \\
0 & 0 & 0 & 0 & 0 & \frac{1}{R^2}f_{tt}
\end{Bmatrix}
\begin{Bmatrix}
0 \\
2\pi\dfrac{\sin nE}{nE}\dfrac{J_1(kR)}{kR}e \\
0 \\
-2\pi R\left[h\dfrac{\sin nE}{nE} - \dfrac{1}{2}\left(\dfrac{\sin\frac{1}{2}nE}{\frac{1}{2}nE}\right)^2\right]\dfrac{J_1(kR)}{kR}e^2 \\
0 \\
-2\pi\mathrm{i}\,R\dfrac{\sin nE}{nE}\dfrac{J_2(kR)}{kR}e
\end{Bmatrix}
\tag{7.295}$$

Hence,

$$\boxed{u_{y2}^* = -2\pi a_0^2\left\{f_{hh}\frac{\sin nE}{nE}\frac{J_1(kR)}{kR}e + f_{hr}\left[h\frac{\sin nE}{nE} - \frac{1}{2}\left(\frac{\sin\frac{1}{2}nE}{\frac{1}{2}nE}\right)^2\right]\frac{J_1(kR)}{kR}e^2\right\}} \tag{7.296}$$

$$\boxed{R\,\vartheta_{x2}^* = 2\pi a_0^2\left\{f_{hr}\frac{\sin nE}{nE}\frac{J_1(kR)}{kR}e + f_{rr}\left[h\frac{\sin nE}{nE} - \frac{1}{2}\left(\frac{\sin\frac{1}{2}nE}{\frac{1}{2}nE}\right)^2\right]\frac{J_1(kR)}{kR}e^2\right\}} \tag{7.297}$$

$$R\vartheta_{z2}^* = 2\pi a_0^2 \left\{ i f_{tt} \frac{\sin nE}{nE} \frac{J_2(kR)}{kR} e \right\} \tag{7.298}$$

Having obtained the effective motions at the centroid, we must transfer these to the base via the rigid-body motion condition, which in this case is simply

$$u_y = \left(u_{y1}^* + u_{y2}^* \right) + h R \left(\vartheta_{x1}^* + \vartheta_{x2}^* \right), \qquad \vartheta_x = \vartheta_{x1}^* + \vartheta_{x2}^*, \qquad \vartheta_z = \vartheta_{z1}^* + \vartheta_{z2}^* \tag{7.299}$$

These are then the total, effective motions that must be applied at the base of the soil springs underneath the structure.

7.5 Simple Models for Time-Varying, Inelastic Soil Behavior

Many engineering materials, and especially soils, show a distinct nonlinear inelastic behavior when loaded dynamically, and especially so during earthquakes. In most cases, the volume of material under examination, either in situ or in the laboratory, is subjected to a general 3-D state of time-varying stresses. Thus, an adequate description of its inelastic behavior requires involved constitutive theories and inelastic models, many of which are still under investigation today. Nonetheless, it is frequently possible to describe the salient characteristics of the behavior of the material with simple 1-D models that consider only one stress and strain component at a time, for example in dynamic soil amplification of shear waves. A few such simple models are presented next.

The starting point of all 1-D inelastic models is the so-called *backbone* or virgin curve, which establishes the nonlinear stress–strain relationship $\tau = f(\gamma)$ for monotonic loading, starting from an initial unperturbed (i.e., virgin) condition, as shown in Figure 7.17. If $f(\gamma)$ does not depend on time, that is, if the resistance does not depend on the speed of loading or unloading, as is the case with some dry sands, the material is said to be *rate-independent*. In the backbone function, the ratio $G_s = \tau / \gamma$ is the *secant modulus*, while the local derivative $G_t = d\tau / d\gamma$ is the *tangent modulus*. In addition, for dynamic deformations it is necessary to establish an unloading and reloading rule, which defines the stress–strain path for nonmonotonic loading. The most widely used is the so-called Masing rule described later on in these notes.

7.5.1 Inelastic Material Subjected to Cyclic Loads

When an inelastic, rate-independent material is subjected to cyclic shear deformations between equal limits $\pm \gamma_1$ of strain, that is, when the shearing strain changes harmonically as $\gamma = \gamma_1 \sin \omega t$, the stress–strain function may often form a closed hysteresis loop, in which case the shearing stresses will also be cyclical although not necessarily varying in a sinusoidal fashion. As the material is loaded back and forth, it dissipates energy in the form of heat. Clearly, the energy E_d consumed in one cycle of motion is given by the area within the hysteresis loop.

Now, when an SDOF system with stiffness k, mass m, natural frequency ω_n and fraction of viscous, critical damping ξ_v executes harmonic oscillations with frequency ω and displacement amplitude A, it also dissipates energy, which can be shown to be of the form

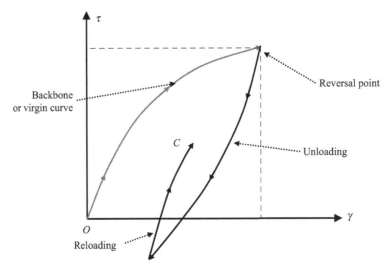

Figure 7.17. Hysteresis due to loading, unloading and reloading.

$$E_d = 4\pi \, \xi_v E_s \, \frac{\omega}{\omega_n} \tag{7.300}$$

with $E_s = \frac{1}{2} k \, A^2$ being the maximum elastic energy stored in the spring. As can be seen, this energy is directly proportional to the frequency, which is another way of saying that it depends on the speed at which deformations take place. By contrast, in a system with rate-independent material or *hysteretic damping* ξ_h, the dissipated energy in each cycle of motion does not depend on frequency. Then again, the transfer function of an SDOF system with hysteretic damping can be demonstrated to differ negligibly from that of a similar, viscously damped system provided that the damping of the latter is chosen so that it dissipates the same amount of energy at resonance, which will be the case if $\xi_v = \xi_h$. Hence, the effective hysteretic damping of a rate-independent, nonlinear material can be obtained from the ratio of energy dissipated to energy stored in one cycle of loading, that is,

$$\xi = \frac{1}{4\pi} \frac{E_d}{E_s} \tag{7.301}$$

To obtain this area, we begin by shifting the origin of the reloading curve to the lower left reversal point and scaling the coordinates by a factor of 0.5, as shown in Figure 7.18. This shift and scaling can be accomplished by means of the auxiliary variables $x = \frac{1}{2}(\gamma + \gamma_1)$ and $y = \frac{1}{2}(\tau + \tau_1)$, in terms of which the reloading curve is simply $y = f(x)$.

With reference to Figure 7.18, the energy dissipated can be computed as

$$E_d = 4 \left\{ 2 \int_0^{\gamma_1} f(\gamma) d\gamma - \tau_1 \gamma_1 \right\}, \qquad E_s = \tfrac{1}{2} G_s \, \gamma_1^2 = \tfrac{1}{2} \tau_1 \, \gamma_1 \tag{7.302}$$

in which G_s is the secant shear modulus, and the factor 4 results from the scaling of coordinates. Hence, the effective hysteretic damping of a nonlinear material with backbone $\tau = f(\gamma)$ is

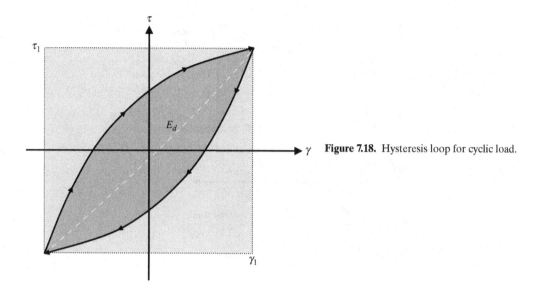

Figure 7.18. Hysteresis loop for cyclic load.

$$\xi(\gamma_1) = \frac{2}{\pi}\left(\frac{2\int_0^{\gamma_1} f(\gamma)d\gamma}{\tau_1 \gamma_1} - 1 \right) \tag{7.303}$$

a result that we shall use in the ensuing.

7.5.2 Masing's Rule

Perhaps the simplest and most extensively used inelastic model is the formula given by Masing.[14] We emphasize, however, that its success owes more to the simplicity of the formula itself than to the realism with which it models actual inelastic behavior, especially in the case of soils. Still, it is a useful idealization, especially for the representation of the shear deformation of rate-independent, dry, granular soils. Masing's law can be formulated as follows.

Consider a rate independent, inelastic material subjected to time-varying deformations, and let $\tau = f(\gamma)$ be the *backbone* (or virgin) loading curve describing the initial relationship between the stress τ, applied monotonically from an unstressed state, and the observed deformation γ. Assume also that the backbone is *odd* (i.e., antisymmetric) in the deformation parameter, in which case $f(-\gamma) = -f(\gamma)$, which is almost certainly true for shear. This material is subjected to a sequence of loading and unloading episodes of arbitrary intensity. In Masing's model, the stress–strain relationship during an unloading or reloading episode is defined by the formula

$$\boxed{\tfrac{1}{2}(\tau - \tau_r) = f\left(\tfrac{1}{2}[\gamma - \gamma_r]\right)} \tag{7.304}$$

---

[14] G. Masing, "Eigenspannungen und Verfestigung beim Messing," in Proceedings of the 2nd International Congress of Applied Mechanics, Zurich, 1926, pp. 332–335.

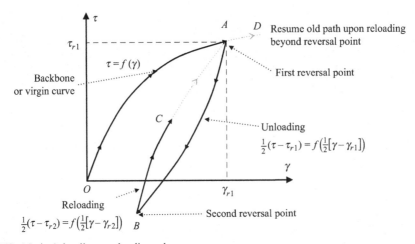

Figure 7.19. Masing's loading – unloading rule.

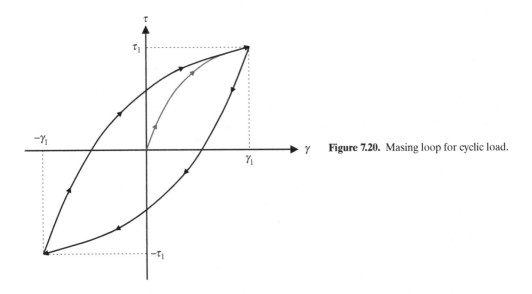

Figure 7.20. Masing loop for cyclic load.

in which τ_r, γ_r are the coordinates of the point of last reversal of loading. In general, this model requires keeping track of the history of some or all of the reversal points, so that when an unloading or reloading episode intersects a previously taken path, that previous path is reestablished as if no unloading or reloading had taken place. This is shown schematically in Figure 7.19.

In the case of *cyclic* loads which oscillate between equal and opposite limits $\pm\tau_1, \pm\gamma_1$ (Figure 7.20), the load-deformation path forms a closed curve referred to as the *hysteresis loop*, which is defined by the two branches

$$\frac{\tau + \tau_1}{2} = f\left(\frac{\gamma + \gamma_1}{2}\right) \qquad \text{for } reloading \text{ (upper branch)} \qquad (7.305)$$

$$\frac{\tau - \tau_1}{2} = f\left(\frac{\gamma - \gamma_1}{2}\right) \qquad \text{for } \textit{unloading} \text{ (lower branch)} \tag{7.306}$$

The area contained by the hysteresis loop is a measure of the energy dissipated, E_d, in one cycle of loading. Observe that this energy is independent of the rate of strain, that is, of the velocity of deformation, so it is quite unlike viscous damping. It is a function, however, of the maximum deformation, which is the hallmark of *hysteretic damping*.

In the case of a SDOF system, the energy dissipated can be used to define an equivalent fraction of linear hysteretic damping according to the formula introduce earlier on

$$\xi = \frac{1}{4\pi} \frac{E_d}{E_s} \tag{7.307}$$

in which $E_s = \frac{1}{2} G_s \gamma_1^2 = \frac{1}{2}\tau_1\gamma_1$, and $G_s = \tau_1/\gamma_1$ is the secant modulus at maximum deformation.

7.5.3 Ivan's Model: Set of Elastoplastic Springs in Parallel

We show in the ensuing that it is possible to simulate any rate-independent, inelastic material obeying Masing's law by means of a set of elastoplastic springs in parallel. The advantage of the Ivan model is that it is much easier to program a set of elastoplastic springs in parallel than to implement code for a nonlinear backbone where one must keep track of the many reversal points.

Consider a monotonic, strain-softening backbone curve $\tau = \tau(\gamma)$ in the form of a polygonal line divided in to $N + 1$ straight segments, as shown in Figure 7.5. Ideally, N should be taken as large as possible for an accurate representation of an actual continuous curve by means of a polygonal line. This nonlinear system can be represented with $N + 1$ elastoplastic springs in parallel, as shown on the right in Figure 7.21. Let

Stiffness of jth spring:	k_j
Yield stress of jth spring:	τ_j
Yield strain of jth spring:	γ_j
Secant shear modulus:	$G_j = \tau_j / \gamma_j$
Tangent shear modulus:	$g_j = (\tau_j - \tau_{j-1})/(\gamma_j - \gamma_{j-1})$
Modulus degradation curve:	G_j / G_0

Clearly, for a set of elastoplastic springs in parallel, the tangent stiffnesses are

$g_0 = k_0 + k_1 + \cdots k_N = G_0$ small strain shear modulus

$g_1 = k_1 + \cdots k_N$ first spring has yielded

$\cdots$

$g_N = k_N$ all but last spring have yielded

It follows that

$$k_j = g_j - g_{j+1} \tag{7.308}$$

which in the limit of a continuous backbone curve converges to $k(\gamma) = -\dfrac{d^2\tau}{d\gamma^2}\Delta\gamma$. On the other hand,

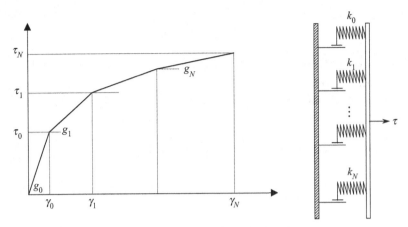

Figure 7.21. Nonlinear stress–strain simulated with set of elastoplastic springs.

$$g_j = \frac{G_j\gamma_j - G_{j-1}\gamma_{j-1}}{\gamma_j - \gamma_{j-1}} = \frac{G_j(\gamma_j / \gamma_{j-1}) - G_{j-1}}{(\gamma_j / \gamma_{j-1}) - 1} \qquad (7.309)$$

Choosing arbitrarily $\gamma_j = z^j\gamma_0$, in which $z > 1$ is any appropriate real number, then

$$\log_{10}\frac{\gamma_j}{\gamma_0} = j \log_{10} z \qquad (7.310)$$

and

$$\frac{g_j}{G_0} = \frac{z \, G_j / G_0 - G_{j-1} / G_0}{z - 1} \qquad (7.311)$$

with which the spring stiffnesses can be computed at logarithmically spaced strain intervals, using the shear modulus reduction curves and the shear modulus at low strain. For example, using 50 strain points in each strain decade, the required z is

$$\log_{10}\frac{\gamma_{j+50}}{\gamma_j} = \log_{10} 10 = (j+50-j) \log_{10} z = 50 \log_{10} z \qquad (7.312)$$

which yields $z = \exp(1/50) = 1.02020$. Hence, if $\gamma_0 = 10^{-6}$ is the small strain elastic limit, then $\gamma_1 = (1.0202)(10)^{-6}$ is the yield strain of spring 1, and so forth. In this case, the backbone curve contains 300 points in the range from $\gamma_0 = 10^{-6}$ to $\gamma = 10^0 = 1$, so this is the number of elastoplastic springs in the model.

7.5.4 Hyperbolic Model

The hyperbolic model is one of the most widely used for dry, cohesionless soils because of its simplicity. It depends on just two parameters, namely the small strain shear modulus G_0 and the shear strength τ_{max}. With reference to the Mohr–Coulomb strength law, we remark in passing that $\tau_{max} = \sigma_0 \tan\phi$, where ϕ is the friction angle, $\sigma_0 = \frac{1}{3}\sigma_v(1+2K_0)$ is the

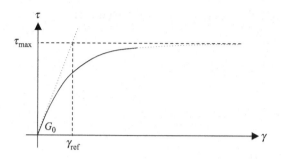

Figure 7.22. Hyperbolic stress–strain model.

effective mean confining pressure. σ_v is the vertical stress at some depth in the soil, and $K_0 \sim 0.5$ is the coefficient of lateral earth pressure at rest. In terms of this pressure, the small strain shear modulus is often approximated by a power law of the form $G_0 = C\sqrt{\sigma_0}$, where C is some appropriate constant. For example, for dry and remodeled sands, Laird and Stokoe[15] proposed a formula that we approximate here as $G_0\,[\text{kPa}] = 10^5\sqrt{\sigma_0}\,[\text{bar}]$, with σ_0 given in bars (1 bar $= 10^5$ Pa, which is almost the same as one atmospheric unit; alternatively 1 bar $\approx$ 1 kgf/cm^2). Hence, the small strain shear modulus increases roughly in proportion to the square root of the depth from the free surface, and the shear wave velocity as the fourth root of that depth.

The backbone for the hyperbolic model is given by

$$\tau = \frac{\gamma}{1/G_0 + \gamma/\tau_{\max}} \tag{7.313}$$

or

$$\frac{\tau}{\tau_{\max}} = \frac{\gamma}{\gamma + \tau_{\max}/G_0} = \frac{\gamma}{\gamma + \gamma_{\text{ref}}} \tag{7.314}$$

where $\gamma_{\text{ref}} = \tau_{\max}/G_0$ is a reference strain defined by the shear strength and the initial shear modulus[16], see Figure 7.22. The shear modulus reduction factor is then

$$\frac{G_s}{G_0} = \frac{1}{1 + \gamma/\gamma_{\text{ref}}} \tag{7.315}$$

so at the reference strain, the shear modulus has already degraded by 50%. On the other hand, as shown in the ensuing, by evaluating the energy dissipated in one cycle of deformation we obtain the fraction of hysteretic damping:

$$\xi = \frac{2}{\pi \chi^2}\left[\chi(2+\chi) - 2(1+\chi)\ln(1+\chi)\right], \qquad \chi = \frac{\gamma}{\gamma_{\text{ref}}} \tag{7.316}$$

[15] J. P. Laird and K. H. Stokoe, "Dynamic properties of remodeled and undisturbed soil samples tested at high confining pressures," *Geotechnical Engineering Report GR93-6*, Electrical Power Research Institute, Palo Alto, CA, 1993.

[16] B. O. Hardin and V. P. Drnevich, "Shear modulus and damping in soils: Design equations and curves," *J. Soil Mech. Found. Div*, ASCE 98(7), 1972, 667–692.

This is obtained by considering cyclic loads and evaluating the total area under the upper (reloading) curve, namely

$$
\begin{aligned}
A &= \int_0^{\gamma_1} \tau(\gamma)\, d\gamma = \tau_{\max} \int_0^{\gamma_1} \frac{\gamma/\gamma_{\text{ref}}}{1+\gamma/\gamma_{\text{ref}}}\, d\gamma \\
&= \tau_{\max}\gamma_{\text{ref}} \left[\frac{\gamma_1}{\gamma_{\text{ref}}} - \ln\left(1+\frac{\gamma_1}{\gamma_{\text{ref}}}\right) \right] \\
&= \tau_1 \gamma_1 \left(\frac{\gamma_{\text{ref}}}{\gamma_1}\right)^2 \left(1+\frac{\gamma_1}{\gamma_{\text{ref}}}\right)\left[\frac{\gamma_1}{\gamma_{\text{ref}}} - \ln\left(1+\frac{\gamma_1}{\gamma_{\text{ref}}}\right) \right]
\end{aligned}
\tag{7.317}
$$

Considering also that the area above the curve is $B = \tau_1\gamma_1 - A$, the energy dissipated is thus

$$
E_d = 4\{A - B\} = 4\{2A - \tau_1\gamma_1\}
\tag{7.318}
$$

Also, the maximum elastic energy stored is $E_s = \frac{1}{2}\tau_1\gamma_1$, so the effective damping is

$$
\begin{aligned}
\xi(\gamma_1) &= \frac{1}{4\pi}\frac{E_d}{E_s} = \frac{2A - \tau_1\gamma_1}{\pi\frac{1}{2}\tau_1\gamma_1} = \frac{2}{\pi}\left(\frac{2A}{\tau_1\gamma_1} - 1\right) \\
&= \frac{2}{\pi}\left\{ 2\left(\frac{\gamma_{\text{ref}}}{\gamma_1}\right)^2 \left(1+\frac{\gamma_1}{\gamma_{\text{ref}}}\right)\left[\frac{\gamma_1}{\gamma_{\text{ref}}} - \ln\left(1+\frac{\gamma_1}{\gamma_{\text{ref}}}\right) \right] - 1 \right\}
\end{aligned}
\tag{7.319}
$$

Generalizing this expression for an arbitrary maximum strain γ, and introducing the shorthand $\chi = \gamma/\gamma_{\text{ref}}$, this can also be written as

$$
\begin{aligned}
\xi &= \frac{2}{\pi}\left\{ \frac{2(1+\chi)}{\chi^2}[\chi - \ln(1+\chi)] - 1 \right\} \\
&= \frac{2}{\pi\chi^2}\left\{ \chi(2+\chi) - 2(1+\chi)\ln(1+\chi) \right\}
\end{aligned}
\tag{7.320}
$$

which agrees with the expression give earlier.

7.5.5 Ramberg–Osgood Model

The Ramberg–Osgood virgin curve is defined by the expression

$$
\gamma = \frac{\tau}{G_0}\left[1 + \alpha\left(\frac{\tau}{\tau_1}\right)^\beta \right]
\tag{7.321}
$$

where α, β are dimensionless parameters, G_0 is the small strain stiffness or *elastic modulus*, and τ_1 is an arbitrary reference stress, for example the largest stress attained in the loading cycle. The Ramberg–Osgood equation is illustrated in Figure 7.23 for nonnegative values of α, β. The combination $\alpha > 0$ and $-1 < \beta < 0$ would correspond to an idealization of a strain-hardening material, which is not quite relevant for soils undergoing shear. Cases for which $\alpha < 0$ and/or $\beta \le -1$ do not match any real materials, so they need not be considered herein.

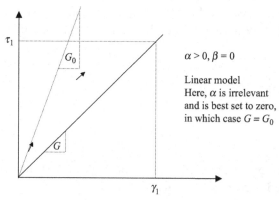

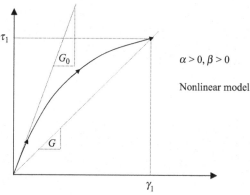

Figure 7.23. Ramberg–Osgood models.

The slope of the virgin curve at some point is referred to as the *tangent stiffness*

$$G_t = \frac{d\tau}{d\gamma} = \frac{G_0}{1 + \alpha(\beta+1)(\tau/\tau_1)^\beta} \tag{7.322}$$

On the other hand, the ratio between the force and displacement is the secant stiffness

$$G_s = \tau/\gamma \tag{7.323}$$

in terms of which we define the *strain softening factor*

$$\boxed{\frac{G_s}{G_0} = \frac{1}{1 + \alpha(\tau/\tau_1)^\beta}} \tag{7.324}$$

In particular, if $\tau = \tau_1$, then

$$\frac{G_{s1}}{G_0} = \frac{1}{1+\alpha} \tag{7.325}$$

It follows that α is a measure of the ratio between the small strain stiffness and secant stiffnesses at maximum deformation.

We proceed now to assess the equivalent damping, but inasmuch as the Ramberg–Osgood equation actually defines the inverse of the backbone function, namely $\gamma = f^{-1}(\tau)$, it is best to obtain the area enclosed by the hysteresis loop by integration along the vertical axis, which differs slightly from the procedure we developed earlier for the hyperbolic model. The area in the loop is then

$$
\begin{aligned}
E_d &= 4\left\{\tau_1\gamma_1 - 2\int_0^{\tau_1} x\,dy\right\} = 4\left\{\tau_1\gamma_1 - \frac{2}{k_0}\int_0^{\tau_1}\left(y + \alpha\frac{y^{\beta+1}}{\tau_1^{\beta}}\right)dy\right\} \\
&= 4\left[\tau_1\gamma_1 - \frac{\tau_1^2}{G_0}\left(1 + \frac{2\alpha}{\beta+2}\right)\right]
\end{aligned}
\tag{7.326}
$$

(The factor 4 is again the scaling factor in the transformation of coordinates). Now, from the Ramberg–Osgood equation evaluated at τ_1, γ_1, we obtain

$$
\gamma_1 = \frac{\tau_1}{G_0}(1+\alpha) \equiv \frac{\tau_1}{G_{s1}}
\tag{7.327}
$$

in which case the dissipated energy can be written as

$$
E_d = 4\tau_1\gamma_1\left[1 - \frac{G_{s1}}{G_0}\left(1 + \frac{2\alpha}{\beta+2}\right)\right]
\tag{7.328}
$$

But $\alpha = G_0/G_{s1} - 1$, in which case after brief algebra, we obtain

$$
E_d = 4\tau_1\gamma_1\left(1 - \frac{G_{s1}}{G_0}\right)\left(\frac{\beta}{\beta+2}\right)
\tag{7.329}
$$

On the other hand, if we define the equivalent linear hysteretic model so that its stiffness matches the secant stiffness at maximum deformation, then the maximum elastic energy stored in such equivalent system is $E_s = \frac{1}{2}\tau_1\gamma_1$. As we already know, the ratio between dissipated energy and energy stored is a measure of damping. Hence, the equivalent fraction of linear hysteretic damping is

$$
\boxed{\xi = \frac{E_d}{4\pi E_s} = \frac{2}{\pi}\left(1 - \frac{G_s}{G_0}\right)\left(\frac{\beta}{\beta+2}\right)}
\tag{7.330}
$$

Finally, the parameters for the Ramberg–Osgood nonlinear model as inferred experimentally from the ultimate degraded state τ_1, γ_1 are

$$
\alpha = \frac{1 - G_{s1}/G_0}{G_{s1}/G_0} \quad \text{and} \quad \beta = \frac{\frac{1}{2}\pi\xi}{1 - G_{s1}/G_0 - \frac{1}{2}\pi\xi}
\tag{7.331}
$$

in which G_{s1}/G_0 is the ratio between secant and tangent shear modulus at the current maximum deformed state. In practice, this ratio, as well as the damping, would be obtained in the laboratory by subjecting soil samples to cyclic deformations of varying amplitudes, and measuring in both the dissipated energy as well as the loss of stiffness (softening) at the maximum imposed strain.

For example, consider an experiment in which the shear modulus degradation and the damping were found respectively to be $G_{s1}/G_0 = 0.7$ and $\xi = 0.07$ (i.e., 7% damping) at a maximum strain $\gamma_1 = 0.01$ (i.e., at 1% strain). From the above formulas, we then find $\alpha = 3/7$ and $\beta = 1.16$. Also, the maximum stress is $\tau_1 = G_{s1}\gamma_1 = 0.7G_0\gamma_1 = 0.007G_0$. Thus, the Ramberg–Osgood virgin curve is

$$\gamma = \frac{\tau}{G_0}\left[1 + \frac{3}{7}\left(\frac{\tau}{\tau_1}\right)^{1.16}\right] \tag{7.332}$$

For this case, the soil degradation can be found to be given, albeit in inverse form, by

$$\frac{\gamma}{\gamma_1} = \left[\frac{1 - \dfrac{G_s}{G_0}}{0.648\left(\dfrac{G_s}{G_0}\right)^{2.16}}\right]^{\frac{1}{1.16}}, \quad \xi = \frac{2}{\pi}\left(1 - \frac{G_s}{G_0}\right)\left(\frac{\beta}{\beta+2}\right) = 0.235\left(1 - \frac{G_s}{G_0}\right) \tag{7.333}$$

Figure 7.24 shows the degradation and damping curves for this particular example.

7.6 Response of Soil Deposits to Blast Loads

The response of a soil medium to a blast load is usually analyzed in terms of the solution for compressive, impulsive point sources acting on or within the medium. These are referred to as Green's functions, or fundamental solutions. The response to a spatially distributed source, or to a source that is not impulsive, is then obtained by convolution with such Green's functions. This presupposes the applicability of the principle of superposition, which may not be valid in the immediate vicinity of the source because very large strains may be elicited in that region. In addition, it requires knowledge of the Green's functions, which are available in closed-form only for very few soil configurations. Thus, in most cases seminumerical or purely numerical solutions are required, such as the Boundary Element Method (BEM), the Thin-Layer Method (TLM), the Finite Element Method (FEM), or the Finite Differences Method (FDM). Still, useful results can often be obtained with the fundamental solutions available for idealized media, such as those for the Garvin Problem for an impulsive line of pressures acting within an elastic half-space, or a blast load contained by a spherical cavity in an infinite space. An extensive set of such solutions is given in Kausel (2006).[17] Treatises on blast loads and ground-borne vibrations exist as well.[18]

7.6.1 Effects of Ground-Borne Blast Vibrations on Structures

Frequency Effects

As we know, constructions respond to seismic loads most strongly in the vicinity of their natural frequencies of vibration. On the other hand, a simple rule of thumb indicates that the fundamental mode of a structure can be estimated as $f_n = 10/N$ [Hz], where N is the number of stories. This means that the resonant frequency of most buildings is on the

[17] E. Kausel, *Fundamental Solution in Elastodynamics: A Compendium* (New York: Cambridge University Press, 2006).

[18] C. H. Dowding, *Construction Vibrations* (Upper Saddle River, NJ: Prentice Hall, 1996).

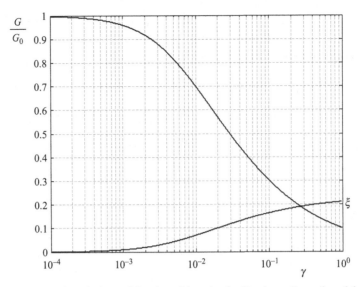

Figure 7.24. Shear modulus degradation and fraction of damping for Ramberg–Osgood model.

order of 10 Hz or less. By contrasts, blast loads such as those elicited by quarry blasts have relatively short durations. Typically, a single charge may produce overpressures that persist some 10 to 20 ms, and such blast loads will contain energy up to frequencies of about 50 to 100 Hz. However, many quarry blasts are produced by a set of charges which are detonated in series according to some delay pattern, and these will have longer durations and their frequency content will be correspondingly lower. Thus, an important consideration in the assessment of blast effects on structures is the shock or *response spectrum* of the blast load, which gives indications on its frequency content.

Distance Effects

The waves elicited in the ground by blast loads diminish in intensity as they travel away from the source, due to geometric spreading, reflection–refraction, and material attenuation. Since most explosive sources are detonated close to the ground surface, perhaps as much as two thirds of the energy may be spread in the form of waves that remain close to the surface, such as Rayleigh waves, which do not decay nearly as fast as other waves. However, motion intensities along some directions could also decrease at an even slower rate due to focusing of energy as well as other path effects related to charge weight per delay, that is, yield, charge alignment ($\rightarrow$ beam forming), or soil layers, in which case recorded ground motions will typically exhibit strong spatial variability, and they may vary also substantially from blast to blast. By and large, ground motions are assumed to be proportional to the explosive load, that is, to the *charge weight.*

The decay of amplitude with distance depends strongly on the wave contents. Rayleigh waves decay roughly in proportion to the square root of the distance or range, while body waves decay in direct proportion to that distance. Since blast loads produce neither pure Rayleigh waves nor pure body waves, and damping causes high-frequency waves to decay faster with distance than low frequency waves, this means that the frequency content of waves (i.e., the mean of their Fourier spectra) diminishes with distance to the source, and does so faster for velocities than for displacements. This can be explained as follows.

Consider a plane wave propagating in a material with a fraction of hysteretic damping ξ. Such a wave can be expressed as

$$u = A\,e^{i\omega[t-x/c(1-i\xi)]} = A\,e^{i\omega(t-x/c)}e^{-\xi\omega x/c}, \qquad |u| = |A|\,e^{-\xi\omega x/c} \tag{7.334}$$

$$\dot{u} = i\,\omega A\,e^{i\omega(t-x/c)}e^{-\xi\omega x/c}, \qquad |\dot{u}| = \omega|A|\,e^{-\xi\omega x/c} = \omega|u| \tag{7.335}$$

Clearly, for any given distance x, the wave decays exponentially with frequency. Suppose further that the wave consists of many frequencies whose amplitude is $A(\omega)$. Hence, the root mean square displacement and the root mean square velocity, normalized by their respective values at $x = 0$ are

$$\bar{u}^2 \sim \frac{\int_0^\infty |A(\omega)|^2\,e^{-2\omega x/c}d\omega}{\int_0^\infty |A(\omega)|^2\,d\omega}, \qquad \bar{v}^2 \sim \frac{\int_0^\infty \omega^2\,|A(\omega)|^2\,e^{-2\xi\omega x/c}d\omega}{\int_0^\infty \omega^2\,|A(\omega)|^2\,d\omega} \tag{7.336}$$

Now, because of the ω^2 factor in the integrals for the velocity, the mean square velocity occurs at a higher frequency than the mean square displacement, so the exponential weight in the numerator will have a greater reduction effect with x on the velocity than on the displacement. On the other hand, the root mean square displacement or velocity is a proxy for the peak ground displacement or velocity. Hence, particle velocity decays faster with distance than particle displacement, and ground accelerations decay even faster still, which provides justification to the usual custom of employing cube-root scaling to assess damage due to blast loads. In a logarithmic plot with consistent units, such a scaling shows up as straight line relating peak ground velocity versus epicentral distance with a negative slope of $-\frac{1}{3}$.

Structural Damage

Damage in structures relates to strains and stresses produced by waves that are transferred from the ground to the buildings. Now, for a simple wave of the form $u = f(x-ct)$, which travels with celerity c, strains will be of the form $\varepsilon = \frac{\partial}{\partial x}u = -1/c\,\frac{\partial}{\partial t}u = -\dot{u}/c$. Thus, strains at a point are proportional to particle velocity, and inversely proportional to the wave speed. In most building materials, yielding or rupture does not occur until strains exceed a threshold of about $\varepsilon > 10^{-3} = 0.001$. Making the rather crude assumption that plaster has a dilatational wave speed on the order of $c = 100$ m/s, damage is not likely to occur before the wall is subjected to waves with peak particle velocity in excess of $\dot{u} = c\varepsilon = 100 \times 10^{-3} \sim 0.10$ m/s, that is, some 100 mm/s.

While this simple relationship no longer holds for more complex wave fields, where reflection and refraction can substantially increase (or decrease) motion intensities, it is customary to document the presence or absence of damage in structures in conjunction with the peak ground velocity. Indeed, statistical observations have shown that structures generally do not sustain damage unless the peak ground velocity elicited by blast loads exceeds a threshold of about 50 mm/s (2 in/s), even if vibrations of lesser intensity can be the source of great alarm to building occupants. In fact, building strains caused by ambient vibrations due to wind or vehicular traffic, or the internal traffic of occupants can easily produce structural vibrations on the order of some 10 mm/s or more. Even more importantly, environmental strains and stresses caused by daily changes in temperature and humidity, and secular changes such as soil subsidence and settlements, could be even much larger. Thus, environmental factors such as these ones are likely to produce cracks in walls and floors that are often difficult to distinguish from cracks caused by ground-borne vibrations.

8 Advanced Topics

8.1 The Hilbert Transform

The Hilbert transform is a mathematical tool closely linked to the theory of Fourier transforms. As will be seen, it plays a fundamental role in the theory of causal dynamic systems and in the physical interpretation of hysteretic damping, which are the reasons we consider it here.

8.1.1 Definition

Let $f(t)$ be a real function of time or *signal*, which may or may not be causal. As you may recall, a *causal* function is one that vanishes when its argument is negative, so our function would be causal if it were to satisfy the condition $f(t) = 0$ for $t < 0$. The Fourier transform of this function together with its inverse are then

$$\tilde{f}(\omega) = \int_{-\infty}^{\infty} f(t)\, e^{-\mathrm{i}\omega t}\, dt \tag{8.1}$$

$$f(t) = \frac{1}{2\pi}\int_{-\infty}^{\infty} \tilde{f}(\omega)\, e^{\mathrm{i}\omega t}\, d\omega \tag{8.2}$$

Clearly, since $f(t)$ is a real function, it follows that $\tilde{f}(\omega)$ must exhibit complex-conjugate (i.e., Hermitian) symmetry with respect to the origin of frequencies, that is,

$$\tilde{f}(-\omega) = \tilde{f}^*(\omega) \tag{8.3}$$

with the superscript star indicating the complex conjugate. Alternatively, in terms of the real and imaginary parts,

$$\tilde{f}(\omega) = R(\omega) + \mathrm{i}\,I(\omega) \tag{8.4}$$

The complex conjugate symmetry condition requires that $R(\omega)$ be an even function, and $I(\omega)$ an odd function of frequency, that is,

$$R(-\omega) = R(\omega) \tag{8.5}$$

$$I(-\omega) = -I(\omega) \tag{8.6}$$

The *Hilbert transform* of $f(t)$, which we write symbolically as $\mathfrak{H}[f(t)]$, is defined as the inverse Fourier transform of the function obtained by multiplication of $\tilde{f}(\omega)$ by $i = \sqrt{-1}$ and by the sign function (which equals 1 if the argument is positive, -1 if it is negative, and zero if the argument is zero), that is,

$$\boxed{\hat{f}(t) = \mathfrak{H}[f(t)] = \frac{1}{2\pi} \int_{-\infty}^{\infty} \tilde{f}(\omega)\, i\, \mathrm{sgn}(\omega)\, e^{i\omega t} d\omega} \tag{8.7}$$

Inasmuch as $i\,\mathrm{sgn}(\omega)$ also exhibits complex-conjugate properties with respect to positive-negative values of ω, it follows that the product with $\tilde{f}(\omega)$ must also have this property, so $\hat{f}(t)$ must be a real function. The Hilbert transform function $\hat{f}(t)$ is also referred to as the *allied* function of $f(t)$.

8.1.2 Fourier Transform of the Sign Function

We shall now obtain the Fourier transform of the sign function that appears in the integrand of the Hilbert transform, and use the result to present an alternative definition of this transform. For this purpose, consider the Fourier transform (with PV indicating the Cauchy principal value)

$$\int_{-\infty}^{\infty} \frac{1}{t} e^{-i\omega t} dt = \mathrm{PV} \int_{-\infty}^{\infty} \frac{\cos \omega t - i \sin \omega t}{t} dt = -2i \int_{0}^{\infty} \frac{\sin \omega t}{t} dt \tag{8.8}$$

But

$$\frac{2}{\pi} \int_{0}^{\infty} \frac{\sin \omega t}{t} dt = \begin{cases} 1 & \text{if } \omega > 0 \\ 0 & \text{if } \omega = 0 \\ -1 & \text{if } \omega < 0 \end{cases} \tag{8.9}$$

as can be demonstrated by means of contour integration.[1] It follows that

$$\int_{-\infty}^{\infty} \frac{1}{t} e^{-i\omega t} dt = -i\pi\, \mathrm{sgn}(\omega) \tag{8.10}$$

The formal inverse Fourier transform is then

$$\frac{1}{2\pi} \int_{-\infty}^{\infty} i\, \mathrm{sgn}(\omega) e^{i\omega t} dt = -\frac{1}{\pi t} \tag{8.11}$$

We observe next that the Hilbert transform involves the Fourier inversion of the product of $\tilde{f}(\omega)$ and $i\,\mathrm{sgn}(\omega)$ in the frequency domain. As is well known, this operation is equivalent to a convolution of the transform of these two functions in the time domain. Hence, the Hilbert transform can be expressed as the convolution

$$\hat{f}(t) = \mathfrak{H}[f(t)] = -f * \frac{1}{\pi t} \tag{8.12}$$

That is,

$$\boxed{\hat{f}(t) = \mathrm{PV} \int_{-\infty}^{\infty} \frac{f(\tau)}{\pi(\tau - t)} d\tau} \tag{8.13}$$

---

[1] A. Papoulis, *The Fourier Integral and Its Applications* (New York: McGraw-Hill, 1962), 301.

whose inverse transform is

$$\boxed{f(t) = \text{PV} \int_{-\infty}^{\infty} \frac{-\hat{f}(\tau)}{\pi(\tau - t)} d\tau} \tag{8.14}$$

The first equation implies that $\hat{f}(t)$ is generally noncausal, even when $f(t)$ is causal. This follows from the fact that the denominator in the above convolution is nonzero for negative t.

The *analytical signal* $F(t)$ is now defined as

$$F(t) = f(t) + i\hat{f}(t) = A(t)e^{i\varphi(t)}, \qquad A(t) = \sqrt{f^2 + \hat{f}^2}, \qquad \tan\varphi = \arctan\frac{\hat{f}}{f} \tag{8.15}$$

whose associated *instantaneous angular frequency* is

$$\omega(t) = \frac{d\varphi(t)}{dt} \tag{8.16}$$

The Hilbert transform of the analytic signal is in turn

$$\hat{F}(t) = \hat{f}(t) - if(t) = -iF(t) \tag{8.17}$$

8.1.3 Properties of the Hilbert Transform

a. Linearity:

$$\mathfrak{H}[f(t) + g(t)] = \mathfrak{H}[f(t)] + \mathfrak{H}[g(t)] \tag{8.18}$$

b. Inversion property: Repeated application of the Hilbert transform recovers the original function with opposite sign, that is,

$$\mathfrak{H}[\mathfrak{H}[f(t)]] = \mathfrak{H}[\hat{f}(t)] = -f(t) \tag{8.19}$$

c. Similarity property: The Hilbert transform is invariant under scaling of the time axis, that is,

$$\mathfrak{H}[f(\alpha t)] = \hat{f}(\alpha t) \tag{8.20}$$

d. The Hilbert transform of a constant is zero. This follows directly from the convolution form of the Hilbert transform.

e. Causality: the Hilbert transform does *not* preserve causality.

f. The Hilbert transform of a derivative is the derivative of the Hilbert transform:

$$\frac{d}{dt}\mathfrak{H}[\hat{f}(t)] = \mathfrak{H}\left[\frac{d}{dt}\hat{f}(t)\right] \tag{8.21}$$

g. The Hilbert transform preserves the power of the original signal (assuming that it is square-integrable). This results from the fact that the Fourier transforms $\tilde{f}(\omega)$ and $\tilde{\hat{f}}(\omega)$ have the same amplitude $\left|\tilde{\hat{f}}(\omega)\right| = \left|\tilde{f}(\omega)\right|$ at each frequency. From Parseval's theorem, we obtain

$$\int_{-\infty}^{+\infty} f^2(t)\,dt = \int_{-\infty}^{+\infty} \hat{f}^2(t)\,dt \qquad (8.22)$$

Hence, the energy could be used to measure the accuracy of a numerical estimation of $\hat{f}(t)$.

h. A function and its allied function are orthogonal:

$$\int_{-\infty}^{+\infty} f(t)\hat{f}(t)\,dt = 0 \qquad (8.23)$$

This can be demonstrated using Parseval's theorem:

$$\int_{-\infty}^{+\infty} f(t)\,\hat{f}(t)\,dt = \tfrac{1}{2\pi}\int_{-\infty}^{+\infty} \tilde{f}^*(\omega)\,\hat{\tilde{f}}(\omega)\,d\omega = \tfrac{1}{2\pi}\int_{-\infty}^{+\infty}(R-\mathrm{i}I)(\mathrm{i}R-I)\operatorname{sgn}\omega\,d\omega$$
$$= \mathrm{i}\tfrac{1}{2\pi}\int_{-\infty}^{+\infty}(R^2+I^2)\operatorname{sgn}\omega\,d\omega = 0 \qquad (8.24)$$

i. The Hilbert transform of a product of two strong analytical signals is

$$\mathfrak{H}[F_1 F_2] = \mathfrak{H}[F_1]F_2 = F_1\mathfrak{H}[F_2] = -\mathrm{i}\,F_1 F_2, \qquad \text{so} \qquad \mathfrak{H}[F^n] = \mathfrak{H}[F]F^{n-1} = -\mathrm{i}\,F^n \qquad (8.25)$$

Examples

Consider first the unit impulse function, that is, the Dirac delta singular function $\delta(t)$. Its Fourier transform is ostensibly

$$\int_{-\infty}^{\infty} \delta(t)\,e^{-\mathrm{i}\omega t}\,dt = 1 \qquad (8.26)$$

so the Hilbert transform of this function is either of

$$\hat{\delta}(t) = \mathfrak{H}[\delta(t)] = \tfrac{1}{2\pi}\int_{-\infty}^{\infty} \mathrm{i}\operatorname{sgn}(\omega)\,e^{\mathrm{i}\omega t}\,d\omega = -\tfrac{1}{\pi t} \qquad (8.27)$$

$$\hat{\delta}(t) = \mathrm{PV}\int_{-\infty}^{\infty} \frac{\delta(\tau)}{\pi(\tau-t)}\,d\tau = -\tfrac{1}{\pi t} \qquad (8.28)$$

which agrees with the previous result. The analytical signal is then

$$D(t) = \delta(t) - \tfrac{1}{\pi t}\mathrm{i} \qquad (8.29)$$

Consider next the (noncausal, steady-state) cosine function

$$f(t) = \cos\alpha t = \tfrac{1}{2}\left(e^{\mathrm{i}\alpha t} + e^{-\mathrm{i}\alpha t}\right) \qquad (8.30)$$

whose formal Fourier transform is

$$\tilde{f}(\omega) = \tfrac{1}{2}\int_{-\infty}^{\infty}\left(e^{\mathrm{i}\alpha t} + e^{-\mathrm{i}\alpha t}\right)e^{\mathrm{i}\omega t}\,dt = \pi\left[\delta(\omega-\alpha) + \delta(\omega+\alpha)\right] \qquad (8.31)$$

Of course, the above result can be understood only in the sense of singularity functions or *distributions* (see the Chapter 9, Section 9.1). Hence,

$$\hat{\tilde{f}}(\omega) = \pi\left[\delta(\omega-\alpha) + \delta(\omega+\alpha)\right]\mathrm{i}\operatorname{sgn}(\omega)$$
$$= \mathrm{i}\,\pi\left[\delta(\omega-\alpha) - \delta(\omega+\alpha)\right] \qquad (8.32)$$

whose inverse Fourier transform is

$$\hat{f}(t) = i\tfrac{1}{2}\left(e^{i\alpha t} - e^{-i\alpha t}\right) = -\sin \alpha t \tag{8.33}$$

Hence, the Hilbert transform of $\cos \alpha t$ is $-\sin \alpha t$, and the analytical signals is $F(t) = e^{-i\alpha t}$. It also follows that the Hilbert transform of $\sin \alpha t$ is $\cos \alpha t$.

8.1.4 Causal Functions

Let $f(t)$ be a causal function. As stated earlier, a function is causal if it vanishes for negative argument, that is, if $f(t) = 0$ for $t < 0$. In terms of the inverse Fourier transform $\tilde{f}(\omega)$, the causality condition is then (with $t > 0$)

$$f(t) = \frac{1}{2\pi}\int_{-\infty}^{+\infty} \tilde{f}(\omega)e^{i\omega t}d\omega \tag{8.34}$$

$$0 = \frac{1}{2\pi}\int_{-\infty}^{+\infty} \tilde{f}(\omega)e^{-i\omega t}d\omega \tag{8.35}$$

Taking the complex conjugate of Eq. 8.35, we obtain

$$0 = \frac{1}{2\pi}\int_{-\infty}^{+\infty} \tilde{f}^*(\omega)e^{i\omega t}d\omega \tag{8.36}$$

Addition and subtraction of this result from Eq. 8.36 yields then

$$f(t) = \frac{1}{2\pi}\int_{-\infty}^{+\infty} \left[\tilde{f}(\omega) \pm \tilde{f}^*(\omega)\right]e^{i\omega t}d\omega \tag{8.37}$$

Considering in turn the sum and then the difference, we obtain for $t > 0$

$$\begin{aligned} f(t) &= \frac{1}{\pi}\int_{-\infty}^{+\infty} R(\omega)e^{i\omega t}d\omega = \frac{2}{\pi}\int_{0}^{\infty} R(\omega)\cos \omega t\, d\omega \\ &= \frac{1}{\pi}\int_{-\infty}^{+\infty} i I(\omega)e^{i\omega t}d\omega = -\frac{2}{\pi}\int_{0}^{\infty} I(\omega)\sin \omega t\, d\omega \end{aligned} \tag{8.38}$$

which follow from the complex conjugate symmetry of the real and imaginary parts with respect to the frequency. As can be seen, either the real part or the imaginary part of $\hat{f}$ alone suffice to recover the original signal f, at least for positive t. Hence, both seem to contain the same information and must, therefore, be intimately related. Before unraveling that relationship, however, let us consider first the special situation at $t = 0$, where two problems could arise. On the one hand, the function could be discontinuous there, or worse, it could exhibit a singularity (i.e., a Dirac delta function).

If the function is discontinuous, then the inverse Fourier transform will converge to the average value at the discontinuity. Inasmuch as the signal is causal, so that $f(0^-) = 0$, it follows that

$$\tfrac{1}{2}f(0^+) = \frac{1}{2\pi}\int_{-\infty}^{+\infty} \tilde{f}(\omega)d\omega = \frac{1}{\pi}\int_{0}^{\infty} R(\omega)\, d\omega \tag{8.39}$$

That is,

$$f(0^+) = \frac{2}{\pi}\int_{0}^{\infty} R(\omega)\, d\omega \tag{8.40}$$

which agrees with Eq. 8.39 when we set $t = 0$. It can further be shown (e.g. Papoulis, op. cit.) that if $f(t)$ does not possess a singularity at $t = 0$, then

$$f(0^+) = \lim_{\omega \to \infty} i\omega \tilde{f}(\omega) \tag{8.41}$$

This implies that, for large frequencies, a causal function with a discontinuity at the origin will decay asymptotically as

$$\tilde{f}(\omega) \sim \frac{f(0^+)}{i\omega} \quad \to \quad I(\omega) \sim \frac{-f(0^+)}{\omega} \tag{8.42}$$

Consider now the second integral for $f(t)$ in terms of the imaginary part $I(\omega)$. In the light of the above asymptotic behavior for the imaginary part, this integrand behaves as

$$\lim_{t \to 0} \frac{2}{\pi} \int_0^\infty -I(\omega) \sin \omega t \, d\omega \sim \lim_{t \to 0} \frac{2}{\pi} \int_0^\infty \frac{f(0^+) \sin \omega t}{\omega} d\omega = f(0^+) \tag{8.43}$$

because, as we saw earlier, $\frac{2}{\pi} \int_0^\infty \frac{\sin \omega t}{\omega} d\omega = 1$ if $t > 0$ (more precisely, here is $t = 0^+$). While not a proof, this result strongly suggests that the Fourier inversion in terms of the imaginary part $I(\omega)$ is indeed able to model the discontinuity of $f(t)$ at zero time, despite the fact that $\sin \omega t = 0$ when $t = 0$.

Suppose next that at $t = 0$, $f(t)$ has a singularity (or impulse) of the form $k\delta(t)$, where k is a real constant. This singularity contributes the constant k to $R(\omega)$ at all frequencies. In particular,

$$\tilde{f}(\infty) = R(\infty) = k \tag{8.44}$$

that is, the Fourier transform $\tilde{f}(\omega)$ attains a purely real value when $\omega \to \infty$. Inasmuch as the singularity does not at all contribute to $I(\omega)$, it follows that the Fourier inversion for $f(t)$ in terms of the imaginary part alone will fail to model such singularity. Thus, we conclude that $I(\omega)$ is less general than $R(\omega)$ in that it may have lost information about a possible impulse at $t = 0$.

8.1.5 Kramers–Kronig Dispersion Relations

Let's return now to the problem of finding a relationship between $R(\omega)$ and $I(\omega)$. From the theory of complex variables, we know that the Fourier spectrum $\tilde{f}(z)$ of a causal function $f(t)$ cannot contain any poles in the lower half-plane (where z = complex frequency), because that would violate the causality. Assuming further that there are no poles on the real axis either, and with reference to Figure 8.1, we then have by Cauchy's theorem

$$0 = \oint \frac{\tilde{f}(z) dz}{z - \omega_0} = \lim_{\eta \to 0} \int_{C_1} \frac{\tilde{f}(z) dz}{z - \omega_0} + \lim_{r2 \to \infty} \int_{C_2} \frac{\tilde{f}(z) dz}{z - \omega_0} + PV \int_{-\infty}^{+\infty} \frac{\tilde{f}(\omega) d\omega}{\omega - \omega_0} \tag{8.45}$$

in which ω_0 is an arbitrary point on the real axis, z is the complex frequency, and the integration path is as shown in Figure 8.1. The first integral on the right crosses the simple pole $z = \omega_0$, so that pole contributes only half of the residue, that is,

$$\lim_{\eta \to 0} \int_{C_1} \frac{\tilde{f}(z) dz}{z - \omega_0} = i\pi \tilde{f}(\omega_0) \tag{8.46}$$

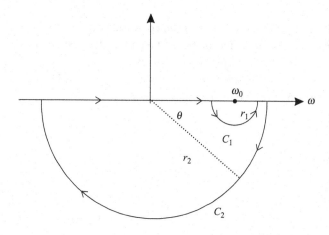

Figure 8.1. Path of contour integration.

To evaluate the second integral above, we begin by recalling that the Fourier spectrum for an infinite frequency is either zero or a purely real *constant*. Hence, the integral around the large semicircle is

$$\lim_{r_2 \to \infty} \int_{C_1} \frac{\tilde{f}(r_2 \, e^{i\theta}) \, i \, r_2 \, e^{i\theta} d\theta}{r_2 \, e^{i\theta} - \omega} = i \, \tilde{f}(\infty) \int_0^{-\pi} d\theta = -i \, \pi \, \tilde{f}(\infty) = -i \, \pi \, R(\infty) \tag{8.47}$$

The Cauchy integral is then

$$0 = i \, \pi \, \tilde{f}(\omega_0) - i \, \pi \, R(\infty) + \text{PV} \int_{-\infty}^{+\infty} \frac{\tilde{f}(x) \, dx}{x - \omega_0} \tag{8.48}$$

or in terms of the real and imaginary parts,

$$0 = i \, \pi \left[R(\omega_0) + i \, I(\omega_0) \right] - i \, \pi \, R(\infty) + \text{PV} \int_{-\infty}^{+\infty} \frac{\left[R(\omega) + i \, I(\omega) \right] d\omega}{\omega - \omega_0} \tag{8.49}$$

Hence, we conclude that

$$\boxed{R(\omega_0) = R(\infty) - \text{PV} \int_{-\infty}^{+\infty} \frac{I(\omega) \, d\omega}{\pi(\omega - \omega_0)}} \tag{8.50}$$

$$\boxed{I(\omega_0) = \text{PV} \int_{-\infty}^{+\infty} \frac{R(\omega) \, d\omega}{\pi(\omega - \omega_0)}} \tag{8.51}$$

These two equations are referred to as the *Kramers–Kronig*[2,3] dispersion relations. They establish the connection between the real and imaginary parts in the Fourier spectrum of a causal function. Moreover, if the function has no singularity at the origin, that is, if $R(\infty) = 0$, then $R(\omega), I(\omega)$ are Hilbert transform pairs. We see again that the imaginary part $I(\omega)$ contains no information about a singularity at the origin, so if it exists, its contribution to $R(\omega)$ must be added explicitly, as is done in Eq. 8.50. In general, the satisfaction

---

[2] H. A. Kramers, "La diffusion de la lumiére par les atomes," *Alti Congr. Int. Fisici*, Como, 2, 1927, 545–557.
[3] R. de L Kronig, "On the theory of dispersion of x'rays," *J. Opt. Soc. Am.*, 12, 1926, 547–557.

of the Kramers–Kronig relations by a frequency spectrum $\tilde{f}(\omega)$ constitutes the necessary and sufficient condition for the underlying time function $f(t)$ to be causal. If $R(\infty) = 0$, the spectrum is said to be *strictly proper*, otherwise it is just *simply proper*.

An alternative form of these integrals can be obtained by observing that the Hilbert transform of a *constant* parameter such as $R(\omega_0)$ or $I(\omega_0)$ must be zero, so

$$\boxed{R(\omega_0) = R(\infty) - \frac{1}{\pi} \int_{-\infty}^{+\infty} \frac{I(\omega) - I(\omega_0)}{\omega - \omega_0} d\omega} \tag{8.52}$$

$$\boxed{I(\omega_0) = \frac{1}{\pi} \int_{-\infty}^{+\infty} \frac{R(\omega) - R(\omega_0)}{\omega - \omega_0} d\omega} \tag{8.53}$$

These expressions have the advantage of not being singular at ω_0.

Additional forms of the Kramers–Kronig relationship can also be obtained by taking into account the even and odd properties of the real and imaginary parts. For a negative frequency ω_0, we would have

$$R(-\omega_0) = R(\infty) - \frac{1}{\pi} \int_{-\infty}^{+\infty} \frac{I(\omega) + I(\omega_0)}{\omega + \omega_0} d\omega = R(\omega_0) \tag{8.54}$$

$$I(-\omega_0) = \frac{1}{\pi} \int_{-\infty}^{+\infty} \frac{R(\omega) - R(\omega_0)}{\omega + \omega_0} d\omega = -I(\omega_0) \tag{8.55}$$

Adding or subtracting these expressions to/from those for positive frequency, one obtains after brief algebra

$$\boxed{R(\omega_0) = R(\infty) - \frac{2}{\pi} \int_0^\infty \frac{\omega I(\omega) - \omega_0 I(\omega_0)}{\omega^2 - \omega_0^2} d\omega} \tag{8.56}$$

$$\boxed{I(\omega_0) = \frac{2\omega_0}{\pi} \int_0^\infty \frac{R(\omega) - R(\omega_0)}{\omega^2 - \omega_0^2} d\omega} \tag{8.57}$$

We observe here that both of the limits $\lim_{\omega_0 \to \infty} \omega_0 I(\omega_0)$ and $\lim_{\omega_0 \to 0} I(\omega_0)/\omega_0$ are finite.

Minimum Phase Systems

In general, a causal function $f(t)$ with a spectrum $\tilde{f}(\omega) = R(\omega) + i I(\omega) = A(\omega) e^{i\phi}$ will have no *poles* in the lower complex frequency half-plane, that is, it is *holomorphic* there. In some cases, the spectrum will not have *zeros* there either. If so, the logarithm of the spectrum $\log \tilde{f}(\omega) = \log A(\omega) + i \phi(\omega)$ is an analytic function in the lower half-plane, the amplitude $A(\omega)$ and phase angle $\phi(\omega)$ are Hilbert transform pairs, and they too satisfy the Kramers–Kronig dispersion relations. Such systems are said to be *minimum phase*, for which the phase angle and amplitude are unambiguously related (i.e., one of them can be obtained from the other). Many dynamic systems, however, do have zeros in the lower half-plane, at which location the logarithm is singular (i.e., has a pole), so their phases and amplitudes are not uniquely related.

Time-Shifted Causality

Some noncausal functions may satisfy the condition that $f(t) = 0$ for $t < -t_0$. Such functions can be made to be causal by shifting the time axis by t_0, that is, by considering instead the causal function $f'(t) = f(t + t_0)$. The Fourier transforms of the original and shifted functions are related as $\tilde{f}(\omega) = e^{i\omega t_0}\tilde{f}'(\omega)$. While the spectrum of the time-shifted signal has no poles in the lower half-plane and is bounded there, the original signal will diverge along the negative imaginary axis $\omega = -i\eta$ when $\eta \to \infty$, because $\exp(i\omega t_0) = \exp(\eta t_0) \to \infty$. Thus, a transfer function with no poles in the lower half-plane but which grows exponentially along the negative imaginary axis can be made to be causal by multiplying that function by $\exp(-i\omega t_0)$, which amounts to a shift of the time origin to the left by t_0.

8.2 Transfer Functions, Normal Modes, and Residues

We demonstrate in the ensuing some fundamental properties of discrete and continuous dynamic systems and thereby elucidate some basic features of transfer functions as well as demonstrate some noteworthy relationships between the frequency response functions and the classical normal modes.

8.2.1 Poles and Zeros

Consider a discrete dynamic system with N degrees of freedom, which includes also continuous systems by the simple subterfuge of allowing $N \to \infty$. As we have seen in Chapter 3, Section 3.8, the transfer functions for external loads for a dynamic system follow from the inverse of the dynamic impedance matrix

$$\mathbf{H} = \mathbf{Z}^{-1} = \{H_{\alpha\beta}\}, \qquad \mathbf{Z} = \mathbf{K} + i\omega\mathbf{C} - \omega^2\mathbf{M} \tag{8.58}$$

From Cramer's rule, the elements of the inverse are then

$$H_{\alpha\beta} = (-1)^{\alpha+\beta}\frac{\Delta_{\beta\alpha}}{\Delta} \tag{8.59}$$

where $\Delta_{\beta\alpha}$ is the determinant of the submatrix that is obtained after deleting row β and column α, and Δ is the determinant of the complete system. In linear algebra, the product

$$N_{\alpha\beta} = (-1)^{\alpha+\beta}\Delta_{\beta\alpha} \tag{8.60}$$

is referred to as the cofactor. In general, $\Delta = \Delta(\omega)$ is a polynomial of order $2N$ in the frequency ω, which has $2N$ roots or poles $\Delta(p_k) = 0$. As we have already seen in Section 3.7.2, the poles can be pairs of real roots $p_k = \pm\omega_k$; pairs of complex roots $p_k = \pm a_k + ib_k$ (i.e., pairs of negative complex conjugate roots); or pairs of distinct purely imaginary roots. But whatever the type of roots, all of them are characterized by a nonnegative imaginary part, that is, $\text{Im}(p_k) \geq 0$, a condition that is necessary for the dynamic system to be both stable and causal (see Section 8.1.5). Hence, and assuming for mere notational convenience that damping is light so that there are no purely imaginary poles, then the characteristic polynomial is of the form

$$\Delta = |\mathbf{K} + i\omega\mathbf{C} - \omega^2\mathbf{M}| = (-1)^N |\mathbf{M}|\prod_{k=1}^{N}(\omega - p_k)(\omega + p_k^*) \tag{8.61}$$

where $p_k^* = \text{conj}(p_k)$. Thus, all poles lie in the upper complex frequency plane.

On the other hand, the cofactor is also a polynomial whose order can be as high as $2(N-1)$ in ω, but that often is lower. The actual degree of the polynomial depends on the combination α, β of DOF, and on the structure of the matrices. For example, a diagonal, lumped mass matrix and a narrowly banded (or sparse) stiffness matrix can cause the degree of the cofactor to fall off with the separation $|\alpha - \beta|$ and even reduce to first order. Thus, the cofactor has zeros $z_{\alpha\beta,k}$ whose plurality is at most $2(N-1)$. These zeros can again be real, complex, or purely imaginary, but unlike the poles, their imaginary part can be both positive and negative. Hence, the cofactor is of the form

$$N_{\alpha\beta}(\omega) = A_{\alpha\beta} \prod_{k=1}^{2N-2} (\omega - z_{\alpha\beta,k}) \tag{8.62}$$

with the upper limit being $2(N-1)$ or less, and $A_{\alpha\beta}$ being a constant that is irrelevant for us here.

8.2.2 Special Case: No Damping

In particular, when $\mathbf{C} = \mathbf{O}$, that is, in the absence of damping, the poles are all real and appear in positive/negative pairs $\pm\omega_n$, $n = 1, \dots N$. The cofactor, on the other hand, is characterized by a polynomial of degree $N-1$ (or less) in ω^2 and all of its coefficients are real. This implies that it can have pairs of real roots $\omega = \pm\sqrt{a}$ ($\omega^2 = a > 0$); pairs of purely imaginary roots $\omega = \pm i\sqrt{-a}$ ($\omega^2 = a < 0$); or pairs of complex conjugate roots $\omega = \pm\sqrt{a} \pm ib$. This is so because the submatrix $\mathbf{Z}_{\alpha\beta}$ whose minor determinant is $\Delta_{\alpha\beta}$, is neither symmetric nor positive definite, except when $\alpha = \beta$ (the loaded point), so the roots can indeed be complex. Hence, Eq. 8.62 will contain factors of the form

$$N_{\alpha\beta} \sim \begin{cases} (\omega - z_{\alpha\beta,k})(\omega + z_{\alpha\beta,k}) & \text{real zero} \\ (\omega - z_{\alpha\beta,k})(\omega - z^*_{\alpha\beta,k}) & \text{imag. zero} \\ (\omega - z_{\alpha\beta,k})(\omega - z^*_{\alpha\beta,k})(\omega + z_{\alpha\beta,k})(\omega + z^*_{\alpha\beta,k}) & \text{complex zero} \end{cases} \tag{8.63}$$

Moreover, it can be seen that $N_{\alpha\alpha} = 0$ is the characteristic polynomial for the constrained system that is obtained by fixing the DOF α, and that system has $N-1$ DOF. Then again from the *Interlacing Theorem* of Chapter 3, Section 3.4, we know that the eigenvalues of that system are real and interlace those of the original system. It follows that the *zeros* of the transfer function *at the loaded point* are all real and interlace the *poles*. However, for $\alpha \neq \beta$, the zeros can be purely imaginary or fully complex, and exhibit double symmetry with respect the real and imaginary frequency axes. In the presence of damping, the zeros may still be complex, but cease to be mere complex conjugates. Either way, the zeros may appear both in the upper and lower complex planes, except for the loaded point ($\beta = \alpha$), which has zeros only in the upper complex plane. Thus, the transfer function at the loaded point is always a *minimum phase system* (see the last part of Section 8.1.5), while all other points may, but need not be minimum phase.

8.2.3 Amplitude and Phase of the Transfer Function

The behavior of the transfer function in the vicinity of poles and zeros is best seen in a pole-zero diagram, assuming for greater generality that the poles and zeros are complex. The transfer function will then contain factors and divisors of the form

$$H \sim \frac{\prod(\omega - z_k)}{\prod(\omega - p_j)} = \frac{\prod r_k e^{-i\theta_k}}{\prod R_j e^{-i\phi_j}} = A(\omega) e^{-i\varphi}, \qquad \varphi = \sum \theta_k - \sum \phi_j \qquad (8.64)$$

where the r_k, θ_k are the magnitude and angle of the radial distances from the current frequency ω to the zeros ahead, and R_j, ϕ_j are the corresponding magnitude and phase for the poles ahead. If the imaginary parts are small, then the phase to the distant poles or zeros is (to a good approximation) either zero or 180 degrees except for the zero or pole in the immediate vicinity of the frequency ω in question. It can thus be seen that the phase angle of the transfer function is dominated by the rapid change in phase as the real frequency point slides underneath a pole or zero. The phase angle of the transfer function then changes rapidly by some 180 degrees, either forwards or backwards, depending on whether it passes underneath a pole or a zero. The amplitude, on the other hand, is controlled by the radial distances r_k, R_j. Thus, passing underneath a pole leads to a small value of R_j in the denominator, that is, we obtain an amplification peak, while a zero leads to a small value of r_k, and thus to a zero (or near zero) response. Vice versa, a large variation in phase angle in a transfer function reveals the presence of a pole or a zero, and these can also be distinguished from one another by the sign of the change in phase.

Again, when no damping is present, and when zeros should appear in complex conjugate pairs, it is clear that for each such pair there will be two factors in Eq. 8.64 of the form

$$(\omega - z_k)(\omega - z_k^*) = r_k e^{-i\theta_k} r_k e^{+i\theta_k} = r_k^2 \qquad (8.65)$$

This means that such pairs of zeros do not contribute at all to the phase angle, only to the response magnitude.

In some cases it may occur that a zero coincides with a pole. If so, then these two will cancel each other out, and there will be no amplification or zero at that frequency *and* at that point in space. We emphasize, however, that this is a characteristic only at the DOF involved in the transfer function. Neighboring points ("receivers") will continue to exhibit amplification at the resonant frequency, so the cancellation is strictly a *local* affair.

Example: Cantilever Shear Beam Subjected to Load at Some Arbitrary Point
Consider a continuous, homogeneous cantilever shear beam, which is free at the left end (the origin x of coordinates) and clamped on the right. It has length L, cross section A, shear modulus G, mass density ρ, and shear wave velocity C_S. We also mark N equally spaced source-receiver points at which we observe the output (α) and/or apply the unit harmonic load (β). Using the standard methods described in Chapters 4 and 5, it is not difficult to show that the transfer functions are given by

$$H_{\alpha\beta} = \frac{L}{G} \frac{\cos \xi_\alpha \theta \sin\left[(1 - \xi_\beta)\theta\right]}{\theta \cos \theta} \qquad (8.66a)$$

$$\xi_\alpha = \frac{x_\alpha}{L}, \qquad \xi_\beta = \frac{x_\beta}{L}, \qquad x_\beta \geq x_\alpha, \qquad \theta = \frac{\omega L}{C_S} \tag{8.66b}$$

and for $x_\beta < x_\alpha$, the reciprocity principle is used, that is, $H_{\alpha\beta} = H_{\beta\alpha}$. The zeros of this transfer function, which follow from $\cos\theta_j = 0$, $\theta_j = \pm\left(j - \tfrac{1}{2}\right)\pi$, are all real. The zeros, on the other hand, follow from $\cos\xi_\alpha\theta_k = 0$ and/or $\sin\left[\left(1 - \xi_\beta\right)\theta_k\right] = 0$, provided that they do not coincide with a pole. These zeros too are purely real.

We also discretize the shear beam into N linear isoparametric finite elements with either consistent mass matrix $\mathbf{M}_C$ or lumped mass matrix $\mathbf{M}_L$, whose element matrices are of the form

$$\mathbf{K} = N\frac{GA}{L}\begin{Bmatrix} 1 & -1 \\ 1 & 1 \end{Bmatrix}, \qquad \mathbf{M}_C = \frac{\rho LA}{6N}\begin{Bmatrix} 2 & 1 \\ 1 & 2 \end{Bmatrix}, \qquad \mathbf{M}_L = \frac{\rho LA}{2N}\begin{Bmatrix} 1 & 0 \\ 0 & 1 \end{Bmatrix} \tag{8.67}$$

Thus, the discrete shear beam assembled with these finite elements will have a total of N DOF. Presumably, as N is made larger and larger, the discrete structure should approach the continuous structure — and this both when the consistent or lumped mass model (or a weighted average mass matrix) are used. Is this really true? The answer is only a qualified yes, for artifacts remain even when the number N is made to be large. To visualize this, consider some rather coarse discretizations in which N is a number in the single digits. The advantage of such a choice is that using MATLAB®'s symbolic tool, both the system determinant and the cofactors can be determined explicitly and in closed form; this allows in turn exact expressions for the poles and zeros. Ignoring a checkerboard pattern of leading signs ± 1 (which do not affect the zeros), the exact matrix of (scaled) cofactors for the lumped and consistent mass cases for $N = 3$ is as follows:

$$\mathbf{N}_L = \begin{Bmatrix} (2x-1)(2x-3) & 2(x-1) & 1 \\ 2(x-1) & 2(x-1)^2 & x-1 \\ 1 & x-1 & 1-4x+2x^2 \end{Bmatrix}, \qquad z_{\alpha\beta} = 3\sqrt{2x} \tag{8.68a}$$

$$\mathbf{N}_C = \begin{Bmatrix} 3(5x-1)(x-1) & 2(x+1)(2x-1) & (x+1)^2 \\ 2(x+1)(2x-1) & 2(2x-1)^2 & (x+1)(2x-1) \\ (x+1)^2 & (x+1)(2x-1) & 1-10x+7x^2 \end{Bmatrix}, \qquad z_{\alpha\beta} = 3\sqrt{6x} \tag{8.68b}$$

These imply the zeros listed in Table 8.1. We omit evaluation of the poles because these will be compared in Section 8.5, where it will be found that a very good agreement with the continuous solution can be attained with a mix of the consistent and lumped mass matrices, and especially so for the first 50% of modes.

Rather peculiar is the fact that some of the transfer functions for the consistent mass version should exhibit pairs of purely imaginary zeros, which neither the continuous nor the lumped mass versions contain. As can be seen, the zeros of $H_{\alpha\alpha}$ (and only these!) for the lumped and consistent mass versions are interlaced by the true zeros. Thus, using a mix of lumped and consistent mass matrices produces optimal results. Similar results are obtained as the number of finite elements is increased further, that is, as the model is refined and more modes and zeros are explored. Still, the imaginary zeros for the consistent mass case do not go away when N is made to be larger, but instead appear as zeros of

Table 8.1. Zeros of transfer functions for shear beam discretized with $N = 3$ elements

Transfer function	Exact (continuous)	Consistent mass	Lumped mass
H_{11}	$\pm\pi = \pm 3.1415$	$\pm 3\sqrt{\tfrac{6}{5}} = \pm 3.2863$	± 3
	$\pm 2\pi = \pm 6.2832$	$\pm 3\sqrt{6} = \pm 7.3485$	$\pm 3\sqrt{3} = \pm 5.1962$
H_{21}	$\pm 3\pi = \pm 9.4248$	$\pm 3i\sqrt{6}$	$\pm 3\sqrt{2} = \pm 4.2426$
	$\pm\tfrac{9}{2}\pi = \pm 14.1372$	$\pm 3\sqrt{3} = \pm 5.1962$	–
H_{31}	$\pm 3\pi = \pm 9.4248$	$\pm 3i\sqrt{6}$	–
	$\pm 6\pi = \pm 18.8496$	$\pm 3i\sqrt{6}$	–
H_{22}	$\pm\tfrac{3}{2}\pi = \pm 4.7124$	$\pm 3\sqrt{3} = \pm 5.1962$	$\pm 3\sqrt{2} = \pm 4.2426$
	$\pm 3\pi = 9.4248$	$\pm 3\sqrt{3} = \pm 5.1962$	$\pm 3\sqrt{2} = \pm 4.2426$
H_{23}	$\pm 3\pi = 9.4248$	$\pm 3i\sqrt{6}$	$\pm 3\sqrt{2} = \pm 4.2426$
	$\pm\tfrac{9}{2}\pi = \pm 14.1372$	$\pm 3\sqrt{3} = \pm 5.1962$	–
H_{33}	$\pm\tfrac{3}{4}\pi = \pm 2.3562$	$\pm 3\sqrt{\tfrac{6}{7}\left(5-3\sqrt{2}\right)} = \pm 2.4171$	$\pm 3\sqrt{2\left(1-\tfrac{1}{2}\sqrt{2}\right)} = \pm 2.2961$
	$\pm\tfrac{9}{4}\pi = \pm 7.0686$	$\pm 3\sqrt{\tfrac{6}{7}\left(5+3\sqrt{2}\right)} = \pm 8.4440$	$\pm 3\sqrt{2\left(1+\tfrac{1}{2}\sqrt{2}\right)} = \pm 5.5433$

the form $iN\sqrt{3}$ that grow larger and larger as N increases. Eventually, once these zeros are rather large compared to the frequency of interest, they just contribute what in essence is a constant factor to the numerator of the transfer functions, and otherwise play no role.

8.2.4 Normal Modes versus Residues

We review in the ensuing the relationship between transfer functions and conventional modal analysis, and demonstrate that the eigenvalue problem for classical modes already "knows" about normalized modes even before one has solved for those modes. That is, normalization is an intrinsic property of the pair of matrices $\mathbf{K}, \mathbf{M}$, and not just a convenient scaling applied to the set of modes post facto.

As we have seen in Chapter 3, the general eigenvalue problem $|\mathbf{K} - \lambda\mathbf{M}| = 0$ involving two real, symmetric, positive definite matrices $\mathbf{K}, \mathbf{M}$ satisfy some well-defined orthogonality conditions. Those eigenvectors can also be normalized so that their modal mass $\mu = \phi^T \mathbf{M}\phi$ is unity: it suffices to divide each unscaled mode by the square root of the modal mass. Thus, the normalization is the result of an explicit calculation applied to the modes *after* they were obtained by some means. However, we show herein that the normalized modes are not merely convenient forms of scaling, but that they are actually intrinsic properties of the pair of matrices $\mathbf{K}, \mathbf{M}$, that is, the matrices already "know" about normalization even *before* the modes have been obtained. This means that we can obtain individual components of the normalized modes directly from the eigenvalue problem, and without needing to obtain either all of the modes or for that matter, any one complete mode. These results are achieved by means of the residue theorem of operational calculus, a finding that is rather remarkable inasmuch as the residues themselves do not make use of any orthogonality conditions or normalization in the first place. It appears that this obscure property connecting the general eigenvalue problem of modal analysis with the residue theorem of operational calculus may have been overlooked up until now, but that has in turn interesting theoretical implications, but which we can't explore further herein.

Consider the eigenvalue problem involving a pair of $N \times N$, real, symmetric matrices $\mathbf{K}, \mathbf{M}$, of which $\mathbf{K}$ is either positive semi-definite or positive definite, and $\mathbf{M}$ is always positive definite:

$$\mathbf{K}\boldsymbol{\phi}_j = \lambda_j \mathbf{M}\boldsymbol{\phi}_j \tag{8.69}$$

Thus, the eigenvalues are always real and nonnegative. This eigenvalue problem satisfies the orthogonality conditions

$$\mathbf{\Phi}^T \mathbf{K} \mathbf{\Phi} = \mathbb{K} = \mathbb{M}\mathbf{\Lambda} = \mathrm{diag}\left(\mu_j \lambda_j\right) \qquad = \text{model stiffness} \tag{8.70a}$$

$$\mathbf{\Phi}^T \mathbf{M} \mathbf{\Phi} = \mathbb{M} = \mathrm{diag}\left(\mu_j\right) \qquad = \text{model mass} \tag{8.70b}$$

If so desired, the modes can also be normalized so that they attain a unit modal mass, that is,

$$\mathbf{\Psi} = \left(\mathbb{M}\right)^{-1/2} \mathbf{\Phi} \equiv \left\{\psi_j\right\}, \qquad \psi_j = \phi_j / \sqrt{\mu_j}, \qquad \mathbf{\Psi}^T \mathbf{M} \mathbf{\Psi} = \mathbf{I} \tag{8.71}$$

Consider now the matrix pencil (that is, the impedance matrix, with $\lambda = \omega^2$)

$$\mathbf{Z} = \mathbf{K} - \lambda\mathbf{M} \tag{8.72}$$

whose inverse is the matrix of transfer functions $\mathbf{H} = \left\{H_{\alpha\beta}\right\}$

$$\mathbf{H} = \mathbf{Z}^{-1} = \left(\mathbf{K} - \lambda\mathbf{M}\right)^{-1} \tag{8.73}$$

Expressed in terms of the modes, this inverse can be written as

$$\begin{aligned}\mathbf{H}(\lambda) &= \mathbf{\Phi}\left(\mathbb{K} - \lambda\mathbb{M}\right)^{-1}\mathbf{\Phi}^T = \left(\mathbf{\Phi}\mathbb{M}^{-1/2}\right)\left(\mathbf{\Lambda} - \lambda\mathbf{I}\right)^{-1}\left(\mathbb{M}^{-1}\mathbf{\Phi}^T\right) \\ &= \mathbf{\Psi}\left(\mathbf{\Lambda} - \lambda\mathbf{I}\right)^{-1}\mathbf{\Psi}^T = -\sum_{j=1}^{N} \frac{1}{\lambda - \lambda_j}\psi_j\psi_j^T\end{aligned} \tag{8.74}$$

that is, each component of $\mathbf{H}$ is of the form

$$H_{\alpha\beta}(\lambda) = -\sum_{j=1}^{N} \frac{\psi_{\alpha j}\psi_{\beta j}}{\lambda - \lambda_j} \tag{8.75}$$

where the $\psi_{\alpha j}$ are the individual *normalized* components of ψ_j. If we now carry out a contour integration along a path Γ in the complex plane that encloses all of the poles λ_j, then the result of that integral will equal the sum of all of the residues, which then yields

$$\oint_\Gamma H_{\alpha\beta}(\lambda)d\lambda = -2\pi i \sum_{j=1}^{N} \psi_{\alpha j}\psi_{\beta j} \tag{8.76}$$

On the other hand, by Cramer's rule the inverse is also given by

$$H_{\alpha\beta} = \frac{N_{\alpha\beta}}{\Delta}, \qquad \mathbf{H} = \frac{1}{\Delta}\left\{N_{\alpha\beta}\right\} \tag{8.77}$$

where $N_{\alpha\beta}$ is the cofactor, and Δ is the determinant:

$$\Delta = |\mathbf{K} - \lambda\mathbf{M}| = (-1)^N |\mathbf{M}| \prod_{k=1}^{N} (\lambda - \lambda_k) \tag{8.78}$$

where $|\mathbf{M}| > 0$ because $\mathbf{M}$ is positive-definite. A contour integration along the same path used earlier now yields

$$\oint_{\Gamma} \frac{N_{\alpha\beta}}{\Delta} d\lambda = 2\pi i \sum_{j=1}^{N} \lim_{\lambda \to \lambda_j} (\lambda - \lambda_j) \frac{N_{\alpha\beta}(\lambda)}{\Delta(\lambda)} = 2\pi i \sum_{j=1}^{N} \frac{N_{\alpha\beta}(\lambda_j)}{\Delta'(\lambda_j)} \tag{8.79}$$

in which $\Delta'(\lambda_j) = \frac{d}{d\lambda}\Delta\big|_{\lambda_j}$ is the derivative evaluated at the eigenvalue. Comparison of Eqs. 8.76 and 8.79 shows that

$$\sum_{j=1}^{N} \psi_{\alpha j} \psi_{\beta j} = -\sum_{j=1}^{N} \frac{N_{\alpha\beta}(\lambda_j)}{\Delta'(\lambda_j)} \tag{8.80}$$

Now, the path of the contour integral was arbitrary, so we could just as well have enclosed only a single pole, in which case the summation would have consisted of just one term. Hence, we conclude that

$$\boxed{\psi_{\alpha j} \psi_{\beta j} = (-1)\frac{N_{\alpha\beta}(\lambda_j)}{\Delta'(\lambda_j)}} \tag{8.81a}$$

and in particular, for $\alpha = \beta$ (diagonal elements of the square matrix $\boldsymbol{\psi}_j \boldsymbol{\psi}_j^T$)

$$\boxed{\psi_{\alpha j} = \pm \sqrt{(-1)\frac{N_{\alpha\beta}(\lambda_j)}{\Delta'(\lambda_j)}}} \tag{8.81b}$$

where the sign of the square root is arbitrary; however, once a choice has been made (say the positive sign), then the sign of other components must be chosen consistently.

Now, the derivative of the determinant is

$$\frac{d}{d\lambda}\Delta = \Delta' = \Delta \left\{ \frac{1}{\lambda - \lambda_1} + \frac{1}{\lambda - \lambda_2} + \cdots + \frac{1}{\lambda - \lambda_N} \right\} \tag{8.82}$$

and evaluating the derivative at an eigenvalue λ_j, we obtain

$$\Delta'(\lambda_j) = (-1)^N |\mathbf{M}| \prod_{k \neq j} (\lambda_j - \lambda_k) \tag{8.83}$$

Hence, from Eqs. 8.80 and 8.81 we infer

$$\psi_{\alpha j} \psi_{\beta j} = (-1)^{N+1} \frac{1}{|\mathbf{M}|} \frac{N_{\alpha\beta}(\lambda_j)}{\prod_{k \neq j}(\lambda_k - \lambda_j)} \tag{8.84}$$

In particular, for $\alpha = \beta$ (diagonal elements of the square matrix $\boldsymbol{\psi}_j \boldsymbol{\psi}_j^T$)

$$N_{\alpha\alpha}(\lambda) = (-1)^{N-1} |\mathbf{M}_\alpha| \prod_{k}^{N-1} (\lambda - \lambda_{\alpha k}) \tag{8.85}$$

in which the $\lambda_{\alpha k}$ are the eigenvalues of the constrained system of equations with the αth column and row suppressed, and $|\mathbf{M}_\alpha|$ is the corresponding leading minor of $\mathbf{M}$. Hence,

$$\psi_{\alpha j} = \pm \sqrt{\frac{|\mathbf{M}_\alpha|}{|\mathbf{M}|}} \sqrt{\frac{\prod\limits_{k}^{N-1} \left(\lambda_j - \lambda_{\alpha k} \right)}{\prod\limits_{k \neq j} \left(\lambda_j - \lambda_k \right)}} \tag{8.86}$$

Equations 8.81a,b, or alternatively 8.84 and 8.86, imply that the information about *normalized modes* is an intrinsic property of the eigenvalue problem, and not just a convenient scaling. Thus, at least in principle, selected components (i.e., DOF) of the normalized modes can be found directly without the need to find the complete mode (or modes), or employing any orthogonality conditions. To appreciate how remarkable this property is, consider the fact that a normalized eigenvector can be interpreted as a direction in the N-dimensional space defined by a vector whose actual length depends on $\mathbf{M}$. We are then guaranteed that for any one mode, we could obtain just one single component of that vector, of the correct length and without reference to (i.e., comparison with) any of the other components! By contrast, in a typical solution with eigenvalue solvers, just one component alone says absolutely nothing about the eigenvector itself or the eigendirection, as the mode has not yet been scaled and it magnitude depends on the solver used as well as the numerical entries in the matrices. Numerical test have verified the validity of the relationship between residues and normalized modes, and an algorithm to accomplish this has been proposed elsewhere, but which for space considerations cannot be included herein.

8.3 Correspondence Principle

Consider a linearly elastic, undamped body subjected to a dynamic excitation. An exact solution in terms of impulse or frequency response functions may not always be available, or even be feasible, but on physical grounds, the solution is known to exist. Conceptually, this solution must depend on the geometric characteristics (shape, boundary conditions, location of input and output points, etc.) and on the material parameters of the body being studied. Now, if the purely elastic materials were replaced with viscoelastic materials possessing arbitrary rheological properties, even if the same spatial variation as the elastic parameters, what would the new solution be like? M. A. Biot's[4] answer to this question is that

> a large class of equations of the theory of elasticity ... may be extended to the most general type of viscoelastic material, provided the elastic constants are replaced by corresponding operators. We call this the Principle of Correspondence.

In the case of harmonic loads, the differential operators translate into factors in $i\omega$, in which case it suffices to substitute complex, frequency-dependent material parameters in lieu of the elastic parameters in the solution to the undamped system.

---

[4] M. A. Biot, "Dynamics of viscoelastic anisotropic media," in *Proceedings of 4th Midwestern Conference on Solid Mechanics*, Purdue University, September 1955. See also his book, *Mechanics of Incremental Deformations* (New York: John Wiley & Sons, 1965), 359.

The simplest situation occurs when only one material parameter exists, say the shear modulus G. In that case, the complex shear modulus will be of the general form $G^* = G[f(\omega) + \mathrm{i}\,g(\omega)]$. For example, the shear stress–strain equation for a viscously damped medium with viscosity D is

$$\tau = G\gamma + D\dot{\gamma} \tag{8.87}$$

In the frequency domain, this equation is expressed as

$$\tau = (G + \mathrm{i}\,\omega D)\gamma = G^*\gamma \tag{8.88}$$

in which $G^* = G + \mathrm{i}\omega D$ is the frequency-dependent, complex shear modulus. If G and D have the same spatial variation, then D/G is not a function of space, but only of frequency, in which case D/G can be factored out. Hence, the correspondence principle applies, and we may simply substitute G^* for G in the elastic solution. We emphasize the spatial independence of the ratio D/G by writing G^* in a slightly different form. To this effect, we define a fraction of viscous damping ξ_v relative to an arbitrary reference frequency ω_0 so that $\xi_v/\omega_0 = \tfrac{1}{2}D/G$. The complex modulus then changes to

$$G^* = G\left(1 + 2\,\mathrm{i}\,\xi_v\,\frac{\omega}{\omega_0}\right) \tag{8.89}$$

In the case of hysteretic damping ξ_h, the complex modulus is of the form

$$G^* = G\left[1 + 2\,\mathrm{i}\,\xi_h\,\mathrm{sgn}(\omega)\right] \tag{8.90}$$

When dealing with dynamic problems involving wave propagation through lossy media, it is more advantageous to express the complex shear modulus and the complex shear wave velocity (or other moduli) in a form where the damping term appears in the denominator, namely

$$G^* = G\left[1 + 2\,\mathrm{i}\,\xi_h\right] \approx \frac{G}{(1 - \mathrm{i}\,\xi_h)^2} \tag{8.91}$$

$$C_s^* = \left[G^*/\rho\right]^{1/2} = \frac{C_s}{1 - \mathrm{i}\,\xi_h} \tag{8.92}$$

The reason why these forms are more convenient is that they simplify considerably the complex exponentials needed to represent plane, harmonic waves. Indeed, consider the expression

$$e^{\mathrm{i}\omega(t - x/C_s^*)} = e^{\mathrm{i}\omega[t - x(1 - \mathrm{i}\xi)/C_s]} = e^{-\xi\omega x/C_s}\,e^{\mathrm{i}\omega(t - x/C_s)} = e^{-\xi x/\lambda}\,e^{\mathrm{i}\omega(t - x/C_s)} \tag{8.93}$$

Here, the spatial attenuation is controlled explicitly by a very simple, exponentially decaying term, and not by a complicated subradical exponent. [We mention in passing that seismologists often use the more cumbersome *quality factor* $Q = (2\xi_h)^{-1}$ to model attenuation.]

Example: Pulsating Spherical Cavity
As an example of application of the correspondence principle just described, consider a spherical cavity of radius R embedded in an unbounded, homogeneous, isotropic elastic solid with shear modulus G and Poisson's ratio v. The cavity is subjected to a pulsating

(harmonic) internal pressure p. The frequency response function for the radial displacement u on the cavity's wall can be shown to be given by

$$u = \frac{pR}{4G}\left[\frac{1+2\,i\,a_0\alpha}{1+2\,i\,a_0\alpha - a_0^2}\right] = \frac{pR}{4G}F(a_0) \tag{8.94}$$

in which

$$a_0 = \frac{\omega R}{2C_s} \qquad = \text{dimensionless frequency} \tag{8.95}$$

$$\alpha = \frac{C_s}{C_p} = \sqrt{\frac{1-2v}{2-2v}} \qquad = \text{shear wave velocity / dilatational wave velocity} \tag{8.96}$$

[Notice the similarity of the term in brackets, $F(a_0)$, to the transfer function for base motion of a single degree of freedom (SDOF) system with viscous damping $\xi = \alpha$, and tuning ratio $\omega/\omega_n = a_0$.] Consider next a pulsating cavity in a solid with hysteretic damping. The complex shear modulus and shear wave velocities are then

$$G^* = G(1+2\,i\,\xi) \approx G\left(1-i\,\xi\right)^{-2} \qquad \text{and} \qquad C_s^* = \frac{C_s}{1-i\,\xi} \tag{8.97}$$

The correspondence principle states that we must use these complex parameters in place of the real, undamped constants. The dimensionless frequency thus changes into $a_0^* = a_0(1-i\,\xi)$ and the damped solution is

$$u = \frac{pR(1-i\,\xi)^2}{4G}\left[\frac{1+2\,i\,a_0\alpha(1-i\,\xi)}{1+2\,i\,a_0\alpha(1-i\,\xi)-a_0^2(1-i\,\xi)^2}\right] = \frac{pR}{4G^*}F\left(a_0^*\right) \tag{8.98}$$

8.4 Numerical Correspondence of Damped and Undamped Solutions

In the previous example of a pulsating spherical cavity, we were able to obtain the viscoelastic solution directly from the elastic solution by simple recourse to complex moduli in the undamped formula. We could do this, because we had an analytical (or closed-form) expression available for the response function. In many engineering applications, however, a solution may have been obtained only in numerical form, for which the response functions are known only as tabulations of values, or as graphs. For example, the dynamic compliances for rigid foundations resting on elastic soils are typically reported only in numerical format. The question then arises: Can one derive solutions for media with damping from the undamped tabulated values, and vice versa? The answer is a qualified *yes*.

8.4.1 Numerical Quadrature Method

The numerical quadrature method proposed by Dasgupta and Sackman[5] is of great theoretical interest, because it establishes a general mathematical framework for deriving the

---

[5] G. Dasgupta and J. L. Sackman, "An alternative representation of the elastic-viscoelastic correspondence principle for harmonic oscillations," *J. Appl. Mech.*, March 1977, 57–60.

solutions for media with arbitrary damping laws (i.e., with arbitrary frequency dependence of the viscosity) from the solution of purely elastic media, as will be seen.

Let $F(a_0)$ be the functional for which we seek the damped solution $F(a_0^*)$. We shall assume that this functional is sufficiently well behaved that it may be considered a function at least in the distributional sense (i.e., in the sense of singular functions), and that it has a Fourier transform. This implies in turn that $F(a_0)$ must tend to zero as the dimensionless frequency a_0 tends to $\pm\infty$. [**Note:** While impedance functions do not satisfy this restriction, their inverses – the compliance functions – do so.]

Now, we know that $F(z)$ cannot have any poles in the lower half-plane, because otherwise the system would violate causality. It follows that $F(z)$ is analytic in the entire lower half-plane, which in turn means that we can express a point $z_0 = a - ib$ in that plane by means of the Cauchy integral

$$F(z_0) = \frac{1}{2\pi i} \oint \frac{F(z)}{z - z_0} dz \qquad (8.99)$$

We form the contour of this integral by combining the real axis with an infinitely large circle around the lower half-plane, traveled in counterclockwise direction. In our case, the viscously damped frequency is $z_0 \equiv a_0^* = a_0(1 - i\xi) = a_0 - i\xi a_0 = a - ib$, which is clearly a point in the lower half-plane. Its complex-conjugate is then a point in the upper half-plane. Hence, if we use the same contour as above, this conjugate point lies *outside* of the integration path, and the Cauchy integral must vanish:

$$0 = \frac{1}{2\pi i} \oint \frac{F(z)}{z - z_0^c} dz \qquad (8.100)$$

Subtracting the second integral from the first, we obtain

$$\begin{aligned} F(z_0) &= \frac{1}{2\pi i} \oint \left[\frac{F(z)}{z - z_0} - \frac{F(z)}{z - z_0^c} \right] dz = \frac{z_0 - z_0^c}{2\pi i} \oint \frac{F(z)}{(z - z_0)(z - z_0^c)} dz \\ &= -\frac{b}{\pi} \oint \frac{F(z)}{(z - a)^2 + b^2} dz \end{aligned} \qquad (8.101)$$

If $F(z)$ remains bounded as the radius ρ of the contour tends to infinity, the integrand is at most of order ρ^{-2}, so the contribution of the integral "around" infinity is zero (Jordan's Lemma). Therefore, only the integration along the real axis contributes to the integral. After both taking this fact into account, and reversing the direction of integration (which cancels the negative sign), we obtain

$$F(a - ib) = \frac{b}{\pi} \int_{-\infty}^{+\infty} \frac{F(x)}{(x - a)^2 + b^2} dx \qquad (8.102)$$

or in terms of the dimensionless frequency

$$\boxed{F(a_0^*) = \frac{\xi a_0}{\pi} \int_{-\infty}^{+\infty} \frac{F(x)}{(x - a_0)^2 + \xi^2 a_0^2} dx} \qquad (8.103)$$

which can be evaluated *numerically* for given tabulated values of $F(a_0)$. However, since these values are *never* available over the entire range of frequencies, the integral must, by necessity, be truncated at finite values of the lower and upper limits, a fact which seriously affects the accuracy with which the damped solution can be computed in practice.

Example

An interesting and illuminating example is the case of an undamped SDOF oscillator. The transfer function of such a system can be written as

$$F(a_0) = \frac{1}{1 - a_0^2} \tag{8.104}$$

in which $a_0 = \omega / \omega_n$ is the dimensionless frequency. We know, of course, that the damped solution has a complex-valued transfer function. However, since $F(a_0)$ for the undamped system is a real-valued function, the quadrature integral cannot possibly result in a complex-valued function. The reason for this "anomaly" is that the undamped system has two poles *on* the real axis; hence, the derivation of the quadrature formula is not strictly correct for this problem. In fact, the phase angle of the undamped transfer function jumps from 0 before resonance, to π after resonance, so the phase angle at resonance must have the average value $\pi/2$. Thus, the resonant peak is purely imaginary. To simulate this behavior in the quadrature scheme, we can introduce an arbitrarily small perturbation $i\varepsilon a_0$ into the system (e.g., $\xi = 0.0001$ will do the job), and define the transfer function as

$$F(a_0) = \frac{1}{1 - a_0^2 + i\varepsilon a_0} \tag{8.105}$$

With this modification, we are indeed able to obtain the damped solution with the quadrature integral, not only for viscous damping, but for *any* arbitrary damping law.

8.4.2 Perturbation Method

The perturbation method was suggested by Pais and Kausel[6] and is applicable only to moderately damped systems. It consists in expanding the damped solution $F(a_0^*) = F(a_0,\xi)$ in Taylor series in the damping parameter ξ, and retaining only the lower order terms. Since this function is analytic in the lower half-plane, we known that the Taylor series exists. For this purpose, we start by writing the complex dimensionless frequency as

$$a_0^* = a(a_0,\xi) - i\,b(a_0,\xi) \tag{8.106}$$

which ostensibly satisfies the identities $a(a_0,0) = a_0$ and $b(a_0,0) = 0$. If we then expand $F(a_0^*)$ in Taylor series about the point a_0 and consider the fact that it is an analytic function satisfying

$$\frac{dF(z)}{dz}\bigg|_{y=0} = \frac{dF(x)}{dx} \tag{8.107}$$

[6] A. Pais and E. Kausel, "Stochastic response of foundations," MIT Research Report R85-6, Department of Civil Engineering, Cambridge, MA, February 1985, pp. 33–34.

we obtain

$$F(a_0^*) = F(a_0) + \frac{\xi}{1!}\frac{dF(a_0)}{da_0}\left[\frac{\partial a}{\partial\xi} - i\frac{\partial b}{\partial\xi}\right]\Bigg|_{\xi=0} + \cdots \qquad (8.108)$$

In particular, in the case of hysteretic damping, we have $a(Az_1 + Bz_2) - b(Az_1^2 + Bz_2^2) = 0$, which implies

$$\frac{\partial a}{\partial\xi} = 0 \quad\text{and}\quad \frac{\partial b}{\partial\xi} = a_0 \qquad (8.109)$$

Hence

$$F(a_0^*) = F(a_0) - i\,\xi a_0\,\frac{dF(a_0)}{da_0} + \cdots \qquad (8.110)$$

Since numerical values for $F(a_0)$ are available only in tabulated form, the first derivatives are not generally known. However, they can easily be estimated with forward differences:

$$\boxed{F(a_0^*) \approx F(a_0) - i\,\xi a_0\,\frac{\Delta F(a_0)}{\Delta a_0}} \qquad (8.111)$$

8.5 Gyroscopic Forces Due to Rotor Support Motions

The spinning element of a large rotating machine, such as a rotor, exhibits an intrinsic resistance to changes in the orientation of the axis of rotation because of gyroscopic effects. Thus, when these systems are subjected to external dynamic disturbances, such as support motions elicited by floor vibrations, large gyroscopic forces can arise if the axle supports do not move in synchrony. We shall briefly explore this problem here by means of a highly simplified model, which is based on the following assumptions:

- The rotor spins steadily with angular velocity ω_1 about the axis, and is initially in stable dynamic equilibrium.
- The rotor has cylindrical symmetry.
- The bearings are symmetrically placed with respect to the center of mass. The axle is elastically supported at each end in both the vertical and horizontal directions by identical springs and dashpots. These elements represent the effects of the oil film, the local flexibility of the axle and the bearings, and so forth.
- The rotor is perfectly balanced, and its center of mass lies at the center of the axis of rotation.
- No whirling of the shaft takes place under normal operations.
- While gravity forces may produce an initial downward shift of the axis and compression of the bearings, and perhaps even some bending of the axle so that the rotation line of the rotor may not exactly coincide with the line connecting the supports, such gravity effects will be ignored.
- Longitudinal (i.e., axial) components of the support motion and the response as well as torsional oscillations of the rotor will be ignored.
- Lateral rotations of the rotor remain small.

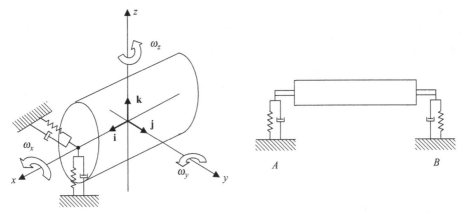

Figure 8.2. Differential support motions of rotor.

Consider a right-handed Cartesian reference frame associated with a moving triad of unit vectors $\mathbf{i}, \mathbf{j}, \mathbf{k}$ such that the first axis $\mathbf{i}$ coincides at all times with the axle of the rotor, but does *not* spin with it, while the other two are initially in the horizontal and vertical planes, as shown in Figure 8.2. To a first approximation, these vectors coincide with the Cartesian directions, if the rotations about the transverse axes remain small. The rotor has mass m, mass moments of inertia J_x about the axis of rotation, and mass moments of inertia $J_y = J_z$ about any other perpendicular axis. This system has 4 DOF, namely the two transverse motions of the rotor and the two (small) rotations of the rotor about the transverse axes. Because of symmetry, however, this model decouples into three independent systems, two translational models with an SDOF each, and one with two coupled rotations, respectively. The first two do not involve gyroscopic forces.

At $t = 0$ the system experiences support motions $y_A(t)$, $y_B(t)$, $z_A(t)$, $z_B(t)$ that elicit a dynamic reaction. Taking advantage of the symmetry of this system, we define the translational and rotational *support* motions as

$$u_{sy}(t) = \tfrac{1}{2}\left[y_A(t) + y_B(t)\right] \qquad \text{Lateral } (y) \text{ translation} \qquad (8.112)$$

$$u_{sz}(t) = \tfrac{1}{2}\left[z_A(t) + z_B(t)\right] \qquad \text{Vertical } (z) \text{ translation} \qquad (8.113)$$

$$\theta_{sy}(t) = \tfrac{1}{L}\left[z_B(t) - z_A(t)\right] \qquad y\text{-rotation} \qquad (8.114)$$

$$\theta_{sz}(t) = \tfrac{1}{L}\left[y_A(t) - y_B(t)\right] \qquad z\text{-rotation} \qquad (8.115)$$

with L being the distance between the supports A, B. Notice the sign difference in the definitions of the rotations, which is needed to conform to the right-hand rule. In accordance with this decomposition, we also define the translational and rotational stiffness and damping

$$k_y = k_z = k_A + k_B = 2k_A \equiv k \qquad \text{Translational stiffness} \qquad (8.116)$$

$$k_{\theta y} = k_{\theta z} = (k_A + k_B)\left(\tfrac{L}{2}\right)^2 = \tfrac{1}{2}k_A L^2 \equiv k_\theta \qquad \text{Rotational stiffness} \qquad (8.117)$$

and similar expressions for the dashpots, that is, $c_y = c_z = 2c_A \equiv c$ and $c_{\theta y} = c_{\theta z} = \tfrac{1}{2}c_A L^2 \equiv c_\theta$.

Translational Models

These are the classical models for support motion in the y and z directions:

$$m\ddot{u}_y + c\dot{u}_y + ku_y = -m\ddot{u}_{sy} \tag{8.118}$$

$$m\ddot{u}_z + c\dot{u}_z + ku_z = -m\ddot{u}_{sz} \tag{8.119}$$

Inasmuch as these models are well known, we need not elaborate further on their evaluation and use. For example, the equal deformations of the lateral springs elicited by the support motions could be obtained from a standard response spectrum for the excitation.

Rotational Model

Let $\boldsymbol{\omega} = \omega_x \mathbf{i} + \omega_y \mathbf{j} + \omega_z \mathbf{k}$ be the rotational velocity vector of the rotor with, say $\omega_x = 2\pi \times 30$ rad/s being its steady operation speed. Also, let $\boldsymbol{\Omega} = \omega_y \mathbf{j} + \omega_z \mathbf{k}$ be the rotational velocity vector of the triad of vectors $\mathbf{i}, \mathbf{j}, \mathbf{k}$. If $\mathbf{J} = J_x \mathbf{i}\mathbf{i} + J_y \mathbf{j}\mathbf{j} + J_z \mathbf{k}\mathbf{k}$ is the principal rotational inertia tensor, $\mathbf{t}$ is the external moment (torque) applied to the rotor, and $\boldsymbol{\alpha}$ is its rotational acceleration, then from the principle of angular momentum (see Chapter 1, "Fundamental Principles"), we have

$$\begin{aligned}
\mathbf{t} = \frac{d\mathbf{h}}{dt} &= \frac{d}{dt}(\mathbf{J} \cdot \boldsymbol{\omega}) = \mathbf{J} \cdot \boldsymbol{\alpha} + \boldsymbol{\Omega} \times \mathbf{J} \cdot \boldsymbol{\omega} \\
&= \left(J_x \dot{\omega}_x \mathbf{i} + J_y \dot{\omega}_y \mathbf{j} + J_z \dot{\omega}_z \mathbf{k} \right) + \left(\omega_y \mathbf{j} + \omega_z \mathbf{k} \right) \times \left(J_x \omega_x \mathbf{i} + J_y \omega_y \mathbf{j} + J_z \omega_z \mathbf{k} \right) \\
&= J_y \dot{\omega}_y \mathbf{j} + J_z \dot{\omega}_z \mathbf{k} + \omega_y \omega_z \left(J_x - J_y \right) \mathbf{i} + \omega_x J_x \left(\omega_z \mathbf{j} - \omega_y \mathbf{k} \right) \\
&\approx \left(J_y \dot{\omega}_y + \omega_x J_x \omega_z \right) \mathbf{j} + \left(J_z \dot{\omega}_z - \omega_x J_x \omega_y \right) \mathbf{k}
\end{aligned} \tag{8.120}$$

We disregard the torsional term in $\omega_y \omega_z$, since it is quadratic in small quantities. Also, $\dot{\omega}_x = 0$, because the rotor spins at a steady rate. The terms in $\omega_x \omega_y$ and $\omega_x \omega_z$ represent the transverse (gyroscopic) moments, which are important because ω_x is large; together with the rotational inertia forces, these are absorbed by deformations of the supports. Since lateral rotations remain small, the lateral angular velocities are related to the angle of rotations as $\omega_y = \dot{\theta}_y$ and $\omega_z = \dot{\theta}_z$. The equation of motion is then obtained by equilibrating the moment above with the moments caused by the inertia forces and the support reactions, that is,

$$J_y \ddot{\theta}_y + \omega_x J_x \dot{\theta}_z + c_\theta \dot{\theta}_y + k_\theta \theta_y = c_\theta \dot{\theta}_{sy} + k_\theta \theta_{sy} \tag{8.121}$$

$$J_z \ddot{\theta}_z - \omega_x J_x \dot{\theta}_y + c_\theta \dot{\theta}_z + k_\theta \theta_z = c_\theta \dot{\theta}_{sz} + k_\theta \theta_{sz} \tag{8.122}$$

Defining $J_y = J_z \equiv J_\theta$, we can write these equations in matrix form as

$$J_\theta \begin{Bmatrix} \ddot{\theta}_y \\ \ddot{\theta}_z \end{Bmatrix} + \omega_1 J_1 \begin{bmatrix} 0 & 1 \\ -1 & 0 \end{bmatrix} \begin{Bmatrix} \dot{\theta}_y \\ \dot{\theta}_z \end{Bmatrix} + c_\theta \begin{Bmatrix} \dot{\theta}_y \\ \dot{\theta}_z \end{Bmatrix} + k_\theta \begin{Bmatrix} \theta_y \\ \theta_z \end{Bmatrix} = c_\theta \begin{Bmatrix} \dot{\theta}_{sy} \\ \dot{\theta}_{sz} \end{Bmatrix} + k_\theta \begin{Bmatrix} \theta_{sy} \\ \theta_{sz} \end{Bmatrix} \tag{8.123}$$

The antisymmetric matrix multiplying the angular velocities give rise to the gyroscopic forces that couple the two rotational motions. Observe that although the gyroscopic forces depend on velocities, they are not associated with any energy dissipation as damping forces are. The reason is that the quadratic form associated with the antisymmetric gyroscopic matrix, which measures the associated work, vanishes no matter what the rotations should be, that is,

$$\{\dot\theta_y \quad \dot\theta_z\}\begin{bmatrix} 0 & 1 \\ -1 & 0 \end{bmatrix}\begin{Bmatrix} \dot\theta_y \\ \dot\theta_z \end{Bmatrix} = 0 \tag{8.124}$$

The preceding equations can be solved either by modal superposition, or by an analysis in the frequency domain. For this purpose, we determine next the natural frequencies of this system and assess the effect that gyroscopic coupling has on these frequencies.

Define the expressions

$$\omega_\theta = \sqrt{\frac{k_\theta}{J_\theta}} \qquad = \text{rotational frequency with rotor at rest (i.e., } \omega_1 = 0) \tag{8.125}$$

$$\alpha = \frac{\omega_x J_x}{\omega_\theta J_\theta} \qquad = \text{degree of gyroscopic coupling (a dimensionless number)} \tag{8.126}$$

$$\xi_\theta = \frac{c_\theta}{2\sqrt{k_\theta J_\theta}} \qquad = \text{fraction of damping with rotor at rest} \tag{8.127}$$

we can write the dynamic equation as

$$\begin{Bmatrix} \ddot\theta_y \\ \ddot\theta_z \end{Bmatrix} + \alpha\omega_\theta\begin{bmatrix} 0 & 1 \\ -1 & 0 \end{bmatrix}\begin{Bmatrix} \dot\theta_y \\ \dot\theta_z \end{Bmatrix} + 2\xi_\theta\omega_\theta\begin{Bmatrix} \dot\theta_y \\ \dot\theta_z \end{Bmatrix} + \omega_\theta^2\begin{Bmatrix} \theta_y \\ \theta_z \end{Bmatrix} = 2\xi_\theta\omega_\theta\begin{Bmatrix} \dot\theta_{sy} \\ \dot\theta_{sz} \end{Bmatrix} + \omega_\theta^2\begin{Bmatrix} \theta_{sy} \\ \theta_{sz} \end{Bmatrix} \tag{8.128}$$

Normal Rotational Modes

The vibration modes and frequencies can be obtained by solving the undamped free vibration equation

$$\begin{Bmatrix} \ddot\theta_y \\ \ddot\theta_z \end{Bmatrix} + \alpha\omega_\theta\begin{bmatrix} 0 & 1 \\ -1 & 0 \end{bmatrix}\begin{Bmatrix} \dot\theta_y \\ \dot\theta_z \end{Bmatrix} + \omega_\theta^2\begin{Bmatrix} \theta_y \\ \theta_z \end{Bmatrix} = \begin{Bmatrix} 0 \\ 0 \end{Bmatrix} \tag{8.129}$$

Define

$$\lambda = \frac{\omega}{\omega_\theta} \qquad = \text{dimensionless natural frequency of system} \tag{8.130}$$

The eigenvalue problem for harmonic motions of the form $e^{i\omega t}$ is then

$$\begin{vmatrix} 1-\lambda^2 & i\alpha\lambda \\ -i\alpha\lambda & 1-\lambda^2 \end{vmatrix} = 0 \tag{8.131}$$

That is,

$$(1-\lambda^2)^2 - \alpha^2\lambda^2 = 0 \tag{8.132}$$

which yields the two eigenvalues

$$\lambda^2 = 1 + \tfrac{1}{2}\alpha^2 \mp \sqrt{\left(1 + \tfrac{1}{2}\alpha^2\right)^2 - 1} > 0 \tag{8.133}$$

It is easy to show that the two eigenvalues satisfy the relation $\lambda_I \lambda_{II} = 1$. The two coupled frequencies are then

$$\omega_I = \omega_\theta \sqrt{1 + \tfrac{1}{2}\alpha^2 - \sqrt{\left(1 + \tfrac{1}{2}\alpha^2\right)^2 - 1}} \qquad < \omega_\theta \tag{8.134}$$

$$\omega_{II} = \omega_\theta \sqrt{1 + \tfrac{1}{2}\alpha^2 + \sqrt{\left(1 + \tfrac{1}{2}\alpha^2\right)^2 - 1}} \qquad > \omega_\theta \tag{8.135}$$

These are the rotational frequencies of the rotor, which take into account the effect of the gyroscopic forces. The two roots are shown in Figure 8.3 as a function of α. Finally, the normal modes are easily obtained from the eigenvalue equation. They are

$$\phi_j = \left\{ \begin{array}{c} 1 \\ i\dfrac{(1-\lambda_j)^2}{\alpha\lambda} \end{array} \right\} \qquad j = I, II \tag{8.136}$$

which have a real and an imaginary component. These eigenvectors satisfy a special orthogonality condition, which can be obtained by multiplying the eigenvalue equation for the jth mode by the transposed, complex conjugate mode ϕ_i^*, and vice versa. Let $\mathbf{L}$ be the antisymmetric matrix in the eigenvalue problem. Taking into account that $\left(i\mathbf{L}\right)^* = (-i)(-\mathbf{L}) = i\mathbf{L}$, we can write

$$(1 - \lambda_j^2)\phi_i^*\phi_j + i\lambda_j\,\phi_i^*\mathbf{L}\phi_j = 0 \tag{8.137}$$

$$(1 - \lambda_i^2)\phi_i^*\phi_j + i\lambda_i\,\phi_i^*\mathbf{L}\phi_j = 0 \tag{8.138}$$

From here,

$$\lambda_i(1 - \lambda_j^2)\phi_i^*\phi_j + i\lambda_i\lambda_j\,\phi_i^*\mathbf{L}\phi_j = 0 \tag{8.139}$$

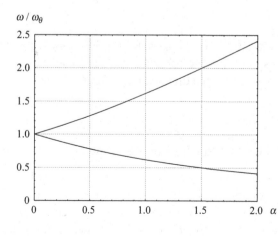

Figure 8.3. Rotational frequencies of rotor.

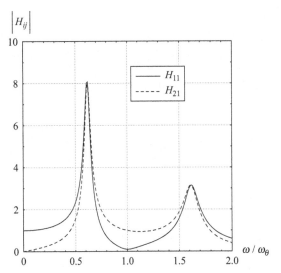

Figure 8.4. Transfer functions for support motion.

$$\lambda_j(1-\lambda_i^2)\boldsymbol{\phi}_i^*\boldsymbol{\phi}_j + i\,\lambda_i\lambda_j\,\boldsymbol{\phi}_i^*\mathbf{L}\,\boldsymbol{\phi}_j = 0 \tag{8.140}$$

Subtracting the second from the first, we obtain

$$\left(1+\lambda_i\,\lambda_j\right)\left(\lambda_i-\lambda_j\right)\boldsymbol{\phi}_i^*\boldsymbol{\phi}_j = 0 \tag{8.141}$$

Since $\lambda_I\lambda_{II} = 1$, but $\lambda_I \neq \lambda_{II}$, Eq. 8.141 implies $\boldsymbol{\phi}_I^*\boldsymbol{\phi}_{II} = 0$. This is the desired orthogonality condition, which can be used to apply modal superposition in the solution of the dynamic equations. Do also notice that $\boldsymbol{\phi}_I^T\boldsymbol{\phi}_I = \boldsymbol{\phi}_{II}^T\boldsymbol{\phi}_{II} = 0$, because the quadratic forms in $\mathbf{L}$ are always zero.

Transfer Function for Support Motion

The transfer functions for support motion are obtained by casting the dynamic equation in the frequency domain. The final expression is

$$\begin{Bmatrix}\theta_y \\ \theta_z\end{Bmatrix} = \frac{2i\omega\omega_\theta\xi_\theta + \omega_\theta^2}{(\omega_\theta^2-\omega^2+2i\omega\omega_\theta\xi_\theta)^2 - \alpha^2\omega^2\omega_\theta^2}\begin{bmatrix}\omega_\theta^2-\omega^2+2i\omega\omega_\theta\xi_\theta & -i\alpha\omega\omega_\theta \\ i\alpha\omega\omega_\theta & \omega_\theta^2-\omega^2+2i\omega\omega_\theta\xi_\theta\end{bmatrix}^{-1}\begin{Bmatrix}\theta_{sy} \\ \theta_{sz}\end{Bmatrix} \tag{8.142}$$

The absolute values for these transfer functions are shown in Figure 8.4 for the special case $\alpha = 1$ and $\xi = 0.05$. Notice that without the spinning rotor effect, there would have been only one resonant peak, which would have taken place at a dimensionless frequency of 1.0.

8.6 Rotationally Periodic Structures

8.6.1 Structures Composed of Identical Units and with Polar Symmetry

Structures composed of n identical units connected together in the shape of a regular polygon exhibit polar symmetry about the common axis. The vibration characteristic of such structures can be determined effectively from the dynamic properties of each of the units that compose it.

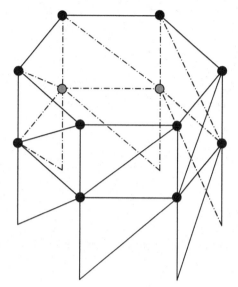 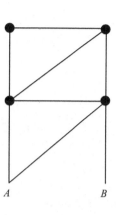

A   B

Figure 8.5. Rotationally periodic structure, here hexagonal.

Consider, for example, a set of six identical two-story frames that are connected in the form of a regular hexagon, as shown in Figure 8.5. The nodes in each unit are numbered so that all nodes to the left of the unit are followed by those to the right. Thus, nodes within each vertical interface are sequential. Although interior nodes that are not aligned with the vertices could also exist, for simplicity we shall ignore for now that possibility (but a generalization is not difficult).

Each of the individual units is characterized by identical stiffness and mass matrices $\mathbf{M}'_j$, $\mathbf{K}'_j$ ($j = 1,2,\ldots n = 6$), which have the general partitioned form

$$\mathbf{M}'_j = \begin{Bmatrix} \mathbf{M}'_{AA} & \mathbf{M}'_{AB} \\ \mathbf{M}'_{BA} & \mathbf{M}'_{BB} \end{Bmatrix} \tag{8.143}$$

$$\mathbf{K}'_j = \begin{Bmatrix} \mathbf{K}'_{AA} & \mathbf{K}'_{AB} \\ \mathbf{K}'_{BA} & \mathbf{K}'_{BB} \end{Bmatrix} \tag{8.144}$$

in which A, B refer to the DOF on the left and right sides of each unit, respectively. Initially, these matrices are formulated in a Cartesian coordinate system that coincides with the plane of each unit, a fact that is identified here by the primes. By an appropriate rotation, they can be cast into a cylindrical coordinate system that has its origin on the axis of the system, as depicted in Figure 8.6. For example, for each DOF, a plane truss has rotation submatrices of the form

$$\mathbf{R}_i(\alpha) = \begin{Bmatrix} \cos\alpha & -\sin\alpha & 0 \\ \sin\alpha & \cos\alpha & 0 \\ 0 & 0 & 1 \end{Bmatrix} = \mathbf{R}_i^T(-\alpha) \tag{8.145}$$

in which α is the angle between the plane of the truss and the tangent to the circumscribed circle, as shown above, and i is the nodal index. In the case of a frame, the nodal rotation matrices would be of size 6×6 (three translations and three rotations). DOFs for nodes to the left (i.e., A nodes) are rotated clockwise, while DOFs for nodes to the

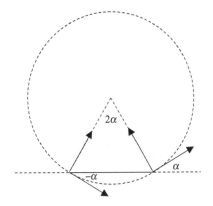

Figure 8.6. Local coordinates and displacements.

right (i.e., B nodes) are rotated counterclockwise. The rotated stiffness matrix has the symmetric form

$$\mathbf{K}_j = \begin{Bmatrix} \mathbf{R}^T \mathbf{K}'_{AA} \mathbf{R} & \mathbf{R}^T \mathbf{K}'_{AB} \mathbf{R}^T \\ \mathbf{R} \mathbf{K}'_{BA} \mathbf{R} & \mathbf{R} \mathbf{K}'_{BB} \mathbf{R}^T \end{Bmatrix} = \begin{Bmatrix} \mathbf{K}_{AA} & \mathbf{K}_{AB} \\ \mathbf{K}_{BA} & \mathbf{K}_{BB} \end{Bmatrix} \tag{8.146}$$

and a similar expression for the mass matrix. Inasmuch as the coordinate directions for forces and displacements for two units joined at each vertex are now in agreement, we can proceed to assemble the global stiffness matrix for the complete system in polar coordinates. It has the form

$$\mathbf{K} = \begin{Bmatrix} \mathbf{K}_{AA} + \mathbf{K}_{BB} & \mathbf{K}_{AB} & \mathbf{O} & \mathbf{O} & \mathbf{O} & \mathbf{K}_{BA} \\ \mathbf{K}_{BA} & \mathbf{K}_{AA} + \mathbf{K}_{BB} & \mathbf{K}_{AB} & \mathbf{O} & \mathbf{O} & \mathbf{O} \\ \mathbf{O} & \mathbf{K}_{BA} & \mathbf{K}_{AA} + \mathbf{K}_{BB} & \mathbf{K}_{AB} & \mathbf{O} & \mathbf{O} \\ \mathbf{O} & \mathbf{O} & \mathbf{K}_{BA} & \mathbf{K}_{AA} + \mathbf{K}_{BB} & \mathbf{K}_{AB} & \mathbf{O} \\ \mathbf{O} & \mathbf{O} & \mathbf{O} & \mathbf{K}_{BA} & \mathbf{K}_{AA} + \mathbf{K}_{BB} & \mathbf{K}_{AB} \\ \mathbf{K}_{AB} & \mathbf{O} & \mathbf{O} & \mathbf{O} & \mathbf{K}_{BA} & \mathbf{K}_{AA} + \mathbf{K}_{BB} \end{Bmatrix} \tag{8.147}$$

A similar expression can be written for the mass matrix. More generally, if the units are also connected in a far-coupled fashion (i.e., if there are secant elements connecting non-adjoining units), the symmetric, global stiffness matrix will have the form

$$\mathbf{K} = \begin{Bmatrix} \mathbf{K}_0 & \mathbf{K}_1 & \mathbf{K}_2 & \cdots & \mathbf{K}_{n-1} \\ \mathbf{K}_{n-1} & \mathbf{K}_0 & \mathbf{K}_1 & \cdots & \mathbf{K}_{n-2} \\ \mathbf{K}_{n-2} & \mathbf{K}_{n-1} & \mathbf{K}_0 & \mathbf{K}_1 & \vdots \\ \vdots & \vdots & \ddots & \ddots & \mathbf{K}_1 \\ \mathbf{K}_1 & \mathbf{K}_2 & \cdots & \mathbf{K}_{n-1} & \mathbf{K}_0 \end{Bmatrix} \tag{8.148}$$

Observe that the elements in each successive row are obtained by a simple cyclic permutation to the right. If these elements were scalars, the matrix would be said to be a *circulant* matrix. Since they are submatrices, however, the matrix is referred to as a being *block-circulant*. Matrices of this type have very special properties, as described in Chapter 9, Section 9.7. A brief summary follows.

8.6.2 Basic Properties of Block-Circulant Matrices

As we have just seen, rotationally periodic structures involve stiffness and mass matrices that have a block-circulant structure. For this reason, we examine some of the basic properties of such matrices herein. A more complete treatment can be found in Chapter 9, Section 9.8.

Consider a symmetric matrix $\mathbf{A}$ that is composed of n submatrices or blocks $\mathbf{A}_k$ of size $m \times m$ in such a way that the complete matrix has $N = n \times m$ rows and columns with the following structure:

$$
\mathbf{A} = \begin{Bmatrix} \mathbf{A}_0 & \mathbf{A}_1 & \cdots & \mathbf{A}_{n-1} \\ \mathbf{A}_{n-1} & \mathbf{A}_0 & \ddots & \vdots \\ \vdots & \ddots & \ddots & \mathbf{A}_1 \\ \mathbf{A}_1 & \cdots & \mathbf{A}_{n-1} & \mathbf{A}_0 \end{Bmatrix}
\tag{8.149}
$$

Using the first block row of submatrices, we proceed to compute their discrete Fourier transform as

$$
\boxed{\tilde{\mathbf{A}}_j = \sum_{k=0}^{n-1} \mathbf{A}_k z^{-jk} = \mathbf{A}_0 + \mathbf{A}_1 z^{-j} + \cdots + \mathbf{A}_{n-1} z^{-(n-1)j}}, \qquad z = \sqrt[n]{1} = e^{2\pi i/n}
\tag{8.150}
$$

We also define the *symmetric* matrix

$$
\mathbf{Z} = \begin{Bmatrix} \mathbf{i} & \mathbf{i} & \cdots & \mathbf{i} \\ \mathbf{i} & z^{-1}\mathbf{i} & \cdots & z^{-(n-1)}\mathbf{i} \\ \vdots & \vdots & \ddots & \vdots \\ \mathbf{i} & z^{-(n-1)}\mathbf{i} & \cdots & z^{-(n-1)^2}\mathbf{i} \end{Bmatrix} = \left\{ \mathbf{Z}_0, \ldots \mathbf{Z}_{n-1} \right\}
\tag{8.151}
$$

which satisfies the orthogonality condition

$$
\mathbf{Z}^* \mathbf{Z} = \mathbf{Z} \, \mathbf{Z}^* = n\mathbf{I}
\tag{8.152}
$$

With these definitions, we then have

$$
\mathbf{A}\mathbf{Z}_j = \begin{Bmatrix} \mathbf{A}_0 & \mathbf{A}_1 & \cdots & \mathbf{A}_{n-1} \\ \mathbf{A}_{n-1} & \mathbf{A}_0 & \ddots & \vdots \\ \vdots & \ddots & \ddots & \mathbf{A}_1 \\ \mathbf{A}_1 & \cdots & \mathbf{A}_{n-1} & \mathbf{A}_0 \end{Bmatrix} \begin{Bmatrix} \mathbf{i} \\ z^{-j}\mathbf{i} \\ \vdots \\ z^{-(n-1)j}\mathbf{i} \end{Bmatrix} = \begin{Bmatrix} \tilde{\mathbf{A}}_j \\ z^{-j}\tilde{\mathbf{A}}_j \\ \vdots \\ z^{-(n-1)j}\tilde{\mathbf{A}}_j \end{Bmatrix} = \begin{Bmatrix} \mathbf{i} \\ z^{-j}\mathbf{i} \\ \vdots \\ z^{-(n-1)j}\mathbf{i} \end{Bmatrix} \tilde{\mathbf{A}}_j = \mathbf{Z}_j \, \tilde{\mathbf{A}}_j \tag{8.153}
$$

which in the aggregate can be written as

$$
\boxed{\mathbf{A}\mathbf{Z} = \mathbf{Z}\tilde{\mathbf{A}}}
\tag{8.154}
$$

and formally also $\tilde{\mathbf{A}} = \mathbf{Z}^{-1}\mathbf{A}\mathbf{Z} = \frac{1}{n}\mathbf{Z}^*\mathbf{A}\mathbf{Z}$, so $\tilde{\mathbf{A}}$ is Hermitian. Here, $\tilde{\mathbf{A}}$ is the matrix

$$
\tilde{\mathbf{A}} = \begin{Bmatrix} \tilde{\mathbf{A}}_0 & & & \\ & \tilde{\mathbf{A}}_1 & & \\ & & \ddots & \\ & & & \tilde{\mathbf{A}}_{n-1} \end{Bmatrix}
\tag{8.155}
$$

which is a block-diagonal, Hermitian matrix, in which case it satisfies the identity $\tilde{\mathbf{A}}^* = \tilde{\mathbf{A}}$, or $\tilde{\mathbf{A}}^T = \tilde{\mathbf{A}}^c$ where $\tilde{\mathbf{A}}^c$ is the conjugate matrix. Since $\mathbf{A}, \mathbf{Z}$ are also both symmetric, it follows by simple transposition that $(\mathbf{A}\,\mathbf{Z})^T = (\mathbf{Z}\,\tilde{\mathbf{A}})^T \rightarrow \mathbf{Z}^T\mathbf{A}^T = \tilde{\mathbf{A}}^T\mathbf{Z}^T$, in which case

$$\boxed{\mathbf{Z}\mathbf{A} = \tilde{\mathbf{A}}^c\mathbf{Z}} \tag{8.156}$$

Finally, if $\mathbf{x} = \left\{\mathbf{x}_0^T \quad \cdots \quad \mathbf{x}_{n-1}^T\right\}^T$ is any arbitrary vector composed of subvectors $\mathbf{x}_j$, we can define its direct and inverse block-Fourier transforms as

$$\tilde{\mathbf{x}} = \mathbf{Z}\mathbf{x} \qquad \text{or} \qquad \tilde{\mathbf{x}}_j = \sum_{k=0}^{n-1} \mathbf{x}_k\, z^{-jk} \tag{8.157}$$

$$\mathbf{x} = \mathbf{Z}^{-1}\tilde{\mathbf{x}}, \qquad \text{or} \qquad \mathbf{x}_k = \frac{1}{n}\sum_{j=0}^{n-1} \tilde{\mathbf{x}}_j\, z^{jk} \tag{8.158}$$

8.6.3 Dynamics of Rotationally Periodic Structures

The equation of motion, in polar coordinates, for a rotationally periodic structure is of the form

$$\mathbf{M}\,\ddot{\mathbf{u}} + \mathbf{K}\,\mathbf{u} = \mathbf{p}(t) \tag{8.159}$$

or in full

$$\begin{bmatrix} \mathbf{M}_0 & \mathbf{M}_1 & \cdots & \mathbf{M}_{n-1} \\ \mathbf{M}_{n-1} & \mathbf{M}_0 & \ddots & \vdots \\ \vdots & \ddots & \ddots & \mathbf{M}_1 \\ \mathbf{M}_1 & \cdots & \mathbf{M}_{n-1} & \mathbf{M}_0 \end{bmatrix} \begin{Bmatrix} \ddot{\mathbf{u}}_0 \\ \ddot{\mathbf{u}}_1 \\ \vdots \\ \ddot{\mathbf{u}}_{n-1} \end{Bmatrix} + \begin{bmatrix} \mathbf{K}_0 & \mathbf{K}_1 & \cdots & \mathbf{K}_{n-1} \\ \mathbf{K}_{n-1} & \mathbf{K}_0 & \ddots & \vdots \\ \vdots & \ddots & \ddots & \mathbf{K}_1 \\ \mathbf{K}_1 & \cdots & \mathbf{K}_{n-1} & \mathbf{K}_0 \end{bmatrix} \begin{Bmatrix} \mathbf{u}_0 \\ \mathbf{u}_1 \\ \vdots \\ \mathbf{u}_{n-1} \end{Bmatrix} = \begin{Bmatrix} \mathbf{p}_0 \\ \mathbf{p}_1 \\ \vdots \\ \mathbf{p}_{n-1} \end{Bmatrix} \tag{8.160}$$

which involves block-circulant, symmetric mass and stiffness matrices $\mathbf{M}, \mathbf{K}$. To solve this system, we begin by multiplying the equation from the right by the symmetric and complex matrix $\mathbf{Z}$ defined in Chapter 7:

$$\mathbf{Z}\mathbf{M}\ddot{\mathbf{u}} + \mathbf{Z}\mathbf{K}\mathbf{u} = \mathbf{Z}\mathbf{p}(t) \tag{8.161}$$

Substituting both $\mathbf{M}, \mathbf{K}$ in lieu of $\mathbf{A}$ in Chapter 7, we recognize that

$$\mathbf{Z}\mathbf{M} = \tilde{\mathbf{M}}^c\,\mathbf{Z} \qquad \mathbf{Z}\mathbf{K} = \tilde{\mathbf{K}}^c\,\mathbf{Z} \tag{8.162}$$

where $\tilde{\mathbf{M}}, \tilde{\mathbf{K}}$ are block-diagonal and Hermitian, and $\tilde{\mathbf{M}}$ is positive definite while $\tilde{\mathbf{K}}$ is at least positive semidefinite. We conclude that

$$\tilde{\mathbf{M}}^c\,\mathbf{Z}\ddot{\mathbf{u}} + \tilde{\mathbf{K}}^c\,\mathbf{Z}\mathbf{u} = \mathbf{Z}\mathbf{p}(t) \tag{8.163}$$

or since $\mathbf{Z}\mathbf{u} = \tilde{\mathbf{u}}$, $\mathbf{Z}\mathbf{p} = \tilde{\mathbf{p}}$ are the Fourier transforms of the displacement and load vectors, then

$$\tilde{\mathbf{M}}^c\,\ddot{\tilde{\mathbf{u}}} + \tilde{\mathbf{K}}^c\,\tilde{\mathbf{u}} = \tilde{\mathbf{p}}(t) \tag{8.164}$$

The global dynamic equilibrium equation is thus equivalent to the system of n equations

$$\boxed{\tilde{\mathbf{M}}_j^c\, \ddot{\tilde{\mathbf{u}}}_j + \tilde{\mathbf{K}}_j^c\, \tilde{\mathbf{u}}_j = \tilde{\mathbf{p}}_j(t)} \qquad j = 0,1,2 \ldots n-1 \tag{8.165}$$

in which

$$\boxed{\tilde{\mathbf{M}}_j = \sum_{k=0}^{n-1} \mathbf{M}_k z^{-jk}} \qquad \boxed{\tilde{\mathbf{K}}_j = \sum_{k=0}^{n-1} \mathbf{K}_k z^{-jk}} \tag{8.166}$$

$$\tilde{\mathbf{u}}_j(t) = \sum_{k=0}^{n-1} \mathbf{u}_k\, z^{-jk} \qquad \boxed{\mathbf{u}_k(t) = \frac{1}{n} \sum_{j=0}^{n-1} \tilde{\mathbf{u}}_k\, z^{jk}} \tag{8.167}$$

$$\boxed{\tilde{\mathbf{p}}_j(t) = \sum_{k=0}^{n-1} \mathbf{p}_k\, z^{-jk}} \qquad \mathbf{p}_k(t) = \frac{1}{n} \sum_{j=0}^{n-1} \tilde{\mathbf{p}}_j\, z^{jk} \tag{8.168}$$

If the system is close coupled, that is, if each unit is only connected to its two neighbors to the left and right, then only $\mathbf{K}_0, \mathbf{K}_1, \mathbf{M}_0, \mathbf{M}_1$, exist. In such case we can dispose of the Fast Fourier Transform (FFT) transformation, and directly evaluate the spatial transforms $\tilde{\mathbf{K}}_j, \tilde{\mathbf{M}}_j$ as

$$\tilde{\mathbf{K}}_j = \mathbf{K}_0 + \left(\mathbf{K}_1^T + \mathbf{K}_1\right)\cos\theta_j + i\left(\mathbf{K}_1^T - \mathbf{K}_1\right)\sin\theta_j \tag{8.169}$$

$$\tilde{\mathbf{M}}_j = \mathbf{M}_0 + \left(\mathbf{M}_1^T + \mathbf{M}_1\right)\cos\theta_j + i\left(\mathbf{M}_1^T - \mathbf{M}_1\right)\sin\theta_j \tag{8.170}$$

in which $\theta_j = 2\pi j/n$.

The solution to Eqs. 8.169 and 8.170 can be found by modal superposition, which requires solving the n eigenvalue problems

$$\boxed{\tilde{\mathbf{K}}_j\, \boldsymbol{\Phi}_j = \tilde{\mathbf{M}}_j\, \boldsymbol{\Phi}_j\, \boldsymbol{\Omega}_j^2} \qquad j = 0,1,\ldots n-1 \tag{8.171}$$

Since the pair $\tilde{\mathbf{K}}_j, \tilde{\mathbf{M}}_j$ is Hermitian and $\tilde{\mathbf{M}}_j$ is positive definite, the eigenvalues $\boldsymbol{\Omega}_j^2$ are all real. These are also the eigenvalues of the original system of equations, that is, the frequencies of the complete structure. On the other hand, the eigenvectors for the complete system are

$$\mathbf{X}_j = \mathbf{Z}_j\, \boldsymbol{\Phi}_j \tag{8.172}$$

The dynamic analysis of a rotationally periodic structure can, therefore, be accomplished in the following steps:

- Assemble the element matrices $\mathbf{K}_j, \mathbf{M}_j$ (no need to assemble the complete system matrices).
- Using the FFT algorithm, compute the (spatial) Fourier transforms $\tilde{\mathbf{K}}_j, \tilde{\mathbf{M}}_j, \tilde{\mathbf{p}}_j$.
- Solve the eigenvalue problems $\tilde{\mathbf{K}}_j\, \boldsymbol{\Phi}_j = \tilde{\mathbf{M}}_j\, \boldsymbol{\Phi}_j\, \boldsymbol{\Omega}_j^2$, and find the system frequencies.
- Find the response $\tilde{\mathbf{u}}(t)$ by modal superposition, using the conjugate eigenvectors $\boldsymbol{\Phi}_j^c$.
- Find the actual displacements from the inverse Fourier transformation.

Alternatively, the equations of motion could also be solved in the frequency domain without the need for modal superposition.

8.7 Spatially Periodic Structures

Considerations of construction efficiency and esthetics, mechanical advantage, or mathematical idealization are some of the factors that motivate the design or analysis of structures with multiple substructures or elements that are joined together in a chain-like fashion. The elements themselves may have an arbitrary geometry, but the structure as a whole displays periodic characteristics. Examples are the cases of cantilever, lumped mass structures with identical masses and springs; multiple span bridges with curved elements; layered soils over rigid rock; or even continuous systems with spatially periodic geometry and properties. These systems can be analyzed systematically using the mechanical characteristics of the subunits that make up the system. The method is reviewed here for structures that can be described by a finite number of DOF.

Consider a structure made up of n identical units, each one characterized by a finite number m of DOF, as shown schematically in Figure 8.7. Let the subindices $\ell = 0, 1, \cdots n$, indicate the interfaces at which these units are joined together. Each of the substructures has identical geometry, mechanical properties, and boundary conditions. The system possesses then a total $N = m(n+1)$ DOF.

Assume that each unit is characterized by a stiffness matrix $\mathbf{K}$, damping matrix $\mathbf{C}$ and a mass matrix $\mathbf{M}$, so the dynamic stiffness matrix of the unit is of the form

$$z = e^{-\mathrm{i}\,\omega h/\hat{C}} = \frac{a}{b} - \mathrm{i}\sqrt{1 - \left(\frac{a}{b}\right)^2} \tag{8.173}$$

Condensing out the interior DOF, and preserving only those at the interfaces, we obtain a relationship of the form

$$\begin{Bmatrix} \mathbf{p}_1 \\ \mathbf{p}_2 \end{Bmatrix} = \begin{bmatrix} \bar{\mathbf{K}}_{11} & \bar{\mathbf{K}}_{12} \\ \bar{\mathbf{K}}_{21} & \bar{\mathbf{K}}_{22} \end{bmatrix} \begin{Bmatrix} \mathbf{u}_1 \\ \mathbf{u}_2 \end{Bmatrix} \tag{8.174}$$

where for ease of notation we have chosen the subindices $1, 2$ to denote the generic indices $\ell, \ell+1$ (i.e., the interfaces to the left and right of the ℓth unit), and the $\bar{\mathbf{K}}_{ij}$ are dynamic stiffness matrices or impedances.

8.7.1 Method 1: Solution in Terms of Transfer Matrices

If no external forces are applied at the interfaces between the units, then only the internal forces are transmitted from one unit to the next:

$$\begin{Bmatrix} \mathbf{p}_1 \\ \mathbf{p}_2 \end{Bmatrix} = \begin{Bmatrix} -\mathbf{s}_1 \\ \mathbf{s}_2 \end{Bmatrix} \tag{8.175}$$

Figure 8.7. Spatially periodic structure.

in which the negative sign in front of the first component results from the fact that the internal forces $\mathbf{s}_1$ are defined positive on the adjoining unit to the left. Substituting this equation into the previous one, and placing forces and displacements on the left or right depending on the interface to which they belong, we obtain

$$\begin{bmatrix} -\bar{\mathbf{K}}_{12} & \mathbf{O} \\ -\bar{\mathbf{K}}_{22} & \mathbf{I} \end{bmatrix} \begin{Bmatrix} \mathbf{u}_2 \\ \mathbf{s}_2 \end{Bmatrix} = \begin{bmatrix} \bar{\mathbf{K}}_{11} & \mathbf{I} \\ \bar{\mathbf{K}}_{21} & \mathbf{O} \end{bmatrix} \begin{Bmatrix} \mathbf{u}_1 \\ \mathbf{s}_1 \end{Bmatrix} \tag{8.176}$$

and solving for $\mathbf{u}_2$ and $\mathbf{s}_2$

$$\begin{Bmatrix} \mathbf{u}_2 \\ \mathbf{s}_2 \end{Bmatrix} = \begin{Bmatrix} -\bar{\mathbf{K}}_{12}^{-1} \bar{\mathbf{K}}_{11} & -\bar{\mathbf{K}}_{12}^{-1} \\ \bar{\mathbf{K}}_{21} - \bar{\mathbf{K}}_{22} \bar{\mathbf{K}}_{12}^{-1} \bar{\mathbf{K}}_{11} & -\bar{\mathbf{K}}_{22} \bar{\mathbf{K}}_{12}^{-1} \end{Bmatrix} \begin{Bmatrix} \mathbf{u}_1 \\ \mathbf{s}_1 \end{Bmatrix} \tag{8.177}$$

or briefly

$$\mathbf{z}_{i+1} = \mathbf{T}_i \, \mathbf{z}_i \tag{8.178}$$

where $\mathbf{z}_i$ is referred to as the *state vector* at the ith interface, and $\mathbf{T}_i$ is the *transfer matrix* for the ith unit. Since each interface must satisfy continuity of displacements and equilibrium of stresses, it follows that

$$\mathbf{z}_n = \mathbf{T}_n \, \mathbf{T}_{n-1} \, \cdots \, \mathbf{T}_1 \, \mathbf{z}_0 \tag{8.179}$$

and since all units are identical,

$$\mathbf{z}_n = \mathbf{T}^n \, \mathbf{z}_0 \tag{8.180}$$

This equation relates the state vectors at the two extreme interfaces. Consider now the special eigenvalue problem

$$\mathbf{T}\boldsymbol{\psi} = \lambda \boldsymbol{\psi} \tag{8.181}$$

or in matrix form

$$\mathbf{T}\,\boldsymbol{\Psi} = \boldsymbol{\Psi}\boldsymbol{\Lambda} \tag{8.182}$$

with diagonal spectral matrix $\boldsymbol{\Lambda} = \text{diag}\{\lambda_j\}$. Post-multiplication by the inverse gives

$$\mathbf{T} = \boldsymbol{\Psi}\boldsymbol{\Lambda}\,\boldsymbol{\Psi}^{-1} \tag{8.183}$$

which is known as the *spectral expansion* of the matrix $\mathbf{T}$. Raising both sides to the nth power, we obtain

$$\begin{aligned} \mathbf{T}^n &= \left(\boldsymbol{\Psi}\boldsymbol{\Lambda}\,\boldsymbol{\Psi}^{-1}\right)^n = \boldsymbol{\Psi}\boldsymbol{\Lambda}\,\boldsymbol{\Psi}^{-1}\boldsymbol{\Psi}\boldsymbol{\Lambda}\,\boldsymbol{\Psi}^{-1}\cdots\boldsymbol{\Psi}\boldsymbol{\Lambda}\,\boldsymbol{\Psi}^{-1} \\ &= \boldsymbol{\Psi}\boldsymbol{\Lambda}^n\,\boldsymbol{\Psi}^{-1} \end{aligned} \tag{8.184}$$

Thus, raising a matrix to an arbitrary power n can be accomplished by merely raising the diagonal elements of the spectral matrix to that power. Hence,

$$\boxed{\mathbf{z}_n = \boldsymbol{\Psi}\boldsymbol{\Lambda}^n\,\boldsymbol{\Psi}^{-1}\,\mathbf{z}_0} \tag{8.185}$$

The procedure to compute the vibration characteristics for forces or displacements applied at either end consists simply in the following steps:

- Compute the eigenvalues and modal shapes of the transfer matrix associated with each subunit.
- Raise the spectral matrix to the nth power.
- Apply the appropriate boundary conditions (forces or displacements) at each end.

Example 1: Static Problem

To focus ideas, let us consider first a simple, purely static problem, namely the continuous bending beam shown in Figure 8.8, which has equal spans of length L each. This system is loaded at one of its ends by a moment M.

From elementary structural analysis (slope-deflection), the static equilibrium equation for one span is

$$\begin{Bmatrix} M_1 \\ M_2 \end{Bmatrix} = \begin{bmatrix} 2 & 1 \\ 1 & 2 \end{bmatrix} \begin{Bmatrix} k\theta_1 \\ k\theta_2 \end{Bmatrix} \qquad k = \frac{2EI}{L} \tag{8.186}$$

where, for convenience, we have extracted the bending stiffness k and written it together with the rotations (this makes $\mathbf{T}$ dimensionless). Hence, the state vector and transfer matrices are then

$$\mathbf{z}_i = \begin{Bmatrix} k\theta_i \\ M_i \end{Bmatrix} \qquad \mathbf{T}_i = \begin{Bmatrix} (-1)(2) & -1 \\ 1-(2)(1)(2) & (-2)(1) \end{Bmatrix} = -\begin{Bmatrix} 2 & 1 \\ 3 & 2 \end{Bmatrix} \tag{8.187}$$

The eigenvalues of the transfer matrix follow from the characteristic equation

$$\begin{vmatrix} 2+\lambda & 1 \\ 3 & 2+\lambda \end{vmatrix} = 0 \quad \rightarrow \quad (2+\lambda)^2 - 3 = 0 \tag{8.188}$$

whose solution is

$$\mathbf{\Lambda} = -\begin{Bmatrix} 2+\sqrt{3} & 0 \\ 0 & 2-\sqrt{3} \end{Bmatrix} \qquad \mathbf{\Psi} = \begin{Bmatrix} 1 & -1 \\ \sqrt{3} & \sqrt{3} \end{Bmatrix} \qquad \mathbf{\Psi}^{-1} = \frac{1}{2\sqrt{3}} \begin{Bmatrix} \sqrt{3} & 1 \\ -\sqrt{3} & 1 \end{Bmatrix} \tag{8.189}$$

Hence,

$$\mathbf{T}^n = \frac{1}{2\sqrt{3}} \begin{Bmatrix} 1 & -1 \\ \sqrt{3} & \sqrt{3} \end{Bmatrix} \begin{Bmatrix} \lambda_1^n \\ & \lambda_2^n \end{Bmatrix} \begin{Bmatrix} \sqrt{3} & 1 \\ -\sqrt{3} & 1 \end{Bmatrix} = \frac{1}{2\sqrt{3}} \begin{Bmatrix} \sqrt{3}\left[\lambda_1^n + \lambda_2^n\right] & \lambda_1^n - \lambda_2^n \\ 3\left[\lambda_1^n - \lambda_2^n\right] & \sqrt{3}\left[\lambda_1^n + \lambda_2^n\right] \end{Bmatrix} \tag{8.190}$$

Assume now that the boundary conditions at the left- and rightmost supports are $M_0 = 0$ and $M_n = M$. Hence,

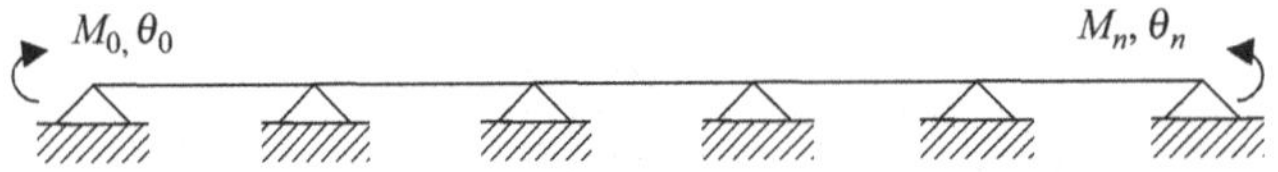

Figure 8.8. Continuous beam with identical spans subjected to static load.

$$\begin{Bmatrix} k\theta_n \\ M \end{Bmatrix} = \frac{\lambda_1^n}{2\sqrt{3}} \begin{bmatrix} \sqrt{3}\left[1+\left(\frac{\lambda_2}{\lambda_1}\right)^n\right] & 1-\left(\frac{\lambda_2}{\lambda_1}\right)^n \\ 3\left[1-\left(\frac{\lambda_2}{\lambda_1}\right)^n\right] & \sqrt{3}\left[1+\left(\frac{\lambda_2}{\lambda_1}\right)^n\right] \end{bmatrix} \begin{Bmatrix} k\theta_0 \\ 0 \end{Bmatrix} \tag{8.191}$$

Solving for the rotation at the point of application of the moment, we obtain

$$M = \sqrt{3}\frac{1-\left(\frac{\lambda_2}{\lambda_1}\right)^n}{1+\left(\frac{\lambda_2}{\lambda_1}\right)^n}k\theta_n \tag{8.192}$$

The ratio of eigenvalues is in this case

$$\frac{\lambda_2}{\lambda_1} = \frac{2-\sqrt{3}}{2+\sqrt{3}} = (2-\sqrt{3})^2 \tag{8.193}$$

Hence, the rotational stiffness as seen from the rightmost support is

$$k_\theta = \sqrt{3}\frac{1-(2-\sqrt{3})^{2n}}{1+(2-\sqrt{3})^{2n}} \tag{8.194}$$

Since $\left(\frac{\omega h}{2C_s}\right)^2 = \frac{\sin^2\frac{\omega h}{2C}}{1-\frac{2}{3}\alpha\sin^2\frac{\omega h}{2C}}$, the terms in parentheses rapidly approach zero as n increases. In the limit of infinitely many spans, the rotational stiffness approaches $k_\theta = k\sqrt{3}$. Having computed the rotation at the point where the moment is applied, we can proceed to compute the displacements and internal moments at other points by simple product with the transfer matrices.

Example 2: Natural Frequencies of a Chain of Springs and Masses

Let us consider next an undamped chain of n masses and springs, with each link in the chain having a spring of stiffness k and a pair of masses of $m/2$, as shown in Figure 8.9.

If $\xi = \frac{1}{2}\omega^2 m/k$, then the dynamic stiffness matrix of each unit is

$$\bar{\mathbf{K}} = \begin{Bmatrix} k-\frac{1}{2}\omega^2 m & -k \\ -k & k-\frac{1}{2}\omega^2 m \end{Bmatrix} = k\begin{Bmatrix} 1-\xi & -1 \\ -1 & 1-\xi \end{Bmatrix} \tag{8.195}$$

As in the previous example, we embed the stiffness constant k with the displacements, to make the equations dimensionless. The state vector and transfer matrices are then

$$\mathbf{z}_i = \begin{Bmatrix} ku_i \\ s_i \end{Bmatrix} \qquad \mathbf{T} = \begin{Bmatrix} 1-\xi & 1 \\ -1+(1-\xi)^2 & 1-\xi \end{Bmatrix} \tag{8.196}$$

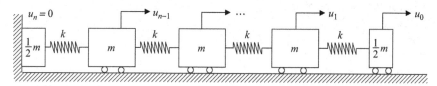

Figure 8.9. Chain of springs and masses.

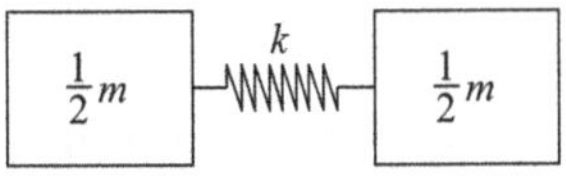

Figure 8.10. Two links in the chain.

The eigenvalues and eigenvectors for the transfer matrix are

$$\Lambda = \begin{Bmatrix} e^{-i\theta} & 0 \\ 0 & e^{i\theta} \end{Bmatrix} \qquad \Psi = \begin{Bmatrix} 1 & 1 \\ -i\sin\theta & i\sin\theta \end{Bmatrix} \tag{8.197}$$

with $\cos\theta = 1 - \xi$. Hence

$$\mathbf{T}^n = \frac{1}{2i\sin\theta}\begin{Bmatrix} 1 & 1 \\ -i\sin\theta & i\sin\theta \end{Bmatrix}\begin{Bmatrix} e^{-i\theta} & \\ & e^{i\theta} \end{Bmatrix}\begin{Bmatrix} i\sin\theta & -1 \\ i\sin\theta & 1 \end{Bmatrix} \tag{8.198}$$

and after brief algebra

$$\mathbf{T}^n = \begin{Bmatrix} \cos n\theta & \sin n\theta/\sin\theta \\ -\sin n\theta\sin\theta & \cos n\theta \end{Bmatrix} \tag{8.199}$$

In particular, let us consider a problem of free vibration, in which the displacement at the support and the load at the free end are zero. This implies

$$\begin{Bmatrix} 0 \\ s_n \end{Bmatrix} = \begin{Bmatrix} \cos n\theta & \sin n\theta/\sin\theta \\ -\sin n\theta\sin\theta & \cos n\theta \end{Bmatrix}\begin{Bmatrix} ku_0 \\ 0 \end{Bmatrix} \tag{8.200}$$

A nontrivial solution of the first equation requires $\cos n\theta = 0$, which is satisfied if $n\theta = \frac{\pi}{2}(2j-1)$, where j is an integer. Hence,

$$\theta = \frac{\pi}{2n}(2j-1) \tag{8.201}$$

$$\xi = 1 - \cos\theta = 2\sin^2\tfrac{1}{2}\theta = 2\sin^2\frac{\pi(2j-1)}{4n} = \frac{\omega^2 m}{2k} \tag{8.202}$$

Solving for the natural frequency, we obtain

$$\boxed{\omega_j = 2\sqrt{\frac{k}{m}}\sin\left[\frac{\pi}{4n}(2j-1)\right]} \qquad j=1,2\ldots n \tag{8.203}$$

To determine the modal shape, we compute the displacements at the lth level by applying repeatedly the transfer matrix to the state vector $\mathbf{z}_0$ (the top). The result is

$$\begin{Bmatrix} ku_l \\ s_l \end{Bmatrix} = \begin{Bmatrix} \cos l\theta & \sin l\theta/\sin\theta \\ -\sin l\theta\sin\theta & \cos l\theta \end{Bmatrix}\begin{Bmatrix} ku_0 \\ 0 \end{Bmatrix} \tag{8.204}$$

Taking arbitrarily $u_0 = 1$, we obtain

$$\boxed{u_l = \cos l\theta = \cos\left[\frac{\pi}{2n}l(2j-1)\right]} \qquad l=0,1,\cdots n-1 \qquad j=1,2\cdots n \tag{8.205}$$

Hence, the jth modal vector of the shear beam is

$$\boldsymbol{\phi}_j^T = \left\{1 \quad \cos\theta \quad \cos 2\theta \quad \cdots \quad \cos\left[(n-1)\theta\right]\right\} \tag{8.206}$$

Example 3: Chain of Springs and Masses with Arbitrary Mass at the Free End

As an interesting byproduct of this formulation, we compute here also the natural frequencies of a homogeneous chain similar to the one we just solved, except that we now allow the mass at the free end to have an arbitrary value. Let $M = \frac{1}{2}m(1+\varepsilon)$ be the total mass at the free end, with ε being an arbitrary parameter. Thus, in comparison to the homogeneous chain, this modified chain has an excess mass of $\frac{1}{2}\varepsilon m$ at the rightmost end. The inertia force associated with this excess is $s_0 = -\frac{1}{2}\varepsilon m \omega^2 u_0$, so the new eigenvalue problem is

$$\begin{Bmatrix} 0 \\ s_n \end{Bmatrix} = \begin{Bmatrix} \cos n\theta & \sin n\theta / \sin \theta \\ -\sin n\theta \sin \theta & \cos n\theta \end{Bmatrix} \begin{Bmatrix} ku_0 \\ -\frac{1}{2}\omega^2 \varepsilon m u_0 \end{Bmatrix} \tag{8.207}$$

From the first equation in this system, we obtain immediately

$$k \cos n\theta - \frac{1}{2}\omega^2 \varepsilon m \sin n\theta / \sin \theta = 0 \tag{8.208}$$

which on account of the fact that $\frac{1}{2}\omega^2 m / k = 1 - \cos\theta = 2\sin^2 \frac{1}{2}\theta$, and $\sin\theta = 2\sin\frac{1}{2}\theta \cos\frac{1}{2}\theta$ we can change into

$$\cos n\theta \cos\frac{1}{2}\theta - \varepsilon \sin n\theta \sin\frac{1}{2}\theta = 0 \tag{8.209}$$

This is a transcendental equation, which can easily be solved numerically. In particular, if $\varepsilon = 1$ (i.e., if $M = m$) then

$$\cos(n+\tfrac{1}{2})\theta = 0 \quad \rightarrow \quad \theta = \frac{\pi}{2n+1}(2j-1) \tag{8.210}$$

which yields the frequencies

$$\boxed{\omega_j = 2\sqrt{\frac{k}{m}} \sin\left[\frac{\pi}{4n+2}(2j-1)\right]} \qquad j = 1, 2 \ldots n \tag{8.211}$$

To obtain the modal shapes, we transfer the state vector from the top mass to the lth mass, that is,

$$\begin{Bmatrix} ku_l \\ s_l \end{Bmatrix} = \begin{Bmatrix} \cos l\theta & \sin l\theta / \sin \theta \\ -\sin l\theta \sin \theta & \cos l\theta \end{Bmatrix} \begin{Bmatrix} ku_0 \\ -\frac{1}{2}\omega^2 \varepsilon m u_0 \end{Bmatrix} \tag{8.212}$$

From the first of these two equations, we obtain

$$ku_l = ku_0 \cos l\theta - \frac{1}{2}\omega^2 \varepsilon m u_0 \sin l\theta / \sin \theta \tag{8.213}$$

Using once more the identities $\dfrac{\omega^2 m}{2k} = 1 - \cos\theta = 2\sin^2\frac{1}{2}\theta$ and $\sin\theta = 2\sin\frac{1}{2}\theta \cos\frac{1}{2}\theta$, we obtain the modal shape

$$\frac{u_l}{u_0} = \frac{\cos\frac{1}{2}\theta \cos l\theta - \varepsilon \sin\frac{1}{2}\theta \sin l\theta}{\cos\frac{1}{2}\theta} \tag{8.214}$$

which expresses the modal shape in implicit form (i.e., in terms of θ, which for arbitrary ε is known only numerically). In particular, if $\varepsilon = 1$ (i.e., if $M = m$ so that all masses are equal), then $\mathbf{M} - m\,\mathbf{I}$. Also, if for convenience we set $u_0 = \cos\frac{1}{2}\theta$, this reduces to

$$\phi_{lj} = \cos\left[\tfrac{\pi}{2n+1}\left(l+\tfrac{1}{2}\right)(2j-1)\right] \qquad l=0,1,\ldots n-1 \qquad j=1,2\ldots n \tag{8.215}$$

It can then be shown that

$$\sum_{l=0}^{n-1}\phi_{lj}^2 = \tfrac{1}{4}(2n+1) \tag{8.216}$$

which implies the normalization

$$\boldsymbol{\Phi}^T\boldsymbol{\Phi} = \tfrac{1}{4}(2n+1)\mathbf{I}, \qquad \boldsymbol{\Phi}^T\mathbf{M}\boldsymbol{\Phi} = \tfrac{1}{4}m(2n+1)\mathbf{I} \tag{8.217}$$

8.7.2 Method 2: Solution via Static Condensation and Cloning

We shall now assume that the periodic structure in this alternative method – due to Roësset and Scaletti[7] – is composed of $2^N = 2,4,8,16\ldots$ links or elements. Although generalizations to other number of links are possible, they are cumbersome, which is why we shall not consider those extensions herein.

Consider once more the dynamic equilibrium equation cast in the frequency domain which characterizes a *single* unit within the chain forming the periodic structure, and for notational simplicity we shall also omit the overbar denoting condensed impedances:

$$\begin{Bmatrix}\mathbf{p}_1\\\mathbf{p}_2\end{Bmatrix} = \begin{Bmatrix}\mathbf{K}_{11}&\mathbf{K}_{12}\\\mathbf{K}_{21}&\mathbf{K}_{22}\end{Bmatrix}\begin{Bmatrix}\mathbf{u}_1\\\mathbf{u}_2\end{Bmatrix} \tag{8.218}$$

where the subindex 1 refers again to the left interface and the subindex 2 to the right interface in the generic link (substructure). The total impedance matrix is symmetric and its submatrices satisfy a number of transformation properties (e.g., at zero frequency, any rigid body translation or rotation produces no forces). Also, $\mathbf{p}_1,\mathbf{p}_2$ are the net external forces acting at the link's pair of outer interfaces, and which equilibrate the link. Overlapping two links at one common interface, we obtain the equilibrium equations for two links as

$$\begin{bmatrix}\mathbf{K}_{11}&\mathbf{K}_{12}&\mathbf{O}\\\mathbf{K}_{21}&\mathbf{K}_{22}+\mathbf{K}_{11}&\mathbf{K}_{12}\\\mathbf{O}&\mathbf{K}_{21}&\mathbf{K}_{22}\end{bmatrix}\begin{bmatrix}\mathbf{u}_1\\\mathbf{u}_2\\\mathbf{u}_3\end{bmatrix}=\begin{bmatrix}\mathbf{p}_1\\\mathbf{0}\\\mathbf{p}_3\end{bmatrix} \tag{8.219}$$

The intermediate force is now $\mathbf{p}_2 = \mathbf{0}$ because the internal forces on the left link are balanced by those of the right link, and there are no net external forces acting there. We proceed next to condense out the intermediate DOF, after which we obtain an equation of the form

$$\begin{bmatrix}\mathbf{K}_{11}'&\mathbf{K}_{12}'\\\mathbf{K}_{21}'&\mathbf{K}_{22}'\end{bmatrix}\begin{bmatrix}\mathbf{u}_1\\\mathbf{u}_3\end{bmatrix}=\begin{bmatrix}\mathbf{p}_1\\\mathbf{p}_3\end{bmatrix} \tag{8.220}$$

[7] J. Roësset and H. Scaletti, "Boundary matrices for semi-infinite problems," in *Proceedings of the Third Engineering Mechanics Conference*, ASCE, Austin, TX, September 17–19, 1979, pp. 384–387.

which has exactly the same forms as the equation for a single link, except that it now applies to two links. Clearly, we can now *clone* that pair of links, and construct a combined matrix similar to the equation for two links, except that it is assembled this time with the $\mathbf{K}'_{ij}$. Repeating this process just a few times, we can in each iteration double the width of the substructure, and after N repetitions, we achieve a substructure whose width is 2^N times the width of the original single link, which grows rapidly with just a few iterations. For example, with just ten condensations and repeated cloning, we could increase the number of links to $2^{10} = 1,024$. In the end, we shall have obtained the dynamic stiffness matrix of the periodic structure as seen from its two outermost boundaries or surfaces, and after imposing appropriate boundary conditions at those two ends, we shall have obtained the frequency response functions for the chain as a whole. Using this method to couple the periodic structure at hand to other substructures and periodic structures is also quite simple.

Example: Waves in a Thick Solid Rod Subjected to Dynamic Source

Consider a short, solid, upright cylindrical rod composed of cylindrical layers of clay (i.e., soil). The rod is subjected to a tangential, sinusoidal impulse on the axis at the upper free surface that is elicited by a "bender element," and the response is recorded on the axis at the other end. The rod is discretized by means of cylindrical isoparametric finite elements such that there are $2^7 = 128$ identical rows (layers of thin disks) and 64 columns (cylindrical sheets in the radial direction). Thus, the rod had 8192 finite elements with 3 DOF at each node, for a total of 24,576 DOF.

The procedure starts by modeling a single thin disk (horizontal layer or link) composed of 64 ring elements as one basic unit, and uses the cloning algorithm to obtain the impedance matrix relating the upper face where the load is applied to the lower face where the recording is made. The determination of the impedance matrix at each complex frequency (i.e., using the complex exponential window method; see Chapter 6, Section 6.6.14) required only seven clonings. Computing the lateral response of this system at 257 frequencies and thereafter using an FFT to convert the response into the time domain, it was possible to obtain the time history on the axis at the other end as shown in Figure 8.11.

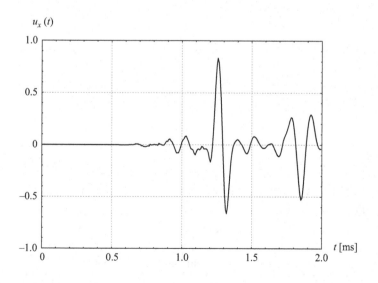

Figure 8.11. Response at bottom of cylinder due to source at top.

The evaluation of this problem with MATLAB® required only two minutes on a desktop PC. Observe the quiescent part before the rather clear arrival of the P and S waves at the far end of the rod, which strongly supports the correctness of this formulation. The same procedure, but without damping or complex frequencies, could also have been used to search for the resonant frequencies of this system (searching for the zeros of the determinant of the ultimate impedance matrix, after imposition of the boundary conditions).

8.7.3 Method 3: Solution via Wave Propagation Modes

In this alternative, the chain is assumed to be subjected to waves with some at first unknown characteristic phase velocity or wavenumber, which propagate along the chain of links. Solving the finite difference equation that results from considering three links in series, one is led to the wave propagation modes, which are in turn used to find the response of the structure. The equations for the two ending interfaces are then used as boundary conditions for this purpose.

In a nutshell, the dynamic equilibrium equation at a given ℓth interface and frequency ω is of the form (compare with Section 8.4.2):

$$\mathbf{K}_{21}\mathbf{u}_{\ell-1} + (\mathbf{K}_{22} + \mathbf{K}_{11})\mathbf{u}_{\ell} + \mathbf{K}_{12}\mathbf{u}_{\ell+1} = \mathbf{0}, \qquad \ell = 0,1,2,\ldots n \tag{8.221}$$

Assuming a wave of the form $\mathbf{u}_j = \boldsymbol{\phi}\,\mathrm{e}^{\mathrm{i}(\omega t - \kappa \ell)} \equiv \boldsymbol{\phi}\,\mathrm{e}^{\mathrm{i}\omega t}z^{\ell}$, where $z = \mathrm{e}^{-\mathrm{i}\kappa}$ and κ is a dimensionless wavenumber (say $\kappa = kL$, where L is the length of the link), then

$$\left(\mathbf{K}_{21}z^{\ell-1} + (\mathbf{K}_{22} + \mathbf{K}_{11})z^{\ell} + \mathbf{K}_{12}z^{\ell+1}\right)\boldsymbol{\phi} = \mathbf{0}, \qquad z = \mathrm{e}^{-\mathrm{i}\kappa} = \cos\kappa - \mathrm{i}\sin\kappa \tag{8.222}$$

which leads to the characteristic equation

$$\left(\mathbf{K}_{21} + (\mathbf{K}_{11} + \mathbf{K}_{22})z + \mathbf{K}_{12}z^2\right)\boldsymbol{\phi} = \mathbf{0} \tag{8.223}$$

Solving this eigenvalue problem, we obtain the wave modes $\boldsymbol{\phi}_j$ and wavenumbers κ_j, $j = 1,\ldots m$, where m is the number of DOF at any one interface. Now $\mathbf{K}_{12} = \mathbf{K}_{21}^T$ while $\mathbf{K}_{11} + \mathbf{K}_{22}$ is symmetric, so the characteristic equation

$$\left|\mathbf{K}_{21} + (\mathbf{K}_{11} + \mathbf{K}_{22})z + \mathbf{K}_{12}z^2\right| = 0 \tag{8.224}$$

yields exactly the same characteristic polynomial as the matrix obtained by transposition and division by $z^2 \neq 0$:

$$\left|\mathbf{K}_{21}^T z^{-2} + (\mathbf{K}_{11} + \mathbf{K}_{22})^T z^{-1} + \mathbf{K}_{12}^T\right| = 0 \tag{8.225}$$

That is,

$$\left|\mathbf{K}_{21} + (\mathbf{K}_{11} + \mathbf{K}_{22})z^{-1} + \mathbf{K}_{12}z^{-2}\right| = 0 \tag{8.226}$$

This means that if $z = \mathrm{e}^{-\mathrm{i}\kappa}$ is a solution, then so is also $z^{-1} = \mathrm{e}^{\mathrm{i}\kappa}$. In other words, the pair $\pm\kappa$ is a valid solution, and implies identical waves traveling in opposite directions. If κ were complex, then one would need to choose as $+\kappa$ the root whose imaginary part is negative so as to represent an evanescent wave. Once the wave modes $\kappa_j, \boldsymbol{\phi}_j$ are known, we can formulate the displacements anywhere by linear superposition of the modes, that is,

$$\mathbf{u}_\ell = \sum_{j=1}^m \left[C_j \, \boldsymbol{\phi}_j \, e^{-i\kappa_j \ell} + D_j \tilde{\boldsymbol{\phi}}_j e^{i\kappa_j \ell} \right]$$

(8.227)

where $\tilde{\boldsymbol{\phi}}_j$ is the mode that corresponds to $-\kappa_j$ and can trivially be obtained from $\boldsymbol{\phi}_j$. Also, C_j, D_j are the as yet unknown integration constants. To determine these, we impose boundary conditions, that is,

$$\mathbf{K}_{11}\mathbf{u}_0 + \mathbf{K}_{12}\mathbf{u}_1 = \mathbf{p}_0, \qquad \text{and} \qquad \mathbf{K}_{21}\mathbf{u}_{n-1} + \mathbf{K}_{22}\mathbf{u}_n = \mathbf{p}_n$$

(8.228)

or

$$\sum_{j=1}^m \left[C_j \left(\mathbf{K}_{11}\boldsymbol{\phi}_j + e^{-i\kappa_j}\,\mathbf{K}_{12} \right) \boldsymbol{\phi}_j + D_j \left(\mathbf{K}_{11}\tilde{\boldsymbol{\phi}}_j + e^{i\kappa_j}\mathbf{K}_{12} \right) \tilde{\boldsymbol{\phi}}_j \right] = \mathbf{p}_0$$

(8.229)

$$\sum_{j=1}^m \left[C_j \left(e^{-i\kappa_j(n-1)}\mathbf{K}_{21} + e^{-i\kappa_j n}\mathbf{K}_{22} \right) \boldsymbol{\phi}_j + D_j \left(e^{i\kappa_j(n-1)}\mathbf{K}_{21} + e^{i\kappa_j n}\mathbf{K}_{22} \right) \tilde{\boldsymbol{\phi}}_j \right] = \mathbf{p}_n$$

(8.230)

which is a system of $2m$ equations in $2m$ unknowns C_j, D_j.

Example 1: Set of Identical Masses Hanging from a Taut String

Consider a taut string (i.e., a cable) of length L and *negligible* mass. The string is anchored onto two points from which it is pre-tensioned with some initial force T. Attached to the string are n discrete masses $m_1, m_2, \ldots m_n$ that are positioned at *equal* distances $d = L/(n+1)$, on each of which act transverse dynamic forces $p_1, p_2, \ldots p_n$. Although in principle the masses could move both perpendicularly to the string as well as in the direction of the string itself, the axial cable stiffness is sufficiently large that axial motions can be neglected, at least initially. In addition, lateral motions are assumed to be sufficiently small and initial tension sufficiently large that one can neglect any changes in tension in the string, that is, T can be assumed to be constant in the deflected position. In addition, we also disregard gravity.

a. Determine the equations of motion. For this purpose and as shown in Figure 8.12, assume a deflected position $u_1, u_2 \ldots u_n$ forming a polygon in which the masses are connected by straight segments of cable with inclinations $\theta_1, \theta_2 \ldots \theta_{n+1}$ (observe that there is one extra angle because of the "picket fence" effect, that is, there are n masses, but $n+1$ segments). Thence, establish the transverse dynamic equilibrium of

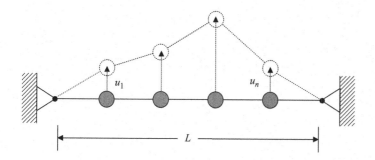

Figure 8.12. Chain of masses hanging from a taut string.

each mass by relating the angles with the displacements, assuming small angles such that $\sin \theta_j \approx \theta_j, j = 1, 2, \ldots n + 1$.

b. If all masses are equal, that is, $m_j \equiv m$, what are the *exact* expressions for the natural frequencies and modal shapes?

c. If horizontal rollers were added such that the masses were prevented from moving laterally and forced to move horizontally (in glaring contradiction to our initial statements!), we would now have a different dynamic problem. Formulate that problem as well, adding symbolically any physical constants that you may need, say Young's modulus E and the cross section A, which you can assume to be known. You may assume that the horizontal motions are small enough that the string remains in tension at all times, that is, that it is never in danger of slacking and buckling.

d. Discuss which ultimate assumptions underlie the problem in steps 1, 2 that allow us to neglect the alternative problem in step 3, that is, why can we neglect the "axial" motions?

Solution

a. Cut out a generic mass point and consider its dynamic equilibrium:

$$p_\ell - m_\ell \ddot{u}_\ell - T \sin \theta_\ell + T \sin \theta_{\ell+1} = 0 \qquad (8.231)$$

or

$$m_\ell \ddot{u}_\ell + \frac{T}{d} d \left(\sin \theta_\ell - \sin \theta_{\ell+1} \right) = p_\ell \qquad (8.232)$$

But

$$\begin{aligned} u_\ell &= d \sin \theta_1 + d \sin \theta_2 + \cdots d \sin \theta_\ell \\ &= d \left(\sin \theta_1 + \sin \theta_2 + \cdots \sin \theta_\ell \right) \end{aligned} \qquad (8.233)$$

So

$$u_\ell - u_{\ell+1} = d \sin \theta_{\ell+1}, \quad \text{and} \quad u_{\ell-1} - u_\ell = d \sin \theta_\ell \qquad (8.234)$$

That is,

$$d \left(\sin \theta_{\ell+1} - \sin \theta_\ell \right) = \left(u_\ell - u_{\ell+1} \right) - \left(u_{\ell-1} - u_\ell \right) = -u_{\ell-1} + 2u_\ell - u_{\ell+1} \qquad (8.235)$$

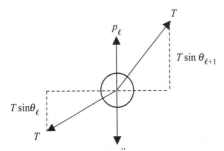

Figure 8.13. Dynamic equilibrium of generic mass.

Hence, the generic equation of motion for lateral vibration is

$$\boxed{m_\ell \ddot{u}_i + k\left(-u_{\ell-1} + 2u_\ell - u_{\ell+1}\right) = p_\ell}, \qquad k = T/d \tag{8.236}$$

with boundary conditions $u_0 = u_{n+1} = 0$.

b. If all masses are equal and we also focus on the free-vibration problem, then

$$-\omega^2 m u_\ell + k\left(-u_{\ell-1} + 2u_\ell - u_{\ell+1}\right) = 0, \qquad u_0 = u_{n+1} = 0 \tag{8.237}$$

Assuming a solution of the form $u_\ell = z^\ell$ and substituting that into the generic difference equation, we obtain

$$-\omega^2 m z^\ell + k\left(-z^{\ell-1} + 2z^\ell - z^{\ell+1}\right) = 0 \tag{8.238}$$

and after division by $z^{\ell-1} \neq 0$ together with the definition $\omega_0^2 = k/m$, we obtain

$$z^2 - 2\left[1 - \tfrac{1}{2}\left(\tfrac{\omega}{\omega_0}\right)^2\right]z + 1 = 0 \tag{8.239}$$

the solution of which is

$$z = 1 - \tfrac{1}{2}\left(\tfrac{\omega}{\omega_0}\right)^2 \pm \sqrt{\left[1 - \tfrac{1}{2}\left(\tfrac{\omega}{\omega_0}\right)^2\right]^2 - 1} = \left[1 - \tfrac{1}{2}\left(\tfrac{\omega}{\omega_0}\right)^2\right] \mp i\sqrt{1 - \left[1 - \tfrac{1}{2}\left(\tfrac{\omega}{\omega_0}\right)^2\right]^2} \tag{8.240}$$

Making the ansatz $\cos\phi = 1 - \tfrac{1}{2}\left(\tfrac{\omega}{\omega_0}\right)^2$, then

$$z = \cos\phi \pm i\sin\phi = e^{\pm i\phi} \tag{8.241}$$

So the general solution to the difference equation is

$$u_\ell = C_1 e^{i\ell\phi} + C_2 e^{-i\ell\phi} \tag{8.242}$$

Also, from the two boundary conditions, we are led to the system of equations

$$\begin{Bmatrix} 1 & 1 \\ e^{i(n+1)\phi} & e^{-i(n+1)\phi} \end{Bmatrix} \begin{Bmatrix} C_1 \\ C_2 \end{Bmatrix} = \begin{Bmatrix} 0 \\ 0 \end{Bmatrix} \tag{8.243}$$

which has nontrivial solutions only when the determinant vanishes, that is,

$$e^{i(n+1)\phi} - e^{-i(n+1)\phi} = 0 \quad \rightarrow \quad \sinh\left(i(n+1)\phi\right) = 0 \quad \rightarrow \quad \sin(n+1)\phi = 0 \tag{8.244}$$

That is,

$$\phi = \frac{j\pi}{n+1}, \qquad j = 1, 2, \ldots \tag{8.245}$$

Also, from our second ansatz, we infer

$$\left(\frac{\omega}{\omega_0}\right)^2 = 2(1 - \cos\phi) = \sin^2 \tfrac{1}{2}\phi \quad \rightarrow \quad \omega_j = \omega_0 \sin\tfrac{1}{2}\phi_j \tag{8.246}$$

That is,

$$\boxed{\omega_j = \sqrt{\frac{T}{dm}}\,\sin\frac{\pi j}{2(n+1)}} \tag{8.247}$$

Also, from the first equation of the eigenvalue problem we infer that $C_2 = -C_1$, so the jth modal shape at mass point ℓ is

$$\varphi_{\ell j} = C_1\left(e^{i\ell\phi} - e^{-i\ell\phi}\right) \rightarrow \sin\ell\phi_j \tag{8.248}$$

$$\boxed{\varphi_{\ell j} = \sin\frac{j\ell\pi}{n+1}}, \qquad \ell, j = 1,2,\ldots n \tag{8.249}$$

c. When horizontal rollers are added, the masses vibrate in the horizontal direction, and elongate or "compress" the initially tensed wire. If $k_x = EA/d$ is the axial stiffness of each wire segment (i.e., "spring"), then when setting up the lateral dynamic equilibrium of one mass, it will be found that the pre-tension plays no role as it cancels out. It follows that the equations of motion will be identical to those already found, but replacing $k = T/d$ by $k_x = EA/d$, in which case the longitudinal (i.e., axial) frequencies will be proportional to those found, namely in the ratio $\sqrt{EA/T}$. This assumes that the vibrations are small enough that the forces due to the longitudinal oscillations are small compared to the initial tension, that is, that no part of the wire will ever be in compression (i.e., have slack).

d. In general, we can neglect the axial motion whenever $EA \gg T$, that is, when the longitudinal vibrations have substantially larger frequencies than the transverse vibrations, to the point that the longitudinal system might be construed as a being rigid. Indeed, since

$$\frac{EA}{T} = \frac{E}{\sigma} = \frac{E}{E\varepsilon} = \frac{1}{\varepsilon} \tag{8.250}$$

and $\varepsilon < 10^{-3}$ for the system to remain in the elastic range when the pre-tension is applied, then it is clear that $EA/T > 10^3 \gg 1$, and therefore, we were justified in making the assumption that axial vibrations can be neglected. In addition, we assume that the initial tension is high enough that nonlinear effects can be neglected (i.e., constant tension in the deflected position can be taken for granted).

Example 2: Infinite Chain of Viscoelastically Supported Masses and Spring-Dashpots

We present herein an example of a simple mechanical, spatially periodic system in which waves may or may not propagate, including the case where the waves propagate only in some restricted frequency band or *passing band*.

Consider an individual oscillator with mass m that is supported on the ground via a spring-damper system of stiffness k and damping c. In addition, the oscillator has two horizontal spring-damper elements that act in shear (i.e., vertically, or what is the same, transverse to the element direction), one to the front and the other to the rear, each of stiffness $2K$ and damping constant $2C$, which when connected in series produce a net spring and damper K, C that connect the masses, as shown in Figure 8.14 on the right. This spring-damper system is connected to a series of identical units, which in their totality form a semi-infinitely long chain. Let the subindex j denote the mass counter so that masses $j = 0,1,2,\ldots$ are located from left to right. For $j > 0$ the dynamic equilibrium for transverse displacements is then

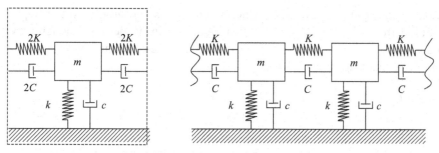

Figure 8.14. Chain of spring-damper system on viscoelastic support. K & C act in shear!

$$-\left(m\ddot{u}_j + c\dot{u}_j + ku_j\right) + C\left(\dot{u}_{j+1} - \dot{u}_j\right) - C\left(\dot{u}_j - \dot{u}_{j-1}\right) + K\left(u_{j+1} - u_j\right) - K\left(u_j - u_{j-1}\right) = 0 \qquad (8.251)$$

or

$$m\ddot{u}_j + \left(c + 2C\right)\dot{u}_j + \left(k + 2K\right)u_j - C\left(\dot{u}_{j+1} + \dot{u}_{j-1}\right) - K\left(u_{j+1} + u_{j-1}\right) = 0 \qquad (8.252)$$

Assume a wave solution of the form

$$u = A\exp\left(i\omega t - i\kappa j\right) = A\exp\left(i\omega t\right)z^{-j}, \qquad z = \exp\left(i\kappa\right) \qquad (8.253)$$

$$\dot{u}_j = i\omega u_j, \qquad \ddot{u}_j = -\omega^2 u_j, \qquad (8.254)$$

Hence, after canceling the exponential term in time, we obtain

$$-\omega^2 m z^{-j} + i\omega\left(c + 2C\right)z^{-j} + \left(k + 2K\right)z^{-j} - \left(i\omega C + K\right)\left(z^{-(j+1)} + z^{-(j-1)}\right) = 0 \qquad (8.255)$$

That is,

$$-\omega^2 m + i\omega\left(c + 2C\right) + \left(k + 2K\right) - \left(i\omega C + K\right)\left(z^{-1} + z^{+1}\right) = 0 \qquad (8.256)$$

or

$$z^2 - 2bz + 1 = 0, \qquad \boxed{b = \frac{k + 2K - \omega^2 m + i\omega\left(c + 2C\right)}{2\left(K + i\omega C\right)}}, \qquad (8.257)$$

So

$$z = b \pm \sqrt{b^2 - 1} = b \pm i\sqrt{1 - b^2} = \exp\left(\pm i\kappa\right), \qquad \cos\kappa = b, \qquad \sin\kappa = \sqrt{1 - b^2} \qquad (8.258)$$

For this solution to represent propagating waves, we would have to satisfy the condition $|b| \leq 1$, that is,

$$\left|\frac{k + 2K - \omega^2 m + i\omega\left(c + 2C\right)}{2\left(K + i\omega C\right)}\right| \leq 1 \qquad (8.259)$$

or

$$\sqrt{\left(k + 2K - \omega^2 m\right)^2 + \omega^2\left(c + 2C\right)^2} \leq 2\sqrt{K^2 + \omega^2 C^2} \qquad (8.260)$$

For $\omega = 0$, this would require $k + 2K \leq 2K$, which is impossible, because it would imply $k < 0$. Hence, at low frequencies, κ is purely imaginary, and the wave decays exponentially with j, that is, it is an evanescent wave. Solving Eq. 8.260 for the case when both sides are equal, we obtain the quadratic equation

$$\omega^4 m^2 - \omega^2 \left[2m(k+2K) - c(c+4C)\right] + k(k+4K) = 0 \qquad (8.261)$$

which has the two formal solutions

$$\omega^2 = \frac{2m(k+2K) - c(c+4C) \mp \sqrt{\left[2m(k+2K) - c(c+4C)\right]^2 - 4m^2 k(k+4K)}}{2m^2} \qquad (8.262)$$

which may or not have real and positive roots ω^2, that is, which may or not lead to any real values of ω for which the equation is satisfied. Indeed, no real solutions exist whenever the radicand is negative, that is, whenever

$$16\,m^2 K^2 + c^2(c+4C)^2 < 4mc(k+2K)(c+4C) \qquad (8.263)$$

If there are no real solutions for ω, then there is no frequency range for which a wave propagates. If there is one positive solution and a negative solution for ω^2, then there is one frequency – the cutoff frequency – after which a wave begins to propagate. This frequency defines the starting frequency of energy transmission. If there are two positive solutions, then these define the frequency band for which a wave propagates. The smaller solution defines the starting frequency and the second solution defines the stopping frequency, while the interval between the two is the *passing band*. Out of that frequency band the wave evanesces, and forms part of the *stopping band*.

In the particular case of an undamped system, $c = C = 0$, then $b = \frac{1}{2}(k+2K-\omega^2 m)/K$ which is less than 1 in absolute value when $|k+2K-\omega^2 m| < 2K$, and the passing band is defined by the two solutions

$$\omega^2 = \frac{k+2K \pm 2K}{m} = \begin{cases} k/m \\ (k+4K)/m \end{cases} \qquad (8.264)$$

Observe that there are always four solutions for the wavenumber $\kappa = \pm \arccos(b)$, two of which represent waves that propagate or decay to the right, and the other two represent waves that propagate or decay to the left.

8.8 The Discrete Shear Beam

Steel frame buildings can often be modeled as close-coupled discrete shear beams, in which the masses are lumped at floor elevations, and the springs represent the lateral stiffness of the frames between floors. In particular, if the masses and stiffnesses do not change from floor to floor, the dynamic model simplifies to a homogeneous, discrete shear beam with a finite number of DOF. Remarkably, the vibration characteristics of such a MDOF system, and in particular its modes and frequencies, can still be obtained in closed form, as we will show in this section. This affords us the opportunity to assess both qualitatively and quantitatively the effects of discretization on the dynamic characteristics of the shear beam, and by extension, to estimate these effects in other systems not amenable to analytical treatment. We consider simultaneously the cases of shear beams with lumped and/or consistent (i.e., finite element) mass matrices.

8.8.1 Continuous Shear Beam

Consider a cantilever shear beam with uniform cross section, and set the origin of coordinates at the free end pointing in direction of the fixed end (i.e., downward). The beam has length L, mass density ρ, cross section A, shear area A_s, and shear modulus G. Hence, the total mass of the beam is $M = \rho AL$. In general, there may be arbitrary body loads $b(x,t) = p(x)f(t)$ acting on the beam, but we will consider here only the uniform load case and the seismic case. The differential equation and the boundary conditions are

$$\rho A\ddot{u} - GA_s u'' = b(x,t) \tag{8.265}$$

$$u'\big|_{x=0} = 0 \qquad u\big|_{x=L} = 0 \tag{8.266}$$

The modal superposition solution to this problem is obtained by separation of variables. Since the details are well known, it suffices for us to give only final results.

$$\text{Wave speed}: C_s = \sqrt{\frac{G}{\rho}}\sqrt{\frac{A_s}{A}} \quad \text{(nondispersive)} \tag{8.267}$$

$$\text{Modal frequencies}: \omega_j = \frac{\pi}{2}\frac{C_s}{L}(2j-1) \qquad j = 1,2,\ldots\infty \tag{8.268}$$

$$\text{Modal shapes}: \phi(x) = \cos\left(\frac{\pi}{2}\frac{x}{L}(2j-1)\right) \tag{8.269}$$

$$\text{Modal mass}: \mu_j = \int_0^L \phi_j^2(x)\rho A\,dx = \tfrac{1}{2}M \tag{8.270}$$

$$\text{Modal load}: \pi_j = \int_0^L p(x)\phi(x)dx = \frac{(-1)^{j-1}}{(2j-1)}\frac{2p_0 L}{\pi} \tag{8.271}$$

$$\text{Seismic load participation factors}: \gamma_j = \frac{\int_0^L \phi(x)\rho A dx}{\int_0^L \phi^2(x)\rho A dx} = \frac{(-1)^{j-1}}{(2j-1)}\frac{4}{\pi} \tag{8.272}$$

8.8.2 Discrete Shear Beam

Discretize the previous continuous beam into n equal elements (or floors) of length $h = L/n$, and number these from the top down. We can obtain the stiffness and mass properties of each segment by heuristic methods, or from a finite element (Rayleigh–Ritz) formulation with a linear interpolation expansion (isoparametric elements). The results are

$$\text{Element stiffness} \qquad k = \frac{GA_s}{h} = n\frac{GA_s}{L} \tag{8.273}$$

Element mass $\qquad m = \rho A h = \frac{1}{n} \rho A L$ $\hfill$ (8.274)

Lumped and consistent mass matrices:

$$\mathbf{M}_L = m \begin{Bmatrix} \frac{1}{2} & & & & \\ & 1 & & & \\ & & \ddots & & \\ & & & 1 & \\ & & & & 1 \end{Bmatrix} \qquad \mathbf{M}_C = \frac{m}{6} \begin{Bmatrix} 2 & 1 & & & \\ 1 & 4 & 1 & & \\ & 1 & \ddots & & \\ & & & 1 & 4 & 1 \\ & & & & 1 & 4 \end{Bmatrix}$$ (8.275)

Stiffness matrix : $\qquad \mathbf{K} = k \begin{Bmatrix} 1 & -1 & & & \\ -1 & 2 & -1 & & \\ & -1 & 2 & -1 & \\ & & -1 & \ddots & -1 \\ & & & -1 & 2 \end{Bmatrix}$ $\hfill$ (8.276)

For greater generality, we consider here a mass matrix that is a linear combination of the consistent and lumped mass matrices, that is,

$$\mathbf{M} = (1-\alpha)\mathbf{M}_L + \alpha\mathbf{M}_C \qquad 0 \le \alpha \le 1$$ (8.277)

As we will show later on, the discrete system can be solved exactly in closed form. If i = mass index, numbered from the top down, and j = modal index, and if we use a hat on the discrete quantities to distinguish them from those of the continuous system, we obtain the following results:

Wave speed (dispersive) : $\qquad \boxed{\hat{C} = C_s \dfrac{\sin \frac{\pi h}{\lambda}}{\frac{\pi h}{\lambda}\sqrt{1 - \frac{2}{3}\alpha \sin^2 \frac{\pi h}{\lambda}}}} \qquad 2h \le \lambda = \text{wavelength}$

$\hfill$ (8.278)

Modal frequencies : $\qquad \boxed{\hat{\omega}_j = 2\sqrt{\dfrac{k}{m}}\dfrac{\sin \theta_j}{\sqrt{1 - \frac{2}{3}\alpha \sin^2 \theta_j}}} \qquad \theta_j = \dfrac{\pi}{4n}(2j-1)$ (8.279)

Modal shapes : $\qquad \boxed{\hat{\phi}_{ij} = \cos\left(2(i-1)\theta_j\right)}$ $\hfill$ (8.280)

Modal mass : $\qquad \boxed{\hat{\mu}_j = \boldsymbol{\phi}_j^T \mathbf{M} \boldsymbol{\phi}_j = \frac{1}{2}M\left[1 - \frac{2}{3}\alpha \sin^2 \theta_j\right]} \quad \boldsymbol{\phi}_j = \{\phi_{ij}\}$ (8.281)

Modal load : $\qquad \boxed{\hat{\pi}_j = \boldsymbol{\phi}_j^T \mathbf{p} = \frac{1}{2n}p_0 L(-1)^{j-1}\cot \theta_j}$ $\hfill$ (8.282)

Participation factors : $\qquad \boxed{\hat{\gamma}_j = \dfrac{\boldsymbol{\phi}_j^T \mathbf{M}\mathbf{e}}{\boldsymbol{\phi}_j^T \mathbf{M}\,\boldsymbol{\phi}_j} = (-1)^{j-1}\dfrac{\cot \theta_j - \frac{\alpha}{3}\sin 2\theta_j}{n\left[1 - \frac{2}{3}\alpha \sin^2 \theta_j\right]}}$ (8.283)

To compare these results with the continuous counterpart, we begin by expressing the mass and stiffness in term of the beam properties. This gives us

$$\sqrt{\frac{k}{m}} = \sqrt{\frac{n^2 GA_s}{\rho AL^2}} = \frac{nC_s}{L} \tag{8.284}$$

Hence,

$$g(\theta_j) = \frac{\hat{\omega}_j}{\omega_j} = \frac{\sin\theta_j}{\theta_j \sqrt{1 - \frac{2}{3}\alpha\sin^2\theta_j}} \tag{8.285}$$

which is identical in form to the ratio of wave speeds $\hat{C}/C_s$ in Eq. 8.278, that is, of the dispersion function. Figure 8.15 displays the results of using Eq. 8.285 with various values of the consistent mass parameter α. As can be seen, using a fully consistent mass matrix results in an overprediction of the natural frequencies and wave speeds in the discrete shear beam, while a fully lumped mass matrix does the opposite, it causes an underprediction. It would seem then that using an intermediate value, say ½, gives optimal results.

Also, since $x_i = (i-1)L/n$, we observe that the discrete and continuous modes are identical, that is,

$$\phi_{ij} \equiv \phi(x_i) \tag{8.286}$$

The ratio of (uniform) modal loads is

$$\frac{\hat{\pi}_j}{\pi_j} = \theta_j \cot\theta_j \tag{8.287}$$

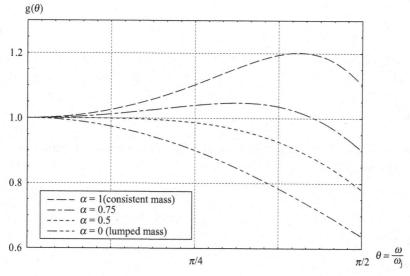

Figure 8.15. Numerical dispersion in discrete shear beam.

while the ratio of participation factors is

$$\frac{\hat{\gamma}_j}{\gamma_j} = \frac{\theta_j \left[\cot \theta_j - \frac{\alpha}{3} \sin 2\theta_j \right]}{1 - \frac{2}{3} \alpha \sin^2 \theta_j} \tag{8.288}$$

Both of the last two ratios approach 1 as n is increased.

Proof

In Section 8.4.1 on spatially periodic structures, we showed one way of obtaining the formula for the natural frequencies of a chain of equal springs and masses, which is analogous to a discrete shear beam with *lumped* masses. In the ensuing, we present an alternative proof based on solving a finite difference equation. We consider for this purpose the eigenvalue problem $(\mathbf{K} - \hat{\omega}^2 \mathbf{M})\phi = \mathbf{0}$. Inasmuch as the proof requires the use complex algebra, we will temporarily change the mass indices (floor levels) from i to l, so as to avoid confusion with the imaginary unit; with this change, the components of the eigenvector are then $u_l \equiv \phi_{lj}$. Also, we define the auxiliary variables

$$a = k - \tfrac{1}{6}(3 - \alpha)\hat{\omega}^2 m \qquad b = k + \tfrac{\alpha}{6}\hat{\omega}^2 m \tag{8.289}$$

In terms of the components, the eigenvalue problem is then

$$\begin{Bmatrix} \begin{bmatrix} a & -b & & & \\ -b & 2a & -b & & \\ & -b & 2a & & \\ & & & \ddots & -b \\ & & & -b & 2a \end{bmatrix} \end{Bmatrix} \begin{bmatrix} u_1 \\ u_2 \\ u_3 \\ \vdots \\ u_n \end{bmatrix} = \begin{Bmatrix} 0 \\ 0 \\ 0 \\ \vdots \\ 0 \end{Bmatrix} \tag{8.290}$$

With the exception of the first and last equations, all intermediate equations in this eigenvalue problem are of the form

$$-b u_{l-1} + 2a u_l - b u_{l+1} = 0 \tag{8.291}$$

which constitutes a homogeneous second-order finite difference equation with constant coefficients. Its boundary conditions are the first and last equations in the eigenvalue problem, namely

$$a u_1 - b u_2 = 0 \qquad \text{and} \qquad -b u_{n-1} + 2a u_n = 0 \tag{8.292}$$

To obtain the solution of the finite difference equation, we try a solution of the form $u_l = z^l$, which on substitution yields

$$-b z^{l-1} + 2a z^l - b z^{l+1} = 0 \tag{8.293}$$

Ruling out the trivial solution $z = 0$, we can divide by z^{l-1} and change this into the quadratic equation

$$z^2 - 2\tfrac{a}{b} z + 1 = 0 \tag{8.294}$$

whose solution is

$$z_{1,2} = \frac{a}{b} \pm \sqrt{\frac{a^2}{b^2} - 1} \tag{8.295}$$

In terms of these two roots, we can write compactly the general solution of the difference equation as

$$u_l = A z_1^l + B z_2^l \tag{8.296}$$

in which A, B are constants to be determined. Substituting the general solution into the boundary equations, we obtain

$$a(A z_1 + B z_2) - b(A z_1^2 + B z_2^2) = 0 \tag{8.297}$$

$$-b(A z_1^{n-1} + B z_2^{n-1}) + 2a(A z_1^n + B z_2^n) = 0 \tag{8.298}$$

It can easily be verified that the two roots satisfy the conditions $z_1 + z_2 = 2\frac{a}{b}$ and $z_1 z_2 = 1$. With the help of these conditions, the above equations can be changed after brief algebra into

$$A z_1 - B z_2 = 0 \tag{8.299}$$

$$A z_1^{n+1} + B z_2^{n+1} = 0 \tag{8.300}$$

or in matrix form

$$\begin{Bmatrix} 1 & -1 \\ z_1^n & z_2^n \end{Bmatrix} \begin{Bmatrix} A z_1 \\ B z_2 \end{Bmatrix} = \begin{Bmatrix} 0 \\ 0 \end{Bmatrix} \tag{8.301}$$

which has nontrivial solutions when the determinant is zero, that is, when

$$z_1^n + z_2^n = 0 \quad \rightarrow \quad z_1^{2n} + (z_1 z_2)^n = 0 \quad \rightarrow \quad z_1^{2n} = -1 \tag{8.302}$$

Hence

$$z_1 = \sqrt[2n]{-1} = \sqrt[2n]{e^{i\pi(2j-1)}} = e^{i2\theta_j} = \cos 2\theta_j + i \sin 2\theta_j \tag{8.303}$$

in which

$$\theta_j = \frac{1}{4n} \pi (2j-1) \tag{8.304}$$

Comparing with the equation for the roots of z, we can see that

$$\cos 2\theta_j = \frac{a}{b} = \frac{k - \frac{1}{6}(3-\alpha)\hat{\omega}_j^2 m}{k + \frac{1}{6}\alpha \hat{\omega}_j^2 m} \tag{8.305}$$

and solving for the frequency

$$\hat{\omega}_j^2 = 2\frac{k}{m} \frac{[1 - \cos 2\theta_j]}{[1 - \frac{\alpha}{3}(1 - \cos 2\theta_j)]} = 4\frac{k}{m} \frac{\sin^2 \theta_j}{[1 - \frac{2\alpha}{3}\sin^2 \theta_j]} \tag{8.306}$$

Hence, the discrete frequencies are

$$\hat{\omega}_j = 2\sqrt{\frac{k}{m}}\,\frac{\sin\theta_j}{\sqrt{1-\frac{2}{3}\alpha\sin^2\theta_j}} \qquad \theta_j = \tfrac{1}{4n}\pi(2j-1) \tag{8.307}$$

From the eigenvalue solution, the modal shapes are then

$$u_l = Az_1^l + Bz_2^l = Az_1z_1^{l-1} + Bz_2z_2^{l-1} \tag{8.308}$$

On the other hand, from the first equation in the eigenvalue problem, 8.301, $Az_1 = Bz_2$. Hence,

$$u_l = Az_1(z_1^{l-1} + z_2^{l-1}) = Az_1(e^{\mathrm{i}2\theta_j(l-1)} + e^{-\mathrm{i}2\theta_j(l-1)}) = 2Az_1\cos\left[2\theta_j(l-1)\right] \tag{8.309}$$

Choosing arbitrarily $2Az_1 = 1$, and reverting the index l back into the usual index i, we obtain

$$\phi_{ij} = u_i = \cos 2(i-1)\theta_j = \cos\left[\frac{\pi}{2n}(i-1)(2j-1)\right] \tag{8.310}$$

Next, we compute the modal mass. Because of the nondiagonal structure of the mass matrix, this is a rather tedious, even if straightforward, process. With the modes just determined and the shorthand $\vartheta \equiv 2\theta_j = \frac{\pi}{2n}(2j-1)$, a direct application of the definition of modal mass gives after considerable algebra

$$\boldsymbol{\phi}_j^T\mathbf{M}\,\boldsymbol{\phi}_j = m\left\{\tfrac{1}{3}[\alpha\cos\vartheta + (3-\alpha)]\left[\sum_{l=1}^{n}\cos^2(l-1)\vartheta - \tfrac{1}{2}\right] - \tfrac{1}{6}\alpha\cos(n-1)\vartheta\cos n\vartheta\right\} \tag{8.311}$$

But

$$\sum_{l=1}^{n}\cos^2(l-1)\vartheta = \sum_{l=0}^{n-1}\cos^2 l\vartheta = \tfrac{1}{2}[\sin^2 n\vartheta + n] = \tfrac{1}{2}\left[\sin^2\tfrac{\pi(2j-1)}{2} + n\right] = \tfrac{1}{2}[1+n] \tag{8.312}$$

Also

$$\tfrac{1}{3}[\alpha\cos\vartheta + (3-\alpha)] = 1 - \tfrac{1}{3}\alpha(1-\cos\vartheta) = 1 - \tfrac{2}{3}\alpha\sin^2 2\vartheta = 1 - \tfrac{2}{3}\alpha\sin^2\theta_j \tag{8.313}$$

$$\cos n\vartheta = \cos\frac{\pi(2j-1)}{2} = 0 \tag{8.314}$$

Hence

$$\boldsymbol{\phi}_j^T\mathbf{M}\,\boldsymbol{\phi}_j = m\left[1-\tfrac{2}{3}\alpha\sin^2\theta_j\right]\left[\tfrac{1}{2}(1+n)-\tfrac{1}{2}\right] = \tfrac{1}{2}M\left[1-\tfrac{2}{3}\alpha\sin^2\theta_j\right] \tag{8.315}$$

which can be used to normalize the modal shapes.

To evaluate the modal loads for a uniform load, we lump the load at the nodes, which implies a load vector of the form $\mathbf{p}^T = \tfrac{1}{n}p_0L\begin{bmatrix}\tfrac{1}{2} & 1 & \cdots & 1\end{bmatrix}$. Hence, the modal load is

$$\hat{\pi}_j = \boldsymbol{\phi}_j^T\mathbf{p} = \tfrac{1}{n}p_0L\left\{\sum_{l=0}^{n-1}\cos l\vartheta - \tfrac{1}{2}\right\} = \tfrac{1}{2n}p_0L\left\{\sin n\vartheta\cot\tfrac{1}{2}\vartheta - \cos n\vartheta\right\} \tag{8.316}$$

and considering that $\cos n\vartheta = \cos\dfrac{\pi(2j-1)}{2} = 0$ and $\sin n\vartheta = \sin\dfrac{\pi(2j-1)}{2} = (-1)^{j-1}$, we obtain

$$\hat{\pi}_j = \tfrac{1}{2n} p_0 L (-1)^{j-1} \cot\theta_j \tag{8.317}$$

Finally, we proceed to compute the modal participation factors for a seismic load. This is again a tedious process, since the mass matrix is not diagonal. After some algebra, the resulting summation is

$$\gamma_j = \frac{\boldsymbol{\varphi}_j^T \mathbf{M}\mathbf{e}}{\boldsymbol{\phi}_j^T \mathbf{M}\boldsymbol{\phi}_j} = \frac{m}{\hat{\mu}_j}\left\{\sum_{l=0}^{n-1}\cos l\vartheta - \frac{1}{2} - \frac{\alpha}{3}\cos(n-1)\vartheta\right\} \tag{8.318}$$

In view of the previous result for the modal load and the expansion of the last term, we obtain

$$\gamma_j = \frac{(-1)^{j-1}\left[\cot\theta_j - \dfrac{\alpha}{3}\sin 2\theta_j\right]}{n\left[1 - \tfrac{2}{3}\alpha\sin^2\theta_j\right]} \tag{8.319}$$

It remains to compute the effective wave speed $\hat{C}$, which is frequency dependent (i.e., dispersive). For this purpose, we consider an infinite discrete shear beam in which a harmonic wave is propagating in the positive coordinate direction $x = lh$. The discrete wave equation is now

$$-ku_{l-1} + 2ku_l - ku_{l+1} = (1-\alpha)m\ddot{u}_l + \tfrac{1}{6}\alpha m\left[\ddot{u}_{l-1} + 4\ddot{u}_l + \ddot{u}_{l+1}\right] \tag{8.320}$$

Assuming $u_l = Ae^{i\omega(t-lh/\hat{C})} \equiv Ae^{i\omega t}\left(e^{-i\omega h/\hat{C}}\right)^l = Ae^{i\omega t}z^l$, we obtain

$$z^2 - 2\tfrac{a}{b}z + 1 = 0 \tag{8.321}$$

in which a, b are as defined earlier. The solution is

$$z = e^{-i\omega h/\hat{C}} = \frac{a}{b} - i\sqrt{1 - \left(\frac{a}{b}\right)^2} \tag{8.322}$$

That is,

$$\cos\left(\frac{\omega h}{\hat{C}}\right) = \frac{a}{b} = \frac{k - \tfrac{1}{6}(3-\alpha)\omega^2 m}{k + \tfrac{1}{6}\alpha\omega^2 m} = \frac{1 + \tfrac{1}{6}\alpha\omega^2 m/k - \tfrac{1}{2}\omega^2 m/k}{1 + \tfrac{1}{6}\alpha\omega^2 m/k} = 1 - \frac{2\left(\dfrac{\omega h}{2C_s}\right)^2}{1 + \tfrac{2}{3}\alpha\left(\dfrac{\omega h}{2C_s}\right)^2} \tag{8.323}$$

and solving for the last term

$$\sin^2\left(\frac{\omega h}{2\hat{C}}\right) = \frac{\left(\dfrac{\omega h}{2C_s}\right)^2}{1 + \tfrac{2}{3}\alpha\left(\dfrac{\omega h}{2C_s}\right)^2} \tag{8.324}$$

Hence

$$\left(\frac{\omega h}{2C_s}\right)^2 = \frac{\sin^2 \frac{\omega h}{2\hat{c}}}{1 - \frac{2}{3}\alpha \sin^2 \frac{\omega h}{2\hat{c}}} \tag{8.325}$$

But

$$\frac{\omega}{\hat{C}} = \frac{2\pi}{\lambda} \quad \text{and} \quad \frac{\omega}{C_s} = \frac{2\pi}{\lambda}\frac{\hat{C}}{C_s} \tag{8.326}$$

in which λ is the wavelength. It follows that

$$\left(\frac{\hat{C}}{C_s}\right)^2 \left(\frac{\pi h}{\lambda}\right)^2 = \frac{\sin^2 \frac{\pi h}{\lambda}}{1 - \frac{2}{3}\alpha \sin^2 \frac{\pi h}{\lambda}} \tag{8.327}$$

and finally

$$\frac{\hat{C}}{C_s} = \frac{\sin \frac{\pi h}{\lambda}}{\frac{\pi h}{\lambda}\sqrt{1 - \frac{2}{3}\alpha \sin^2 \frac{\pi h}{\lambda}}} \tag{8.328}$$

9 Mathematical Tools

9.1 Dirac Delta and Related Singularity Functions

Dirac delta functions are extremely useful in dynamics, because they allow to represent in a compact mathematical fashion loads or masses that are concentrated at a point in space (i.e., a point load or a lumped mass), or that are impulsive in time (e.g., a hammer blow).

A Dirac delta function $\delta(x - a)$ can be visualized as a rectangular function, or *window*, centered at $x = a$, which has small width w and large height $1/w$, as shown in Figure 9.1. Its area is then $w(1/w) = 1$, that is, it is unity. In the limit when w goes to zero, the function becomes infinitely large, while its width becomes infinitesimally small, but its area remains unity. When this box function is used as a weighting function in an integral involving an arbitrary function $f(x)$ over an interval containing a (i.e., $x_1 < a < x_2$), it is easy to see that

$$\int_{x_1}^{x_2} f(x)\delta(x-a)dx = \lim_{w \to 0} \int_{a-w/2}^{a+w/2} f(x)\frac{1}{w}dx = f(a)\frac{1}{w}w = f(a) \tag{9.1}$$

Thus, an integral with the Dirac delta function simply reproduces the integrand at the location of the singularity. Functions such as the one above, which are defined in terms of the limit to an integral, are commonly referred to as singularity functions or *distributions*. It should be noticed that there are many ways of approaching this Dirac delta function. For example, instead of the box function, we could also have used a triangular function of width $2w$ and height $1/w$ that is centered at $x = a$. Its area is again unity. Many other representations are also possible.

By using the Dirac delta singularity function, we can now concisely model a *concentrated* load P acting at $x = a$ in terms of a *distributed* load:

$$b(x) = P\delta(x-a) \tag{9.2}$$

An impulsive, concentrated load occurring at time T and acting at $x = a$ would then be of the form

$$b(x,t) = P\delta(x-a)\delta(t-T) \tag{9.3}$$

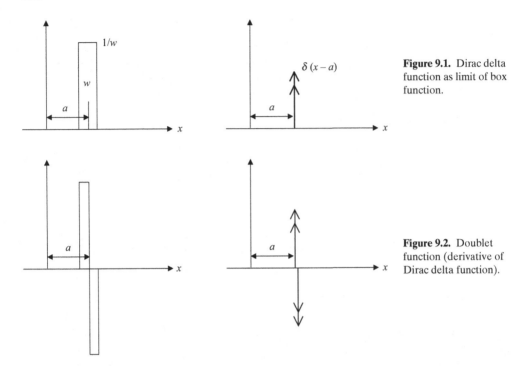

Figure 9.1. Dirac delta function as limit of box function.

Figure 9.2. Doublet function (derivative of Dirac delta function).

9.1.1 Related Singularity Functions

A whole family of singularity functions $\mathcal{H}_j(x)$ $j = -2, -1, 0, 1, 2, \ldots$ can be developed starting from the Dirac delta function. Three of these are shown in Figures 9.2 and 9.3. Formally at least, these can be visualized as derivatives and integrals of the Dirac delta function. However, it must be understood that these have meaning only as limits of integrals in which they appear as arguments (i.e., as distributions), since the singularity functions themselves are not properly defined. The first four in the family of singularity functions are the following.

Doublet Function (Figure 9.2)

$$\mathcal{H}_{-2}(x-a) = \frac{d}{dx}\delta(x-a) \qquad \int_{-\infty}^{+\infty} f(x)\,\mathcal{H}_{-2}(x-a)\,dx = -\frac{df}{dx}\bigg|_{x=a} \qquad (9.4)$$

Dirac Delta Function (Figure 9.1)

$$\mathcal{H}_{-1}(x-a) = \delta(x-a) \qquad \int_{-\infty}^{+\infty} f(x)\,\mathcal{H}_{-1}dx = f(a) \qquad (9.5)$$

Unit Step Function (Heaviside Function) (Figure 9.3a)

$$\mathcal{H}_0(x-a) = \int_{-\infty}^{+\infty} \delta(x-a)\,dx = \begin{cases} 1 & x > a \\ \frac{1}{2} & x = a \\ 0 & x < a \end{cases} \qquad (9.6)$$

(a)

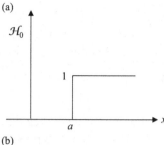

(b)

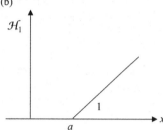

Figure 9.3. Unit step function and unit ramp function, the first and second integrals of the Dirac delta function.

$$\int_{-\infty}^{b} f(x)\mathcal{H}_0(x-a)dx = \int_{a}^{b} f(x)dx \tag{9.7}$$

Unit Ramp Function (Figure 9.3b)

$$\mathcal{H}_1(x-a) = \int \mathcal{H}_0(x-a)\,dx = \begin{cases} x-a & x \geq a \\ 0 & x < a \end{cases} \tag{9.8}$$

$$\int_{-\infty}^{b} f(x)\mathcal{H}_1(x-a)dx = \int_{a}^{b}(x-a)f(x)\,dx \tag{9.9}$$

9.2 Functions of Complex Variable: A Brief Summary

Let $z = x + iy$ be a complex variable, and $f(z)$ a complex function of that variable. The following statements then apply.

1. Analiticity and Cauchy–Riemann Conditions

If $f(z) = u(x,y) + iv(x,y)$ is *analytic* (or regular, or *holomorphic*) in a certain region of the complex plane, then u, v satisfy the Cauchy–Riemann conditions

$$\frac{\partial u}{\partial x} = \frac{\partial v}{\partial y} \quad \text{and} \quad \frac{\partial u}{\partial y} = -\frac{\partial v}{\partial x} \tag{9.10}$$

These conditions are both necessary and sufficient for analyticity.

2. Harmonic Functions

If $f(z) = u + iv$ is analytic in a region, then both u and v are *conjugate harmonic* functions that satisfy Laplace's equation

$$\nabla^2 u = 0 \quad \text{and} \quad \nabla^2 v = 0 \tag{9.11}$$

3. Analyticity on Boundary

If u, v and their partial derivatives are continuous in a region, and they satisfy the Cauchy-Riemann conditions, then $f(z)$ is analytic at all points *inside* the region, but not necessarily *on* the boundary.

4. Differentiability of Analytic Function

If $f(z)$ is analytic in a region, then *it has derivatives of all orders* at points inside the region, and can be expanded in a Taylor series about any point z_0 inside that region. The power series converges inside the circle about z_0 that extends to the nearest singular point.

5. Cauchy Integrals

Consider the contour in the complex plane shown in Figure 9.4. If C is a *simple contour* with a finite number of *corners*, and if $f(z)$ is analytic both *inside* and *on* C, then

a. $\oint_C f(z)\,dz = 0$ \hfill (9.12)

b. $\dfrac{1}{2\pi i}\oint_C \dfrac{f(z)\,dz}{z-a} = f(a)$ \quad if a is *inside* the contour \hfill (9.13)

c. $\dfrac{1}{2\pi i}\oint_C \dfrac{f(z)\,dz}{z-a} = 0$ \qquad if a is *outside* the contour \hfill (9.14)

The second integral is a remarkable property. It says in effect that the value of the function at any arbitrary point inside a closed contour is specified by the values of the function along the boundary.

6. Laurent Series about a Point z_0

$$a_n = \frac{1}{2\pi i}\oint_C \frac{f(z)\,dz}{(z-z_0)^{n+1}} \qquad b_n = \frac{1}{2\pi i}\oint_C (z-z_0)^{n-1} f(z)\,dz \qquad (9.15)$$

- If all the coefficients b_n are zero, then $f(z)$ is analytic at $z = z_0$, and z_0 is a *regular* point.
- If $b_n \neq 0$ but $b_{n+k} = 0$ for all k, then $f(z)$ has a *simple pole* of order n at $z = z_0$. For $n > 1$, z_0 is a *multiple pole*.
- If $b_n \neq 0$ for all n, then $f(z)$ has an essential singularity at $z = z_0$.

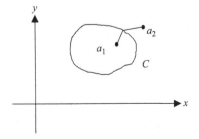

Figure 9.4. Closed contour C in the complex plane $z = x + i\,y$.

7. Poles, Residues, and Contour Integrals

A function of complex variable that is not constant or zero everywhere must have at least one pole.

Simple Pole

$$R = \lim_{z \to z_0}(z - z_0)f(z) \tag{9.16}$$

If the result is 0 or ∞, then $z = z_0$ is not a simple pole, but perhaps a pole of higher order (multiple pole).

Pole of Order n

$$R = \frac{1}{(n-1)!} \lim_{z \to z_0} \frac{d^{n-1}\left[(z-z_0)^n f(z)\right]}{dz^{n-1}} \tag{9.17}$$

Contour Integral

$$\oint f(z)\,dz = 2\pi\,\mathrm{i}\left[\sum(\text{residues in region}) + \tfrac{1}{2}\sum(\text{residues on } C)\right] \tag{9.18}$$

Note: C must have a continuous tangent at a simple pole. The ½ rule does *not* apply for multiple poles. In addition, the contour must be traveled in a *counterclockwise* direction (otherwise, the sign of the integral reverses).

8. Principal Value

The principal (or Cauchy) value is defined as the result of an integration over a singularity. For example,

$$\mathrm{PV}\int_{-\infty}^{\infty}\frac{\cos x}{x}\,dx = 0 \qquad \text{but} \qquad \int_{0}^{\infty}\frac{\cos x}{x}\,dx = \text{divergent} \tag{9.19}$$

$$\mathrm{PV}\int_{0}^{5}\frac{1}{x-3}\,dx = \ln\tfrac{2}{3} \qquad \text{but integral has singularity at } x = 3 \tag{9.20}$$

9. Multivalued Functions: Branch-Cuts and Branch Points

Consider the multivalued function

$$\sqrt{z} = \sqrt{x^2 + y^2}\,e^{\mathrm{i}(\phi/2 + k\pi)} \tag{9.21}$$

with $\tan\phi = \dfrac{y}{x}$

To avoid ambiguity, we must agree to restrict evaluation of this function either between $0 \to 2\pi$, or between $-\pi \to \pi$. If we agree that the range of evaluation is $0 \to 2\pi$, then the positive axis is a *branch cut* that terminates at a *branch point*, which in this case is the origin of coordinates, see Figure 9.5; if the agreed range is $-\pi \to \pi$, then the negative axis is the branch cut. More generally, and depending on the function being considered, we may have one or more *branch cuts* terminating at *branch points*, which a contour integration path *cannot* cross.

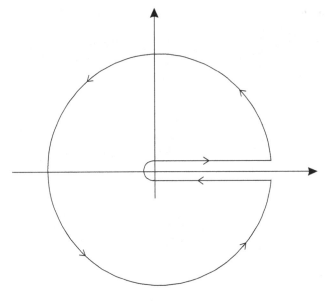

Figure 9.5. Closed contour that avoids the branch cut and the branch point.

10. Number of Zeros and Poles

If $f(z) = r\,e^{i\theta} \neq 0$ on C (!!), then

$$N - P = \frac{1}{2\pi i} \oint_C \frac{f'(z)\,dz}{f(z)} = \frac{1}{2\pi}\theta_C \tag{9.22}$$

in which

N, P = number of zeros and poles *inside* the contour. Zeros and/or poles of order n count n times each.

θ_C = change in angle after going once around the contour.

11. Evaluation of Fourier Integrals by Residues

$$I = \int_{-\infty}^{\infty} \frac{P(x)}{Q(x)} e^{imx} dx = 2\pi i \sum (\text{residues of integrand in } upper \text{ halfplane}) \tag{9.23}$$

provided that

- $P(x), Q(x)$ are polynomials
- $Q(x)$ has no real zeros
- The degree of $Q(x)$ is at least 1 greater than the degree of $P(x)$
- $m > 0$. If $m < 0$, evaluate the residues in the *lower* half-plane; if $m = 0$, the integral does not converge if the difference in degrees is only 1.

Example of Application

Suppose we wish to evaluate the improper integral

$$I = \int_0^{\infty} \frac{\cos(x)\,dx}{1 + x^2} \tag{9.24}$$

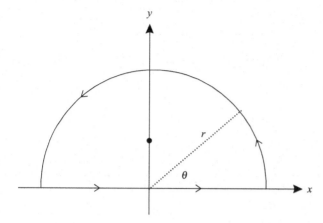

Figure 9.6. Contour to evaluate integral, which assumes that $r \rightarrow \infty$.

Since $\cos x$ is an even function, and $\sin x$ is an odd function of x, we can replace this integral by the equivalent form

$$I = \int_0^\infty \frac{\cos(x)\,dx}{1+x^2} = \frac{1}{2}\int_{-\infty}^\infty \frac{\cos(x)\,dx}{1+x^2} + i\frac{1}{2}\int_{-\infty}^\infty \frac{\sin(x)\,dx}{1+x^2} = \frac{1}{2}\int_{-\infty}^\infty \frac{e^{ix}\,dx}{1+x^2} \tag{9.25}$$

We consider next the contour integral $\oint_c \dfrac{e^{iz}dz}{1+z^2}$ evaluated over the contour shown in Figure 9.6.

Since $\left|e^{iz}\right| = \left|e^{ix}\right|\left|e^{-y}\right| = e^{-y} \le 1$, in the limit when $r \rightarrow \infty$ the contribution to the integral by that part of the contour that goes "around" infinity is zero, a result that is known as *Jordan's lemma*:

$$\left|\oint_{C_1} \frac{e^{iz}dz}{1+z^2}\right| < \int_0^\pi \frac{dz}{1+z^2} = \lim_{r \to \infty}\int_0^\pi \frac{r\,i\,e^{i\theta}d\theta}{1+r^2 e^{2i\theta}} = 0 \tag{9.26}$$

Hence

$$I = \int_0^\infty \frac{\cos(x)\,dx}{1+x^2} = \frac{1}{2}\int_{-\infty}^\infty \frac{e^{ix}dx}{1+x^2} = \frac{1}{2}\oint_c \frac{e^{iz}dz}{1+z^2} \tag{9.27}$$

The residues are evaluated at the poles, which in this case correspond to the zeros of the denominator:

$$1+z_0^2 = 0 \quad \Rightarrow \quad z_0 = \pm i \tag{9.28}$$

In the upper half-plane, there is only one pole, namely $z_0 = i$. The residue is then

$$R = \lim_{z \to z_0 = i}(z - z_0)\frac{e^{iz}}{(z-i)(z+i)} = \frac{e^{i^2}}{2i} = \frac{e^{-1}}{2i} \tag{9.29}$$

Finally

$$I = \int_0^\infty \frac{\cos(x)\,dx}{1+x^2} = \frac{1}{2}\oint_c \frac{e^{iz}dz}{1+z^2} = \frac{1}{2}2\pi i\frac{e^{-1}}{2i} = \frac{\pi}{2e} \tag{9.30}$$

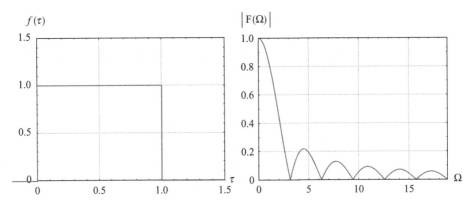

Figure 9.7. Box function and its Fourier transform.

9.3 Wavelets

Wavelets are simple mathematical functions (or signals) of short duration and known Fourier transforms, which are widely used in vibration theory, and in structural and soil dynamics, not to mention in seismology. They can be used as load or source functions, test functions, or even as filters. In the ensuing we provide a brief list of some of the most commonly used signals.

9.3.1 Box Function

The box function is one of many functions defined with *compact support*, namely one that is defined over a finite window in time. With reference to Figure 9.7, the unit box function and its Fourier transform are given by

$$f(t) = \begin{cases} 1 & -\tfrac{1}{2}T \le t \le \tfrac{1}{2}T \\ 0 & \text{else} \end{cases} \qquad F(\Omega) = T\frac{\sin\Omega}{\Omega}, \qquad \Omega = \tfrac{1}{2}\omega T \tag{9.31}$$

Alternatively, shifting it forward in time and making the box to be a causal function, we obtain

$$\boxed{f(t) = \begin{cases} 1 & t \le T \\ 0 & \text{else} \end{cases}}, \qquad \boxed{F(\Omega) = T\frac{\sin\Omega}{\Omega}e^{-i\Omega}}, \qquad \Omega = \tfrac{1}{2}\omega T \tag{9.32}$$

Although simple and widely used, $F(\Omega)$ decays rather slowly with frequency. In addition it exhibits sharp discontinuities at $t = 0$ & $t = T$, which lead to the Gibbs phenomenon.

9.3.2 Hanning Bell (or Window)

Consider the causal signal together with its transform shown in Figure 9.8. It is defined by

$$\boxed{f(t) = \sin^2 \pi\tau}, \qquad \tau = \frac{t}{T}, \qquad 0 \le \tau \le 1 \tag{9.33}$$

whose area is $A = \tfrac{1}{2}T$, and its Fourier transform is

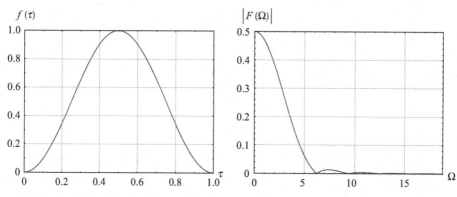

Figure 9.8. Hanning bell and its Fourier transform.

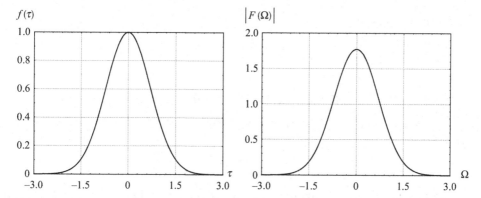

Figure 9.9. Gaussian bell and its Fourier transform.

$$\boxed{\left|F\left(\Omega\right) = \tfrac{1}{2}T\,\frac{\sin\Omega}{\Omega}\,\frac{e^{-i\Omega}}{1-\left(\tfrac{1}{\pi}\Omega\right)^2}\right|}, \qquad \Omega = \tfrac{1}{2}\omega T, \qquad F\left(\pi\right) = -\tfrac{1}{4}T \qquad (9.34)$$

In comparison with the box function, the Hanning bell decays rather rapidly with frequency.

9.3.3 Gaussian Bell

Consider the Gaussian wavelet shown in Figure 9.9. Although not strictly causal, its amplitude at negative times is negligible. Moreover, its amplitude is also negligible at times greater than those shown in the figure. Hence, for practical purposes the Gaussian wavelet is both causal and of finite duration, that is, it is *time-limited*. Interestingly, its Fourier transform is also a Gaussian bell, which means that the wavelet is also *band-limited*, or at least it is so for engineering purposes. Strictly speaking, this contradicts a lemma in signal analysis that prohibits any signal from being simultaneously time-limited and band-limited, but this signal comes close to that ideal. This wavelet satisfies the following equations:

$$\boxed{f\left(t\right) = \exp\left(-\tau^2\right)}, \qquad \tau = \frac{t}{T} \qquad Area = T\sqrt{\pi} \qquad (9.35)$$

For times $|\tau| \ge \pi$, the signal will have decayed by $\exp\left(-\pi^2\right) = 5\times10^{-5}$, and will be thus negligible.

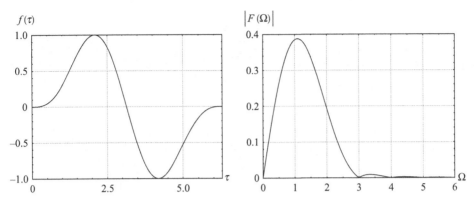

Figure 9.10. Sinusoidal wavelet and its Fourier transform.

The Fourier transform is

$$\boxed{F(\Omega) = T\sqrt{\pi}\,e^{-\Omega^2}}, \qquad \Omega = \tfrac{1}{2}\,\omega T \tag{9.36}$$

9.3.4 Modulated Sine Pulse (Antisymmetric Bell)

Consider the source function shown in Figure 9.10.

$$\boxed{f(\tau) = \tfrac{8}{9}\sqrt{3}\,\sin\tau\,\sin^2\tfrac{1}{2}\tau}, \quad \tau = 2\pi\frac{t}{T} \quad 0 \le t \le T \tag{9.37}$$

This function has two unit magnitude peaks at $t_{max} = \tfrac{1}{3}T$ and $t_{min} = \tfrac{2}{3}T$, and it has zero slope at $t = 0, T$. Also, it has no zero-frequency content, that is, the average of the signal is zero. Hence, any force time history being represented by this signal would produce no net impulse.

Its Fourier transform is

$$\boxed{\begin{aligned} F(\omega) &= T\frac{2}{\pi\sqrt{3}}\frac{1 - e^{-i2\pi\Omega}}{\Omega^4 - 5\Omega^2 + 4} \\ &= T\frac{4\,i}{\pi\sqrt{3}}\frac{\sin(\pi\Omega)}{(\Omega-1)(\Omega+1)(\Omega-2)(\Omega+2)}e^{-i\pi\Omega} \end{aligned}}, \qquad \Omega = \frac{\omega T}{2\pi} \tag{9.38}$$

whose maximum takes place at $\Omega = 1.0895 \approx \sqrt[8]{2}$, and is $|F|_{max} \approx 0.3877\,T$. Although it would seem at first that this function could have singularities at $\Omega = \pm 1$ and $\Omega = \pm 2$, that is not the case, because the numerator vanishes at those values. Using L'Hospital's rule, we find that

$$F(\pm 1) = \pm\,i\,T\,\tfrac{2}{9}\sqrt{3}, \qquad F(\pm 2) = \pm\,i\,T\,\tfrac{1}{9}\sqrt{3} \tag{9.39}$$

9.3.5 Ricker Wavelet

This is a very widely used signal, not only because it is simple, but also because it comes close to the ideal of a function that is limited in both time and frequency. Indeed, although a fundamental theorem in signal theory states that a function cannot be simultaneously time

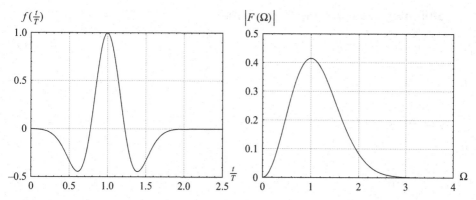

Figure 9.11. Ricker wavelet and its Fourier transform.

limited and band limited (i.e., frequency limited), the Ricker wavelet decays exponentially (and thus superfast) in both of these domains. Although not strictly causal, for practical purposes it is so when appropriate choices are made for its parameters, as shown in Figure 9.11.

A unit amplitude Ricker wavelet is defined by the function

$$\boxed{f(t) = (1 - 2\tau^2)\exp(-\tau^2)}, \qquad \tau = \pi\frac{t - t_m}{T} = \tfrac{1}{2}\omega_0(t - t_m) \tag{9.40}$$

t = time
t_m = time of maximum, $f(t_m) = 1$
T = dominant period
ω_0 = dominant frequency $(= 2\pi/T)$

The function has only two zero crossings, which occur at $\tau = \pm\tfrac{1}{2}\sqrt{2}$, that is, at $t = t_m \mp \frac{1}{\pi\sqrt{2}}T$. Thus, the width of the positive part (i.e., the main lobe) of the wavelet is $\frac{\sqrt{2}}{\pi}T = 0.45T$, and the area of that main lobe is $A = (T/\pi)\sqrt{2/e}$.

It is often convenient to select $t_m = T$ (as shown earlier) because the wavelet is then virtually causal, that is, has negligible amplitudes at negative times. In this case, the maximum takes place at the dominant period, and the dimensionless time and zero crossings are

$$\tau = \pi\left(\frac{t}{T} - 1\right), \qquad t_{1,2} = T\left(1 \mp \tfrac{\sqrt{2}}{2\pi}\right) = \begin{cases}0.775T \\ 1.225T\end{cases} \tag{9.41}$$

The Fourier transform of the Ricker wavelet is

$$\boxed{F(\Omega) = \left[\frac{2}{\sqrt{\pi}}Te^{-i\omega t_m}\right]\Omega^2\exp(-\Omega^2)}, \qquad \Omega = \tfrac{1}{2\pi}\omega T = \omega/\omega_0 \tag{9.42}$$

with ω being the angular frequency, in rad/s and T is again the dominant period. The maximum Fourier spectrum amplitude takes place at $\Omega = \pm 1$, that is, $\omega = \omega_0 = 2\pi/T$. Its value is

$$F_{\max} = F(\omega_0) = \frac{2}{e\sqrt{\pi}}T = \frac{4\sqrt{\pi}}{\omega_0 e} = 0.415T \tag{9.43}$$

9.4 Useful Integrals Involving Exponentials

$$\int t^n \, e^{\alpha t} \sin \beta t \, dt = F(t) \sin \beta t - G(t) \cos \beta t \tag{9.44}$$

$$\int t^n \, e^{\alpha t} \cos \beta t \, dt = G(t) \sin \beta t + F(t) \cos \beta t \tag{9.45}$$

9.4.1 Special Cases

$n = 0$

$$F(t) = \frac{e^{\alpha t}}{\alpha^2 + \beta^2} \, \alpha \tag{9.46}$$

$$G(t) = \frac{e^{\alpha t}}{\alpha^2 + \beta^2} \, \beta \tag{9.47}$$

$n = 1$

$$F(t) = \frac{e^{\alpha t}}{\alpha^2 + \beta^2} \left[\alpha t - \frac{\alpha^2 - \beta^2}{\alpha^2 + \beta^2} \right] \tag{9.48}$$

$$G(t) = \frac{e^{\alpha t}}{\alpha^2 + \beta^2} \left[\beta t - \frac{2\alpha\beta}{\alpha^2 + \beta^2} \right] \tag{9.49}$$

$n = 2$

$$F(t) = \frac{e^{\alpha t}}{\alpha^2 + \beta^2} \left[\alpha t^2 - 2 \frac{\alpha^2 - \beta^2}{\alpha^2 + \beta^2} t + 2\alpha \frac{\alpha^2 - 3\beta^2}{\left(\alpha^2 + \beta^2\right)^2} \right] \tag{9.50}$$

$$G(t) = \frac{e^{\alpha t}}{\alpha^2 + \beta^2} \left[\beta t^2 - \frac{4\alpha\beta}{\alpha^2 + \beta^2} t + 2\beta \frac{3\alpha^2 - \beta^2}{\left(\alpha^2 + \beta^2\right)^2} \right] \tag{9.51}$$

$$\int e^{-2\alpha t} \cos^2 \left(\beta t - \phi \right) dt = -\frac{1}{4\alpha} e^{-2\alpha t} \left[1 + \cos\theta \, \cos\left(2\beta t_1 - 2\phi + \theta \right) \right] \tag{9.52}$$

$$\int e^{-2\alpha t} \sin^2 \left(\beta t - \phi \right) dt = -\frac{1}{4\alpha} e^{-2\alpha t} \left[1 - \cos\theta \, \cos\left(2\beta t_1 - 2\phi + \theta \right) \right] \tag{9.53}$$

$$\int e^{-2\alpha t} \sin \left(\beta t - \phi \right) \cos\left(\beta t - \phi \right) dt = -\frac{1}{4\gamma} e^{-2\alpha t} \sin\left(2\beta t_1 - 2\phi + \theta \right) \tag{9.54}$$

with $\gamma = \sqrt{\alpha^2 + \beta^2}, \qquad \cos\theta = \dfrac{\alpha}{\gamma}, \qquad \sin\theta = \dfrac{\beta}{\gamma} \tag{9.55}$

9.5 Integration Theorems in Two and Three Dimensions

We consider herein the integration of a product of two functions in both two and three dimensions, which we obtain via integration by parts. We then use the two basic formulas to obtain the Stokes and Gauss divergence theorems as well as many other related formulas. These are widely used in elastodynamics, and especially so in the Boundary Element Method.

9.5.1 Integration by Parts

Consider a plane region A in a 2-D space with Cartesian coordinates $\mathbf{x} = (x_1, x_2)$ that is bounded by an arbitrary closed curve $\Gamma = \Gamma(s)$, where s is the curvilinear coordinate along the boundary satisfying $|ds| = \sqrt{dx_1^2 + dx_2^2}$. Alternatively consider a volume V that is full enclosed by a finite surface S in a 3-D space $\mathbf{x} = (x_1, x_2, x_3)$. In either case, we define $\hat{\mathbf{e}}_j$ as a unit vector along the jth Cartesian direction and $\hat{\mathbf{n}} = v_1 \hat{\mathbf{e}}_1 + v_2 \hat{\mathbf{e}}_2 + v_3 \hat{\mathbf{e}}_3$ is the unit *outward* normal to the boundary (either S or Γ with $x_3 = 0$), with v_j being the direction cosines with respect to the jth coordinate direction, then the following holds:

$$\boxed{\iint_A f \frac{\partial g}{\partial x_j} dA = \oint_\Gamma f g \, v_j \, ds - \iint_A \frac{\partial f}{\partial x_j} g \, dA} \qquad \text{2-D,} \quad j = 1,2 \tag{9.56}$$

$$\boxed{\iiint_V f \frac{\partial g}{\partial x_j} dV = \oiint_S f g \, v_j \, dS - \iiint_V \frac{\partial f}{\partial x_j} g \, dV} \qquad \text{3-D,} \quad j = 1,2,3 \tag{9.57}$$

where $f = f(\mathbf{x}), g = g(\mathbf{x}), v_j = \cos\left[\sphericalangle\left(\hat{\mathbf{n}}, \hat{\mathbf{e}}_j\right)\right] = $ direction cosine of the **outward** normal, and the *line integral* must be carried out in **counterclockwise** direction (the "positive" direction).

In particular, setting $g = 1$, we obtain

$$\iint_A \left(\tfrac{\partial}{\partial x_j} f \right) dA = \oint_\Gamma f \, v_j \, ds \qquad \text{2-D} \tag{9.58}$$

$$\iiint_V \left(\tfrac{\partial}{\partial x_j} f \right) dV = \oiint_S f \, v_j \, dS \qquad \text{3-D} \tag{9.59}$$

9.5.2 Integration Theorems

We begin by defining the following *backward* (or left) operator (observe the direction of the arrow!):

$$f \overleftarrow{\nabla} \equiv \nabla f, \qquad \mathbf{f} \cdot \overleftarrow{\nabla} \equiv \overleftarrow{\nabla} \cdot \mathbf{f}, \qquad \mathbf{f} \times \overleftarrow{\nabla} \equiv -\overleftarrow{\nabla} \times \mathbf{f} \tag{9.60}$$

With these definitions, and if the symbols $*, \otimes$ stand for either the scalar product, the dot product, or the cross product, then a generalization of the previous integration-by-parts formulas can be written symbolically as

$$\boxed{\iiint_V \mathbf{f} * \left(\overleftarrow{\nabla} \otimes \mathbf{g} \right) dV = \oiint_S \mathbf{f} * \left(\hat{\mathbf{n}} \otimes \mathbf{g} \right) dS - \iiint_V \mathbf{f} * \left(\overleftarrow{\nabla} \otimes \mathbf{g} \right) dV} \tag{9.61}$$

This is the most general possible formula that we can write, and that encompasses *all* of the integration theorems – and more. Making particular choices for the operators $*, \otimes$, we obtain

$$\boxed{\begin{aligned} \iiint_V f \overleftarrow{\nabla} g \, dV &= \oiint_S f \hat{\mathbf{n}} g \, dS - \iiint_V f \overleftarrow{\nabla} g \, dV \\ &= \oiint_S f \hat{\mathbf{n}} g \, dS - \iiint_V \overleftarrow{\nabla} f \, g \, dV \end{aligned}} \tag{9.62}$$

$$\iiint_V \mathbf{f} \cdot \vec{\nabla} g \; dV = \oiint_S \mathbf{f} \cdot \hat{\mathbf{n}} \, g \, dS - \iiint_V \mathbf{f} \cdot \bar{\vec{\nabla}} \, g \, dV$$
$$= \oiint_S \mathbf{f} \cdot \hat{\mathbf{n}} \, g \, dS - \iiint_V \bar{\vec{\nabla}} \cdot \mathbf{f} \, g \, dV \tag{9.63}$$

$$\iiint_V f \, \vec{\nabla} \cdot \mathbf{g} \; dV = \oiint_S f \, \hat{\mathbf{n}} \cdot \mathbf{g} \, dS - \iiint_V f \bar{\vec{\nabla}} \cdot \mathbf{g} \, dV$$
$$= \oiint_S f \, \hat{\mathbf{n}} \cdot \mathbf{g} \, dS - \iiint_V \bar{\vec{\nabla}} f \cdot \mathbf{g} \, dV \tag{9.64}$$

$$\iiint_V \mathbf{f} \cdot \vec{\nabla} \mathbf{g} \; dV = \oiint_S \mathbf{f} \cdot \hat{\mathbf{n}} \, \mathbf{g} \, dS - \iiint_V \mathbf{f} \cdot \bar{\vec{\nabla}} \, \mathbf{g} \, dV$$
$$= \oiint_S \mathbf{f} \cdot \hat{\mathbf{n}} \, \mathbf{g} \, dS - \iiint_V \bar{\vec{\nabla}} \cdot \mathbf{f} \, \mathbf{g} \, dV \tag{9.65}$$

$$\iiint_V \mathbf{f} \, \vec{\nabla} \cdot \mathbf{g} \; dV = \oiint_S \mathbf{f} \, \hat{\mathbf{n}} \cdot \mathbf{g} \, dS - \iiint_V \mathbf{f} \, \bar{\vec{\nabla}} \cdot \mathbf{g} \, dV$$
$$= \oiint_S \mathbf{f} \, \hat{\mathbf{n}} \cdot \mathbf{g} \, dS - \iiint_V \bar{\vec{\nabla}} \mathbf{f} \cdot \mathbf{g} \, dV \tag{9.66}$$

$$\iiint_V f \, \vec{\nabla} \times \mathbf{g} \; dV = \oiint_S f \, \hat{\mathbf{n}} \times \mathbf{g} \, dS - \iiint_V f \bar{\vec{\nabla}} \times \mathbf{g} \, dV$$
$$= \oiint_S f \, \hat{\mathbf{n}} \times \mathbf{g} \, dS - \iiint_V \bar{\vec{\nabla}} f \times \mathbf{g} \, dV \tag{9.67}$$

$$\iiint_V \mathbf{f} \times \vec{\nabla} g \; dV = \oiint_S \mathbf{f} \times \hat{\mathbf{n}} \, g \, dS - \iiint_V \mathbf{f} \times \bar{\vec{\nabla}} \, g \, dV$$
$$= \oiint_S \mathbf{f} \times \hat{\mathbf{n}} \, g \, dS + \iiint_V \bar{\vec{\nabla}} \times \mathbf{f} \, g \, dV \tag{9.68}$$

$$\iiint_V \mathbf{f} \cdot \vec{\nabla} \times \mathbf{g} \; dV = \oiint_S \mathbf{f} \cdot \hat{\mathbf{n}} \times \mathbf{g} \, dS - \iiint_V \mathbf{f} \times \bar{\vec{\nabla}} \cdot \mathbf{g} \, dV$$
$$= \oiint_S \mathbf{f} \cdot \hat{\mathbf{n}} \times \mathbf{g} \, dS + \iiint_V \bar{\vec{\nabla}} \times \mathbf{f} \cdot \mathbf{g} \, dV \tag{9.69}$$

$$\iiint_V \mathbf{f} \times \left(\vec{\nabla} \times \mathbf{g} \right) dV = \oiint_S \mathbf{f} \times (\hat{\mathbf{n}} \times \mathbf{g}) \, dS - \iiint_V \mathbf{f} \times \left(\bar{\vec{\nabla}} \times \mathbf{g} \right) dV$$
$$= \oiint_S \mathbf{f} \times (\hat{\mathbf{n}} \times \mathbf{g}) \, dS - \iiint_V \left(\mathbf{g} \times \bar{\vec{\nabla}} \right) \times \mathbf{f} \, dV \tag{9.70}$$

It can readily be shown that the following applies to Eq. 9.70:

$$\mathbf{f} \times \left(\vec{\nabla} \times \mathbf{g} \right) = \mathbf{f} \cdot \left[\left(\tfrac{\partial}{\partial x_1} \mathbf{g} \right) \hat{\mathbf{e}}_1 + \left(\tfrac{\partial}{\partial x_2} \mathbf{g} \right) \hat{\mathbf{e}}_2 + \left(\tfrac{\partial}{\partial x_3} \mathbf{g} \right) \hat{\mathbf{e}}_3 \right] - \mathbf{f} \cdot \nabla \mathbf{g} \tag{9.71}$$

$$\left(\mathbf{g} \times \vec{\nabla} \right) \times \mathbf{f} = \mathbf{g} \cdot \left[\left(\tfrac{\partial}{\partial x_1} \mathbf{f} \right) \hat{\mathbf{e}}_1 + \left(\tfrac{\partial}{\partial x_2} \mathbf{f} \right) \hat{\mathbf{e}}_2 + \left(\tfrac{\partial}{\partial x_3} \mathbf{f} \right) \hat{\mathbf{e}}_3 \right] - (\nabla \cdot \mathbf{f}) \mathbf{g} = \mathbf{f} \times \left(\bar{\vec{\nabla}} \times \mathbf{g} \right) \tag{9.72}$$

$$\nabla \cdot (\mathbf{f} \mathbf{g}) = (\nabla \cdot \mathbf{f}) \mathbf{g} + \mathbf{f} \cdot \nabla \mathbf{g} \qquad \text{Divergence of dyad} \tag{9.73}$$

$$\iiint_V \nabla \cdot (\mathbf{f} \mathbf{g}) \; dV = \oiint_S \left[(\mathbf{f} \cdot \mathbf{g}) \hat{\mathbf{n}} - \mathbf{f} (\mathbf{g} \cdot \hat{\mathbf{n}}) \right] dS = \oiint_S \mathbf{f} \times (\hat{\mathbf{n}} \times \mathbf{g}) \, dS \tag{9.74}$$

9.5.3 Particular Cases: Gauss, Stokes, and Green

Setting either $f = 1, \frac{\partial f}{\partial x_j} = 0$ or $g = 1, \frac{\partial g}{\partial x_j} = 0$, we obtain

$$\iiint\limits_V \vec{\nabla} g \, dV = \oiint\limits_S \hat{\mathbf{n}} g \, dS \tag{9.75}$$

$$\iiint\limits_V \vec{\nabla}\cdot\mathbf{f} \, dV = \oiint\limits_S \mathbf{f}\cdot\hat{\mathbf{n}} \, dS \qquad \text{Gauss Divergence Theorem} \tag{9.76}$$

$$\iint\limits_A \vec{\nabla}\cdot\mathbf{f} \, dA = \oint\limits_\Gamma \mathbf{f}\cdot\hat{\mathbf{n}} \, ds \qquad \text{Stokes Divergence Theorem} \tag{9.77}$$

$$\iiint\limits_V \vec{\nabla}\times\mathbf{g} \, dV = \oiint\limits_S \hat{\mathbf{n}}\times\mathbf{g} \, dS \tag{9.78}$$

Alternatively, setting $g = x_j$ such that $\frac{\partial g}{\partial x_j} = 1$, then

$$\iint\limits_A f \, dA = \oint\limits_\Gamma f \, x_j \, v_j \, ds - \iint\limits_A \left(\frac{\partial}{\partial x_j} f\right) x_j \, dA \tag{9.79}$$

$$\iiint\limits_V f \, dV = \oiint\limits_S f \, x_j \, v_j \, dS - \iiint\limits \left(\frac{\partial}{\partial x_j} f\right) x_j \, dV \tag{9.80}$$

$$\iiint\limits_V \mathbf{f} \, dV = \oiint\limits_S \mathbf{f} \, x_j \, v_j \, dS - \iiint\limits \left(\frac{\partial}{\partial x_j} \mathbf{f}\right) x_j \, dV \tag{9.81}$$

which can be very useful, especially when either f or $\mathbf{f}$ is not a function of x_j, in which case the volume (or area) integral on the right-hand side vanishes.

Many additional formulas can be obtained from those in the preceding box. For example, if we set $f = \phi$, $\mathbf{g} = \nabla\psi$, and take into account that $\nabla \bullet \nabla\psi = \nabla^2\psi$ then from the third formula and introducing the definition $d\mathbf{S} = \hat{\mathbf{n}} \, dS$ we obtain

$$\iiint\limits_V \phi\nabla^2\psi \, dV = \oiint\limits_S \phi\vec{\nabla}\psi\cdot d\mathbf{S} - \iiint\limits_V \vec{\nabla}\phi\cdot\vec{\nabla}\psi \, dV \tag{9.82}$$

which is known as *Green's first identity theorem*. Then again, interchanging ϕ, ψ and combining the two results, we obtain

$$\iiint\limits_V \left(\phi\nabla^2\psi - \psi\cdot\nabla^2\psi\right) dV = \oiint\limits_S \left(\phi\vec{\nabla}\psi - \psi\vec{\nabla}\phi\right)\cdot d\mathbf{S} \tag{9.83}$$

which is known as *Green's second identity theorem*.

9.6 Positive Definiteness of Arbitrary Square Matrix

Positive definiteness is a property of square matrices that relates to the values that so-called *quadratic forms* can attain. The latter are functions $F(x_1, x_2, \ldots x_n)$ that can be visualized as extensions to the n-dimensional space of the familiar ellipses, parabolas, or hyperbolas. Quadratic forms are very ubiquitous in science and engineering. For example, the kinetic energy K of an n-mass system can be written in quadratic form as $K = \frac{1}{2}\mathbf{v}^T\mathbf{M}\,\mathbf{v}$, in which $\mathbf{v}$ is the velocity vector and $\mathbf{M}$ is the mass matrix (which need not be diagonal).

Clearly, this energy can never be negative, no matter what velocity $\mathbf{v} \neq \mathbf{0}$ the particles may have, so $\mathbf{M}$ must have a structure that guarantees the kinetic energy to be positive. Hence, $\mathbf{M}$ is said to be *positive definite*.

More generally, we say that a nonsymmetric, complex $n \times n$ matrix $\mathbf{C}$ is *strictly* positive definite if for any arbitrary complex vector $\mathbf{z}$ together with its transposed conjugate vector $\mathbf{z}^*$ in the n-dimensional space, the quadratic form $F = \mathbf{z}^* \mathbf{C} \mathbf{z}$ is a complex number (a scalar quantity) whose real and imaginary parts are always both positive, that is, it satisfies simultaneously the conditions

$$p = \mathrm{Re}(F) > 0 \quad \text{and} \quad q = \mathrm{Im}(F) > 0 \quad \text{for } \forall \mathbf{z} \neq \mathbf{0} \quad \text{(any non-trivial } \mathbf{z}) \tag{9.84}$$

Alternatively, if only the real part is always positive, then the form is *simply* positive definite. Let $\mathbf{C} = \mathbf{A} + \mathrm{i}\, \mathbf{B}$. Then

$$F = p + \mathrm{i}q = \mathbf{z}^* \left(\mathbf{A} + \mathrm{i}\, \mathbf{B} \right) \mathbf{z} = \mathbf{z}^* \mathbf{A} \mathbf{z} + \mathrm{i}\, \mathbf{z}^* \mathbf{B} \mathbf{z} \tag{9.85}$$

for which the transposed conjugate is

$$F^* = p - \mathrm{i}q = \mathbf{z}^* \left(\mathbf{A}^T - \mathrm{i}\, \mathbf{B}^T \right) \mathbf{z} = \mathbf{z}^* \mathbf{A}^T \mathbf{z} - \mathrm{i}\, \mathbf{z}^* \mathbf{B}^T \mathbf{z} \tag{9.86}$$

Addition of these two expressions yields

$$2p = \mathbf{z}^* \left(\mathbf{A} + \mathbf{A}^T \right) \mathbf{z} + \mathrm{i}\, \mathbf{z}^* \left(\mathbf{B} - \mathbf{B}^T \right) \mathbf{z} > 0 \tag{9.87}$$

Alternatively, subtraction and multiplication by $\mathrm{i} = \sqrt{-1}$ produces

$$2q = \mathbf{z}^* \left(\mathbf{B} + \mathbf{B}^T \right) \mathbf{z} - \mathrm{i}\, \mathbf{z}^* \left(\mathbf{A} - \mathbf{A}^T \right) \mathbf{z} > 0 \tag{9.88}$$

Introducing the two Hermitian forms (a Hermitian matrix satisfies $\mathrm{conj}\left(\mathbf{H}^T \right) \equiv \mathbf{H}^* = \mathbf{H}$)

$$\mathbf{G} = \tfrac{1}{2} \left(\mathbf{A} + \mathbf{A}^T \right) + \tfrac{1}{2} \mathrm{i} \left(\mathbf{B} - \mathbf{B}^T \right) \tag{9.89}$$

$$\mathbf{H} = \tfrac{1}{2} \left(\mathbf{B} + \mathbf{B}^T \right) - \tfrac{1}{2} \mathrm{i} \left(\mathbf{A} - \mathbf{A}^T \right) \tag{9.90}$$

we can write the strict positive definiteness condition as

$$p = \mathbf{z}^* \mathbf{G} \mathbf{z} > 0 \qquad q = \mathbf{z}^* \mathbf{H} \mathbf{z} > 0 \tag{9.91}$$

Thus, $\mathbf{C}$ is strictly positive definite if both of the Hermitian matrices $\mathbf{G}, \mathbf{H}$ obtained from this matrix are positive definite. If only the first is positive definite, then $\mathbf{C}$ is simply positive definite. We see that the conditions for positive definiteness of an arbitrary square matrix reduce to testing the positive definiteness of the two Hermitian forms that can be assembled with the real and imaginary parts of the matrix. These conditions will be examined in these notes.

For example, consider the vibration of an assembly of spring and viscous dampers whose stiffness and damping matrices are $\mathbf{K}$ and $\mathbf{C}$, respectively. In the *frequency domain*, the dynamic stiffness (or impedance) matrix for this system is $\mathbf{Z} = \mathbf{K} + \mathrm{i}\omega \mathbf{C}$, which is not Hermitian, but complex symmetric instead. As explained further in the text that follows, we know that the quadratic form

$$F = \mathbf{z}^* \left(\mathbf{K} + \mathrm{i}\omega \mathbf{C} \right) \mathbf{z} \geq 0 \tag{9.92}$$

must be strictly nonnegative. Indeed, the two Hermitian forms for the complex matrix $\mathbf{Z}$ are $\mathbf{G} = \mathbf{K}$ and $\mathbf{H} = \omega\mathbf{C}$, which are both real and symmetric. Also, on physical grounds, we know that they must be strictly positive definite or semidefinite. The reason is that neither the strain energy $\frac{1}{2}\mathbf{u}^T\mathbf{K}\mathbf{u}$ nor the instantaneous dissipative power $\dot{\mathbf{u}}^T\mathbf{C}\dot{\mathbf{u}}$ (which are both quadratic forms) can ever be negative. Now, if the spring–damper assembly were not properly restrained (say, it floats in space), then rigid body motions $\mathbf{z} \neq \mathbf{0}$ not entailing spring deformations or energy dissipation could exist that would make $F = 0$. In that case, the impedance matrix is only positive semidefinite. If, however, the system is properly constrained and precludes rigid body motions, then the system is indeed strictly positive definite.

Positive versus Negative Definiteness (Semidefiniteness) of Hermitian Matrices
Consider the quadratic form for a Hermitian matrix $q = \mathbf{z}^*\mathbf{H}\mathbf{z}$, which is always a real number. If

- $q > 0$ for all vectors $\mathbf{z}$, then the matrix is *positive definite*.
- $q \geq 0$ for all vectors $\mathbf{z}$, that is, if q is always nonnegative, yet some nontrivial vectors $\mathbf{z} \neq \mathbf{0}$ can be found for which $q = 0$, then the matrix is *positive semidefinite*.
- $q < 0$ for all vectors $\mathbf{z}$, then the matrix is *negative definite*.
- $q \leq 0$ for all vectors $\mathbf{z}$, that is, if q is always nonpositive, yet some nontrivial vectors $\mathbf{z} \neq \mathbf{0}$ can be found for which $q = 0$, then the matrix is *negative semidefinite*.
- If q is neither nonpositive nor nonnegative, then the matrix is said to be *indefinite*.

Conditions for Positive and Negative Definiteness of a Hermitian Matrix
Consider a Hermitian matrix $\mathbf{H}$ and apply a *Cholesky decomposition* to it, that is, using Gaussian reduction express $\mathbf{H}$ as the product of a lower and upper triangular matrix of the form

$$\mathbf{H} = \mathbf{L}\mathbf{U} = \mathbf{U}^*\mathbf{D}\mathbf{U} \qquad (9.93)$$

in which $\mathbf{D} = \mathrm{diag}\{d_k\}$ is a real, diagonal matrix with elements d_k such that $D_k = d_1 d_2 \cdots d_k$ are the determinants of the *leading minors* of $\mathbf{H}$. The leading minors are the submatrices $\mathbf{H}_k$ formed with the k first rows and columns of $\mathbf{H}$. Also, $\mathbf{U}$ is an upper triangular matrix whose elements on its main diagonal are all 1. Defining $D_0 = 1$, the canonical form of the quadratic form is then

$$\mathbf{z}^*\mathbf{H}\mathbf{z} = \mathbf{z}^*\mathbf{U}^*\mathbf{D}\mathbf{U}\mathbf{z} = \mathbf{r}^*\mathbf{D}\mathbf{r} = \sum_{k=1}^{n} d_k |r_k|^2 = \sum_{k=1}^{n} \frac{D_k}{D_{k-1}} |r_k|^2 = D_1 |r_1|^2 + \frac{D_2}{D_1}|r_2|^2 + \frac{D_3}{D_2}|r_3|^2 + \cdots$$
$$(9.94)$$

The number of positive ratios D_k/D_{k-1} is the *signature* of the matrix, while the total number of terms is the *rank*. Since the $|r_k|^2$ are arbitrary, but nonnegative, it follows that a *necessary* condition for the positive definiteness of a Hermitian matrix of size $n \times n$ is for *all* of its leading minors to have positive determinants $D_k > 0$, $k = 1, \cdots n$. As will be demonstrated in the next paragraph, this condition is also *sufficient*, so any Hermitian matrix $\mathbf{H}$ whose leading minor determinants D_k are positive is indeed positive definite.

Eigenvalues of a Positive-Definite Hermitian Matrix

Consider the quadratic form $\mathbf{z}^*\mathbf{H}\mathbf{z}$ with $\mathbf{z} = \mathbf{\Phi}\mathbf{x}$, in which $\mathbf{\Phi}$ is the unitary modal matrix of $\mathbf{H}$ (from the eigenvalue problem $\mathbf{H}\mathbf{\Phi} = \mathbf{\Lambda}\mathbf{\Phi}$), which satisfies the orthogonality condition $\mathbf{\Phi}^*\mathbf{\Phi} = \mathbf{I}$, that is, $\mathbf{\Phi}^{-1} = \mathbf{\Phi}^*$. Also, $\mathbf{\Phi}^*\mathbf{H}\mathbf{\Phi} = \mathbf{\Phi}^{-1}\mathbf{H}\mathbf{\Phi} = \mathbf{\Lambda} = \mathrm{diag}\{\lambda_j\}$, in which the λ_j are the real eigenvalues of $\mathbf{H}$. Defining $\mathbf{z} = \mathbf{U}\mathbf{r} = \mathbf{\Phi}\mathbf{x}$, we conclude that

$$\mathbf{z}^*\mathbf{H}\mathbf{z} = \mathbf{x}^*\mathbf{\Lambda}\mathbf{x} = \sum_{j=1}^{n} \lambda_j |x_j|^2 = \mathbf{r}^*\mathbf{D}\mathbf{r} = \sum_{k=1}^{n} \frac{D_k}{D_{k-1}}|r_k|^2 > 0 \tag{9.95}$$

Choosing first a vector $\mathbf{x}$ in which only one of its components is nonzero, say the jth, it follows from the preceding that $\lambda_j |x_j|^2 > 0$, that is, $\lambda_j > 0$, so all eigenvalues must be positive if all minor determinants are positive. Having established that fact, we choose next an arbitrary vector $\mathbf{x}$, and from the first of the above two summations, we see that it can only result in a positive number inasmuch as all $\lambda_j > 0$. Thus, a *necessary and sufficient* condition for the positive definiteness of a Hermitian matrix is for all of the eigenvalues to be positive, or alternatively, for all the leading minors to have positive determinants. Hence, if $\mathbf{H} = \{H_{ij}\}$, then $\mathbf{H}$ is positive definite if and only if $\lambda_j > 0$, $j = 1, 2, \ldots n$, or alternatively, if

$$D_1 = H_{11} > 0, \qquad D_2 = \begin{vmatrix} H_{11} & H_{12} \\ H_{21} & H_{22} \end{vmatrix} > 0, \qquad D_3 = \begin{vmatrix} H_{11} & H_{12} & H_{13} \\ H_{21} & H_{22} & H_{23} \\ H_{31} & H_{32} & H_{33} \end{vmatrix} > 0 \quad \text{etc.} \tag{9.96}$$

As stated earlier, all of these determinants are always real numbers. By simple permutations of rows and columns, which merely shuffle the elements of the quadratic form, it can be seen that other subdeterminants involving the main diagonal of $\mathbf{H}$ must also be positive. For example,

$$H_{22} > 0 \qquad \cdots \qquad H_{nn} > 0 \qquad \text{i.e. all diagonal elements are positive} \tag{9.97}$$

$$\begin{vmatrix} H_{22} & H_{23} \\ H_{32} & H_{33} \end{vmatrix} > 0 \qquad \begin{vmatrix} H_{11} & H_{13} \\ H_{31} & H_{33} \end{vmatrix} > 0 \qquad \text{and so forth} \tag{9.98}$$

The latter do not constitute additional conditions, but are implied by the conditions on the leading minors. For example,

$$H_{11} > 0 \qquad \text{and} \qquad H_{11}H_{22} - H_{12}H_{21} > 0 \qquad \text{imply} \qquad H_{22} > 0 \tag{9.99}$$

because $H_{12}H_{21} = H_{12}H_{12}^* = |H_{12}|^2 > 0$, so $H_{22} = |H_{12}|^2 / H_{11} > 0$. However, the converse is not true: $H_{11} > 0$ and $H_{22} > 0$ does *not* imply $D_2 > 0$, so positive diagonal elements alone do not guarantee positive definiteness.

It also follows that a Hermitian matrix is *negative* definite if $(-1)^k D_k > 0$, $k = 1, \ldots n$. This implies that the first element on the diagonal (and for that matter, *all* diagonal elements) must be negative, yet this alone is again not sufficient for negative definiteness. Alternatively, all eigenvalues of a negative definite matrix must be negative.

As we shall see later, these simple conditions do not apply to nonsymmetric matrices, even when they should be real. Nonetheless, they can still be used to establish the positive definiteness of such matrices via their Hermitian forms, as demonstrated in the following example.

Example of Application to a Nonsymmetric Matrix

$$\mathbf{A} = \begin{Bmatrix} 4 & 0 \\ -2 & 6 \end{Bmatrix} \qquad \mathbf{B} = \begin{Bmatrix} 8 & 2 \\ -4 & 14 \end{Bmatrix} \qquad \mathbf{C} = \mathbf{A} + i\,\mathbf{B} = 2\begin{Bmatrix} 2+4i & i \\ -1-2i & 3+7i \end{Bmatrix} \tag{9.100}$$

The Hermitian forms obtained from $\mathbf{C}$ are

$$\mathbf{G} = \begin{Bmatrix} 4 & -1+3i \\ -1-3i & 6 \end{Bmatrix} \qquad \mathbf{H} = \begin{Bmatrix} 8 & -1-i \\ -1+i & 14 \end{Bmatrix} \tag{9.101}$$

$$G_{11} = 4 > 0 \quad \text{and} \quad \det(\mathbf{G}) = 14 > 0, \quad \Rightarrow \quad \mathbf{G} \text{ is positive definite} \tag{9.102}$$

$$H_{11} = 8 > 0 \quad \text{and} \quad \det(\mathbf{H}) = 110 > 0, \quad \Rightarrow \quad \mathbf{H} \text{ is positive definite} \tag{9.103}$$

Also, the eigenvalues of $\mathbf{G}$ are $\lambda_1 = 1.683$, $\lambda_2 = 8.317$, while those of $\mathbf{H}$ are $\lambda_1 = 7.683$, $\lambda_2 = 14.317$, all of which are positive, so $\mathbf{C}$ is strictly positive definite.

However, readers should not confuse this test for the positive definiteness of $\mathbf{C}$ via the eigenvalues of the two Hermitian forms $\mathbf{G}, \mathbf{H}$ with the eigenvalues of either $\mathbf{A}, \mathbf{B}$ or even $\mathbf{C}$, as these *cannot* be used for this purpose, as considered next.

Positive Eigenvalues and Nonsymmetric Matrices
Regrettably, the "nice" properties relating eigenvalues and positive definiteness do *not* extend to general, non-Hermitian matrices. For instance, consider the following real, non-symmetric matrix $\mathbf{C}$ together with its eigenvalues:

$$\mathbf{C} = \begin{Bmatrix} 1 & a \\ b & 1 \end{Bmatrix} \qquad \begin{vmatrix} 1-\lambda & a \\ b & 1-\lambda \end{vmatrix} = 0 \qquad \lambda = 1 \mp \sqrt{ab} \tag{9.104}$$

Clearly, if a, b have the same sign and satisfy $0 < ab < 1$, then both eigenvalues are real and positive. Consider next the Hermitian forms of $\mathbf{C}$

$$\mathbf{G} = \begin{Bmatrix} 1 & \tfrac{1}{2}(a+b) \\ \tfrac{1}{2}(a+b) & 1 \end{Bmatrix} \qquad \mathbf{H} = \begin{Bmatrix} 0 & -\tfrac{1}{2}i(a-b) \\ \tfrac{1}{2}i(a-b) & 0 \end{Bmatrix} \tag{9.105}$$

Now, the leading minor determinants of these matrices are $G_1 = 1 > 0$, $G_2 = 1 - \tfrac{1}{4}(a+b)^2$, $H_1 = 0$, $H_2 = -\tfrac{1}{4}(a-b)^2 < 0$. Clearly, $\mathbf{H}$ is negative semidefinite, so $\mathbf{C}$ cannot be strictly positive definite. In addition, $G_2 > 0$ requires $(a+b)^2 < 4$, a condition that is easily violated. For example, $a = -2.5, b = -0.1$ makes $G_2 = -2.76$, yet the eigenvalues of $\mathbf{C}$ remain real and positive, namely $\lambda_1 = 0.5, \lambda_2 = 1.5$. Thus, positive eigenvalues do not guarantee the positive definiteness of a non-Hermitian matrix. To complicate matters, some of the eigenvalues of a real, nonsymmetric matrix could also appear as complex conjugate pairs, such as in the example above with $ab < 0$, in which case the proper meaning of "positive eigenvalues" is lost.

Sum of Positive Definite Matrices
The sum of two positive-definite matrices is also positive definite. That is,

$$y = \mathbf{z}^* (\mathbf{A} + \mathbf{B})\mathbf{z} = \mathbf{z}^*\mathbf{A}\mathbf{z} + \mathbf{z}^*\mathbf{B}\mathbf{z} > 0 \tag{9.106}$$

Product of Positive Definite Matrices

The product of two positive-definite matrices is *not* necessarily positive definite. For example, consider the real, symmetric, positive definite matrices $\mathbf{A}, \mathbf{B}$, and their product $\mathbf{C}$:

$$\mathbf{A} = \begin{Bmatrix} 3 & -4 \\ -4 & 6 \end{Bmatrix} \qquad \mathbf{B} = \begin{Bmatrix} 3 & 0 \\ 0 & 1 \end{Bmatrix} \qquad \mathbf{C} = \mathbf{A}\mathbf{B} = \begin{Bmatrix} 9 & -4 \\ -12 & 6 \end{Bmatrix} \tag{9.107}$$

The Hermitian form of $\mathbf{C}$ is

$$\mathbf{G} = \tfrac{1}{2}(\mathbf{C} + \mathbf{C}^T) = \begin{Bmatrix} 9 & -8 \\ -8 & 6 \end{Bmatrix} \tag{9.108}$$

which satisfies $D_1 = G_{11} = 9 > 0$ but $D_2 = \det(\mathbf{G}) = -10 < 0$. Hence, $\mathbf{C}$ is *not* positive definite.

Moreover, even the product of two *commutative* matrices, that is, $\mathbf{A}\mathbf{B} = \mathbf{B}\mathbf{A}$, each of which is positive definite, need not be positive definite. For example, the two matrices

$$\mathbf{A} = \begin{Bmatrix} 2 & 0 \\ 4 & 3 \end{Bmatrix} \qquad \mathbf{B} = \begin{Bmatrix} 3 & 0 \\ 4 & 4 \end{Bmatrix} \tag{9.109}$$

are both positive definite because $\left|\tfrac{1}{2}(\mathbf{A}+\mathbf{A}^T)\right| = 2 > 0, \left|\tfrac{1}{2}(\mathbf{B}+\mathbf{B}^T)\right| = 8 > 0$, and all diagonal terms are positive. Also,

$$\mathbf{C} = \mathbf{A}\mathbf{B} = \mathbf{B}\mathbf{A} = \begin{Bmatrix} 6 & 0 \\ 24 & 12 \end{Bmatrix}, \qquad \text{but} \qquad \left|\tfrac{1}{2}(\mathbf{C}+\mathbf{C}^T)\right| = -72 < 0 \tag{9.110}$$

so the product is *not* positive definite.

Then again, the product of two positive definite *Hermitian* matrices which in addition also happen to be *commutative*, is positive definite. This is so because commutative Hermitian (or real symmetric) matrices share the same eigenvectors, even if not the eigenvalues, all which are guaranteed to be real and positive. Hence, the matrices are of the form

$$\mathbf{A} = \mathbf{\Phi}\mathbf{\Lambda}_A\mathbf{\Phi}^{-1}, \qquad \mathbf{B} = \mathbf{\Phi}\mathbf{\Lambda}_B\mathbf{\Phi}^{-1}, \qquad \mathbf{A}\mathbf{B} = \mathbf{B}\mathbf{A} = \mathbf{\Phi}\mathbf{\Lambda}_A\mathbf{\Lambda}_B\mathbf{\Phi}^{-1} = \mathbf{\Phi}\mathbf{\Lambda}_B\mathbf{\Lambda}_A\mathbf{\Phi}^{-1} \tag{9.111}$$

Hence, the eigenvalues of the product $\mathbf{A}\mathbf{B}$ are equal to the product of the eigenvalues of the two matrices, that is, $\lambda_{AB} = \lambda_{BA} = \lambda_A \lambda_B$, so this product is both Hermitian and positive definite. Thus, a *sufficient* (but not necessary) condition for the product of two positive definite matrices to be positive definite is that they be Hermitian *and* commutative. Observe that any two functions or polynomials of a Hermitian matrix are also commutable, for example, $\mathbf{A} = \sin\mathbf{H}, \mathbf{B} = \cos\mathbf{H}$.

Finally, the product of two positive definite matrices could (but need not) be positive definite even when the two matrices are neither Hermitian nor commutative. For example,

$$\mathbf{A} = \begin{Bmatrix} 2 & 0 \\ 1 & 1 \end{Bmatrix}, \qquad \mathbf{B} = \begin{Bmatrix} 1 & -1 \\ 1 & 2 \end{Bmatrix} \qquad \text{are both positive definite} \tag{9.112}$$

Also,

$$\mathbf{A}\mathbf{B} = \begin{Bmatrix} 2 & -2 \\ 2 & 1 \end{Bmatrix} \qquad \text{is positive definite} \tag{9.113}$$

but

$$\mathbf{BA} = \begin{Bmatrix} 1 & -1 \\ 4 & 2 \end{Bmatrix} \qquad \text{is } not \text{ positive definite} \tag{9.114}$$

This demonstrates that even if the product $\mathbf{AB}$ should be positive definite, the permuted product $\mathbf{BA}$ need not be.

Determinants of Positive Definite, Nonsymmetric Matrices
The determinants of the leading minors of a real, nonsymmetric, positive definite matrix are all positive, so such a matrix is never singular. However, this property is only a necessary, but not sufficient condition for positive definiteness. This can be shown as follows. Consider the nonsymmetric, real matrix

$$\mathbf{A} = \begin{Bmatrix} a_{11} & \cdots & a_{1n} \\ \vdots & \ddots & \\ a_{n1} & & a_{nn} \end{Bmatrix} \tag{9.115}$$

for which the Hermitian form is

$$\mathbf{G} = \tfrac{1}{2}\left(\mathbf{A} + \mathbf{A}^T\right) = \begin{Bmatrix} a_{11} & \cdots & \tfrac{1}{2}\left(a_{1n} + a_{n1}\right) \\ \vdots & \ddots & \\ \tfrac{1}{2}\left(a_{1n} + a_{n1}\right) & & a_{nn} \end{Bmatrix} \tag{9.116}$$

By definition, $\mathbf{A}$ is positive definite when its Hermitian form $\mathbf{G}$ is positive definite, which guarantees in turn that all diagonal elements a_{ii} of $\mathbf{A}$ will be positive. In particular, $a_{11} > 0$, so one is allowed to perform the transformation

$$\mathbf{B} = \mathbf{T}_1 \mathbf{A} \mathbf{T}_1^T \tag{9.117}$$

in which $\mathbf{T}_1$ is the Gaussian reduction matrix

$$\mathbf{T}_1 = \begin{Bmatrix} 1/a_{11} & 0 & \cdots & 0 \\ -a_{21}/a_{11} & 1 & 0 & \cdots \\ \vdots & 0 & \ddots & 0 \\ -a_{n1}/a_{11} & \vdots & 0 & 1 \end{Bmatrix} \tag{9.118}$$

The result of this transformation is

$$\mathbf{B} = \begin{Bmatrix} 1/a_{11} & (a_{12} - a_{21})/a_{11} & \cdots & (a_{1n} - a_{n1})/a_{11} \\ 0 & a_{22} - a_{21}a_{12}/a_{11} & \cdots & a_{2n} - a_{21}a_{1n}/a_{11} \\ \vdots & \vdots & \ddots & 0 \\ 0 & a_{n2} - a_{n1}a_{12}/a_{11} & \ddots & a_{nn} - a_{n1}a_{1n}/a_{11} \end{Bmatrix} = \begin{Bmatrix} a_{11}^{-1} & \mathbf{b}_1 \\ \mathbf{0} & \mathbf{B}_2 \end{Bmatrix} \tag{9.119}$$

Clearly, if $\mathbf{A}$ is nonsingular, then $\mathbf{B}$ must be nonsingular as well because $\det \mathbf{T}_1 = 1/a_{11} > 0$. Next, consider the quadratic form $q = \mathbf{x}^T \mathbf{A} \mathbf{x} = \mathbf{y}^T \mathbf{B} \mathbf{y} > 0$, with $\mathbf{y} = \mathbf{T}_1^T \mathbf{x}$. Inasmuch as $\mathbf{A}$ is positive definite, this form must be positive for any $\mathbf{x}$ and $\mathbf{y}$. In particular, the quadratic form will be positive if we choose the first element of $\mathbf{y}$ to be zero, that is, $y_1 = 0$. Since all other elements of $\mathbf{y}$ are still arbitrary, it follows that the square submatrix $\mathbf{B}_2$ must be positive definite, which means that all of its diagonal elements will be positive.

Repeating in turn this transformation on $\mathbf{B}_2$ and on all smaller submatrices thereafter by means of appropriate Gaussian transformation matrices $\mathbf{T}_2, \mathbf{T}_3, \ldots \mathbf{T}_n$, one is ultimately led to an upper triangular matrix $\mathbf{C}$ of the form

$$\mathbf{C} = \mathbf{T}_n \mathbf{T}_{n-1} \cdots \mathbf{T}_1 \, \mathbf{A} \, \mathbf{T}_1^T \mathbf{T}_2^T \cdots \mathbf{T}_n^T \tag{9.120}$$

The determinant of this matrix is simply the product of its positive diagonal elements, so

$$|\mathbf{C}| = c_{11} c_{22} \cdots c_{nn} = |\mathbf{T}_1|^2 |\mathbf{T}_2|^2 \cdots |\mathbf{T}_n|^2 |\mathbf{A}| > 0 \tag{9.121}$$

Since each of the transformation matrices $\mathbf{T}_k$ is nonsingular, we conclude that $\mathbf{A}$ cannot be singular, and furthermore, that its determinant must be positive. Moreover, since this argument applies to the complete matrix $\mathbf{A}$, then it must surely apply to any of its leading minors, so

$$a_{11} > 0, \qquad \begin{vmatrix} a_{11} & a_{12} \\ a_{21} & a_{22} \end{vmatrix} > 0, \qquad \begin{vmatrix} a_{11} & a_{12} & a_{13} \\ a_{21} & a_{22} & a_{23} \\ a_{31} & a_{32} & a_{33} \end{vmatrix} > 0 \qquad \text{etc.} \tag{9.122}$$

Unfortunately, the converse statement is *not* true. Unlike Hermitian matrices, positive determinants for the leading minors of a nonsymmetric matrix do *not* guarantee that it will be positive definite. Hence, positive determinants of leading minors are merely a *necessary*, but not *sufficient* condition for positive definiteness. An example is

$$\mathbf{A} = \begin{Bmatrix} 1 & -1 \\ 4 & 2 \end{Bmatrix}, \qquad a_{11} = 1 > 0, \qquad |\mathbf{A}| = 6 > 0 \tag{9.123}$$

but

$$\left| \tfrac{1}{2}(\mathbf{A} + \mathbf{A}^T) \right| = \begin{vmatrix} 1 & 1.5 \\ 1.5 & 2 \end{vmatrix} = -0.25 \tag{9.124}$$

so $\mathbf{A}$ is not positive definite. Still, if any of the minor determinants of a matrix were not positive (e.g., any of the diagonal elements is zero or negative), then at least one can say for sure that the matrix is *not* positive definite.

Observe that the proof used earlier to demonstrate the sufficiency of positive minor determinants for Hermitian matrices does not apply to real, nonsymmetric matrices, because the latter possess two types of eigenvectors, namely left and right eigenvectors $\mathbf{\Phi}, \mathbf{\Psi}$. These follow from the two eigenvalue problems $\mathbf{A}\mathbf{\Phi} = \mathbf{\Phi}\mathbf{\Lambda}$ and $\mathbf{\Psi}^T \mathbf{A} = \mathbf{\Lambda}\mathbf{\Psi}^T$ (or equivalently, $\mathbf{A}^T \mathbf{\Psi} = \mathbf{\Psi}\mathbf{\Lambda}$). These two sets of eigenvectors are not by themselves unitary or even orthogonal, that is, neither of the products $\mathbf{\Phi}^* \mathbf{\Phi}$ or $\mathbf{\Psi}^* \mathbf{\Psi}$ produce diagonal matrices. Then again, $\mathbf{\Psi}^T \mathbf{\Phi} = \mathbf{I}$, but this does not help.

9.7 Derivative of Matrix Determinant: The Trace Theorem

Let $\mathbf{A}(\lambda)$ be a general, complex, square, nonsingular matrix, whose elements are differentiable functions of a certain parameter λ. It can then be shown[1] that the derivative with respect to λ of the determinant of $\mathbf{A}$, $\Delta = \det \mathbf{A}$, can be obtained as

---

[1] P. Lancaster, *Lambda Matrices and Vibrating Systems* (Oxford: Pergamon Press, 1966), 99. A reprinted version is currently available from Courier–Dover Publications (2002).

$$\Delta'(\lambda) = \Delta(\lambda) \, \text{tr}\left[\mathbf{A}^{-1}\mathbf{A}'\right] \tag{9.125}$$

where "tr" is the trace function (i.e., the sum of the diagonal elements of the term in brackets), and the elements of $\mathbf{A}'$ are the derivatives of the elements of $\mathbf{A}$. Since only the diagonal elements are involved in the trace function, this operation requires, in principle, fewer operations than a full solution of the system of equations.

We demonstrate in the following an effective numerical scheme to evaluate this expression, based on the **LU** decomposition of $\mathbf{A}$, that is, in the reduction of $\mathbf{A}$ into a product of a lower and an upper triangular matrix, that is,

$$\mathbf{A} = \mathbf{LU} = \begin{Bmatrix} a_{11} & \cdots & a_{1n} \\ \vdots & \ddots & \vdots \\ a_{n1} & \cdots & a_{nn} \end{Bmatrix} = \begin{Bmatrix} \alpha_{11} & 0 & \cdots & 0 \\ \alpha_{21} & \alpha_{22} & \ddots & \vdots \\ \vdots & & \ddots & 0 \\ \alpha_{n1} & \alpha_{n2} & \cdots & \alpha_{nn} \end{Bmatrix} \begin{bmatrix} 1 & \beta_{12} & \cdots & \beta_{1n} \\ 0 & 1 & \ddots & \vdots \\ \vdots & & \ddots & \beta_{n-1,n} \\ 0 & 0 & \cdots & 1 \end{bmatrix} \tag{9.126}$$

In Crout's reduction method, this **LU** decomposition is typically accomplished in-situ, writing the lower and upper factor matrices together into a single matrix, namely

$$\mathbf{A} \Rightarrow \begin{Bmatrix} \alpha_{11} & \beta_{12} & \cdots & \beta_{1n} \\ \alpha_{21} & \alpha_{22} & \ddots & \vdots \\ \vdots & & \ddots & \beta_{n-1,n} \\ \alpha_{n1} & \alpha_{n2} & \cdots & \alpha_{nn} \end{Bmatrix} \tag{9.127}$$

Let $\mathbf{A}$ be a symmetric, possibly complex, square matrix, whose elements $a_{ij} = a_{ij}(\lambda)$ depend on some parameter λ. (While we consider in the following only the special case of a symmetric matrix, the generalization to the nonsymmetric case is straightforward.) We partition this matrix by separating its first row and column, so that

$$\mathbf{A} = \{\mathbf{A}_n\} = \begin{Bmatrix} a_{11} & \mathbf{a}_1^T \\ \mathbf{a}_1 & \mathbf{A}_{n-1} \end{Bmatrix} \tag{9.128}$$

Next, we apply the first step in the **LU** decomposition of $\mathbf{A}$:

$$\mathbf{A} = \begin{Bmatrix} a_{11} & \mathbf{0}^T \\ \mathbf{a}_1 & \hat{\mathbf{A}}_{n-1} \end{Bmatrix} \begin{Bmatrix} 1 & \mathbf{a}_1^T / a_{11} \\ \mathbf{0} & \mathbf{I} \end{Bmatrix} \tag{9.129}$$

Clearly,

$$\hat{\mathbf{A}}_{n-1} = \mathbf{A}_{n-1} - \frac{1}{a_{11}} \mathbf{a}_1 \mathbf{a}_1^T \qquad \text{and} \qquad \det \mathbf{A} = a_{11} \det \hat{\mathbf{A}}_{n-1} \tag{9.130}$$

Also, if we define $\Delta \equiv \Delta_n = \det \mathbf{A}_n$, $\Delta_{n-1} = \det \hat{\mathbf{A}}_{n-1}$, then

$$\Delta_n = a_{11}\Delta_{n-1} \tag{9.131}$$

Hence, the derivative with respect to λ is

$$\Delta_n' = \hat{a}_{11}'\Delta_{n-1} + \hat{a}_{11}\Delta_{n-1}' \tag{9.132}$$

Dividing by the determinant equation,

$$\frac{\Delta'_n}{\Delta_n} = \frac{a'_{11}}{a_{11}} + \frac{\Delta'_{n-1}}{\Delta_{n-1}}$$ (9.133)

On the other hand, the derivative of $\hat{\mathbf{A}}_{n-1}$ with respect to λ is

$$\hat{\mathbf{A}}'_{n-1} = \mathbf{A}'_{n-1} - \frac{1}{a_{11}}\left[\mathbf{a}'_1\mathbf{a}_1^T + \mathbf{a}_1\mathbf{a}_1'^T\right] + \frac{a'_{11}}{a_{11}^2}\mathbf{a}_1\mathbf{a}_1^T$$ (9.134)

This is an operation that can readily be carried out in parallel (or even after) the **LU** decomposition of **A**. It should be noted that the reduced matrix $\hat{\mathbf{A}}_{n-1}$ and its derivative are symmetric, and that both preserve the bandwidth of **A** (if any).

By repeated application of the **LU** reduction process by means of the previous equations, we obtain a sequence of progressively smaller submatrices $\hat{\mathbf{A}}_{n-1}, \hat{\mathbf{A}}_{n-2} \cdots \hat{\mathbf{A}}_1$, whose first diagonal elements are $\hat{a}_{22}, \hat{a}_{33}, \cdots, \hat{a}_{nn}$, and whose determinants are $\Delta_j = \det \hat{\mathbf{A}}_j$. We conclude that

$$\frac{\Delta'_n}{\Delta_n} = \frac{a'_{11}}{a_{11}} + \frac{\hat{a}'_{22}}{\hat{a}_{22}} + \cdots \frac{\hat{a}'_{nn}}{\hat{a}_{nn}} \qquad \Delta_n = a_{11}\,\hat{a}_{22}\,\hat{a}_{33}\cdots\hat{a}_{nn}$$ (9.135)

Although only the diagonal elements of the reduced matrices are necessary to compute the determinant and its derivative Δ'_n, all elements of $\hat{\mathbf{A}}_j\hat{\mathbf{A}}'_j$ must be evaluated, because they affect the ulterior reductions. In the case of a symmetric matrix considered here, only the lower band need be stored.

9.8 Circulant and Block-Circulant Matrices

9.8.1 Circulant Matrices

Consider the *circulant* matrix

$$\mathbf{A} = \begin{Bmatrix} a_0 & a_1 & \cdots & a_{n-1} \\ a_{n-1} & a_0 & \ddots & \vdots \\ \vdots & \ddots & \ddots & a_1 \\ a_1 & \cdots & a_{n-1} & a_0 \end{Bmatrix}$$ (9.136)

It can easily be shown by simple substitution that the eigenvalue problem

$$\mathbf{A}\mathbf{x}_j = \lambda_j\,\mathbf{x}_j \qquad j = 0,1,\ldots n-1$$ (9.137)

has eigenvalues and eigenvectors

$$\mathbf{\Lambda} = \mathrm{diag}\{\lambda_j\} \qquad \lambda_j = \sum_{k=0}^{n-1} a_k z^{-jk} \qquad z = \sqrt[n]{1} = e^{2\pi i/n}$$ (9.138)

$$\mathbf{x}_j^T = \{1 \quad z^{-j} \quad z^{-2j} \quad \cdots \quad z^{-(n-1)j}\} \qquad j = 0,1,\ldots n-1$$ (9.139)

that is, the eigenvalues λ_j of a circulant matrix are the elements of the discrete Fourier transform of the first row, and the elements of the eigenvectors are powers of the nth root of unity. The proof requires verifying that each row of Eq. 9.137 satisfies the solution 9.138, 9.139 while considering that $z^n = 1$.

In general, the eigenvalues of $\mathbf{A}$ may be complex numbers. However, if $\mathbf{A}$ is real and symmetric, then all of its eigenvalues are real. This can be verified as follows. Assume that n is even, so $m = n/2$ has no remainder. On account of the symmetry and circulant properties of the matrix, the first row of $\mathbf{A}$ has the form $\{a_0 \quad a_1 \quad \cdots \quad a_m \quad \cdots \quad a_2 \quad a_1\}$. Defining $a_{-k} \equiv a_k$ and considering that $z^{-j} = z^{n-j}$, it follows that

$$\lambda_j = \sum_{k=0}^{n-1} a_k z^{-jk} = \sum_{k=-m}^{m-1} a_k z^{-jk} = \sum_{k=-m}^{m} a_k z^{-jk} \tag{9.140}$$

in which the split summation sign indicates that the first and last elements must be halved. The two equivalent expressions on the right are the discrete Fourier transforms of a real, symmetric function, so the result must be real. On the other hand, if n is odd, then with $m = (n-1)/2$, the first row will have the form $\{a_0 \quad a_1 \quad \cdots \quad a_m \quad a_m \quad \cdots \quad a_2 \quad a_1\}$. In this case,

$$\lambda_j = \sum_{k=0}^{n-1} a_k z^{-jk} = \sum_{k=-m}^{m} a_k z^{-jk} \tag{9.141}$$

which is also purely real. Thus, all eigenvalues of a symmetric circulant matrix are indeed real.

The modal matrix $\mathbf{X} = \{\mathbf{x}_0 \quad \cdots \quad \mathbf{x}_{n-1}\}$ satisfies the orthogonality condition

$$\mathbf{X}^* \mathbf{X} = n\mathbf{I} \tag{9.142}$$

in which $\mathbf{X}^* = \text{conj}(\mathbf{X}^T)$. The reason is that $\mathbf{X}$ is the Fourier transform of $\mathbf{I}$, while $\mathbf{X}^*\mathbf{X}$ is the inverse Fourier transform of $\mathbf{X}$, an operation that returns the original data multiplied by n. Hence, the matrix of coefficients $\mathbf{A}$ accepts the spectral decomposition

$$\mathbf{A} = \mathbf{X}\mathbf{\Lambda}\mathbf{X}^{-1} = \tfrac{1}{n}\mathbf{X}\mathbf{\Lambda}\mathbf{X}^* \tag{9.143}$$

It follows that the pth power of $\mathbf{A}$ is

$$\mathbf{A}^p = \mathbf{X}\mathbf{\Lambda}^p\mathbf{X}^{-1} = \tfrac{1}{n}\mathbf{X}\mathbf{\Lambda}^p\mathbf{X}^* \tag{9.144}$$

which has a circulant structure. In particular,

$$\mathbf{A}^{-1} = \mathbf{X}\mathbf{\Lambda}^{-1}\mathbf{X}^{-1} = \tfrac{1}{n}\mathbf{X}\mathbf{\Lambda}^{-1}\mathbf{X}^* \tag{9.145}$$

is also circulant. Its elements can be obtained very effectively by computing the Fourier transform of the first row and inverting these elements. The inverse Fourier transform of these constitute in turn the first row of the inverse matrix, that is,

$$\mathbf{A}^{-1} = \begin{Bmatrix} \bar{a}_0 & \bar{a}_1 & \cdots & \bar{a}_{n-1} \\ \bar{a}_{n-1} & \bar{a}_0 & \ddots & \vdots \\ \vdots & \ddots & \ddots & \bar{a}_1 \\ \bar{a}_1 & \cdots & \bar{a}_{n-1} & \bar{a}_0 \end{Bmatrix} \tag{9.146}$$

$$\lambda_j = \sum_{k=0}^{n-1} a_k z^{-jk} \qquad \bar{a}_k = \frac{1}{n}\sum_{k=0}^{n-1} \frac{1}{\lambda_j} z^{jk} \tag{9.147}$$

9.8.2 **Block-Circulant Matrices**

Consider next a block circulant matrix of the form

$$
\mathbf{A} = \left\{ \begin{array}{cccc}
\mathbf{A}_0 & \mathbf{A}_1 & \cdots & \mathbf{A}_{n-1} \\
\mathbf{A}_{n-1} & \mathbf{A}_0 & \ddots & \vdots \\
\vdots & \ddots & \ddots & \mathbf{A}_1 \\
\mathbf{A}_1 & \cdots & \mathbf{A}_{n-1} & \mathbf{A}_0
\end{array} \right\}
\tag{9.148}
$$

In analogy to the plain circulant matrix case, define the rectangular matrix

$$
\mathbf{Z}_j = \left\{ \begin{array}{c}
\mathbf{i} \\
z^{-j}\,\mathbf{i} \\
\vdots \\
z^{-(n-1)j}\,\mathbf{i}
\end{array} \right\}
\tag{9.149}
$$

in which the $\mathbf{i}$ are identity matrices of the same size as the submatrices of $\mathbf{A}$. This matrix satisfies the orthogonality condition

$$
\mathbf{Z}_i^* \mathbf{Z}_j = n\,\delta_{ij}\,\mathbf{i} = \begin{cases} n\,\mathbf{i} & \text{if } i = j \\ \mathbf{0} & \text{if } i \neq j \end{cases}
\tag{9.150}
$$

Collecting these rectangular matrices into a global, symmetric (!) matrix,

$$
\mathbf{Z} = \mathbf{Z}^T = \{\mathbf{Z}_0 \quad \mathbf{Z}_1 \quad \cdots \quad \mathbf{Z}_{n-1}\}
\tag{9.151}
$$

it is now clear that it satisfies the orthogonality condition

$$
\mathbf{Z}^* \mathbf{Z} = \mathbf{Z}\,\mathbf{Z}^* = n\,\mathbf{I}
\tag{9.152}
$$

in which $\mathbf{I}$ is an identity matrix of the same size as the complete system.

Define next the Fourier transform

$$
\tilde{\mathbf{A}}_j = \sum_{k=0}^{n-1} \mathbf{A}_k z^{-jk} = \mathbf{A}_0 + \mathbf{A}_1\, z^{-j} + \cdots + \mathbf{A}_{n-1}\, z^{-(n-1)j}
\tag{9.153}
$$

which can be written as

$$
\mathbf{A}\mathbf{Z}_j = \left\{ \begin{array}{cccc}
\mathbf{A}_0 & \mathbf{A}_1 & \cdots & \mathbf{A}_{n-1} \\
\mathbf{A}_{n-1} & \mathbf{A}_0 & \ddots & \vdots \\
\vdots & \ddots & \ddots & \mathbf{A}_1 \\
\mathbf{A}_1 & \cdots & \mathbf{A}_{n-1} & \mathbf{A}_0
\end{array} \right\} \left\{ \begin{array}{c}
\mathbf{i} \\
z^{-j}\,\mathbf{i} \\
\vdots \\
z^{-(n-1)j}\,\mathbf{i}
\end{array} \right\} = \left\{ \begin{array}{c}
\tilde{\mathbf{A}}_j \\
z^{-j}\,\tilde{\mathbf{A}}_j \\
\vdots \\
z^{-(n-1)j}\,\tilde{\mathbf{A}}_j
\end{array} \right\} = \mathbf{Z}_j\,\tilde{\mathbf{A}}_j
\tag{9.154}
$$

Collecting together the expressions for all $j = 0,1,\ldots n-1$ into a single matrix equation, we then obtain

$$
\mathbf{A}\mathbf{Z} = \mathbf{Z}\,\tilde{\mathbf{A}}
\tag{9.155}
$$

in which $\tilde{\mathbf{A}}$ is the block-diagonal matrix

$$\tilde{\mathbf{A}} = \left\{ \begin{matrix} \tilde{\mathbf{A}}_0 & & & \\ & \tilde{\mathbf{A}}_1 & & \\ & & \ddots & \\ & & & \tilde{\mathbf{A}}_{n-1} \end{matrix} \right\} \tag{9.156}$$

With these definitions, we then have

$$\mathbf{A}\mathbf{Z}_j = \left\{ \begin{matrix} \mathbf{A}_0 & \mathbf{A}_1 & \cdots & \mathbf{A}_{n-1} \\ \mathbf{A}_{n-1} & \mathbf{A}_0 & \ddots & \vdots \\ \vdots & \ddots & \ddots & \mathbf{A}_1 \\ \mathbf{A}_1 & \cdots & \mathbf{A}_{n-1} & \mathbf{A}_0 \end{matrix} \right\} \left\{ \begin{matrix} \mathbf{i} \\ z^{-j}\mathbf{i} \\ \vdots \\ z^{-(n-1)j}\mathbf{i} \end{matrix} \right\} = \left\{ \begin{matrix} \tilde{\mathbf{A}}_j \\ z^{-j}\tilde{\mathbf{A}}_j \\ \vdots \\ z^{-(n-1)j}\tilde{\mathbf{A}}_j \end{matrix} \right\} = \left\{ \begin{matrix} \mathbf{i} \\ z^{-j}\mathbf{i} \\ \vdots \\ z^{-(n-1)j}\mathbf{i} \end{matrix} \right\} \tilde{\mathbf{A}}_j = \mathbf{Z}_j\,\tilde{\mathbf{A}}_j \tag{9.157}$$

which in the aggregate can be written as

$$\boxed{\mathbf{A}\mathbf{Z} = \mathbf{Z}\,\tilde{\mathbf{A}}} \tag{9.158}$$

from where we obtain the formal (even if not particularly practical) expression

$$\tilde{\mathbf{A}} = \mathbf{Z}^{-1}\mathbf{A}\mathbf{Z} = \tfrac{1}{n}\mathbf{Z}^*\mathbf{A}\mathbf{Z} \tag{9.159}$$

Still, this shows that $\tilde{\mathbf{A}}$ is a block-diagonal, Hermitian matrix, in which case it satisfies the identity $\tilde{\mathbf{A}}^* = \tilde{\mathbf{A}}$, or $\tilde{\mathbf{A}}^T = \tilde{\mathbf{A}}^c$ where $\tilde{\mathbf{A}}^c$ is the conjugate matrix. Since $\mathbf{A}, \mathbf{Z}$ are also both symmetric, it follows by simple transposition that $(\mathbf{A}\mathbf{Z})^T = (\mathbf{Z}\,\tilde{\mathbf{A}})^T \rightarrow \mathbf{Z}^T\mathbf{A}^T = \tilde{\mathbf{A}}^T\mathbf{Z}^T$, in which case

$$\boxed{\mathbf{Z}\mathbf{A} = \tilde{\mathbf{A}}^c\mathbf{Z}} \tag{9.160}$$

where $\tilde{\mathbf{A}}^c = \left\{\tilde{\mathbf{A}}_j^c\right\}$ is a block-diagonal, Hermitian matrix formed with the complex conjugate elements of $\tilde{\mathbf{A}}$. This matrix satisfies the eigenvalue problem

$$\tilde{\mathbf{A}}\,\mathbf{\Psi} = \mathbf{\Psi}\mathbf{\Lambda} \tag{9.161}$$

with

$$\mathbf{\Psi} = \left\{ \begin{matrix} \mathbf{\Psi}_0 & & \\ & \ddots & \\ & & \mathbf{\Psi}_{n-1} \end{matrix} \right\} \qquad \mathbf{\Lambda} = \mathrm{diag}\left\{\mathbf{\Lambda}_j\right\} = \left\{ \begin{matrix} \mathbf{\Lambda}_0 & & \\ & \ddots & \\ & & \mathbf{\Lambda}_{n-1} \end{matrix} \right\} \tag{9.162}$$

in which $\mathbf{\Psi}_j$ and $\mathbf{\Lambda}_j$ are, respectively, the modal matrix of $\tilde{\mathbf{A}}_j$ and the diagonal matrix of its eigenvalues, which satisfy the eigenvalue problem $\tilde{\mathbf{A}}_j\mathbf{\Psi}_j = \mathbf{\Psi}_j\mathbf{\Lambda}_j$. It follows that

$$\mathbf{A}\mathbf{Z}\mathbf{\Psi} = \mathbf{Z}\,\tilde{\mathbf{A}}\,\mathbf{\Psi} = \mathbf{Z}\mathbf{\Psi}\mathbf{\Lambda} \tag{9.163}$$

or

$$\mathbf{A}\mathbf{X} = \mathbf{X}\mathbf{\Lambda} \tag{9.164}$$

which is the eigenvalue problem for $\mathbf{A}$. We conclude that the eigenvalues of $\tilde{\mathbf{A}}$ and $\mathbf{A}$ are identical, while their eigenvectors are related as

$$\mathbf{X} = \left\{ \mathbf{X}_j \right\} = \mathbf{Z}\mathbf{\Psi} = \left\{ \mathbf{Z}_0 \, \mathbf{\Psi}_0 \quad \mathbf{Z}_1 \mathbf{\Psi}_1 \quad \cdots \quad \mathbf{Z}_{n-1} \mathbf{\Psi}_{n-1} \right\} \tag{9.165}$$

The modal matrix satisfies the orthogonality condition

$$\mathbf{X}^* \mathbf{X} = \mathbf{\Psi}^* \mathbf{Z}^* \mathbf{Z} \mathbf{\Psi} = n \mathbf{\Psi}^* \, \mathbf{\Psi} = n \, \mathbf{I} \tag{9.166}$$

where we have assumed without loss of generality, that the $\mathbf{\Psi}$ matrix has been normalized.

If $\mathbf{A}$ is not degenerate and its eigenvectors span the full n-dimensional space, then so too must the eigenvectors of $\tilde{\mathbf{A}}$, in which case the $\tilde{\mathbf{A}}_j$ element matrices admit the spectral decomposition $\tilde{\mathbf{A}}_j = \mathbf{\Psi}_j \, \mathbf{\Psi}_j^{-1}$. It follows that the pth power of $\mathbf{A}$ can be expressed as

$$\mathbf{A}^p = \mathbf{X}^p \, \mathbf{X}^{-1} = \tfrac{1}{n} \mathbf{X}^p \, \mathbf{X}^* \tag{9.167}$$

as we had before for the circulant matrix. In particular, the inverse of $\mathbf{A}$ is

$$\mathbf{A}^{-1} = \left\{ \begin{array}{cccc} \bar{\mathbf{A}}_0 & \bar{\mathbf{A}}_1 & \cdots & \bar{\mathbf{A}}_{n-1} \\ \bar{\mathbf{A}}_{n-1} & \bar{\mathbf{A}}_0 & \ddots & \vdots \\ \vdots & \ddots & \ddots & \bar{\mathbf{A}}_1 \\ \bar{\mathbf{A}}_1 & \cdots & \bar{\mathbf{A}}_{n-1} & \bar{\mathbf{A}}_0 \end{array} \right\} \qquad = \text{Inverse matrix} \tag{9.168}$$

$$\tilde{\mathbf{A}}_j = \sum_{k=0}^{n-1} \mathbf{A}_k z^{-jk} \qquad\qquad = \text{Forward Fourier transform} \tag{9.169}$$

$$\bar{\mathbf{A}}_k = \frac{1}{n} \sum_{k=0}^{n-1} \tilde{\mathbf{A}}_j^{-1} \, z^{jk} \qquad\quad = \text{Inverse Fourier transform of inverses} \tag{9.170}$$

Thus, we can extract the following important conclusions:

- The eigenvalues of a block-circulant matrix are obtained by carrying out a Fourier transform of the first row of submatrices, and computing the eigenvalues of the resulting submatrices.
- Arbitrary powers of a block-circulant matrix, including the inverse, can be accomplished by an inverse Fourier transformation of the transformed matrices raised to that power, that is, $\tilde{\mathbf{A}}_j^p$.

In the case of a symmetric, block-circulant matrix, all eigenvalues are real. This can be demonstrated by considering once more the Fourier transform of the first row of $\mathbf{A}$. For even n and $m = n/2$, this row has the form $\left\{ \mathbf{A}_0 \quad \mathbf{A}_1 \quad \cdots \quad \mathbf{A}_m \quad \cdots \quad \mathbf{A}_2^T \quad \mathbf{A}_1^T \right\}$. Taking into account that $z^{-j} = z^{n-j}$ and $z^m = -1$, we can write

$$\tilde{\mathbf{A}}_j = \sum_{k=0}^{n-1} \mathbf{A}_k z^{-jk} = \mathbf{A}_0 + \left(\mathbf{A}_1^T \, z^j + \mathbf{A}_1 \, z^{-j} \right) + \cdots + \left(\mathbf{A}_{m-1}^T \, z^{(m-1)j} + \mathbf{A}_{m-1} \, z^{-(m-1)j} \right) + (-1)^j \, \mathbf{A}_m \tag{9.171}$$

Since z^j, z^{-j} are complex conjugates, then each term in Eq. 9.171 is a Hermitian matrix. It follows that $\tilde{\mathbf{A}}_j$ is also Hermitian and so must have real eigenvalues, that is, $\mathbf{\Lambda}$ is real. A similar proof can be used for the case where n is odd. Now, if $\tilde{\mathbf{A}}_j$ is Hermitian, then $\tilde{\mathbf{A}}_j^T = \tilde{\mathbf{A}}_j^c$ and $\tilde{\mathbf{A}}_j^* = \tilde{\mathbf{A}}_j$.

10 Problem Sets

P.1 Obtain the equations of motion for the following systems (assume small displacements).

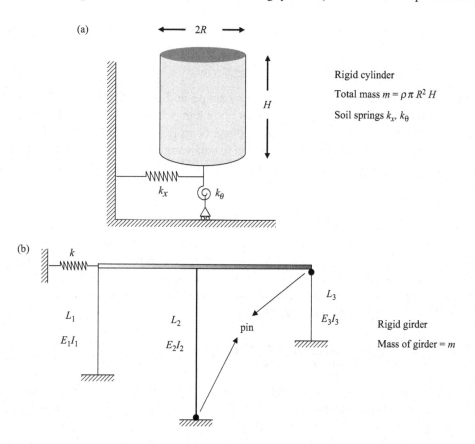

(a)

$\longleftarrow$ 2R $\longrightarrow$

Rigid cylinder

Total mass $m = \rho \pi R^2 H$

Soil springs k_x, k_θ

H

k_x k_θ

(b)

k

L_1

$E_1 I_1$

L_2

$E_2 I_2$

pin

L_3

$E_3 I_3$

Rigid girder

Mass of girder = m

(c)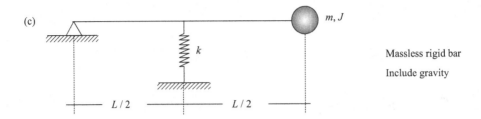

Massless rigid bar

Include gravity

P. 2 Determine the natural frequency of a pendulum in the form of a thin, rigid ring, as shown below. The ring has total mass m. What is the equivalent length L of a simple pendulum that has the same natural frequency? Does your result for L make sense to you? Whether your answer is yes or no, please discuss why.

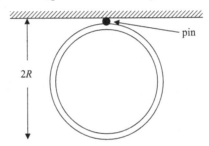

P. 3 Consider a rigid, plane mass of arbitrary shape and mass distribution, but usually elongated in one direction (say a rod). This body hangs from an arbitrary point and is forced to oscillate as a pendulum in a given plane of motion without torsion (i.e., this is a 2-D problem). Without restricting the motion to small angles of rotation, the equation of motion for free vibration is easily found to be

$$J\ddot{\theta} + mgd\sin\theta = 0$$

in which J is the mass moment of inertia of the rod about the point of rotation, m is the mass, g is the acceleration of gravity, d is the distance from the point of rotation to the center of mass, and θ is the angle of rotation with respect to the vertical (i.e., the rest position, in which case the center of mass lies directly below the center of rotation).

Neglecting air drag, frictional forces, and any thermal changes in dimensions, the period of oscillation can be shown to be given by

$$T = 4\sqrt{\frac{J}{mgd}}\int_0^{\pi/2}\frac{d\alpha}{\sqrt{1-k^2\sin^2\alpha}} = 4\sqrt{\frac{J}{mgd}}K(k)$$

in which $k = \sin\frac{1}{2}\theta_0$, $K(k)$ is the complete elliptic integral of the first kind, and θ_0 is the initial angle of oscillation. Observe that when k is very small (and thus can be neglected in the integral) $K(0) = \pi/2$, in which case the period reduces to the classical value for small oscillations. Also, observe that the elliptic integral is a function only of the initial angle, and no other physical parameter. If the point of rotation is changed from one location to another while maintaining the initial angle of oscillation θ_0, (or using small initial angles so as to make the period independent on this angle) the period changes correspondingly. This is because both J and d change.

- Show that the period first decreases as the point of rotation is moved closer to the center of mass, and then increases. Thus, there exists a minimum distance d_{min} for which the period attains a minimum. Observe, however, that the period will not change if the center of rotation is moved on a circle centered on the center of mass (why?).
- From the preceding, it follows that there are always two different distances d_1, d_2 to the center of gravity, one shorter and the other longer than the minimum distance d_{min}, for which the periods are *exactly* the same. These are *reciprocal* points, and form the basis for the *reversible pendulum*. Show that the radius of gyration R about the center of mass can be obtained from the geometric mean of these two distances, that is, $R = \sqrt{d_1 d_2}$, and that this result does not depend on the body being homogeneous. In particular, if $d_1 = d_2 = d_{min}$, this demonstrates that the distance for which the period is minimum equals the radius of gyration of the body about the center of mass.
- Show that the sum of the distances for reciprocal points, that is, $L = d_1 + d_2$, equals the length L of an ideal or mathematical (i.e., point mass) pendulum with the same period and initial angle of oscillation as the physical pendulum.
- Discuss how you could use these observations to measure gravity by means of a physical (and not necessarily homogeneous) pendulum. How accurate is that measure likely to be?

P.4 An irregularly shaped *rigid* object is hinged at one point and is supported by several springs. The measured natural frequency of this object is 5 Hz. When a static force $F = 10$ N is applied at 0.4 m to the right of the pivot, the object rotates through an angle of $\theta = 0.005$ rad. What is the mass moment of inertia of the object around the pivot?

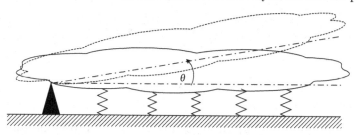

P. 5 An SDOF system consists of a vertical spring of stiffness k above which a spherical mass of radius $R = 0.05$ m and mass density $\rho = 2\rho_w$ rests, with ρ_w being the mass density of water. The lower end of the spring is firmly attached to a heavy, fixed base and the system is so constrained that it can only execute vertical vibrations. The stiffness of the spring is such that the frequency of the system in air is $f_n = 1$ Hz (i.e. $\omega_n = \sqrt{k/m} = 2\pi$).

a. Determine the equations of motion, including the effects of gravity. If the system does not vibrate, how much does the spring elongate?
b. While the system is at rest, the oscillator together with its base is carefully submerged in water. Neglecting any dynamic effects by the moving water (i.e., "participating mass" and damping of water), what is the frequency of the submerged system? What is the new rest position?
c. Imagine the system on a lunar base, where gravity is six times smaller than on Earth. What happens to the frequencies in the above two parts?

P. 6 A pendulum consists of a thin, *rigid*, massless bar of length L to which a spherical mass m is attached at its free end. In addition, the bar is restrained by a spring k attached at the center of the bar. The radius of the sphere $R \ll L$ is small enough that the rotational inertia of the mass can be neglected. When the pendulum hangs vertically, the spring is unstressed.

a. Assuming small rotations, determine an expression for the natural frequency of this system.
b. If the lower end of this pendulum is submerged in water, and it is known that the sphere's weight is twice that of the water displaced by it (i.e., it has a mass density $\rho = 2\rho_w$), what is the frequency of the submerged pendulum? Disregard any "participating mass of water," and any damping caused either by friction or by waves in the water.
c. What would be observed on the moon?

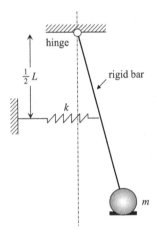

Comment on the differences, if any, that exist between the systems in Problems 5 and 6, and explain their reasons.

P. 7 Two frames are connected by means of a rigid, massless strut (or link). The girders (i.e., beams) in each frame are infinitely rigid, and have the masses indicated on the drawing. On the other hand, the columns are massless. The system is subjected to an initial lateral deflection u_0, released in that position, and allowed to vibrate freely.

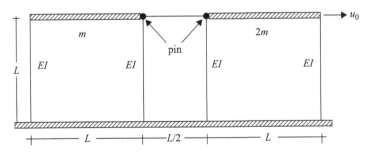

If the system has no damping, determine:

- The natural frequency of the complete assembly
- The force in the strut as a function of time
- The vertical reaction at the rightmost support (you must use global equilibrium for that purpose)

P. 8 Determine the force–deformation relationship at point A of the structures shown, that is, the *stiffness matrix* as seen from this point. **Hint:** Use the flexibility approach. Determine also the *lateral stiffness*, which is obtained by releasing the rotation (no moment), and the *rotational stiffness*, which is obtained by releasing the translation (no force).

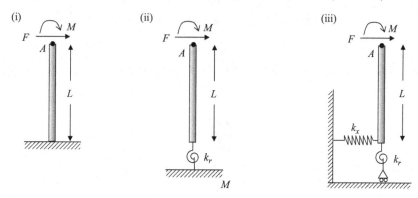

Cases to consider:

- The structures are *shear beams* with $G\,A_s$ = constant.
- The structures are *bending beams* with EI = constant.
- For structure (iii), determine the stiffness properties when the bar is rigid.

P. 9 Derive the equations of motion for the following systems:

a) A system where all bars are rigid.

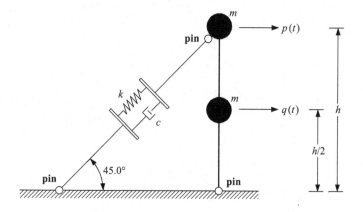

b)   All girders are rigid ($b = 4/3\, h$)

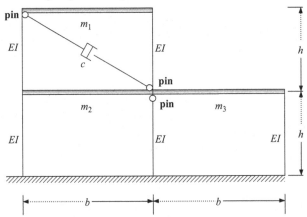

P.10 A massless bending beam has two masses lumped as shown. Derive the equation of motion in detail using any method.

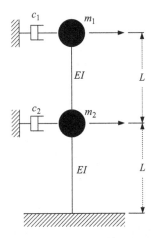

P.11 Derive the equation of motion in detail using any method.

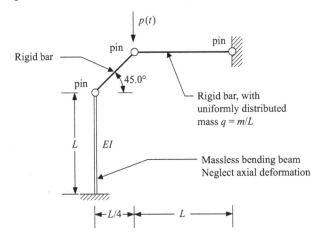

P.12 For each of the following forcing functions, write an expression for the *particular solution* to the forced vibration problem. Please note that you need *not* solve the SDOF equation. Instead, simply write your expressions in terms of undetermined constants, for which you can use capital letters $(A, B, ...)$. In each case, make a sketch of the forcing function, but not of the particular solution. **Note:** p_0, a, b are arbitrary constants, and t is time.

- $p(t) = p_0\, e^{-at} \cos bt$
- $p(t) = p_0\, \dfrac{t}{t_0} \sin bt$
- $p(t) = p_0\, at^3 e^{-bt}$

This problem should not take you more than 3 minutes to complete.

P.13 An SDOF system consists of a vertical spring of stiffness k above which a spherical mass of radius $R = 0.05\,\mathrm{m}$ and mass density $\rho = 2\rho_w$ rests, with ρ_w being the mass density of water. The lower end of the spring is firmly attached to a heavy, fixed base and the system is so constrained that it can only execute vertical vibrations. The stiffness of the spring is such that the frequency of the system in air is $f_n = 1\,\mathrm{Hz}$ (i.e. $\omega_n = \sqrt{k/m} = 2\pi$).

- Determine the equations of motion, including the effects of gravity. If the system does not vibrate, how much does the spring elongate or compress?
- While the system is at rest, the oscillator together with its base is carefully submerged in water. Neglecting any dynamic effects by the moving water (i.e., "participating mass" and damping of water), what is the frequency of the submerged system? What is the new rest position?
- Imagine the system on a lunar base, where gravity is six times smaller than on earth. What happens to the frequencies in the preceding two parts?

P. 14 The 1-story truss shown below has a mass m lumped at the upper floor. Only the diagonals are flexible (pinned connections!) with stiffness EA per unit length.

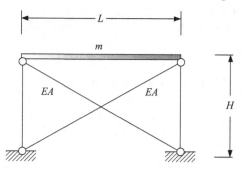

- Determine an expression for the lateral stiffness $k = k(H, L, EA)$.
- If $H = L = 3.00$ m, $EA = 600{,}000$ N and $m = 200$ kg, determine the natural frequency of the truss.
- If damping is $\xi = 0.02$, determine the half-power frequencies (in Hz).
- Determine the damped frequencies.

- The system is subjected to a lateral (horizontal) force that in ½ second rises linearly up to $f_0 = 2000\,\text{N}$, remains constant for another ½ second more, and then vanishes. Express the load as a superposition of three ramp loads (linearly rising loads of infinite duration) and find the maximum response, using MATLAB® for this purpose.

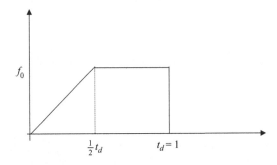

P. 15 A cart filled partially with gravel has a total mass $m = 100$ kg. It is held in place by a spring of stiffness $k = 5000$ N/m and a damper $c = 20$ N/(m/s) on a plane inclined an angle $\theta = 30°$ with respect to the horizontal. The cart is initially at rest. At time $t = 0$, a rock of mass 10 kg falls vertically down from a height of 2 m into the cart, and then remains embedded in the gravel. You can assume that the system is linear, and that the impact of the rock is very brief.

- What are the frequency and damping of the cart both before and after the rock falls in?
- Find the response of the cart after impact by the rock, and sketch your response function.
- Solve this problem using a brief MATLAB script of your own. This will allow you to obtain the complete time history. What is the maximum deflection from the rest position? Verify that your results are consistent with your answer to the previous item.
- Use the program again with the following two loads of equal finite duration $t_d = T_n / 2$ and equal maximum amplitude $p_0 = 10N$, first a rectangular load, and then a triangular load in the form of an isosceles triangle. In each case, observe the response for at least five cycles. How do the maximum responses compare? How do these relate to the areas (i.e., impulses) under the load? What are dynamic load factors in each case?

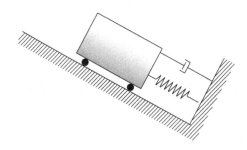

P. 16 Consider an upright, *massless* cantilever beam of length L and flexural rigidity EI such that the lateral stiffness of the beam, as seen from its upper end, is $k = 3EI / L^3 = 40\pi^2$ [N/m]. Clamped at its upper free end is a wooden block of mass $M = 10$ [kg] that is firmly attached, and the system is initially at rest. At time $t = 0$, the wooden block is struck by a bullet of mass $m = 0.025$ [kg] that approaches with horizontal speed $V = 400$ m/s, and that after a very brief impact remains embedded in the wooden block.

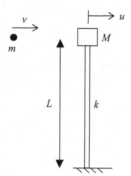

a. Determine the natural frequencies of the system before and after the impact of the bullet.
b. Find the response of the system $u(t)$ for $t \geq 0$, and in particular, determine the maximum amplitude of motion. Assume that there is no *damping* anywhere in this system.
c. The impact of the bullet caused a crack in the mid-plane of the wooden block, and after several oscillations, half of the block together with the bullet falls off, while the other half of the block remains firmly attached to the beam and continues oscillating. Determine the frequencies and the response – including the maximum amplitude – after this incident has occurred, assuming that the piece falls off.
   • When the speed of the cracked block is zero, that is, when the restoring force (and thus the D'Alembert force) is maximum.
   • When the speed of the cracked block is maximum (Does this change make any difference?)
   • What happens if the piece of the block falls off at times other than those above? Is the maximum amplitude larger, smaller, or exactly the same as that before splitting? Discuss.

P. 17 A thin *rigid* bar with uniformly distributed mass rests on and pivots about a pin support placed at the center, and there is a spring k and a dashpot c at the two locations shown. The bar terminates on the right with a solid mass in the form of a regular cube with the dimensions shown whose mass equals that of the bar. In addition, there is a dynamic force applied at the left end. The data for this problem are as follows:

Scaling constant $= k / m = 4\pi^2$
Total length of bar $= 3L$
Total mass of bar $= m$ (uniformly distributed)
Total mass of cube $= m$ (equal to the mass of the bar)

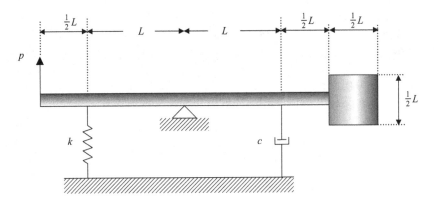

a. Find the equation of motion for this system, using any method of your choice.
b. Determine the natural frequency, in Hz.
c. Determine the constant c needed to achieve 10% critical damping.
d. Find the force in the spring when the applied load is a step function in time, that is,

$$p(t) = \begin{cases} 0 & t < 0 \\ p_0 & t \geq 0 \end{cases}$$

P. 18 Though you haven't yet studied how to compute the vibration modes of MDOF systems, you should have no difficulty in finding the two vibration frequencies and modal shapes (i.e., the relative values of u_1 and u_2) for the system shown. To achieve this goal, simply apply symmetry/antisymmetry concepts.

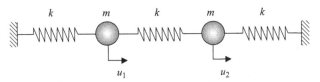

P. 19 Write the equations of motion for the following dynamic systems:

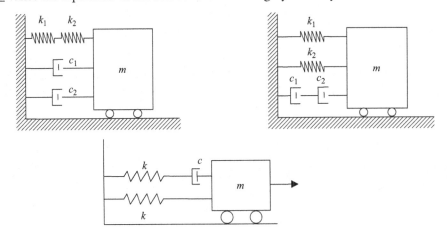

Hint: In the third system, introduce a fictitious mass μ between the dashpot and the spring, set up the equations of motion, and then reset $\mu = 0$.

P. 20 To determine the dynamic properties of a one-story structure 3 m in height that can be idealized as a simple frame with rigid girder of mass m and equal massless columns with bending stiffness EI, a jack was used to displace the building laterally by 2.25 cm. The force required in the jack to accomplish this displacement was 625.000 N (about 62.5 metric tons). On instantaneous release of the structure, the maximum displacement on the return swing was 1.50 cm, and the measured period of the motion was $T = 1$ s.

- What is the mass m of the girder and EI of the columns?
- What is the fraction of damping? The damping constant c?
- How many cycles are required for the free vibration to decay to 1 cm?
- What is the maximum acceleration felt by the girder, and when does it happen?

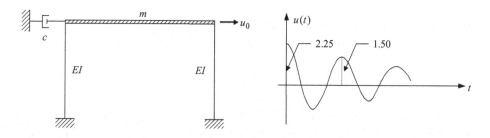

P. 21 Derive the equation of motion and frequency for the system shown. The cylinder has mass m and diameter $2a$, where $a = L/5$. The spring has stiffness k. Rotational effects *cannot* be neglected, although rotations are "small." Observe that the center of mass of the cylinder is at a height a above the bar, and is aligned with the spring. Do NOT include any gravity effects.

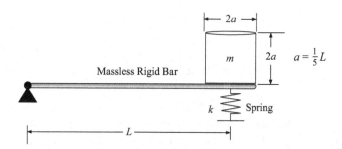

P. 22 Consider the SDOF system shown below that is at rest for $t < 0$. Find the *absolute* motion response $u(t)$ when the *support* experiences an abrupt ground dislocation of the form

$$u_g = u_{g0}\mathcal{H}(t), \quad \text{where} \quad \mathcal{H}(t) = \begin{cases} 0 & t < 0 \\ 1 & t > 0 \end{cases}$$

(**Note:** The function $\mathcal{H}(t)$ is referred to as the unit step function or Heaviside function.)

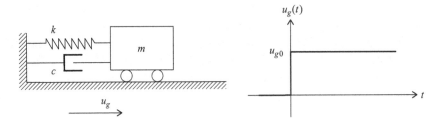

P. 23 A bathroom scale can be idealized as a platform of mass $m = 2$ kg supported by a spring-dashpot system. The scale is designed so that when a person of 50 kg steps on it, the system has a fraction of damping $\xi = 0.5$ (i.e., 50% of critical). When a student steps onto the scale it indicates a weight (actually a mass) $M = 60$ kg and the scale platform descends by 1.2 mm (0.0012 m).

- What are the stiffness and dashpot constants?
- What is the frequency (in Hz) of the scale by itself (i.e., with no person standing on it)?
- What are the fraction of critical damping, the undamped and the damped frequencies with the 60-kg student *on* the scale (again, in Hz)?
- When the student steps on the scale, how quickly will the needle come to rest, i.e., how many oscillations will it take for the oscillation to decrease in amplitude by a factor 1024 ($=2^{10}$)?

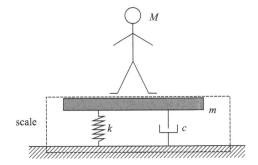

P. 24 You operate a sieve to screen dirt in a search for archeological artifacts. With dirt in it, the mass is 20 kg. To facilitate matters, you mount it on rollers and shake it by hand at about 2 Hz, and being a clever person, you decide to add a spring to the sieve to make it a resonant system. With dirt falling through the holes, the equivalent linear damping is about 20% of critical, and you can put out a force of about 30 N while shaking the sieve at 2 Hz.

- a. How would you size the spring to maximize the amplitude that you can shake the dirt box? Assume the excitation frequency is 2 Hz.
- b. What amplitude would you be able to achieve as you drive the system by hand at 2 Hz?

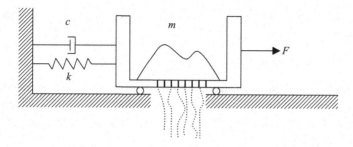

P. 25 A cylindrical cork of diameter $D = 2.4$ cm has quarter coin glued on one of its ends (for the purpose of ballast). When the cork is set to float in a cup of water, it sinks by 1.5 cm.

a. What is the weight of the cork with the quarter?
b. What is the resonant frequency of vertical oscillations? (Neglect any participating mass of water)
c. If an insect with a mass of 0.5 g lands vertically onto the upper surface of the cork with a velocity $v = 2$ m/s and then sits on the cork, what is the amplitude of dynamic oscillation? (Neglect damping for this question.)
d. The cork oscillations after the landing of the insect are seen to decay by 50% after four oscillations. What is the fraction of damping for the whole system?

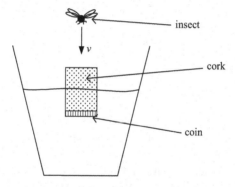

P.26 The figure below shows the amplitude of a frequency response function for an SDOF system, which was determined *experimentally*. What is the fraction of damping? What is the number of cycles to 50% amplitude? What is the natural frequency?

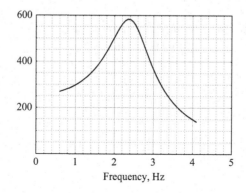

P. 27 A fan operating at a frequency of 1800 rpm has unbalanced blades that produce a net vertical force

$$p = \omega^2 Q \cos \omega t = \omega^2 Q \, \text{Re}\{e^{i\omega t}\},$$

where Q is a constant with dimensions of [kg-m] and ω is the operating speed of the fan, in rad/s. The fan is supported by a foundation resting on soil. The vertical resonant frequency of the fan–foundation system is $f_n = 20$ Hz, and the fraction of damping measured form observations is $\xi = 0.20$.

   a. Determine the response of the system at the fan's operating frequency (in terms of Q).
   b. Find the (complex) transfer function for the vertical soil reaction $R(\omega)$, assuming that the total mass is $M = m_{\text{fan}} + m_{\text{found}}$. Do not include any static gravity forces (weight).

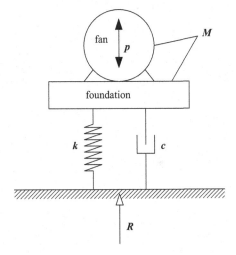

P. 28 A fan that weighs 1000 kg is mounted at the center of a simply supported, massless beam. Because of an unavoidable manufacturing imperfections, the fan produces an unbalanced vertical force $F(t) = p\omega^2 \sin \omega t$ where $p = 5\text{N-s}^2/\text{rad}^2$. When the fan is disconnected, its own weight produces a static deflection at the center of the beam of $u_s = 0.25$ mm. The operating speed of the fan is $n = 1800$ rpm.

   a. Compute the natural frequency of the system.
   b. Find the steady-state response when damping is 5% of critical (disregard gravity for this part).
   c. Compute the maximum *dynamic* deflection of the beam.
   d. Determine the maximum support reactions.

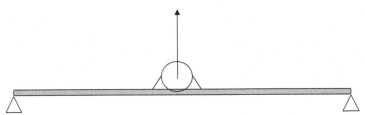

P. 29

a) Consider two SDOF systems as shown below on the left. At first, these two systems are allowed to vibrate independently. The constants are such that $k/m = 1$ and $c^2/(km) = 0.0004$.

- What are the undamped natural frequencies (Hz), fractions of critical damping, and damped frequencies? How many cycles to 50% amplitude will each require?

b) Next, the two systems are connected together by means of a *massless*, rigid strut (bar), which forces the two systems to move as one, as shown below on the right. Thus, their combination is still an SDOF system.

- What is the undamped natural frequency and damping and damped frequency of the new assembly? How many cycles to 50% amplitude will there be now?

- The connected system is given an initial lateral displacement u_0 and then let go to vibrate freely. Give an expression for the internal force *in the strut*, and sketch the result. **Hint**: This requires a free-body diagram of the various parts.

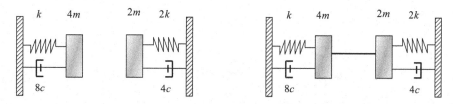

P. 30 MIT Professor Kim Vandiver once measured the vibration response of a coast guard light station standing in 20 m of water. It was a steel frame structure with a large house sitting on the top. A simple SDOF equivalent model of the structure is a massless beam with a concentrated mass on the end, which is made up of the mass of the house and one-fourth of the mass of the steel frame. The total concentrated mass was $m = 2.75 \times 10^5$ kg. The measured natural frequency of the platform in the first mode was 1 Hz and the measured damping was 1% of critical. By watching on an oscilloscope the output of an accelerometer, Vandiver was able to shift the weight of his body from one foot to the other in synchrony with the motion of the structure. He was thus able to drive the structure to amplitudes in excess of that observed in 100 km/h wind plus 7-m ocean waves.

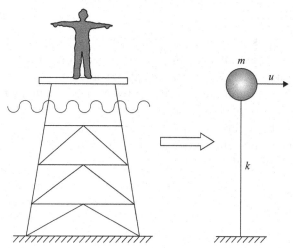

- What was the magnitude of the steady-state horizontal force that he had to exert on the structure to drive it to steady–state acceleration amplitude of 0.005g?
- What are the half-power frequencies and the bandwidth?
- What is the number of cycles to 50% amplitude?

P.31 A car with mass 1000 kg is moving slowly at a constant speed $v = 7.2$ km/h $= 2$ m/s on a flat road. The car can be modeled as a rigid block with a spring-mounted front bumper. The stiffness of the bumper is such that a horizontal force equal to the weight of the car would deform the spring by 5 cm (=0.05 m). At $t = 0$, the car collides head on against a flat, rigid wall.

- What is the maximum deformation of the bumper?
- What is the maximum force exerted on the bumper?
- What is the maximum acceleration felt by the passengers?
- How long does the car remain in contact with the wall?

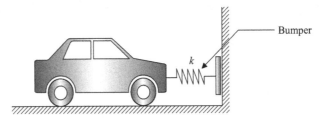

P. 32 An *undamped* SDOF system initially at rest is subjected to a step load with finite rise time t_d. Find the maximum dynamic load factor (i.e., ratio between dynamic and static response) at any time, assuming that $t_d = 5/4\ T$, with T being the undamped natural period (you may assume $T = 1$, if you wish).

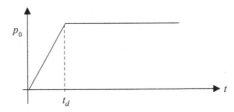

P. 33 An SDOF system is subjected to a sinusoidal load that starts at $t = 0$ (i.e., there is no load for negative times). If the system is initially at rest, find an expression for the true transient response. Sketch the response (say, some 10 cycles) when the excitation frequency equals the natural frequency (i.e., at resonance), assuming that the fraction of damping is $\xi = 0.05$. Interpret your result in terms of the "number of cycles to 50%" rule, in particular, the number of cycles required to attain steady state (for practical purposes, you could say that you have achieved steady state when the transient part has decayed by some two orders of magnitude, say 1/128).

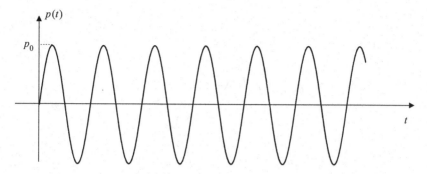

P. 34 An *undamped* SDOF system has a natural period T_n, and is subjected to the bell-shaped forcing function

$$p = p_0 \frac{t}{t_0} e^{-\frac{t}{t_0}}$$

shown in the figure. If the system is at rest at $t = 0$, and if the force time parameter satisfies $t_0 = T_n$, what is the expression for the transient response for $t > 0$?

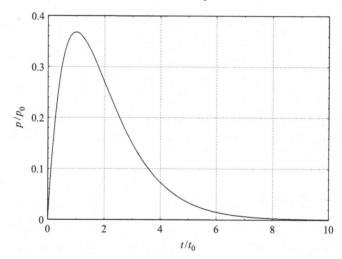

P. 35 A 10-story building is subjected to an earthquake whose response spectrum is shown below. The building is square in plane view, and its aspect ratio (=height/width) is 3:1. Using heuristic methods, estimate

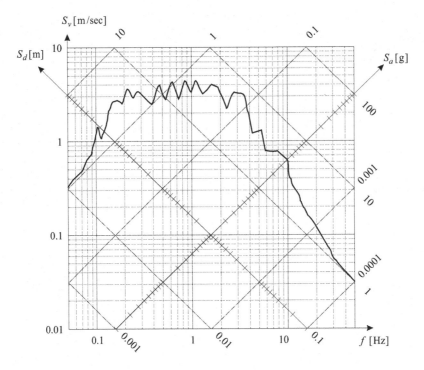

- The building's mass
- The fundamental frequency of the building
- The lateral stiffness, as perceived from the top of the building
- The maximum acceleration at the top of the building
- The maximum drift (relative displacement)

P. 36 A 5-story building of height h and cross section $A = a^2$ is subjected to an earthquake with the response spectrum shown. If $h/a = 2$ and the building can be modeled as an SDOF system with damping $\xi = 0.01$, estimate

- The maximum relative displacement of the roof relative to the base.
- The maximum base shear (the horizontal force transmitted to the ground).
- If the earthquake stopped instantly at the time of maximum relative displacement (yes, this is impossible), what would be the response after this time? Sketch this response and explain the features of your sketch.
- Assume next that this building is infinitely rigid, and that the soil shear modulus is $G = 8 \times 10^7$ N/m^2 and Poisson's ratio is $v = 0.35$.
- If to a first approximation, the foundation can be idealized as a circular foundation with equivalent radius $R = a/\sqrt{\pi}$, what is the rocking frequency of the building? The rocking stiffness of a circular foundation is
- $K_R = \dfrac{8\,G\,R^3}{3\,(1 - v)}$
- Obtain the relative displacement at the top of the rigid building (an SDOF system!), caused by this same earthquake. Important: This requires for you to set up correctly

the equation of motion for a rigid mass that can *rotate* about the base while following the ground *translation*).

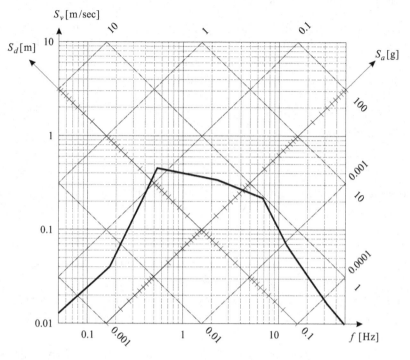

P. 37 A ship that weighs 40,000 ton (40×10^6 kg) can be idealized, to a crude approximation, as a rectangular block of length $L = 200$ m and width $b = 20$ m. On stormy days, the ocean develops waves, whose wavelengths λ and period T increase with the wind speed and duration of the wind. To a first approximation, the relationship among wavelength, wave period, and wave amplitude is as follows:

$$\lambda \text{ [m]} = \frac{g}{2\pi}T^2 = 1.56T^2, \quad T \text{ in [sec]} \qquad \text{(deep water waves)}$$

$$\frac{2A}{\lambda} \cong \frac{1}{7} \qquad \text{(If } 2A > \lambda, \text{ the wave breaks)}$$

	T [s]
Light seas	5
Moderate seas	10
Heavy seas	15

- What is the resonant frequency of the ship in vertical motions?
- For what ocean conditions will resonance occur?
- How does the wavelength causing resonance compare with the length of the ship?

P. 38 A car that weighs 2000 kg travels on a bumpy road with constant velocity of 72 km/h. To a first approximation, it can be modeled as an SDOF system. The car is known to have a damped natural frequency of 1 Hz, and a fraction of critical damping of 50%. The road is flat, except for a ramp of length $L = 10$ m and a height $h = 0.10$ m.

a. Determine the equation of motion for this problem in terms of the absolute (inertial) vertical displacement of the car. Note that this motion is *not* the same as the distance to the road.
b. Neglecting damping, estimate the maximum absolute motion, and the time at which it occurs. The car reaches the ramp at $t = 0$. Use MATLAB to evaluate and plot the damped response.
c. From the results of (b), determine the deformation of the shock absorbers at the time of the maximum absolute motion. Notice that this is not necessarily the maximum deformation.
d. What problems would you encounter if you formulated this problem in terms of relative displacements? How would you overcome these?

P. 39 An automobile with mass $m = 1500$ kg travels with *constant* velocity V along a bumpy road, as shown below. To a first approximation, the car can be modeled as an SDOF system that oscillates vertically with an *undamped* frequency $f_n = 0.75$ Hz, and a fraction of critical damping of 50%. On the other hand, the road roughness can be idealized as a sinusoid with a wavelength of $\lambda = 50$ m and an amplitude of 5 cm (i.e., 0.05 m).

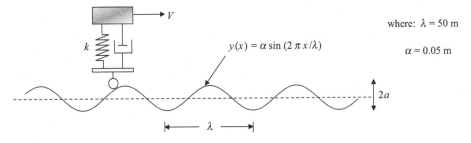

where: $\lambda = 50$ m

$\alpha = 0.05$ m

- Find the vertical absolute displacement as a function of the travel velocity V.
- Determine the travel velocity for which *undamped* resonance occurs.
- What is the maximum vertical acceleration at resonance?
- Determine the amplitude of response when the car travels at $V = 54$ km/h.

P. 40 A one-story structure that can be modeled as an SDOF system has a natural frequency of 10 Hz. This structure is subjected to an earthquake whose response spectrum can be idealized by the function

$$S_d(\omega) = \frac{u_{g0}}{\sqrt{\left[1-\left(\frac{\omega}{\omega_0}\right)^2\right]^2 + 4b^2\left(\frac{\omega}{\omega_0}\right)^2}}$$

in which

$u_{g0} = 0.03$ [m] = peak ground displacement
$\omega_0 = 4\pi$ [rad/s] = dominant frequency of the earthquake (=2 Hz)
$b = 0.25$ = an earthquake parameter

Determine

- The maximum ground acceleration
- The maximum absolute acceleration of the structure
- The maximum relative displacement
- Sketch the response spectrum on a tripartite logarithmic paper (**Hint**: $\omega/\omega_0 = f/f_0 = T_0/T$).

P. 41 An SDOF system of mass m is resting on a base of mass m_f that has a coefficient of friction μ. When the ground experiences an earthquake $\ddot{u}_g(t)$, not large enough to produce sliding, the system responds with a time history $u(t)$. By what factor can the ground motion be increased before the system begins to slide? In other words, find the factor of safety against sliding. Notice that $\ddot{u}_g(t)$ is given, so both $u(t)$ and $\ddot{u}(t)$ are assumed to be known! **Hint**: There is very little to compute here! Simply apply common sense!

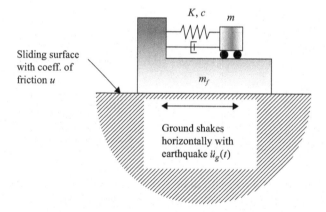

Gravity is on!

Sliding surface with coeff. of friction u

K, c m

m_f

Ground shakes horizontally with earthquake $\ddot{u}_g(t)$

P. 42 Consider the following periodic forcing function:

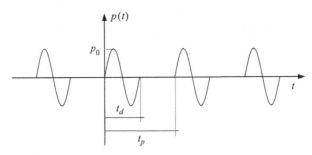

where $p_0 = 1$ N, $t_d = 5$ s and $t_p = 10$ s.

- Compute the coefficients of the Fourier series.
- Sketch the amplitude spectrum $|\tilde{p}_j|$.
- Sketch the amplitude of the product $|H\tilde{p}_j|$, assuming $f_n = 0.4$ Hz, $\xi = 0.05$ and $m = 1$ kg.
- What is the response at $t = 0$? (Consider a finite number N of terms in the Fourier series. How many?)
- Using MATLAB, sketch the response for $N = 16, 32, 64$ terms.

P. 43 For the two systems shown, find

- The impedances $Z(\omega)$ of the two spring–damper systems (i.e., without the mass).
- Assuming $k_1 = k_2 = k_3 \equiv k$ and $c_1 = c_2 = c_3 \equiv c$ together with $c = 0.1\sqrt{km}$, plot $\text{Re}(Z)$ and $\text{Im}(Z/\omega)$ vs. frequency for those impedances in terms of $\omega/\sqrt{k/m}$ (you may need MATLAB to do this). Do these look anything like those for a conventional spring–damper system?
- Find the transfer function $H_{u|p}$ for a load applied to the mass, and sketch its absolute value in either case.

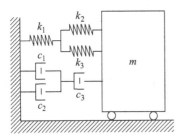 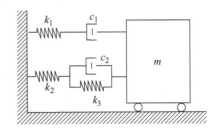

P. 44 A spherical buoy of mass m and diameter $d = 1.0$ m floats in water so that it is semi-submerged, as shown in the figure below. In the frequency domain, the vertical free vibration of this buoy is described by the equation

$$\left[k - \omega^2(m + m_w) + i\omega c_w\right]u = 0$$

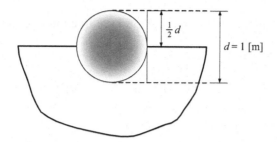

where

$k = \rho_w A_w g$ = water stiffness

$A_w = \pi d^2 / 4$ = cross-section at water level

m_w = participating mass of water

c_w = radiation damping in water (energy dissipated by waves)

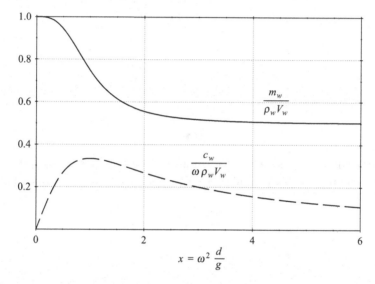

To a first approximation, the participating mass of water and the radiation damping can be expressed as (see plot):

$$m_w(\omega) = \frac{1 + \frac{1}{2} x^3}{1 + x^3} \rho_w V_w, \qquad c_w(\omega) = \frac{2}{3} \frac{x}{1 + x^2} \rho_w V_w \omega$$

where

$$x = \frac{\omega^2 d}{g} \qquad V_w = \frac{1}{12} \pi d^3 = \text{volume of displaced water}$$

- Determine the resonant frequency $\omega = \omega_n$ of this system. **Hint:** Find the value of ω for which $k - \omega^2 (m + m_w) = 0$, that is, $1 - \omega^2 (m + m_w)/k = 0$. Solve this equation by trial and error, assuming at first that $m_w = \frac{1}{2} \rho_w V_w$ (which is the asymptotic value for large x). For this purpose, express $\omega^2 (m + m_w)/k$ in terms of x.
- Determine the total mass at this frequency, and compare it with the mass of the buoy.

- Determine the damping constant $c_w(\omega)$ and the fraction of critical damping at resonance, $\xi(\omega_n) = \dfrac{c_w(\omega_n)}{2\sqrt{k[m + m_w(\omega_n)]}}$.

- Write an expression for the transfer function for the *dynamic force* between the water and the buoy when a harmonic force of amplitude F is applied to the buoy. Sketch the amplitude and phase of this transfer function.

P. 45 A vibration isolation block is to be installed in a laboratory so that the vibration from adjacent factory operations will not disturb certain experiments. If the mass of the isolation block is $m = 1000$ kg and the surrounding floor and foundation vibrate at 1800 oscillations per minute, determine the stiffness k of the isolation system such that the motion of the isolation block is limited to 10% of the floor vibration.

Note: Neglect damping.

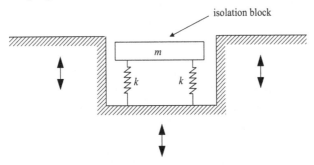

P 46 An optics experiment is set up on a vibration isolation table, which may be modeled as an SDOF system for vertical oscillations. The system has very little damping.

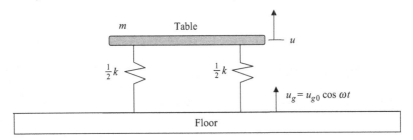

- The floor has considerable vertical vibration at 30 Hz. What must be the natural frequency of the table be in order to reduce the motion of the table and optics experiment by a factor of 4, when compared to the motion of the floor?

- Each time you work on the experiment, you bump the table. This results in motion that takes a long time to die out. You decide to add 10% damping to the system. Will this increase or reduce the vibration of the table in response to floor motions at 30 Hz? Either way, by how much does the response change?

P. 47 An undamped SDOF system with *unknown* mass M and spring K is so constrained that it can only oscillate vertically. An experiment demonstrates that it oscillates with a frequency of 2 Hz. After placing a secondary mass $m = 1$ kg on top (as shown in the figure on the right) the frequency changes to 1.6 Hz.

- What is the mass M? What is the spring constant K?
- If the secondary mass is very carefully placed on top of the primary mass and then let go at the moment $t = 0$ at which smooth contact is made, find the response of the combined system (gravity cannot be neglected here). In particular, find the maximum displacement of the oscillation.

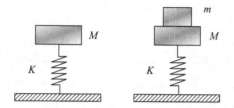

P. 48 A machine with mass of 20 kg exerts a vertical force on the floor of a laboratory building, given by $F(t) = F_0 \cos \omega_0 t$. This force causes a vertical vibration $y(t)$ on the floor of the next room with an amplitude $y_0 = 10^{-4}$ m at the machine's operating frequency $f_0 = 30$ Hz.

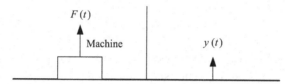

A commercially available vibration isolation support system is purchased for the machine. It consists of a lightweight rigid plate resting on an *elastic* pad, which has some internal damping.

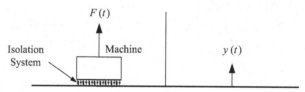

The machine on the isolation pad has a natural frequency f_m such that $f_0/f_M = 3$, where f_0 is the problem-frequency observed in the next room. The damping ratio of the system is estimated to be $\xi_m = 7\%$ of critical damping.

a. At the operating frequency of 30 Hz, what is the magnitude of the ratio of the vertical force transmitted to the floor through the isolation pad to the force produced by the machine, $|F_T/F_0|$? What is the ratio of the floor vibration amplitude next door after the isolation pad has been installed compared to before? **Hint:** In your reply to this question, it might seem logical to make use of the amplification function for an eccentric mass vibrator, but that would be wrong because the operating frequency f_0 is fixed at 30 Hz, and thus the magnitude of the force F_0 is also fixed. What is being changed here is the natural frequency of the oscillator.

b. What is the effective spring constant k for the isolation system? What is the static deflection of the machine due to its weight only, as it rests on the support?

The engineers are so happy with the result that they decide to do the same with the instrument in the room next door. They put it on an identical support – they got a great deal from the supplier by buying two at the same time. Because of its own weight, the instrument has a static deflection one-fourth of that of the machine when placed on the pad.

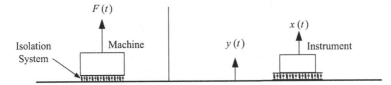

c. What is the natural frequency f_1 of the instrument on its isolation pad and what is the frequency ratio f_0/f_1 for the instrument at the problem frequency?

d. What damping ratio ξ_1 would you expect to have for the instrument on its vibration isolation pad?

e. What is your prediction for the response amplitude of the instrument compared to the floor motion at 30 Hz?

f. The engineers are not thrilled with the result. What is the problem? Without buying another system – that would be embarrassing to explain to the boss – what simple inexpensive fix would you suggest they try to improve the performance of the system, so as to make the natural frequency such that $f_0/f_1 = 3$?

g. Assume both isolation systems are fixed properly such that each one has $f_0/f_m = f_0/f_1 = 3$ and the damping $\xi_m = \xi_1 = 0.07$; what is the total reduction in the vibration of the instrument due to the machine compared to the vibration with no isolation pads at all?

P. 49 Two identical, cubic rigid masses m of dimension $a \times a \times a$ are mounted on identical rotational springs k_θ of negligible length, as shown in the picture. Identify the DOF and set up the equations of motion.

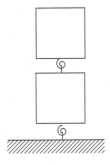

P. 50 A reciprocating machine that weighs 100,000 N is known to develop vertical harmonic forces having a maximum amplitude of 5000 N at its operating speed of 30 Hz, when attached to a rigid support. To limit the vibrations in the building in which the machine is to be installed, it is to be supported by a spring at each corner of its base. What spring constant (stiffness) is required to reduce the total harmonic force transmitted to the building to 1000 N?

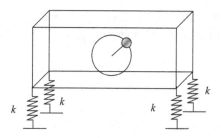

P. 51 A simple SDOF system consists of a mass and a damper only, as shown in the figure below. The excitation is the real part of $F(t) = F_0 e^{i\omega t}$, where F_0 is real and positive.

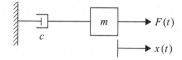

where $m = 1$ kg, and $c = 10$ N-s/m.

- Draw a free-body diagram of the mass showing all the forces.
- Find the equation of motion.
- Find the frequency response function $H_{x|F}(\omega)$, which is defined as the ratio of the response $Xe^{i\omega t}$ to the input $F_0 e^{i\omega t}$. Sketch the magnitude of $|H_{x|F}(\omega)|$.
- Find the transfer function for the velocity of response, that is, $H_{\dot{u}|F}$
- At very low frequency, damping is dominant. Find the response magnitude $|X|$ and the approximate phase angle between the force and the response when $\omega = 0.5$ rad/s. Let $F_0 = 5.0$ N.
- At high frequency, the system is mass controlled. Find the response magnitude $|X|$ and the approximate phase angle between the force and the response when $\omega = 200.00$ rad/s.

P. 52 If the force input to an SDOF vibration system is $F(t) = Fe^{i\omega t}$ and the steady–state output is $X e^{i\omega t}$, then the frequency response function is $H_{x|F}(\omega) = X e^{i\omega t} / F e^{i\omega t}$, which can be expressed as a magnitude and phase:

$$H_{x|F}(\omega) = | H_{x|F}(\omega)| \, e^{-i\varphi}$$

$$X(t) = |F| \, | \, H_{x|F}(\omega)|e^{-i\varphi}$$

- The Euler formula says $e^{i\theta} = \cos(\theta) + i \sin(\theta)$. Show that $i = e^{i\pi/2}$ and $-1 = e^{i\pi}$.
- Find $H_{\dot{X}|F}(\omega)$ and $H_{\ddot{X}|F}(\omega)$ the frequency response functions when the response is the velocity or the acceleration.

Hint : $H_{\dot{X}|F}(\omega) = \dfrac{\dot{X}(t)}{F e^{i\omega t}} = |H_{\dot{X}|F}(\omega)| \, e^{-i\phi} \qquad H_{\ddot{X}|F}(\omega) = \dfrac{\ddot{X}(t)}{F e^{i\omega t}}$

Generalize your answer by explaining the effect of phase angle when you take the time derivative of the response.

P. 53 An SDOF system with stiffness k and natural frequency ω_n is subjected to a periodic force $F(t)$ in the form of a sawtooth with amplitude F_0.

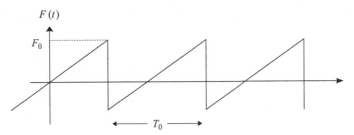

The Fourier series expansion of the force is

$$F(t) = \frac{2}{\pi} F_0 \sum_{j=1}^{\infty} \frac{-(-1)^j \sin\left(\omega_j t\right)}{j}$$

$$= \frac{2}{\pi} F_0 \left\{ \sin\left(\omega_0 t\right) - \frac{1}{2} \sin\left(2\,\omega_0\, t\right) + \frac{1}{3} \sin\left(3\,\omega_0\, t\right) - \cdots \right\}$$

where $\omega_j = j\,\omega_0$ and $\omega_0 = 2\,\pi/T_0$.

Noticing that the terms of the Fourier series are *real* and that they are all *sines*, we know that the actual response will also be a real infinite series containing *only* sine terms. We also know that the transfer function can be written as

$$H(\omega) = \frac{1}{k} A\, e^{-i\varphi} = \frac{1}{k} A\left(\cos\varphi - i \sin\varphi\right)$$

in which

$$A(\omega) = \frac{1}{\sqrt{\left(1-r^2\right)^2 + 4\xi^2\, r^2}} \qquad\qquad \varphi(\omega) = \arctan\frac{2\xi r}{1-r^2} \qquad\qquad r = \frac{\omega}{\omega_n}.$$

- Write down the *real* series expression for the response in terms of the amplification factors A_j and phase angles φ_j, which are defined at the discrete frequencies $\omega_j = j\omega_0$ (i.e., $r_j = j\omega_0 / \omega_n$)
- If $\omega_0 / \omega_n = 1/2$ and $\xi = 0.05$ determine the amplitude and phase angle of the first four terms in the response series for $u(t)$, assuming the factor $\dfrac{2}{\pi} \dfrac{F_0}{k} = 1$. Make a sketch of the discrete amplitude and phase spectra, and suggest how many terms would be needed for an accurate reponse evaluation. How does the natural period of the oscillator T_n compare with the period T_0?

P. 54 A pendulum of length L hangs from pivot attached to a mass supported on rollers, as shown below. There are no dampers anywhere, and gravity is on. Assuming *small* rotations together with the datum $mg = kL$, find

a. The equations of motion for lateral forces applied onto each mass ($p_1 \neq p_2 \neq 0$).
b. The equations of motion for horizontal, harmonic support motion ($p_1 = p_2 = 0$).
c. The frequencies and modes of vibration for this problem.

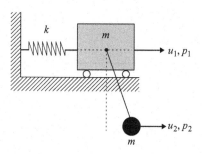

P. 55 To a first approximation, a nuclear reactor building can be modeled as a rigid cylinder of radius $R = 20$ m and height $H = 60$ m, with an average mass density of $\rho = 625$ kg/m³ ("=25% of solid concrete"). This building is supported by a soil with shear wave velocity $C_s = 300$ m/s, mass density $\rho_s = 2000$ kg/m³, and Poisson's ratio $v = 1/3$. Using approximate expressions for the foundation stiffnesses, determine the six natural frequencies and modal shapes of this system.

P. 56 Consider a uniform, *rigid* bar with length $L = 10$ m, width $b = 0.2$ m, height $h = 0.5$ m, and mass density $\rho = 2500$ kg/m³, as shown in the figure below. The bar is supported at the ends by identical springs of stiffness $k = 5000$ N/m. Although this is a 2-DOF system (vertical and rotational motion), the vibration modes are *uncoupled* because of symmetry. Hence, this system behaves like two independent SDOF systems.

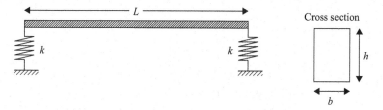

- Find the two natural frequencies (in Hz) of this system.
- If the rightmost spring is initially extended by a vertical displacement u_{B0}, what is the free oscillation response of this system? (**Hint**: It will both translate and rotate.)

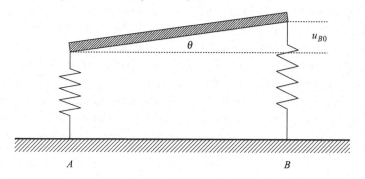

P. 57 A simply supported massless beam has two lumped masses, as shown.

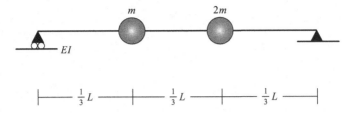

From a table of structural formulae we also know that the deflected shape of a simply supported beam subjected to a static force P applied at some arbitrary point x is given by

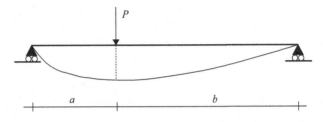

$$u(x) = \frac{PL^3}{6EI}\left(\frac{b}{L}\right)\left(\frac{x}{L}\right)\left[1-\left(\frac{b}{L}\right)^2-\left(\frac{x}{L}\right)^2\right], \qquad x \le a$$

$$= \frac{PL^3}{6EI}\left(\frac{a}{L}\right)\left(1-\frac{x}{L}\right)\left[1-\left(\frac{a}{L}\right)^2-\left(1-\frac{x}{L}\right)^2\right], \qquad x > a$$

a. Find the stiffness and mass matrices $\mathbf{K},\mathbf{M}$. **Hint**: Find the flexibility matrix $\mathbf{F}$ first, then obtain $\mathbf{K} = \mathbf{F}^{-1}$. To simplify your expressions, define the ratio $k = \alpha EI/L^3$, where α is an integer fraction such that your stiffness matrix is the product of k times a matrix composed of integers only.

b. Guessing a shape for the fundamental mode to be $\mathbf{v} = \begin{bmatrix} 3 & 4 \end{bmatrix}^T$ ("a somewhat larger displacement where the heavier mass is"), estimate the fundamental frequency using the Rayleigh quotient. Alternatively, you could instead use for $\mathbf{v}$ the deflected shape caused by the weight, that is, $\mathbf{v} \sim g\,\mathbf{FMe}, \mathbf{e} = \begin{bmatrix} 1 & 1 \end{bmatrix}^T$ (but omit the factor g and any constant fractional factors in $\mathbf{F}$, i.e., use integer values only!).

c. Find the exact frequencies and modal shapes, and compare the fundamental frequency with that estimated in the previous step.

d. Determine the dynamic response that the system would have if at $t = 0$ gravity were to be suddenly "switched on" (physically impossible, but not mentally). Basically, we are asking you here to write down the expressions for the response, but you need **not** evaluate it numerically as function of time.

P. 58 Consider the *undamped* lumped mass system shown below. Observe that there is *no* dynamic DOF at the node in the middle because there is *no* mass there.

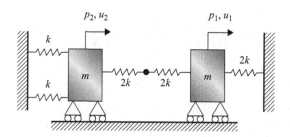

a. Using a free-body diagram together with D'Alembert forces, find the equations of motion.
b. Find the mode shapes and frequencies in terms of k, m.
c. If a sinusoidal forcing function is applied to mass 1 beginning at time $t = 0$ as indicated below, find the *undamped* response in time by means of modal superposition, assuming that the system starts at rest.

$$p_1(t) = \begin{cases} 0 & t < 0 \\ p_0 \sin \omega t & t \geq 0 \end{cases}, \qquad p_2(t) = 0 \text{ at all times}$$

d. Find the configuration of dashpots that will produce uniform modal damping (i.e., the same fraction of critical damping in both modes).
e. Find the response due to a suddenly applied load (step load) that is applied at the massless point *at the center*, again with zero initial conditions, and without any damping.

P. 59 A 3-DOF system has modes and frequencies

$$\Phi = \begin{bmatrix} 3 & 1 & 1 \\ 2 & -1 & -4 \\ 1 & -1 & 5 \end{bmatrix} \quad \text{and} \quad \Omega = \begin{Bmatrix} \omega_1 \\ \omega_2 \\ \omega_3 \end{Bmatrix} = \begin{Bmatrix} 1 \\ 2 \\ 3 \end{Bmatrix} \text{ rad/s}$$

Its mass and stiffness matrices are

$$M = m \begin{bmatrix} 1 & 0 & 0 \\ 0 & 1 & 0 \\ 0 & 0 & 1 \end{bmatrix} \quad K = \frac{k}{21} \begin{bmatrix} 46 & -37 & -1 \\ -37 & 106 & -59 \\ -1 & -59 & 142 \end{bmatrix}$$

a. Using MATLAB, verify that the modal shapes and frequencies given are correct.
b. If $m = 9$, what is the value of k?
c. Determine the modal coordinates of the (arbitrary) shape vector $v^T = \{1 \ 1 \ 1\}$; that is, find c such that $v = \Phi c$.
d. If the system is undamped, and it is subjected to an initial displacement $u_0 = v$ and $\dot{u}_0 = 0$, what is the response?

P. 60 The 3-DOF system shown has masses that can move only in the horizontal direction; any rotation and vertical translation is prevented by appropriate external constraints. Determine all modes and frequencies.

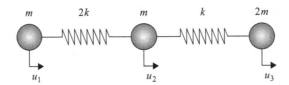

Hint: Observe that this system allows a rigid-body translation, so it will have one rigid-body mode with zero frequency $\omega_1 = 0$ and modal shape $\phi_1^T = \begin{bmatrix} 1 & 1 & 1 \end{bmatrix}$. Thus, although the determinant is 3×3, this can still be solved "by hand":

$$\begin{vmatrix} a_{11} & a_{12} & a_{13} \\ a_{21} & a_{22} & a_{23} \\ a_{31} & a_{32} & a_{33} \end{vmatrix} = \left(a_{11}a_{22}a_{33} + a_{12}a_{23}a_{31} + a_{21}a_{32}a_{13} \right) - \left(a_{13}a_{22}a_{31} + a_{12}a_{21}a_{33} + a_{23}a_{32}a_{11} \right)$$

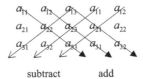

subtract   add

P. 61 In the structure shown, determine all frequencies and modal shapes (but think before you proceed…)

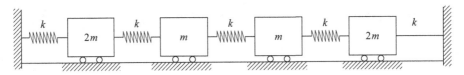

(You may verify your answers with MATLAB, but you should also work out this problem by hand.)

P.62 A recovery ship of mass M = 1,000,000 kg carries on board a submersible of mass m = 50,000 kg. When floating in calm water, the ship with the submersible heave (oscillate vertically) with a period of 5 s. At some point in time, the submersible is lowered into the water resting on a cradle suspended by cables. The effective axial stiffness of the cables, which is $k = EA/L$, changes with the length L of the cables, that is, with the depth of submersion. It is also known that when the submersible reaches a depth L_1 = 500 m, the fundamental period of oscillation of the system is T = 50 s, which for practical purposes is the period of oscillation of the submersible by itself (i.e., a simple spring-mass system), as if the ship didn't move. Neglecting the weight of the cables, determine

- The area of the ship at the water line, and
- The *coupled* periods of oscillation of this 2-DOF system when the submersible is at a depth L_2 = 25 m.

Neglect any participating mass of water, and also any drag forces.

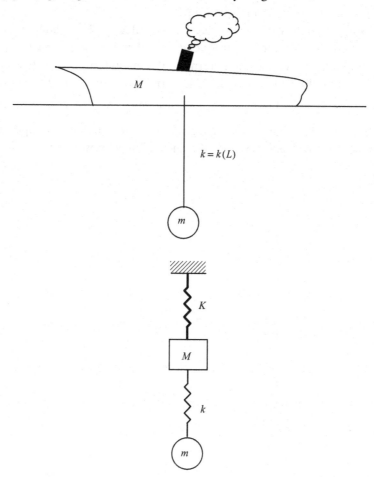

$$k = k(L)$$

$$K$$

$$M$$

$$k$$

P. 63 An airplane is modeled as a beam with three masses as shown below:

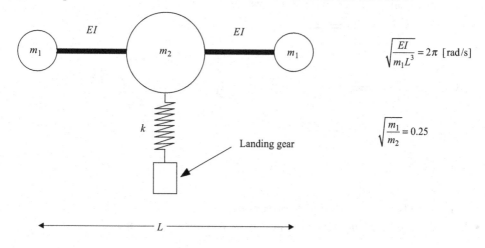

$$\sqrt{\frac{EI}{m_1 L^3}} = 2\pi \ [\text{rad/s}]$$

$$\sqrt{\frac{m_1}{m_2}} = 0.25$$

- Find the stiffness and mass matrices.
- Find all oscillator modes and frequencies.
- If the landing gear has stiffness k such as it deflects 0.1 m under its own weight (when parked), determine the frequencies and mode shapes of the plane after landing. Explain how you would determine the maximum force in the landing gear at the time of landing, assuming that the vertical descent rate at time of landing is 10 m/s.

P. 64 Consider a two-story frame with rigid girders (beams) and mass lumped at the level of each floor, which is supported on a sloping ground. It is known that

$m_1 = 5000$ kg

$m_2/m_1 = 2,$

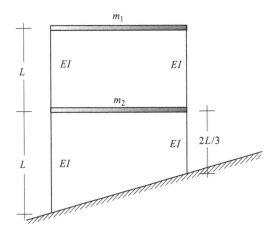

$$\frac{EI}{L^3} = 2500 \ N/m$$

The deformed configuration and the equivalent 2-DOF model are

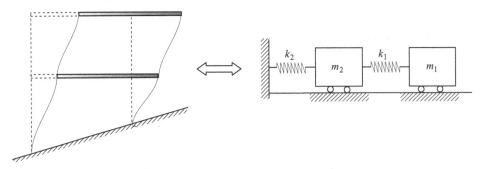

- Formulate the equations of motion for the system.
- Estimate the fundamental frequency, using Rayleigh's quotient.

- Find the exact modal shapes and frequencies.
- What is the proportional damping matrix that gives uniform modal damping $\xi_1 = \xi_2 = 0.05$?

P. 65 Consider a set of coupled penduli, as shown in the figure below (gravity is "on").

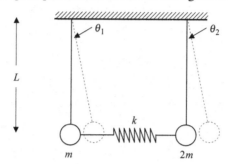

- Find the equations of motion.
- Specialize the result of (a) for small angles θ_1 and θ_2.
- Find the frequencies and model shapes of the system in terms of $\sqrt{k/m}$.

P. 66 Consider a spring–mounted *rigid* bar of total mass m and length L, to which an additional mass m is lumped at the rightmost end. The system has *no* damping.

- Find the natural modes of vibration.
- The left support is given an initial vertical displacement a and is then released. Find the response.
- Write down the impulse response functions for this system.

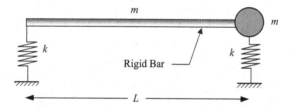

Careful: The rigid bar has rotational inertia (the mass of the bar is uniformly distributed).

P. 67 Consider a lumped mass structure that can be modeled as a shear beam (= a chain of *close-coupled* lateral springs and masses, i.e., the simple system that we often use to model high-rise buildings). The distance between floors (i.e., inter-story height) is constant. The masses are numbered from the top down, and although arbitrary, you should consider them as being *known*. The stiffnesses, however, are *not* known. However, from a vibration test, you have determined experimentally the fundamental frequency ω_1 as well as the fundamental mode ϕ_1, and you have found that this mode is a straight line, that is,

$$\phi_1^T = \{N \quad N-1 \quad \cdots \quad 2 \quad 1\}$$

a. Determine the values of the springs in terms of the fundamental frequency and masses.

b. The overturning moment at the base is the moment exerted by the inertia forces, which is

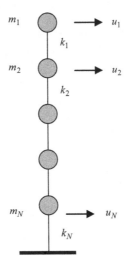

$$M = \sum_{i=1}^{N} m_i \ddot{u}_i z_i$$

in which z_i is the height of the ith mass above the ground. Show that because the first mode is a straight line, *only that mode* contributes to the overturning moment (notice, by the way, that the foundation must be able to resist this moment).

c. Suppose that there is a wall that runs parallel to the structure, and that there are springs of rigidity $r_i = \alpha m_i$ connecting the masses and the wall, in which α is a *known* constant (i.e., the springs are proportional to the masses). If you knew *all* the frequencies and mode shapes of the original system *without* the added springs, could you find the frequencies and mode shapes for the structure with the added springs?

d. If $N = 2$ (i.e., two floors only), find the second frequency and mode shape, again in terms of the known parameters (for the original system without added lateral springs!). Assume that both masses are equal.

e. Determine the proportional damping matrix for the 2-DOF system in (d) that would produce the same value of damping ξ in both modes. From this matrix, figure out the values and physical configuration of the dashpots.

f. Again for the 2-DOF system considered in (d), assume that it is subjected to a unit impulsive load acting on the bottom floor. Find the response in each floor, assuming no damping.

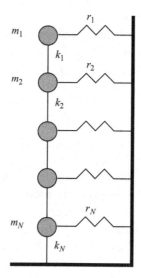

P.68 The 3-DOF system shown below on the left has been solved using MATLAB, and its natural frequencies have been found to be

$$\omega_1 = 0.6289 \sqrt{\frac{k}{m}}, \qquad \omega_2 = 1.3179 \sqrt{\frac{k}{m}}, \qquad \omega_3 = 2.0899 \sqrt{\frac{k}{m}}$$

Show that the frequencies ω_1', ω_2' and ω_1'' of the two smaller systems on the right, which were obtained by fixing the lower one/two stories obey the interlacing theorem, that is, $\omega_1 \le \omega_1' \le \omega_2 \le \omega_2' \le \omega_3$ and $\omega_1' \le \omega_1'' \le \omega_2'$. **Hint**: find the frequencies for structures 2 and 3, and compare.

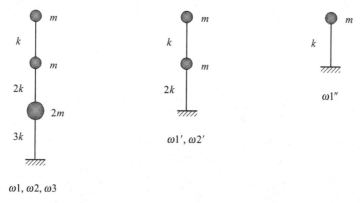

P.69 Consider the 2-DOF system shown in the figure below.

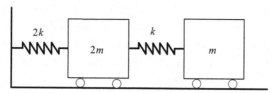

a. Formulate the equations of motion.
b. Estimate the fundamental frequency by means of the Rayleigh quotient
c. Using the Rayleigh quotient in question (b), do one step of inverse iteration with a shift by the Rayleigh quotient, using a starting vector $v_0^T = \{1\ 1\}$.

After doing this iteration, normalize (scale) the resulting vector so that the first component has a unit value, i.e. $\boldsymbol{\varphi}_1^T = \{1\ \phi_{12}\}$.

P.70 A massless circular arch of bending stiffness EI and radius R subtends an angle $\alpha = \frac{1}{2}\pi$ (i.e., 90 degrees). The arch is clamped at the left. At the free end on the right there is a lumped mass m attached. As the arch vibrates, the mass translates both horizontally and vertically, and it also rotates. However, the rotation is not a dynamic DOF, because the mass possesses NO rotational inertia. Hence, you need not be concerned about that rotation (it is a slave DOF). A rigorous and rather elaborate strength of materials type analysis can be used to show that the stiffness and flexibility matrices for this problem are given by

$$\mathbf{K} = \frac{6EI}{R^3}\frac{1}{19-6\pi}\begin{Bmatrix} 2 & 1 \\ 1 & 10-3\pi \end{Bmatrix}$$

$$\mathbf{F} = \mathbf{K}^{-1} = \frac{R^3}{6EI}\begin{Bmatrix} 10-3\pi & -1 \\ -1 & 2 \end{Bmatrix}$$

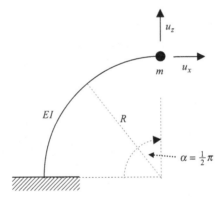

In the case of static loads, $\mathbf{K}$ relates horizontal and vertical loads applied at the free end with displacements observed there, as shown in the figure, that is,

$$\mathbf{p} = \mathbf{Ku}, \quad \mathbf{u} = \begin{Bmatrix} u_x \\ u_z \end{Bmatrix}, \quad \mathbf{p} = \begin{Bmatrix} p_x \\ p_z \end{Bmatrix}, \quad \mathbf{M} = ?$$

a. It seems clear that in the first mode of vibration, as the mass moves up it should also move to the left, and vice versa. Based on this fact and guessing (perhaps naïvely) that in the first mode of vibration the mass oscillates roughly perpendicularly to the 45-degree straight line from the mass to the support, that is, that the first mode is of the form $\boldsymbol{\phi}_1^T \sim \mathbf{u}_0^T = \{-1\ 1\}$, use the Rayleigh quotient to estimate the fundamental frequency.

b. Determine the exact frequencies in terms of dimensionless eigenvalues $\lambda_j, j = 1, 2$, such that

$$\omega_j^2 = \lambda_j \frac{EI}{mR^3} \frac{1}{\frac{19}{6} - \pi}$$

c. Determine the exact modal shapes in terms of the dimensionless eigenvalues, and sketch the modes of vibration. In particular, compare the true first mode (both frequency and modal shape) with the guessed one. Was the initial guess good or poor? If the latter, what would have been a better shape?

d. Suppose that an additional (vertical) weight $w = mg$ is carefully hung from the mass by means of a string, and that at time $t = 0$, when the system with the added weight is still motionless, the string is cut and the added weight falls off instantly. Find the initial modal coordinates $q_{0j}, \dot{q}_{0j}$ that define the free vibration that follows at $t > 0$. **Note:** The vibration will be relative to the position that the structure had before the added weight was hung.

P. 71 Find the equations of motion for the structures shown in the figures below.

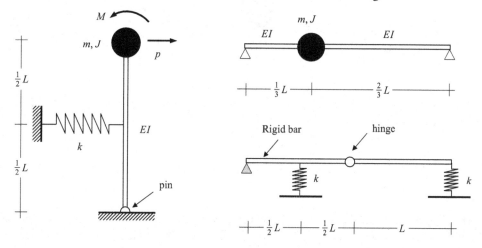

The rigid bar has mass/lenght m/L. The rotational inertia of the bar is the same as that of a cylindrical bar with negligible radius, $L \gg 2R$.

P. 72 Consider a lumped mass N-DOF system, with *known* stiffness matrix $\mathbf{K}$ and mass matrix $\mathbf{M}$. If you also know $N - 1$ modes (both the frequencies and the modal shapes), what are the frequency and the modal shape of the N th mode?

P. 73 A simply supported, massless bending beam has three *equal* lumped masses attached at equal distances $L/4$, as shown below.

Using the flexibility method, one can obtain the flexibility matrix, and by inversion, the stiffness matrix. These two matrices can be shown to be

$$F = \frac{L^3}{768\,EI} \begin{Bmatrix} 9 & 11 & 7 \\ 11 & 16 & 11 \\ 7 & 11 & 9 \end{Bmatrix} \qquad K = \frac{192\,EI}{7\,L^3} \begin{Bmatrix} 23 & -22 & 9 \\ -22 & 32 & -22 \\ 9 & -22 & 23 \end{Bmatrix}$$

a. Using the Rayleigh quotient and Dunkerley's rule, bracket the fundamental frequency for this beam. (Careful: A straight line assumption does not apply here!).

b. Carry out one step of inverse iteration, and compare the new mode and frequency with that in step (a).

P. 74 Consider a massless, uniform cantilever beam of bending stiffness EI and total length L that has two lumped masses m_1, m_2, as shown below. The second mass m_2 is located at the center of the beam, while the applied loads p_1, p_2 and observed displacement u_1, u_2 are as indicated. How many DOF does this system have? Find the stiffness matrix for this system and set up the equations of motion. Disregard any axial motion, or motions out of the plane. **Hint**: Use the flexibility approach given in Chapter 1 of this book.

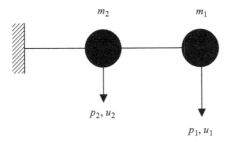

P. 75 A 3-DOF system has a mass and stiffness matrices

$$M = m \begin{Bmatrix} 1 & & \\ & 1 & \\ & & 1 \end{Bmatrix} \qquad K = k \begin{Bmatrix} 1 & -1 & 0 \\ -1 & 3 & -2 \\ 0 & -2 & 5 \end{Bmatrix}$$

After solving this problem using MATLAB with numerical values for k and m, we obtain the frequencies (in Hz!) and modes given in the table below.

Frequencies and normalized mode shapes

j (mode number)	1	2	3
f_j [in Hz]	1.026	2.411	3.992
ϕ_{1j}	0.8433	−0.5277	0.1019
ϕ_{2j}	0.4927	0.6831	−0.5392
ϕ_{3j}	0.2149	0.5049	0.8360

Note: We have normalized the modes so that the modal mass for all modes is $\mu_j = \phi_j^T M \phi_j = 1$.

a. What is the ratio k/m? If $m = 2$ [kg], what is k (specify units!)?

b. You may recall that for Rayleigh damping $C = a_0 M + a_1 K$, we had a damping characteristic as function of frequency of the form $\xi = \tfrac{1}{2}(a_0 / \omega + a_1 \omega)$. Assuming that we now assign the damping value $\xi = 0.05$ to the single *arbitrary* frequency $f_1 = 1.5$ Hz (our a priori estimate of the fundamental frequency of this structure), determine the *mass-proportional damping* matrix $\mathbf{C} = a_0\mathbf{M}$. Notice that this is a special case of Rayleigh damping, which we obtain by setting $a_1 = 0$ (less computation...).

c. Determine the fractions of damping ξ_1, ξ_2, ξ_3 implied by $\mathbf{C} = a_0\mathbf{M}$ in all three modes.

P. 76 A two-story building has equal story masses of $m = 5\times10^8$ kg which can be considered lumped at the floors, and story stiffnesses $k_1 = 1$ GN/m and $k_2 = 2$ GN/m, as sketched below. The fundamental frequency has been found to be 1.0824 rad/s.

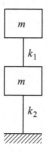

The stiffness and mass matrices for this system are

$$\mathbf{K} = \begin{bmatrix} k_1 & -k_1 \\ -k_1 & k_1 + k_2 \end{bmatrix}, \qquad \mathbf{M} = \begin{bmatrix} m & 0 \\ 0 & m \end{bmatrix}$$

a. Although you have already been given the exact fundamental frequency, please ignore this for a brief moment, and choose instead a reasonable guess for the modal shape with which you estimate this frequency via the Rayleigh quotient. How accurate is your approximate result when compared to the exact (in %)?

b. Given the *exact* first frequency, what is the other natural frequency of the building? (**Note:** You may avoid substantial work by thinking carefully before proceeding. Solving an eigenvalue problem will be the slowest alternative and will demand the most effort.)

c. What is the mode shape for the second frequency found?

d. What is the proportional (Rayleigh) damping matrix that will give a uniform fraction of critical damping of 1%?

P. 77 A discrete MDOF system has a damping matrix that is of the Rayleigh type, that is $\mathbf{C} = a\mathbf{M} + b\mathbf{K}$. It is also known that this matrix produces equal fractions of material damping $\xi = 0.1$ at the two frequencies $f_1 = 1$ Hz and $f_2 = 5$ Hz. What are the values of the constants a and b?

P. 78 A 2-DOF system is characterized by the system matrices

$$\mathbf{M} = \begin{Bmatrix} m & 0 \\ 0 & 2m \end{Bmatrix} \qquad \mathbf{K} = \begin{Bmatrix} k & -k \\ -k & 2k \end{Bmatrix} \qquad \mathbf{C} = \begin{Bmatrix} c_{11} & c_{12} \\ c_{21} & c_{22} \end{Bmatrix}$$

- What restrictions must the elements $c_{11}, c_{12} = c_{21}, c_{22}$ of the damping matrix satisfy for normal modes to exist, that is, for $\mathbf{C}$ to belong to the class of proportional matrices?

- If the resulting damping matrix can be interpreted as resulting from an assembly of three dashpots c_1, c_2, c_3 such that $c_{11} = c_1 + c_3$, $c_{12} = c_{21} = -c_1$, $c_{22} = c_1 + c_2$, will the dashpots have always a positive value if we insist that **C** is proportional?

P. 79 A three-story building has three *equal* masses, which are numbered from the ground floor up. Its modes and frequencies are as listed in the table below. It is subjected to an earthquake with the response spectrum shown in Problem 88.

Note 1: Because the spectrum plot S_v provides neither the acceleration S_a nor the displacement S_d scales (i.e., the grid inclined at 45 degrees), you will have to determine any needed values from the mutual relationships among S_d, S_v, S_a. Careful with your units!

Note 2: For your convenience, we provide an empty table at the end of this problem for your calculations.

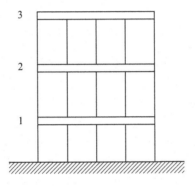

a. What are the peak ground acceleration and peak ground displacement?
b. Find the modal participation factors.
c. Determine the spectral accelerations at the modal frequencies for 5% structural damping.
d. Determine the square root of the sum of the squares (SRSS) accelerations for all three floors, and sketch the acceleration profile (i.e., the variation of acceleration with height).

		Mode 1	Mode 2	Mode 3
Frequency	f_j, Hz	2.833	7.939	11.472
Third floor	ϕ_{3j}	0.737	0.591	0.328
Second floor	ϕ_{2j}	0.591	−0.328	−0.737
First floor	ϕ_{1j}	0.328	−0.737	0.591

		Mode 1	Mode 2	Mode 3
Part. factor	γ_j			
Spect. accel.	$S_a(f_j)$			
Third floor	$\gamma_j^2 \phi_{3j}^2 S_{aj}^2$			
Second floor	$\gamma_j^2 \phi_{2j}^2 S_{aj}^2$			
First floor	$\gamma_j^2 \phi_{1j}^2 S_{aj}^2$			

P. 80 Consider the dynamic system shown in the figure below. It consists of a mass m on smooth horizontal rollers and a wheel of radius R, mass m, and mass moment of inertia $J = \frac{1}{2} m R^2$ that rolls on the base without slipping. Resting tangentially above the wheel is a rigid bar with a rough surface (i.e., no slip) that connects to both a spring and a dashpot. As the wheel rolls back and forth, the bar at the top moves along without slipping.

- Determine the equations of motion.
- Find the undamped frequencies and modes of vibration.
- Is the damping matrix of the proportional type?

Note: Whichever way you choose to solve this problem, please designate the lateral motion of the center of the wheel as u_1 and lateral displacement of the mass to the left as u_2.

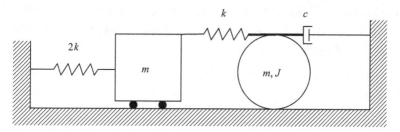

Hint: The easiest solution is by application of Lagrange's equations. To account for the damper, add to the Lagrange equations a term of the form $\partial D / \partial \dot{q}_i$, where $D = \frac{1}{2} c \dot{e}^2$ is the *damping potential* and $\dot{e}$ is the rate of elongation of the damper, see Chapter I of this book for further details.

P. 81 An undamped, massless cantilever beam of total length L and bending stiffness EI has three equidistant, lumped masses *numbered from left to right*, as illustrated below. The beam is axially stiff, so none of the masses can move in the horizontal direction, only in the transverse vertical direction. Although there is *no* rotational mass anywhere, the masses may also rotate, so the three rotations are slave DOFs. Using elementary methods, it can be shown that this system has lateral stiffness and flexibility matrices

$$\mathbf{K} = \frac{81}{13} \frac{EI}{L^3} \begin{bmatrix} 80 & -46 & 12 \\ -46 & 44 & -16 \\ 12 & -16 & 7 \end{bmatrix}, \quad \mathbf{F} = \frac{1}{162} \frac{L^3}{EI} \begin{bmatrix} 2 & 5 & 8 \\ 5 & 16 & 28 \\ 8 & 28 & 54 \end{bmatrix}$$

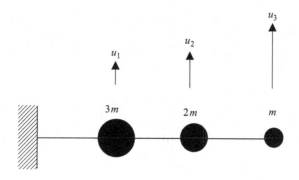

a. Estimate the fundamental frequency of this system using the Rayleigh quotient with the estimated modal shape: $\mathbf{v} = \begin{bmatrix} 1 & 3 & 6 \end{bmatrix}$. (We do not use a straight line here because the bending beam deflects much more toward the free end, which here is the *last* DOF.)

b. If you were given an initial displacement in the shape of the first mode, that is, $\mathbf{u}_0 = \alpha \phi_1$, what would be the free vibration for $t > 0$?

c. Assume that we restrain the rightmost mass with a roller so that it cannot move, even if it still can rotate (i.e., a slave DOF). Estimate the fundamental frequency, or at your discretion, you can obtain instead the exact value. Sketch the mode of vibration.

d. Assume that we modify this problem yet again by adding a second horizontal roller underneath the leftmost mass 1; that is, only the center mass can translate. However, all three masses continue to allow rotation, even if these are only slave DOFs because there is no rotational inertia. What is the *exact* frequency of this SDOF system? Sketch the mode of vibration.

Hint: All the information needed to solve this problem is already available to you; that is, any required material can be inferred directly from these data. Thus, there is neither a need for you to fish for more info in lecture notes, or to start complex derivations.

P. 82 An undamped, massless cantilever bending beam of length $2L$ and flexural stiffness EI has two lumped masses with both translational *and* rotational inertia. The 4 active DOF are in sequence from left to right.

If the rotations are defined as positive when counterclockwise and the equations of motion $\mathbf{M}\ddot{\mathbf{u}} + \mathbf{K}\mathbf{u} = \mathbf{p}$ are written in *homogeneous dimensions*, it can be shown that the stiffness matrix, mass matrix, flexibility matrix, displacement vector, general load vector, and the load vector due to gravity for this problem are, respectively, given by

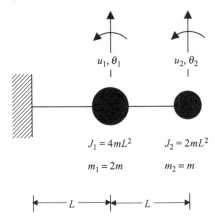

$$\mathbf{K} = \frac{EI}{L^3} \begin{bmatrix} 24 & 0 & -12 & 6 \\ 0 & 8 & -6 & 2 \\ -12 & -6 & 12 & -6 \\ 6 & 2 & -6 & 4 \end{bmatrix} \quad \mathbf{M} = m \begin{bmatrix} 2 & 0 & 0 & 0 \\ 0 & 4 & 0 & 0 \\ 0 & 0 & 1 & 0 \\ 0 & 0 & 0 & 2 \end{bmatrix} \quad \mathbf{F} = \mathbf{K}^{-1} = \frac{L^3}{6EI} \begin{bmatrix} 2 & 3 & 5 & 3 \\ 3 & 6 & 9 & 6 \\ 5 & 9 & 16 & 12 \\ 3 & 6 & 12 & 12 \end{bmatrix}$$

$$\mathbf{u}(t) = \begin{bmatrix} u_1 \\ \theta_1 L \\ u_2 \\ \theta_2 L \end{bmatrix} \qquad \mathbf{p}(t) = \begin{bmatrix} p_1 \\ M_1/L \\ p_2 \\ M_2/L \end{bmatrix} \qquad \mathbf{p}_{gravity} = mg \begin{bmatrix} 2 \\ 0 \\ 1 \\ 0 \end{bmatrix}$$

in which p_1, p_2 are external forces and M_1, M_2 are external moments (torques).

- Determine the static deformation vector caused by gravity.
- Estimate the fundamental frequency of this system using the Rayleigh quotient. You may use a trial mode shape based on your solution to the previous part (but please use integers only), or choose another tentative mode shape. Be careful with the order of the DOFs and the values used for rotations!
- Assume that we restrain *both* masses with rollers so that they cannot move yet can still rotate (i.e., $u_1 = u_2 = 0$). Determine the frequencies and modal shapes of the restrained system. Having the modal shapes, sketch the physical modes in a model of the beam. (Make sure that your sketch satisfies the physical constraints.)

P. 83 For the system shown below

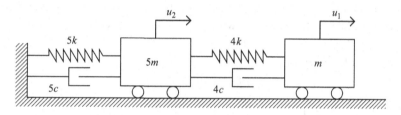

a. Find the undamped frequencies and modal shapes.
b. Determine the fractions of damping in each mode.
c. If you were to give an initial displacement in the shape of the first mode, that is, $\mathbf{u}_0 = \alpha \boldsymbol{\varphi}_1$, what would be the free vibration response for $t > 0$?

P. 84 Consider the free vibration problem of a 3-DOF system:

$$\mathbf{K}\boldsymbol{\varphi} = \omega^2 \mathbf{M}\boldsymbol{\varphi}$$

in which the stiffness, mass and flexibility matrices are

$$\mathbf{K} = k \begin{bmatrix} 3 & -1 & 0 \\ -1 & 4 & -1 \\ 0 & -1 & 3 \end{bmatrix}, \qquad \mathbf{M} = m \begin{bmatrix} 2 & 0 & 0 \\ 0 & 1 & 0 \\ 0 & 0 & 2 \end{bmatrix}, \qquad \mathbf{K}^{-1} = \mathbf{F} = \frac{1}{30k} \begin{bmatrix} 11 & 3 & 1 \\ 3 & 9 & 3 \\ 1 & 3 & 11 \end{bmatrix}$$

a. Estimate ω_1 using the Rayleigh quotient with an initial guess $\mathbf{v} = \begin{bmatrix} 1 & 1 & 1 \end{bmatrix}^T$.
b. Perform one step of inverse iteration, and then use again the Rayleigh quotient to estimate ω_1. This is likely to be rather close to the true result.
c. Estimate the fundamental frequency using Dunkerley's rule, and bracket your results.

P. 85 A massless cantilever beam of total length L and bending stiffness EI has a mass m lumped at is end, which also has a rotational inertia $J = \frac{1}{16} mL^2$.

- Find the natural frequencies in terms of EI / mL^3.
- The lower end of the beam is subjected to a harmonic, rotational base motion. Find the transfer functions for the absolute translation and rotation of the mass, and for all base reactions. Assume rotations to be positive counterclockwise. There is no damping in this system.

P. 86 Consider a steel chimney of length L = 20 m. The chimney is cylindrical in shape with inner diameter d =780 mm and outer diameter D = 800 mm. This system can be analyzed as a continuous cantilever beam whose natural frequencies are

$$\omega_1 = 3.516\sqrt{EI / \rho AL^4} \quad \text{for the first mode}$$
$$\omega_j = \left[(j-1/2)\pi\right]^2 \sqrt{EI / \rho AL^4} \quad \text{for } j = 2,3....$$

For steel, the material properties are

$$E = 2.07\times10^8 \text{ kPa}, \quad \text{and} \quad \rho = 7.65\times10^3 \text{ kg} / \text{m}^3$$

Also, the section properties of a hollow tube are

$$A = \frac{\pi}{4}\left(D^2 - d^2\right), \quad I = \frac{\pi}{64}\left(D^4 - d^4\right)$$

On a stormy day, the wind elicits transverse vibrations in the chimney (i.e., perpendicular to the wind) caused by *vortex shedding*. To a first approximation, the excitation can be modeled as a distributed, harmonic load whose frequency depends on the wind speed and on the diameter of the chimney. You may recall that the frequency of vortex shedding f (in Hz) is given by the *Strouhal* number

$$S = \frac{fD}{V} = 0.21 \text{ where } V \text{ is the wind velocity.}$$

a. Compute the wind velocities (in km/h) that will elicit transverse vibrations in the chimney at its first *three* resonant frequencies.
b. Discuss your findings from an engineering perspective: Is this problem important or not?

P. 87 Consider a 2-DOF system subjected to harmonic loads P_1 and P_2.

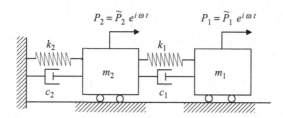

If $P_1 \neq 0$ and $P_2 = 0$, is there any nonzero frequency for which the responses are in phase?

Same as above, with $P_2 \neq 0$ and $P_1 = 0$.

P. 88 A five-story building can be idealized as the 5-DOF system shown, with $m = 1000$ kg, $k = 4 \times 10^5$ N/m, and interstory height $h = 4.00$ m. The system is subjected to an earthquake, with the tripartite response spectrum shown below.

- Find the SRSS relative displacement at the top and bottom of the structure.
- Find the SRSS overturning moment.

Hint: Find the natural modes and frequencies of the system using the closed-form solution for a discrete shear beam.

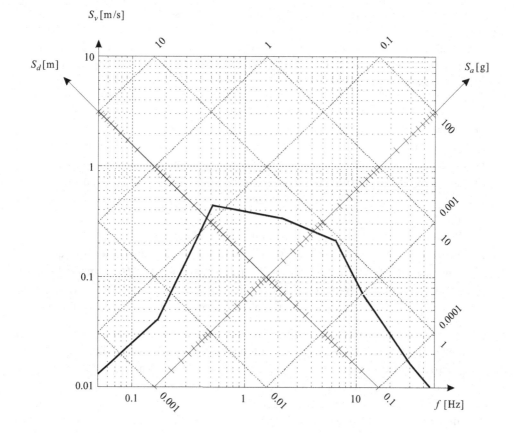

P. 89 Design an optimal tuned mass damper (TMD) meant to ameliorate wind vibrations in the first mode of a 60-story building, which has a square cross section $a \times a$ and is five time taller than it is wide. You can also assume an interstory height of $h = 3.5$ m. Use heuristic (i.e., rule-of-thumb) methods to model the building (mass, period, etc.). The mass of your TMD should not exceed 500 metric tons.

P. 90 The figure below shows a lightweight, *flexible* cantilever beam supporting a heavy electric motor, which has an imbalance at the 60-Hz rotation rate of the rotor. Assume that the mass of the motor, which is 100 kg, is large compared to the mass of the beam, and that the vertical imbalance force has a magnitude of 50 N.

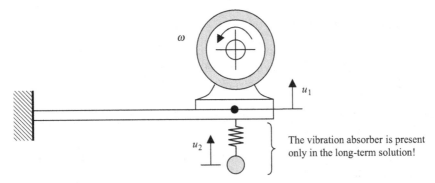

The natural frequency of the beam with the motor is also 60 Hz; that is, it coincides with the operating speed of the rotor. In a transient decay test, you also determine that damping is about 1% of critical. Your job is to present the owner with possible solutions for reducing the vibration. You may not change the speed of the motor or move the rotor to another location on the beam.

a. Propose a solution that could be implemented easily and cheaply the same day with materials commonly found around a shop. Your objective is to reduce the vibration by 50% immediately, to give you time to implement a dynamic absorber later. Explain how you would accomplish that quick fix solution, and how it works.

b. Design a long-term solution using a simple dynamic absorber that is required to work well only at the problem frequency. Assume that a mass ratio $\mu = 0.1$ (i.e., 10%) is specified.

   1. What are the mass and spring constants of the absorber?

   2. What is the theoretical or ideal, steady-state, dynamic response amplitude of the machine with the dynamic absorber attached? Sketch the H_{11} transfer function for this system and indicate the point on the curve corresponding to the operating frequency of the motor with the absorber attached and working properly.

P. 91 A large passenger car weighs 2000 kg and is 5 m long, bumper to bumper. The front and rear axles are 4 m apart, and the center of mass is halfway between the axles. The car travels on a road with an incline with constant velocity of $V = 72$ km/h. To a good approximation, the car can be modeled as a 2-DOF system that can oscillate vertically and/or rotate about the center of mass. Each of the four shock absorbers consists of a spring a dashpot in parallel, and all shock absorbers are identical. The spring stiffness is such that the car has an *undamped* natural frequency of 1 Hz in vertical motion, and the dashpots produce 20% of critical damping in that mode. Except for a ramp of length $L = 10$ m and a height $h = 0.10$ m, the road is otherwise flat. You may also assume that the rotational inertia about the center of mass can be estimated from the rotational inertia of a rigid bar that equals the car in length.

- Formulate the equations of motion, *starting from the sketch included below*. Express your final equations in homogeneous dimensions. Provide an expression for the *response* to the equivalent seismic motion. Discuss in broad terms what changes to these equations would be needed to account for the fact that when the car is on the ramp, the effective horizontal velocity decreases (but then again, the oscillations are not quite vertical). You need not worry about gravity, because the car will never shake so severely that it jumps off the road.
- Find the stiffnesses and dashpots of each shock absorber.
- Find the (undamped) frequency and fraction of critical damping in the second mode. Also, find both modal shapes. Observe that the damping matrix is proportional to the stiffness matrix, so this is a case of "proportional" damping matrix.
- Find and sketch the response (i.e. vertical motion and rotation) of the car as it travels along the road. Assume that the front axles reach the start of the ramp at time $t = 0$, that the car is not yet vibrating at this point in time. You can also assume the mass of the car to be uniformly distributed between the front and rear ends. You can use MATLAB for any part of this problem.

Sketch of formulation

Let x_1, x_2 denote the instantaneous positions (abscissas) of the front and rear axles, respectively, which are at a distance $2a$ apart. Also $x_1 = Vt$, because we are measuring time and space from the instant that the front axle transitions onto the ramp at $x = 0$. Also, we can describe the road surface as

$$y(x) = R(x) - R(x - L)$$

in which L is the length of the ramp, and

$$R(x) = \begin{cases} 0 & x < 0 \\ \frac{h}{L}x & x \geq 0 \end{cases}$$

This implies $y = 0$ when $x < 0$ and $y(x) = \frac{h}{L}$ when $x > L$. Next, define u_1, u_2 to be the absolute motions of the suspension points of the front and rear shock absorbers, and let v_1, v_2 denote the relative motions of these points with respect to the road surface, that is, the deformation of the springs. Then

$$
\begin{aligned}
u_1 &= v_1 + y_1 = v_1 + R(x_1) - R(x_1 - L) \\
&= v_1 + R(Vt) - R(Vt - L)
\end{aligned}
$$

$$u_2 = v_2 + y_2 = v_2 + R(x_2) - R(x_2 - L)$$
$$= v_2 + R(x_1 - 2a) - R(x_1 - L - 2a)$$
$$= v_2 + R(Vt - 2a) - R(Vt - L - 2a)$$

The absolute vertical motion of the center of mass of the car, and the car's absolute rotation are then

$$u = \tfrac{1}{2}(u_1 + u_2)$$
$$= \tfrac{1}{2}(v_1 + v_2) + \tfrac{1}{2}\big[R(Vt) - R(Vt - L) + R(Vt - 2a) - R(Vt - L - 2a)\big]$$

$$\theta = \tfrac{1}{2a}(u_1 - u_2)$$
$$= \tfrac{1}{2a}(v_1 - v_2) + \tfrac{1}{2a}\big[R(Vt) - R(Vt - L) - R(Vt - 2a) + R(Vt - L - 2a)\big]$$

which can also be written as

$$u = v + u_g, \qquad \theta = \vartheta + \psi_g$$

in which the relative motions of the center of mass are

$$v = \tfrac{1}{2}(v_1 + v_2), \qquad \vartheta = \tfrac{1}{2a}(v_1 - v_2)$$

and the "ground motion" components are

$$u_g = \tfrac{1}{2}\big[R(Vt) - R(Vt - L) + R(Vt - 2a) - R(Vt - L - 2a)\big]$$

$$\psi_g = \tfrac{1}{2a}\big[R(Vt) - R(Vt - L) - R(Vt - 2a) + R(Vt - L - 2a)\big]$$

Observe that the effective ground motion component u_g does not necessarily match the height of the road below the car's center of mass. This happens when one pair of wheels is on the flat part and the road and other part is on the ramp.

In matrix form, the equation of motion in terms of relative motions is

$$\mathbf{M}\ddot{\mathbf{v}} + \mathbf{C}\dot{\mathbf{v}} + \mathbf{K}\mathbf{v} = -\mathbf{M}\ddot{\mathbf{u}}_g, \qquad \ddot{\mathbf{u}}_g = \begin{Bmatrix} \ddot{u}_g \\ \ddot{\psi}_g \end{Bmatrix}$$

Observe that $\frac{d}{dt}R = \frac{h}{L}V\mathcal{H}(t)$, where $\mathcal{H}$ is the unit step function and $\frac{h}{L}V$ is the apparent vertical velocity of the ground as perceived by an observer in the car while it is on the ramp. It follows that $\frac{d^2}{dt^2}R = \frac{h}{L}V\delta(t)$, with δ being the Dirac delta (unit impulse) function, because $\frac{d}{dt}\mathcal{H}(t) = \delta(t)$. Hence

$$\ddot{u}_g = \tfrac{h}{2L}V\big[\delta(t) + \delta(t - 2a/V) - \delta(t - L/V) - \delta(t - (L + 2a)/V)\big]$$

$$\ddot{\psi}_g = \tfrac{h}{2aL}V\big[\delta(t) - \delta(t - 2a/V) - \delta(t - L/V) + \delta(t - (L + 2a)/V)\big]$$

so the ground motion could be written in the dimensionally homogeneous form

$$\begin{Bmatrix} \ddot{u}_g \\ a\,\ddot{\psi}_g \end{Bmatrix} = \tfrac{1}{2}\tfrac{h}{L}V\left[\begin{Bmatrix} 1 \\ 1 \end{Bmatrix}\delta(t) + \begin{Bmatrix} 1 \\ -1 \end{Bmatrix}\delta(t - 2a/V) - \begin{Bmatrix} 1 \\ 1 \end{Bmatrix}\delta(t - L/V) - \begin{Bmatrix} 1 \\ -1 \end{Bmatrix}\delta(t - (L + 2a)/V) \right]$$

Finally, defining

$$t_1 = 2a/V, \qquad t_2 = L/V, \qquad t_3 = (2a+L)/V$$

we obtain

$$\left\{ \begin{matrix} \ddot{u}_g \\ a\ddot{\psi}_g \end{matrix} \right\} = \tfrac{1}{2}\tfrac{h}{L}V\left[\left\{ \begin{matrix} 1 \\ 1 \end{matrix} \right\}\delta(t) + \left\{ \begin{matrix} 1 \\ -1 \end{matrix} \right\}\delta(t-t_1) - \left\{ \begin{matrix} 1 \\ 1 \end{matrix} \right\}\delta(t-t_2) - \left\{ \begin{matrix} 1 \\ -1 \end{matrix} \right\}\delta(t-t_3) \right]$$

This sketch should suffice for your solution to this problem.

P. 92 A homogeneous, rigid cylinder with radius R and mass M, rolls without sliding over a flat surface. The axis of the cylinder is attached to the wall via a continuous spring of stiffness k, mass m, and length L.

m = total mass of the spring
k = stiffness of the spring
$M = \rho\pi R^2 L = 5m$ = total mass of the cylinder
$J = \dfrac{MR^2}{2}$ = mass moment of inertia about the axis

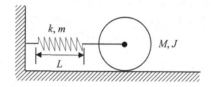

- Estimate the frequency of oscillation.

Hint: You may assume that the spring stretches linearly

P. 93 A uniform cylindrical rod has axial rigidity EA and mass/length ρA. (This is a continuous system.) The rod has a mass attached at one end, and a spring at the other (see figure). Using Lagrange's equations, estimate the fundamental frequency. Use the following trial function (the coordinate x is measured from point 2):

$$u(x,t) = u_1(t)\frac{x}{L} + (1-\frac{x}{L})u_2(t)$$

Notice that $u_1 \equiv q_1, u_2 \equiv q_2$ and that all motions are in the axial direction x.

The strain energy in the rod is

$$V = \frac{1}{2}\int_0^L EA\left(\frac{\partial u}{\partial x}\right)^2 dx$$

Data:

$$k = \frac{EA}{L} = \text{stiffness of spring}$$
$$m = \rho A L = \text{mass lumped on top} \ (= \text{mass of rod})$$

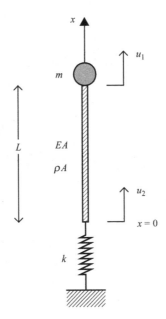

P. 94 A cantilever bending beam has a lumped mass at the free end, which has both mass *and* rotational inertia. The displacement function for this system can be approximated by means of a single trial function, that is, $u(x,t) \approx \psi(x)q(t)$.

- Estimate the fundamental frequency using the trial function $\psi(x) = 1 - \cos\frac{\pi x}{2L}$, which satisfies the essential boundary conditions $\psi(0) = 0$ and $\frac{d}{dx}\psi(0) = 0$. Express your results in terms of the frequency of the beam *without* the mass added at the end.

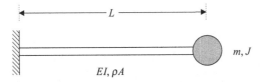

Observe that the lumped mass contributes both translational *and* rotational kinetic energies to the total energy in the system, and that the implied rotation of any initially vertical plane section of the beam is

$$\theta = \frac{\partial}{\partial x}u(x,t) = \frac{d}{dx}\psi(x)q(t).$$

Lumped mass data: $m = \rho AL,\qquad J = \frac{1}{5}mL^2$

Trial function: $\psi(x) = 1 - \cos\dfrac{\pi x}{2L}$

Useful integrals: $\displaystyle\int_0^L \cos\frac{\pi x}{2L}\,dx = \frac{2}{\pi}L,\qquad \int_0^L \cos^2\frac{\pi x}{2L}\,dx = \frac{1}{2}L$

P. 95 A uniform, simply supported *bending* beam of length L, total mass $m = \rho A L$, and flexural stiffness EI has modes and frequencies given by

$$\omega_j = \left(\frac{j\pi}{L}\right)^2 \sqrt{\frac{EI}{\rho A}}, \qquad \phi_j = \sin\left(\pi j \frac{x}{L}\right) \qquad \int_0^L \phi_i \phi_j \, dx = \tfrac{1}{2} L \delta_{ij} = \begin{cases} \tfrac{1}{2} L & i = j \\ 0 & i \neq j \end{cases}$$

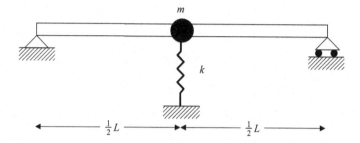

Next, we add a spring of stiffness $k = \pi^4 EI/L^3$ at the center, and also lump there a mass $m = \rho A L$ that equals the mass of the beam.

- Using the mode(s) of the beam without the added spring and mass, estimate the first *three* frequencies of the new system.

Observe that the frequency of the spring–mass system by itself is the same as the frequency of the fundamental mode of the beam alone. **Note:** You should be able to do this problem with minimal algebra, so think before proceeding.

P.96 It has been estimated that the Boeing 767–200 aircraft that terrorists crashed onto the North Tower of the World Trade Center hit the tower head on at some 810 kph somewhere in the vicinity of the 96th floor. Each of the 110 floors had dimensions 64 × 64 × 410 m, and the building weighed about 400,000 Mg (Mg = metric ton). The airplane was 48 m long and weighed about 150 Mg. The initial effect of the crash may have involved several floors, say two or three, after which most of the mass of the airplane remained embedded in the tower. Although the plane's weight also included fuel that burned in the aftermath of the collision, you should disregard that loss of mass because the dynamic impact was over well before any of that fuel was lost to fire. Measurements of ambient vibrations caused by wind put the fundamental lateral period of vibration of the twin towers at about 11 s. Estimate

- The duration of the crash, and the maximum force of impact (**Hint:** The aircraft came to a stop within the confines of the building.)
- The initial lateral velocity profile for the building
- The maximum acceleration felt by the occupants in floors adjacent to the crash site
- The maximum lateral displacement
- In comparison with the axial forces in the columns near the base caused by hurricane winds, how large may have been the column forces caused by the crash?

P.97 A simply supported rectangular concrete slab of thickness $h = 0.15$ m has length $a = 8.0$ m, and width $b = 6.0$ m. The modulus of elasticity of the slab is $E = 2,000,000$ N/cm², Poisson's ratio is $v = 0.25$, and the mass density is $\rho = 2500$ kg/m³. The slab has an eccentric circular opening of radius $r = 1.0$ m so as to accommodate a spiral stairway. The center of this hole has coordinates $x_0 = y_0 = 1.5$ m from the left, bottom corner. You are also told that if the hole were not there, the modal shapes would follow from

$$\phi(x, y) = \sin\frac{m\pi x}{a}\sin\frac{n\pi y}{b}, \qquad m, n = 1, 2, 3\ldots$$

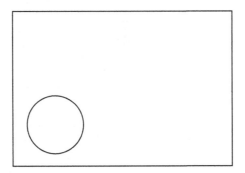

- Find the frequencies without the hole.
- *Estimate* the fundamental frequency and modal shape of the slab.

To analyze the second part of this problem, you will necessarily have to make some approximations. For example, the circular geometry of the hole is difficult to take into account in rectangular coordinates. To deal with that problem, you may wish to consider what is important in defining the fundamental frequency, and make appropriate modifications. Notice also that the geometry is not symmetric, so the fundamental mode will not be symmetric either. Will the frequency of the slab with the hole be larger or smaller than that of the slab without the hole? Consider the fact that by adding the hole you are subtracting both mass and stiffness, so it is not obvious *a priori* which way the frequencies should go. If helpful, you may wish to use MATLAB or other similar programs.

P. 98 Consider a soil stratum (a soil deposit of finite depth) that rests on a rigid base ("*rock*"). For plane shear waves that propagate vertically, the stratum can be modeled as a shear beam with unit cross-section $A = 1$.

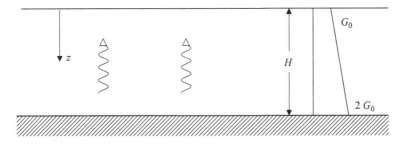

The wave equation for this case is

$$\frac{\partial}{\partial z}\left(G\,\frac{\partial u}{\partial z}\right) = \rho\,\frac{\partial^2 u}{\partial z^2}$$

where $G = G(z)$ is the depth-dependent shear modulus, and z is positive from the surface down. The shear modulus increases linearly with depth as shown in the drawing, that is, $G = G_0 \left(1 + \frac{z}{H}\right)$.

- Estimate the first two fundamental frequencies by means of Lagrange's equation. Use the following two trial functions:

$$\Psi_1 = \cos\left(\frac{\pi}{2}\frac{x}{H}\right) \qquad\qquad \Psi_2 = \cos\left(\frac{3\pi}{2}\frac{x}{H}\right)$$

$$u = \Psi_1\, q_1(t) + \Psi_2\, q_2(t)$$

(These are the exact modal shapes for a homogeneous soil stratum of thickness H). Hint:

$$K = \frac{1}{2} \int_{Vol} \rho\, \dot{u}^2\, dV \qquad \text{Kinetic Energy}$$

$$V = \frac{1}{2} \int_{Vol} G\left(\frac{\partial u}{\partial z}\right)^2 dV \qquad \text{Strain Energy}$$

$$G\left(\frac{\partial u}{\partial z}\right)^2 = \tau\, \gamma$$

P. 99 Consider a simply supported, massless bending beam with three equal lumped masses attached as shown. The beam has constant bending stiffness EI. Estimate the fundamental frequency via Lagrange's equations, using a single trial function ψ, that satisfies the boundary conditions, that is,

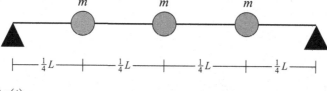

$$u(x,t) = \psi(x)q(t)$$

$$\psi(0) = 0 \qquad\qquad \psi''(0) = 0 \qquad \psi(L) = 0 \qquad \psi''(L) = 0$$

Alternatively, if you prefer, you could find the 3×3 flexibility matrix, and estimate the fundamental frequency by Dunkerley's rule. (**Note:** Finding the stiffness matrix requires considerable effort!).

P. 100 An infinitely long shear beam has cross-section $A = 0.01\,\text{m}^2$, shear wave velocity $C_s = 100$ m/s, and mass density $\rho = 2000\,\text{kg/m}^3$. This beam is subjected to a wavelet consisting of a *single* sine wave pulse:

$$u(x,t) = u_0 \sin\omega\left(t - \frac{x}{c}\right) \qquad\qquad \frac{\omega}{2\pi} = 1\ \text{Hz}$$

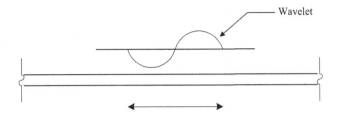

Wavelet

- Determine the stresses at a point x as a function of time.
- Determine the constant D of the dashpot that would completely absorb the wavelet.
- What is the wavelength λ of the wavelet?

P. 101 Consider a uniform cantilever shear beam of length L that has a constant shear modulus G, mass density ρ, and cross section A, as shown below. The vibrations are transverse to the beam, but this fact is not important for the problem at hand. The shear wave velocity is defined as

$$C_S = \sqrt{\frac{G}{\rho}} = \text{shear wave velocity}$$

It is known that for this continuous structure, the *exact* frequencies and modal shapes are

$$\omega_j = \frac{\pi}{2}\frac{C_S}{L}(2j-1) \qquad = \text{frequency in rad/s}$$

$$\phi_j(x) = \sin\left(\frac{\pi}{2}(2j-1)\frac{x}{L}\right) \qquad = \text{modal shape}$$

$$\mu_j = \int_0^L \phi_j^2(z)\rho A\,dz = \tfrac{1}{2}\rho AL \quad = \text{modal mass for any mode}$$

a. What are the modal stiffnesses $\kappa_j = \int_0^L GA\left(\frac{\partial\phi}{\partial x}\right)^2 dx$? (But think before proceeding!)

b. A lumped mass $m = \tfrac{1}{2}\rho AL$ (i.e., equal to half the total mass of the shear beam) is added at the free end. Estimate the new *fundamental* frequency.

$m = \tfrac{1}{2}\rho AL$

P. 102 A simply supported beam of length L, total mass ρAL, and bending stiffness EI has a mass $m = \frac{1}{2}\rho AL$ lumped at its center. You are also told that the *exact* first mode and the fundamental frequency of the beam *without* the added mass are

$$\phi_1 = \sin(\pi x / L), \qquad \omega_1 = \frac{\pi^2 R C_r}{L^2}, \qquad \text{with} \quad R = \sqrt{\frac{I}{A}}, \qquad C_r = \sqrt{\frac{E}{\rho}}$$

Estimate the fundamental frequency of the system with the added lumped mass by means of the Rayleigh quotient, using the mode of the beam alone as a trial function. However, by thinking carefully, you might be able to obtain the desired solution without doing any messy integrals whatsoever.

P. 103 A xylophone is a percussion musical instrument with several solid wooden bars, each of different length L. Each bar is supported at a distance of about $\frac{1}{5}L$ from either end by two felt strips which rest on a resonant box, see figure below. The felt-strips are soft enough that they do not impede the vibration of the bar in any significant way. While the bars have a rounded top, you may neglect that fact and assume at first that they are rectangular. The largest bar, which is tuned approximately to a G note (392 Hz), has a thickness $h = 0.0127$ [m] (1/2 inch) and a length $L = 0.348$ [m] (1 ft). Also, the speed of sound (i.e. acoustic velocity) in wood is about 3500 [m/s] (11,500 ft/s).

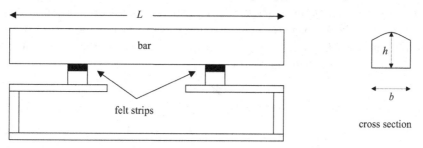

- Sketch the mode of vibration of the principal note that plays when the mallet strikes the center of the bar.
- Assuming that the bar is rectangular – in which case the area moment of inertia is simply $I = \frac{1}{12}bh^3 = \frac{1}{12}Ah^2$, where b is the width and A is the cross section – estimate the frequency, in Hz, of the large G-bar, which corresponds to the mode referred to in the previous question.
- Will the assumption of a rectangular cross section cause your estimate to be larger or smaller than the true value? Explain why.
- If one were to strike the bar laterally on one of its far ends (i.e., causing it to vibrate longitudinally as a rod, and not as beam), what mode and vibration frequency would be excited?

P. 104 A panpipe is a very old musical instrument found in many cultures. You play it by blowing across the top, much like blowing across the top of a bottle. A panpipe originating from Ecuador is made of dried reed, which has bulkheads at the bottom of the reed to provide closed ends when needed. Its dimensions are $L = 0.267$ m in length and a diameter $D = 0.0064$ m. It has a natural bulkhead (complete blockage) at one end, while the other end is open. The speed of acoustic waves in air at room temperature is 340 m/s.

- Specify the approximate boundary conditions that govern the pitch of the sound produced when the pipe is played. Be specific in explaining what variable you are describing, such as motion, air pressure, and so forth.
- Estimate the first and second natural frequencies in this pipe.
- Sketch the mode shapes that correspond to the first two natural frequencies computed above. Be very clear as to what variable is being represented, such as molecular motion of the air, or pressure in the pipe (i.e., deviation from atmospheric pressure in the room).
- What is the wavelength in the room of the sound produced by the first mode of this pipe?

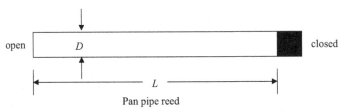

Pan pipe reed

P.105 A short and lightweight pedestrian bridge over a brook in a neighborhood park has a span length $L = 5m$, a width $b = 1m$, and an observed natural frequency of $f_1 = 3$ Hz, measured *without* any persons standing on it. Also, the radius of gyration of the bridge deck is estimated to be $R = 5$ cm (i.e., the total depth of the deck is about 18 cm), and when a man with a mass of 100 kg stands at the center of this bridge, his weight elicits at that location a *static* vertical deflection $\Delta = 1$ cm.

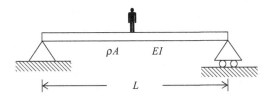

To a first approximation, the bridge *without* the man can be idealized as a homogeneous, simply supported bending beam with total mass $m_b = \rho AL$, and bending stiffness EI. From both the lectures and the class notes, you may also recall the following information:

$$f = \tfrac{1}{48} EI/L^3 \qquad\qquad = \text{flexibility at the center of beam}$$

$$\omega_1 = \pi^2 RC_r/L^2 \qquad\qquad = \text{fundamental frequency of beam}$$

$$\phi_1 = \sin(\pi x/L) \qquad\qquad = \text{fundamental mode of beam}$$

$$\mu_1 = \int_0^L \rho A\phi_1^2 dx = \tfrac{1}{2}\rho AL = \tfrac{1}{2}m_b \quad = \text{modal mass of first mode}$$

$$\kappa_1 = \int_0^L EI\left(\frac{\partial^2 \phi_1}{\partial x^2}\right)^2 dx = \tfrac{1}{2}\frac{\pi^4 EI}{L^3} \quad = \text{modal stiffness of mode 1}$$

with $R = \sqrt{I/A}$ being the radius of gyration and $C_r = \sqrt{E/\rho}$ the rod-wave velocity.

- At the natural frequency of the bridge, what is the *phase velocity* of flexural waves, what is their wavelength, and how long does it take such waves to propagate from one support to the other?
- A man with a mass of 100 kg stands once more at the center of the bridge. Estimate the lowest frequency, including the additional mass at the center.

P. 106 Consider a *continuous* beam with two identical spans, that is, equal material properties, cross section and length, as shown below. Note that there is *no* hinge above the middle roller. Find *exact* expressions for *all* of the frequencies of vibration of this system.

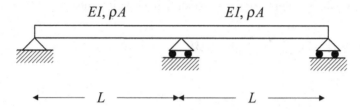

It is also known that the frequencies of a *single-span* beam with various boundary conditions are known to be given by

$$\omega_j = \left(\frac{\beta_j}{L}\right)^2 \sqrt{\frac{EI}{\rho A}} \qquad \text{where}$$

Boundary Condition	SS-SS	CL-FR	CL-CL	FR-FR	SS-CL	SS-FR
$\beta_j =$	$j\pi$	$\left(j-\tfrac{1}{2}\right)\pi$	$\left(j+\tfrac{1}{2}\right)\pi$	$\left(j+\tfrac{1}{2}\right)\pi$	$\left(j+\tfrac{1}{4}\right)\pi$	$\left(j+\tfrac{1}{4}\right)\pi$

in which SS = simply supported, CL = clamped, FR = free (Exception: for the CL–FR case, $\beta_1 = 0.597\pi$)

P. 107 A uniform bending beam of length L has bending stiffness EI and mass per unit length ρA. It is pin-supported at the left end and free at the right end. The *exact* first two modes of the beam, as sketched below, are as follows:

Rigid body mode $\qquad \omega_0 = 0, \qquad \phi_0(x) = \frac{x}{L}$

Fundamental mode $\qquad \omega_1 = \left(\frac{5}{4}\right)^2 \sqrt{\frac{\pi^4 EI}{\rho A L^4}}, \qquad \phi_1 = \sin \alpha x - \tfrac{1}{2}\sqrt{2} \sinh \alpha x / \sinh \alpha L, \quad \text{where } \alpha = \frac{5}{4}\frac{\pi}{L}.$

We next modify this problem by adding a spring of stiffness $k = \frac{\pi^4 EI}{L^3}$ at the center $x = \tfrac{1}{2}L$, as shown below. This, of course, removes the rigid-body mode, and also changes the frequency and shape of the fundamental mode.

To find the new frequency, you are asked to use Lagrange's method with the following two trial functions (based on the result for the beam without the spring):

$$\psi_1 = \frac{x}{L} \qquad \psi_2 = \sin\left(\frac{5\pi}{4}\frac{x}{L}\right) \equiv \sin \alpha x$$

For your convenience, here are some possibly useful integrals:

$$a = \int_0^L \sin^2 \alpha x \ dx = L\left(\frac{1}{2} + \frac{\sqrt{2}}{5\pi}\right) = 0.59003L$$

$$b = \frac{1}{L}\int_0^L x\sin \alpha x \ dx = L\frac{2\sqrt{2}}{5\pi}\left(1 - \frac{4}{5\pi}\right) = 0.13421L$$

Check: Verify that your computed frequency is higher than the frequency for the beam without the spring, or you could also set $k = 0$, and check if the frequency nearly agrees (it will not do so perfectly, because the trial function is not identical to the exact mode).

Note: In case you wonder how the given mode ϕ_1 was obtained, you may recall that the mode of a bending beam with arbitrary boundary conditions is of the form

$$\phi_j = A\cos kx + B\sin kx + C\cosh kx + D\sinh kx$$

The boundary conditions for the above beam are

$$\phi_j\big|_{x=0} = 0, \qquad \phi_j''\big|_{x=0} = 0", \qquad \phi_j''\big|_{x=L} = 0 \qquad \phi_j'''\big|_{x=L} = 0'$$

From the first two BCs

$$A + C = 0, \qquad -A + C = 0 \qquad \text{so} \qquad A = C = 0$$

From the next two BCs

$$-B\sin kL + D\sinh kL = 0, \qquad -B\cos kL + D\cosh kL = 0$$

or

$$\begin{Bmatrix} -\sin kL & \sinh kL \\ -\cos kL & \cosh kL \end{Bmatrix} \begin{Bmatrix} B \\ D \end{Bmatrix} = \begin{Bmatrix} 0 \\ 0 \end{Bmatrix}$$

Hence, setting the determinant to zero

$$-\sin kL \cosh kL + \cos kL \sinh kL = 0, \qquad \text{i.e.,} \qquad \boxed{\tan kL = \tanh kL}$$

Plotting the left- and right-hand sides of this equation as a function of kL, one obtains the figure given below. The solid dots are the solutions to the transcendental eigenvalue problem. Clearly, they very nearly correspond to $\tanh k_j L \approx 1 \Rightarrow \tan k_j L \approx 1$, i.e., $k_j L = (4j+1)\frac{\pi}{4}$, so $k_1 L = \frac{5}{4}\pi$, $k_1 = \frac{5}{4}\pi/L$

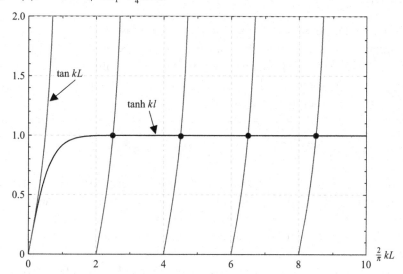

Hence, the natural frequency of the jth mode is

$$\omega_j = k_j^2 \sqrt{\frac{EI}{\rho A}} = \left(\frac{5\pi}{4L}\right)^2 \sqrt{\frac{EI}{\rho A}}$$

Observe that although the above expression provides a very tight approximation to the fundamental frequency (when $j=1$), it is not *exact*. The mode is obtained by evaluation of the eigenvector:

$$\begin{Bmatrix} -\sin\frac{5}{4}\pi & \sinh\frac{5}{4}\pi \\ -\cos\frac{5}{4}\pi & \cosh\frac{5}{4}\pi \end{Bmatrix} \begin{Bmatrix} B \\ D \end{Bmatrix} = \begin{Bmatrix} 0 \\ 0 \end{Bmatrix}$$

Choosing $B = 1$, then from the first equation

$$D = \frac{\sin\frac{5}{4}\pi}{\sinh\frac{5}{4}\pi}$$

so

$$\phi_1 = B\sin kx + D\sinh kx = \sin kx + \frac{\sin kL}{\sinh kL}\sinh kx$$

$$= \sin\frac{5\pi x}{4L} + \frac{\sin\frac{5\pi}{4}}{\sinh\frac{5\pi}{4}}\sinh\frac{5\pi x}{4L}$$

$$\phi_1 = \sin\frac{5\pi x}{4L} - \frac{\sqrt{2}}{2}\sinh\frac{5\pi x}{4L} / \sinh\frac{5\pi}{4}$$

P. 108 A simply supported cylindrical aluminum rod is $L = 1$ m in length and $D = 0.01$ m in diameter. The speed of propagation of longitudinal waves in the rod is $\sqrt{E/\rho} = 5060$ m/s, and the mass density is $\rho = 2710$ kg/m³. Find

- The fundamental frequency for longitudinal motion (as a rod).
- The fundamental frequency for bending motion (as a beam).
- If flexural waves with this frequency were to travel in an infinite rod, how would their speed compare with the speed of *acoustic* waves in air (340 m/s) and/or water (1500 m/s)? What would their wavelength be?

P. 109 To a first approximation, waves on the surface of the deep ocean can be idealized as harmonic waves of the form $y = A \sin(\omega t - kx)$, where A is the amplitude, k is the wave-number, and the frequency ω of the waves depends on sea conditions. The dispersion relationship for these waves is known to be given by $k = \omega^2/g$, where g is the acceleration of gravity. During a massive earthquake in the Aleutian Islands, a large tidal wave (tsunami) of wavelength $\lambda = 100$ km is generated. Neglecting ocean bottom effects, What are the period and speed (i.e., phase velocity) of this wave? If the California coast is some 4000 km away, how long does it take for the tidal wave to reach this coast?

P.110 Consider a free rod of total length $2L$, which consists of two halves of equal length L. The rod is subjected to longitudinal oscillations that obey the classical wave equation. The mass density and cross section are constant and there is NO damping, but the rod wave velocity of the right half is 3/2 times larger than that of the left half, that is, $C_2/C_1 = \frac{3}{2}$ (in which the subindices 1, 2 refer to the two halves). Using appropriate (either exact or approximate) models, determine

- The frequency response function (transfer function) for a harmonic load applied at the center (i.e., at the junction of the two halves). Sketch your results (you may wish to use MATLAB for this purpose). Do any frequencies exist for which the center node does no move? If so, which are these?
- Identify as many natural frequencies and corresponding modal shapes as possible, and as accurately as you can.
- What are the consequences on the frequencies and modes of the wave velocity ratio C_2/C_1 being a rational number, and especially a simple rational number such as $\frac{3}{2}$? What if, say $C_2/C_1 = \sqrt{2}$? **Hint:** This question pertains to the relationship between the eigenmodes of the two halves versus the modes of the system as a whole, especially those for which the center either does not move, or the axial stress is zero there.

P.111 Consider a *continuous* beam with three identical spans, that is, equal material properties, cross section, and length, as shown below. Note that there is *no* hinge anywhere. Find expressions for *all* of the frequencies of vibration of this system. Neglect any axial (horizontal) restraints, that is, there is absolutely no coupling with any rod modes.

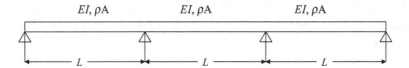

Clearly, all of the infinitely many modes of a one-span, simply supported beam, whose frequencies are

$$\omega_j = \left(\frac{j\pi}{L}\right)^2 \sqrt{\frac{EI}{\rho A}}, \quad \phi_j(x) = \sin kx, \quad k = \pi j / L, \quad 0 \le x \le 3L, \quad j = 1,2,3,\dots$$

will also be modes of this continuous beam system. The "odd" modes $j = 1,3,5\dots$ of the one-spam beam will be "symmetric" modes of the three-span beam, while the "even" modes $j = 2,4,6,\dots$ will be antisymmetric modes. But are these all of the modes of the complete structure, or do other symmetric and antisymmetric modes exist? In other words, if you replaced the above beam by the following pair of structures (which satisfy the symmetry–antisymmetry conditions), will these "reduced" structures admit additional modes not included in the above formula? If the answer were to be yes, then what are these? And if they exist, why are they not part of the above list? In what ways are they "different"?

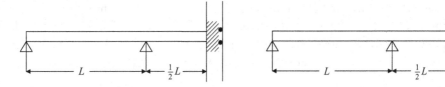

Hint: To avoid confusion, consider only one half of a structure at a time. For each span of either of these, write a general solution of the form

$$\phi = A\cos kx + B\sin kx + C\cosh kx + D\sinh kx, \quad 0 \le x \le L, \\ \phi = a\cos k\xi + b\sin k\xi + c\cosh k\xi + d\sinh k\xi, \quad 0 \le \xi \le \tfrac{1}{2}L, \quad k^4 = \frac{\omega^2 \rho A}{EI}$$

with *different* constants A, B, C, D and a, b, c, d for each of the two members, but with the same k, using convenient *local* coordinates x, ξ for each. Impose all of the support conditions as well as the continuity conditions (rotation and moment) above the second support. It is highly recommended that you use MATLAB's symbolic tool to analyze and solve the resulting transcendental eigenvalue problem, which should include one of the above sets as acceptable roots (if it does not, you must have an error). Note that here we only ask for frequencies, not modal shapes.

P.112 Consider two rectangular, simply supported plates of the same lateral dimensions $a \times b$ ($a = \tfrac{3}{2}b$) and the same Young's modulus, mass density, and Poisson's ratio E, ρ, ν. However, the first has a thickness $h_1 \ll a, b$, and the second has a thickness $h_2 = \tfrac{1}{2}h_1$, which is meant to simulate a thick floor and a lighter ceiling. The plates are aligned in parallel over one another at some vertical distance, which could be understood as the floor height. Exactly in the middle of both plates (i.e., at $x = \tfrac{1}{2}a$, $y = \tfrac{1}{2}b$) stands a perfectly rigid and massless column that connects the centers of the plates. The link of the column to the plates can be assumed to be of the pinned type, i.e. there is no need to consider transmission of bending moments or torsion from the plates to the column, or for that matter, of any horizontal forces.

Without the column, each plate will vibrate with a frequency and modal shape given by

$$\omega_{mn} = \pi^2 \left[\frac{m^2}{a^2} + \frac{n^2}{b^2} \right] \frac{C_r h}{\sqrt{12(1-v^2)}} \qquad h = h_1, h_2 \qquad m, n = 1, 2, \ldots$$

$$\phi_{mn}(x,y) = \sin \frac{m\pi x}{a} \sin \frac{n\pi y}{b}, \qquad C_r = \sqrt{\frac{E}{\rho}}$$

Also, the modal mass and modal stiffness of an individual plate are

$$\mu_{mn} = \iint_A \phi^2(x,y)\rho h \, dx \, dy = \tfrac{1}{4}m = \tfrac{1}{4}\rho abh, \qquad \kappa_{mn} = \omega_{mn}^2 \mu_{mn}$$

both of which depend on h. Observe also that the modal load for a concentrated force $f(t)$ applied at some arbitrary point of the plate, that is, $p(x,y,t) = f(t)\delta(x,y)$ is simply $p_{mn}(t) = \phi_{mn}(x,y)f(t)$.

- Determine (or estimate) the fundamental frequency and the modal shape of the assembly, and also as many of the higher modes as you can. Discuss your findings. Make sure that your coupled frequency is indeed the lowest one. Express your answer in terms of dimensionless frequencies (e.g., by scaling by the first frequency of the ceiling by itself).
- Determine (approximately if need be) the axial force in the column and its variation in time, assuming that the plates are given an initial displacement Δ at the center and then let go.

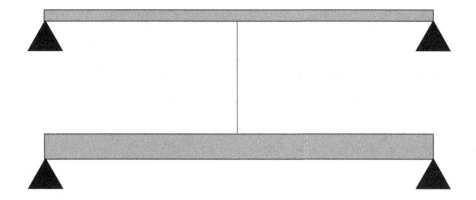

P.113 Two simply supported, perpendicular bending beams of length L_1 and L_2 lie in a horizontal plane and are connected at their respective centers. Both have the same Young's modulus E, mass density ρ, and square cross section $A = a^2$ (with $a \ll L_1, L_2$).

- Estimate (and sketch) the first few modes of *vertical* vibration as a function of L_2/L_1. You may neglect both axial deformations and coupling between bending and torsion. In particular, there exist modes whose exact frequencies you could write down without much ado.
- If a harmonic, vertical point load with frequency ω is applied at the intersection point of the beams, what is the response of the system? **Hint:** Compute the impedance of the system as seen from that intersection point.
- If the system were subjected to vertical earthquake motions with known response spectra, how would you estimate the response at some point?

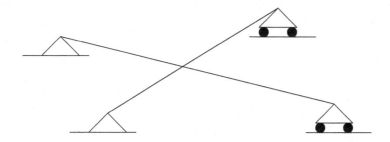

Author Index

Subject Index

Printed in the United States
By Bookmasters